NCCER's Contren Instructo[r]

This Web site is passcode-protected and is only meant for instructor use. The access code below provides you with access to all levels of the craft area. In order for you to access other craft areas, you will need to have an access code that is specific to that craft.

This site contains the following resources for instructor use!

TestGen – This product contains software to create the secure paper-based module exams that trainees must pass in order to acquire NCCER credentials. Instructors may use this software to scramble the order of exam questions and create additional questions as desired. Note that according to the NCCER Accreditation Policies and Guidelines, existing questions may not be removed or edited from any module examination. Doing so may seriously jeopardize either the accreditation status of the training program sponsor or any recognition of trainees, instructors, and trainers through the NCCER National Registry.

Performance Profile Sheets – Included on the Web site is a downloadable document containing all the hands-on tasks that are necessary for trainees to complete to receive credentials from NCCER.

Getting started is easy!

1. Visit **www.NCCERContrenIRC.com**. You will need a valid email address that you check on a regular basis as well as the access code that is on this sheet. You are allowed to redeem this access code only once.
2. Select the appropriate craft area by either using the drop down menu or by selecting "Login" next to the book cover.
3. Login using your username and password every time you return to **www.NCCERContrenIRC.com**.

Note: Additional resources may be added to this site as they become available.

Access Code: HVAC 2

PEEL HERE →

STORE IN A SECURE AREA!

HVAC
Level Two

Annotated Instructor's Guide
Third Edition

PEARSON

Upper Saddle River, New Jersey
Columbus, Ohio

National Center for Construction Education and Research

President: Don Whyte
Director of Curriculum Revision and Development: Daniele Stacey
HVAC Project Manager: Carla Sly
Production Manager: Tim Davis
Quality Assurance Coordinator: Debie Ness
Editors: Rob Richardson and Matt Tischler
Desktop Publishing Coordinator: James McKay

NCCER would like to acknowledge the contract service provider for this curriculum:
Topaz Publications, Liverpool, New York.

This information is general in nature and intended for training purposes only. Actual performance of activities described in this manual requires compliance with all applicable operating, service, maintenance, and safety procedures under the direction of qualified personnel. References in this manual to patented or proprietary devices do not constitute a recommendation of their use.

Copyright © 2007, 2001, 1995 National Center for Construction Education and Research, Gainesville, FL 32606. No part of this work may be reproduced in any form or by any means, including photocopying, without written permission of the publisher. Developed by the National Center for Construction Education and Research. Published by Pearson Education, Inc., Upper Saddle River, NJ 07458. All rights reserved. Printed in the United States of America. This publication is protected by Copyright and permission should be obtained from the NCCER prior to any prohibited reproduction, storage in a retrieval system, or transmission in any form or by any means, electronic, mechanical, photocopying, recording, or likewise. For information regarding permission(s), write to: NCCER Product Development, 3600 NW 43rd St, Bldg G, Gainesville, FL 32606.

10 9 8 7 6 5 4
ISBN 0-13-614387-3

PREFACE

TO THE INSTRUCTOR

Heating and air-conditioning systems (HVAC) regulate the temperature, humidity, and the total air quality in residential, commercial, industrial, and other buildings. This also extends to refrigeration systems used to transport food, medicine, and other perishable items. Other systems may include hydronics (water-based heating systems), solar panels, or commercial refrigeration. HVAC technicians and installers set up, maintain, and repair such systems. As a technician, you must be able to maintain, diagnose, and correct problems throughout the entire system. Diversity of skills and tasks is also significant to this field. You must know how to follow blueprints or other specifications to install any system. You may also need working knowledge of sheet metal practices for the installation of ducts, welding, basic pipefitting, and electrical practices.

Think about it! Nearly all buildings and homes in the United States alone use forms of heating, cooling and/or ventilation. The increasing development of HVAC technology causes employers to recognize the importance of continuous education and keeping up to speed with the latest equipment and skills. Hence, technical school training or apprenticeship programs often provide an advantage and a higher qualification for employment. NCCER's program has been designed by highly-qualified subject matter experts with this in mind. Our four levels present an apprentice approach to the HVAC field, including theoretical and practical skills essential to your success as an HVAC installer or technician.

As the population and the number of buildings grow in the near future, so will the demand for HVAC technicians. According to the U.S. Bureau of Labor Statistics, employment of HVAC technicians and installers is projected to increase 18 to 26 percent by 2014. We wish you the best as you begin an exciting and promising career.

WHAT'S NEW IN HVAC LEVEL TWO?

HVAC Level Two has three new modules: Introduction to Hydronic Systems, Sheet Metal Duct Systems, and Fiberglass and Flexible Duct Systems. Two modules that were previously in Level Three — Commercial Airside Systems, Troubleshooting Gas Heating, and Troubleshooting Cooling — have moved to Level Two.

The first two levels of the HVAC curriculum now provide you with a broad knowledge of installation and service requirements for residential and commercial heating and cooling systems, both forced-air and hydronic. Through this new course design, you will enter the workforce with the knowledge and skills needed to perform productively in either the residential or commercial market.

CONTREN® LEARNING SERIES

The National Center for Construction Education and Research (NCCER) is a not-for-profit 501(c)(3) education foundation established in 1995 by the world's largest and most progressive construction companies and national construction associations. It was founded to address the severe workforce shortage facing the industry and to develop a standardized training process and curricula. Today, NCCER is supported by hundreds of leading construction and maintenance companies, manufacturers, and national associations. The Contren® Learning Series was developed by NCCER in partnership with Pearson Education, Inc., the world's largest educational publisher.

Some features of NCCER's Contren® Learning Series are as follows:

- An industry-proven record of success
- Curricula developed by the industry for the industry
- National standardization providing portability of learned job skills and educational credits
- Compliance with the Office of Apprenticeship requirements for related classroom training (CFR 29:29)
- Well-illustrated, up-to-date, and practical information

NCCER also maintains a National Registry that provides transcripts, certificates, and wallet cards to individuals who have successfully completed modules of NCCER's Contren® Learning Series. *Training programs must be delivered by an NCCER Accredited Training Sponsor in order to receive these credentials.*

Contents

03201-07 Commercial Airside Systems 1.i
Describes the systems, equipment, and operating sequences used in a variety of commercial airside system configurations, such as constant volume single-zone and multi-zone, VVT, VAV, and dual-duct VAV. **(12.5 Hours)**

03202-07 Chimneys, Vents, and Flues.......... 2.i
Covers the principles of venting fossil-fuel furnaces and the proper methods for selecting and installing vent systems for gas-fired heating equipment. **(5 Hours)**

03203-07 Introduction to Hydronic Systems 3.i
Introduces hot water heating systems, focusing on safe operation of the low-pressure boilers and piping systems commonly used in residential applications. **(10 Hours)**

03204-07 Air Quality Equipment.............. 4.i
Covers the basic principles, processes, and devices used to control humidity and air clean-lines, as well as devices used to conserve energy in HVAC systems. **(5 Hours)**

03205-07 Leak Detection, Evacuation, Recovery, and Charging 5.i
Covers the basic refrigerant handling and equipment servicing procedures to service HVAC systems in an environmentally safe manner. **(20 Hours)**

03206-07 Alternating Current................. 6.i
Covers transformers, single-phase and three-phase power distribution, capacitors, the theory and operation of induction motors, and the instruments and techniques used in testing AC circuits and components. Also reviews electrical safety. **(7.5 Hours)**

03207-07 Basic Electronics................... 7.i
Explains the theory of solid-state electronics, as well as the operation, use, and testing of the various electronic components used in HVAC equipment. Includes an introduction to computers. **(5 Hours)**

03208-07 Introduction to Control Circuit Troubleshooting............ 8.i

Covers the operation, testing, and adjustment of conventional and electronic thermostats, as well as the operation of common electrical, electronic, and pneumatic circuits used to control HVAC systems. Also explains how to analyze circuit diagrams for electronic and microprocessor-based controls used in comfort heating and cooling equipment and how to troubleshoot systems that use these controls. **(30 Hours)**

03209-07 Troubleshooting Gas Heating........ 9.i

Covers tools, instruments, and techniques used in troubleshooting gas heating appliances, including how to isolate and correct faults. **(12.5 Hours)**

03210-07 Troubleshooting Cooling.......... 10.i

Covers the basic techniques and equipment used in troubleshooting cooling equipment, focusing on analyzing system temperatures and pressures in order to isolate faults. **(20 Hours)**

03211-07 Heat Pumps..................... 11.i

Covers the principles of reverse cycle heating, describes the operation of the various types of heat pumps, and describes how to analyze heat pump control circuits. Includes heat pump installation and service procedures. **(20 Hours)**

03212-07 Basic Installation and Maintenance Practices............. 12.i

Covers the application and installation of various types of fasteners, gaskets, seals, and lubricants, as well as the installation and adjustment of different types of belt drives, bearings, and couplings. Includes job documentation and customer relations. **(17.5 hours)**

03213-07 Sheet Metal Duct Systems......... 13.i

Covers layout, fabrication, installation, and insulating of sheet metal ductwork. Also includes selection and installation of registers, diffusers, dampers, and other duct accessories. **(5 Hours)**

03214-07 Fiberglass and Flexible Duct Systems..................... 14.i

Covers the layout, fabrication, installation, and joining of fiberglass ductwork and fittings. Describes the proper methods for attaching and supporting flex duct. **(5 Hours)**

Glossary of Trade Terms................... G.1

Index..................................... I.1

Contren® Curricula

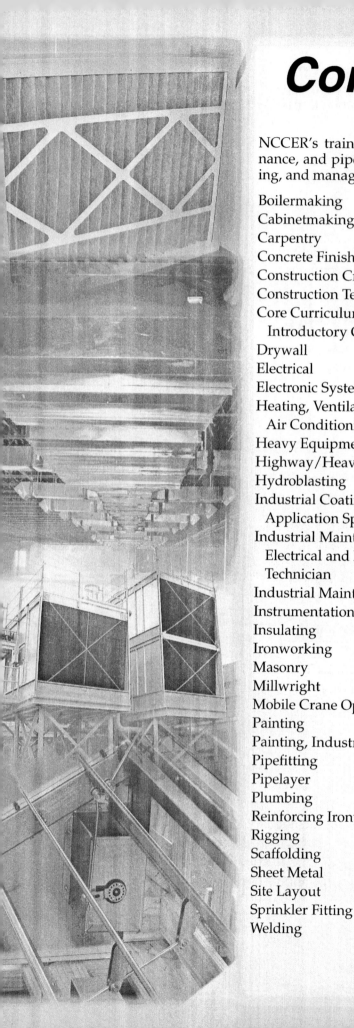

NCCER's training programs comprise over 50 construction, maintenance, and pipeline areas and include skills assessments, safety training, and management education.

Boilermaking
Cabinetmaking
Carpentry
Concrete Finishing
Construction Craft Laborer
Construction Technology
Core Curriculum:
   Introductory Craft Skills
Drywall
Electrical
Electronic Systems Technician
Heating, Ventilating, and
   Air Conditioning
Heavy Equipment Operations
Highway/Heavy Construction
Hydroblasting
Industrial Coating and Lining
   Application Specialist
Industrial Maintenance
   Electrical and Instrumentation
   Technician
Industrial Maintenance Mechanic
Instrumentation
Insulating
Ironworking
Masonry
Millwright
Mobile Crane Operations
Painting
Painting, Industrial
Pipefitting
Pipelayer
Plumbing
Reinforcing Ironwork
Rigging
Scaffolding
Sheet Metal
Site Layout
Sprinkler Fitting
Welding

Pipeline
Control Center Operations,
   Liquid
Corrosion Control
Electrical and Instrumentation
Field Operations, Liquid
Field Operations, Gas
Maintenance
Mechanical

Safety
Field Safety
Safety Orientation
Safety Technology

Management
Introductory Skills for the
   Crew Leader
Project Management
Project Supervision

Spanish Translations
Andamios
Currículo Básico
   Habilidades Introductorias
   del Oficio
Instalación de Rociadores
   Nivel Uno
Orientación de Seguridad
Seguridad de Campo

Supplemental Titles
Applied Construction Math
Careers in Construction

Acknowledgments

This curriculum was revised as a result of the
farsightedness and leadership of the following sponsors:

ABC of Wisconsin

Lincoln Technical Institute

W. B. Guimarin & Co., Inc.

This curriculum would not exist were it not for the dedication
and unselfish energy of those volunteers who served on the Authoring Team.
A sincere thanks is extended to the following:

Frank Kendall

Daniel Kerkman

Troy Staton

NCCER PARTNERING ASSOCIATIONS

American Fire Sprinkler Association
Associated Builders and Contractors, Inc.
Associated General Contractors of America
Association for Career and Technical Education
Association for Skilled and Technical Sciences
Carolinas AGC, Inc.
Carolinas Electrical Contractors Association
Center for the Improvement of Construction Management and Processes
Construction Industry Institute
Construction Users Roundtable
Design-Build Institute of America
Electronic Systems Industry Consortium
Merit Contractors Association of Canada
Metal Building Manufacturers Association
NACE International
National Association of Minority Contractors
National Association of Women in Construction
National Insulation Association
National Ready Mixed Concrete Association
National Systems Contractors Association
National Technical Honor Society
National Utility Contractors Association
NAWIC Education Foundation
North American Crane Bureau
North American Technician Excellence
Painting and Decorating Contractors of America
Portland Cement Association
SkillsUSA
Steel Erectors Association of America
Texas Gulf Coast Chapter ABC
U.S. Army Corps of Engineers
University of Florida
Women Construction Owners and Executives, USA

Product Supplements

Windows/Macintosh-Based
TestGen
ISBN 0-13-614779-8 $30

Ensure test security with NCCER's computerized testing software. This software allows instructors to scramble the module exam questions and answer keys in order to print multiple versions of the same test, customize tests to suit the needs of their training units*, add questions, or easily create a final exam.

Due to NCCER's Accreditation Guidelines, instructors may not delete existing questions from exams. Doing so may seriously jeopardize either the accreditation status or the training program sponsor or any recognition of trainees, instructors, and trainers through the NCCER National Registry.

Transparency Masters
ISBN 0-13-614795-X $25

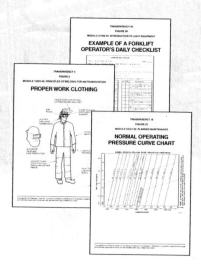

Spend more time training and less time at the copier. In response to instructor feedback, NCCER offers loose, reproducible copies of the overhead transparencies referenced in the Instructor's Guides. The transparency masters package includes most of the Trainee Guide graphics, enlarged for projection and printed on loose sheets* for easy copying onto transparency film using your photocopier.

* *Transparency masters are provided on regular, loose sheets of paper, not acetates.*

To order any of these supplements,
contact Prentice Hall Publishing at

800-922-0579

Module 03201-07

Commercial Airside Systems

NCCER STANDARDIZED CRAFT TRAINING PROGRAM

The National Center for Construction Education and Research (NCCER) provides a standardized national program of accredited craft training. Key features of the program include instructor certification, competency-based training, and performance testing. The program provides trainees, instructors, and companies with a standard form of recognition through a National Craft Training Registry. The program is described in full in the *Guidelines for Accreditation*, published by NCCER. For more information on standardized craft training, contact the NCCER by writing us at 3600 NW 43rd St., Bldg. G, Gainesville, FL 32606; calling 352-334-0911; or emailing info@nccer.org. More information may be found at our website, www.nccer.org.

HOW TO USE THIS ANNOTATED INSTRUCTOR'S GUIDE

Each page presents two sections of information. The larger section displays each page exactly as it appears in the Trainee Module. The narrow column ties suggested trainee and instructor actions to each page and provides icons (detailed below) to call your attention to material, safety, audiovisual, or testing requirements. The bottom of each page includes space for your notes.

The **Audiovisual** icon indicates an appropriate time to show a transparency or other audiovisual aid.

The **Classroom** icon prompts you to define a term, stress a point, ask trainees to explain a concept, or give examples.

The **Demonstration** icon directs you to show trainees how to perform tasks.

The **Examination** icon tells you to administer the written module examination.

The **Homework** icon is placed where you may wish to assign reading for the next class, assign a project, or advise trainees to prepare for an examination.

The **Laboratory** icon is used when trainees are to practice performing tasks.

The **Materials** icon is a reminder for you to gather materials needed for classes, labs, and testing.

The **Performance Testing** icon tells you to administer a performance test or a portion thereof.

The **Safety** icon is used to emphasize safety issues. It is often keyed to *Caution* and *Warning!* statements in the Trainee Module.

The **Teaching Tip** icon indicates additional guidance is available, such as how to conduct an exercise, get the most educational value from a field trip, or encourage class participation. Teaching Tips may expand on a feature (*Think About It*, *Did You Know?*) or provide *Quick Quizzes* or similar exercises. You will be referred to the Teaching Tips section at the back of the module if there is additional material.

The **Combination** icon indicates that the laboratory listed corresponds with a performance task. If desired, you can note the proficiency of the trainees during the laboratory, and use it to satisfy performance testing requirements.

PREPARATION

Before teaching this module, you should review the Objectives, Performance Tasks, Materials and Equipment List, and Module Outline. Be sure to allow ample time to prepare your own training or lesson plan and gather all required materials and equipment.

Commercial Airside Systems
Annotated Instructor's Guide

Module 03201-07

MODULE OVERVIEW

This module introduces trainees to the various types of all-air systems used in commercial buildings.

PREREQUISITES

Prior to training with this module, it is recommended that the trainee shall have successfully completed *Core Curriculum* and *HVAC Level One*.

OBJECTIVES

Upon completion of this module, the trainee will be able to do the following:

1. Identify the differences in various types of commercial all-air systems.
2. Identify the type of building in which a particular type of system is used.
3. Explain the typical range of capacities for a commercial air system.

PERFORMANCE TASKS

Under the supervision of the instructor, the trainee should be able to do the following:

1. Through observation of the equipment, identify the types of commercial air systems installed in selected buildings.
2. Given a list of several commercial-type buildings, identify the type of airside system(s) commonly used in each application. Describe the reason why.

MATERIALS AND EQUIPMENT LIST

Overhead projector and screen
Transparencies
Blank acetate sheets
Transparency pens
Whiteboard/chalkboard
Markers/chalk
Pencils and scratch paper
Appropriate personal protective equipment
Ducts in different shapes and materials
Duct seams

Grilles and registers
Manufacturers' literature on air terminals
Manufacturers' literature on packaged equipment
Fan rating curves
Filters used with packaged equipment
Manufacturers' literature on packaged air handlers
Copies of the Quick Quiz*
Module Examinations**
Performance Profile Sheets**

* Located at the back of this module.
**Located in the Test Booklet.

SAFETY CONSIDERATIONS

Ensure that the trainees are equipped with appropriate personal protective equipment and know how to use it properly. Emphasize basic site safety. This module require that trainees visit commercial buildings and observe HVAC systems. Ensure all trainees are properly briefed on site safety procedures.

ADDITIONAL RESOURCES

This module is intended to present thorough resources for task training. The following reference works are suggested for both instructors and motivated trainees interested in further study. These are optional materials for continued education rather than for task training.

HVAC Systems, 1992. Samuel C. Monger. Englewood Cliffs, NJ: Prentice Hall.

HVAC Systems and Equipment Handbook, 2000. Atlanta, GA: American Society of Heating, Refrigeration, and Air Conditioning Engineers (ASHRAE)

TEACHING TIME FOR THIS MODULE

An outline for use in developing your lesson plan is presented below. Note that each Roman numeral in the outline equates to one session of instruction. Each session has a suggested time period of 2½ hours. This includes 10 minutes at the beginning of each session for administrative tasks and one 10-minute break during the session. Approximately 12½ hours are suggested to cover *Commercial Airside Systems*. You will need to adjust the time required for hands-on activity and testing based on your class size and resources. Because laboratories often correspond to Performance Tasks, the proficiency of the trainees may be noted during these exercises for Performance Testing purposes.

Topic	Planned Time
Session I. Introduction	
A. Introduction	
B. Zoning	_____
C. Commercial Systems	_____
D. Outdoor Air and Air Systems	_____
Session II. Types of All-Air Systems and Duct Systems	_____
A. Types of All-Air Systems	
B. Laboratory	_____
Trainees practice identifying the types of airside systems commonly used in each application. This laboratory corresponds to Performance Task 2.	
Session III. Air Terminals, Source Equipment, and Handlers	
A. Duct Systems	
B. Air Terminals	_____
C. Air Source Equipment	_____
Session IV. Identification of Commercial Air Systems	_____
A. Air Handlers	
B. Laboratory	_____
Trainees practice identifying the types of commercial air systems installed in selected buildings. This laboratory corresponds to Performance Task 1.	
Session V. Review and Testing	
A. Review	
B. Module Examination	_____
1. Trainees must score 70% or higher to receive recognition from NCCER.	_____
2. Record the testing results on Craft Training Report Form 200, and submit the results to the Training Program Sponsor.	
C. Performance Testing	_____
1. Trainees must perform each task to the satisfaction of the instructor to receive recognition from NCCER. If applicable, proficiency noted during laboratory exercises can be used to satisfy the Performance Testing requirements.	
2. Record the testing results on Craft Training Report Form 200, and submit the results to the Training Program Sponsor.	

HVAC Level Two

03201-07

Commercial Airside Systems

Assign reading of Module 03201-07.

03201-07
Commercial Airside Systems

Topics to be presented in this module include:

1.0.0	Introduction	1.2
2.0.0	Zoning	1.2
3.0.0	Commercial Systems	1.2
4.0.0	Outdoor Air and Air Systems	1.5
5.0.0	Types of All-Air Systems	1.10
6.0.0	Duct Systems	1.19
7.0.0	Air Terminals	1.22
8.0.0	Air Source Equipment	1.30
9.0.0	Air Handlers	1.39

Overview

HVAC Level One focused on the types of systems and equipment used in residential and small commercial applications. Larger buildings use much different systems. You may have seen office buildings and shopping malls with many air conditioning units on the roof. What you can't see is that, unlike residential systems where refrigerant lines penetrate the building, these are likely to be packaged units in which the ductwork penetrates the building. Large buildings have different needs than small buildings. For example, there are cases where some zones in the building will be calling for heat, while others are calling for cooling, and still others are making no demands. The system design must be able to accommodate these differences. In commercial systems, it is more common to find single systems serving multiple zones, with each zone having its own comfort control device. These applications require special air distribution equipment not commonly found in residential applications.

Instructor's Notes:

Objectives

When you have completed this module, you will be able to do the following:

1. Identify the differences between types of commercial air systems.
2. Identify the type of building in which a particular type of system is used.
3. Explain the typical range of capacities for a commercial air system.

Trade Terms

Adiabatic
Air handler
Constant-volume system
Control zone
Dual-duct systems
Isothermal
Packaged air conditioner (PAC)
Packaged rooftop unit
Packaged unit
Sick building syndrome
Variable air volume (VAV) system
Variable volume, variable temperature (VVT) system
Year-round air conditioner (YAC)
Zoned system

Required Trainee Materials

1. Pencil and paper
2. Appropriate personal protective equipment

Prerequisites

Before you begin this module, it is recommended that you successfully complete *Core Curriculum* and *HVAC Level One*.

This course map shows all of the modules in the second level of the HVAC curriculum. The suggested training order begins at the bottom and proceeds up. Skill levels increase as you advance on the course map. The local Training Program Sponsor may adjust the training order.

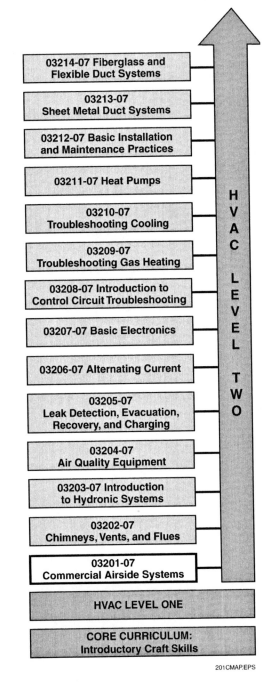

Ensure that you have everything required to teach the course. Check the Materials and Equipment list at the front of this module.

See the general Teaching Tip at the end of this module.

Explain that terms shown in bold are defined in the Glossary at the back of this module.

Show Transparency 1, Objectives, and Transparency 2, Performance Tasks. Review the goals of the module, and explain what will be expected of the trainee.

Review the modules covered in Level Two and explain how this module fits in.

Explain that this module covers commercial air systems.

Show Transparency 3 (Figure 1). Describe a typical control zone.

Explain that a control zone has its own thermostat. Discuss the use of zoned systems in commercial buildings.

Explain that most systems in commercial buildings are all-air systems.

Identify the control zone in the room where you are holding class.

See the Teaching Tip for Sections 1.0.0–9.0.0 at the end of this module.

1.0.0 ♦ INTRODUCTION

This module will familiarize you with commercial air systems, the type of equipment and applications associated with each system, and how each system operates. Commercial air systems provide for complete sensible and latent cooling, preheating, and humidification of the air supplied by the system. Heating may be provided by the same airstream, either in the central system or at a particular zone. In some applications, heating is provided by a separate system.

The ultimate purpose of any HVAC system is to provide comfort. Comfort is commonly associated with two aspects of the air that surrounds us in a building: temperature and relative humidity. However, the broader definition of comfort has come to also involve air cleanliness, odor level, and air motion.

Relative humidity is a measure of the moisture in the air around us, and is measured in percentages. When it is raining, the relative humidity is 100 percent; the air cannot hold any more moisture. Temperature is also called dry-bulb temperature because it is measured by a standard thermometer that has a dry sensing bulb. Because comfort is commonly associated with the control of temperature, and temperature control is normally associated with the use of a thermostat, we will begin our discussion of commercial air systems with the use of a thermostat.

2.0.0 ♦ ZONING

Figure 1 shows a typical office where people are at work. Heat flows into the office from people, lights, equipment, and from outdoors during warm weather. The office must be cooled in order to prevent an uncomfortable rise in room temperature. A thermostat located in the office measures the rise in room temperature and controls the cooling capacity supplied to the room. Because this office area contains a thermostat, we call the area a **control zone**.

Each air-conditioned building must have at least one control zone. Each zone is controlled by its own thermostat. The control zone may be a single room, several rooms, part of a building, or an entire building. As the size of a building increases, the number of control zones also tends to increase. Many commercial buildings use **zoned systems** in which a single system controls multiple zones.

3.0.0 ♦ COMMERCIAL SYSTEMS

Most systems in commercial buildings are all-air systems with a condensing unit and an air handler that provides cooling and heating to one or more zones. These systems can be self-contained (packaged) or split systems. Large buildings may use a chilled-water system in place of the condensing unit to transfer heat to the outdoors.

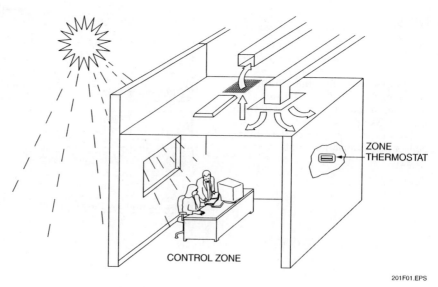

Figure 1 ♦ Typical control zone.

1.2   HVAC ♦ LEVEL TWO

3.1.0 Typical All-Air System

Figure 2 shows a typical commercial all-air system, and identifies its major components (items 1 through 10). The heart of any all-air system is a cooling coil (1) and a fan (10). They are part of a factory-assembled package unit called an air handler unit. As an airstream comes in contact with the coil's cold surface temperature, it is cooled and dehumidified. A cold fluid is passed through the coil's tubes to induce heat flow from the warmer airstream. The cold fluid may be either water or refrigerant, depending on the type of air conditioning system. For the system shown in *Figure 2*, the cold fluid is a refrigerant. It is routed to and from the cooling coil via the refrigerant lines (2) connected between the cooling coil and condensing unit. Moisture is removed from the air as it condenses on the cold coil surface. This condensing process is similar to what happens on the surface of a glass filled with a cold beverage on a hot, humid summer day. The condensed moisture is called condensate. It is collected in a drain pan under the coil and drained away.

The cold, dry air leaves the cooling coil and travels through a supply air duct (3) to the entrance of the control zone. The supply duct is insulated to prevent the supply air from absorbing heat before it reaches the zone. The insulation also prevents moisture in the surrounding air from condensing on the cold outer surface of the duct. A central supply air fan (10) moves the air around the building.

The cool air is distributed into the control zone through a supply air terminal (4), or diffuser. This device is most important in providing good room air motion, which is one of the five aspects of comfort. Poor selection or location of the diffuser(s) causes drafts, hot spots in the zone, and dumping of cold air directly on the occupants.

Show Transparency 4 (Figure 2). Identify the components of a typical all-air system. Describe the airflow in an all-air system.

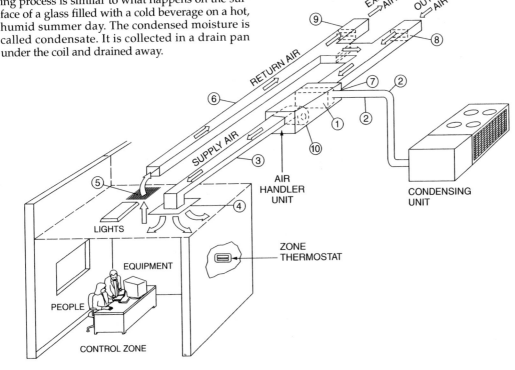

LEGEND
1. COOLING (DX) COIL
2. REFRIGERANT LINES
3. SUPPLY AIR DUCT
4. SUPPLY AIR TERMINAL
5. RETURN GRILLE
6. RETURN AIR DUCT
7. AIR FILTERS
8. OUTDOOR AIR DAMPER(S)
9. EXHAUST AIR DAMPER(S)
10. FAN

Figure 2 ♦ Typical all-air refrigerant system.

Keep Transparency 4 (Figure 2) showing. Continue a discussion of the air flow in an all-air system.

Explain the role of the filter and point out how it is positioned to filter return air as well as outdoor air entering the air handler. Discuss odor control.

Identify the two types of cooling coil systems. Explain how a direct expansion (DX) coil functions.

Show Transparency 5 (Figure 3). Describe the operation of a typical all-air chilled water system.

Explain that most commercial systems with loads less than 100 tons use DX coils. Larger projects use chilled-water systems, in which the DX cooling coil in the air handler is replaced with a chilled-water coil.

Point out that activated carbon or potassium permanganate filters are often used to improve indoor air quality in commercial applications.

Discomfort caused by the results of poor diffuser selection or location is a common complaint made by building occupants.

After the air enters the zone, it absorbs heat and moisture and establishes the room's temperature and relative humidity. The temperature of the zone is usually maintained between 72°F and 76°F, and the relative humidity is maintained between 50 and 60 percent. The air supplied to the zone is typically between 55°F and 60°F.

Without a means for the supply air to escape from the zone, the zone pressure would rise and the airflow into the zone would cease. Thus a return grille (5) is provided in the zone. Specially designed fluorescent lights known as troffers can also act as return grilles. In some instances, dummy supply air terminals are used as return grilles.

Air from the return grille is drawn back to the coil through a return air duct (6). At the coil, the air can once again be cooled, dehumidified, and recycled back to the control zone.

Filters (7), typically located upstream of the coil in the air handler, provide another of the five aspects of comfort; they clean the air. The filters remove dirt, pollen, and other impurities from the circulating air.

Bringing fresh outdoor air (ventilation air) into the zone controls another aspect of comfort—odor. A set of dampers (8) is used to bring outdoor air into the system, and another set of dampers (9) is used to exhaust odor-laden return air back outside. Therefore, the air handler mixes outdoor air with return air to reduce odors before sending the conditioned air to the control zone. Not all air conditioning systems do a good job of odor dilution. This is particularly true in residential systems. However, one of the reasons that commercial all-air systems are so popular in the commercial building market is that most systems provide good odor dilution.

As mentioned previously, the cold fluid flowing through the coil is the cooling source for the recirculated air. The type of fluid flowing through the coil is either refrigerant or chilled water. Thus, the coil can be either a direct-expansion (DX) or a chilled-water coil, respectively. In the system shown in *Figure 2*, the coil (1) is a DX coil.

In *Figure 2*, the DX coil (1) is shown connected by refrigerant lines (2) to an outdoor unit called a condensing unit. The condensing unit contains the compressor(s) and air-cooled condenser portions of a typical mechanical refrigeration system. The metering device and cooling coil located in the air handler make up the rest of the refrigeration system. The metering device can be found in the liquid line soldered to the inlet of the cooling coil. Because liquid refrigerant evaporates inside the tubes of the DX coil, the DX coil is also referred to as the evaporator.

The compressor, condenser, evaporator, and metering device can be combined into two pieces of manufactured equipment, as shown here, or they may be combined into a single unit (a rooftop packaged unit, for instance). In either case, the refrigeration system absorbs building heat from the air system at the cooling coil, and rejects that heat to the outside air through the condenser.

The means of rejecting heat back outside the building can be handled by chilled water instead of a DX refrigeration system. *Figure 3* shows a typical all-air chilled-water system and identifies its components (items 1 through 15). A chilled water coil (1) replaces the DX coil. Chilled water coils are typically used in larger multi-story buildings that use systems above 100 tons in size. DX coils are popular in buildings of less than 100 tons in capacity. Buildings of less than 100 tons in capacity represent some 80 percent to 90 percent of all the commercial buildings in the United States.

A chilled water pump (11) circulates 40°F to 48°F chilled water through the coil. The water absorbs heat from the air passing over the coil. The water is then pumped to a water chiller containing the basic components of the refrigeration cycle. The water chiller absorbs heat from the water in its evaporator (13). The evaporator is commonly referred to as a cooler when used in a chiller.

Activated Carbon Filters

Odor control is of particular concern today in commercial buildings because of tight buildings, new building materials, and outside air pollution. Filter devices used to control odors and other contaminants in all-air systems typically use activated carbon or potassium permanganate as the basic filter medium.

1.4 HVAC ♦ LEVEL TWO

Instructor's Notes:

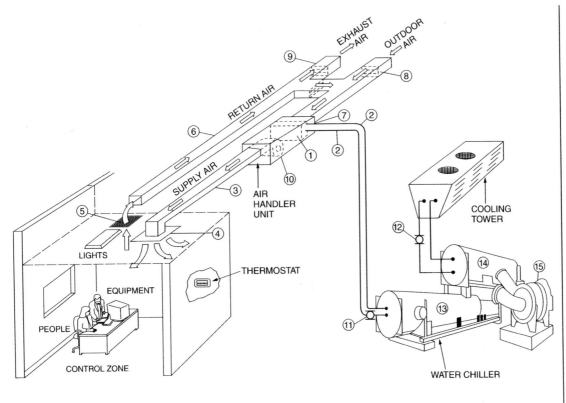

Figure 3 ♦ Typical all-air chilled-water system.

The heat absorbed by the chiller's cooler is rejected from the chiller's condenser (14). In this diagram, the chiller is located indoors, so the condenser is a water-cooled condenser. A condenser pump (12) circulates water through the chiller's condenser where it absorbs building heat, leaving at a temperature of about 95°F. The water is then circulated through a cooling tower where the building heat is rejected to the outdoor air. The water returns to the condenser at a temperature of about 85°F.

The chiller can also be located outside the building (typically on the roof) and equipped with an air-cooled condenser. Using an air-cooled chiller eliminates the need for a cooling tower and the associated condenser water pump and piping. Condenser circuit water treatment is also eliminated.

4.0.0 ♦ OUTDOOR AIR AND AIR SYSTEMS

All-air systems easily incorporate the use of outdoor ventilation air by virtue of their design. This is one of the reasons commercial air systems dominate the marketplace today. Outdoor air (ventilation air) is introduced into buildings for four reasons.

First, ventilation air is needed to dilute gaseous contaminants released within the conditioned space. Chemicals used in furniture, draperies, carpeting, wall coverings, and other furnishings release toxic gases for several years after being manufactured. Without dilution, these toxins accumulate and create a condition that can contribute to **sick building syndrome**.

Keep Transparency 5 (Figure 3) showing. Continue explaining the all-air chilled-water system.

Take a tour of the air handling system in the building in which you are holding class. Identify various system components and identify their functions.

Discuss the importance of outdoor air in commercial applications. Explain how outdoor air dilutes gaseous contaminants. Define sick building syndrome.

See the Teaching Tip for Section 4.0.0 at the end of this module.

Explain how the airspace above a suspended ceiling is used as a return plenum.

If your classroom has a suspended ceiling and a plenum, have trainees examine the plenum.

Discuss the ways outdoor air is used in the ventilation system. Explain how ventilation air replaces the oxygen used by people who occupy the building.

Explain how outdoor air can be used as the first stage of cooling. Discuss pressure control.

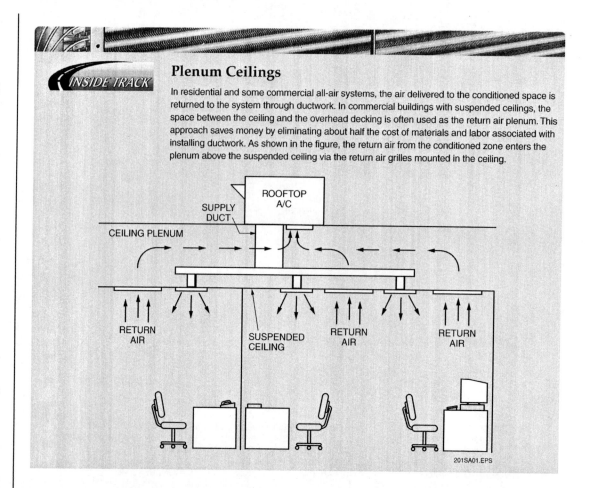

Plenum Ceilings

In residential and some commercial all-air systems, the air delivered to the conditioned space is returned to the system through ductwork. In commercial buildings with suspended ceilings, the space between the ceiling and the overhead decking is often used as the return air plenum. This approach saves money by eliminating about half the cost of materials and labor associated with installing ductwork. As shown in the figure, the return air from the conditioned zone enters the plenum above the suspended ceiling via the return air grilles mounted in the ceiling.

In addition, people give off carbon dioxide (CO_2) as they breathe. While not toxic, concentrated accumulations of CO_2 can cause drowsiness and other unpleasant symptoms. The maximum safe concentration of carbon dioxide is about 1,000 parts per million (ppm).

Second, the ventilation air replenishes the oxygen used by people who occupy the conditioned space and removes smoke and odors. The amount of air supplied per person varies with the activity and location of the people. For example, normal design practice is to provide about 20 cfm/person of ventilation air for office workers. In contrast, 30 cfm/person of ventilation air is recommended for people working in a dry cleaning plant. *ASHRAE Standard 62* specifies minimum ventilation rates for residential, commercial, and public buildings. Because ventilation rates vary with application and locality, you should always check the local code requirements.

The third purpose for using ventilation air is to save operating costs. When the outdoor air is cool and dry enough compared to the system return air, it can be used as the first stage of cooling instead of mechanical cooling. The set of dampers in the central equipment that introduces outdoor air in this manner is called an economizer. An economizer only requires fan energy to cool the building; it is used on jobs where the climate favors it.

The fourth purpose is to control the pressure within the building. A slight positive pressure in the conditioned space prevents the unintentional inward leakage of untreated outdoor air called infiltration. Introducing more ventilation air than is exhausted from a space, and maintaining a space pressure from 0.05 to 0.10 inches water gauge (in wg), is normally sufficient to offset infiltration.

Instructor's Notes:

4.1.0 Ventilation and Exhaust

The ventilation air brought in through the central cooling equipment displaces an equal amount of air from the conditioned space, and must be exhausted from the building. Exhaust takes place through a combination of three methods (*Figure 4*):

- Exhausting at or near the central cooling equipment
- Direct exhaust from the conditioned zone by local fans (for example, bathroom exhausts and kitchen hood exhausts)
- Exfiltration (the outward leakage through cracks around windows and doors)

The first method usually provides the majority of exhaust airflow. For most commercial applications, more ventilation air is brought into a building than can be exhausted through the combined effects of exfiltration and direct exhaust.

4.2.0 Exhausting at the Central Equipment

Air is exhausted at or near the central equipment in three ways (*Figure 5*). As stated earlier, the objective is to operate one of these exhaust methods in order to maintain a positive space pressure of about 0.05 to 0.10 in wg. Overpressurization of the space causes doors to blow open when released. A negatively pressurized space works in the opposite manner, making it difficult to open doors. A gravity relief damper located in the roof or wall lets air spill outside when the pressure in the space becomes too positive. Weights added to the relief damper establish its opening point. When the space pressure drops back to or below the desired level, gravity closes the weighted relief damper and prevents indoor air from leaving.

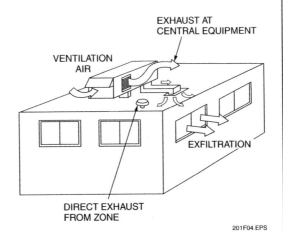

Figure 4 ♦ Ventilation and exhaust.

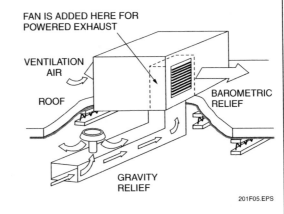

Figure 5 ♦ Central exhaust.

Poor Indoor Air Quality

Two terms commonly used when describing the results of poor indoor air quality in a building are sick building syndrome and building-related illness. Sick building syndrome is defined as a condition when more than 20 percent of a building's occupants complain during a two-week period of a set of symptoms, including headaches, fatigue, nausea, eye irritation, and throat irritation, that are alleviated by leaving the building and are not known to be caused by any specific contaminants. A building-related illness is defined as the condition in which the symptoms of a specific illness can be traced directly to airborne building contaminants.

Explain how powered exhaust maintains building pressurization.

Explain how the powered exhaust is coordinated with the fire control panel; first to keep smoke from circulating, and then to exhaust it from the building.

Show Transparency 8 (Figure 6). Discuss the use of return fans to overcome static pressure loss in the return duct runs of larger systems.

Show Transparency 9 (Figure 7). Explain how an economizer is used to introduce outdoor air for cooling when the temperature and humidity conditions are appropriate.

Identify the three modes of economizer operation.

Provide literature on return fans and economizers for trainees to examine.

A barometric relief damper operates on the same principle. Some have a spring that determines the opening point, while most use their own weight. This device can be roof- or wall-mounted. Manufacturers also offer them as standard options on **packaged rooftop units**. The damper is usually sized to handle up to 75 percent of the ventilation air. The other 25 percent leaves the building through exfiltration and direct exhaust.

A third option for handling exhaust at the central apparatus is powered exhaust. Powered exhaust is an option on most packaged rooftop units. The powered exhaust maintains proper building pressure by exhausting slightly less air than is brought into the building through the economizer. It typically consists of a centrifugal exhaust fan located in the return air section of the unit. The fan exhausts air directly outside through a grille located in the side of the unit. A backdraft damper located near the grille prevents outdoor air from blowing backward into the return duct when the powered exhaust fan is not running.

One power exhaust control method controls the fan to maintain static pressure in the conditioned space. In another method, the powered exhaust fan is controlled by economizer position. A switch on the outdoor air damper activates when the damper opens beyond a preset point, typically 50 percent.

Powered exhaust can also be coordinated with a fire control panel. When activated by a fire marshal, signals from the fire panel can place the rooftop unit in one of two operating modes:

- *Building pressurization mode* – In this mode, the exhaust fan is shut down, the supply fan runs, and the outside air damper is opened 100 percent. This allows outside air to pressurize the zone, and prevents smoke from entering from adjacent areas.
- *Smoke exhaust mode* – In this mode, the supply fan is shut down, the outside air dampers are 100 percent open, and the exhaust fan fan runs. This causes infiltration of smoke-free air from surrounding spaces and the exhausting of smoke-laden air from the controlled space.

4.3.0 Return Fans

One problem mentioned earlier is the overpressurization of the conditioned space. This can be caused by bringing in too much ventilation air compared to that exhausted. Another cause could be an excessive resistance to airflow back through the return duct system. This may be encountered on large jobs with long return ducts; however, it is rarely encountered in jobs using packaged rooftop units or split systems below 20 tons. The return fan (*Figure 6*) is generally sized to overcome the static pressure loss of the return duct system, and to move air from the space back to the mixing chamber of the central unit.

Return fans may be found as accessories in large-capacity **packaged units** and in built-up air handling units. When used on **variable air volume (VAV) systems**, the return fan is set to maintain a fixed cubic feet per minute (cfm) differential below the amount of supply air being introduced into the conditioned space. This requires velocity measuring stations in both the supply and return ducts and can be quite expensive.

4.4.0 Economizers

An outside air economizer (*Figure 7*) is simply a means for bringing outside air into the building to provide free cooling. Using fan energy, the economizer excludes or minimizes the use of compression energy to cool the conditioned space. This can be done whenever the outside air temperature and humidity conditions are favorable.

The economizer contains dampers that increase the entrance of outdoor air as the return airflow is decreased. The dampers are usually driven by a bi-directional motor, and are placed in the central unit so that the outside air flows into the unit upstream of the evaporator coil. This arrangement allows the negative fan inlet pressure to induce the inward flow of outside air.

There are three types of economizers:

- Dry-bulb economizer
- Enthalpy economizer
- Integrated economizer

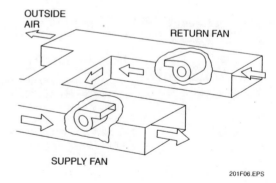

Figure 6 ◆ Return fan.

1.8 HVAC ◆ LEVEL TWO

Instructor's Notes:

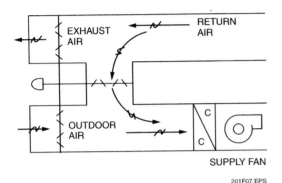

Figure 7 ♦ Economizer.

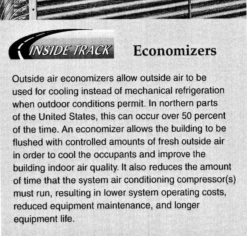

Outside air economizers allow outside air to be used for cooling instead of mechanical refrigeration when outdoor conditions permit. In northern parts of the United States, this can occur over 50 percent of the time. An economizer allows the building to be flushed with controlled amounts of fresh outside air in order to cool the occupants and improve the building indoor air quality. It also reduces the amount of time that the system air conditioning compressor(s) must run, resulting in lower system operating costs, reduced equipment maintenance, and longer equipment life.

Explain the operation of the dry-bulb economizer. Point out that a system that operates with outdoor temperatures below 55°F is a candidate for a dry-bulb economizer.

Explain how an enthalpy economizer permits the use of outdoor air at higher temperatures than the dry-bulb economizer.

Discuss the integrated economizer and point out the differences between it and the other two types.

Manufacturers offer all three types of economizers as accessories on packaged equipment and built-up air handlers.

4.4.1 Dry-Bulb Economizer

The air temperature normally supplied by the central apparatus to cool the conditioned space is between 55°F and 60°F. Therefore, when the outside air is 55°F or less, it can be used directly as the supply air to the control zone. Using outdoor air for cooling comes essentially free and eliminates the need to run the compression equipment. Any location with a significant number of operating hours below 55°F is a good candidate for a dry-bulb economizer.

4.4.2 Enthalpy Economizer

The enthalpy economizer measures both dry-bulb and wet-bulb temperatures. To understand this economizer we must first define the word enthalpy. Enthalpy measures total heat; in other words, it measures the heat contained in the outdoor air due not only to its temperature, but also to its moisture content. Moist, cool air could have the same heat content as dry, warm air.

The maximum allowable room relative humidity (RH) for comfort purposes is 60 percent. With a little psychrometric work, this 60 percent RH can be translated into a maximum outdoor air heat content of 24.7 Btuh/lb and a maximum allowable outdoor air temperature of 60°F. The enthalpy economizer can be operated without the compression equipment during all hours when the outside air is between 55°F and 60°F. The dry-bulb economizer could not operate at outdoor air temperatures above 55°F. This higher temperature range (55°F to 60°F) can occur frequently when the system is installed in a temperate climate, allowing much greater energy savings.

4.4.3 Integrated Economizer

The integrated economizer is more sophisticated than the other types. With the enthalpy economizer, outside air cannot be used for cooling when its temperature is above 60°F. With an integrated economizer, however, if the air is dry enough, it can be used to reduce the amount of cooling and dehumidification required by the compression equipment. That is, if the outdoor air enthalpy (outdoor total heat and humidity) is lower than the return air enthalpy, then the outdoor air can be used for cooling.

Explain that the integrated economizer design allows the system to use a combination of compression cooling and outdoor air.

Have the trainees review Sections 5.0.0–5.5.2.

Ensure that you have everything required for teaching this session.

Identify the five types of all-air systems.

Show Transparency 10 (Figure 8). Describe a single-zone constant volume system. Explain that these systems can be used as whole-building systems or in multi-zone applications.

Explain how the thermostat is used to control the central unit. Discuss occupied cooling and occupied heating.

The integrated economizer is a compression-assisted economizer. When the total cooling capacity of the outdoor air is not adequate to control the space-cooling load, the compression equipment makes up the difference. The integrated economizer can be used during all those operating hours when the outside air temperature is above 60°F. This can represent a significant number of hours, depending on the location of the building.

4.5.0 Demand-Controlled Ventilation

Traditionally, the amount of ventilation air required for a building, as defined by *ASHRAE Standard 62*, was based on the peak design occupancy rate for a building. This ventilation rate was maintained by the central air source equipment at a constant cfm as long as the building was occupied. With recent advancements in carbon dioxide sensing equipment and direct digital controls, this traditional method of providing a constant cfm of ventilation air to a building has changed.

Because all humans exhale carbon dioxide (CO_2) at a predictable rate, the CO_2 level in a space can be used as a reliable indicator of the number of people in a conditioned space. Demand-controlled ventilation (DCV) is an approved *ASHRAE 62* building control strategy that uses CO_2 sensors and direct digital controls to match a building's ventilation rate to a building's actual occupancy pattern.

The DCV strategy is achieved by mounting digital controls on the HVAC system's central air source and air terminal equipment. Each building control zone is provided with temperature and CO_2 sensors. Using direct digital control communication capability, space CO_2 levels, and thus the number of people in the building, are monitored and transmitted back to the central air source's controller. The central air source then compares outside air CO_2 levels to space CO_2 levels and adjusts the building ventilation air quantities to meet space needs.

5.0.0 ♦ TYPES OF ALL-AIR SYSTEMS

There are five basic types of all-air systems in use today. Each has been created to satisfy a building's zoning needs at a price affordable for the type of building involved. The five types are as follows:

- Single-zone constant volume
- Multi-zone constant volume
- Variable volume, variable temperature (VVT)
- Variable air volume (VAV)
- Dual-duct variable air volume

5.1.0 Single-Zone Constant Volume

Figure 8 shows a single-zone **constant-volume system**. Each system serves a single control zone and consists of six major components. The zone is equipped with a supply air diffuser(s) (1), and a return grille (2). A central unit (3) provides cooling and heating to the zone through supply air ductwork (4). Air is returned to the central unit through return ductwork (5). A single thermostat (6) located in each zone controls the central unit's cooling or heating capacity. The central unit supplies a constant volume of air to its assigned control zone whenever the thermostat calls for heating or cooling.

Two types of equipment are typically used in single-zone constant-volume systems: rooftop units and vertical packaged units. Both types of units are equipped to cool or heat the single control zone.

5.1.1 Control Sequence

A mechanical thermostat starts or stops the central unit when the zone temperature is above the cooling setpoint or below the heating setpoint. An alternative is to use a programmable thermostat with a time clock and start the central unit based on time. The zone occupant can then define occupied and unoccupied times for each day of the week, weekend days, and holidays.

Occupied cooling – When signaled to start by the thermostat, the central unit starts the supply fan and opens the outside air dampers to the minimum ventilation position. With a programmable thermostat, the fan typically remains on as long as the thermostat determines the zone is occupied. The fan provides a constant volume of air to the space. If the space temperature rises above the zone cooling setpoint, the central unit activates the economizer (if used) as the first stage of cooling. If the zone temperature continues to rise, capacity control is provided by cycling the compressor(s) on and off to maintain the zone setpoint. Larger central units (above 7.5 tons) are typically provided with compressor capacity unloaders to increase the number of cooling capacity stages that are available.

Occupied heating – Heating sources for rooftop units include electric resistance, natural gas, liquid propane gas, and hot water. Vertical packaged products use electric resistance heaters, hot water, or steam. The heating device is cycled as needed when the zone temperature falls below the heating setpoint. Hot water or steam coils usually provide capacity control with a two-way or three-way control valve. Electric resistance

Instructor's Notes:

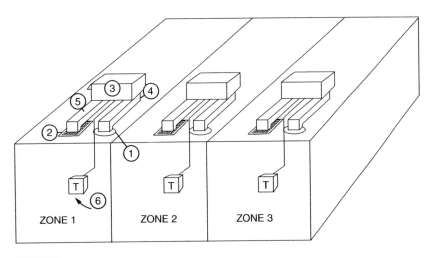

LEGEND
1. SUPPLY AIR DIFFUSER
2. RETURN GRILLE
3. CENTRAL UNIT
4. SUPPLY AIR DUCT
5. RETURN AIR DUCT
6. THERMOSTAT

Figure 8 ♦ Single-zone constant-volume system.

Discuss the unoccupied period, demand-controlled ventilation, and simultaneous heating and cooling.

Describe typical applications for a single-zone constant volume system.

Show Transparency 11 (Figure 9). Explain how several single-zone packaged units are used in a multiple-zone system.

heat is usually stepped by the thermostat in 5kW increments. Gas heaters are typically staged between low and high burners.

Unoccupied period – When the thermostat determines that the space is unoccupied, the compressors and supply air fan stop, and the outside air damper is closed. If the thermostat is equipped with setback temperatures, the central unit operates just like the occupied period, except that the outside air damper remains closed.

If the central unit is equipped with direct digital controls, the unit may operate in a nighttime free-cooling mode. In this mode, the central unit flushes the building with 100 percent outdoor air between 2 a.m. and 7 a.m. to pre-cool the building prior to occupancy.

Demand-controlled ventilation – If each central unit is equipped with direct digital controls and a CO_2 sensor is provided in each space, each central unit's outdoor air damper may be varied to match the building's ventilation rate to the corresponding space occupancy pattern.

Simultaneous cooling and heating – Because the central unit serves a single zone, there is never a need for the single zone system to provide both cooling and heating at the same time.

5.1.2 Typical Applications

For small buildings (less than 5 tons) it is normal to see only one single-zone unit providing cooling and heating for the building. The building may have multiple rooms, but one room will have the thermostat and control the unit operation. Buildings of this size are very cost sensitive and typically not able to bear the added cost of multiple control zones.

An approach to improving the zoning of buildings, however, is to use multiple single-zone packaged units. The layout in *Figure 9* shows a building with two zones. Each zone has a single-zone constant-volume system (1) serving it. The system consists of the unit itself, supply ductwork (3), supply diffusers (4), return grille (5), and a return duct (6) open to the plenum area above the ceiling. Because each zone has its own thermostat (2), the overall building comfort level has been increased. This design is used extensively in buildings up to 40 tons in size with multiple 3.5-, 5-, 7.5-, or 10-ton packaged units.

Explain how a system is designed to handle commercial building requirements for simultaneous cooling and heating.

Show Transparency 12 (Figure 10). Explain how multiple single-zone units can be used in shopping malls. Point out that the modular approach with multiple packaged units is common in applications of up to 40 tons.

Show Transparency 13 (Figure 11). Describe a multi-zone constant volume system. Compare multi-zone constant volume systems with the modular approach using multiple packaged systems.

Explain that dampers are used to control the airflow to the individual zones.

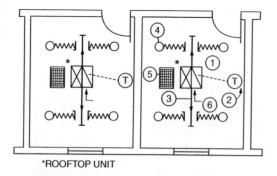

*ROOFTOP UNIT

LEGEND
1. SINGLE-ZONE CONSTANT VOLUME UNIT
2. THERMOSTAT
3. SUPPLY DUCT
4. SUPPLY DIFFUSERS
5. RETURN GRILLE
6. RETURN DUCT

Figure 9 ◆ Single-zone system layout.

Commercial buildings have a tendency to require simultaneous cooling and heating. For example, in the morning during the fall, the east face might require cooling, while the west face requires heating. That same day in the afternoon, with the western sun pouring through the west glass, the west zone would need cooling. The east zone, now in the shade, would require heating. The building's simultaneous cooling and heating needs can be satisfied with two separate single-zone systems.

Shopping malls are excellent examples of the use of multiple single-zone units (*Figure 10*). Each small store has its own unit and is responsible for its own associated utility bill. Larger stores are provided with multiple units sized to match the mall's typical store size. If the larger store leaves, the mall can easily break the store space back down into smaller rental stores with minimal startup costs. Likewise, the mall can easily combine multiple small store systems into one larger store with minimal startup costs.

5.2.0 Multi-Zone Constant Volume

The layout in *Figure 11* shows a double-deck multi-zone constant-volume system. The system consists of a factory-assembled, blow-through, central air handler or rooftop packaged type unit (1). The central unit is equipped with a supply fan (2) that blows air in parallel through a cooling coil (3), and through a heating coil (4) into a discharge damper section with multiple outlet plenums (5). The central unit may have up to 12 individual outlet plenums.

Each outlet plenum is equipped with a set of mixing dampers (6), and provides a constant volume of supply air through a supply duct (10), then through a supply diffuser (7), and on into an individual control zone. The air is returned from each control zone through a return air grille (8), and flows back to the central unit. At the central unit, an economizer (9) brings fresh ventilation air into the supply airstream while stale return air is exhausted. A thermostat (11) located in each control zone modulates a set of dampers (6) located in the discharge plenum (5) to provide cooling, or activates a heater located in the zone's supply duct (12) to supply heating to the zone.

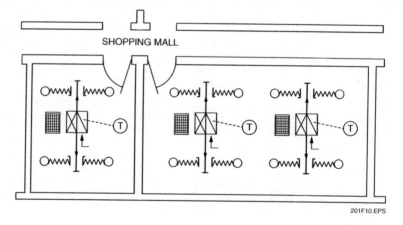

Figure 10 ◆ Modular approach to zoning.

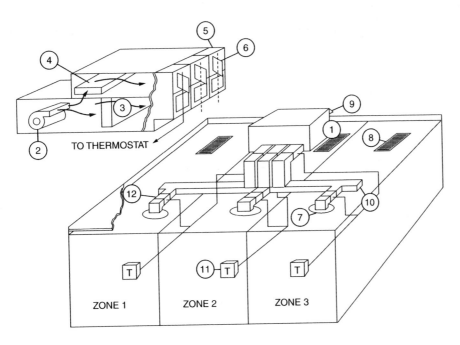

LEGEND
1. ROOFTOP PACKAGED UNIT
2. SUPPLY FAN
3. COOLING COIL
4. HEATING COIL
5. DISCHARGE DAMPER WITH MULTIPLE PLENUMS
6. MIXING DAMPER(S)
7. SUPPLY DIFFUSER
8. RETURN AIR GRILLE
9. ECONOMIZER
10. SUPPLY DUCT
11. THERMOSTAT
12. ZONE HEATER

201F11.EPS

Figure 11 ♦ Multi-zone constant-volume system.

Identify the two types of equipment used in multi-zone constant volume systems.

Explain the control sequence for a multi-zone constant volume system in the occupied cooling mode.

Two types of equipment are typically used in multi-zone constant-volume systems. They are rooftop multi-zone units and indoor multi-zone air handling units. Both types of units are equipped with blow-through fans and multiple sets of discharge dampers to cool or heat multiple control zones.

5.2.1 Control Sequence

An external time clock or energy-management system starts or stops the unit though the unit's external start contacts. As an alternative, the unit's electronic control panel may contain its own programmable time clock. The external or internal time clock can be programmed to define occupied and unoccupied times for each day of the week, weekend days, and holidays. If this is a multi-zone air handler, the associated condensing unit or water chiller (cooling source) is also started at this time.

Occupied cooling – When signaled to start by the thermostat, the central unit starts the multi-zone unit's supply fan and opens the outside air dampers to minimum ventilation position. The supply fan typically remains on as long as the space is occupied. The fan provides a constant volume (in cfm) of air to the control zones. The air passes in parallel through the unit's heating and cooling coils.

If the space temperature in any control zone rises above the zone cooling setpoint, the thermostat modulates its set of mixing dampers toward the wide-open position. When one of the cold-deck dampers is fully open, the central unit activates the economizer (if used) as the first stage of cooling.

If the zone temperature continues to rise, compression cooling capacity is added. If this is a multi-zone air handler, the associated chiller or condensing unit provides more cooling to the multi-zone unit's cooling coil. If this is a pack-

Explain the control sequence for a multi-zone constant volume system in the occupied heating, unoccupied period, demand-controlled ventilation, and simultaneous cooling and heating modes.

Explain the night-time free-cooling mode in which the building is flushed with 100 percent outdoor air during the night.

Provide some examples when simultaneous heating and cooling might be required. Explain that multi-zone constant volume systems are preferred for buildings with large, permanent control zones, such as schools.

Define reheat and explain why it is undesirable in constant volume systems with multiple zones.

Show Transparency 14 (Figure 12). Describe the layout of a multi-zone system.

aged rooftop multi-zone unit, the first stage of compression is activated. As more zone dampers call for cooling, more compression capacity stages are activated and staged to meet the cooling demand. In either case, cold air leaving the cooling coil is mixed by each zone's damper with untreated air that has bypassed through the inactive heating coil to maintain the zone thermostat cooling setpoint.

Occupied heating – The original intent of the two-deck multi-zone design was to use the unit's central heat source to provide zone heat. In this mode of operation, the zone thermostat modulates the zone damper, mixing hot and cold air streams to achieve the zone thermostat heating setpoint. This is using reheat. Reheat is an expensive way to operate and is not allowed by *ASHRAE Standard 90.1* due to the high energy consumption involved. Central multi-zone unit heat sources include electric resistance, gas, hot water, or steam coils.

To eliminate reheat with a two-deck multi-zone unit, heating for each zone is accomplished with a heater located in the supply air duct leading to each control zone. This prevents the mixing of cold and hot air streams to maintain the zone setpoint. The duct heater is typically an electric or hot water coil. The duct heater is cycled as needed when the zone temperature falls below the heating setpoint and the zone's heating damper is in the fully open position. The outside air dampers are set to the minimum ventilation position. In this arrangement, the central multi-zone unit's heat source is used for morning warm-up purposes. Once the building is occupied, the duct heaters are used and central heat locked out.

Unoccupied period – When the thermostat determines the zone is unoccupied, the cooling source and supply air fan stop, and the outside air damper is closed. If the thermostat is equipped with setback temperatures, the central unit operates just like the occupied period except that the outside air damper remains closed.

If the central unit is equipped with electronic controls, the unit may operate in a nighttime free cooling mode. In this mode, the central unit flushes the building with 100 percent outdoor air between 2 a.m. and 7 a.m. to pre-cool the building prior to occupancy.

Demand-controlled ventilation (DCV) – The DCV control strategy may be employed if the central air handling unit is equipped with direct digital controls and the ability to measure outdoor air intake velocity, and a CO_2 sensor is provided in each zone. By monitoring each zone's CO_2 level, the direct digital controls can calculate the required outdoor air cfm and modulate the central unit's outdoor air damper to match the building's occupancy pattern.

Simultaneous cooling and heating – Because the central unit serves multiple zones, and each zone has its own thermostat, the multi-zone system can provide both heating and cooling to a building at the same time. For those zones requiring cooling, the thermostat will modulate the individual zone damper and provide cooling capacity to the zone. For those zones requiring heating, the thermostat will position the zone damper to the wide open heating position (closing off cooling), and stage the duct heater to provide heating capacity to the zone.

5.2.2 Typical Applications

As we have seen with packaged equipment, one way to provide multiple control zones is to provide a single packaged unit for each zone. There are buildings, however, in which this approach does not work well. Providing multiple packaged units requires multiple roof penetrations and multiple individual power connections, and is not the best way to air condition such a building. For buildings like this, with large permanent control zones, the multi-zone unit is well suited.

Multi-zone systems (*Figure 12*) may be found in a variety of buildings where the control zones are permanent and large. Permanent zones are favored because it is very expensive to change control zones once a multi-zone system is installed. Typical applications include schools, libraries, museums, banks, laboratories, medical

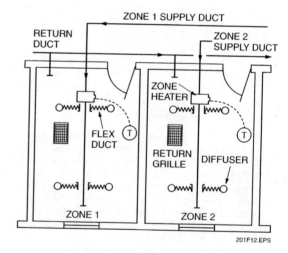

Figure 12 ♦ Multi-zone system layout.

clinics, radio and television studios, department stores, and industrial facilities. Schools with permanent classrooms and the need for individual classroom temperature control have traditionally been a major user of multi-zone systems.

Furthermore, multi-zone systems are best used with buildings requiring a somewhat limited number of large zones (4 to 12) that require a supply air quantity greater than 500 to 1,000 cfm per zone. Thus, buildings that range in capacity from 10 to 40 tons in size, or buildings that can be broken down into 10- to 40-ton divisions, would be likely candidates for this system.

5.3.0 Variable Volume, Variable Temperature (VVT)

Figure 13 shows a **variable volume, variable temperature (VVT) system**. A single-zone constant-volume packaged unit (1) equipped with an air source control module provides conditioned air through a supply duct to VVT air terminals (2). Each VVT terminal is equipped with a zone control module and supplies air to a single control zone. Each zone contains a space sensor (3) that is wired back to an appropriate zone controller. The zone controller monitors zone temperature and modulates an air terminal damper to maintain the zone temperature setpoint. After absorbing room loads, the circulated air leaves the zone through a return air grille (4), and returns to the central packaged unit. A heating coil is located in the central unit to provide heating to the zones. A bypass damper equipped with a bypass controller (5) maintains constant airflow across the packaged unit DX cooling coil when the air terminals throttle air to the zones. A user interface module (sometimes referred to as a system pilot), which also contains a space sensor (6), allows the operator to interface with the control system. Information is exchanged between the zones and central packaged unit through a communication bus (dashed line) that runs between the user interface, zone controllers, bypass controller, and packaged unit control module.

Older versions of this system used zone thermostats instead of zone sensors and did not have a user interface module. In this system the communication bus ran between zone thermostats, a bypass controller, and a relay pack. A master thermostat provided the functions of the user interface. The relay pack provided communication between the zone controllers and the central packaged unit.

Identify applications for a multi-zone system.

Show Transparency 15 (Figure 13). Describe a variable volume, variable temperature (VVT) system and compare it to a multi-zone constant volume system.

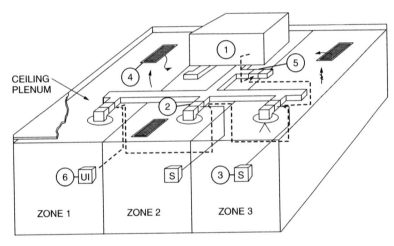

LEGEND
1. SINGLE-ZONE CONSTANT VOLUME PACKAGED UNIT WITH CONTROLLER
2. VVT AIR TERMINAL WITH ZONE CONTROLLER
3. SENSOR(S)
4. RETURN GRILLE
5. BYPASS DAMPER WITH BYPASS CONTROLLER
6. USER INTERFACE (UI)

Figure 13 ♦ VVT system.

Identify equipment commonly used in VVT systems.

Show Transparency 16 (Figure 14). Explain that a bypass damper is used to maintain a constant air volume.

Show Transparency 17 (Figure 15). Discuss common VVT applications. Explain that although 20 tons is generally viewed as the capacity limit for VVT, a larger building can be divided into 20-ton blocks.

Three types of constant-volume central equipment are typically used in VVT systems. They are single-zone rooftop units, vertical packaged units, and split systems. All unit types are equipped to cool or heat the control zones with an airstream provided from the central air source. In addition, air terminals equipped with modulating dampers and zone controllers are required for each zone.

Each VVT system requires a bypass system. As shown in *Figure 14*, the bypass system consists of a bypass damper, a static pressure sensor, a bypass controller, and a backdraft damper. In a typical job, the bypass controller is mounted right on the bypass damper.

5.3.1 Typical Applications

For small buildings from 5 tons to 20 tons that have small control zones, or a mixture of small and large control zones, VVT (*Figure 15*) is one of the most popular ways to achieve comfort control. In fact, larger capacity buildings can be broken down into 20-ton blocks served by multiple VVT systems.

Typical applications include offices, banks, schools, and medical clinics, all of which are typically in the 20 tons or smaller capacity range and have some small control zones (less than 500 cfm). VVT air terminals range in capacity from 200 to 1,400 cfm (0.5 to 5.5 tons), and can cover a wide range of zone sizes.

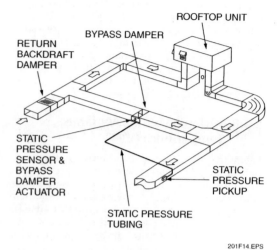

Figure 14 ◆ VVT system bypass.

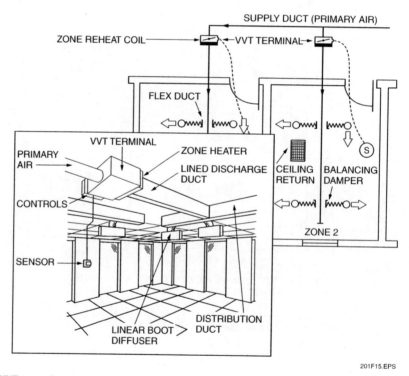

Figure 15 ◆ VVT system layout.

1.16 HVAC ◆ LEVEL TWO

Instructor's Notes:

Unit Design Variations

In order to eliminate the need for reheat, two popular variations of the traditional two-deck multi-zone unit have become available from manufacturers. One variation is the use of a packaged unit with a major change in the way the cooling and heating coil is arranged internally. In this design there is not just one cooling and heating coil, there is one cooling and heating coil for each control zone served by the unit. Thus, when any zone needs cooling or heating, the appropriate coil is activated and neutral air passes through the other coil. The zone thermostat is always mixing neutral air with cold or hot air to maintain zone temperature.

The second arrangement involves the use of three sets of dampers instead of two. In this arrangement, during cooling, the hot deck damper is closed while the thermostat mixes neutral air with cold air to maintain zone temperature. During heating, the cold damper is closed, and the thermostat mixes neutral air with hot air to maintain zone temperature.

5.4.0 Variable Air Volume (VAV)

Figure 16 shows a variable air volume (VAV) system. A central air source unit (1) equipped with an air source control module provides conditioned air through a supply duct (2) to VAV air terminals (3). Each VAV terminal is equipped with a zone control module and supplies air to a single control zone. Each zone contains a space sensor (4) that is wired back to an appropriate zone controller. The zone controller monitors zone temperature and modulates an air terminal damper to maintain the zone temperature setpoint. After absorbing room loads, the circulated air leaves the zone through a return air grille (5) and returns to the central packaged unit. A communication bus (dashed line) runs between the zone controllers and the air source control module. One zone controller (the master controller) polls all other zones and regularly transmits zone needs to the central air source control module. Based on zone needs, the central air source controller chooses the operating mode of the system and transmits the mode to the zone controllers.

Three types of central equipment are typically used in VAV systems. They are VAV rooftop units, central station VAV air handling units, and VAV vertical packaged units. Regardless of the

Show Transparency 18 (Figure 16). Discuss variable air volume (VAV) systems. Explain the significant differences between it and the VVT system, such as the use of modular air terminals and inlet guide vanes or variable speed drives.

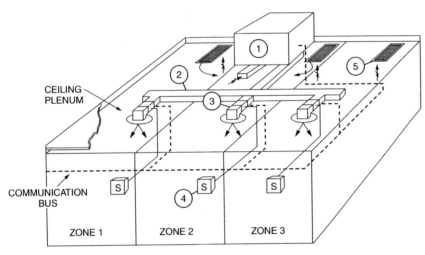

LEGEND
1. CENTRAL AIR SOURCE WITH AIR SOURCE CONTROLLER
2. SUPPLY DUCT
3. VARIABLE AIR VOLUME (VAV) TERMINALS WITH ZONE CONTROLLERS
4. SENSOR
5. RETURN AIR GRILLE

Figure 16 ♦ Variable air volume system.

Discuss VAV applications. Explain that VAV is used in buildings with demands greater than 20 tons.

Explain that large buildings generally have core areas with year-round cooling loads, even in heating season. Perimeter areas require cooling in summer and heating in winter.

Explain that in multi-story buildings, it is common to have one or two VAV air handlers per floor.

Show Transparency 19 (Figure 17). Discuss the differences between a VAV system and a dual-duct VAV system.

type of central equipment used, the indoor supply air fan must be equipped with some form of airflow modulation to track airflow changes caused by the VAV air terminals. Airflow modulation is typically achieved with inlet guide vanes or variable speed drives.

Several types of VAV air terminals are also available. Depending on the building cooling and heating needs, the VAV system may use VAV single duct boxes, series fan-powered mixing boxes, and parallel fan-powered mixing boxes. All are typically mounted in the ceiling as shown and require a low-pressure downstream duct system with diffusers.

5.4.1 Typical Applications

VAV systems are used in building sizes starting at 20 tons. As buildings tend to become larger, they also tend to have internal core areas with cooling loads based on people, lights, and equipment. Such cooling loads occur year-round and are not tied to outside climate conditions. Thus, core areas always require cooling, while the perimeter areas of the building require cooling in the summer and heating during other seasons. The need for simultaneous cooling and heating makes VAV an ideal choice for buildings with internal core loads.

In summary, VAV systems are applicable to perimeter or interior spaces in offices, banks, medical clinics, libraries, schools, hotels, motels, and public buildings. VAV systems are not used with central equipment below 20 tons in size. Thus, VAV tends to be used in larger multi-story buildings. In multi-story buildings it is common to have one or two VAV air handlers per floor. VAV air terminals come in sizes from 200 to 3,000 cfm (0.5 to 7.5 tons) and are used in small or large control zones. Recent trends in the market are toward using VAV rooftop equipment in sizes up to 100 tons per unit. Units of this size serve up to four-story buildings. For buildings larger than this, central VAV air handlers with air-cooled or water-cooled chillers are used.

5.5.0 Dual-Duct VAV System

Figure 17 shows a dual-duct variable air volume system. Each **dual-duct system** serves multiple control zones. Each zone is equipped with a supply air diffuser(s) (1) and a return grille (2). A central blow-through double-duct unit (3) provides all control zones with cold air through a cold air duct (4) and neutral bypassed air through a neutral air duct (5). A single sensor (6) located in each zone controls the cold air damper of the zone's

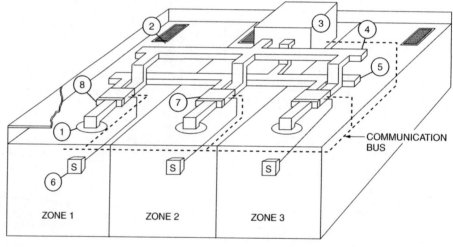

LEGEND
1. SUPPLY AIR DIFFUSER
2. RETURN GRILLE
3. CENTRAL DOUBLE DUCT UNIT WITH AIR SOURCE CONTROLLER
4. COLD AIR DUCT
5. NEUTRAL AIR DUCT
6. SENSOR
7. DUAL-DUCT AIR TERMINAL WITH ZONE CONTROLLER
8. ZONE HEATER

Figure 17 ◆ Dual-duct VAV system.

Instructor's Notes:

dual-duct air terminal (7) for cooling. A zone heater (8) located in the zone or on the air terminal is used for heating. A communication bus connects all zone controllers to the central unit's air source controller. Air is returned to the central unit through a return duct connected to the central unit. The central unit supplies varying quantities of cold and neutral air to each of its assigned control zones.

The central air source in a dual-duct VAV system is a built-up blow-through double-duct unit. No packaged equipment is available to serve the central unit function. Each zone is served by a special dual-duct VAV air terminal. A zone heater for each zone is mounted on the air terminal or located directly in the zone.

5.5.1 Typical Applications

As described for multi-zone equipment, one way to provide multiple control zones is to provide a single multi-zone central unit with a single duct serving each control zone. The system worked well when serving up to about 12 large zones. In the past, a modified version of the multi-zone system was developed for larger multi-story buildings that had many small zones. It was called the double-duct system. In this system, the mixing dampers were removed from the central unit, and two ducts—a cold duct and a warm or neutral duct—were run around the building. At each zone, a double-duct air terminal was added to mix the cold and neutral air streams to achieve zone temperature control.

The double-duct system allowed the expanded use of the multi-zone system principle with many small zones. This double-duct system circulated a constant volume of air all the time, just like the multi-zone system. However, in larger buildings, this approach resulted in a significant amount of fan energy use.

The dual-duct VAV system came about as a means of improving the fan energy consumption of existing double-duct systems. Old double-duct systems were retrofitted with air terminals, providing variable damper control. Rather than be a constant-volume system, the dual-duct VAV system provides fan energy savings by operating like a VAV system during cooling. You are not likely to see any new construction using a dual-duct VAV system. Rather, you may come upon this system in older multi-story buildings that have been modified to improve energy use. The use of direct digital controls (DDC) allows for the control sequence just described, or other variations of it.

6.0.0 ♦ DUCT SYSTEMS

The purpose of a duct system is to transmit air from the central air source to the air diffusers located in the building control zones. This section describes the common components of a duct system, the classification of duct systems, and the common layout and construction techniques. With this information you should be able to look at the duct system in a commercial building and quickly recognize potential causes of poor delivery of air to the building control zones.

6.1.0 Classification

In the past, duct systems were classified in terms of their application, velocity, and pressure. These older terms were rather vague and have been replaced by pressure classification values. *Table 1* shows a numerical pressure classification table according to Sheet Metal and Air Conditioning Contractor's National Association, Inc. (SMACNA) standards. The table shows that static pressure classes from ½ inch to 3 inches can be designated as either (+) positive or (–) negative. All static pressure classes at or above 4 inches are for positive pressure systems.

Static pressure classifications are much more useful than the older classification system that used terms such as high velocity or low pressure. The static pressure class levels may be used directly to establish appropriate duct construction materials, metal gauge, duct dimensions, and duct reinforcement.

If a duct drawing refers to a SMACNA pressure class of +3, according to the table, the duct static pressure is greater than +2 in wg below or equal to +3 in wg, and has a maximum allowable air velocity of 2,500 feet per minute (fpm).

From an HVAC system design point of view, however, some guidelines should be followed with regard to maximum duct velocities to avoid unwanted noise in various applications.

Table 1 Pressure-Velocity Duct Classification

Static Pressure Class (in wg)	Pressure Range (in wg)	Maximum Velocity (fpm)
±0.5	0 to 0.5	2,000
±1	>0.5 to 1	2,500
±2	>1 to 2	2,500
±3	>2 to 3	4,000
±4	>3 to 4	4,000
+6	>4 to 6	*
+10	>6 to 10	*

201T01.EPS

Explain that the dual-duct system requires a special built-up unit, and cannot be served by a standard packaged unit.

Describe the double-duct system.

Provide a list of several commercial-type buildings. Show trainees how to identify the type of airside system(s) commonly used in each application, and explain why.

Have trainees practice identifying the type of airside system(s) on the list provided, and have them describe the reason why each is used. Note the proficiency of each trainee. This laboratory corresponds to Performance Task 2.

Have the trainees review Sections 6.0.0–8.4.0.

Ensure that you have everything required for teaching this session.

Discuss types of duct systems.

Show Transparency 20 (Table 1). Explain how commercial duct systems are classified by velocity and pressure.

Identify guidelines for determining the maximum airflow velocity for use in selected applications.

Show Transparency 21 (Figure 18). Describe the three different types of duct layouts. Explain that the H pattern is the most common.

Show Transparency 22 (Figure 19). Explain that trunk ducts are usually installed above corridors to minimize noise in occupied spaces.

Discuss the use of flexible duct. Identify its limitations, such as using straight runs, and limiting run lengths to 6 feet.

The following are some guidelines for determining the maximum airflow velocity to use in selected applications of low-velocity systems:

- *Residences* – 600 fpm
- *Theaters, churches, auditoriums* – 800 fpm
- *Apartments, hotels rooms* – 1,000 fpm
- *Offices, libraries* – 1,200 fpm
- *Stores, restaurants, banks* – 1,500 fpm
- *Cafeterias* – 1,800 fpm

6.2.0 Duct Layout

All duct systems are designed around one simple principle: deliver the air from the central air handling unit to the air terminals in the simplest direct route possible while following the line of maximum duct clearance. Following this principle usually results in one of three basic symmetrical layouts for the main (trunk) ducts (*Figure 18*):

- The spine
- The H pattern
- The loop or doughnut

The H pattern is the most popular method and is seen in most buildings. The spine method is likely to be found in narrow buildings with a width of 50 to 75 feet. The loop layout is found in larger multi-story buildings. The loop is a modification of the H pattern and results in smooth, even changes in duct static pressure at the supply fan, particularly on VAV type systems.

The main (trunk) ducts are usually located above corridors in the cavity above the ceiling to minimize noise transmission to the conditioned zones and allow easy access without disturbing the building occupants. Ducts that connect the trunk ducts to the air diffusers are called runout ducts (*Figure 19*). Ducts that connect air terminals to diffusers are called header ducts. Runout/header ducts extend outward from the trunk ducts for distances up to 25 feet. Thus, a trunk duct with runout/header ducts on either side will cover a 50-foot path through the building.

For constant-volume systems, the runout/header ducts are made of sheet metal or fiberglass and come within 6 feet of the air diffusers. The final 6 feet are usually made of flexible material to facilitate alignment and to minimize noise. Good duct designs will not have any more than 6 feet of flexible duct, and the flexible duct will be as straight as possible. A good duct design will never have the flexible duct coiled, bent, or collapsed. These bad practices lead to excessive pressure drop and the inability to deliver sufficient air to the control zone. For VVT, VAV, and dual-duct type systems, an air terminal with a variable damper (as shown in *Figure 19*) is located between the trunk duct and the header duct. In all systems, manual balancing dampers are located just before the flexible duct connections to the air diffusers.

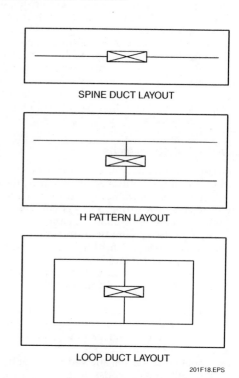

Figure 18 ◆ Main (trunk) duct layouts.

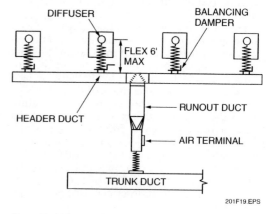

Figure 19 ◆ Runout duct layouts.

Providing a Smooth Exit for Supply Air

Good duct design provides a smooth transition from the central air handler fan discharge to the initial section of the main duct. Failure to provide a smooth transition will result in excessive loss of fan pressure and lack of air at the control zones. This is where a lot of jobs run into trouble, particularly with small packaged units that have limited fan pressure capability to begin with.

Trying to achieve this objective with rooftop equipment may present problems on some jobs. The rooftop unit discharge may have a long dimension that will not fit within the roof joist spacing (see diagram). A good job will turn and locate the rooftop unit in such a way as to allow the supply duct to leave the unit, run down between the roof bar joists, and then transition through a long-radius elbow into the ceiling cavity. After making the turn, the supply duct should then transition smoothly to the initial section of the main duct.

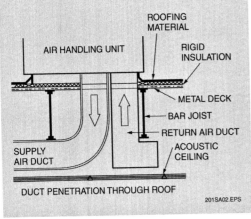

6.3.0 Duct Condensation

Ducts may sweat when their surface temperature is below the dew point of the surrounding air. This is the same phenomenon that occurs when moisture condenses on a glass filled with a cold beverage in the summer. Condensation can lead to water damage and microbial growth within the building. To prevent this from happening, rectangular supply ducts are typically insulated on the inside. Round and oval ducts are insulated on the outside due to the cost and difficulty involved in providing inside insulation.

6.4.0 Duct Materials

A wide variety of materials are used in HVAC system duct designs. *Table 2* shows a few of the common duct materials and their applications. As the table shows, the most common material used is galvanized steel. The other materials are used for more specific applications such as kitchen exhausts, moist air, and so on. Airshafts, typically found in multi-story buildings, are usually made of concrete or gypsum board.

6.5.0 Duct Construction

Ducts come in round, flat oval, and rectangular cross sectional shapes (*Figure 20*). Because the ducts must usually fit in a building's ceiling cavity or between or under roof joists, space is a premium.

Table 2 Duct Materials and Their Applications

Material	Applications
Galvanized steel	Widely used for most HVAC systems
Aluminum	For systems with high moisture-laden air, or special exhaust systems
Stainless steel	For kitchen exhaust, fume exhaust, or high moisture-laden air
Concrete	Underground ducts and air shafts
Rigid fibrous glass	Interior, low-pressure HVAC systems
Gypsum board	Ceiling plenums, corridor ducts, and air shafts

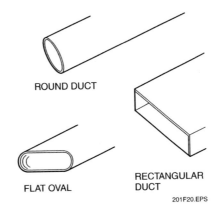

Figure 20 ♦ Duct cross-section shape.

Discuss the effects of duct condensation. Explain that ducts are insulated to prevent condensation (rectangular ducts insulated on the inside, round and oval ducts insulated on the outside).

Show Transparency 23 (Table 2). Discuss different duct materials and their applications.

Show Transparency 24 (Figure 20). Identify common duct shapes.

Provide a variety of ducts in different shapes and materials for trainees to examine.

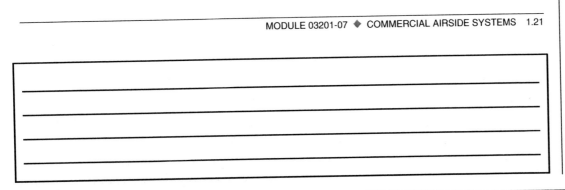

Discuss common applications for different duct shapes.

Show Transparencies 25 and 26 (Figures 21 and 22). Explain the different methods for seaming and joining ductwork.

Provide various examples of duct seams for trainees to examine.

Describe air terminals and identify the functions controlled by air terminals.

When the space allows, round metal duct is preferred. Round metal duct is strong and rigid, offers the least frictional pressure loss, and is economical to install. It also is best suited to withstand the physical abuse caused by handling during installation. The flat, oval metal duct is the next most efficient type of duct, fitting in areas where round metal duct cannot. Rectangular metal duct is the most flexible when it comes to fitting within limited spaces. It is used in many duct systems with pressures under 2.0 in wg, particularly when the installing contractor has a sheet metal shop.

In systems with pressures less than 2.0 in wg, it is common to find fiberglass ducts instead of metal. Fiberglass ducts consist of a rigid fiber board and a reinforced aluminum applied facing. Fiberglass ducts are available in both round and rectangular construction.

In medium- and high-pressure systems (above 4.0 in wg) it is common to find round and flat, oval ducts. These ducts are easier to seal and prevent leaks than rectangular ones. Leaks with medium- and high-pressure systems can create considerable noise. Sealing materials include gaskets, pressure-sensitive tapes, embedded fabrics, mastics, and liquids.

Depending on the duct pressure and type, joints and seams are made in different ways. *Figure 21* shows the various types of slip joints made for a low-pressure rectangular duct system. This is the type of seam you are likely to see in runout ducts because runout ducts are low-pressure ducts. Noise caused by leakage in this type of duct is minimal due to the low pressure involved.

When round duct is used, the joints and seams are typically joined with sheet metal screws and then sealed as shown in *Figure 22*. They are usually covered with an outer layer of insulation. You are likely to see this type of construction in supply mains and trunk ducts for medium- or high-pressure applications.

7.0.0 ♦ AIR TERMINALS

Air terminals are devices that control the distribution and volume of conditioned air introduced into a space or zone from the supply air duct. Depending on their design, air terminals can control one or more of the following functions:

- Pressure
- Airflow rate
- Temperature
- Air blending

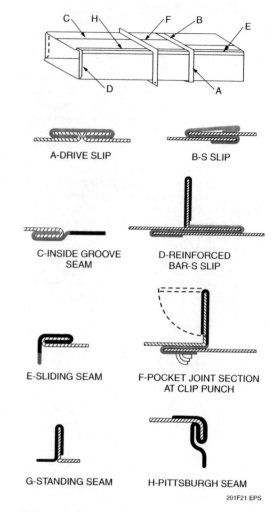

Figure 21 ♦ Rectangular duct seams.

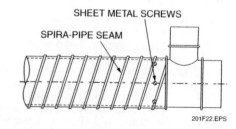

Figure 22 ♦ Round duct seams.

7.1.0 Supply Outlets

All air systems require some type of device to shape the pattern of airflow as it enters the conditioned control zone. In a well-designed and installed air system, the patterned airflow within the control zone will mix supply air with room air, satisfy zone cooling or heating loads, and do so without the awareness of the occupants. Air conditioning systems may have costly sophisticated central equipment and well-designed and installed ductwork; however, the installation will be perceived as poor if a good job is not done with the selection and location of the supply air outlet devices. Selection of the supply outlets that perform this important air conditioning function depends on the size of the space being supplied and the type of air system supplying it, constant or variable air volume.

Normally, supply outlets for cooling are located in the center of the ceiling of the control zone (*Figure 23*) to take advantage of the natural phenomenon that dense cool air tends to drop. With this tendency, a good mix of supply air with room air can be accomplished during cooling. When used for heating, however, a ceiling location tends to result in the warm air staying at the ceiling with resultant cool stagnant regions in the occupied area of the control zone.

The best location for heating outlets is in the floor next to the outside wall of a perimeter control zone (*Figure 24*). This is because less dense warm air rises, and can mix with and absorb cold air drifting downward along the outside wall.

Outlets that provide both heating and cooling result in a compromise location. Because most commercial jobs locate the outlets in the ceiling, outlet models can be purchased that direct warm air downward along the outside wall during heating, and in two directions along the ceiling during cooling. This is a reasonable compromise for an outlet used for both purposes.

Supply outlets fall into four common types, each with its own individual physical construction and performance characteristics:

- Grilles
- Ceiling diffusers
- Slot diffusers
- Perforated ceiling panels

7.1.1 Grilles

Grilles are typically used with constant-volume air systems and are located in high sidewall, ceiling, or floor applications. They may also be

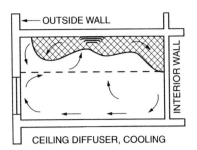
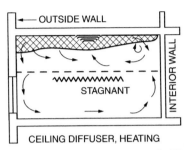

Figure 23 ♦ Ideal supply outlet location for cooling.

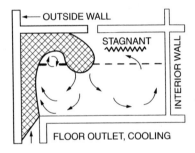
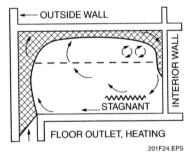

Figure 24 ♦ Ideal supply outlet location for heating.

Discuss the function of supply outlets.

Show Transparencies 27 and 28 (Figures 23 and 24). Explain how the location of supply ducts is determined by heating and cooling applications.

Identify the four common types of supply outlets.

Explain that grilles are used with constant volume systems.

Explain that grilles equipped with adjustable dampers are known as registers.

Discuss the different types of diffusers. Explain that an optional damper may be added to make the diffuser adjustable.

Discuss the use of perforated panel diffusers.

Discuss the use of slot diffusers. Explain that slot diffusers are a popular choice for VAV systems.

Provide grilles and registers for trainees to examine.

located in the windowsill of a perimeter control zone. Normally, grilles are equipped with vertical and horizontal vanes that direct the air into the control zone as desired by the occupant. When supplied with a second set of vanes (a damper) the grille becomes a register. The damper is installed behind, and at right angles to the face vanes and can be used by the occupant to manually adjust the quantity of air entering the control zone. *Figure 25* shows a pole-operated louvered supply grille typical of those used in sidewall applications.

7.1.2 Ceiling Diffusers

The most common type of diffuser in commercial applications is the ceiling diffuser (*Figure 26*). Normally, this diffuser is used with constant-volume air systems and sometimes with variable air volume systems. Typically, multiple diffusers are located symmetrically in square or rectangular ceiling areas to provide complete coverage of the control zone. The diffuser may be used for both cooling and heating.

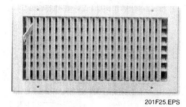

Figure 25 ◆ Typical pole-operated supply grille.

Figure 26 ◆ Ceiling diffusers.

Ceiling diffusers are available in round, square, and rectangular shapes. They consist of an outer cone and inner assembly made of a series of flaring louvers. The louvers form a series of concentric air passages that discharge air simultaneously in a variety of patterns. The branch air duct connects directly to the diffuser's duct collar. An optional damper assembly may be added above the inner louver assembly to provide manual adjustment of air quantity.

7.1.3 Perforated Panels

Much like ceiling diffusers, the perforated panel diffuser (*Figure 27*) is located in the ceiling, and more specifically replaces a ceiling tile in a T-bar type ceiling. The diffuser can be square or rectangular and is typically used with constant-volume air systems. A directional deflector adapter may be installed in the inlet duct collar to distribute the supply air quantity into the desired horizontal airflow patterns.

7.1.4 Slot Diffusers

Slot diffusers (*Figure 28*) perform well in both constant-volume and VAV systems. They are used

Figure 27 ◆ Perforated panels.

Instructor's Notes:

in ceilings, high sidewall, floor, or windowsill locations. Slot diffusers are also used as structural members in special ceilings and recessed lighting fixtures. They are the diffusers of choice in variable air volume systems due to their excellent part-load performance, preventing the cool supply airstream from dumping down on the heads of occupants. The slot diffuser is very useful in all-air systems that heat and cool from overhead. Heat from the central unit travels down the same duct system and is delivered to the control zone through the same supply outlet as cooling from the central unit. The central unit switches from heat to cool mode as needed by the control zone. As stated earlier, the location of the diffuser must be a compromise between the ideal cooling and ideal heating location in the space. With a slot diffuser, the location is closer to the outside wall than the interior wall. In addition, the diffuser is provided with a power element that is capable of switching the flow of air based on its temperature. During cooling, the power element causes air to flow in both directions, providing good room air motion. During heating, the power element forces all warm supply air along the outside wall. The high-velocity warm airstream absorbs any downdrafts and provides good heating performance.

7.1.5 Fan Coil Heating System

A constant-volume overhead heating system is used in some commercial buildings. This heating system is separate from the building's cooling system. The heating system (*Figure 29*) consists of a heating fan coil unit, low-pressure constant-volume ductwork, and multiple diffusers for each control zone. A constant volume of warm supply air is projected vertically downward along the outside wall through either slot diffusers or round nozzles. The warm projected airstreams mix with and absorb any downdrafts from the cool outside wall and keep the zone warm. When cooling is needed in the zone, this system is inactive and a separate parallel cooling system serves the control zone.

PLENUM SLOT DIFFUSER

LINEAR SLOT DIFFUSER

Figure 28 ◆ Slot diffusers.

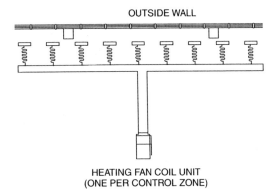

HEATING FAN COIL UNIT
(ONE PER CONTROL ZONE)

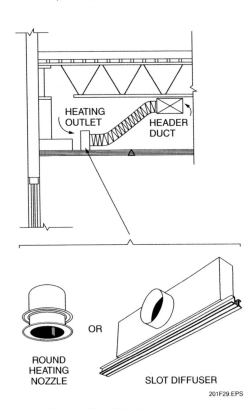

Figure 29 ◆ Separate fan coil heating system.

Discuss the use of the concentric duct diffuser.

Show Transparency 30 (Figure 30). Explain that this design is generally used with single-zone systems of 7.5 tons or less, although it has been used with larger systems.

Discuss VVT air terminals. Identify common types of air terminals.

Show Transparency 31 (Figure 31). Explain that round VVT terminals are used in systems with terminal inlet duct pressure of 1.0 in wg or less.

Provide manufacturers' literature on air terminals.

7.1.6 Concentric Duct Diffuser

The simplest air distribution system used in commercial buildings employs a concentric ductwork diffuser. This system is used mostly with single-zone rooftop units up to 7.5 tons in capacity, although it has been used with rooftop units up to 20 tons in size.

The system (*Figure 30*) consists of a rooftop unit mounted on a roof curb, a field-supplied duct transition, a small vertical duct section, and a concentric duct diffuser. A constant quantity of supply air is discharged through the perimeter of the diffuser, while the return air enters the center of the diffuser.

Because of the tendency to discharge the air downward, this system is often used in high-bay ceilings where the supply air has a chance to mix and warm before coming in contact with the occupants below. Home improvement stores and volume discount stores are common places to find such systems, although fast food and small strip mall stores may also use them.

7.2.0 Air Terminals

In constant-volume air systems, supply air travels down the supply duct from the central unit to the diffuser and is delivered in a constant quantity to the control zone. A balancing damper is located upstream of the diffuser, or is located as an integral part of the diffuser. The balancing damper is manually adjusted at installation to evenly distribute the supply air quantity between diffusers and assure an even air pattern throughout the control zone. In variable air volume systems, a device called a controllable air terminal is inserted between the diffusers and the central unit. The air terminal provides a modulating damper that can vary airflow to the zone in response to a zone thermostat. Generally, the air terminal also contains a discharge heater to provide zone heat.

There are several types of air terminals available to match the type of variable airflow system installed. As described earlier, these systems include VVT, VAV, and dual-duct VAV systems. The types of air terminals used with these systems include:

- Standard VVT
- Single-duct VAV
- Parallel fan-powered mixing box
- Series fan-powered mixing box
- Dual-duct VAV boxes

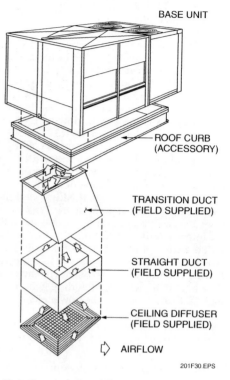

Figure 30 ♦ Concentric duct diffuser system.

7.2.1 Standard VVT Air Terminal

Air terminals (*Figure 31*) with both round and rectangular dampers are available for VVT. The most economical installation cost is achieved when the damper size and shape match those of the branch duct in which it is installed, and when the number of dampers is minimized.

Round air terminals are made in several sizes, typically ranging from 160 to 1,125 cfm (0.4 to 2.8 tons). They have dampers that range from 6 to 16 inches in diameter in order to match typical runout duct sizes. Round dampers come from the factory without insulation. Insulation and vapor barriers are applied externally during installation. Round dampers are normally installed on jobs where the duct pressure at the terminal inlet is 1.0 in wg or less.

Rectangular air terminals are typically made in sizes ranging from 8" × 10" to 8" × 24", where 8" represents the height of the terminal. Rectangular air terminals are available from the factory with an insulated duct sleeve designed for new duct systems. The sheet metal sleeve has internal insu-

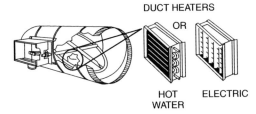

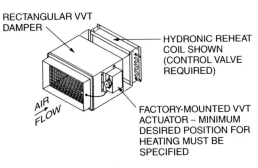

Figure 31 ♦ VVT air terminal.

lation that complies with the *National Fire Protection Association (NFPA) 90A* fire protection code. The internal damper is made of multiple opposing blades and is powered by a high torque actuator. The actuator delivers sufficient torque to handle higher velocity, higher static inlet pressures, and more turbulence than can be handled by round dampers.

When a control zone needs heat in a VVT system, and the central unit is in the cooling mode, a zone heater is activated to keep the zone above the heating setpoint. Two types of zone heaters can be used with a round air terminal. A hot water or electric heater may be installed in the branch duct downstream of the air terminal. An alternative is to install a hot water or electric baseboard heater in the zone along the outside wall. If a rectangular air terminal is used, the hot water or electric duct heater is mounted directly on the discharge of the terminal. Otherwise, a baseboard heater may be used as an alternate heat source.

7.2.2 VAV Air Terminals

There are three types of air terminals typically used with variable air volume (VAV) systems:

- Single-duct VAV terminal
- Parallel fan-powered mixing box
- Series fan-powered mixing box

7.2.3 Single-Duct VAV Air Terminal

The single-duct VAV air terminal (*Figure 32*) consists of an primary air inlet connection, a set of opposed blade dampers, an inlet velocity sensor, and a damper (zone) controller mounted on the side of the box. A sensor located in the control zone monitors temperature and is wired to the zone controller. Round branch ductwork typically connects to the primary air inlet. Low-pressure rectangular ductwork is usually installed downstream between the air terminal and the supply air diffusers.

The single-duct air terminal is made in several sizes, typically ranging from 200 to 4,000 cfm (0.5 to 10 tons). Depending on the model and size of the terminal, the input duct connection can be round, oval, or rectangular. All connection shapes and sizes facilitate the location of the unit in a ceiling cavity located above a hallway, utility closet, restroom, or other area where the space below is less sensitive to noise.

A zone controller is mounted on the side of the unit and responds to a room sensor. Available control types include pneumatic, electric, and direct digital control (DDC). The zone controller monitors the inlet air velocity with the velocity sensor. It operates to adjust the damper position and factor out any increase or decrease in supply cfm caused by varying inlet static pressure. This action is called pressure-independent control.

Figure 32 ♦ Single-duct VAV air terminal.

Explain that a zone controller mounted on the side of the terminal responds to the room sensor.

Describe a parallel fan-powered mixing box. Explain that this unit is designed for perimeter zones in buildings with a plenum ceiling return.

Show Transparency 32 (Figure 34). Explain the operation of a parallel fan-powered mixing box.

The single-duct VAV air terminal is used as part of a conventional VAV system where the central air source is in the cooling mode when the building is occupied. Any control zone served by a single-duct air terminal that requires occupied heat is satisfied by a factory-installed heater connected to the air terminal discharge. A hot water or electric heater may be used.

7.2.4 Parallel Fan-Powered Mixing Box Air Terminal

The parallel fan-powered mixing box air terminal (*Figure 33*) typically consists of a primary air inlet connection, a set of opposed-blade primary air dampers, an inlet velocity sensor, a heating fan, a backdraft damper, a heater, and a damper (zone) controller mounted on the side of the box. A sensor located in the control zone monitors temperature and is wired to the zone controller. Round branch ductwork typically connects to the primary air inlet. Low-pressure rectangular ductwork is usually installed downstream between the air terminal and the supply air diffusers.

The parallel fan-powered mixing box air terminal is made in several sizes typically ranging from 500 to 4,000 cfm (1.25 to 10 tons). Depending on the model and size of the terminal, the input duct connection can be round, oval, or rectangular. All connection shapes and sizes facilitate the location of the unit in a ceiling cavity located above a hallway, utility closet, restroom, or other area where the space below is less sensitive to noise.

The parallel fan-powered mixing box is designed for perimeter zones in a building with a ceiling plenum return system. The cavity above the zone's ceiling is used as a return duct to bring zone return air back to the air terminal.

A zone controller is mounted on the side of the unit and responds to a room sensor. Available control types include pneumatic, electric, or direct digital control. The controller monitors the primary air inlet velocity with the velocity sensor. Thus the controller can adjust the damper position and factor out any increase or decrease in supply cfm caused by varying inlet static pressure. This action provides pressure-independent control.

The parallel fan-powered mixing box is used as part of a conventional VAV system where the central air source is in the cooling mode when the building is occupied. Any control zone served by a parallel fan-powered mixing box that requires heat is satisfied by a fan and factory-installed heater connected to the air terminal discharge. The heater may be hot water or electric. The fan runs only during the heating mode.

Parallel fan-powered mixing boxes made by different manufacturers can operate in different ways. One common way is described here (*Figure 34*). When a rise in zone temperature above the cooling setpoint (typically 74°F) is monitored by the zone sensor, the zone controller modulates the primary air damper between full open and the minimum ventilation position to satisfy the zone cooling load. The terminal fan is not running, and the pressure of the air being discharged from the primary air damper closes the backdraft damper.

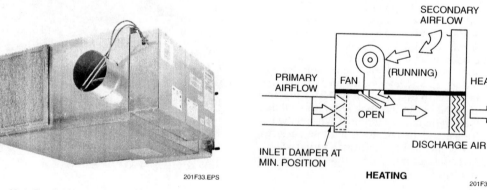

Figure 33 ♦ Parallel fan-powered mixing box.

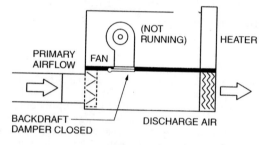

Figure 34 ♦ Parallel fan-powered mixing box control.

When a drop in zone temperature below the cooling setpoint (typically 74°F) is monitored by the zone sensor, the zone controller positions the primary air damper at the minimum cooling (ventilation) cfm position. With a further drop in zone temperature below the zone heating setpoint (typically 70°F), the zone controller activates the first stage of heat. The first stage of heat is the activation of the unit's heating fan. The fan draws warm ceiling plenum air into the unit. Thus heat is reclaimed from the building and reused for heating purposes. The fan discharges this warm plenum air into the primary airstream, opening the backdraft damper in the process. The warm plenum air heats the primary airstream containing ventilation air, and the heated airstream is directed to the diffusers in the control zone. The heating cfm delivered to the space is determined by the fan setting, and is typically 50 to 60 percent of the design primary air cooling cfm. With a further drop in zone temperature, the zone controller activates the unit discharge heater as necessary to maintain the zone heating setpoint.

7.2.5 Series Fan-Powered Mixing Box Air Terminal

A second type of fan-powered mixing box has the air terminal fan located in series with the primary air damper (*Figure 35*). Other than this change, the series fan-powered mixing box components, with two exceptions, are identical to those of the parallel fan-powered mixing box. With the fan in series there is no need for a backdraft damper.

The second difference is that the unit is supplied with variable plenum air dampers located upstream of the unit fan.

The series fan-powered mixing box is characterized by the delivery of a constant volume of air to the control zone as the primary airstream varies. Thus it is suited for applications that have difficult room air motion problems. Hotel atriums or public areas with tall ceilings are good applications for the series unit. Like the parallel unit, the series fan-powered mixing box uses the ceiling cavity as a return duct to bring zone return air back to the air terminal, and operates in a pressure-independent manner.

The series fan-powered mixing box is used as part of a conventional VAV system where the central air source is in the cooling mode when the building is occupied. Any control zone served by a series fan-powered mixing box that requires heat is satisfied by a factory-installed heater connected to the air terminal discharge. The heater may be hot water or electric. The unit fan runs during cooling and heating in a continuous manner as long as the control zone is occupied.

7.2.6 Dual-Duct VAV Air Terminals

The dual-duct VAV air terminal is used as part of a dual-duct VAV system (*Figure 36*). It consists of two primary air inlet connections, two primary air dampers, inlet velocity sensors, and a zone controller mounted on the side of the box.

Describe how mixing boxes use reclaimed heat.

Show Transparency 33 (Figure 35). Explain how the series mixing box differs from the parallel mixing box.

Explain that the series mixing box is used in building areas with air circulation problems, such as hotel atriums and public areas with tall ceilings.

Show Transparency 34 (Figure 36). Describe the air terminals used in dual-duct VAV systems.

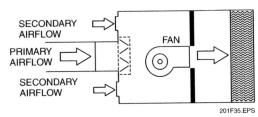

Figure 35 ♦ Series fan-powered mixing box.

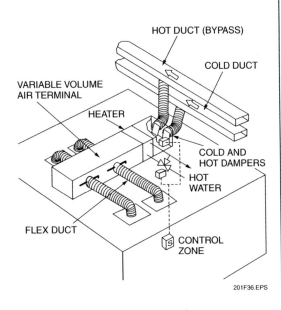

Figure 36 ♦ Dual-duct VAV air terminal.

Explain that heat is provided by a factory-installed heater connected to the air terminal discharge.

Show Transparency 35 (Figure 37). Describe the layout of a dual-duct system.

Explain that the central core of an all-air system is the fan and a coil, which are contained in packaged equipment or built-up air handling equipment.

A thermostat located in the control zone monitors the temperature and is wired to the zone controller and hot-water valve. Round branch ductwork typically connects to the primary air inlets and low-pressure round ductwork is usually installed downstream between the air terminal and the supply air diffusers.

The dual-duct VAV air terminal is made in several sizes, typically ranging from 300 to 2,400 cfm (0.75 to 6 tons). Depending on the model and size of the terminal, the input duct connection can be round, oval, or rectangular. All connection shapes and sizes facilitate the location of the unit in a ceiling cavity located above a hallway, utility closet, restroom, or other locations where the space below is less sensitive to noise.

A zone controller is mounted on the side of the unit and responds to a room sensor. Available control types include pneumatic, electric, and digital. The zone controller monitors the inlet air velocity with the velocity sensor. Thus the controller can adjust the damper position and factor out any increase or decrease in supply cfm caused by varying inlet static pressure. This action provides pressure-independent control.

In the dual-duct VAV system, the central air handler is in the cooling mode when the building is occupied. Any control zone served by a double-duct air terminal that requires heat is satisfied by a factory-installed heater connected to the air terminal discharge. The heater may be hot water or electric. Essentially, this terminal allows older constant-volume double-duct systems to be upgraded to variable air volume-type operation with the associated savings in central fan energy.

When a rise in zone temperature above the cooling setpoint (typically 74°F) is sensed by the zone thermostat, the zone controller modulates the primary cold air damper between full open and the minimum ventilation position to satisfy the zone cooling load. This minimum position represents 50 to 60 percent of the primary cold airflow. The neutral primary air damper remains closed until the primary cold air damper has reached its minimum position. When zone cooling loads fall below this, the cold and neutral air dampers are modulated to match the zone cooling load.

When a drop in zone temperature below the heating setpoint (typically 70°F) is sensed by the zone thermostat, the zone controller positions the primary cold air damper at the minimum ventilation position and activates the unit discharge heater to satisfy the zone heating load. The hot or neutral primary air damper remains closed.

When the central unit is in heating, any zone requiring heat has its primary cold air damper positioned at the minimum ventilation position and its primary hot air damper modulated to meet the zone heat load.

A typical dual-duct VAV system layout is shown in *Figure 37*. Notice that each control zone has a sensor (S) wired to the zone controller located on the side of each air terminal. Each thermostat contains cooling and heating setpoints for its zone. All dual-duct VAV air terminals are equipped with a supplementary hot-water heater.

The central air handler starts based on its own time clock and provides cooling during the occupied period. Cooling supply air of the required temperature and quantity is supplied through the primary cold air duct to the zone air terminals. Any zone requiring heat is satisfied by the zone thermostat activating its own heater.

8.0.0 ◆ AIR SOURCE EQUIPMENT

The central core of all-air systems is a fan and a coil. The coil provides the means to cool and dehumidify the circulated air, and the fan provides the energy to circulate the air around the building. Two broad categories of equipment contain the central fan and coil. They are packaged equipment and built-up air handling equipment.

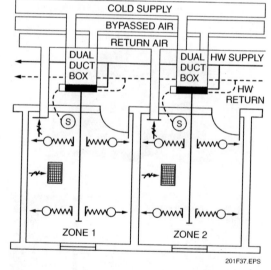

Figure 37 ◆ Double-duct VAV layout.

8.1.0 Packaged Equipment

Packaged equipment covers a wide range of factory-assembled equipment, from room air conditioners and central residential equipment through large tonnage absorption and centrifugal water chillers. Because this module covers all-air systems, we will focus on the equipment containing a central fan and coil that apply to commercial buildings.

A packaged air conditioner (PAC) consists of one or more factory-made assemblies that normally include an evaporator or cooling coil, an air-moving device or fan, and a compressor and condenser combination. It may include a heating function as well.

Packaged units are further categorized by the method they use to provide heating:

- Packaged air conditioners (PACs)
- **Year-round air conditioners (YACs)**
- Heat pumps (HPs)
- Vertical packaged air conditioners (VPACs)

8.1.1 Packaged Air Conditioner (PAC)

A packaged air conditioner (PAC) is a unit that provides cooling capacity only, or cooling capacity with electric heat. A PAC unit (*Figure 38*) consists of a DX cooling coil, filters, indoor fan, compressors, air condenser, condenser fans, and controls. Because these units are usually installed on the roof, they are commonly called rooftop units. In some instances, however, these units are mounted on a slab and the ductwork penetrates the wall of the building.

The primary function of the PAC is cooling. Electric heaters are provided as an accessory. Sometimes, PACs are called all-electric or electric-electric because the source of heat is an electric coil. Some manufacturers also offer steam or hot-water coils as accessories. Generally, these units come with either horizontal or vertical discharge.

Buildings requiring small rooftop units in the 1.5- to 10-ton range offer the widest range of applications for PAC units. In some areas, rooftop units are used in residential as well as commercial buildings.

Popular applications for PAC units include fast food restaurants, hotel atriums, and public spaces, with the units mounted on the roof. A single concentric diffuser is used. Supply air exits downward through the perimeter of the diffuser, and return air enters upward through the center of the diffuser. This eliminates some ductwork.

Figure 38 ◆ Packaged air conditioner (PAC).

On larger fast food restaurants, these units are installed in multiples, typically with one in the kitchen area and another serving the dining area. A two-unit arrangement is also frequently used in larger retail stores.

Single-story office buildings also use small rooftop units. One unit might handle each exposure, with perhaps another to handle the core load of the building that is nearly constant throughout the day. The load in a typical classroom is about three tons. Thus, a school might use a small PAC unit to handle each classroom. These may be located on the roof, or on a slab outside the room.

Typically, stores in a strip mall have less than a 10-ton load, so this is another common application for small units. As tenants move into a shopping center, they may be required to pay fit-up or build-out costs, which may include air conditioning. Using PACs is a very flexible and economical way to accomplish these tenant improvements.

Buildings requiring rooftop units in the 10- to 20-ton range are almost exclusively commercial. These include larger fast food restaurants, convenience stores, and medium-sized office buildings. Single, rather than multiple units installed with ductwork are most commonly used. Each unit handles an area or floor of the building. These units tend to have more application flexibility than smaller units and they also have a greater number of available options and accessories.

Identify the common components of a packaged air conditioner (PAC) for commercial buildings. Explain that packaged units are categorized by the methods they use to provide heating.

Describe large and small packaged air conditioners. Explain that their primary function is cooling.

Discuss common applications for packaged air conditioners.

Provide manufacturers' literature on packaged equipment.

See the Teaching Tip for Section 8.1.0 at the end of this module.

Describe a year-round air conditioner (YAC) and identify applications.

Describe vertical packaged air conditioners (VPAC).

Buildings requiring rooftop units in the 20- to 100-ton range are entirely commercial. These larger size PACs (*Figure 39*) offer the greatest application flexibility of any packaged equipment.

Perhaps the largest commercial use is office buildings. These units can be used as singles or in multiples to handle either single-story, or more commonly, multiple-story buildings (up to six stories). Most commercial applications include VAV control because the buildings are large enough to have more than one zone.

Large rooftop units are also used in industrial applications with large open areas that require cooling, or where machines and processes give off vast quantities of heat that must be removed.

8.1.2 Year-Round Air Conditioner (YAC)

A year-round air conditioner (YAC) is a packaged rooftop unit with a gas heating section (*Figure 40*). It is called year-round because heating and cooling are integral unit functions. Gas heating sections are normally designed for natural gas, but propane models are available and conversion kits are typically available as unit accessories. YACs are also known as gas-electric units. YACs are typically available in capacities from 1.5 to 130 tons. YACs can be found in the same type of buildings as PAC units, because they are simply an alternative heating energy choice to electricity.

8.1.3 Vertical Packaged Air Conditioner (VPAC)

Figure 41 shows a vertical packaged air conditioner (VPAC). VPACs consist of a DX cooling coil, filters, indoor fan, compressors, and a condenser. Hot water or steam heating coils are available as heating accessories for most models. VPACs made with a water-cooled condenser are used with a cooling tower that is typically located on the roof. VPACs made with an air-cooled condenser are commonly used when the unit can be mounted against an outside wall or window.

A typical VPAC has the condensers and compressors enclosed in the bottom compartment. The evaporator coil filters and the fan(s) are located in the top. The control panel is mounted at the end of the unit. Fans discharge a constant volume of air freely into the conditioned space or are arranged for either horizontal or vertical discharge into a supply air duct system. Return air ductwork is not required if the unit is installed in an equipment room or conditioned space that serves as a return air plenum.

Figure 39 ♦ Larger size PAC (20 to 45 tons).

Figure 40 ♦ Year-round air conditioner (YAC).

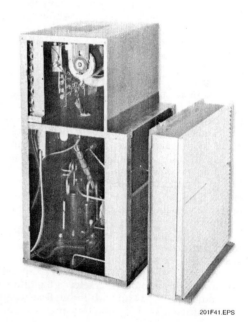

Figure 41 ♦ Vertical packaged air conditioner (VPAC).

1.32   HVAC ♦ LEVEL TWO

Instructor's Notes:

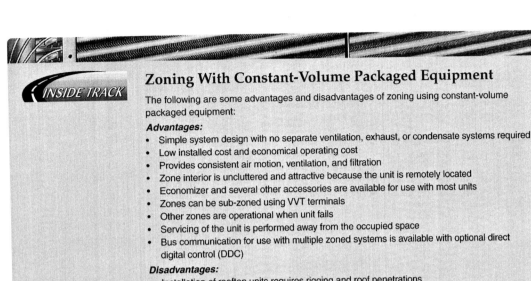

Zoning With Constant-Volume Packaged Equipment

The following are some advantages and disadvantages of zoning using constant-volume packaged equipment:

Advantages:
- Simple system design with no separate ventilation, exhaust, or condensate systems required
- Low installed cost and economical operating cost
- Provides consistent air motion, ventilation, and filtration
- Zone interior is uncluttered and attractive because the unit is remotely located
- Economizer and several other accessories are available for use with most units
- Zones can be sub-zoned using VVT terminals
- Other zones are operational when unit fails
- Servicing of the unit is performed away from the occupied space
- Bus communication for use with multiple zoned systems is available with optional direct digital control (DDC)

Disadvantages:
- Installation of rooftop units requires rigging and roof penetrations
- Each vertical packaged unit requires an equipment room or the loss of interior floor space in the zone
- Separate electrical power wiring must be installed and maintained for each unit
- Visible rooftop units can be unattractive without an aesthetic roof treatment scheme
- Rumble is a common problem with rooftop units and machine noise is a common problem in or near the occupied space with vertical package units

Discuss the advantages and disadvantages of zoning with constant volume equipment.

Identify typical applications for VPAC units.

Describe a packaged heat pump and identify typical applications.

Discuss the types of compressors used with packaged equipment. Explain that part load efficiency with hermetic compressors is achieved by using more than one compressor, while cylinder unloaders are used with semi-hermetic compressors.

With water-cooled units, the condensers are piped to the cooling tower where condenser heat is rejected to the tower water. The tower water circulates to the cooling tower where heat is rejected to the outdoor air. As an alternative, city water may be used for water-cooled condensing purposes where local codes and cost allow.

VPACs are typically made in the 3- to 60-ton range. They are often used in interior floor-by-floor office building installations. A major use of VPACs is in renovation work. A floor-by-floor renovation can be completed in stages without shutting down the entire building. VPACs are often used with VAV airside systems to provide comfort control at each control zone.

8.1.4 Packaged Heat Pumps

A heat pump is any packaged unit with the ability to reverse its internal refrigeration cycle. A heat pump uses the refrigeration cycle to provide cooling or heating. In the heating mode, the refrigerant flow is reversed using a four-way valve. The cycle extracts heat from the outdoor air and transfers it to the indoor air. Physically, the packaged unit heat pump looks just like a PAC.

The heat pump is more than an air conditioner with a four-way valve. The evaporator and condenser are larger and the metering device arrangement is different to allow for the dual role. Because the heat pump does not heat efficiently at very low outdoor temperatures, the unit often contains accessory electric heaters that supplement the reverse cycle heat.

Heat pump versions of packaged equipment are available in capacities from 1.5 to 130 tons per unit. Heat pumps are usually rooftop units, mounted on a roof curb, with either a vertical or horizontal discharge.

8.2.0 Packaged Equipment Components

As described earlier, every packaged unit has a compressor, evaporator (cooling coil), filters, condenser, indoor fan, condenser fan, and control system. These components are briefly described in the sections that follow.

8.2.1 Compressors

Reciprocating and scroll compressors (*Figure 42*) are used in packaged units. Welded-shell hermetic compressors are generally used in 10- to 15-ton units; semi-hermetic compressors are optional in units of 20 tons and larger. However, semi-hermetic reciprocating compressors are becoming less common in new equipment. They

Explain that the largest systems use multiple semi-hermetic compressors.

Discuss the types of coils used in packaged equipment.

Explain that evaporator coils are usually installed in a slanted position to obtain more coil surface area and enhance condensate drainage.

Figure 42 ◆ Compressors.

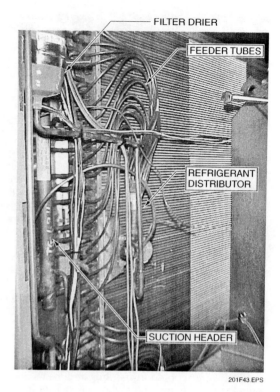

Figure 43 ◆ Commercial unit evaporator coil.

are being replaced with multiple hermetic compressors. Semi-hermetic or multiple hermetics are used in the larger sizes because nearly all of them require capacity control. Capacity must be matched to airflow for proper cooling, and semi-hermetic and multiple hermetic compressors offer better capacity controls for larger space-cooling requirements. VAV systems in particular require significant capacity control in order to match the refrigerant-side capacity to the changing airflow volume.

8.2.2 Coils and Fans

Figure 43 shows a typical evaporator coil in a packaged unit. Evaporator (cooling) coils are generally made of aluminum-plate fins pressed on aluminum or copper tubing, and usually are two or three rows deep. Small rooftop units usually have one refrigeration circuit, and the expansion device may be a capillary tube, a fixed orifice, or a thermal expansion valve (TXV).

Large rooftop units typically have two separate refrigeration circuits. The expansion device is usually a TXV, one per circuit. In order to keep the full coil face area active when one circuit is shut down at part load, two-circuit coils often have their headers arranged with special circuiting. The continually active coil face area provides good part-load humidity control.

Evaporator coils are usually installed in a slanted position so more coil surface can be put into a given package. The slanted position also helps condensate to drain properly. An evaporator fan is usually installed downstream of the coil in a draw-through configuration so that the air passing over the coil is better distributed and of a uniform velocity.

Like evaporator coils, condenser coils are typically made of aluminum plate fins pressed on aluminum or copper tubing. For corrosive atmospheres, copper fins are sometimes offered as an option. The condenser coil is usually two or three rows deep. It is likely to be internally grooved for

Instructor's Notes:

higher efficiency, and headered to provide multiple parallel circuits. Each circuit feeds hot gas refrigerant to the leaving airside. The refrigerant leaves the entering airside of the coil as a liquid. This counterflow arrangement provides for maximum subcooling of 20°F or so for a typical air-cooled condenser.

PAC condenser coils are most often positioned vertically, where they become part of the wall of the unit. Direct-drive propeller fans draw outdoor air through the coils and discharge it upward.

Indoor vertical PACs are typically offered with water-cooled condensers that are mounted in the base of the unit near the compressors. These condensers may be of the shell-and-coil type, with a water coil inside a steel shell. They may also be the tube-in-tube type in which the water and refrigerant make a single pass down a concentric pair of tubes.

Evaporator fans, sometimes called indoor fans or supply fans, are generally forward-curved centrifugal blowers so that sufficient static pressure can be generated for good medium- to low-pressure air distribution systems. These fans are usually belt-driven on units 10 tons or larger so the speed can be adjusted for proper flow and pressure (*Figure 44*). Rooftop units sometimes have variable pitch pulleys on the fan-drive motor shaft to adjust the fan speed. Units smaller than 10 tons usually have a direct-drive fan with multiple motor taps for fan speed selection. Normal fan speeds generally fall in the 600 to 1,200 rpm range.

Manufacturers provide fan rating curves or tables showing the rpm and power requirements for various cfm and external static pressure conditions. Sometimes, two fans are used rather than a single fan in order to reduce the unit size. However, the static capabilities and flow control are not as good with multiple fans as with a single fan. Fans are dynamically balanced at the factory and mounted with vibration isolators to limit transmission of vibration to the unit chassis, and ultimately to the building.

Condenser fans are usually direct-driven propeller fans. In rooftop units, several fans are generally used so that the air volume can be varied by switching individual fans on and off to maintain head-pressure control.

8.2.3 Heat Options

Electric heat is usually a factory-supplied option for rooftop units, and is sometimes available for vertical PACs as well. The heaters are generally resistance elements installed as close as possible to the outlet of the supply fan. They are usually controlled by contactors or relays, arranged so that heat is available in two stages as activated by a standard two-stage commercial heat/cool thermostat. The first stage of heating is brought on at a temperature set by the customer, and the second stage is activated if the temperature drops below a lower preset temperature. For example, the first stage may activate at 70°F, and if the temperature continues to drop, the second stage activates at 68.5°F.

Gas units (YACs) are usually larger than electric units (PACs) because of the additional space needed for the gas controls, burners, and heat exchangers. Supply air discharges into the heat exchanger section and flows to the supply duct after passing over the heat exchangers. Though most YACs are built for natural gas, manufacturers usually offer the option of converting for use with propane. Early gas-fired units used clamshell heat exchangers; however, current units use more efficient tubular heat exchangers. A burner

CONDENSER FANS

EVAPORATOR FAN

201F44.EPS

Figure 44 ♦ Evaporator and condenser fans.

Describe condenser coils and discuss their positioning.

Explain that multiple condenser fans are often used to maintain head pressure control.

Discuss fan rating curves.

Provide fan rating curves for trainees to examine.

Describe the operations of electric and gas heating units.

Discuss roof curbs. Explain that most curbs are designed for flat roofs, but curbs are also available for mounting units on pitched roofs. Packaged units installed on the roofs of houses are common in some parts of the country.

Describe the different types of filters used with packaged equipment and discuss their relative efficiency.

Provide various types of filters for trainees to examine.

Discuss options for outdoor air and exhaust on packaged systems.

at the entrance of the tube shoots a flame into the first section of the tube. Gas is fed to the burner by a two-stage combination gas valve. Pilots are usually the spark-ignition type with flame rectification sensing when the pilot is fully lit and ready for gas from the gas valve. Induced-draft blowers draw in combustion air and exhaust flue gas to the atmosphere.

8.3.0 Packaged Unit Accessories

Many accessories are available for installation on packaged units. The more common accessories are described here.

8.3.1 Roof Curbs

Rooftop units are generally mounted on the roof deck. Because the roofs on most commercial buildings have a 20-year guarantee, the method used to penetrate the roof is extremely important. In most cases, standard roof curbs (*Figure 45*) are flashed-in or made part of the roof during the roof installation. This ensures a weatherproof connection to the building compliant with the roof guarantee.

Usually all electrical, duct, and piping connections penetrate the roof inside the curb. Duct connections are usually made directly to the curb. Most roof curbs are 14 inches high; however, in areas of significant snow accumulation, 24-inch curbs are used. The higher curb ensures the ventilation intake is not covered with snow and that snow is not drawn into the building.

Most commercial applications are on a flat or nearly flat roof and use the standard curb. However, rooftop units can be installed on a sloped roof using a sloped roof curb. These usually are not available directly from the unit manufacturer, but are available from companies that specialize in making accessories for rooftop units.

In sound-sensitive applications (sound studios, for example), where it is absolutely necessary that no noise or vibration be transmitted to the building, special vibration-isolation curbs are used. These special curbs are typically obtained from companies specializing in rooftop accessories.

8.3.2 Filters

Most packaged units come with either 1- or 2-inch throwaway filters. They filter particles visible with the naked eye from the air, but in terms of filtering efficiency, they do not remove much else. Several filter types are available to overcome this problem. The most common are as follows:

- *1, 2, and 4-inch cartridge filters* – A cartridge filter (a mechanical filter) is an optional accessory with packaged units because they fit in the same filter track as the standard throwaway filters. Cartridge filters do a much better filtration job than throwaway filters; however, they are not truly high-efficiency filters. For high efficiency, an electrostatic or bag filter must be used.
- *12, 24, and 36-inch bag filters* – Bag filters are mechanical filters that are available in 12-, 24-, or 36-inch lengths. They are more expensive than the other filters but do a much better job of filtration. Bag filters require a large amount of unit space. For this reason, they are typically available only with packaged units above 20 tons in capacity.
- *High-efficiency particulate arresting (HEPA) filters* – HEPA filters are highly efficient filters originally designed during the Manhattan Project in the 1940s to prevent the spread of radioactive contaminants. HEPA filters are comprised of a mat of randomly arranged fibers, and they target pollutants as small as 0.3 micron.

8.3.3 Outdoor Air and Exhaust Options

With the exception of the smallest packaged units, one or more types of economizer options are available. As explained earlier, an economizer provides for cooling the control zone with free cooling when the conditions are favorable. This allows for substantial energy savings.

Because air is being introduced into the building through the packaged unit, some means of relieving the resultant rise in building pressure must be provided. As discussed earlier, packaged equipment is available with either barometric relief dampers, separate building exhaust, or unit powered exhaust options (*Figure 46*).

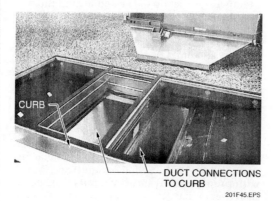

Figure 45 ♦ Roof curb.

Instructor's Notes:

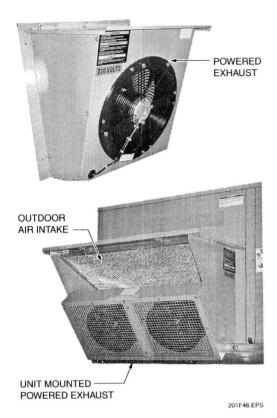

Figure 46 ♦ Power exhaust.

Figure 47 ♦ Low-ambient condenser fan speed controls.

Finally, when return duct pressure losses are significant, a return fan is used to overcome this problem. Because this return pressure loss is likely to be encountered in larger buildings, the return fan is available as a factory option on larger packaged units.

8.3.4 Low-Ambient Control

Standard rooftop units must have a method of controlling head pressure so they can operate properly in cold weather. Head pressure is the difference between the condenser and evaporator pressures, and it must be high enough to assure proper operation of the metering device. Depending on how many condenser fans there are and how they are controlled, most units can operate down to about 45°F outdoor air temperature by cycling fans on/off with a thermostat or pressurestat (*Figure 47*).

If a unit is equipped with an economizer that handles cooling below 55°F, it is not necessary to add optional low-temperature control. Normal fan cycling will maintain sufficient head pressure to ensure proper operation of the expansion device, and thus the system.

If an economizer is not used, and the building requires cooling when air temperatures are below 45°F, one of two available accessory low-temperature operation methods should be considered. The simplest way is to let the condenser fan cycle on a pressurestat. As head pressure decreases, the fan cycles off. This causes head pressure to rise, making the fan cycle on again. This method works well if the fan motor and pressurestat are designed for this kind of duty.

A more refined method involves modulating the speed of the lead condenser fan. This is more expensive than fan cycling, but provides more system stability and allows operation down to –20°F. Solid-state fan speed controllers are available for both single-phase and three-phase fan motors. Fan speed can be made to vary from 0 percent to 100 percent of full rpm in direct response to the head pressure.

NOTE

The fan motor controlled by a solid-state fan speed controller must be designed specifically for this duty. Controller input is typically a thermal sensor mounted on a tube in the saturated liquid portion of the condenser. This controller varies the fan speed to maintain the desired saturated liquid temperature, usually around 100°F.

8.3.5 Coil Guards and Check-Filter Switch

Two popular accessories commonly used in packaged equipment are the coil guard and the check-filter switch.

Explain how return fans are used to overcome duct pressure losses.

Discuss the operation of low-temperature head pressure control. Discuss the use of economizers with these units.

Explain that units with economizers that provide cooling at temperatures below 55°F do not require head pressure control.

Explain that an alternative to cycling fans on and off is a control that modulates condenser fan speed.

Explain that two popular accessories used with packaged equipment are coil guards and check-filter switches.

Show Transparency 36 (Figure 48). Describe coil guards and check-filter switches. Explain that a pressure-sensing switch mounted across the filter bank is used to monitor filter status and signal the need for filter cleaning or replacement.

Discuss crankcase heaters. Explain that they are used when outdoor temperatures are lower than those of the conditioned space.

Explain how a supply air reheat option can maintain the relative humidity at equipment part load.

Identify the differences between packaged VAV units and constant-volume units.

When packaged rooftop units are slab-mounted on the ground, they are more easily accessible to the public. In these cases, vandals can bend the condenser coil fins thereby altering the flow of condenser air. To protect the coils from such vandalism or unintentional damage, most unit manufacturers offer accessory coil guards (*Figure 48*). They can also be field-fabricated, but must be built to permit proper airflow across the condenser coils. Coil guards can also prevent damage caused by hailstones.

All air conditioning systems require maintenance, especially filter changes, which are often overlooked by customers. A check-filter switch can solve this problem. This desirable option is simply a pressure sensor connected across the filter bank. When the pressure exceeds a pre-set pressure drop, a light signals the customer that it is time to change the filters. Although this option is not offered with all units, it is usually installed when available.

8.3.6 Crankcase Heaters

Rooftop units above 10 tons are equipped with crankcase heaters. However, they may not be included on smaller units under 10 tons.

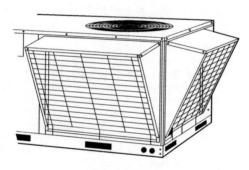

COIL GUARDS

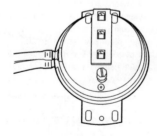

CHECK-FILTER SWITCH

Figure 48 ♦ Coil guards and check-filter switch.

The purpose of a crankcase heater is to keep the compressor crankcase oil warm. Refrigerant always migrates to the coldest part of a system, which may be the compressor oil sump. At cold temperatures oil has a tendency to absorb refrigerant. This tendency diminishes at higher temperatures. The compressor oil can become saturated with refrigerant if the compressor oil sump becomes cold. At startup, this refrigerant will boil from the crankcase oil and enter the discharge line, taking oil with it and leaving insufficient oil in the compressor for proper lubrication. This can cause compressor failure.

To prevent this, a crankcase heater is added to the compressor. The heater raises the oil temperature and minimizes any absorption of the refrigerant by the oil. There are two common types of heaters. One is an insertion heater, and the other is a bellyband heater that straps around the base of the compressor. In most cases, the heater activates only when the compressor is off, assuring that the refrigerant does not migrate to the oil in the crankcase. Typical crankcase heaters draw about 100 watts. Crankcase heaters are used if unit operation is likely when outdoor air temperatures are lower than those in the conditioned space.

8.3.7 Supply Air Reheat Option

Constant-volume rooftop units are designed to provide humidity control at full load conditions by dehumidifying the supply air as it crosses the cooling coil. At full load, the room percent RH can easily be maintained between 50 and 60 percent in most applications. Controlling space humidity when the building cooling load has dropped presents more of a problem. Most 2- to 25-ton rooftop units use one or two scroll compressors, and therefore have one or two steps of unloading. Thus, these units will spend a lot of time in the off cycle where no dehumidification of the supply air to the space is being done. Under these conditions, the room percent RH will rise. Rooftop equipment can be ordered with a supply air reheat option to maintain control of space relative humidity at equipment part load.

8.4.0 VAV Unit Modifications

A VAV packaged unit has several features that are different from a constant-volume unit. Three major issues must be addressed to provide proper unit operation when providing air to variable air volume terminals. First, the unit must be able to modulate airflow. Second, the unit refrigeration system controls must be able to handle a varying flow rate without damaging the compressor(s). Finally, a morning warmup heating cycle must be included.

There are two methods for varying the unit's supply airflow: using inlet guide vanes and modulating the fan motor speed. Inlet guide vanes (*Figure 49*) are located at the entrance to the supply air fan and give a spin (pre-whirl) to the air that affects fan performance. On fans designed for variable inlet guide vanes, airflow can be varied from 100 percent to 25 percent of design. Guide vanes are controlled from a static pressure sensor usually located two-thirds of the distance down the main supply duct. The duct pressure increases as the air terminals throttle airflow. The added duct resistance causes the fan to decrease the amount of supply air. Inlet guide vanes alter fan performance to reduce duct pressure and deliver the quantity of air required by the air terminals.

Another method of controlling airflow is to slow the speed of the fan motor in response to static pressure. The variable-speed controller is an electronic device that controls the power supplied to the fan motor. These devices are more efficient than variable inlet vanes and can save even more in fan operating costs. However, they can be relatively expensive.

If a slipping belt or dirty filter causes the airflow to a constant-volume unit to be reduced, the coil can freeze. If it does, the refrigerant in the coil cannot evaporate properly. This results in liquid refrigerant flowing back to the compressor. Liquid will damage the compressor because compressors are designed to handle refrigerant vapor only.

Precautions are taken to prevent coil freeze-up and compressor damage on units with variable airflow. Airflow to the space varies as the sensible load changes. The unit maintains a constant supply air temperature (usually about 55°F) and is controlled by a discharge air sensor. Supply air temperature is held constant, so the compressors on VAV units are controlled by electric unloaders rather than by standard suction pressure unloaders. The latter are not sensitive enough for this duty.

In addition, hot gas bypass is available on VAV units to keep the units on the line when the building load falls below the unit's minimum step of unloading.

9.0.0 ♦ AIR HANDLERS

As discussed earlier, packaged equipment can serve as the central source of air in an all-air system. The packaged unit contains the cooling coil, supply fan, the refrigeration system components, all in a single-unit housing. It is typically mounted on the roof.

In many jobs, the packaged unit is split into two or more components to meet the physical constraints of the building and facilitate the transmission of cooling from the refrigeration unit to the cooling coil.

When the central packaged unit is split, one of the divided components contains the supply fan and the cooling coil. The combination of a fan and cooling coil is called an air handler. Depending on the building size, the air handler comes in two varieties: packaged air handler and central station air handler.

9.1.0 Packaged Air Handler

In many applications, particularly when the building is taller than three or four floors, the space required to run ductwork vertically in the building becomes too costly. In these situations split systems are typically used.

A split system consists of a condensing unit and a constant-volume packaged DX air handler connected together with field-installed refrigerant piping lines. A suction and liquid line are required for each refrigerant circuit (usually two) in the condensing unit.

The condensing unit consists of the compressor, air-cooled condenser, condenser fans, and control system. Typically, the condensing unit is mounted on the roof of a building or on a slab alongside the building.

Figure 49 ♦ Inlet guide vanes.

Describe the types of packaged DX air handlers available.

Provide manufacturers' literature on packaged air handlers.

Show Transparency 37 (Figure 51). Explain that central station air handlers are configured at the factory to meet requirements specified by the designer.

Explain that central station air handlers are typically available in capacity ranges from 2,500 to 63,000 cfm.

The constant-volume packaged DX air handler consists of a DX cooling coil, filters, and a supply air fan (*Figure 50*). It is located inside the building, typically one on each floor.

Packaged DX air handlers are available in capacities ranging from 5 to 30 tons and they come from the manufacturer completely assembled in one piece. Air handler heating sources include hot water coil, steam coil, or electric heat.

Figure 50 ◆ Packaged air handler.

Water-cooled condensing units are also available although they are not used very often. They are just like the air-cooled condensing units, except the compressor is mounted on top of a water-cooled condenser.

9.2.0 Central Station Air Handler

Central station air handlers (*Figure 51*) typically have selectable sections for air mixing, filtration, heating coils, cooling coils, and the fan. Some are equipped with an accessories section. The selectable sections are chosen by the design engineer and assembled by the manufacturer in a building-block fashion.

Central station air handling equipment is made in several capacity ranges, typically from 2,500 cfm up to 63,000 cfm. Units can be floor-mounted or hung from a ceiling inside the building. Increasing use is being made of units that are properly treated for outdoor applications in an effort to save valuable interior space. Some manufacturers even offer an outdoor unit modified for rooftop installation on a factory-furnished roof curb.

With the exception of curb-mounted units, air discharged from the units can be in almost any direction, front, top, bottom, or back.

9.3.0 Basic Air Handler Makeup

This section describes the basic makeup of an air handler. Included are descriptions of the different fan locations within an air handler. Also covered are the locations and functions of preheat and reheat coils used with air handlers.

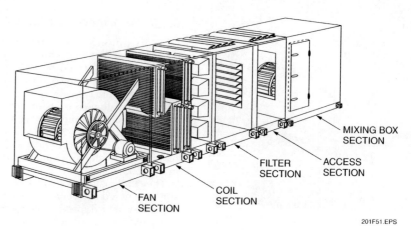

Figure 51 ◆ Central station air handler.

1.40   HVAC ◆ LEVEL TWO

9.3.1 Fan Location

The position of the coil within the air handler varies depending on the air handler style. Its position becomes part of the description of the air handler. All air handling units fall into one of three categories (*Figure 52*):

- Horizontal draw-through
- Horizontal blow-through
- Vertical draw-through

If the fan is placed downstream of the coil section, the unit is called a draw-through unit. If the fan is placed upstream of the coil section, the unit is called a blow-through unit. In addition, the relative height of the fan to the coil also contributes to the description of the air handler.

If the fan is located above the cooling coil, the unit is said to be a vertical unit. If the fan and coil are on the same level, the unit is said to be a horizontal air handler.

9.3.2 Heating Coil Location

The cooling and heating coils are located within the coil section of an air handler. The relative position of the heating coil to the cooling coil in the coil section depends on the use of the heating coil in the HVAC system.

In northern climates, when cold outdoor air is being drawn into the coil section, there is a possibility that the chilled water coil can freeze. To protect the chilled water coil, a heating coil is placed upstream of the cooling coil as a preheat coil (*Figure 53*). The preheat coil heats the outside air above the freezing point before allowing it to enter the chilled water coil.

Another common practice is to place a preheat coil in the outdoor air inlet section of the air handler instead of the coil section. The preheat coil heats the outdoor air above the freezing temperature of the chilled water.

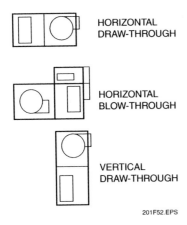

Figure 52 ♦ Fan-coil locations within an air handler.

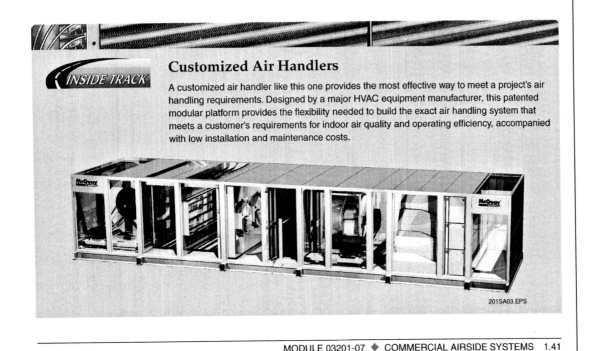

Customized Air Handlers

A customized air handler like this one provides the most effective way to meet a project's air handling requirements. Designed by a major HVAC equipment manufacturer, this patented modular platform provides the flexibility needed to build the exact air handling system that meets a customer's requirements for indoor air quality and operating efficiency, accompanied with low installation and maintenance costs.

Explain that some units have two preheat coils, one in the outdoor air duct and the other in the coil station.

Discuss how reheat is accomplished by placing a heating coil downstream of the cooling unit.

Identify the components in a basic air handler.

Describe the construction of an air handler.

Discuss different types of cooling and heating coils used with air handlers.

Explain that a drip pan with a trapped condensate drain is usually required under the coil section.

See the Teaching Tip for Section 9.4.0 at the end of this module.

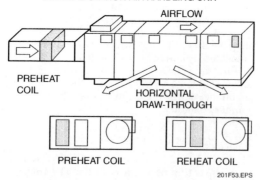

Figure 53 ◆ Coil placement within an air handler.

Some applications provide two preheat coils in the air handler. One coil is placed in the outdoor air duct. The second is located in the coil section and heats the outdoor air to the desired leaving temperature.

When the heating coil is placed downstream of the cooling coil, it is called a reheat coil and may serve two functions. When the chilled water coil is controlling humidity it will overcool the air in an attempt to reduce the air's humidity level. The reheat coil adds heat to the supply air in order to prevent overcooling the conditioned zones. The coil may also serve as a heating source for the conditioned space during cold weather when heating is required.

9.4.0 Basic Air Handler Component Descriptions

This section describes the components used in a basic air handler. The components include the following:

- Casing
- Coils
- Fans
- Filters
- Mixing box
- Humidifiers

9.4.1 Casing

Air handler unit casings are typically made of sheet metal panels held in place by angles or channels. The casings may also consist of self-supporting panels that do not require a separate support frame. Most air handlers used for comfort applications are made of steel and are factory-painted to control corrosion. Units made of heavy-gauge galvanized sheet steel are already protected from corrosion so they are not normally painted.

Fan and coil sections are normally insulated internally with one inch of fire-resistant insulation. All casings on the leaving or downstream side of the cooling coil are also insulated to prevent condensation on the panels.

9.4.2 Coils

Cooling coils are designed for use with chilled water or with an evaporating liquid refrigerant (*Figure 54*). The refrigerant coil is a DX coil. Both coil types are available in a variety of row depths, fin spacing, fin design, fin material, circuits, and in the case of DX coils, a variety of coil split arrangements. Coil row offerings may vary from 2 to 12 rows with 4, 6, and 8 being used predominantly in the comfort air conditioning industry.

Fins are mechanically bonded to the tubes of a coil to increase its effective heat-transfer surface area. Common spacing intervals are 8 and 14 fins per inch of tube length. Coil tubes are inserted through sheets of fin material, which are appropriately spaced. Each coil tube is then expanded to achieve a mechanical bond with the fins.

Aluminum and copper are used almost exclusively for both the tubes and fins of cooling coils. Copper tubes are used for the vast majority of commercial comfort applications. Aluminum fins are used most extensively for air conditioning duty. Copper fins are more expensive, and have limited use. However, copper is used as a precaution against corrosion.

The air velocity through the cooling coil is typically limited from 400 to 550 fpm. In non-latent applications, the velocity may go as high as 700 fpm. The number of rows and the fin spacing are determined by the temperature rise required for the job.

The air handler is usually equipped with a condensate drip pan located under the coil section. The drip pan is usually coated with ½ inch of waterproof insulation. This insulation is needed to prevent exterior sweating due to the cold condensate collected from the cooling coil. A liquid leg and condensate trap (liquid seal) are required to prevent trapping of condensate in the pan and to eliminate leakage of air through the condensate drain into the unit.

Heating coils that look just like chilled water coils are located either within the air handler or the ductwork. The energy sources used in heat-

1.42 HVAC ◆ LEVEL TWO

Discuss the types of fans used with air handlers.

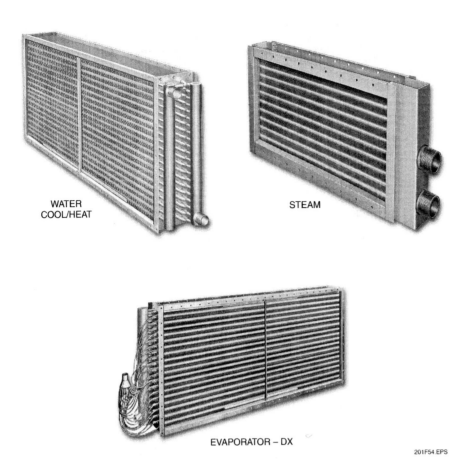

Figure 54 ♦ Typical cooling and heating coils.

ing coils include steam, hot water, and electricity. These coils are basically used for preheating (heating upstream of the cooling coil), and for tempering or reheating (heating downstream of the cooling coil with the cooling coil active).

9.4.3 Fans

Comfort air conditioning air handling units almost exclusively use centrifugal fans (*Figure 55*). Centrifugal fans come in three basic impeller designs that make up four basic centrifugal fan types. The first type of fan is the forward-curved fan. Forward-curved fans have a large number of blades that are curved in the direction of fan rotation. They provide large volumes of air at or below static pressures around 5.0 in wg. These fans are used in most air handler applications due to their lightweight and fairly inexpensive construction. The second type is called backward-inclined because the tips of the fan blades point in a direction opposite to the direction of fan rotation. The impeller blades have either single thickness straight or curved-type construction. Backward-inclined fans are noted for their production of medium to high static pressures.

The airfoil fan is a refinement of the backward-inclined fan. The airfoil fan uses a double thickness metal blade shaped like the cross-section of a curved airplane wing. The airfoil fan is designed to improve the static efficiency of airflow through the wheel. The overall construction of the fan is much heavier to support the high pressures it produces. Because of its heavier construction, it tends to be used in high-pressure applications (6.0 to 12.0 in wg).

Discuss fan design and fan volume control.

Explain the operation of a mixing box.

Identify applications where humidifiers are commonly used with air handlers.

Figure 55 ♦ Centrifugal fan.

Figure 56 ♦ Fan mounted in the fan section.

The fourth type of fan is the radial fan with blades that are straight, single-thickness metal with the tips pointing in a radial direction. This type of fan is not used in commercial air conditioning.

Fan section designs vary from manufacturer to manufacturer. The motor and drive assembly is typically mounted within the fan section to reduce the equipment room space required at installation (*Figure 56*). With this internal drive and fan support arrangement, the fan shaft is relatively short and the motor and bearings are within the airstream.

Another approach involves a motor and drive assembly mounted outside the fan casing. In this situation, the fan wheel is fastened to a longer fan shaft that extends through the fan section casing. The fan motor is typically mounted on the fan casing and drives the fan through a series of belts and pulleys.

Air handlers above 20 tons in capacity can be applied to VAV systems. In these systems, fan capacity reduction is necessary to reduce duct pressure rise caused by throttling the air terminals. It is also necessary to take advantage of fan energy savings. Inlet guide vanes and variable speed drives (inverters) are two common air handler accessories that make this possible. Inlet guide vanes were covered earlier in the module.

Another popular method of fan volume control is the use of variable-speed drives. An inverter receives a signal from a pressure sensor mounted in the supply duct and varies the drive motor speed accordingly. Lowering the fan speed lowers the fan volume and the fan horsepower.

9.4.4 Mixing Box

The mixing box is a plenum with two air inlets. Each inlet is provided with a set of dampers. The mixing box is installed at the entering air end of the air handler and mixes return and outside air. The damper sets are mechanically linked, so as one damper closes the other damper is opening. An operator is used to open the dampers to a minimum ventilation position, or to modulate them for economizer control. In addition, the dampers may be controlled by an indoor air quality (IAQ) routine that modulates the dampers to achieve a maximum carbon dioxide level in the control zones.

9.4.5 Humidifiers

Air handlers commonly contain humidification equipment to maintain consistent humidity levels in commercial and industrial facilities. Textile facilities, printing plants, museums, data processing rooms, and food processing areas are a few examples of applications where humidity levels must be maintained to ensure that manufactured or stored product quality standards are met. Of course, humidity levels play a key role in human comfort as well.

Several types of humidifiers are generally applied in commercial air handling systems.

These include spray humidifiers, ultrasonic humidifiers, and steam humidifiers. The final selection of a humidification approach depends a great deal on the quality of the available water supply, capacity requirements, and energy costs.

Most humidifiers can easily be categorized by the process used to create minute water particles. Units using an **isothermal** process generally use electricity or gas as an external heat source to convert liquid water to steam. The energy consumed in the process (added to the water) is approximately 1,000 Btuh/lb of water. Systems that use an **adiabatic** process employ mechanical energy to create water particles or evaporate water from a medium. This process allows the water to remove heat from the airstream, again at the rate of roughly 1,000 Btuh/lb of water. The adiabatic process is generally the most energy-efficient choice overall, but other factors, such as water quality requirements, can significantly affect final operating and maintenance costs. The positive or negative heat gain resulting from humidification should be considered in the heating and cooling system load calculations and design.

Spray humidifiers (*Figure 57*) discharge atomized water particles into the airstream, using the adiabatic process. Although some models rely solely on water pressure to atomize the water, models that provide the smallest particles for quick absorption into the airstream rely on compressed air to create an atomized fog. Each nozzle provides a given amount of flow, depending on both water pressure and air pressure. In larger applications, multiple spray units are mounted to a single manifold spanning the width of the air handling unit. Although the output of the individual nozzles remains constant, capacity is controlled by the number of spray heads activated, based on demand. With spray units, minerals and other solids carried in the water supply are also added to the airstream. As a result, proper water filtration is required to reduce airborne dust. Where a dust-free environment is required, a de-mineralized water supply must be provided, which also reduces spray apparatus maintenance due to the buildup of mineral deposits.

Ultrasonic units use high-frequency, inaudible sound waves to break liquid water into extremely small particles approximately 1 micron (0.001 mm) in diameter. An ultrasonic unit can be placed directly inside an air handling unit or duct system, downstream of the air handling unit. Electronic transducers, which convert electronic signals into high frequency sound waves, are

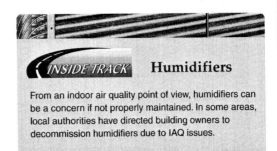

INSIDE TRACK — Humidifiers

From an indoor air quality point of view, humidifiers can be a concern if not properly maintained. In some areas, local authorities have directed building owners to decommission humidifiers due to IAQ issues.

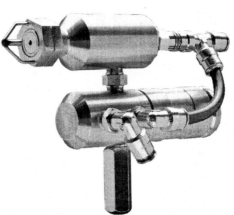

Figure 57 ♦ Ultrasonic atomizing spray humidifier assembly.

immersed in the water. Due to its mass, the water is unable to keep up with the speed of the sound wave, creating momentary vacuums and high compressions near the transducer. In the vacuum stage, water cavitates and boils into a gas. The compression stage forces the newly formed gas bubble through the water, creating an airborne mist. Since this is an adiabatic process, heat is not added to the water and a cooling effect in the airstream results. Most units are of modular construction. This allows units of specific capacity to be built by using multiple transducers, along with individual control by demand. Advantages include low power consumption, instantaneous response to demand, and low water consumption, since little or no water is lost to flushing cycles. Operation with a de-mineralized water supply to eliminate mineral deposits and airborne dust is generally required and always recommended.

Discuss the different types of humidifiers used in air handlers.

Explain the importance of properly maintaining humidifiers.

Describe spray and ultrasonic humidifiers.

Describe steam and electrode-type humidifiers.

Show trainees how to identify the types of commercial air systems installed in selected buildings through observation of the equipment.

Have trainees practice identifying the types of commercial air systems installed in selected buildings through observation of the equipment. Note the proficiency of each trainee. This laboratory corresponds to Performance Task 1.

Steam humidifiers rely on the isothermal process to create steam, with electricity, gas, or central steam supply used as the external heat source. Heat contained in the steam is added to the airstream as a result. In many industrial applications, where a plentiful supply of steam might exist, a humidifier using a steam-to-steam heat exchanger provides a cost-effective solution. Steam from the main plant can also be injected directly into the airstream. Such units are relatively simple in design and construction, and take advantage of an existing energy source. Electric units are more commonly used in commercial applications. The steam-generating apparatus is generally installed outside of the air handling unit and piped to the airstream. Electric heating elements placed in water reservoirs provide the needed heat to create the steam. This process, much like that used in water distillation units, creates water vapor free of the minerals and other particulates. These solids are left behind in the reservoir and as hard deposits on the heating elements. Periodic flushing cycles allow the unit to rinse away many of the remaining solids. These humidifiers can be used with most water sources, although maintenance is significantly reduced and reliability enhanced with the use of de-mineralized or de-ionized water.

Electrode-type electric units *(Figure 58)* pass electrical current through water, inside a sealed plastic canister, to create heat and generate steam. One significant advantage of electrode-type humidifiers over many other types is their ability to modulate steam production to match environmental requirements by varying current flow. Although most types of humidifiers can use de-mineralized or distilled water to reduce mineral deposits, the electrode-type humidifier depends on some mineral content to create sufficiently high conductivity in the water. Distilled or de-mineralized water cannot be used. As a result, water conditions play a major role in the life of

Figure 58 ◆ Electrode-type electric humidifier.

the electrodes. The vast majority of canisters are hermetically sealed and not serviceable. Once the electrodes become fouled with mineral deposits and efficiency is substantially reduced, the canister is replaced as a unit. On larger models, the canister can be opened and the electrode assembly replaced periodically.

Dispersion of the steam into the air handling unit or duct can be done through one or more tubes and/or panels, depending on the desired capacity and the distance available for the steam to be fully absorbed by the airstream. With steam humidification, it is important to prevent any condensed water in the steam supply from reaching the nozzles. This causes spitting or dripping. Water droplets of this size are not readily absorbed into the airstream and fall to the bottom of the unit cabinet or duct, potentially causing water damage or IAQ problems from microbial growth. Ultraviolet lamps that prevent or destroy microbial growth are often good partners for systems that incorporate humidification.

Review Questions

1. All-air systems equipped with direct expansion (DX) coils are commonly used in buildings requiring a cooling capacity below _____ tons.
   a. 50
   b. 100
   c. 150
   d. 200

2. An internal building air pressure normally sufficient to prevent infiltration is between _____.
   a. 0.05 and 0.10 in wg
   b. 0.10 and 0.15 in wg
   c. 0.15 and 0.20 in wg
   d. 0.20 and 0.25 in wg

3. Which of the following accessories depends on the outdoor temperature being below 55°F?
   a. Dry-bulb economizer
   b. Powered exhaust
   c. Enthalpy economizer
   d. Integrated economizer

4. In a variable volume, variable temperature (VVT) system, the source of heat for the zones is _____.
   a. air terminal reheat
   b. heating coil
   c. perimeter baseboard heating
   d. separate overhead fan coil heating system

5. A building requiring a cooling capacity of under 20 tons with small control zones typically would use a _____ system.
   a. single-zone constant volume
   b. multi-zone constant volume
   c. variable volume, variable temperature
   d. variable air volume

6. In the past, duct systems were classified in terms of their application, velocity, and pressure. These older terms were rather vague and have been replaced by SMACNA _____ classification values.
   a. weight
   b. pressure
   c. insulation
   d. velocity

7. The best location to install a cooling diffuser delivering cool air is in the _____.
   a. floor along the outside wall
   b. center of the ceiling
   c. ceiling close to the outside wall
   d. floor along the inside wall

8. A diffuser that delivers supply air through the perimeter of the diffuser and returns the room air through the center of the same diffuser is called a _____ diffuser.
   a. liner slot
   b. perforated panel
   c. concentric ductwork
   d. ceiling

9. When a parallel fan-powered mixing box is used in a VAV system, the air terminal fan is activated _____.
   a. when the zone calls for cooling
   b. during morning warmup
   c. when the zone requires heating
   d. during optimal start

10. Which of the following statements is correct?
    a. The parallel fan-powered mixing box is designed for use in perimeter zones.
    b. The fan in a series-powered mixing box operates only during the heating mode.
    c. VAV systems are typically used in buildings with a cooling capacity of less than 20 tons.
    d. Multi-zone systems are typically used in buildings with more than 30 control zones.

11. Any control zone served by a dual-duct air terminal that requires heat is satisfied by _____.
    a. perimeter baseboard hot water
    b. a heater connected to the air terminal discharge
    c. a central air handler heating coil
    d. the fan being activated in the air terminal

Review Questions

12. If it is equipped with heat, the heat source in a packaged air conditioner is _____.
    a. electric heat
    b. LP gas
    c. hot water
    d. a heat pump

13. A unit equipped with a four-way reversing valve is the _____.
    a. packaged air conditioner
    b. year-round air conditioner
    c. vertical packaged air conditioner
    d. packaged heat pump

14. An air-cooled condenser in a packaged rooftop unit provides a maximum subcooling of _____.
    a. 5°F
    b. 10°F
    c. 20°F
    d. 30°F

15. In order to improve part load humidity control with small sized rooftop units, a(n) _____ option may be used that can raise the unit's discharge air temperature up to about 75°F at part load.
    a. energy recovery heat pump
    b. flat plate heat exchanger
    c. run around loop
    d. supply air reheat coil

Summary

This module has introduced you to all-air systems used in commercial buildings. All-air systems provide for complete sensible and latent cooling, preheating, and humidifying of the air supplied by the system. Heating may be provided by the same airstream, either in the central system or at a particular zone. In some applications, heating is provided by a separate system.

The basic components used in all-air systems consist of packaged air conditioning units, air-handling units, air terminals, duct systems, and related controls. Depending on the type of building, its commercial application, and the number of controlled zones, these components are arranged into one of five different types of all-air systems to meet the building's requirements. The five types of all-air systems are as follows:

- Single-zone constant-volume systems
- Multi-zone constant-volume systems
- Variable volume, variable temperature (VVT) systems
- Variable air volume (VAV) systems
- Dual-duct variable air volume systems

Based on the information covered in this module, the next time you walk into a commercial building you will be equipped to recommend the type of system that should be installed, recognize what type of system actually is installed, and know how the system should operate.

Notes

Trade Terms Introduced in This Module

Adiabatic: A term used to describe a thermodynamic process that happens without loss or gain of heat.

Air handler: A commercial air handler is a packaged unit containing a cooling coil and usually several other components such as a filter, heating coil or element, dampers, and fans connected to ducts. Conditioned air from the air handler leaves through ducts and is delivered to air terminals for distribution around the conditioned spaces.

Constant-volume system: A constant-volume system maintains a constant airflow while varying the air temperature in response to the space load.

Control zone: In HVAC, a control zone is a building, group of rooms, single room, or part of a room controlled by its own thermostat.

Dual-duct systems: Dual-duct systems condition all the air in a central unit and distribute it to the conditioned spaces through two parallel ducts, one duct carrying cold air and the other warm air. In each conditioned zone, a device mixes the warm and cold air in proper proportions to satisfy the load of the zone.

Isothermal: The relationship between variables, especially pressure and volume, at a constant temperature.

Packaged air conditioner (PAC): Packaged air conditioners provide cooling only or can provide both cooling and heating when equipped with electric resistance heaters.

Packaged rooftop unit: A self-contained air conditioning unit that is installed outdoors on a rooftop with connections through a roof opening to the internal duct system. They are commonly used on flat roof commercial structures such as office buildings and shopping malls.

Packaged unit: Packaged units are factory-assembled units that contain all the components needed to support an HVAC function, such as cooling, heating, or air handling.

Sick building syndrome: A combination of symptoms (headache, nausea, eye, nose and throat irritation) that are attributed to flaws in the HVAC systems. Symptoms can be cured by boosting the overall turn-over rate in fresh air exchange with the outside air. Other causes have been attributed to contaminants produced by out-gassing of some types of building materials, or improper exhaust ventilation of light industrial chemicals.

Variable air volume (VAV) system: A VAV system is one that controls the temperature within a control zone by varying the quantity of supply air rather than by varying the supply air temperature. Dual-duct VAV systems blend cold and warm air in various volume combinations.

Variable volume, variable temperature (VVT) system: A VVT system is one that delivers a variable volume of air to each controlled zone, as the load dictates. The temperature of the air supplied by the central unit varies with time.

Year-round air conditioner (YAC): The year-round air conditioner provides both cooling and heating. It differs from a packaged air conditioner in that its heating capability is provided by a natural or LP gas heating section, instead of electric resistance heaters.

Zoned system: A system that has more than one thermostat used to control the areas (zones) it conditions.

Additional Resources and References

Additional Resources

This module is intended to be a thorough resource for task training. The following reference works are suggested for further study. These are optional materials for continued education rather than for task training.

HVAC Systems, 1992. Samuel C. Monger. Englewood Cliffs, NJ: Prentice Hall.

HVAC Systems and Equipment Handbook, 2000. Atlanta, GA: American Society of Heating, Refrigeration, and Air Conditioning Engineers (ASHRAE).

Figure Credits

Carrier Corporation, 201F07 (photo), 201F14, 201F15, 201T01, 201F21, 201F30–201F33, 201F38–201F40, 201F43–201F47, 201F49–201F51, 201F56

Titus, 201F25, 201F27

Selkirk Corporation, 201F26

Hart & Cooley, Inc., 201F28

Nailor Industries Inc., 201F35 (photo)

ECR International – Dunkirk Boilers, 201F41

McQuay International, 201SA03, 201F54

CML Northern Blower, Inc., 201F55

Axair Nortec, 201F57

Topaz Publications, Inc., 201F42, 201F58

MODULE 03201-07 — TEACHING TIPS

The following are suggested activities or instructional methods to help you teach the material in this module.

General

When you call on someone to answer a question, the rest of the class relaxes or even tunes out because they expect that the question and answer will take place only between you and the trainee you called on. Instead, use this technique to involve more trainees in answering questions and to keep them on their toes.

1. Ask trainees to define a term or explain a concept.
2. After one trainee has answered, ask a trainee seated nearby if the answer is right. Then ask whether a trainee in the back of the room agrees.
3. Ask trainees to explain why they think an answer is right or wrong.
4. Use the session to clear up incorrect ideas and encourage trainees to learn from their mistakes.

Sections 1.0.0 through 9.0.0

Trade Terms

This Quick Quiz will familiarize trainees with trade terms commonly used by HVAC specialists. You will need photocopies of the quiz provided on the following page. Trainees will need pencils. If you allow trainees to use the Trainee Module, decrease the amount of time you give them to complete the quiz.

1. Make a photocopy of the quiz for each trainee.
2. Give trainees between 5 and 10 minutes to complete the quiz.
3. Go over the answers to the quiz.
4. Ask trainees if they have questions.

Answers to Quick Quiz

1. c
2. f
3. g
4. b
5. d
6. h
7. j
8. a
9. e
10. i

Quick Quiz   *Trade Terms*

For each description listed, identify the term that the text best describes. Write the corresponding letter in the blank provided.

_____ 1. A building, group of rooms, single room, or part of a room controlled by its own thermostat is called a(n) _____.

_____ 2. Without dilution, toxins can accumulate in a building and cause a condition that can contribute to _____.

_____ 3. A system that controls the temperature within a room by varying the quantity of supply air is called a(n) _____.

_____ 4. A small building of less than 5 tons would have a single-zone _____.

_____ 5. In older buildings that have been modified to improve energy use, you may find a(n) _____.

_____ 6. One of the most popular ways to achieve comfort control in small buildings from 5 to 20 tons is a(n) _____.

_____ 7. A unit with integral heating and cooling functions is called a(n) _____.

_____ 8. The most energy efficient choice for dehumidification is _____.

_____ 9. A unit that provides cooling capacity only, or cooling capacity with electric heat, is known as a(n) _____.

_____ 10. Many commercial building have a single system that controls multiple zones called a(n) _____.

a. adiabatic
b. constant-volume system
c. control zone
d. dual-duct VAV system
e. packaged air conditioner (PAC)
f. sick building syndrome
g. variable air volume (VAV)
h. variable volume, variable temperature (VVT)
i. year-round air conditioner (YAC)
j. zoned system

Section 4.0.0 *Sick Building Syndrome*

This exercise will familiarize trainees with sick building syndrome. Trainees will need pencils and paper. Arrange for a speaker to give a presentation to the class on sick building syndrome. Allow 30 to 45 minutes for this exercise.

Alternately, obtain fact sheets from the US EPA or other sources on sick building syndrome. A detailed analysis of the causes and solutions is contained in the US EPA Office of Indoor Air Quality Fact Sheet #4: Sick Building Syndrome, which is available online at www.epa.gov/iaq/pubs/sbs.html.

Additionally, trainees can use the internet and other sources to identify articles on the causes and solutions for sick building syndrome. Have each trainee summarize the information they located for the class.

1. Before the presentation, discuss sick building syndrome. Brainstorm with the trainees for questions to ask the presenter.
2. Introduce the speaker who will give a presentation on the causes, investigation procedures, and solutions to sick building syndrome.
3. After the presentation, answer any questions the trainees may have.

Section 8.1.0 *Packaged Equipment*

This exercise familiarizes trainees with packaged equipment. Trainees will need pencils and paper. You will need to arrange for a manufacturer's representative or environmental specialist to give a presentation on various types of packaged equipment available. Allow 30 to 45 minutes for this exercise.

1. Before the exercise, review the topics to be covered and brainstorm possible questions with the trainees.
2. Introduce the speaker who will give a presentation on various types of packaged systems and their applications.
3. Have the trainees take notes and write down additional questions during the presentation.
4. After the presentation, answer any questions trainees may have.

Section 9.4.0 *Air Handler Components*

This exercise will familiarize trainees with the components of air handlers. Trainees will need appropriate personal protective equipment, pencils, and paper. Arrange for an opportunity to examine an air handler. Allow 30 to 45 minutes for this exercise.

Alternately, have trainees use the internet and other sources to obtain information on air handler components. Have each trainee summarize the information they located for the class.

1. Before the tour, review the components of an air handler and their function.
2. Have trainees examine an air handler. Point out the various components and explain their functions in the system.
3. After the tour, answer any questions the trainees may have.

MODULE 03201-07 — ANSWERS TO REVIEW QUESTIONS

Answer	Section Reference
1. b	3.1.0
2. a	4.0.0
3. a	4.4.1
4. b	5.3.0
5. c	5.3.1
6. b	6.1.0
7. b	7.1.0; Figure 23
8. c	7.1.6
9. c	7.2.4
10. a	7.2.4
11. b	7.2.6
12. a	8.1.1
13. d	8.1.4
14. c	8.2.2
15. d	8.3.7

CONTREN® LEARNING SERIES — USER UPDATE

NCCER makes every effort to keep these textbooks up-to-date and free of technical errors. We appreciate your help in this process. If you have an idea for improving this textbook, or if you find an error, a typographical mistake, or an inaccuracy in NCCER's Contren® textbooks, please write us, using this form or a photocopy. Be sure to include the exact module number, page number, a detailed description, and the correction, if applicable. Your input will be brought to the attention of the Technical Review Committee. Thank you for your assistance.

Instructors – If you found that additional materials were necessary in order to teach this module effectively, please let us know so that we may include them in the Equipment/Materials list in the Annotated Instructor's Guide.

Write: Product Development and Revision
National Center for Construction Education and Research
3600 NW 43rd St, Bldg G, Gainesville, FL 32606

Fax: 352-334-0932

E-mail: curriculum@nccer.org

Craft _____ Module Name _____

Copyright Date _____ Module Number _____ Page Number(s) _____

Description

_____

_____

_____

_____

(Optional) Correction

_____

_____

_____

(Optional) Your Name and Address

_____

_____

_____

Module 03202-07

Chimneys, Vents, and Flues

NCCER STANDARDIZED CRAFT TRAINING PROGRAM

The National Center for Construction Education and Research (NCCER) provides a standardized national program of accredited craft training. Key features of the program include instructor certification, competency-based training, and performance testing. The program provides trainees, instructors, and companies with a standard form of recognition through a National Craft Training Registry. The program is described in full in the *Guidelines for Accreditation*, published by NCCER. For more information on standardized craft training, contact the NCCER by writing us at 3600 NW 43rd St., Bldg. G, Gainesville, FL 32606; calling 352-334-0911; or emailing info@nccer.org. More information may be found at our website, www.nccer.org.

HOW TO USE THIS ANNOTATED INSTRUCTOR'S GUIDE

Each page presents two sections of information. The larger section displays each page exactly as it appears in the Trainee Module. The narrow column ties suggested trainee and instructor actions to each page and provides icons (detailed below) to call your attention to material, safety, audiovisual, or testing requirements. The bottom of each page includes space for your notes.

The **Audiovisual** icon indicates an appropriate time to show a transparency or other audiovisual aid.

The **Classroom** icon prompts you to define a term, stress a point, ask trainees to explain a concept, or give examples.

The **Demonstration** icon directs you to show trainees how to perform tasks.

The **Examination** icon tells you to administer the written module examination.

The **Homework** icon is placed where you may wish to assign reading for the next class, assign a project, or advise trainees to prepare for an examination.

The **Laboratory** icon is used when trainees are to practice performing tasks.

The **Materials** icon is a reminder for you to gather materials needed for classes, labs, and testing.

The **Performance Testing** icon tells you to administer a performance test or a portion thereof.

The **Safety** icon is used to emphasize safety issues. It is often keyed to *Caution* and *Warning!* statements in the Trainee Module.

The **Teaching Tip** icon indicates additional guidance is available, such as how to conduct an exercise, get the most educational value from a field trip, or encourage class participation. Teaching Tips may expand on a feature (*Think About It, Did You Know?*) or provide *Quick Quizzes* or similar exercises. You will be referred to the Teaching Tips section at the back of the module if there is additional material.

The **Combination** icon indicates that the laboratory listed corresponds with a performance task. If desired, you can note the proficiency of the trainees during the laboratory, and use it to satisfy performance testing requirements.

PREPARATION

Before teaching this module, you should review the Objectives, Performance Tasks, Materials and Equipment List, and Module Outline. Be sure to allow ample time to prepare your own training or lesson plan and gather all required materials and equipment.

Chimneys, Vents, and Flues
Annotated Instructor's Guide

Module 03202-07

MODULE OVERVIEW

This module covers proper venting of fossil-fuel furnaces and the procedures for selecting and installing vents in all types of gas furnaces.

PREREQUISITES

Prior to training with this module, it is recommended that the trainee shall have successfully completed *Core Curriculum*; *HVAC Level One*; and *HVAC Level Two*, Module 03201-07.

OBJECTIVES

Upon completion of this module, the trainee will be able to do the following:

1. Describe the principles of combustion and explain complete and incomplete combustion.
2. Describe the content of flue gas and explain how it is vented.
3. Identify the components of a furnace vent system.
4. Describe how to select and install a vent system.
5. Perform the adjustments necessary to achieve proper combustion in a gas furnace.
6. Describe the techniques for venting different types of furnaces.
7. Explain the various draft control devices used with natural-draft furnaces.
8. Calculate the size of a vent required for a given application.
9. Adjust a thermostat heat anticipator.

PERFORMANCE TASKS

Under the supervision of the instructor, the trainee should be able to do the following:

1. Measure supply and return temperature and determine the temperature rise of a furnace.
2. Adjust a thermostat heat anticipator.
3. Calculate the correct size and type of PVC pipe using manufacturer's instructions or *National Fuel Gas Code* or American Gas Association specifications.
4. Calculate the correct size and type of furnace vent connector and metal vent using manufacturer's instructions or *National Fuel Gas Code* or American Gas Association specifications.

MATERIALS AND EQUIPMENT LIST

Overhead projector and screen
Transparencies
Blank acetate sheets
Transparency pens
Whiteboard/chalkboard
Markers/chalk
Pencils and scratch paper
Copy of latest edition of the *National Fuel Gas Code* or American Gas Association specifications
Various vent manufacturers' product data and catalogs
Videotape (optional) *Principles of Gas Combustion*
Videotape (optional) *Ventinox Chimney Solution*
TV/VCR/DVD player
Thermometer

Selection of vent piping:
    Double wall (Types B, L, and B-W)
    Single wall
    Schedule 40 PVC
High-temperature plastic
PVC and metal tubes
Smoke source
Flame source
Concentric vent termination
Temperature probes
Operating gas-fired furnace
Copies of the Quick Quiz*
Module Examinations**
Performance Profile Sheets**

* Located in the back of this module.
**Located in the Test Booklet.

SAFETY CONSIDERATIONS

Ensure that the trainees are equipped with appropriate personal protective equipment and know how to use it properly. The module requires that trainees work with operating gas-fired furnaces. Ensure all trainees are briefed on fire safety procedures.

ADDITIONAL RESOURCES

This module is intended to present thorough resources for task training. The following reference works are suggested for both instructors and motivated trainees interested in further study. These are optional materials for continued education rather than for task training.

Mid-Efficiency Furnace Installation Awareness. Latest Edition. Syracuse, NY: Carrier Corporation.

National Fuel Gas Code (NFPA 54/ANSI/Z223.1). Latest Edition. Quincy, MA: National Fire Protection Association.

TEACHING TIME FOR THIS MODULE

An outline for use in developing your lesson plan is presented below. Note that each Roman numeral in the outline equates to one session of instruction. Each session has a suggested time period of 2½ hours. This includes 10 minutes at the beginning of each session for administrative tasks and one 10-minute break during the session. Approximately 5 hours are suggested to cover *Chimneys, Vents, and Flues*. You will need to adjust the time required for hands-on activity and testing based on your class size and resources. Because laboratories often correspond to Performance Tasks, the proficiency of the trainees may be noted during these exercises for Performance Testing purposes.

Topic	Planned Time
Session I. Introduction to Chimneys and Venting Requirements	
A. Introduction	
B. Combustion	_____
C. Flue Gases	_____
D. Furnace Venting	_____
E. Vent System Components	_____
F. Natural-Draft Furnaces	_____
G. Induced-Draft Furnaces	_____
H. Laboratory	_____
Trainees practice measuring the temperature and determining the temperature rise. This laboratory corresponds to Performance Task 1.	
I. Laboratory	_____
Trainees practice adjusting the thermostat anticipator. This laboratory corresponds to Performance Task 2.	
Session II. Vent Calculations, Review, and Testing	
A. Venting Considerations	_____
B. Laboratory	_____
Trainees practice calculating the correct size and type of vent connector and metal vent. This laboratory corresponds to Performance Task 4.	
C. Condensing Gas Furnaces	_____
D. Laboratory	_____
Trainees practice calculating the correct size and type of PVC pipe. This laboratory corresponds to Performance Task 3.	
E. Draft Controls	_____
F. Review	_____

G. Module Examination
   1. Trainees must score 70% or higher to receive recognition from NCCER.
   2. Record the testing results on Craft Training Report Form 200, and submit the results to the Training Program Sponsor.
H. Performance Testing
   1. Trainees must perform each task to the satisfaction of the instructor to receive recognition from NCCER. If applicable, proficiency noted during laboratory exercises can be used to satisfy the Performance Testing requirements.
   2. Record the testing results on Craft Training Report Form 200, and submit the results to the Training Program Sponsor.

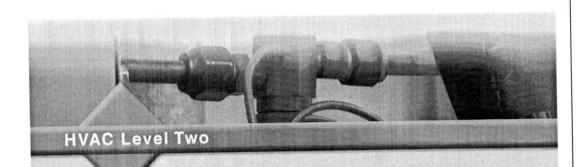

HVAC Level Two

03202-07

Chimneys, Vents, and Flues

Assign reading of Module 03202-07.

03202-07
Chimneys, Vents, and Flues

Topics to be presented in this module include:

1.0.0	Introduction	2.2
2.0.0	Combustion	2.2
3.0.0	Flue Gases	2.4
4.0.0	Furnace Venting	2.5
5.0.0	Vent System Components	2.7
6.0.0	Natural-Draft Furnaces	2.9
7.0.0	Induced-Draft Gas Furnaces	2.9
8.0.0	Condensing Gas Furnaces	2.14
9.0.0	Draft Controls	2.17

Overview

In order to operate efficiently and safely, furnaces and other fuel-burning appliances must have an adequate supply of air to support combustion, as well as proper venting of the gases created as byproducts of combustion. There are two specific concerns when selecting and installing a furnace vent. The first is to make sure that toxic by-products of combustion are vented to the outdoors. The second is to avoid condensation that could corrode the vent and the furnace heat exchangers. There are well-defined standards covering the materials and methods used to vent furnaces. These standards are reflected in the furnace manufacturer's installation instructions. Failure to follow these instructions could create a deadly environment for building occupants.

Instructor's Notes:

Objectives

When you have completed this module, you will be able to do the following:

1. Describe the principles of combustion and explain complete and incomplete combustion.
2. Describe the content of flue gas and explain how it is vented.
3. Identify the components of a furnace vent system.
4. Describe how to select and install a vent system.
5. Perform the adjustments necessary to achieve proper combustion in a gas furnace.
6. Describe the techniques for venting different types of furnaces.
7. Explain the various draft control devices used with natural-draft furnaces.
8. Calculate the size of a vent required for a given application.
9. Adjust a thermostat heat anticipator.

Trade Terms

Complete combustion
Condensing furnace
Dilution air
Heat anticipator
Incomplete combustion
Induced-draft furnace
Natural-draft furnace
Primary air
Secondary air
Vent
Vent connector

Required Trainee Materials

1. Pencil and paper
2. Appropriate personal protective equipment

Prerequisites

Before you begin this module, it is recommended that you successfully complete *Core Curriculum; HVAC Level One;* and *HVAC Level Two,* Module 03202-07.

This course map shows all of the modules in the second level of the HVAC curriculum. The suggested training order begins at the bottom and proceeds up. Skill levels increase as you advance on the course map. The local Training Program Sponsor may adjust the training order.

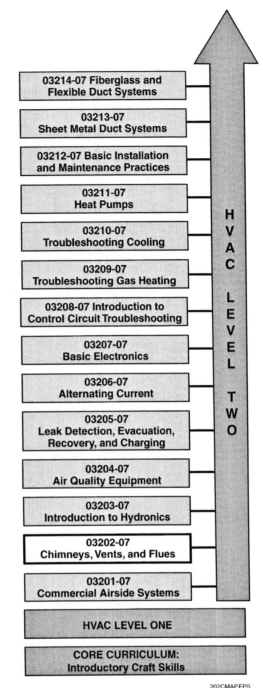

Ensure that you have everything required to teach the course. Check the Materials and Equipment list at the front of this module.

See the general Teaching Tip at the end of this module.

Explain that terms shown in bold are defined in the Glossary at the back of this module.

Show Transparency 1, Objectives, and Transparency 2, Performance Tasks. Review the goals of the module, and explain what will be expected of the trainee.

Review the modules covered in Level Two and explain how this module fits in.

Show Transparency 3 (Figure 1). Discuss the function of a vent system. Identify venting requirements for different types of furnaces.

List the conditions necessary for combustion.

Explain the difference between complete combustion and incomplete combustion.

See the Teaching Tip for Sections 2.0.0–9.0.0 at the end of this module.

1.0.0 ♦ INTRODUCTION

All fossil-fuel furnaces produce flue gases as a byproduct of burning fuel. These gases contain materials that are dangerous. In addition, they contain moisture, soot, and acids that can damage equipment. Flue gases must be vented to the outdoors in order for occupants to avoid their harmful effects (*Figure 1*). The design of the **vent** system depends on the building construction, the type of furnace, and the temperature of the flue gases.

Proper venting is especially important in **induced-draft furnaces**. These are furnaces with an Annual Fuel Utilization Efficiency (AFUE) rating of 78 to 85 percent. Because of their low flue gas temperatures, these furnaces need to be designed in a way that will prevent the formation of condensation. Moisture can damage the furnace and vent.

Natural-draft furnaces are no longer made in large numbers because they cannot meet the minimum AFUE standard of 78 percent without the use of special accessories. Although you may service them, it is unlikely that you will have to install one.

High-efficiency **condensing furnaces** (AFUE of 90 percent and higher) are fairly easy to vent. The condensing coil removes much of the moisture before the flue gases reach the vent. The furnace is also equipped to capture and dispose of any condensation that forms.

2.0.0 ♦ COMBUSTION

During combustion, oxygen combines with fuel to release stored energy in the form of heat. There are three conditions necessary for combustion to take place:

- First, there must be fuel. The fuel can be gas, such as natural gas; liquid, such as fuel oil; or solid, such as coal. Two elements that all fuels have in common are carbon and hydrogen.
- Second, fuel must be heated in order to burn or to reach the kindling temperature. A pilot burner or electronic ignition is used to ignite a gas burner, an electric spark is used to ignite fuel oil, and a wood-burning fire is used to ignite coal.
- Third, oxygen must be present for burning to take place.

There are two types of combustion: **complete combustion** and **incomplete combustion**. Incomplete combustion is dangerous; therefore, complete combustion must be obtained in all fuel-burning systems.

2.1.0 Complete Combustion

Complete combustion takes place when carbon combines with oxygen to form carbon dioxide. Carbon dioxide is nontoxic and can be exhausted to the outdoors. Hydrogen combines with oxygen to form water vapor, which is also harmless.

2.2.0 Incomplete Combustion

Incomplete combustion results from too little oxygen and causes the formation of undesirable products, such as carbon monoxide, pure carbon or soot, and aldeheydes (highly reactive compounds). Both carbon monoxide and aldehydes are toxic. Soot coats the heating surfaces of the furnace and reduces heat transfer.

Enough air must be provided to allow for proper combustion to take place, and to avoid

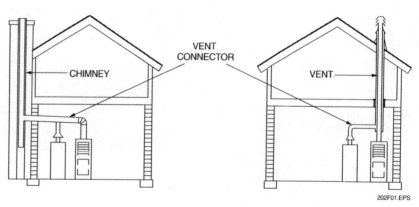

Figure 1 ♦ Furnace venting.

incomplete combustion. In practice, 15 to 30 percent excess air has been found to provide satisfactory combustion without seriously lowering burner efficiency. Operating burners at a lower percentage of excess air is not practical, because the small improvement in efficiency may not offset the hazards that may be created.

2.3.0 Combustion Efficiency

When fuel is burned in a furnace, a certain amount of heat is lost in the hot gases that go out through the vent. This heat loss is necessary to establish a draft in the chimney or vent, but should be minimized to allow the furnace to operate at its peak efficiency. For example, if the amount of heat lost is 20 percent, the furnace efficiency would be 80 percent.

Air entering the furnace at room temperature or lower is heated to flue gas temperatures that range from 100°F to 600°F, depending upon the design and adjustments of the furnace. The flue gas temperature in a natural-draft furnace ranges from 350°F to 600°F; from 275°F to 400°F in an induced-draft furnace; and from 100°F to 125°F in a high-efficiency furnace.

The acceptable minimum amount of carbon dioxide should be about 8.5 percent for natural gas, with no carbon monoxide. For oil, it should be about 10 percent without any smoke.

2.4.0 Flames

The type of flame and the intensity with which it burns affects the efficiency of the heating unit. Pressure-type oil burners burn with a yellow flame. Bunsen-type gas burners burn with a blue flame. The difference is mainly due to the manner in which air is mixed with the fuel. The color of a gas flame indicates the amount of air being supplied for combustion. A yellow flame is produced when gas is burned by igniting it as it flows out of the open end of a pipe. A blue flame is produced when about 50 percent of the required air is mixed with the gas prior to ignition. This mix is called **primary air** (*Figure 2*). The balance of air,

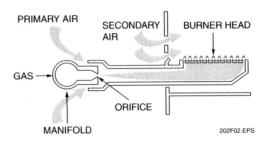

Figure 2 ◆ Combustion air.

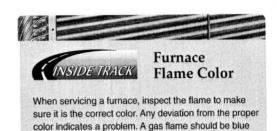

Furnace Flame Color

When servicing a furnace, inspect the flame to make sure it is the correct color. Any deviation from the proper color indicates a problem. A gas flame should be blue with an orange tip. An oil flame should be solid yellow.

Discuss combustion efficiency and flue gas temperatures.

Describe the flame characteristics of oil and gas burners.

Show Transparency 4 (Figure 2). Explain the importance of the proper balance of primary and secondary air.

Discuss carbon monoxide hazards.

Show the video *Principles of Gas Combustion*.

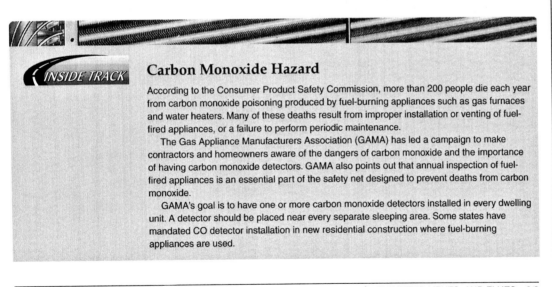

Carbon Monoxide Hazard

According to the Consumer Product Safety Commission, more than 200 people die each year from carbon monoxide poisoning produced by fuel-burning appliances such as gas furnaces and water heaters. Many of these deaths result from improper installation or venting of fuel-fired appliances, or a failure to perform periodic maintenance.

The Gas Appliance Manufacturers Association (GAMA) has led a campaign to make contractors and homeowners aware of the dangers of carbon monoxide and the importance of having carbon monoxide detectors. GAMA also points out that annual inspection of fuel-fired appliances is an essential part of the safety net designed to prevent deaths from carbon monoxide.

GAMA's goal is to have one or more carbon monoxide detectors installed in every dwelling unit. A detector should be placed near every separate sleeping area. Some states have mandated CO detector installation in new residential construction where fuel-burning appliances are used.

Discuss the components of flue gas.

Explain how the temperature of the flue gas affects condensation in the vent system.

List the features of a good vent system.

called **secondary air**, is supplied during combustion to the exterior of the flame. The primary air is drawn by the negative pressure of the gas. The secondary air is drawn by the vacuum created by the combustion process. These air mixtures are adjustable. Improper gas flames are the result of inefficient or incomplete combustion and can be caused by too much primary air, too little secondary air, or by the flame touching a cool surface.

3.0.0 ♦ FLUE GASES

Both gas and oil furnaces rely on the combustion of fuel to generate heat. In the process, they also produce wastes in the form of vent gases. The bulk of these waste gases are carbon dioxide, water vapor, excess air, and small amounts of other elements. If incomplete combustion occurs, these gases may also include carbon monoxide, aldehydes, and soot, all of which are potentially dangerous to people. Venting of these gases to the outdoors is an important part of a heating system. Proper gas venting is the removal of all products of combustion, together with excess air and **dilution air**, to the outside of the building. In most furnaces, venting is done through a chimney flue or vertical vent which leads from the furnace area up through the roof. A horizontal metal vent pipe (**vent connector**) is used to connect the furnace to the chimney or a metal flue, which vents to the outdoors. In condensing furnaces, venting is done with plastic pipe through an outside wall. This is possible because the flue gases from these furnaces are much cooler than those of other furnaces.

There are problems related to the removal of flue gases that must be considered when sizing vents. Often, the true volume of the flue gases or products of combustion is underestimated. Flue gas volume is many times greater than the volume of gas burned, so the inside of the chimney or vent must be large enough to handle the large volume of flue gases. Also, in all types of furnaces, water vapor produced by combustion can be troublesome if allowed to condense into a liquid.

In burning 100 cubic feet of natural gas, a furnace can produce 200 cubic feet of water vapor (about one gallon of water). This water vapor must be prevented from condensing in the vent system. In natural-draft and induced-draft furnaces, the vent temperature stays well above the dew point. In condensing furnaces, where the vent temperature is much closer to the dew point, the condensing coil removes moisture from the flue gases. In addition, a system to collect and remove condensation is included in the design.

When coal furnaces were widely used, the constant heat from the glowing coals, plus high-temperature flue gases (about 1,000°F) helped push the products of combustion up the chimney. Gas furnaces are much different. They usually operate intermittently and produce flue gas temperatures much lower than those of coal. This creates a greater potential for condensation.

For perfect combustion, natural gas is united with oxygen to form one part of carbon dioxide and two parts of water vapor, plus heat. The oxygen needed for combustion comes from air. If 10 cubic feet of air is divided into its elements, it contains about eight cubic feet of nitrogen and other inert gases, and slightly less than two cubic feet of oxygen. Thus, there is roughly 20 percent oxygen and 80 percent nitrogen in a given quantity of air. Theoretically then, when one cubic foot of natural gas is burned, its carbon and hydrogen combine with the oxygen present in 10 cubic feet of air to form one cubic foot of carbon dioxide, plus two cubic feet of water vapor, plus heat. The eight cubic feet of nitrogen remain unchanged. Therefore, it takes 10 cubic feet of air to burn one cubic foot of natural gas. If that one cubic foot of natural gas is burned in the presence of less than 10 cubic feet of air, incomplete combustion results.

Incomplete combustion produces carbon monoxide instead of carbon dioxide in the flue gas. Thus, in order to ensure complete combustion, an extra five cubic feet of air are generally supplied for each cubic foot of gas. This extra air is usually termed excess air.

4.0.0 ♦ FURNACE VENTING

Gas-fired appliances produce flue gases in quantities of about 30 times the volume of gas burned, at temperatures that affect both venting power and moisture condensation. Vents should include the following features:

- Low resistance to flue gas flow
- Small mass to enhance quick warm-up
- Insulating properties to maintain flue gas temperature
- Exact-size availability so that they can be matched to fit specific appliances

Installing gas vents requires the same technical understanding and early-stage planning as the installation of an air system. Nothing should be left to chance. It is necessary to understand the basic principles of vent operation and the factors that interfere with vent action. It is also important to know the rules that apply to proper installation and operation of gas vents.

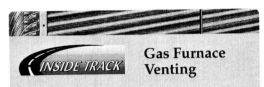

Gas Furnace Venting

The *National Fuel Gas Code*, which is published jointly by the American Gas Association and the National Fire Protection Association, allows fan-assisted furnaces to use indoor air for combustion in certain circumstances. However, it is important to keep in mind that the environment in which the furnace is installed may change over time. Consider the following examples:

- Occupants may remodel a basement where the furnace is installed, adding walls and doors that reduce the amount of available open space from which the furnace can draw combustion air.
- Gas appliances such as stoves, water heaters, and clothes dryers might be added, increasing the demand for combustion air.
- Occupants may add insulation and caulk around windows and doors, reducing the amount of infiltration air available.

Many dealers, concerned with the long-term safety of their customers as well as liability issues, require that combustion air be drawn from outside the building, even if existing conditions would allow the use of indoor air. Local governments have also tightened requirements for furnace venting. Be sure to check local codes and your employer's policies before undertaking a gas furnace installation. Also, encourage homeowners to install carbon monoxide detectors.

For example, a 100,000-Btuh gas furnace consumes about 100 cubic feet of gas during each hour of constant operation. Because of the air/gas ratio, about 3,000 cubic feet of air is also consumed during that period. Assume this furnace is in a house with an area of 1,250 square feet and a volume of about 10,000 cubic feet. With an average infiltration rate of one air change per hour, 10,000 cubic feet of outside air will move through the structure every hour. This exceeds the 3,000 cubic feet of air required by the furnace and vent system.

In the past, normal air infiltration was enough to satisfy the furnace needs. However, modern building construction has become tighter. In addition, slab floors have replaced basements, and more dampered exhaust fans have been built into kitchens and bathrooms. Thus, the air leakage rate (air supply to the furnace) has become a critical design factor. Now it is sometimes necessary to deliver outside combustion air to fan-assisted furnaces. Condensing furnaces must use outdoor air for combustion.

4.1.0 Requirements

A service technician should know the local codes and regulations that govern vent systems. If local codes or manufacturer's instructions do not cover vent piping, refer to the *National Fuel Gas Code (NFPA 54/ANSI Z223.1)* published by the American Gas Association (AGA) and the National Fire Protection Association (NFPA). All wiring and connections should be made in accordance with the *National Electrical Code®* and with any local codes that may apply. Supply gas pipe sizing should be made in accordance with the standards of the AGA.

In general, the vent system should meet the following minimum requirements as defined by *The National Fuel Gas Code*:

- The vent must not be smaller in diameter than the vent collar on the furnace.
- The combination of the vent and vent connector must not exceed a specified length.
- The installation must not have more than a specified number of elbows.

4.2.0 Clearances

Local codes and manufacturers' installation instructions usually specify the minimum distance between the furnace and combustible materials.

 WARNING!
Flammable materials must not come into contact with the heat exchangers, burners, or any other hot surfaces, such as the flue vent.

Accessibility clearances take precedence over minimum fire protection clearances. Allow at least 24" at the front of the furnace if all parts can be reached from the front. Otherwise, allow 24" on three sides of the furnace if the back must be reached for servicing. When the installation is made in a utility room or closet, the door must be big enough to allow replacement of the appliance. Consult local codes and manufacturer's installation instructions for allowable clearances.

4.3.0 Air Supply

Return air plenums should be lined with an acoustical duct liner to reduce fan noise. This is of particular importance when the return air grille is close to the furnace. All duct connections to the furnace must extend outside the furnace closet.

Discuss flue gas supply.

Discuss the requirements for vent systems.

Emphasize the importance of maintaining the proper clearance between the furnace and combustible materials.

Explain that the minimum clearance for service access is typically 24 inches.

Explain that return air plenums should be lined to reduce fan noise.

Provide a copy of the *National Fuel Gas Code* for trainees to examine.

List the requirements of the return air system.

Show Transparency 5 (Figure 3). Describe the requirements for furnace room venting.

Discuss the four categories of gas appliances.

See the Teaching Tip for Section 5.0.0 at the end of this module.

Combustion Air Contamination

Do not install furnaces that use indoor air for combustion near sources of air contamination such as cleaning solvents, aerosol sprays, detergents, bleaches, air fresheners, etc. Some manufacturers will not honor warranties on their heat exchangers unless the combustion air is drawn from outside the building.

Return air must not be taken from the furnace room or closet. Adequate return air duct height must be provided to allow filters to be removed and replaced. All return air must pass through the filter after it enters the return air plenum. Air required for combustion, draft hood dilution, and ventilation differs somewhat for a furnace in a confined space as opposed to a furnace in an open space. As a general rule, there must be two permanent openings: one within 12" of the ceiling, and one within 12" of the floor. Each opening must have a free area of at least one square inch per 1,000 Btuh of the total input rating of all the gas-fired appliances in the enclosure (see *Figure 3*).

5.0.0 ◆ VENT SYSTEM COMPONENTS

The type of vent system used is based on the type of furnace and the construction of the building. The *National Fuel Gas Code* identifies vented appliance categories and describes them as follows:

- *Category I* – An appliance that operates with a non-positive vent static pressure and with a vent gas temperature that avoids excessive condensate production in the vent.
- *Category II* – An appliance that operates with a non-positive vent static pressure and with a vent gas temperature that may cause excessive condensate production in the vent.
- *Category III* – An appliance that operates with a positive vent static pressure and with a vent gas temperature that avoids excessive condensate production in the vent.
- *Category IV* – An appliance that operates with a positive vent static pressure and with a vent gas temperature that may cause excessive condensate production in the vent.

This module focuses on Category I (natural-draft and induced-draft furnaces) and Category IV (condensing furnaces). Category II furnaces are natural-draft condensing furnaces. Category III furnaces are sidewall-vented 80 percent

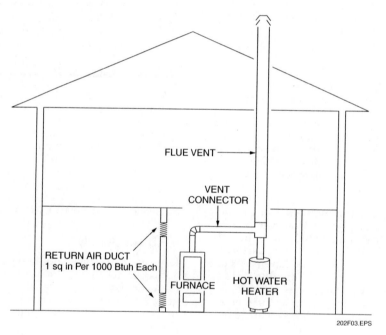

Figure 3 ◆ Furnace room venting.

AFUE induced-draft furnaces vented with high-temperature plastic pipe. Neither Category II nor Category III is in common use at this time.

Masonry and factory-built chimneys, along with metal and plastic vents, comprise the basic venting systems for coal, gas, oil, and wood-burning appliances. Factory-built chimneys with an inner wall of stainless steel that are in compliance with Underwriters Laboratories (UL) Standard No. 959 are suitable for all of these fuels.

Figure 4 shows the components of a typical factory-built chimney. The flue-gas temperature-rise limit for these applications is set at 1,730°F.

There are several types of vent construction approved for use with gas appliances.

Type B vents have inner and outer walls made of corrosion-resistant material. They are round and are available in diameters to suit all uses. The double-wall construction of Type B vents helps conserve heat and therefore promotes better draft and reduced condensation. They are often used

Identify the components of a factory-built chimney.

Provide manufacturers' literature on factory-built chimneys for trainees to examine.

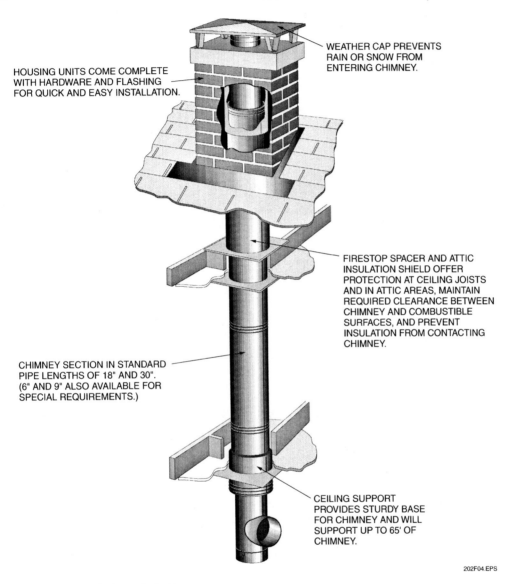

Figure 4 ◆ Components of a factory-built chimney.

List the various types of vents and their applications.

Provide examples of various vents for trainees to examine.

Show Transparencies 6 and 7 (Figures 5 and 6). Discuss the airflow through the vent of a natural-draft furnace.

Demonstrate natural draft using metal tubes and matches.

Chimney-Related Fires

A major cause of chimney-related fires is the failure to maintain the required clearances or air spaces between the chimney and adjacent combustible materials. For this reason, it is essential that a chimney be installed in strict accordance with local codes as well as the manufacturer's instructions.

in new construction because they are cheaper than a lined masonry chimney.

Type B-W vents have the same type of double-wall construction as the B-vent, but are oval. They are designed for venting in-the-wall gas heaters.

Type L vents are also double-wall vents. They are similar to Type B, but are made of materials that are more resistant to heat and corrosion. In general, Type L can be used in any application where Type B is suitable. The reverse is not true, however. Type L vents are also used in venting combination gas/oil appliances, as well as residential incinerators and certain appliances equipped with draft hoods.

Category III furnaces, which are relatively few in number, were once vented using special high-temperature plastics designed to withstand the condensate that might be produced, as well as the heat. Due to problems associated with these plastics, they are no longer used. Instead, most Category III products now use single-wall stainless steel vents. These vents are able to handle the corrosion and condensate production in the flue, while easily resisting heat. Because the flue is under positive pressure, most of these vents use proprietary clamping and gasket systems that ensure a sound, leak-free installation.

Schedule 40 PVC pipe is used in venting Category IV (condensing) furnaces. Because these vents are used in positive-pressure applications, they must also be carefully sealed to eliminate vent gas and condensate leakage.

6.0.0 ◆ NATURAL-DRAFT FURNACES

Sixteen cubic feet of combustion gases result from burning one cubic foot of natural gas. This does not equal the total gas volume passing up through the vent to the atmosphere. Before leaving the furnace through the vent, the 16 cubic feet of combustion gas is joined by 14 cubic feet of air at the furnace draft hood (*Figure 5*). This additional air is called dilution air. Therefore, the total volume in a properly operating vent is 30 cubic feet of vent gas for every cubic foot of natural gas burned.

In natural-draft furnaces, vent gases are not forced out through the vent pipe and chimney, but are drawn out instead. The chimney does this by producing a suction or drawing action called a draft. This principle is shown in *Figure 6*. When no heat is applied to the air or gas, the temperatures in and around the pipe are the same, and no movement occurs. When a fire is lit, however, it heats the air around it. This heated air expands in volume and becomes lighter in weight (less dense). Due to its lighter weight, the warm air rises, creating a draft in the chimney or vent.

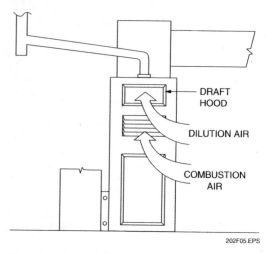

Figure 5 ◆ Natural-draft furnace.

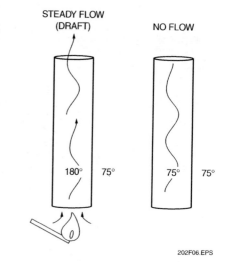

Figure 6 ◆ Natural drawing action.

2.8 HVAC ◆ LEVEL TWO

When confined within a vertical pipe, the warm air cannot mix with the surrounding air and cool down. When it is within the pipe, it retains its heat and therefore rises at a faster rate. In the process of rising, it draws fresh air in behind to replace it. As this new air is in turn heated, the process is continued and a constant flow of air moves through the pipe as long as heat is applied. The volume of gas the chimney will move, and the amount of draft it will create, depend on two factors: the temperature of the vent gas, and the diameter and height of the chimney.

Poor draft results from a chimney that is too small or too short, or vent gas temperatures that are too low. Increasing the diameter or the height of the chimney will increase the draft. The diameter of the chimney or vent pipe is important because of friction. If the pipe is too small, it will restrict the flow of gases.

Sufficient draft is essential for the proper operation of natural-draft furnaces. Draft allows the products of combustion to be removed safely. It also provides the oxygen used for combustion. Sufficient draft is also necessary with fan-assisted furnaces, but it is less critical because the flue gases receive some push from the inducer fan.

Because natural-draft furnaces are no longer manufactured, it is unlikely that you will be involved in the installation of a vent system for one of these furnaces. However, you may encounter existing installations in which something has been done to reduce the amount of draft air available to the furnace. Reduced draft can result in a condition known as depressurization, which can cause spillage, or backdrafting, in which some of the products of combustion that would normally be vented to the outdoors are drawn into the indoors. Some of the changes that can cause backdrafting are the addition of exhaust vents; the addition of a vented combustion appliance such as a gas fireplace or wood stove; and renovations that reduce air infiltration, such as new windows, insulation, or siding. Depressurization can also affect induced-draft furnaces, but does not affect furnaces that are direct-vented or power-vented.

7.0.0 ♦ INDUCED-DRAFT GAS FURNACES

In induced-draft furnaces, condensation can be a major problem because the flue gases are cooler than those of a natural-draft furnace. As mentioned earlier, natural-draft furnaces have flue gas temperatures of 350°F to 600°F. They are not likely to have condensation problems unless the furnace was greatly oversized or the vent system was improperly designed or installed. On the other hand, the flue gases of induced-draft furnaces run in the range of 275°F to 400°F. Because of that, the moisture in the flue gases condenses more readily.

To avoid condensation that can damage both the vent system and the furnace, there are five key tasks that must be completed when installing an induced-draft furnace:

- The furnace must be properly sized.
- The burners must be firing as close as possible to 100 percent of their rated input.
- The correct amount of air must be flowing over the heat exchangers in order to maintain the temperature rise at the correct level.
- The thermostat **heat anticipator** must be adjusted correctly.
- The furnace must be properly vented.

7.1.0 Furnace Sizing

A furnace must not be severely oversized. However, it is better to slightly oversize than it is to undersize. Generally, the furnace heat output rating should be from 95 to 120 percent of the heating load. The load is established by an analysis that considers building construction, type and amount of insulation, amount of glass, and other factors. *Manual J*, published by the Air Conditioning Contractors of America (ACCA), provides a method for properly estimating heating and cooling loads.

WARNING!
If a furnace is oversized, the only effective way to correct the problem is to replace it. Do not attempt to derate the furnace by drilling or changing orifices, or by reducing manifold pressure. It will make the condensation problem worse, and may create a safety hazard. Always follow the installation instructions supplied with the unit.

A correctly sized furnace will have a long operating cycle, particularly when the outdoor temperature approaches the heat loss design temperature. This keeps the furnace and vent warm and prevents condensation. A severely oversized furnace will deliver a large amount of heat in a very short time. During its long off cycle, the furnace and vent will cool, allowing condensation to form when the warm flue gas hits the vent during the next on cycle.

Discuss poor draft in natural-draft furnaces.

Explain that induced-draft furnaces are prone to condensation in the vent system due to low flue gas temperatures.

Emphasize the importance of proper furnace sizing.

Explain that attempting to derate a furnace may create a safety hazard.

List the furnace adjustments that are critical for preventing condensation.

Explain how to adjust the burner input.

Show Transparency 8 (Figure 7). Explain how to measure the temperature of the supply and return air.

Show trainees how to measure the supply and return air temperatures and determine the temperature rise of a furnace.

Have trainees practice measuring the supply and return air temperatures and determining the temperature rise of a furnace. Note the proficiency of each trainee. This laboratory corresponds to Performance Task 1.

Show Transparency 9 (Figure 8). Explain how to adjust the thermostat anticipator.

7.2.0 Burner Input Adjustment

The burners must be operated at as close to 100 percent of their rated input as possible. In most cases, this can be done by adjusting the manifold pressure at the gas valve. This adjustment was covered in the HVAC Level One module, *Introduction to Heating*.

7.3.0 Temperature Rise Adjustment

Temperature rise is the temperature difference between the supply air and the return air. The amount of air flowing over the heat exchangers determines the temperature rise. If there is too much airflow, the heat exchangers will not be able to reach normal operating temperature. Moisture will condense on the heat exchangers and in the vent. If there is too little air, the heat exchangers will become overheated.

The furnace nameplate will specify the correct temperature rise range; 45°F to 75°F is common. The actual rise is determined by drilling holes in the supply and return ducts and measuring the temperatures (*Figure 7*). The difference between the two temperatures must be within the specified range. Ideally, it should be just above the mid-point of the range. If it is not, the blower speed can be changed to compensate. The blower speed is increased for too much rise and decreased for too little rise.

7.4.0 Thermostat Heat Anticipator Adjustment

A heat anticipator is a small resistive element that heats up as current flows through the thermostat. It is usually adjustable (*Figure 8*), and the adjustment is required at the time of installation. The heat from the anticipator causes the thermostat to turn off just before the setpoint is reached. This prevents the system from exceeding the desired temperature and allows some of the residual heat from the heat exchangers to be dissipated by the circulating air.

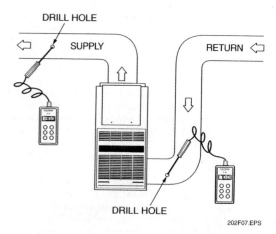

Figure 7 ♦ Measuring supply and return air temperature.

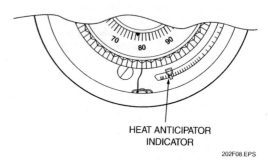

Figure 8 ♦ Thermostat heat anticipator adjustment.

If the anticipator is set incorrectly, it can cause the furnace to short-cycle and allow moisture to condense. This can occur if the furnace cycles more than six times an hour.

The heat anticipator should be set to the same current that is flowing through the thermostat contacts, which is usually in the range of 0.15 to 1 amp.

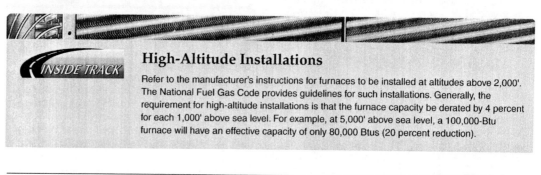

High-Altitude Installations

Refer to the manufacturer's instructions for furnaces to be installed at altitudes above 2,000'. The National Fuel Gas Code provides guidelines for such installations. Generally, the requirement for high-altitude installations is that the furnace capacity be derated by 4 percent for each 1,000' above sea level. For example, at 5,000' above sea level, a 100,000-Btu furnace will have an effective capacity of only 80,000 Btus (20 percent reduction).

Instructor's Notes:

Electronic thermostats do not have heat anticipators. If an electronic thermostat is used, set the cycle rate for three cycles per hour or consult the furnace manufacturer's literature for the correct cycle rate.

7.5.0 Venting Considerations

Proper furnace sizing and adjustment is meaningless if the furnace is not vented correctly. Vents must be carefully sized. A vent that is too small may not be able to handle the volume of flue gas, while a vent that is too large may not be able to establish a proper draft. This is true of both induced-draft and natural-draft furnaces. Flue gases from gas appliances can cool too quickly in a large vent. This causes condensation. It is important to follow the manufacturer's instructions, along with local and national codes, when selecting and installing furnace vents.

NOTE

Control of condensation in the induced-draft furnace is critical. If condensation occurs, it could not only corrode the vent, it could trickle down into the heat exchangers and cause corrosion there. This corrosion can eat holes in the heat exchangers, allowing toxic byproducts of combustion to enter occupied areas. This is one of the reasons that furnaces must be periodically inspected. These inspections must include a thorough examination of the heat exchangers.

The *National Fuel Gas Code* contains tables and instructions for selecting the diameter and type of metal vents and vent connectors for induced-draft furnaces and other gas appliances. Part of this data is provided in the installation instructions for most furnaces. The following sections summarize some of the important furnace venting requirements specified by the *National Fuel Gas Code*.

7.5.1 General Guidelines for Metal Vents and Vent Connectors

Gas appliances may be vented through a metal pipe or lined chimney (*Figure 9*). The vent is the vertical section of the vent system. The vent connector is the horizontal section that connects the appliance(s) to the vent.

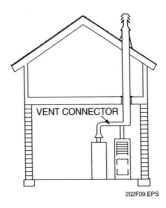

Figure 9 ◆ Vent connector.

Because of the potential for condensation, metal vents must be of double-wall (Type B) construction. Vent connectors for induced-draft furnaces should also be the double-wall type, which heats up faster, thereby limiting the risk of condensation. A single-wall vent connector can sometimes be used with induced-draft furnaces, but only under the following conditions:

- The furnace must be common-vented with another gas appliance, such as a hot water heater. This helps to keep the vent connector from cooling down.
- The length of the vent connector (in feet) cannot exceed 1½ times its diameter (in inches). For example, a 4" diameter vent connector can be no longer than 6'. The selection tables are used to determine the diameter of the vent connector, based on the furnace input (in Btuh) and the height of the vent.

Single-wall pipe has a high heat loss, which allows combustion products to cool rapidly. In general, a single-wall vent connector is suitable only for natural-draft furnaces because their high flue gas temperatures and the use of dilution air prevent condensation. A double-wall vent connector increases the initial cost of the installation, but that is far outweighed by the potential cost of major repairs or furnace replacement due to condensation. Using a single-wall vent connector also hampers installation flexibility because of the limited length of the vent connector.

Vent connectors should be pitched upward toward the vent at a slope of no less than ¼" per foot and should be as short as possible.

Show trainees how to adjust a thermostat heat anticipator.

Have trainees practice adjusting a thermostat heat anticipator. Note the proficiency of each trainee. This laboratory corresponds to Performance Task 2.

Have the trainees review Sections 7.5.0–9.3.0.

Ensure that you have everything required for teaching this session.

Explain the importance of proper vent sizing.

Show Transparency 10 (Figure 9). Provide an overview of the general guidelines for metal vents and vent connectors.

MODULE 03202-07 ◆ CHIMNEYS, VENTS, AND FLUES 2.11

Provide an overview of the general guidelines for venting through a masonry chimney.

Describe a flexible chimney liner kit.

Show the video *Ventinox Chimney Solution* on how to install a flexible liner in a masonry chimney.

Show trainees how to calculate the correct size and type of furnace vent connector and metal vent.

Have trainees practice calculating the correct size and type of furnace vent connector and metal vent. Note the proficiency of each trainee. This laboratory corresponds to Performance Task 4.

Avoid elbows, because they create more resistance. Also avoid sharp turns—two 45-degree connections are better than one 90-degree connection. If single-wall vent connectors are used with more than two 90-degree elbows, a 10 percent reduction in the maximum length of the vent connector must be made for each extra elbow. If the furnace is common-vented with another gas appliance, neither appliance should have a vent damper because dampers can cause condensation. Vent dampers are covered later in this module.

When working with metal vents, the selection tables assume that the vertical run will not be exposed to outdoor air, except above the point where it penetrates the roof. In many locales, exposed metal vents will cause serious condensation problems. The selection tables are not intended for these applications in these areas.

7.5.2 Venting Through a Masonry Chimney

When a tile-lined masonry chimney is available, it can be used to vent gas appliances. However, a fan-assisted furnace cannot be vented through a masonry chimney unless it is common-vented with another gas appliance or the chimney is suitably lined. In all cases, follow the local codes.

An unlined chimney must not be used in any circumstance. Every chimney must have a liner, because corrosive substances in the flue gases will attack mortar and cause the chimney to deteriorate. This could cause falling mortar and debris to block the vent opening, creating a hazard for occupants. At best, the unlined chimney will be seriously damaged over time. In addition, unlined chimneys are prone to condensation, which can damage the furnace and vent system. Under no circumstances may a furnace or other gas appliance be vented through a chimney that serves a fireplace or other wood- or coal-burning device.

Ideally, a chimney used for furnace venting should run inside the building. If the chimney is exposed below the roofline, it will probably need a metal liner. In some cases, the metal liner will have to be insulated. A chimney with a metal liner is treated the same as an unexposed chimney.

The vent selection tables will identify the chimney size (in square inches) required to match the furnace capacity. An oversized chimney could be dangerous because it will have difficulty in establishing the proper draft, making it hazardous for occupants. Also, it will be more likely to develop condensation. If the opening is too large, a liner must be installed. A double-wall metal vent may be used for this purpose; however, it may be difficult to install, especially if there are offsets in the chimney. As an alternative, flexible chimney liners are available (see *Figure 10*). They are expandable; a 30'-long vent might be shipped in a 3'-long box.

Flexible liners are not sized in the same way as solid metal vents. This is due to the added resistance to flue gas flow created by the corrugations that allow the material to be flexible and compressible. This series of ridges inhibits the flow of flue gas more than a smooth pipe of equal size would. When flexible liners are used, a larger diameter vent than the one called for by the table may be required because the corrugated structure of the flexible chimney liner acts as a restriction. For example, if a 6" diameter metal vent is suitable for a particular furnace, the same furnace may need an 8" flexible liner.

INSIDE TRACK

Flexible Chimney Liners

Flexible chimney liner kits have made lining chimneys relatively easy. The most difficult part of the job is working on high chimneys or steeply pitched roofs. In those cases, scaffolding or a motorized lift should be used. To install the liner, drop a weighted rope down the chimney. Attach the flexible liner to the rope and have someone pull the liner up or down the chimney. Use the vent cap and all other components of the installation kit to ensure a safe installation.

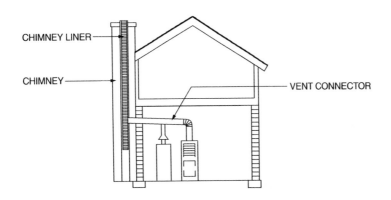

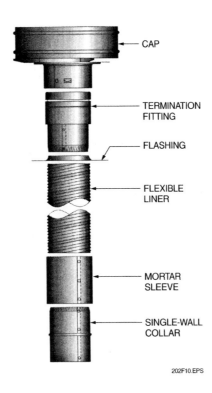

Figure 10 ♦ Flexible chimney liner kit.

8.0.0 ♦ CONDENSING GAS FURNACES

Because condensing furnaces produce low-temperature flue gases, they can be vented with Schedule 40 PVC pipe. These furnaces also require a lot of combustion air, which may be drawn from outdoors or from inside the building. The diameter of the pipe depends on the furnace input, the length of the pipe run, and the number and type of elbows used. If 45-degree elbows are used, each one increases the length of the run by an equivalent of 5'. A 90-degree elbow is equivalent to an increase of 10'. *Table 1* shows the type of pipe selection chart you might see in the installation instructions for a condensing furnace.

Explain that condensing furnaces can be vented with PVC pipe.

Show Transparency 11 (Table 1). Explain how to size PVC pipe for use in venting.

Provide PVC pipe for trainees to examine.

Show Transparencies 12 and 13 (Figures 11 and 12). Discuss the requirements for venting condensing furnaces.

Show trainees how to calculate the correct size and type of PVC pipe.

Have trainees practice calculating the correct size and type of PVC pipe. Note the proficiency of each trainee. This laboratory corresponds to Performance Task 3.

Table 1 PVC Selection Chart

Pipe Length (Feet)	SCHEDULE 40 PVC DIAMETER Number of 90° Elbows				
	0	2	4	6	8
5	2	2	2	2	2
10	2	2	2	2	2
20	2	2	2	2	2½
30	2	2	2	2½	2½
40	2	2	2½	2½	2½
50	2	2½	2½	2½	2½
60	2½	2½	2½	2½	3
70	2½	2½	2½	3	3
80	2½	2½	3	3	3
90	2½	3	3	3	3

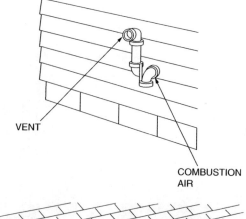

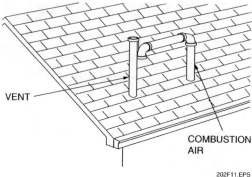

Figure 11 ♦ Terminating PVC vent and combustion air pipes.

In a direct-vent, two-pipe system, combustion air is drawn in through one pipe (outdoors) and the products of combustion are vented through another pipe to the outdoors. Both pipes must be in the same pressure zone, meaning the intake and vent pipes are located close to each other. In a non-direct venting situation, the products of combustion are still vented to the outdoors. However, the combustion air can be taken in from a well ventilated area such as an attic or crawlspace instead of from directly outdoors.

The following are good practices to consider when installing vents and intake piping:

- The vent and intake pipes should always terminate in the same pressure zone, whether they run through the roof or through an exterior wall. They should also be as close together as possible; separations of 3" on a roof and 6" on a sidewall are standard.
- Piping should be sloped back toward the furnace at least ¼" per foot and supported. The slope allows condensate to drain back to the furnace and into the condensate trap for disposal.
- Keep the termination well above expected snow levels. In cold climates, the outdoor portion and any part of the exhaust pipe that runs through an unconditioned space should be insulated.
- As shown in *Figure 11*, the combustion air intake should be bent downward to prevent dirt and moisture from entering the system. The exhaust vent must be straight up (roof) or straight out (sidewall). Rooftop terminations are usually best because there is less chance of the pipes being damaged or blocked and less chance of receiving contaminated combustion air.
- When venting through a sidewall, avoid terminating the pipes in a corner, under a deck, or near shrubs or trees, in order to prevent the recirculation of moist vent gases. Also avoid terminating pipes near doors and windows. This can cause flue gases to enter the building and condense on glass.

Special concentric termination devices (*Figure 12*) are available from some manufacturers. These devices allow combustion and exhaust gases to be carried through the same hole in the roof or exterior wall.

NOTE
The *National Fuel Gas Code* will reference the manufacturer's instructions for venting condensing furnaces. Always check the prevailing codes and the furnace installation instructions for proper venting requirements before venting a furnace with PVC pipe or through a sidewall. It may not be permitted in all jurisdictions.

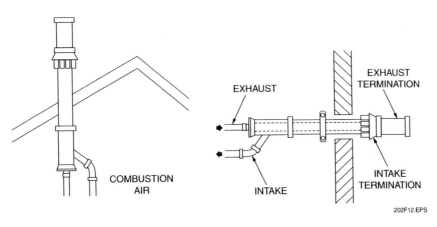

Figure 12 ◆ Concentric termination.

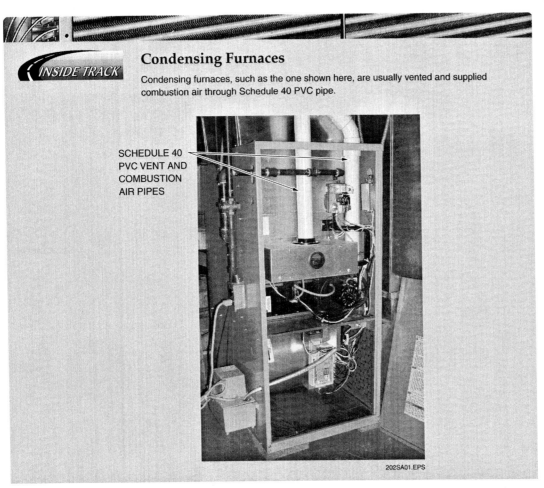

Condensing Furnaces

Condensing furnaces, such as the one shown here, are usually vented and supplied combustion air through Schedule 40 PVC pipe.

Explain the purpose of draft controls.

Explain that a draft regulator keeps a constant draft over the fire.

Explain that vent dampers are energy-saving devices.

9.0.0 ♦ DRAFT CONTROLS

Draft controls regulate the amount of air feeding a fire. They are used in the vent systems of natural-draft, gas-fired furnaces as well as oil and solid-fuel furnaces.

9.1.0 Draft Regulator

A draft regulator (*Figure 13*) keeps a constant draft over the fire, usually 0.01 to 0.03 inches water column (in. w.c.) on oil-burning furnaces. Too high a draft causes undue loss of heat through the chimney. Too little draft causes incomplete combustion.

A draft regulator consists of a small door in the side of the flue pipe. The door is hinged near the center and controlled by adjustable weights. Basement air is admitted to the flue pipe as required to maintain a proper draft over the fire.

9.2.0 Vent Dampers

Some vent dampers (*Figure 14*) are energy-saving devices that can be added to in-service furnaces.

Figure 13 ♦ Draft regulator.

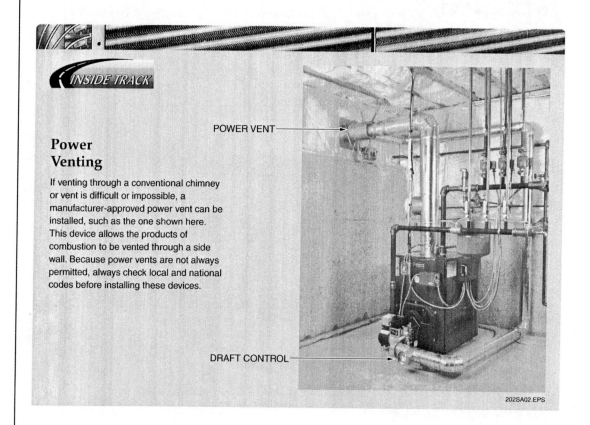

Power Venting

If venting through a conventional chimney or vent is difficult or impossible, a manufacturer-approved power vent can be installed, such as the one shown here. This device allows the products of combustion to be vented through a side wall. Because power vents are not always permitted, always check local and national codes before installing these devices.

Figure 14 ♦ Vent damper.

They are designed to stay open while the burner is operating in order to vent combustion gases.

When the burner shuts off, vent dampers are designed to close and stop the heat from escaping up the flue vent or chimney. They are relatively easy to install, but vent dampers can be both health and fire hazards if they fail to open when the furnace is operating. In addition, some furnace warranties are voided if vent dampers are added. Consult local codes. Some manufacturers are building furnaces with control wiring installed for adding a vent damper. Vent dampers are most effective when combustion air is being drawn from within the house.

9.3.0 Draft Diverters

The draft diverter or draft hood (*Figure 15*), is designed to provide a balanced draft (slightly negative) over the flame in a gas-fired, natural-draft furnace.

The bottom of the diverter is open to allow air from the furnace room to blend with the products of combustion. The hot vent gases from the furnace normally pass into the draft diverter and then into the vent pipe without any spilling out of the bottom opening. This is because the hot gases tend to stay toward the top of the draft diverter and are removed by the draw from the vent pipe and chimney. The chimney draft is greater than required to remove the vent gases; therefore, additional air from the furnace room is drawn into the bottom opening and passes up the vent along with the gases.

Figure 15 ♦ Draft diverter.

If not enough draft is available to remove the vent gases due to a restriction or a downdraft in the chimney, the draft diverter acts as a relief valve. Since the vent gases are prevented from going up the chimney, they go out the bottom opening in the draft hood. This relief factor prevents combustion from being upset in the furnace. The discharge of gases from the bottom opening in the draft hood or diverter is called spillage.

> **WARNING!**
> When spillage occurs, vent gases are discharged into the structure. The result is that all the water vapor from combustion passes into the conditioned space, causing high humidity. If combustion is complete, relatively harmless carbon dioxide also passes into the space. If combustion is not complete, the vent gases will also contain deadly carbon monoxide. For this reason, the cause of spillage must be found and corrected immediately. Homeowners should be strongly encouraged to install carbon monoxide detectors.

Explain the operation of a vent damper.

Explain the operation of a draft diverter.

Emphasize the importance of avoiding vent spillage.

Have the trainees complete the Review Questions, and go over the answers prior to administering the Module Examination.

Review Questions

1. If there are indications that incomplete combustion is occurring, the problem is solved by _____.
   a. increasing the flow of fuel
   b. increasing the amount of air
   c. changing the size of the flue vent
   d. adding carbon dioxide

2. The temperature range of the flue gases in an induced-draft furnace is _____.
   a. 100°F to 125°F
   b. 100°F to 600°F
   c. 275°F to 400°F
   d. 350°F to 600°F

3. The horizontal pipe that connects the furnace to the chimney is called the _____.
   a. flex duct
   b. vent connector
   c. vent
   d. flue pipe

4. All _____ furnaces require combustion air to be piped in from outdoors.
   a. natural-draft
   b. induced-draft
   c. condensing
   d. wood-burning

5. Which type of vent does *not* have double-wall construction?
   a. PVC
   b. Type B-W
   c. Type L
   d. Type B

6. If an induced-draft furnace is oversized and is *not* venting properly, the problem can be solved by changing to a _____.
   a. smaller burner orifice
   b. larger burner orifice
   c. smaller flue vent
   d. different furnace

7. Temperature rise is a term that describes the _____.
   a. temperature difference between supply air and return air
   b. natural venting of flue gases
   c. amount of heat in the flue gases
   d. heat remaining in the heat exchanger when the burner is shut off

8. Exterior chimneys may be used to vent gas furnaces as long as they have a _____ liner.
   a. masonry
   b. plastic
   c. suitable
   d. PVC

9. A gas furnace may be vented through an unlined chimney _____.
   a. when it is a natural-draft furnace
   b. when the climate is not too cold
   c. when a double-wall vent connector is used
   d. under no circumstances

10. How many equivalent feet does a 45-degree elbow add to the length of a PVC vent?
    a. 0
    b. 5
    c. 10
    d. 20

Instructor's Notes:

Summary

There are two important reasons for properly venting a furnace. The first, and most important, is to make sure that harmful flue gases are exhausted from the building and thus do not create a hazard for building occupants. The other reason is to prevent moisture from condensing out of the flue gases and causing serious damage to the vent system and possibly the furnace.

Although proper venting is important with all types of fossil-fuel heating systems, there is a special concern with induced-draft, gas-fired furnaces. The flue gases of these furnaces are cooler than those of natural-draft devices. For that reason, it is very important to select the correct size and type of vent and to install it in accordance with applicable codes and instructions.

Summarize the major concepts presented in the module.

Administer the Module Examination. Record the results on Craft Training Report Form 200, and submit the results to the Training Program Sponsor.

Administer the Performance Test, and fill out Performance Profile Sheets for each trainee. If desired, trainee proficiency noted during laboratory sessions may be used to complete the Performance Test. Record the results on Craft Training Report Form 200, and submit the results to the Training Program Sponsor.

Notes

Trade Terms Introduced in This Module

Complete combustion: Burning in which there is enough oxygen to prevent the formation of carbon monoxide.

Condensing furnace: A high-efficiency furnace containing a secondary heat exchanger that extracts additional heat from the flue gases.

Dilution air: Air added to the flue gases in a natural-draft furnace to aid flue gas removal.

Heat anticipator: A resistive heating element in a thermostat that shuts off the furnace before the space temperature reaches the setpoint. It prevents the system from exceeding the desired temperature.

Incomplete combustion: Burning in which there is not enough oxygen to prevent the formation of carbon monoxide.

Induced-draft furnace: A fan-assisted furnace with an AFUE rating of 78 to 85 percent.

Natural-draft furnace: A furnace that depends on the pressure created by the heat in the flue gases to force them out through the vent system.

Primary air: Air that is added to the fuel before it goes to the burner.

Secondary air: Air that is added during combustion.

Vent: The vertical section of the vent pipe.

Vent connector: The horizontal section of the vent system that connects the appliance(s) to the vent pipe or chimney.

Additional Resources and References

Additional Resources

This module is intended to be a thorough resource for task training. The following reference works are suggested for further study. These are optional materials for continued education rather than for task training.

Mid-Efficiency Furnace Installation Awareness. Latest Edition. Syracuse, NY: Carrier Corporation.

National Fuel Gas Code (NFPA 54/ANSI Z223.1), Latest Edition. Quincy, MA: National Fire Protection Association.

Figure Credits

Hart & Cooley, Inc., 202F04, 202F10 (bottom)

Topaz Publications, Inc., 202SA01

Field Controls LLC, 202F13, 202SA02, 202F14

Burnham Hydronics, 202F15

MODULE 03202-07 — TEACHING TIPS

The following are suggested activities or instructional methods to help you teach the material in this module.

General

When you call on someone to answer a question, the rest of the class relaxes or even tunes out because they expect that the question and answer will take place only between you and the trainee you called on. Instead, use this technique to involve more trainees in answering questions and to keep them on their toes.

1. Ask trainees to define a term or explain a concept.
2. After one trainee has answered, ask a trainee seated nearby if the answer is right. Then ask whether a trainee in the back of the room agrees.
3. Ask trainees to explain why they think an answer is right or wrong.
4. Use the session to clear up incorrect ideas and encourage trainees to learn from their mistakes.

Sections 2.0.0 through 9.0.0

Trade Terms Quiz

This Trade Terms Quiz will familiarize trainees with the terms and definitions that are commonly used in venting. You will need photocopies of the quiz provided in the following page. Trainees will need pencils. If you allow trainees to use the Trainee Guide, decrease the amount of time you give them to complete the quiz.

1. Make a photocopy of the quiz for each trainee.
2. Give trainees between 5 and 10 minutes to complete the quiz.
3. Go over the answers to the quiz.
4. Ask trainees if they have questions.

Answers to Quick Quiz

1. d
2. k
3. h
4. c
5. f
6. a
7. e
8. j
9. i
10. g
11. b

Quick Quiz

Trade Terms

For each description listed, identify the term that the text best describes. Write the corresponding letter in the blank provided.

_____ 1. The heat from the _____ causes the thermostat to turn off just before the setpoint is reached.

_____ 2. The horizontal section of pipe that connects the furnace to the chimney is called the _____.

_____ 3. The _____ is drawn by the negative pressure of the gas.

_____ 4. Venting is the removal of all products of combustion and _____ to the outside of the building.

_____ 5. In a(n) _____, vent gases are drawn out of the pipe and chimney.

_____ 6. When carbon combines with oxygen to form carbon dioxide, _____ takes place.

_____ 7. When there is too little oxygen in the flame, _____ produces undesirable products like carbon monoxide.

_____ 8. A(n) _____ that is too large may not be able to establish a proper draft.

_____ 9. The _____ is drawn by the vacuum created by the combustion process.

_____ 10. Condensation is a factor in _____ because the flue gases are cooler than those of a natural-draft furnace.

_____ 11. A(n) _____ produces low-temperature flue gases.

a. complete combustion
b. condensing furnace
c. dilution air
d. heat anticipator
e. incomplete combustion
f. induced-draft furnace
g. natural-draft furnace
h. primary air
i. secondary air
j. vent
k. vent connector

Section 5.0.0 *Vent System Components*

This exercise familiarizes trainees with various parts of a vent system. Trainees will need appropriate personal protective equipment, pencils, and paper. You will need to arrange for an opportunity to examine an existing furnace and vent system in a commercial building or the building in which you are conducting class. Allow 20 to 30 minutes for this exercise.

1. Discuss the components of vent systems with trainees.
2. Show trainees typical vent systems. Have trainees identify different components of the system.
3. After the tour, answer any questions trainees may have.

MODULE 03202-07 — ANSWERS TO REVIEW QUESTIONS

Answer	Section Reference
1. b	2.2.0
2. c	2.3.0
3. b	3.0.0
4. c	4.0.0
5. a	5.0.0
6. d	7.1.0
7. a	7.3.0
8. c	7.5.2
9. d	7.5.2
10. b	8.0.0

CONTREN® LEARNING SERIES — USER UPDATE

NCCER makes every effort to keep these textbooks up-to-date and free of technical errors. We appreciate your help in this process. If you have an idea for improving this textbook, or if you find an error, a typographical mistake, or an inaccuracy in NCCER's Contren® textbooks, please write us, using this form or a photocopy. Be sure to include the exact module number, page number, a detailed description, and the correction, if applicable. Your input will be brought to the attention of the Technical Review Committee. Thank you for your assistance.

Instructors – If you found that additional materials were necessary in order to teach this module effectively, please let us know so that we may include them in the Equipment/Materials list in the Annotated Instructor's Guide.

Write: Product Development and Revision
National Center for Construction Education and Research
3600 NW 43rd St, Bldg G, Gainesville, FL 32606

Fax: 352-334-0932

E-mail: curriculum@nccer.org

Craft

Module Name

Copyright Date

Module Number

Page Number(s)

Description

(Optional) Correction

(Optional) Your Name and Address

Module 03203-07

Introduction to Hydronic Systems

NCCER STANDARDIZED CRAFT TRAINING PROGRAM

The National Center for Construction Education and Research (NCCER) provides a standardized national program of accredited craft training. Key features of the program include instructor certification, competency-based training, and performance testing. The program provides trainees, instructors, and companies with a standard form of recognition through a National Craft Training Registry. The program is described in full in the *Guidelines for Accreditation*, published by NCCER. For more information on standardized craft training, contact the NCCER by writing us at 3600 NW 43rd St., Bldg. G, Gainesville, FL 32606; calling 352-334-0911; or emailing info@nccer.org. More information may be found at our website, www.nccer.org.

HOW TO USE THIS ANNOTATED INSTRUCTOR'S GUIDE

Each page presents two sections of information. The larger section displays each page exactly as it appears in the Trainee Module. The narrow column ties suggested trainee and instructor actions to each page and provides icons (detailed below) to call your attention to material, safety, audiovisual, or testing requirements. The bottom of each page includes space for your notes.

The **Audiovisual** icon indicates an appropriate time to show a transparency or other audiovisual aid.

The **Classroom** icon prompts you to define a term, stress a point, ask trainees to explain a concept, or give examples.

The **Demonstration** icon directs you to show trainees how to perform tasks.

The **Examination** icon tells you to administer the written module examination.

The **Homework** icon is placed where you may wish to assign reading for the next class, assign a project, or advise trainees to prepare for an examination.

The **Laboratory** icon is used when trainees are to practice performing tasks.

The **Materials** icon is a reminder for you to gather materials needed for classes, labs, and testing.

The **Performance Testing** icon tells you to administer a performance test or a portion thereof.

The **Safety** icon is used to emphasize safety issues. It is often keyed to *Caution* and *Warning!* statements in the Trainee Module.

The **Teaching Tip** icon indicates additional guidance is available, such as how to conduct an exercise, get the most educational value from a field trip, or encourage class participation. Teaching Tips may expand on a feature (*Think About It, Did You Know?*) or provide *Quick Quizzes* or similar exercises. You will be referred to the Teaching Tips section at the back of the module if there is additional material.

The **Combination** icon indicates that the laboratory listed corresponds with a performance task. If desired, you can note the proficiency of the trainees during the laboratory, and use it to satisfy performance testing requirements.

PREPARATION

Before teaching this module, you should review the Objectives, Performance Tasks, Materials and Equipment List, and Module Outline. Be sure to allow ample time to prepare your own training or lesson plan and gather all required materials and equipment.

Introduction to Hydronic Systems
Annotated Instructor's Guide

Module 03203-07

MODULE OVERVIEW

This module introduces hydronic systems. It covers the types of systems available and the various system components, including boilers, valves, radiators, and piping. Radiant floor heating systems are also covered.

PREREQUISITES

Prior to training with this module, it is recommended that the trainee shall have successfully completed *Core Curriculum*; *HVAC Level One*; and *HVAC Level Two*, Modules 03201-07 and 03202-07.

OBJECTIVES

Upon completion of this module, the trainee will be able to do the following:

1. Explain the terms and concepts used when working with hot-water heating.
2. Identify the major components of hot-water heating.
3. Explain the purpose of each component of hot-water heating.
4. Demonstrate the safety precautions used when working with hot-water systems.
5. Demonstrate how to operate selected hot-water systems.
6. Demonstrate how to safely perform selected operating procedures on low-pressure systems.
7. Identify the common piping configurations used with hot-water heating.
8. Read the pressure across a water system circulating pump.
9. Calculate heating water flow rates.
10. Select a pump for a given application.

PERFORMANCE TASKS

Under the supervision of the instructor, the trainee should be able to do the following:

1. Demonstrate the safety precautions used when working on hot-water systems.
2. Identify the major components of a selected hot-water heating system.
3. Demonstrate how to safely perform selected operating procedures on hot-water boilers.
4. Identify the types of common piping configurations used with hot-water systems.
5. Calculate heating water gpm requirements from base information provided by the instructor.
6. Select a pump from manufacturer's data given the friction loss of a piping system and the gpm requirements from the previous performance task.

MATERIALS AND EQUIPMENT LIST

Overhead projector and screen
Transparencies
Blank acetate sheets
Transparency pens
Whiteboard/chalkboard
Markers/chalk
Pencils and scratch paper
Appropriate personal protective equipment
Hot-water heating system(s) and assorted system components, including:
    Cast-iron and steel boiler parts
    Electric boiler heating elements and other parts
    Differential pressure gauges
    Pump curve chart

Pressure-temperature gauges
Pressure relief valves
Low water controls
Aquastats
Electronic-type water level controls
Expansion/compression tanks
Air control devices
Circulating pumps
Assorted gate, ball, globe, and angle valves
Pressure-reducing valves
Backflow preventer valves
Zone control valves
Multipurpose valves
Balancing and flow control valves
Venturi and orifice-type flow meters
Two-way and three-way valves

continued

Assorted hot-water terminals, including:
  Convectors
  Baseboard and finned-tube radiators
  Unit heaters
  Heating coils
  Shell-and-tube plate heat exchangers
Tankless water heaters
Indirect water heaters

Manufacturers' instructions for safety relief valves
Manufacturers' literature on expansion/compression tanks
Manufacturers' literature on circulating pumps
Copies of the Quick Quizzes*
Module Examinations**
Performance Profile Sheets**

* Located at the back of this module.
**Located in the Test Booklet.

SAFETY CONSIDERATIONS

Ensure that the trainees are equipped with appropriate personal protective equipment and know how to use it properly. The module requires that trainees work with operating hot-water boilers. Ensure all trainees are briefed on appropriate safety procedures.

ADDITIONAL RESOURCES

This module is intended to present thorough resources for task training. The following reference works are suggested for both instructors and motivated trainees interested in further study. These are optional materials for continued education rather than for task training.

ASHRAE Handbook — HVAC Systems and Equipment, 2004. Atlanta, GA: American Society of Heating and Air Conditioning Engineers, Inc.

ASHRAE Handbook — HVAC Applications, 2007. Atlanta, GA: American Society of Heating and Air Conditioning Engineers, Inc.

HVAC Systems, 1992. Samuel C. Monger. Englewood Cliffs, NJ: Prentice Hall.

TEACHING TIME FOR THIS MODULE

An outline for use in developing your lesson plan is presented below. Note that each Roman numeral in the outline equates to one session of instruction. Each session has a suggested time period of 2½ hours. This includes 10 minutes at the beginning of each session for administrative tasks and one 10-minute break during the session. Approximately 12½ hours are suggested to cover *Introduction to Hydronic Systems*. You will need to adjust the time required for hands-on activity and testing based on your class size and resources. Because laboratories often correspond to Performance Tasks, the proficiency of the trainees may be noted during these exercises for Performance Testing purposes.

Topic   **Planned Time**

Session I. Introduction to Hot-Water Heating System Components and Boilers
    A. Introduction
    B. Water System Terms   _____
    C. Hot-Water Heating Systems   _____
    D. Hot-Water Boilers   _____
    E. Safety Controls   _____
    F. Laboratory   _____

        Trainees practice demonstrating safety precautions when working on a boiler. This laboratory corresponds to Performance Task 1.

    G. Laboratory   _____

        Trainees practice safely performing selected operating procedures on a boiler. This laboratory corresponds to Performance Task 3.

Session II. Hot-Water Heating System Components
   A. Expansion/Compression Tanks
   B. System Air Control Devices
   C. Pumps and Valves
   D. Terminals
   E. Tankless and Indirect Water Heaters
   F. Laboratory
      Trainees practice identifying the major components of a hot-water heating system. This laboratory corresponds to Performance Task 2.
   G. Radiant Floor Heating Systems

Session III. Piping
   A. One-Pipe Systems
   B. Two-Pipe Systems
   C. Hot-Water Zoning
   D. Laboratory
      Trainees practice identifying common piping configurations. This laboratory corresponds to Performance Task 4.
   E. Dual-Temperature Water Systems

Session IV. Water Balance
   A. Water Flow Measuring Devices and Flow-Control Devices
   B. Laboratory
      Trainees practice calculating the water gpm requirements. This laboratory corresponds to Performance Task 5.
   C. Friction Losses
   D. Laboratory
      Trainees practice selecting a pump given gpm requirements and piping system friction loss. This laboratory corresponds to Performance Task 6.

Session V. Review and Testing
   A. Review
   B. Module Examination
      1. Trainees must score 70% or higher to receive recognition from NCCER.
      2. Record the testing results on Craft Training Report Form 200, and submit the results to the Training Program Sponsor.
   C. Performance Testing
      1. Trainees must perform each task to the satisfaction of the instructor to receive recognition from NCCER. If applicable, proficiency noted during laboratory exercises can be used to satisfy the Performance Testing requirements.
      2. Record the testing results on Craft Training Report Form 200, and submit the results to the Training Program Sponsor.

HVAC Level Two

03203-07

Introduction to Hydronic Systems

Assign reading of Module 03203-07.

03203-07
Introduction to Hydronic Systems

Topics to be presented in this module include:

1.0.0	Introduction	3.2
2.0.0	Water System Terms	3.3
3.0.0	Hot-Water Heating Systems	3.5
4.0.0	Hot-Water Heating System Components	3.6
5.0.0	Water Piping Systems	3.26
6.0.0	Dual-Temperature Water Systems	3.31
7.0.0	Water Balance	3.31

Overview

A hydronic system uses water, rather than air, as a medium for providing heating and cooling. In hydronic systems, instead of conditioned air being carried in ductwork, conditioned water is carried through pipes to heat exchangers. Circulating pumps are used to move the water. Hydronic systems also use different types of control systems and safety devices than forced-air systems.

In a hydronic heating system, water is heated in a boiler and piped to heating terminals where the heat is transferred to the conditioned space. The water can be heated using gas or oil as a fuel source. Residential and small commercial systems, which are the type covered in this introductory module, use low-temperature, low-pressure water. Large commercial systems, which are covered in a later module, use high-temperature, high-pressure steam systems.

Instructor's Notes:

Objectives

When you have completed this module, you will be able to do the following:

1. Explain the terms and concepts used when working with hot-water heating.
2. Identify the major components of hot-water heating.
3. Explain the purpose of each component of hot-water heating.
4. Demonstrate the safety precautions used when working with hot-water systems.
5. Demonstrate how to operate selected hot-water systems.
6. Demonstrate how to safely perform selected operating procedures on low-pressure systems.
7. Identify the common piping configurations used with hot-water heating.
8. Read the pressure across a water system circulating pump.
9. Calculate heating water flow rates
10. Select a pump for a given application.

Trade Terms

Aquastat
Cavitation
Corrosion
Gravity hot-water system
Head pressure
High/low pump head
Hydronic system
MBh
Pressure drop
Redundancy
Specific heat
Static pressure

Required Trainee Materials

1. Pencil and paper
2. Appropriate personal protective equipment

Prerequisites

Before you begin this module, it is recommended that you successfully complete *Core Curriculum*; *HVAC Level One*; and *HVAC Level Two*, Modules 03201-07 and 03202-07.

This course map shows all of the modules in the second level of the HVAC curriculum. The suggested training order begins at the bottom and proceeds up. Skill levels increase as you advance on the course map. The local Training Program Sponsor may adjust the training order.

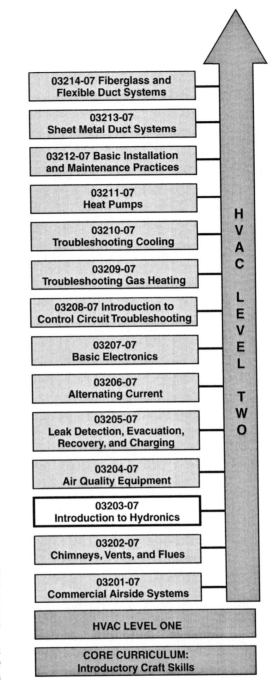

Ensure that you have everything required to teach the course. Check the Materials and Equipment list at the front of this module.

See the general Teaching Tip at the end of this module.

Explain that terms shown in bold are defined in the Glossary at the back of this module.

Show Transparency 1, Objectives, and Transparency 2, Performance Tasks. Review the goals of the module, and explain what will be expected of the trainee.

Review the modules covered in Level Two and explain how this module fits in.

See the Teaching Tip for Sections 1.0.0–7.0.0.

Explain that the term hydronic refers to all water systems and that a hydronic system uses pipes to transport heated fluid (water or steam) from the source to where it is needed.

Compare and contrast forced-air and hydronic heating systems.

Describe steam heating systems.

Ask trainees what some of the advantages are for using a steam heating system instead of a hot-water heating system.

Discuss the advantages of steam heating over hot-water heating. Explain that hot-water systems will be covered in this module and steam systems will be covered in Level Three.

1.0.0 ♦ INTRODUCTION

There are two basic ways to deliver mechanically generated heat and cold from its source of generation to the point where it is needed. A forced-air system uses a fan to deliver heated and/or cooled air through a ductwork system for distribution to the conditioned space. Similarly, a **hydronic system** uses pipes to transport heated or cooled fluid. The fluid used for heating can be either hot water or steam. For cooling, it is chilled water. The term hydronic refers to all water systems. However, the name hydronic heating or heating hot-water system is commonly used when referring to hot-water and steam heating systems (*Figure 1*). The phrase chilled-water system is commonly used when referring to an air-conditioning system that circulates chilled water to cooling coils.

Some advantages of hydronic systems over forced-air systems are:

- Water systems usually provide more comfortable heating and/or cooling throughout the building.
- Water has a greater **specific heat** than air, so it can carry a large amount of heat per pound. Specific heat is the amount of heat required to raise the temperature of one pound of a substance 1°F. One pound of water can hold one Btu for every degree that the temperature of the water is raised. In comparison, one pound of air can hold an average of 0.24 Btu for every degree the temperature is raised.

Figure 1 ♦ Gas-fired hot-water boiler.

- Hydronic systems may be more economical to use than forced-air systems and they may use less energy. This is because it is generally less expensive to move heated or cooled water through piping than it is to move heated or cooled air through a duct. This is true mainly if there is a great distance from the source of heating/cooling to where it is needed.

Steam systems generate and distribute steam used for comfort heating as well as commercial and industrial processes. Steam heating systems can be separate systems used only to heat buildings, or they can be part of a combined-process steam and heating system. This occurs mainly in locations where steam heat is required for use with other processes, such as in hospitals and industrial plants. In these systems, the process steam can be used directly to heat the building, or as the heat source for a heat exchanger (converter) to produce hot water for comfort heating.

Some advantages of steam systems relative to hot-water heating systems are:

- Steam flows through a system as a result of the natural pressures produced in the boiler, eliminating the need for circulating pumps. After condensing in the system terminals, the liquid condensate flows by gravity back to the boiler. In some systems, it flows back to a receiver tank from which it is pumped back to the boiler.
- The density of steam is low, resulting in system pressures that are suitable for use in tall buildings. If denser hot water were used in the same building, excessive system pressures would result.
- Steam depends both on temperature and pressure. This allows the system temperature to be controlled by changing either the steam temperature or pressure.
- Temperature changes are relatively small as steam flows throughout the heating system. This makes steam heat a good choice when using a central boiler to provide heat for two or more buildings in a given location. Such systems are encountered in campus or institutional facilities, as well as industrial environments.

Although a thorough understanding of steam and chilled water systems may be necessary as you advance through your career, this module will focus on introducing significant concepts related to hot-water systems only. Steam and chilled-water systems will be covered in much greater detail in *HVAC Level Three*.

Instructor's Notes:

2.0.0 ♦ WATER SYSTEM TERMS

Before you can study how water systems work, it is first necessary that you understand some concepts and know the meaning of some of the common terms used when describing water systems.

2.1.0 Water

Water is a chemical compound of two elements: oxygen and hydrogen. It can exist as ice, water, or steam due to changes in temperature. At sea level, it freezes at 32°F and boils at 212°F. Water changes weight with a change in temperature; that is, the higher the temperature of water, the less it weighs. This change in weight is due to the expansion and reduction in water volume. For example, one cubic foot of water weighs 62.41 pounds at 32°F and 59.82 pounds at 212°F. Different water weights resulting from different temperature levels cause natural gravity circulation of water in a system to occur. This natural circulation in a water system is referred to as thermal circulation.

Water itself is non-corrosive to most materials used in hydronic systems. However, a variable amount of air, air consisting primarily of nitrogen and oxygen, is also dissolved in fresh water. Although nitrogen is relatively inert, the dissolved oxygen will cause **corrosion** and rusting. For this reason, closed systems are used with minimal amounts of fresh water admitted after the initial fill. Open systems (meaning they are open to the atmosphere at one or more points in the system) allow the water to absorb additional oxygen continuously. These systems are no longer used. Each time fresh water is admitted in a closed system, following a leak repair for example, additional oxygen is also admitted and fresh corrosion begins until all free oxygen has been used in new chemical bonds such as iron oxide, commonly known as rust. If the admission of fresh water continues consistently, the piping system and its components will potentially suffer irreparable damage and failure.

Because most systems use automatic valves to admit fresh water when needed, and leaks can occur in unmonitored areas, hydronic systems should be inspected regularly for leaks. On larger systems, water meters may be installed to monitor the volume of fresh water being added over a given period of time.

As water passes through the boiler and is heated, air dissolved in the water is liberated in the form of bubbles. Eventually, most of the air is liberated and contributes to the air that must be removed from the system for it to function properly. Methods of air removal are covered later in this module.

Because water contains minerals, scale can form in the piping system. Reducing mineral scale build-up within the system, which impedes both heat transfer and flow dramatically, is another important reason to minimize the admittance of fresh water. Special chemicals are often added to systems, especially larger ones, which combine with minerals to form harmless, non-scale compounds. Such chemicals can also work to reduce free oxygen, again using the oxygen to produce harmless materials and prevent the oxygen from bonding with iron to form rust.

2.2.0 Pressure Drop

Pressure drop is the difference in pressure between two points. Pressure drop is a result of power being consumed as the water moves through pipes, heating units, and fittings. It is caused by the friction created between the inner walls of the pipe or device and the moving water. In a horizontal pipe in which there is no flow, the pressure is equal at all points. Once flow starts, friction is encountered which increases in direct proportion to the velocity of flow. The change in pressure drop when there is an increase or decrease in the flow in gallons per minute (gpm) can be calculated using the following formula:

(Final gpm ÷ initial gpm)^2 × initial pressure drop = final pressure drop

For example, assume that the water flow through a system or device is increased from 5 to 10 gpm and that the initial pressure drop was 5 psi. The new or final pressure drop will be 20 psi, as shown in *Figure 2*, and calculated as follows:

(10 gpm ÷ 5 gpm)^2 × 5 psi = final pressure drop
(2 gpm)^2 × 5 psi = final pressure drop
4 gpm × 5 psi = 20 psi

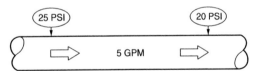
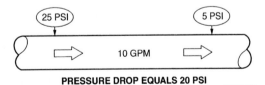

Figure 2 ♦ Pressure drop.

Explain that in a hot-water system, enough pressure must be available to overcome the effects of pressure drop in the system.

Identify the different units used for expressing pressure drop.

Show Transparency 4 (Figure 3). Define head pressure and explain how it is measured.

Discuss head pressure and its relationship to the capacity of a pump. Show the formulas for converting head pressure of water in feet to pressure in psi, and vice versa.

Discuss head pressure relationships. See the answer to the "Think About It" at the end of this module.

Give trainees some problems that require them to convert head pressure of water given in feet to pressure in psi, and vice versa.

In a water system, enough power must be available to overcome the effects of pressure drop to make sure the total system provides the desired heating/cooling. This means there must be enough power to overcome the total pressure drop (which creates the need for power consumption) of all the system piping and components. In a forced hot- and/or chilled-water system this power is normally provided by a circulating pump. Manufacturers will provide pressure drop information about their equipment. It can be expressed in pounds per square inch (psi), in feet of water, or in milli-inches. These units are interchangeable as follows:

1 pound per square inch = 2.31 feet of water
1 foot of water = 0.43 pounds per square inch
1 foot of water = 12,000 milli-inches
1 inch of water = 1,000 milli-inches

2.3.0 Head Pressure

Head pressure (or head) is another measure of pressure, expressed in feet of water. It is known that a column of water that is 1' high produces a pressure of 0.43 psi. It is also known that a column of water 2.31' high will produce a pressure of 1 psi (*Figure 3*). The formulas used to convert head pressure in feet to the equivalent psi or to convert psi to the equivalent head pressure in feet are shown in *Figure 3*.

Head pressure in feet is normally used in describing the capacity of a circulating pump. In this regard, it indicates the height of a column of water being lifted by the pump, without considering friction losses in the piping. The maximum head pressure of a pump is actually the maximum pressure drop against which the pump can cause a flow of water. For a water pump, the head pressure in feet can be divided by 2.31 to get the equivalent pump pressure in psi.

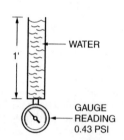

FORMULA FOR CONVERTING HEAD PRESSURE OF WATER IN FEET TO PRESSURE IN PSI
WHERE: P = PRESSURE (PSI)   H = HEAD PRESSURE (FEET OF WATER)

$$P = H \times 0.43$$

FORMULA FOR CONVERTING PRESSURE IN PSI TO HEAD PRESSURE OF WATER IN FEET
WHERE: H = HEAD PRESSURE (FEET OF WATER)   P = PRESSURE (PSI)

$$H = 2.31 \times P$$

Figure 3 ♦ Relationship of head pressure in feet to pounds per square inch.

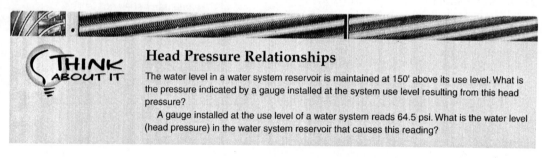

Head Pressure Relationships

The water level in a water system reservoir is maintained at 150' above its use level. What is the pressure indicated by a gauge installed at the system use level resulting from this head pressure?

A gauge installed at the use level of a water system reads 64.5 psi. What is the water level (head pressure) in the water system reservoir that causes this reading?

2.4.0 Static Pressure

Static pressure is created by the weight of the water in the system. We know that a column of water that is 1' high produces a pressure of 0.43 psi. Therefore, static pressure is equal to 0.43 psi for each foot of height above the system gauge. Consider a simple water system to be an upright loop of water confined in a pipe. As such, the static pressure in one of the vertical pipes of the loop is identical to the static pressure at the same level in the opposite vertical pipe. Static pressure levels in a system decrease as we move from the lowest point in the system toward the highest point in the system. At the highest point, the static pressure is 0 psi. For example, if the highest point holding water in a heating system is 30' above the hot-water boiler, the ascending static pressure levels within the system are as follows:

30' (highest point in system) = 0 psi
20' (10' below highest point) = 4.3 psi (10 × 0.43)
10' (20' below highest point) = 8.6 psi (20 × 0.43)
0' (30' below highest point) = 12.9 psi (30 × 0.43)

3.0.0 ♦ HOT-WATER HEATING SYSTEMS

Hot-water heating systems are used mainly to provide comfort heating. They can be separate systems used to heat buildings that have separate or no cooling systems. They can also be used to heat domestic water in dual-temperature systems. This section describes separate hot-water heating systems. Hot-water heating in a dual-temperature system is covered in *HVAC Level Three*.

Hot-water heating systems are of two types: **gravity hot-water systems** and forced hot-water systems.

3.1.0 Gravity Hot-Water Systems

In a gravity hot-water system (*Figure 4*), water contained in the boiler is heated to the operating temperature needed for use in the system. The circulation of hot water occurs as a result of thermal circulation or buoyancy. This is because of the water temperature and weight differences that exist in the system. Water, when hot, is

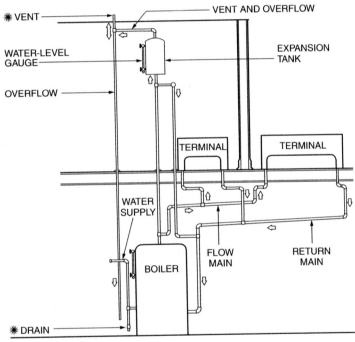

Figure 4 ♦ Gravity hot-water system.

▷ Direction of water flow

✱ Vents and drains must be piped to provide a safe place for the discharge of overflow water. Discharges can occur when the system is operating at the maximum temperature or a fill valve has malfunctioned.

Explain how a gravity hot-water system functions. Point out the disadvantages of gravity hot-water systems.

Show Transparency 6 (Figure 5). Show the configuration and explain the operation of a forced hot-water system. Explain that pumps are used to satisfy system flow and pressure requirements.

Identify the components of a hot-water heating system and explain their purposes.

See the Teaching Tip for Section 4.0.0 at the end of this module.

lighter than when it is cold. This is due to the expansion of the water as its temperature is increased. In operation, the temperature of the water in the boiler rises and its volume per pound increases. This disturbs the equilibrium of the system. As the warmer and lighter water rises upward in the supply main, the cooler and heavier water in the return main starts to flow downward. This movement starts the circulation of the water in the system. The heat contained in the circulating water is transferred via the terminal devices to the cooler air in the conditioned space. This heat transfer takes place both by radiation and convection. Cooled water leaving the various system terminals is then returned to the boiler, where it is reheated and recirculated.

Because the gravity hot-water system relies on thermal circulation, it responds slowly to changes in load conditions. For the same application, the gravity hot-water system is less efficient than a forced hot-water system. Since water flow depends on gravity, in order to achieve the required heating levels, it requires the use of larger pipes to reduce pressure drop and larger terminal devices to accommodate the reduced heated water flow rates than a pump-driven system. Because no pump is used, the system operates with little or no noise.

3.2.0 Forced Hot-Water Systems

Figure 5 shows the major components of a typical forced hot-water heating system. Please note that a variety of piping arrangements are used, depending on the application. Other practical piping layouts will be presented later in this module. Depending on the type of boiler, the water is heated to operating temperature using gas, oil, or an electric heat source. The burners or heating elements used in boilers are basically the same as those used in gas, oil, and electric furnaces. The boiler combustion or electric heating process is controlled by temperature and/or water level sensing elements located in the boiler. These controls start and stop the process based on the temperature and/or level of the water in the boiler.

Forced hot-water systems use one or more pumps to force the water through the system piping. The pump or pumps are sized to satisfy the system flow and pressure requirements. Piping networks commonly used with hot-water systems are described in detail later in this module. Buildings that use forced hot-water systems are normally divided into several zones controlled by zone thermostats. When a zone thermostat calls for heat, the system circulating pump turns on. This causes the heated water from the boiler to circulate through a zone control valve to terminal devices located in the zone. The heat contained in the circulating water is transferred via the terminal devices to the cooler air in the conditioned space. Normally, this heat transfer takes place both by radiation and convection. Cooled water leaving the terminal devices is then routed by the return line to the boiler, where the water is reheated and recirculated.

The check valve located in the return line stops any flow of water by gravity when the circulating pump is turned off. The balancing valve is used to adjust the resistance to water flow through the zone to meet the system design requirements.

4.0.0 ♦ HOT-WATER HEATING SYSTEM COMPONENTS

Typical components of a hot-water heating system include the following:

- Hot-water boiler
- Boiler operating/safety controls and accessories
- Expansion/compression tank
- System air control devices
- Circulating pump(s)
- Valves
- Terminal heating equipment (radiators)
- Hot-water heat exchangers/converters
- Tankless and indirect water heaters

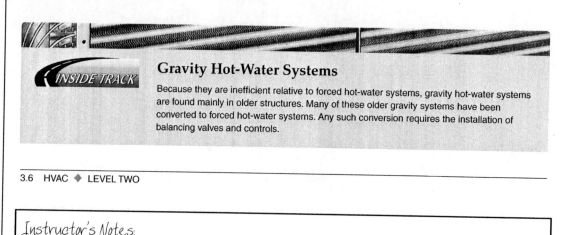

Gravity Hot-Water Systems

Because they are inefficient relative to forced hot-water systems, gravity hot-water systems are found mainly in older structures. Many of these older gravity systems have been converted to forced hot-water systems. Any such conversion requires the installation of balancing valves and controls.

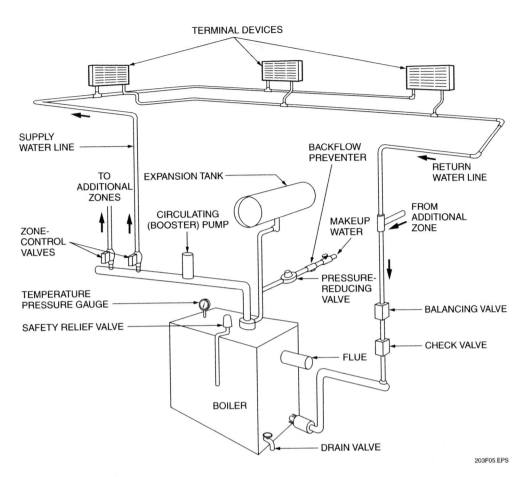

Figure 5 ♦ Simplified zoned forced hot-water heating system.

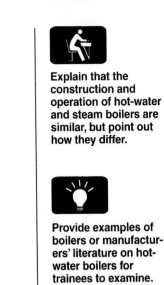

4.1.0 Hot-Water Boilers

A boiler heats water using gaseous fuels, oil fuels, solid fuels, or electricity as the heat source. Geothermal heat pumps are sometimes used to produce hot water for comfort heating. Some boilers can be fired with more than one fuel. This is done by burner conversion or by using dual-fuel burners. Gaseous fuels include natural gas, manufactured gas, and liquefied petroleum (propane and butane). Oil fuels include both lightweight and heavy oils. Solid fuels include coal and wood. However, the residential market is dominated by natural gas in urban and suburban areas, while lightweight fuel oil (grade #2) and propane are the primary fuels in rural areas.

Combustion in a boiler occurs by combining the fuel with oxygen and igniting the mixture. The methods used for combustion in gas-fired and oil-fired boilers are basically the same as those used with furnaces. The ignition of the gas or oil can be achieved by the use of a standing gas pilot, electric spark-type intermittent ignitor, or by a hot surface ignitor. The presence of a pilot or flame is proven by a protective device before the main gas or oil valve is allowed to open. Most boiler combustion systems have a purge and pre-purge sequence to make sure that there is not a combustible mixture in the boiler which might cause an explosion during start-up. The methods used to vent boilers are similar to those used with furnaces.

The construction and operation of hot-water and steam boilers are similar, with two exceptions. The operating and safety controls used with hot-water boilers are different than those

Compare and contrast different types of boilers.

Discuss the hazards and safety precautions needed when working with large boilers.

Describe copper-finned tube and stainless steel boilers.

used with steam boilers. Also, hot-water boilers are entirely filled with water, while steam boilers are not. Low-temperature boilers are the most widely used type of boiler. They are used for residential, apartment, and commercial buildings. Low-pressure hot-water boilers can be constructed to have working pressures of up to 160 psi. Normally, they are designed for a 30 psi maximum working pressure, but are frequently operated below that pressure level, with 12 to 15 psig being common. Low-pressure hot-water boilers are limited to a maximum operating temperature of 250°F. Above this temperature, even water under low pressure will begin to boil and begin the change of state to steam, creating a dangerous condition.

Medium- and high-pressure hot-water boilers are built to operate at pressures above 160 psig and/or temperatures well above 250°F. In extremely large systems, such high pressures and temperatures allow the water to be circulated great distances, from a central heating plant to other buildings for example, without losing all heating capacity before arriving at the intended destination. It should be noted that such systems can also be extremely dangerous.

WARNING!
Larger hydronic heating systems often use high-temperature, high-pressure water, especially in central plant applications. Water at such extreme temperatures—in excess of 450°F in some cases—and under high pressure, instantly flashes to steam when even a small fracture or tiny leak develops in piping or components. Quite often, evaporation of the superheated steam occurs so rapidly, no steam plume is seen. This is due to its sudden exposure to atmospheric pressures, where water boils at 212°F. Serious physical harm can result by simply walking near or moving a hand across the path of the escaping steam. Use extreme caution when servicing or working in the presence of such systems.

Most boilers are made of cast iron or steel. Many of today's boilers also incorporate copper-finned tubes to further extract heat from the combustion process. As is the case with warm-air furnaces, many levels of efficiency are also available, including condensing boilers with efficiencies above 95 percent. Today's boilers even offer the appearance of home appliances. Stylish cabinets, some made of stainless steel, better reflect the tastes of the modern homeowner.

4.1.1 Copper-Finned Tube and Stainless Steel Boilers

Small, compact boilers with heat exchangers constructed of copper-finned tubes and/or stainless steel are growing in popularity. Older boilers with reduced heat transfer characteristics required that a significant amount of heated water remain in storage to provide for the instantaneous demand for warmth. With the introduction of copper-finned tubes, the speed of recovery eliminates the need for large volumes of heated water ready for circulation, and further improves efficiency of the products by reducing heat losses from the water while in stand-by. Units using copper tubes are often made with cast-iron headers and/or sidewalls, as shown in *Figure 6*. The use of such materials, which are lightweight and transfer heat quickly and effectively, has allowed units to shrink remarkably in both size and weight (*Figure 7*). Although capacities range high enough to easily satisfy the needs of a standard home, large residential or commercial applications may require the use of two or more boilers piped in parallel, providing **redundancy**, increasing total system heating capacity, and providing further control of active heating capacity to match heat losses (*Figure 8*). For smaller applications, gas-fired boilers designed for wall mounting and using direct through-wall venting packages are readily available (*Figure 9*).

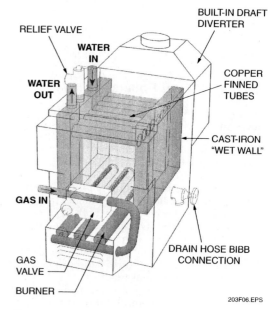

Figure 6 ♦ Cast-iron/copper-finned tube boiler.

3.8   HVAC ♦ LEVEL TWO

Instructor's Notes:

Figure 7 ♦ Lightweight boilers simplify installation.

Figure 9 ♦ Wall-mounted boiler with direct venting.

Figure 8 ♦ Two boilers, piped in parallel configuration.

Condensing models, primarily made with stainless steel heat exchangers and reaching efficiencies above 95 percent, are also popular (*Figure 10*). Such units can generally be vented using Schedule 40 PVC pipe, due to the low temperature of the exiting flue gas. As is the case with condensing furnaces, cooling the by-products of combustion to this degree causes a rather acidic moisture to condense out of the flue gases. This condensate must be collected and removed from the unit.

Although gas burner designs vary widely among manufacturers, the basic operation remains the same as that used in today's warm-air furnaces. Gas valves can be either single stage, two-stage, or fully modulating. Ignition is generally accomplished through direct spark or hot surface ignitors. Sealed combustion burners use

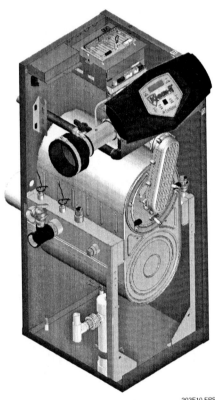

Figure 10 ♦ Condensing boiler with stainless steel heat exchanger.

Describe condensing boilers.

Describe the operation of gas boilers.

Describe on-board controls for gas boilers.

Describe cast-iron boilers.

Show Transparency 7 (Figure 12). Identify the controls and accessories of hot-water boilers.

Explain that boilers are equipped with safety controls.

See the Teaching Tip for Section 4.2.0 at the end of this module.

combustion air from outdoors, rather than using indoor air which has been heated at some cost. The use of draft inducer fans also allows for sidewall venting and other, more convenient venting options. These vents are routed along with the combustion air intake piping, terminating in special manifolds installed at the outside wall for easy installation.

Some gas boilers are equipped with sophisticated on-board controls, using sensors placed as necessary to monitor boiler water conditions, space conditions and outdoor conditions simultaneously. The firing rate and run time of the burner are then controlled based on a constant comparison of these conditions. For maximum fuel efficiency, these boilers provide only the required water temperature and/or flow rate required to meet the needs of the heated space, and no more. Controls providing night setback can also be incorporated directly into the boiler controls.

4.1.2 Cast-Iron Boilers

Many of today's boilers are still made of cast iron, due to its durability, resistance to thermal shock, and heat transfer characteristics. Cast-iron boilers are formed by assembling individual cast-iron heat exchanger sections together (*Figure 11*). Each section is basically a separate boiler. The number of sections used determines the size of the boiler and its energy rating. Cast-iron boiler capacities range from those required for small residences up to large commercial systems of 13,000 **MBh**. (One MBh is equal to 1,000 Btuh.) In a cast-iron boiler, the water circulates inside the cast sections with the flue gases on the outside of the sections. The cast sections are usually mounted vertically, but they can be mounted horizontally. In both arrangements, the heating surface is large relative to the volume of water. This allows the water to heat up quickly.

The boiler depicted in *Figure 12* uses modular construction. This allows different combinations

Figure 11 ♦ Cast-iron sections in a packaged boiler.

of boiler sections, bases, and flue collectors to be assembled to match heating requirements. Cast-iron boilers used for residential and smaller commercial jobs usually are supplied completely assembled (packaged). A packaged boiler is one that includes the burner, boiler, controls, and auxiliary equipment. Larger boilers can be packaged units, or they can be assembled on the job site.

4.2.0 Boiler Operating/Safety Controls and Accessories

In a properly designed hot-water system, the boiler is equipped with several operating/safety controls to guarantee safe and proper operation of the boiler (*Figure 12*). Accessories are also used to improve operation or make the system more efficient. The following controls and accessories are used with hot-water boilers:

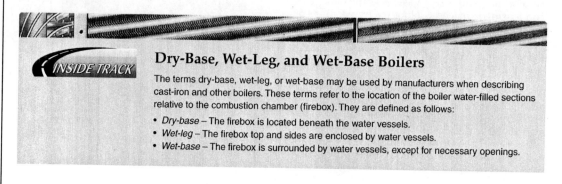

Dry-Base, Wet-Leg, and Wet-Base Boilers

The terms dry-base, wet-leg, or wet-base may be used by manufacturers when describing cast-iron and other boilers. These terms refer to the location of the boiler water-filled sections relative to the combustion chamber (firebox). They are defined as follows:

- *Dry-base* – The firebox is located beneath the water vessels.
- *Wet-leg* – The firebox top and sides are enclosed by water vessels.
- *Wet-base* – The firebox is surrounded by water vessels, except for necessary openings.

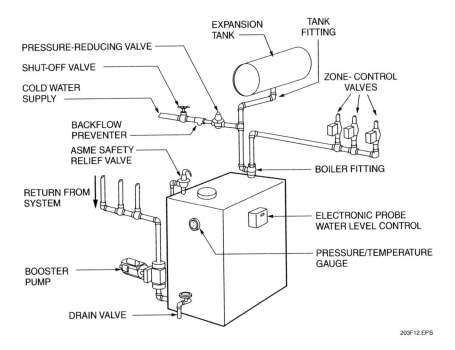

Figure 12 ♦ Hot-water boiler controls and accessories.

- Pressure/temperature gauge
- Pressure (safety) relief valve
- Thermal/electronic probe operating/safety controls
- Backflow preventer
- Drain valve

4.2.1 Pressure/Temperature Gauge

Pressure and temperature gauges (*Figure 13*) are usually combined into one gauge to show the pressure and temperature of the water in the boiler. Typically, these gauges provide water-pressure readings from 0 to 50 psi, corresponding to altitudes from 0' to 70', and temperatures from 60°F to 320°F.

4.2.2 Pressure (Safety) Relief Valve

A pressure (safety) relief valve (*Figure 14*) is used to protect the boiler and the system from high pressures caused by either water thermal conditions or steam pressure conditions in the boiler. It does not operate unless an overpressure condition exists. The typical hot-water boiler is constructed for a maximum working pressure of less than 30 psi, at which point the valve is designed to be fully open.

Figure 13 ♦ Pressure/temperature gauge.

It should be noted that this is significantly less than the average household domestic water pressure. As a result, water fed to the boiler is admitted through a water pressure reducing valve, set at an appropriate pressure to accommodate proper system operation while protecting the boiler from damage. They are generally factory-set to 12 psig, but most are adjustable to match actual system and equipment requirements.

Identify the various operating and safety controls and accessories used with a hot-water boiler.

Explain the operation of a pressure/temperature gauge and a pressure (safety) relief valve. Point out that they are normally set to 12 psig.

Provide pressure/temperature gauges and safety relief valves for trainees to examine.

Explain how the safety relief valve is used and installed.

Discuss the importance of maintaining proper boiler water pressure. Emphasize that the pressure setting of a safety relief valve is not field-adjustable and must not be altered.

Explain how to check a safety relief valve.

Provide examples of safety relief valves and/or manufacturers' instructions for safety relief valves for trainees to examine.

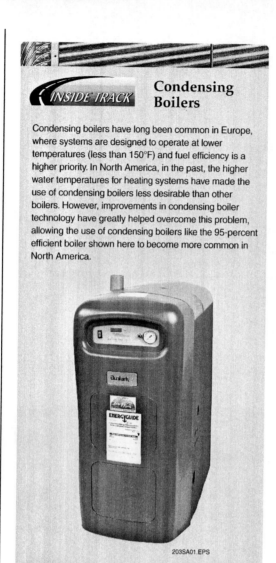

INSIDE TRACK

Condensing Boilers

Condensing boilers have long been common in Europe, where systems are designed to operate at lower temperatures (less than 150°F) and fuel efficiency is a higher priority. In North America, in the past, the higher water temperatures for heating systems have made the use of condensing boilers less desirable than other boilers. However, improvements in condensing boiler technology have greatly helped overcome this problem, allowing the use of condensing boilers like the 95-percent efficient boiler shown here to become more common in North America.

Figure 14 ♦ Pressure relief valve.

Safety relief valves should always be installed as directed by the manufacturer. They are usually installed in an upright position at the top of the boiler. Discharge is piped toward the floor or to a safe outside location. Shutoff valves must never be installed anywhere in relief valve piping.

WARNING!

Maintaining proper boiler water pressure is crucial to the safety of anyone nearby, as well as to the boiler itself. Boiler pressure ratings vary widely due to differences in construction, with many units having a maximum pressure rating of 30 psig. Excessive pressures can cause unexpected explosive failure of the unit and serious personal injury. For this reason, properly selected and installed safety relief valves are essential.

The pressure setting of a safety relief valve is not field-adjustable and must not be altered. Misadjustment will result in an explosion, which can result in death or personal injury. Replace a safety relief valve if it does not open when its pressure rating is exceeded or if it leaks when the boiler is operating at normal pressures.

The safety relief valve is set to open at the same pressure as the boiler's specified maximum operating pressure, and discharge hot water via a waste pipe attached to its outlet port. This waste pipe must be the same size as the relief valve's outlet port, never smaller. Should a runaway over-firing of the boiler burners occur, temperatures and pressures will rise in the boiler. This is because the system cannot dissipate the heat energy as fast as it is developed. With this condition, the safety relief valve will discharge both high-temperature hot water and steam. Note that a hazardous low-water condition can result in this situation if no makeup water is supplied to replace the water lost through the relief valve.

Safety relief valves should be checked according to the manufacturer's instructions intermittently during operation or after long periods of inactivity. Large relief valves may need to be removed and sent to a certified testing site for reconditioning and testing.

3.12 HVAC ♦ LEVEL TWO

Instructor's Notes:

Inside Track — Reusing Boiler Parts

When a boiler is replaced, some of its components may be safely reused. However, the pressure-relief valve should not be reused. Relief valves may develop leaks over time, and scale buildup on a used valve may cause it to stick.

Relief valves are inexpensive, but they are vital safety components. When replacing a boiler, install a new relief valve as specified by the boiler manufacturer. Depending on the manufacturer, the new boiler may come with a new relief valve installed, or it may be shipped loose to be installed at the location.

4.2.3 Thermal/Electronic Probe Operating/Safety Controls

Thermally operated switches called **aquastats** (*Figures 15* and *16*) can be used to control hot-water boilers. An aquastat works basically the same way as a thermostat with the exception that it is designed to control water temperatures instead of air temperatures. Aquastats may be surface-mounted. This type is typically strapped to a pipe. Because they are strapped to a pipe, it is important that clean contact be made between the mating surfaces of the aquastat and the pipe. Aquastats can also be of an immersion type. This type has a sensing probe that is immersed in the system water by inserting it into a well on the boiler.

Aquastats are used as both operating controls and as high-limit safety controls. Direct-acting aquastats open their contacts on a temperature rise. This type is used mainly as a high-limit control. Reverse-acting aquastats open their contacts on a temperature drop. This type is used to stop pump or fan operation when the system water temperature falls below its setpoint. Reverse-acting aquastats also can be set to switch on the burner if the water temperature gets too low.

Electronic probe-type controls are also common (*Figure 17*). This device uses an electrode placed in the water. The electrode uses the conductivity of the water to complete a circuit to ground. If the probe is in the water, the circuit to ground is completed; if not, the circuit to ground is open. In relation to the water level as sensed by the control, the absence or presence of the circuit to ground can be used to turn on visual and/or audible alarms at a remote control panel, energize/de-energize relays, and turn pumps on and off. A common use for this control is as a low-water safety device that prevents firing of the boiler burners whenever the water level drops below a safe level. Low-water safety devices are required by most codes.

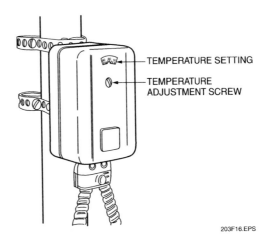

Figure 16 ♦ Surface-mounted aquastat.

Figure 15 ♦ Aquastat.

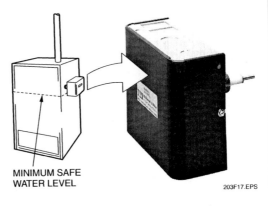

Figure 17 ♦ Electronic probe water level control.

Discuss the reuse of boiler parts.

Show Transparency 8 (Figure 16). Show examples of thermal surface-mounted and immersion-type aquastats. Explain their purposes, how they operate, and the differences between them.

Show Transparency 9 (Figure 17). Explain how an aquastat can be used as either an operating control or as a high-limit safety control.

Show trainees safety precautions necessary when working on a boiler and how to safely perform selected operating procedures on a boiler.

Have trainees practice safety precautions necessary when working on a boiler and safely performing selected operating procedures on a boiler. Note the proficiency of each trainee. This laboratory corresponds to Performance Tasks 1 and 3.

Have the trainees review Sections 4.3.0–4.9.0.

Ensure that you have everything required for teaching this session.

Show Transparency 10 (Figure 18). Describe an expansion tank and explain its purpose in a hot-water system.

Explain that when a heating system is filled with water, the expansion tank normally will be from one-third to one-half full of water, with the remainder of the tank containing air that acts as a cushion for thermal expansion.

Show Transparency 11 (Figure 20). Explain the operation of a pressurized expansion tank.

Provide an expansion/compression tank or manufacturers' literature on expansion/compression tanks for trainees to examine.

4.3.0 Expansion/Compression Tanks

An expansion tank (*Figure 18*), also called a compression tank, is used to maintain system pressures. It allows for the safe expansion of water as it is heated. The volume of water in a piping system varies with temperature changes. When water is heated or cooled, it expands or contracts. When the system water is at maximum operating temperature, the resulting excess water volume caused by expansion is stored in the expansion tank. When the system water temperature drops and the water volume in the tank contracts, water is returned to the system. An expansion tank must be able to store the maximum volume of excess water that exists when the system is operating at maximum temperature. It must do this without exceeding the maximum system operating pressure. It also must maintain a required minimum level of system pressure when the system is cold.

When a heating system is filled with water, the expansion tank will normally be about one-third to one-half full of water. The rest of the tank will contain air, which acts as a cushion for thermal expansion. The air compresses and expands as the volume of water in the system expands and contracts. This increases or decreases the pressure in the tank and serves to maintain the system pressure within established design limits.

The traditional hot-water system uses an airtight expansion tank to provide room for expansion, as shown in *Figure 18*. Standard tanks have no separation between the air and water. They are simply empty vessels, with a variety of fittings for piping installation and air controls. Larger systems often use two or more tanks. These tanks are normally installed near the boiler. They are equipped with various air-control devices used to continuously separate free air from the water in the system and route it to the expansion tank. These air-control devices are covered later in this section. If an air leak occurs in the tank, the air will escape from the tank and be replaced by water. Eventually, all the air will leak out and the tank will become filled with water (waterlogged). An air-control tank fitting is used to prevent this problem.

Pressurized, or diaphragm, expansion tanks (*Figure 19*) are also in common use in closed systems. They contain a flexible diaphragm, or bladder, that separates the system water from the air. *Figure 20* shows the action of the diaphragm as the water expands. It should be noted that tank pressure must be adjusted and checked with the tank isolated from the water system—tank and system pressure will change as thermal expansion takes place. Pressurized expansion tanks often come pre-charged with air, but the pressure can be adjusted through the charge valve, if needed, to fit system design conditions.

Figure 19 ◆ Pressurized diaphragm expansion tanks.

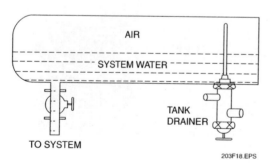

Figure 18 ◆ Standard expansion tank.

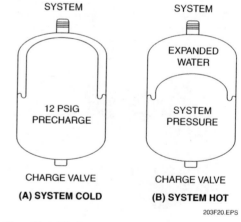

Figure 20 ◆ Diaphragm tank operation.

Waterlogged Expansion Tanks

You should suspect a waterlogged expansion tank if the boiler relief valve discharges whenever the boiler water is being heated. If this happens, check the expansion tank and any related air-control components for leaks.

Leaking Air Valves or Bladders/Diaphragms

If a pressurized tank fails to hold its pre-charge of air, you should suspect a leaking air valve or bladder/diaphragm. If a diaphragm-type expansion tank has a leaking diaphragm, the tank must be replaced. If a bladder-type expansion tank has a leaking bladder, the tank can usually be repaired by replacing the damaged bladder. Bladder-type expansion tanks can be recognized by their large flanged opening, which allows access to the bladder. When removing the bladder, the system pressure must be reduced before opening the system.

Pressurized expansion tanks must be installed according to the manufacturer's instructions. This is because the piping between the system and tank differs from that used with a standard expansion tank. It is also important that any air in the system be purged to the atmosphere and not allowed to enter the tank(s).

4.4.0 System Air-Control Devices

In a closed hot-water heating system, the only air in the system should be the air in the expansion tank. Because air is absorbed into water, air-control devices must be used to continuously separate free air from the water in the system and route it to the expansion tank. At the same time, these devices must prevent the flow of water by gravity circulation back to the boiler. Various air-control fittings for installation on the expansion tank, boiler, and/or in-line with the piping are available for this purpose. *Figure 21* shows some of the air-control devices commonly used in hot-water systems.

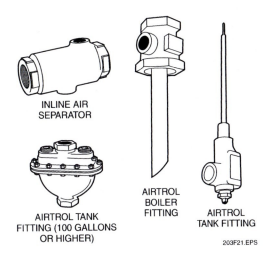

Figure 21 ◆ Air-control devices.

Makeup Water and Corrosion

Air contains oxygen, which corrodes metal. A typical closed hot-water heating system contains some air, but the constant heating of the water eventually drives it out.

If there are leaks or other problems in the system that permit fresh water to enter, corrosion can build up because fresh water contains oxygen. A properly designed and properly operating closed hot-water heating system should require very little makeup water during operation. This helps prevent corrosion.

Explain that free air in an open system can cause noise and interfere with water circulation.

Describe a circulating pump and explain its function in the hot-water system. Explain that the pump can be installed in the supply side of the system or on the return side. Give examples of each application.

Describe pump installation.

Provide circulating pumps or manufacturers' literature on circulating pumps for trainees to examine.

Free air in an open system, or in a system that uses a pressurized expansion tank, can cause noise and also interfere with water circulation. On startup of a heating system, the air which is forced to the high points must be removed or purged from the system. This can be done with manually operated or automatic air vents installed at the high points in the system. When automatic air vents are used, drain lines must be installed to carry any vented water from the vent to a suitable receptor.

4.5.0 Circulating Pumps

The circulating pump forces the hot water from the boiler through the piping to the terminal (radiation) units and back to the boiler. It is often referred to as a booster pump. Most circulating pumps are centrifugal-type pumps. In a centrifugal pump, the rotating action of an impeller in a spiral housing generates a pressure that forces the water through the piping system. The pressure and volume developed is a function of the pump size, pump motor horsepower, and rotational speed. In small systems, in-line pumps are commonly used, while base-mounted pumps are used in large systems (*Figure 22*). Shutoff valves are normally installed on the inlet and outlet piping of the pump so that the pump can be isolated for repair or replacement. These valves also act to hold the water in the system, thus avoiding air problems that can arise if a zone or system needed refilling.

A circulating pump creates circulation in a piping system by establishing a pressure differential between its suction and discharge openings. Water flows in the system in an attempt to equalize this difference. As long as the pump

Figure 22 ♦ In-line circulating pump.

runs, the difference remains and the water keeps flowing. The pump can be installed either in the supply side or return side of the system. On larger systems with a high pump head, it is recommended that the pump(s) be installed in the supply side of the system so that the pump is discharging away from the boiler and expansion tank. The exception is in small systems or systems in low-rise buildings. In these systems, the pump(s) may be installed in the return side of the system so that the pump is discharging into the boiler and expansion tank. This is possible because the pumps encounter low operating head pressures and pumping into the boiler does not cause any problems because the pressure drops are small.

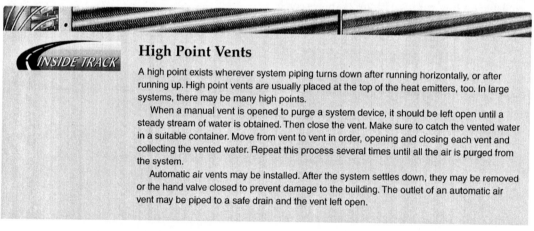

High Point Vents

A high point exists wherever system piping turns down after running horizontally, or after running up. High point vents are usually placed at the top of the heat emitters, too. In large systems, there may be many high points.

When a manual vent is opened to purge a system device, it should be left open until a steady stream of water is obtained. Then close the vent. Make sure to catch the vented water in a suitable container. Move from vent to vent in order, opening and closing each vent and collecting the vented water. Repeat this process several times until all the air is purged from the system.

Automatic air vents may be installed. After the system settles down, they may be removed or the hand valve closed to prevent damage to the building. The outlet of an automatic air vent may be piped to a safe drain and the vent left open.

Instructor's Notes:

The pump's location in the system, relative to the expansion tank, determines whether the pump's pressure is added to or subtracted from the system static pressure. This is because the point where the expansion tank is connected to the system is the point of no pressure change when the pump is started or stopped. It is recommended that this point be on the suction side of the pump in order to minimize total pressure. If the connection is made on the discharge side of the pump, the total pressure must be greater in order to prevent **cavitation** at the pump inlet. Cavitation is a result of vapor pockets formed when the pressure on a liquid drops below its vapor pressure in a pumping system. It can result from incorrect sizing or installation of the pump.

The circulating pump is sized to overcome the system pressure drop and to supply the necessary amount of water in gpm at the proper temperature to each terminal device. Pressure drop results from the friction caused by the system piping, fittings, boiler, terminals, and other heating components. The term head pressure is used to give the capacity of a circulating pump. It is just another way of expressing pressure drop.

The maximum head of a pump is actually the maximum pressure drop against which the pump can cause water to flow. It is usually expressed in feet of water. **High/low pump head** are terms often used to indicate the relative magnitude of the height of a column of water that a circulating pump is moving, or must move, in a water system.

Manufacturers use various kinds of curves to show pump performance. The exact curve used depends on the pump characteristic of interest. *Figure 23* shows a set of typical pump curves for a pump operating at a speed of 1,750 rpm. It should be noted that for the same pump operating at a different speed, a different set of curves must be used. The bottom horizontal scale shows the pump delivery in gpm. The vertical scale shows the head pressure in feet. At 0 gpm, there is no delivery on the curve. At this point, the power of the pump is exactly equal to the pressure drop opposed to it. Since a difference in pressure between two points is necessary before flow can occur, the pump will not deliver water at this point. If the pressure drop (head) is lowered, the pump can deliver water.

Explain how the location of the pump in the system determines whether the pump's pressure is added or subtracted from the system's static pressure.

Define cavitation and explain that it is caused from incorrect pump sizing or installation.

Discuss factors involved in properly sizing a circulation pump.

Show Transparency 13 (Figure 23). Describe pump curves.

Provide pump curves for trainees to examine.

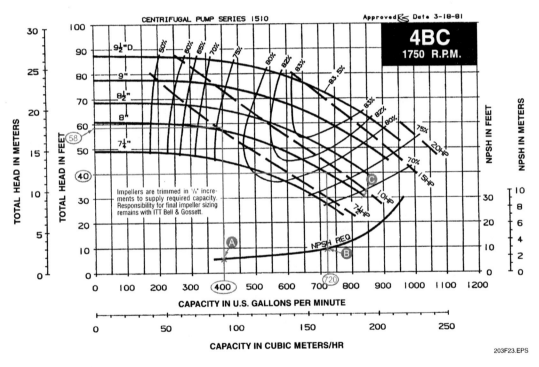

Figure 23 ◆ Typical pump curves.

Keep Transparency 13 (Figure 23) showing. Explain how to read pump curves.

Show trainees how to interpret a set of circulating pump curves. Explain how to interpret the efficiency, horsepower, and NPSHR curves on the chart.

Explain how a zone valve can control a circulating pump.

> ### Pump Cavitation
>
> Cavitation can occur in a water circulating pump when the pressure at the inlet of the impeller falls below the vapor pressure of the water (a function of water temperature and pressure), causing the water to vaporize and form bubbles. These bubbles are carried through the pump impeller inlet to an area of higher pressure where they implode (burst inward) with terrific force that can cause the impeller vane tips or inlet to be damaged due to pitting or erosion. The problem of cavitation can be eliminated by maintaining a minimum suction pressure at the pump inlet to overcome the pump's internal losses. This minimum suction pressure is called the net positive suction head required (NPSHR). Here are some symptoms that indicate cavitation is occurring in a pump:
>
> - Snapping and crackling noises at the pump inlet
> - Pump vibration
> - A reduction or no water flow
> - A drop in pressure

Examination of the chart capacity curves for various impeller sizes shows that:

- The lower the head pressure (pressure drop), the more water the pump can deliver in gpm, and vice versa.
- The lower the volume of water delivered by the pump in gpm, the higher the head pressure against which the pump can deliver water, and vice versa.

These points are illustrated by the examples shown on the chart in *Figure 23*. It shows that a pump with an 8" diameter impeller will deliver 400 gpm, when operating with a head pressure of 58' (point A on the chart). However, with a decrease in head pressure to 40', the same pump will deliver about 720 gpm (point B on the chart). Further examination of the curve for the 8" impeller shows that the maximum flow this pump can deliver is about 840 to 845 gpm (point C).

The chart in *Figure 23* also shows curves for the pump performance characteristics of efficiency and motor horsepower. Examination of these curves shows that for the example shown at point A, the pump is operating at an efficiency of between 75 percent and 80 percent (about 76 percent) and at a non-overloading motor horsepower of 7½ hp. When operating at point B, the same pump is operating at a slightly higher efficiency (about 78 percent), with an increase of the non-overloading motor horsepower from 7½ hp to about 9 hp.

The NPSHR curve shown on the chart is used to determine the minimal total head in feet that must be maintained above the pump suction inlet center line in order to prevent cavitation and provide proper pump operation. Referring to the NPSHR curve for our examples, the chart shows that when the pump is delivering 400 gpm (point A), the NPSHR is 6'. When the pump is delivering 720 gpm (point B), the NPSHR is 10'.

> ### How a Zone Valve Controls a Circulating Pump
>
> Here's how a zone valve controls a circulating pump:
>
> - A room thermostat in a calling zone energizes a motor-driven zone valve, causing it to open.
> - The zone valve is equipped with normally open switch contacts that close when the valve is fully open.
> - The closure of the switch contacts completes the path to energize the circulating pump.

Instructor's Notes:

In some zoned systems, a single circulating pump may be used to provide water flow to all the zones. In this case, the pump is wired such that it will start when any one of the zone thermostats or control valves calls for heat. The pump is sized to handle the total system flow rate at the required head pressure of the longest zone.

In other zoned systems, a circulating pump is used as the zone control for each zone. In this case, a circulating pump is installed in each zone, with each pump being under the control of its zone thermostat. When used in separate zones, the pump is sized to handle the flow rate and the head pressure of its zone piping circuit only.

4.6.0 Valves

This section describes the many different kinds of valves used in hydronic systems. In addition to being used in hot-water systems, many of the valves described here are also used in chilled-water and/or steam systems. Valves can be separated into two groups: common valves and specialty valves. Common valves are typically used to provide for component isolation or flow regulation. Common valves include the following:

- Gate valves
- Ball valves
- Globe valves
- Angle valves
- Check valves

Specialty valves are typically used to provide safe operation or enhance the operation of the system. These must be installed in accordance with the manufacturer's instructions. Specialty valves include the following:

- Pressure-reducing valves
- Backflow-preventer valves
- Flow-control valves
- Multi-purpose valves
- Balancing valves and flow meters
- Two-way and three-way valves

4.6.1 Gate, Ball, Globe, and Angle Valves

Gate valves (*Figure 24A*) are used to turn on or shut off the flow of water or steam. They have a gate-like disc that slides across the path of flow. In its fully open position, the gate is out of the way and resistance to water or steam flow is minimal. In its fully closed position, the gate seats tightly and flow is stopped. Gate valves have a straight-through flow pattern. There is no damming effect that might trap particles or sediment. Gate valves

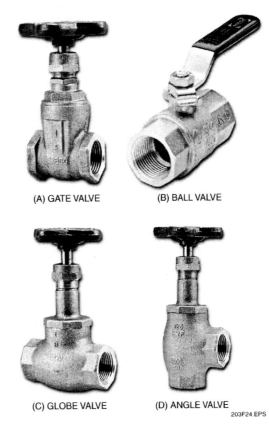

Figure 24 ♦ Typical gate, ball, globe, and angle valves.

should be operated either fully open or fully closed. If opened partway, the throttling that results from the water or steam being obstructed wears the seating surfaces of the valve.

Ball valves (*Figure 24B*), like gate valves, are used mainly to turn on or shut off the flow of water or steam. As the name implies, a ball valve has a ball-shaped flow control element with a hole in its center. As the ball is rotated through 90 degrees of arc, the hole moves from being fully aligned with the output valve port in the fully opened position, to being at a right angle and blocking the output valve port in its fully closed position. Ball valves can be easily recognized because they have a lever-type handle instead of a handwheel. The position of the lever relative to the center line of the valve indicates the position of the hole in the ball.

Globe valves (*Figure 24C*) are used to adjust water or steam flow within limits that depend on system pressure variations. They are also called

Describe check valves and explain how they are used in hot-water systems.

Explain the purpose and function of pressure-reducing valves.

Describe backflow-preventer valves and explain their purpose in a hot-water system.

Provide various types of specialty valves, including check, pressure-reducing, and backflow-preventer valves for trainees to examine.

compression valves. Globe valves come in many configurations, but all have a plug that allows gradual throttling of the water or steam. Flow through a globe valve moves in a Z-pattern that allows throttling without excessive wear on the valve seat. An angle valve (*Figure 24D*) is a form of globe valve, and is made with its outlet port rotated at an angle, typically 90 degrees relative to its input port. Angle valves are commonly used in various applications in place of a standard globe valve and elbow combination.

4.6.2 Check Valves

Check valves (*Figure 25*) come in several types and are used to allow water or steam to flow through a pipe in one direction, but not in the other. The most common type of check valve has a swinging gate or flapper that swings open to allow flow in one direction, but closes if the flow is reversed. This type of valve must be installed horizontally with the hinge at the top of the valve. Some check valves are spring-loaded to help in closing the flapper. Another type, called a lift check valve, works so that the flapper lifts off the seat to allow flow.

Check valves are used anywhere in a system where it is necessary to prevent the reverse flow of water or steam. For example, when installed in the circulating pump line of a hot-water system, the check valve allows hot water to flow only when the circulating pump is running. When the pump is off, it prevents any backflow of water caused by gravity to be returned to the boiler. In hot-water and steam heating systems, check valves are used in the domestic water supply lines to the boiler. They are also commonly used in the condensate return lines of steam systems.

4.6.3 Pressure-Reducing Valves

Pressure-reducing valves are used in hydronic systems anywhere it is necessary to reduce a higher water or steam pressure to a lower one for input to a device. All systems have a feed water pressure-reducing valve installed in the boiler water makeup line (*Figure 26*). This device automatically replenishes any water lost through leaks in the system. It also reduces the pressure of the cold water supplied from the water utility to a pressure suitable for use with the boiler. The valve maintains the water supplied to the boiler at a pressure less than that of the boiler relief valve. Feed water pressure-reducing valves have a built-in strainer and low inlet pressure check valve. The pressure-reducing valve typically used as a boiler feed water line valve controls the output pressure and flow by keeping a balance

Figure 25 ♦ Check valve.

Figure 26 ♦ Feed water pressure-reducing valve.

between the pressure of an internal spring applied to one side of a diaphragm and the pressure of the delivered water applied to the other side of the diaphragm. It allows water to pass through the valve whenever the pressure at its outlet side drops below its pressure setting. The spring pressure is manually adjusted to set the desired output pressure level. Typically, it is factory set at 12 psi, but is adjustable between 10 and 25 psi. When adjusted, care must be taken to make sure it provides the required pressure at all points in the system.

4.6.4 Backflow-Preventer Valves

A backflow-preventer valve (*Figure 27*) is used to protect the domestic water supply from boiler back-siphonage and back pressure, which would contaminate the water supply and render it unfit for drinking and cooking. The valve consists of

> ### Inside Track
> **Factory Settings for Pressure-Relief Valves**
>
> Normally, a pressure-relief valve comes factory set for a specific pressure. A feed water pressure-reducing valve is often combined with a pressure-relief valve. This is done to make sure that the system pressure does not exceed the boiler's pressure-relief setting. Although the setting of a pressure-relief valve can be adjusted over a range of pressures, it is factory adjusted and should not be changed.

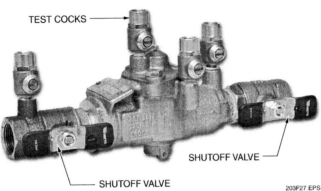

Emphasize that work on backflow-preventer valves is governed by many state and local codes, and that some states require health department certification to work on these valves.

Describe zone control valves and explain how they are used in hot-water systems.

Figure 27 ◆ Backflow-preventer valve.

two in-line independent check valves with an intermediate relief valve. Most states also require that it be equipped with a strainer to capture small particles that can prevent proper operation of the discharge vent. When a back-siphonage condition occurs, the relief valve opens to permit air to enter and break the siphon. In the event of back pressure and a fouled second check valve, leakage is vented through the relief valve. Installation and maintenance of backflow-preventer valves are governed by many state and local codes. This includes the skilled trade(s) authorized to perform these tasks. Some states require health-department certification to work on these valves. Always check the state and local codes before testing, installing, or working on a backflow-preventer valve.

4.6.5 Zone Control Valves

Zone control valves manage the flow of water in each zone of a zoned system (*Figure 28*). They are two-position, thermostatically controlled valves. Zone control valves can be operated by heat or with an electric motor. In a heat-operated valve, a resistance wire around the valve heats the valve's bimetal element when the zone thermostat calls for heat. This causes the bimetal element to

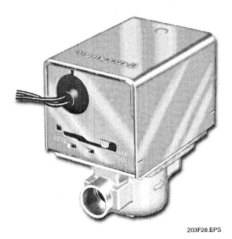

Figure 28 ◆ Zone control valve.

expand and slowly open the valve. When the zone thermostat is satisfied, the bimetal element cools, allowing the valve to slowly close. Electric-motor operated valves also operate to open and close the valve slowly, under control of the zone thermostat. They may contain a switch to operate the circulating pump. Slow opening and closing of zone control valves is necessary to reduce

Show Transparency 14 (Figure 29). Describe two-way and three-way valves and their applications. Explain typical piping applications for these valves.

Discuss the function of hot-water heating terminals.

Identify terminals commonly used in heating-only systems.

Show Transparency 15 (Figure 30). Describe convectors.

Provide two-way and three-way valves for trainees to examine.

expansion noise and prevent water hammer. The time required for a typical zone control valve to open and/or close ranges from about 15 to 30 seconds. Some zone control valves have a feature that enables the valve to be opened manually in the event of a power failure. This feature is also useful when troubleshooting heating problems in a zoned system. Some zone control valves have a flow indicator dial that aids in system balancing.

4.6.6 Two-Way and Three-Way Valves

Two-way and three-way valves are used to control the flow and/or temperature of hot water, chilled water, or both. Their operation is normally controlled by a signal applied to the valve's actuator. Actuators used with these valves may be thermostatic, pneumatic, electric, or electronically controlled. In two-way valves, the water enters the input port and exits the output port, as shown in *Figure 29(A)*. Flow at the output can be at full volume or at a reduced volume, depending on the valve's position.

The three-way valve is often used as a mixing (blending) or diverting valve. As a mixing valve, it has two inputs and one output. It mixes the two water streams into one, based on the position of the valve's plug in relation to its upper and lower valve seats. Typically, it is used to mix or blend hot water and chilled water inputs so that the single water stream leaving the valve has a controlled temperature.

When the three-way valve is used as a diverting valve, it has one input and two outputs. In this application, it splits the single input water stream into two smaller output streams to achieve temperature control. Typically, the flows of the two output water streams are routed in the following manner: one travels through a heat-transfer coil, and the other travels through a bypass pipe around the transfer coil, as shown in *Figure 29(B)*.

4.7.0 Heating System Terminals

Hot-water heating terminals transfer the heat carried by the system hot water to the conditioned space. This can be done in many ways and using many kinds of devices. In combined heating and cooling systems (dual-temperature systems) different kinds of terminals are used than in heating-only systems. The focus of this section is on terminals used in heating-only systems. The terminals used in dual-temperature systems are covered later in this module. In heating-only systems, the transfer of the heat from the terminals results from a combination of radiation to the space and convection to the air in the space. Normally, these terminals are installed at the points of greatest heat loss, such as under windows, along exposed walls, and near door openings. The following terminals are commonly used in heating-only systems:

- Convectors
- Baseboard and finned-tube units
- Radiators
- Unit heaters, unit ventilators, and fan coils
- Heating coils

4.7.1 Convectors

A convector (*Figure 30*) is a heating device that depends mainly on gravity conductive heat transfer. The heating element is a finned-tube coil

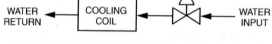

(A) TWO-WAY VALVE CONTROLLING THE FLOW OF WATER THROUGH A COOLING COIL

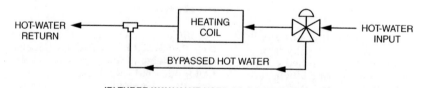

(B) THREE-WAY VALVE USED AS A DIVERTING VALVE

203F29.EPS

Figure 29 ♦ Two-way and three-way valves.

Instructor's Notes:

or coils, mounted in an enclosure designed to increase the convective flow. The enclosure can have many shapes. The room air enters the enclosure below the heating element, is heated in passing through the element, and leaves the enclosure through the grill at the top. Convectors are usually mounted at or near the floor on an outside wall of the room.

4.7.2 Baseboard and Finned-Tube Units

Baseboard units are mounted on the wall in place of the usual baseboard. They can be either a finned-tube system, similar to a convector but much smaller, or a cast-iron section with convective heat channels. Heat transfer takes place by convection. Baseboard radiation is usually installed in a continuous run along the outside walls of a room. *Figure 31* shows baseboard and finned-tube units.

Self-contained hydronic baseboard heaters made in various lengths can also be used. These have a thermostat-controlled, electric heating element. When heated by this element, water contained in a closed-loop finned-tube system circulates in a continuous cycle. The control thermostat may be built into the baseboard unit, or it can be wall-mounted. This type of unit can also be floor-mounted, and is normally used in commercial and special residential applications.

Finned-tube units are room heaters similar to baseboard units. They are composed of larger tubing or pipe, typically 1¼" to 2", with fins bonded to the pipe. Fins can be 3½" to 4½" square. The finned-tube assembly can be housed in a variety of enclosures depending on where it is installed. Finned-tube units are used mostly for perimeter heating, especially in large glassed-in areas.

4.7.3 Radiators

As the name suggests, radiators transfer heat mainly by radiation. *Figure 32* shows an example of a sectional cast-iron radiator. This is the most common type of radiator in use. However, the use of radiators is limited mainly to some older systems. In newer systems, convector radiators or baseboard radiation units are being used instead of radiators. Steel-tube radiators are also available and are used in newer systems.

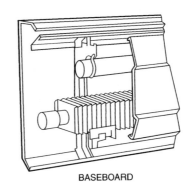
BASEBOARD

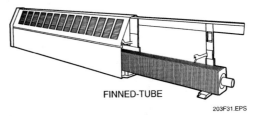
FINNED-TUBE

Figure 31 ◆ Baseboard and finned-tube heating terminals.

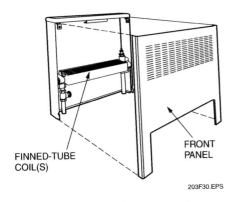
Figure 30 ◆ Convector-type heating terminal.

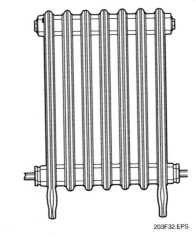
Figure 32 ◆ Radiator.

MODULE 03203-07 ◆ INTRODUCTION TO HYDRONIC SYSTEMS 3.23

Describe unit heaters and explain their method of heat transfer. Point out that unit heaters are used mainly in commercial systems.

Show Transparency 18 (Figure 34). Describe tankless heaters. Explain how they are used to produce domestic hot water.

Provide examples of tankless heaters and indirect water heaters for trainees to examine.

4.7.4 Unit Heaters and Unit Ventilators

A unit heater consists of heating elements and a circulating fan in an enclosure. The fan blows room air across the heating elements. In hydronic unit heaters, the heating element is made from seamless copper tubing with bonded aluminum fins through which the water flows. Unit heaters have a relatively large heating capacity for their size and can project heated air over a considerable distance. They are used mainly in commercial systems, such as those in garages, factories, and warehouses.

Both horizontal and vertical units are used (*Figure 33*). Horizontal units blow heated air in a horizontal direction. They are used in areas with low to medium ceiling heights. Vertical units blow air in a vertical (downward) direction and are normally used in areas with high ceilings. They are also used when the unit must be installed in an out-of-the-way location. This may be necessary because of floor or space limitations.

Both types of unit heaters usually have an adjustable diffuser used to vary their air discharge pattern. Unit ventilators are used for both heating and cooling. Unit ventilators are covered in the section on chilled-water system terminals, presented later in this module.

4.8.0 Tankless and Indirect Water Heaters

Many cast-iron boilers used for residential and commercial hot-water systems are equipped with an internal heater coil called a tankless heater (*Figure 34*). It is a water-to-water heat exchanger used to supply domestic hot water. The domestic hot water is contained in the coil and is quickly heated by the boiler water. In most cases, this eliminates the need for a hot-water storage tank. This type of system requires a special aquastat that maintains boiler water temperature at a certain level. The boiler runs all year long because it heats domestic water.

Figure 33 ◆ Unit heaters.

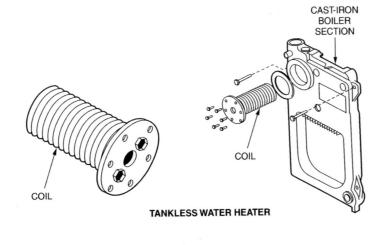
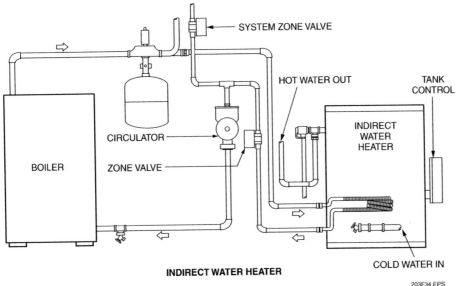

Figure 34 ♦ Tankless and indirect water heaters.

Another method of producing domestic hot water is by use of an indirect water heater. An indirect water heater is also a water-to-water heat exchanger. It is usually a standalone unit connected to the boiler. It is connected as a separate zone with a circulator or zone valve that starts the boiler burners only when necessary. As shown, heated boiler water flows through the heat exchanging coil at the bottom of the indirect water heater tank, resulting in the heating of the domestic water contained in the tank. Like tankless heaters, the boiler runs all year long in this system.

4.9.0 Radiant Floor Heating Systems

Radiant floor heating is a method used for heating the rooms in a building from a heat source in or under a floor. Radiant energy transmitted from the floor material into a room travels in straight lines and at the speed of light. These energy rays do not heat the air through which they travel but they can be reflected by surfaces and objects in the room. Radiant heat reduces the heat loss from the body.

In modern radiant floor heating systems (*Figure 35*), heated water is pumped from a boiler (or

Describe indirect water heaters.

Show trainees how to identify the major components of a hot-water system.

Have trainees practice identifying the major components of a hot-water system. Note the proficiency of each trainee. This laboratory corresponds to Performance Task 2.

Show Transparency 19 (Figure 35). Discuss the method of heat transfer in a radiant floor heating system.

See the Teaching Tip for Section 4.9.0 at the end of this module.

MODULE 03203-07 ♦ INTRODUCTION TO HYDRONIC SYSTEMS 3.25

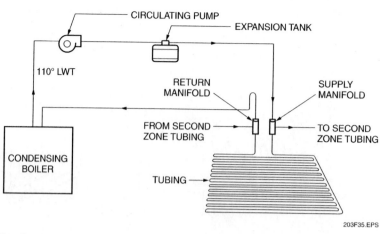

Figure 35 ♦ Basic radiant floor heating system.

other water heating source) through continuous tubing loops embedded in or fastened under the floor. The heat from the tubing loops is absorbed by the surrounding floor materials, and then slowly radiated into the rooms. The result is that the warmth stays down around the floor level in an evenly distributed pattern that is free of hot and cold spots.

Heating with a floor radiant system is energy efficient. Because people's feet are in direct contact with the heat source, comfortable room temperatures are achieved at lower thermostat settings, typically 65°F. Depending on the application, the radiant floor system water typically only needs to be heated to between 90°F and 110°F, thereby reducing heating costs.

The tubing currently used in radiant floor heating systems is made from a cross-linked polyethylene material (PEX) or similar polymer-based material. The tubing is incorporated into floors using one of several methods of installation. Typically, these methods include embedding the tubing in floor slabs on-grade, attaching the tubing to the surface of the floor and embedding it in a layer of concrete or gypsum, mounting the tubing in or below the subfloor, or attaching the tubing directly to the underside of the subfloor. As with other types of hot-water heating systems, a floor radiant-heat system can be divided into several independent room heating zones.

Because the tubing for a floor heating system must be built into the building's flooring systems, the installation costs for radiant floor heating systems tend to be higher than for other types of heating systems, such as a forced-air or baseboard heating system.

Radiant Floor Heating Systems

Radiant floor heating systems have been used for centuries. Historical evidence shows that around 1300 BC a radiant floor heating system was used in a Turkish king's palace. Over a period ranging from about 80 BC to 324 AD, the Romans used improved versions of floor radiant heat systems in upper-class houses and public baths throughout their empire. These systems typically consisted of under-floor chambers or tile flues built into a stone floor. The floors were heated by routing the heated combustion gases, produced by a fire enclosed in a chamber located at one end of the floor, through the floor chambers or tile flues, to exhaust vents placed in the outside walls at the other end of the floor.

5.0.0 ♦ WATER PIPING SYSTEMS

Water piping systems provide for the routing and distribution of hot water, chilled water, or both. There are four general classifications of piping systems. Combinations of all four types can exist in any particular water system. The four classes are as follows:

- One-pipe systems
- Two-pipe systems
- Three-pipe systems
- Four-pipe systems

5.1.0 One-Pipe Systems

One-pipe water systems are used primarily for hot-water heating systems. One-pipe systems can be series-loop one-pipe systems or single-loop one-pipe systems.

5.1.1 Series-Loop One-Pipe Systems

A series-loop system has a continuous run of pipe from the supply connection to the return connection. In a series-loop system, all the hot water flows through all of the heating terminals. Neither the flow nor the temperature of the water can be changed without affecting the whole loop. When the system thermostat calls for heat, the hot water flows from one terminal in the loop to the next.

In doing so, it gives up more and more heat as it flows through each terminal. While this design is relatively inexpensive to install, its disadvantage is that rooms receiving the heat first tend to overheat, while those at the end of the loop tend to be cold. One or more series loops may be used with a complete system. They may connect to mains, or all loops can run directly to and from the boiler. The pipe diameter size used in a series loop is the same for the entire loop. The length of the series loop determines the water flow rate, pressure drop, and temperature drop. *Figure 36* shows a two-circuit (two-loop) series-loop system. Series-loop systems are commonly used in residential hot-water systems.

5.1.2 Single-Loop One-Pipe Systems

Single-loop one-pipe systems have a single loop (one pipe) main supply with branches to each of the terminals. The piping in the branches is smaller than the main. To route hot water through each terminal, a supply (input) and return tee are used to connect the terminal to the main. The supply tee is a special tee called a diverting tee. It is also referred to as a one-pipe fitting or Monoflow® fitting. This tee creates a pressure drop in the main flow that allows some of the water flowing in the main to be diverted through the terminal. The rest of the water continues flowing in the main supply line. The terminals and diverting tees are matched in design for a low pressure drop. Some single-loop systems also use diverting tees on the return at each terminal to overcome very high resistance in pipes and terminals. Since the water returning to the main from each of the terminals is cooler than the water in the main, the water in the main gets progressively cooler as more terminals empty into the main. Also, water entering each successive terminal in the system will be cooler than the water in the terminal before it. However, in a well-designed system, this temperature drop is not large enough to cause comfort problems because the volume of water flowing through each terminal is much less than the total amount of water flowing in the main. *Figure 37* shows a two-circuit, single-loop system. Single-loop systems are commonly used in residential, commercial, and industrial hot-water systems.

5.2.0 Two-Pipe Systems

Two-pipe systems have terminals connected across separate supply and return main piping. Two-pipe systems are classified as either direct-return or reverse-return systems.

5.2.1 Two-Pipe, Direct-Return Systems

In the direct-return system, the supply water and return water flow in opposite directions (*Figure 38*). Also, the return water from each terminal takes the shortest path back to the boiler. This

Show Transparency 20 (Figure 30). Describe water flow in a series-loop, one-pipe water system.

Show Transparency 21 (Figure 37). Describe water flow in a single-loop, one-pipe water system.

Show Transparency 22 (Figure 38). Describe water flow in a two-pipe, direct-return water system.

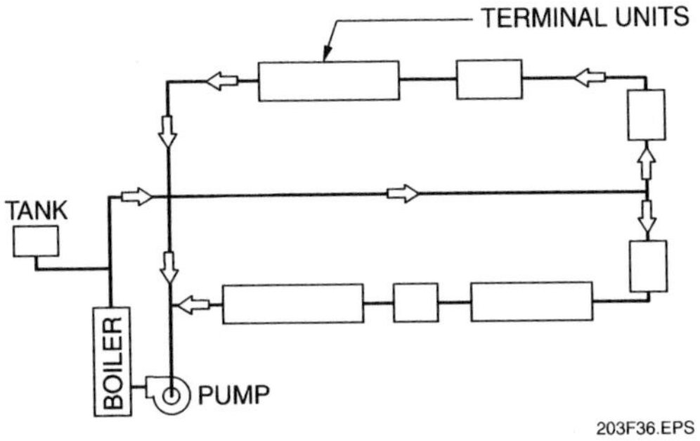

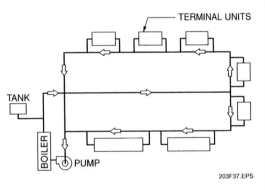

Figure 36 ◆ Series-loop, two-circuit hot-water system.

Figure 37 ◆ Single-pipe, two-circuit hot-water system.

MODULE 03203-07 ◆ INTRODUCTION TO HYDRONIC SYSTEMS 3.27

Show Transparency 23 (Figure 39). Describe water flow in a two-pipe, reverse-return water system.

Show Transparency 24 (Figure 40). Explain how hot water zoning is accomplished using zone valves.

Show Transparency 25 (Figure 41). Explain how hot water zoning is accomplished using circulators.

Show trainees how to identify types of common piping configurations used with hot-water systems.

Have trainees practice identifying types of common piping configurations used with hot-water systems. Note the proficiency of each trainee. This laboratory corresponds to Performance Task 4.

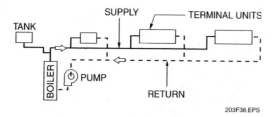

Figure 38 ♦ Two-pipe, direct-return system.

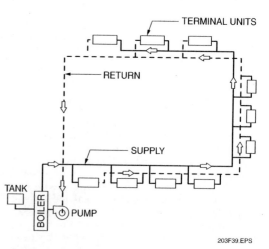

Figure 39 ♦ Two-pipe, reverse-return system.

means the water flowing through the terminal nearest to the boiler is the first back to the boiler, because it has the shortest run. The water flowing to the terminal farthest away from the boiler has the longest run and is the last to return. This arrangement results in an unequal length of travel for water flow through the terminals. This causes an imbalance in the system, with the result being poor heat distribution. Circuit balancing valves are normally used with direct-return systems to aid in balancing the system.

5.2.2 Two-Pipe, Reverse-Return Systems

In the reverse-return system, the supply water and return water flow in the same direction through parallel pipe runs (*Figure 39*). In this system, the supply water going to the first terminal is the last to be returned to the boiler. The supply water going to the last terminal is the first to be returned to the boiler. This arrangement equalizes the distance of the water flow in the system, thus providing a balanced system. Because the reverse-return system is more easily balanced, it is used more often than the direct-return system.

5.3.0 Hot Water Zoning

As is often necessary in air systems, a method of zoning hot water systems is needed to accommodate different loads and comfort requirements throughout a structure. Systems can be easily zoned using multiple circulators or zone valves.

Systems using zone valves (*Figure 40*) generally have a single pump. In the figure, the installation of an indirect domestic water heater that often uses its own dedicated circulating pump is also shown. As individual zone thermostats call for heat, the zone valves open and allow water to flow through to the zone. To prevent the pump from dead-heading or attempting to pump with the flow fully restricted, a differential bypass valve or three-way zone valves can be used. This provides a path for some water flow when all zone valves are closed. Occasionally, this is necessary when only a very small zone is in operation as well. Otherwise, the circulating pump should operate only when one or more zone valves are in their heating position. The pump is sized to accommodate total system flow at the required head of the longest, or most resistant, zone.

Other systems use multiple circulating pumps, and often in addition to a boiler circulator pump (*Figure 41*). Each zone is fitted with its own pump, sized to accommodate the flow rate and head pressure through its own zone piping only. The boiler circulating pump has only the responsibility of maintaining flow through the boiler loop, rather than through individual zone piping as well. Both the boiler circulator and the zone circulator pump must operate for the system to function properly on a call for heating.

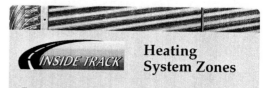

Heating System Zones

For improved indoor comfort, zones can be incorporated in a residential heating system. Each zone is a series loop. These series loop zones are typically piped in parallel with each other. A zone valve and room thermostat controls each zone. Using zones provides a greater level of control over temperatures in different areas of a building.

3.28 HVAC ♦ LEVEL TWO

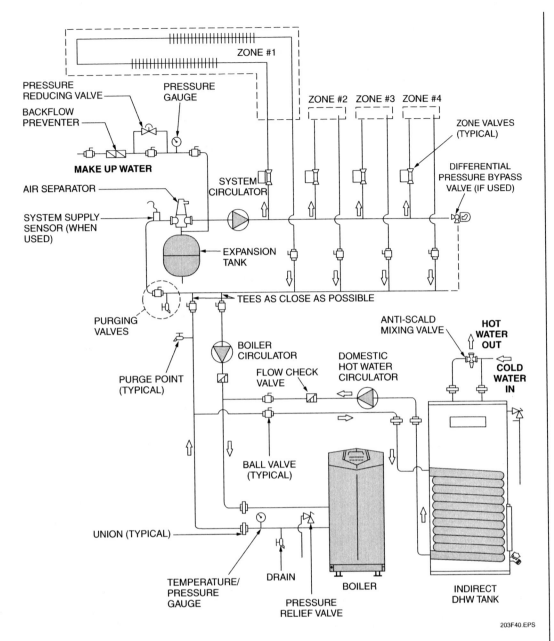

Figure 40 ♦ Boiler piping zoned with valves.

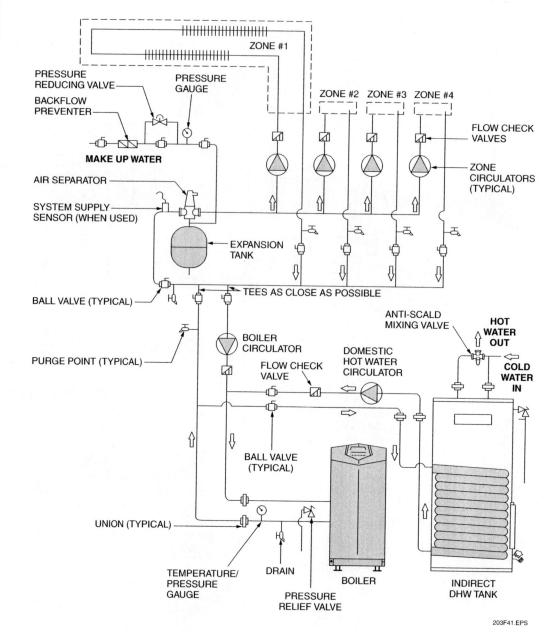

Figure 41 ♦ Boiler piping zoned with circulators.

6.0.0 ◆ DUAL-TEMPERATURE WATER SYSTEMS

CAUTION

There is an inherent risk of damage to boilers and chillers used with some dual-temperature water systems because of the common piping and valve arrangements involved. Never open a valve that will permit hot water to enter a chiller as this will cause the maximum chiller pressure of 15 pounds to be exceeded, causing the chiller safety relief valve disc to rupture, and the chiller to lose all its refrigerant to the atmosphere. Also, never open a valve that will permit the flow of cold water into a boiler. The application of low-temperature water to a boiler will subject it to shocking, possibly resulting in the cracking of the heat exchangers.

A dual-temperature system is a water system that circulates hot and chilled water to heat or cool with common piping and heat transfer terminals. It operates within the pressure and temperature limits of low-temperature water systems with usual winter supply temperatures of 100°F to 150°F and summer supply water temperatures of 40°F to 55°F. *Figure 42* shows a typical dual-temperature system.

As shown, almost all of the components used in a dual-temperature system are the same as those used in the individual heating or chilled-water systems. The main differences are in the system piping and the valves or other system controls used to select heating, cooling, or both.

Dual-temperature water systems have been added herein as an introduction, as they are generally used only in large commercial and industrial applications. This type of system and other advanced hydronic applications are covered in much greater detail in future modules.

7.0.0 ◆ WATER BALANCE

In water comfort systems, water balancing means the proper delivery of hot water, chilled water, or both in the correct amounts to each of the areas in a structure being conditioned. The satisfactory distribution of conditioned water depends upon a well-designed piping system and properly chosen water system components. Water balancing also means that the correct amount of water is being returned to the boiler or chiller unit. Once a water system is installed, it must be balanced to make sure that it meets design conditions. Balancing the water system requires the adjustment of the flow controls so that the right amount of hot and/or chilled water is circulated in the required spaces at the proper velocity to provide satisfactory heating and/or cooling.

7.1.0 Water Flow Measuring and Flow-Control Devices

Balancing the flow of water throughout a system is done by adjusting the various multi-purpose valves and/or other types of flow-control valves installed in the system. Flow measurements that must be made in order to balance a system are done by measuring the water flow and/or pressure drops at specific locations throughout the system. Venturi tubes, orifice plates, multi-purpose valves, and flow-control valves equipped with pressure readout taps are commonly installed in the system for this purpose. The differential pressure drop across each of these devices is measured using a manometer or differential pressure gauge (*Figure 43*), the readout from which can be easily converted into actual flow in gallons per minute (gpm) using a conversion chart. Note that many instrument manufacturers commonly refer to their water differential pressure gauges as readout meters. In some systems, the flow-control devices also have the capability to provide a direct readout of water flow in gpm to aid in balancing a system. Venturi tubes, orifice plates, multi-purpose valves, and flow-control valves commonly used to balance water systems were covered in detail earlier in this module.

Some considerations in the design of water systems that have a relationship to system balancing include the system flow rate, pump selection, and system friction losses. Some background information about these factors is given here.

7.1.1 System Rate of Flow and Pump Selection

The system pump must be able to provide the rate of flow required by the system. The rate of flow in gallons per minute (gpm) for the system can be calculated using the formula:

$$gpm = \frac{\text{hourly heat loss (HL) in Btuh}}{TD \times 8.33 \times 60}$$

Where:

TD = temperature differential across terminals
8.33 = weight of one gallon of water in pounds
60 = number of minutes in one hour

MODULE 03203-07 ◆ INTRODUCTION TO HYDRONIC SYSTEMS 3.31

Show Transparency 26 (Figure 42). Discuss dual-temperature water systems. Point out the inherent risk of damage to boilers and chillers in some systems because of the common piping and valve arrangements used.

Have the trainees review Sections 7.0.0–7.3.0.

Ensure that you have everything required for teaching this session.

Discuss the reasons why it is important to balance a water system.

Explain that the rate of water flow within a system can be determined by measuring water flow directly or by measuring pressure drops across pressure measurement devices installed at specific locations within the system.

Explain that the pump used in a water system must be able to deliver the rate of flow required by the system. Show the formula and explain how to calculate the required rate of a system water flow for a given set of parameters.

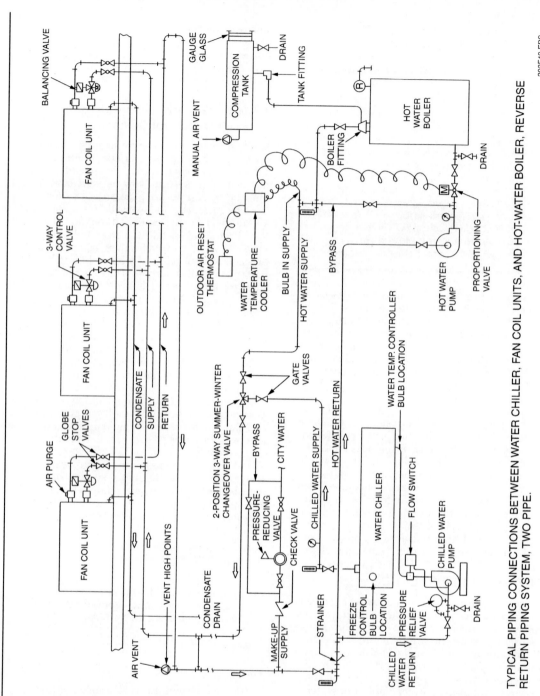

Figure 42 ◆ Dual-temperature water system.

Figure 43 ◆ Differential pressure gauge (readout meter).

Study Example

Assume the total heat loss per hour of a building is 110,000 Btuh. Calculate the conditioned water flow in gallons per minute with a temperature drop across the coils of 20°F.

$$gpm = \frac{\text{hourly heat loss (HL) in Btuh}}{TD \times 8.33 \times 60}$$

$$gpm = \frac{110,000}{20 \times 8.33 \times 60}$$

$$gpm = 11.0$$

Once the required system flow rate is known, a pump can be selected that will provide the needed flow. Using the rate of flow of 11 gpm from our example, a pump is selected that will give the rate of flow required. As marked on the pump curve chart in *Figure 44*, a pump with the permissible head of 9.2' at its output is needed to provide the flow rate of 11 gpm.

Information on how to interpret the various curves shown on a pump curve chart was given earlier in this module.

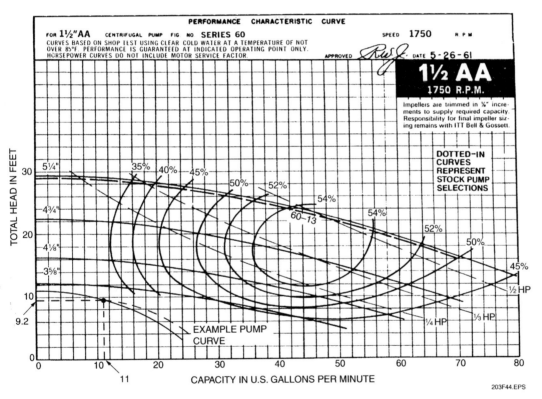

Figure 44 ◆ Pump curve chart.

MODULE 03203-07 ◆ INTRODUCTION TO HYDRONIC SYSTEMS 3.33

Show trainees how to calculate the flow rate using the sample problem.

Give trainees sample problems to practice calculating rate of flow.

Explain that once the required rate of flow is determined, then the pump can be selected.

Show Transparency 27 (Figure 44). Explain how to use a curve chart that will deliver a given rate of flow.

Give trainees sample problems that require them to calculate the system rate of flow. For each rate of flow calculated, have them use pump curve charts to select the appropriate pump.

Show trainees how to calculate gpm requirements.

Have trainees practice calculating gpm requirements. Note the proficiency of each trainee. This laboratory corresponds to Performance Task 5.

Discuss friction loss in a hydronic system.

Explain how to calculate friction loss for a pipe circuit for a given set of parameters. Work the sample problem and answer any questions trainees may have.

Have trainees calculate the friction loss for one or more pipe circuits.

Explain how to measure the added head pressure by a circulating pump by measuring the pressure differential.

Describe how the flow rate and head loss can be used in conjunction with published tables to determine pipe sizes needed to deliver a given flow.

7.2.0 Friction Losses

When water flows, there is a friction loss caused by its motion. As in air distribution systems, the components of the water system resist the motion of the water, and this resistance must be accounted for. The higher the velocity of the water and the longer the piping, the higher the friction loss.

Pipe fittings produce more friction loss than straight lengths of pipe due to the turbulence created by changes in direction and the edges and ridges found at points of connection. Pipe fittings are assigned an equivalent feet of loss value. When sizing a pipe run, the equivalent feet of loss value for each fitting in the run is added to the straight physical length of pipe to find the total equivalent length in feet. By using the correct loss in feet per 100 equivalent feet of pipe, the total loss in feet of head pressure can be calculated. The individual friction losses (resistances) of the other system components are determined in feet of pressure drop from the manufacturer's data and are added to the system piping losses.

Using the pump head pressure at a known flow rate, and the total equivalent length of pipe for the longest run in the system, the friction loss for the circuit can be calculated as follows:

Friction loss (per 100') =

$$\frac{\text{permissible head}}{\text{equivalent length of pipe}} \times 100$$

or

Friction loss (mill-in/ft) =

$$\frac{\text{permissible head}}{\text{equivalent length of pipe}} \times 12{,}000$$

Study Example

Assume a head pressure of 9.2 and an equivalent length of pipe of 250'. Calculate the friction loss per 100' of pipe.

Friction loss (per 100')

$$= \frac{\text{permissible head}}{\text{equivalent length of pipe}} \times 100$$

$$= \frac{9.2}{250} \times 100$$

$$= 3.68$$

The flow rate and head loss (pressure drop per 100' or milli-inches per foot) can be used in conjunction with published tables to determine pipe sizes needed to serve a given flow.

If the total friction loss for the circuit exceeds the head pressure that the pump can handle, some changes in pipe size may have to be made. If the total friction loss is too small, pipe sizes in some sections may have to be reduced. If necessary, corrections can be made by changing some pipe sizes or by adding adjustable flow controls or balancing devices.

7.3.0 Determining Circulating Pump Pressure, Head Pressure, and Flow

The amount of head pressure added by a circulating pump at a given flow rate can be determined by measuring the pressure differential between the pump's suction and discharge ports. Pressure taps used for making such measurements should be located as close to the pump flanges as possible. Both pressure taps should be at the same height with respect to the pump center line and there should be no gate valves or check valves installed between the readout taps and the pump flanges. The pressure differential measured across a pump can be converted to head using the following formula:

$$\text{Head} = \frac{(\Delta P)144}{D}$$

Where:

ΔP = pressure differential in psi between the suction and discharge ports of the pump

D = density of the fluid being pumped in lb/ft^3

NOTE

The density of water and other fluids is dependent on temperature. As the temperature of water increases, its density decreases. The standard density of water measured at sea level and at 68°F is 62.4 lb/ft^3. To find the density of water at other than the standard temperature, you can refer to tables and/or curves typically given in most HVAC reference books and texts concerning hydronic systems.

The calculated total pump head pressure in feet and the pump impeller size can be used to determine the amount of water flow through the pump in gpm. This is done by plotting the calculated total pump head on the pump's performance characteristic curve chart at zero flow.

Following this, draw a line horizontally to the right to intercept the appropriate pump impeller curve. Then, at this intersection point, draw a line vertically to the bottom of the chart to read the flow in gpm.

Study Example

Calculate the pump head pressure developed for a measured pump pressure differential of 10.8 psi when pumping water at the standard density of 62.4 lb/ft³.

$$\text{Head} = \frac{10.8 \times 144}{62.4} = 24.92' \text{ of head}$$

Using the pump curve in *Figure 45*, determine the flow through a pump with a 5¼" impeller for the calculated total head pressure of 24.92'. As shown, circled in the figure, the intersection of the 24.92' total head horizontal line with the 5¼" pump curve shows that the water flow through the pump is 40 gpm.

Show Transparency 28 (Figure 45). Work the sample problem and answer any questions trainees may have.

Show trainees how to select a pump from the manufacturer's data given the friction loss of a piping system and gpm requirements.

Have trainees practice selecting a pump from the manufacturer's data given the friction loss of a piping system and gpm requirements. Note the proficiency of each trainee. This laboratory corresponds to Performance Task 6.

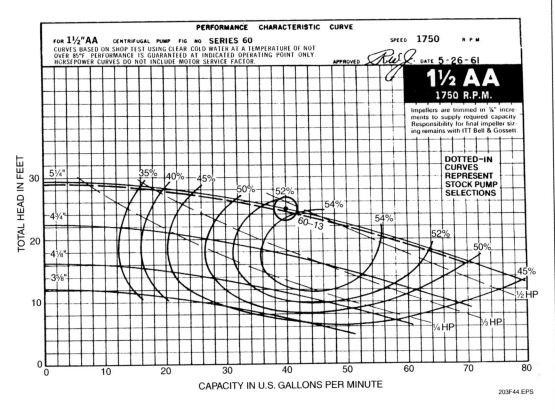

Figure 45 ♦ Pump curve for study example.

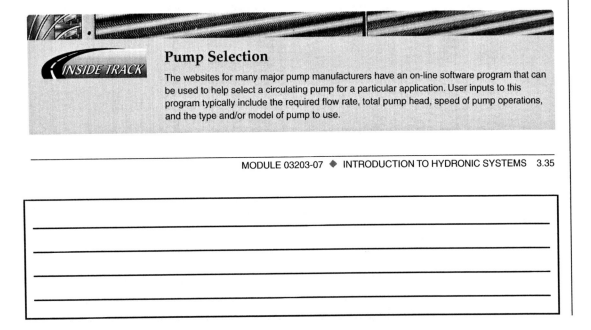

Pump Selection

The websites for many major pump manufacturers have an on-line software program that can be used to help select a circulating pump for a particular application. User inputs to this program typically include the required flow rate, total pump head, speed of pump operations, and the type and/or model of pump to use.

Have the trainees complete the Review Questions, and go over the answers prior to administering the Module Examination.

Review Questions

1. The element primarily responsible for the corrosion of metals in hot-water systems is _____.
   a. water
   b. nitrogen
   c. oxygen
   d. sulphur

2. Chemical treatment of hydronic systems can help to _____.
   a. increase the heating capacity of the boiler
   b. decrease pump operating noise
   c. increase the specific heat of water
   d. prevent corrosion and scale in the boiler and piping system

3. A column of water that is 25' high produces a static pressure of _____ psi.
   a. 4.3
   b. 8.6
   c. 10.75
   d. 12.9

4. Thermal circulation is the method used to cause water flow in _____ systems.
   a. forced hot-water
   b. chilled-water
   c. gravity hot-water
   d. dual-temperature water

5. In a forced hot-water heating system, what happens first when the zone thermostat calls for heat?
   a. Makeup water enters the system.
   b. The circulating pump turns on.
   c. Water from the boiler circulates to zone control valves.
   d. The check valve closes.

6. A boiler design which allows the capacity to be increased by adding boiler sections is called a _____ boiler.
   a. firetube
   b. watertube
   c. cast-iron
   d. vertical tubeless

7. The pressure-relief valve used with a typical hot-water boiler will discharge hot water at a pressure of about _____ psi.
   a. 10
   b. 20
   c. 30
   d. 35

8. Electronic probe-type low water safety controls operate by _____.
   a. using optic sensors to determine water level
   b. sensing the level of an internal float
   c. using the conductivity of water to complete a circuit to ground
   d. using sound waves to sense the height of the water

9. The expansion tank used in a closed forced hot-water system is normally installed _____.
   a. near the boiler
   b. anywhere space allows
   c. at the highest point in the system
   d. on the system pump

10. A component that is pre-charged with air before filling the hot water system with water is called a _____.
    a. standard expansion tank
    b. pressurized diaphragm expansion tank
    c. pressure reducing valve
    d. air separator

11. On larger forced hot-water systems with high pump head pressures, the circulating pump should be installed _____.
    a. in the return side of the system
    b. in the supply side of the system
    c. so that the pump discharges into the expansion tank
    d. so that the pump discharges into the boiler

12. A common type of valve used to turn on or shut off the flow of water is a _____ valve.
    a. butterfly
    b. gate or ball
    c. pressure-reducing
    d. compression

Instructor's Notes:

Review Questions

13. A heating device that depends mainly on gravity conductive heat transfer is a _____.
    a. unit ventilator
    b. radiator
    c. finned-tube unit
    d. convector

14. In a zoning system that uses zone valves, the circulating pump must be sized to _____.
    a. provide sufficient water flow for the entire system
    b. provide water flow to only one zone at a time
    c. match the electrical power available
    d. pump around the clock

15. Assume the total heat loss per hour of a building is 120,000 Btuh. The required water flow in gallons per minute with a temperature drop across the coils of 20°F is _____ gpm.
    a. 11
    b. 12
    c. 14
    d. 60

Summarize the major concepts presented in the module.

Administer the Module Examination. Record the results on Craft Training Report Form 200, and submit the results to the Training Program Sponsor.

Administer the Performance Test, and fill out Performance Profile Sheets for each trainee. If desired, trainee proficiency noted during laboratory sessions may be used to complete the Performance Test. Record the results on Craft Training Report Form 200, and submit the results to the Training Program Sponsor.

Summary

Residential and commercial water systems (hydronic systems) use pipes to transport heated or cooled fluid from its source to where it is needed. The fluid used for heating can be either hot water or steam. For cooling, it is chilled water. The term hydronic refers to all water systems. However, it is commonly used when referring to hot-water and steam heating systems. An air conditioning system that circulates chilled water to cooling coils is known as a chilled-water system.

Hot-water heating systems are used mainly to provide comfort heating. They can be separate systems used to heat buildings that have separate or no cooling systems. Many very old systems were the gravity hot-water type, but forced hot-water systems in use today are far more effective. A boiler is basically a heat exchanger used to transfer heat to water. Dissolved oxygen in water can cause damaging corrosion to boilers and piping systems. For that reason, it is important to minimize the admittance of fresh water and/or chemically treat the water circuit. Although commercial/industrial boilers can operate at high pressures and water temperatures, residential systems most often operate at pressures below 30 psig and temperatures well below 250°F.

Cast-iron boilers remain popular for their durability and heat transfer characteristics. Copper-finned tube heat exchangers are also common, sometimes in combination with cast-iron sections. Today's boilers incorporate a number of features for more efficient operation, such as fuel modulation, sealed combustion systems, and condensing heat exchangers.

Common boiler system accessories and controls include pressure/temperature gauges, pressure safety relief valves, and both float-operated and electronic probe-style low water cut-off devices. Expansion tanks are also important parts of hydronics systems because water expands when heated. Expansion tanks provide the required space for water to occupy in its expanded state. A variety of air elimination devices are also needed to remove air as it becomes separated from the water.

Circulating pumps provide the necessary force to move water through the hydronic circuit, including the boiler and terminal devices used to deliver heat to the occupied space. Such pumps operate by creating a differential pressure between the suction and discharge openings. Pumps must be sized to accommodate the total system pressure drop that occurs as a result of flow, and to provide the appropriate amount of flow volume. Hydronic systems can also be zoned, using a single pump for all zones or a separate pump for each zone. Pressure-reducing valves reduce domestic water pressure to the proper pressure for filling the hydronic circuit, often 12 psig, to avoid over-pressurizing the boiler, which could result in damage.

Terminal devices are the heat transfer devices used to deliver heat to the occupied space. They include finned-tube baseboard units, unit heaters and ventilators, hot water coils placed in air ducts, and even radiators in older systems. Radiant-floor heating systems, with hot water circulated through loops of tubing that is either embedded in or fastened to the floor, are especially popular in colder climates and offer excellent comfort.

Trade Terms Introduced in This Module

Aquastat: A control that works basically the same way as a thermostat with the exception that it is designed to control water temperature instead of air temperature.

Cavitation: The result of air formed due to a drop in pressure in a pumping system.

Corrosion: The breaking down or destruction of a material, especially a metal, through chemical reactions. The most common form of corrosion is rusting, which occurs when iron combines with oxygen and water.

Gravity hot-water system: A hot-water heating system in which the circulation of the hot water through the system results from thermal conduction. No system circulating pump is used.

Head pressure: A measure of pressure drop, expressed in feet of water or psig. It is normally used to describe the capacity of circulating pumps. It indicates the height of a column of water that can be lifted by the pump, neglecting friction losses in piping. Commonly referred to as head.

High/low pump head: Trade terms used to indicate the relative magnitude of the height of a column of water that a circulating pump is moving, or must move, in a water system. See head pressure.

Hydronic system: A system that uses water or water-based solutions as the medium to transport heat or cold from the point of generation to the point of use.

MBh: One MBh equals 1,000 Btus per hour.

Pressure drop: The difference in pressure between two points. In a water system, it is the result of power being consumed as the water moves through pipes, heating units, and fittings. It is caused by the friction created between the inner walls of the pipe or device and the moving water.

Redundancy: In HVAC systems, designs that provide a back-up of primary equipment such as boilers or pumps, allowing for system operation to continue in spite of a failed unit. With 100 percent redundancy, for example, a system may have two boilers installed, each sized to handle the complete heating needs of the structure alone

Specific heat: The amount of heat required to raise the temperature of one pound of a substance one degree Fahrenheit. Expressed as Btu/lb/°F. At sea level, water has a specific heat of 1 Btu/lb/°F. At sea level, air has a specific heat of 0.24 Btu/lb/°F.

Static pressure: In a water system, static pressure is created by the weight of the water in the system. It is referenced to a point such as a boiler gauge. Static pressure is equal to 0.43 pounds per square inch, per foot of water height.

Additional Resources and References

Additional Resources

This module is intended to be a thorough resource for task training. The following reference works are suggested for further study. These are optional materials for continued education rather than for task training.

ASHRAE Handbook – HVAC Systems and Equipment, 2004. Atlanta, GA: American Society of Heating and Air Conditioning Engineers, Inc.

ASHRAE Handbook – HVAC Applications, 2007. Atlanta, GA: American Society of Heating and Air Conditioning Engineers, Inc.

HVAC Systems, 1992. Samuel C. Monger. Englewood Cliffs, NJ: Prentice Hall.

Figure Credits

Utica Boilers, 203F01, 203F11

LAARS Heating Systems Company, 203F06

KNIGHT Heating Boiler by Lochinvar Corporation, 203F07, 203F08, 203F10, 203F40, 203F41

Eric Legacy – Buderus Heating Systems, 203F09

Carrier Corporation, 203F13

ECR International – Dunkirk Boilers, 203SA01

Courtesy ITT, 203F14, 203F17 (photo), 203F19, 203F23, 203F44, 203F45

Courtesy of Honeywell International Inc., 203F15

Taco, Inc., 203F22

NIBCO INC., 203F24, 203F25

Watts Regulator Company, 203F26, 203F27, 203F28

Topaz Publications, Inc., 203F33

Trane, 203F42

Ashcroft, 203F43

Instructor's Notes:

MODULE 03203-07 — TEACHING TIPS

The following are suggested activities or instructional methods to help you teach the material in this module.

General When you call on someone to answer a question, the rest of the class relaxes or even tunes out because they expect that the question and answer will take place only between you and the trainee you called on. Instead, use this technique to involve more trainees in answering questions and to keep them on their toes.

1. Ask trainees to define a term or explain a concept.
2. After one trainee has answered, ask a trainee seated nearby if the answer is right. Then ask whether a trainee in the back of the room agrees.
3. Ask trainees to explain why they think an answer is right or wrong.
4. Use the session to clear up incorrect ideas and encourage trainees to learn from their mistakes.

Sections 1.0.0 through 7.0.0 *Trade Terms Quiz*

This Trade Terms Quiz will familiarize trainees with the terms and definitions that are commonly used in hydronic systems. You will need photocopies of the quiz provided in the following page. Trainees will need pencils. If you allow trainees to use the Trainee Guide, decrease the amount of time you give them to complete the quiz.

1. Make a photocopy of the quiz for each trainee.
2. Give trainees between 5 and 10 minutes to complete the quiz.
3. Go over the answers to the quiz.
4. Ask trainees if they have questions.

Answers to Quick Quiz

1. g
2. d
3. k
4. j
5. f
6. b
7. h
8. c
9. a
10. i
11. e

Quick Quiz *Trade Terms*

For each description listed, identify the term that the text best describes. Write the corresponding letter in the blank provided.

_____ 1. A(n) _____ uses pipes to transport heated or cooled fluid.

_____ 2. There are two types of hot-water heating systems: forced hot-water systems and _____.

_____ 3. The weight of the water in the system creates _____.

_____ 4. Water has a greater _____ than air.

_____ 5. The capacity of a circulating pump is typically expressed in feet of _____.

_____ 6. In a pumping system, when the pressure on a liquid drops below its vapor pressure it can result in _____.

_____ 7. Large commercial boilers have a capacity of 13,000 _____.

_____ 8. Dissolved oxygen in air will cause _____ and rusting in pipes.

_____ 9. Thermally operated switches called _____ can be used to control boilers.

_____ 10. The difference in water pressure between two points is known as the _____.

_____ 11. In a water system, the relative magnitude of the height of a column of water that a circulating pump is moving is indicated by the _____.

a. aquastats
b. cavitation
c. corrosion
d. gravity hot-water systems
f. head pressure
e. high/low pump head
g. hydronic system
h. MBh
i. pressure drop
j. specific heat
k. static pressure

Section 2.3.0 *Think About It – Head Pressure Relationships*

A gauge installed at the use level in a water system reservoir maintained at 150 feet above the use level will indicate 64.5 psi calculated as follows:

$P = H \times 0.43$
$P = 150 \text{ feet} \times 0.43$
$P = 64.5 \text{ psi}$

A gauge installed at the use level of a water system that reads 64.5 psi corresponds to a water level (head pressure) in the water system reservoir of 149 feet calculated as follows:

$H = 2.31 \times P$
$H = 2.31 \times 64.5 \text{ psi}$
$H = 148.99$ (149 feet rounded up)

Section 4.0.0 *Hot-Water Systems Components*

This exercise familiarizes trainees with various parts of a hot-water system. Trainees will need appropriate personal protective equipment, pencils, and paper. You will need to arrange for an opportunity to inspect a hot-water heating system in a commercial building or the building in which you are conducting class. Allow 20 to 30 minutes for this exercise.

1. Discuss the components of a hot-water heating system with trainees.
2. Show trainees a typical hot-water system. Have trainees identify different components of the system, including boiler, valves, pumps, and piping connections.
3. After the tour, answer any questions trainees may have.

Section 4.2.0 *Boiler Components and Controls*

This Quick Quiz will familiarize trainees with boiler components and controls. You will need photocopies of the quiz provided on the following page. Trainees will need pencils. If you allow trainees to use the Trainee Module, decrease the amount of time you give them to complete the quiz.

1. Make a photocopy of the quiz for each trainee.
2. Give trainees between 5 and 10 minutes to complete the quiz.
3. Go over the answers to the quiz.
4. Ask trainees if they have questions.

Answers to Quick Quiz

1. j
2. k
3. d
4. h
5. l
6. g
7. e
8. a
9. n
10. f
11. i
12. m
13. b
14. c

Quick Quiz   *Identifying Hot-Water Boiler Controls and Accessories*

Write the letter corresponding to the boiler component in the blank provided.

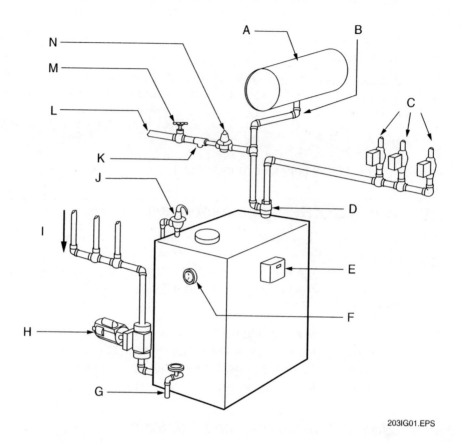

_____ 1. ASME safety relief valve
_____ 2. Backflow preventer
_____ 3. Boiler fitting
_____ 4. Booster pump
_____ 5. Cold water supply
_____ 6. Drain valve
_____ 7. Electronic probe water level control
_____ 8. Expansion tank
_____ 9. Pressure-reducing valve
_____ 10. Pressure/temperature gauge
_____ 11. Return from system
_____ 12. Shut-off valve
_____ 13. Tank fitting
_____ 14. Zone-control valves

Section 4.6.0 *Identifying Valves*

This exercise will familiarize trainees with various types of valves. Trainees will need pencils and paper. You will need to arrange workstations with a different type of valve at each station. Allow 20 to 30 minutes for this exercise.

1. Discuss various types of valves and their applications.
2. Have trainees rotate between the various workstations and identify the type of valve at each station. If desired, have each trainee write down one fact about each valve; for example, its function in a hot-water system.
3. After all trainees have visited each station, have one trainee identify the type of valve at their station. Answer any questions trainees may have.

Section 4.9.0 *Radiant Floor Heating Systems*

This exercise familiarizes trainees with radiant floor heating systems. Trainees will need pencils and paper. You will need to arrange for manufacturer's representative to give a presentation on radiant floor heating systems. Allow 20 to 30 minutes for this exercise.

1. Before the presentation, brainstorm questions with trainees.
2. Have the representative give a presentation on radiant floor heating systems, including installation, troubleshooting, and different types of systems available.
3. After the presentation, answer any questions trainees may have.

MODULE 03203-07 — ANSWERS TO REVIEW QUESTIONS

Answer	Section Reference
1. c	2.1.0
2. d	2.1.0
3. c	2.4.0
4. c	3.1.0
5. b	3.2.0
6. c	4.1.2
7. c	4.2.2
8. c	4.2.3
9. a	4.3.0
10. b	4.3.0
11. b	4.5.0
12. b	4.6.1
13. d	4.7.1
14. a	5.3.0
15. b	7.1.1

CONTREN® LEARNING SERIES — USER UPDATE

NCCER makes every effort to keep these textbooks up-to-date and free of technical errors. We appreciate your help in this process. If you have an idea for improving this textbook, or if you find an error, a typographical mistake, or an inaccuracy in NCCER's Contren® textbooks, please write us, using this form or a photocopy. Be sure to include the exact module number, page number, a detailed description, and the correction, if applicable. Your input will be brought to the attention of the Technical Review Committee. Thank you for your assistance.

Instructors – If you found that additional materials were necessary in order to teach this module effectively, please let us know so that we may include them in the Equipment/Materials list in the Annotated Instructor's Guide.

Write: Product Development and Revision
National Center for Construction Education and Research
3600 NW 43rd St, Bldg G, Gainesville, FL 32606

Fax: 352-334-0932

E-mail: curriculum@nccer.org

Craft _____ Module Name _____

Copyright Date _____ Module Number _____ Page Number(s) _____

Description

_____

_____

_____

_____

(Optional) Correction

_____

_____

_____

(Optional) Your Name and Address

_____

_____

_____

Module 03204-07

Air Quality Equipment

NCCER STANDARDIZED CRAFT TRAINING PROGRAM

The National Center for Construction Education and Research (NCCER) provides a standardized national program of accredited craft training. Key features of the program include instructor certification, competency-based training, and performance testing. The program provides trainees, instructors, and companies with a standard form of recognition through a National Craft Training Registry. The program is described in full in the *Guidelines for Accreditation*, published by NCCER. For more information on standardized craft training, contact the NCCER by writing us at 3600 NW 43rd St., Bldg. G, Gainesville, FL 32606; calling 352-334-0911; or emailing info@nccer.org. More information may be found at our website, www.nccer.org.

HOW TO USE THIS ANNOTATED INSTRUCTOR'S GUIDE

Each page presents two sections of information. The larger section displays each page exactly as it appears in the Trainee Module. The narrow column ties suggested trainee and instructor actions to each page and provides icons (detailed below) to call your attention to material, safety, audiovisual, or testing requirements. The bottom of each page includes space for your notes.

 The **Audiovisual** icon indicates an appropriate time to show a transparency or other audiovisual aid.

 The **Classroom** icon prompts you to define a term, stress a point, ask trainees to explain a concept, or give examples.

 The **Demonstration** icon directs you to show trainees how to perform tasks.

 The **Examination** icon tells you to administer the written module examination.

 The **Homework** icon is placed where you may wish to assign reading for the next class, assign a project, or advise trainees to prepare for an examination.

 The **Laboratory** icon is used when trainees are to practice performing tasks.

 The **Materials** icon is a reminder for you to gather materials needed for classes, labs, and testing.

 The **Performance Testing** icon tells you to administer a performance test or a portion thereof.

 The **Safety** icon is used to emphasize safety issues. It is often keyed to *Caution* and *Warning!* statements in the Trainee Module.

 The **Teaching Tip** icon indicates additional guidance is available, such as how to conduct an exercise, get the most educational value from a field trip, or encourage class participation. Teaching Tips may expand on a feature (*Think About It, Did You Know?*) or provide *Quick Quizzes* or similar exercises. You will be referred to the Teaching Tips section at the back of the module if there is additional material.

 The **Combination** icon indicates that the laboratory listed corresponds with a performance task. If desired, you can note the proficiency of the trainees during the laboratory, and use it to satisfy performance testing requirements.

PREPARATION

Before teaching this module, you should review the Objectives, Performance Tasks, Materials and Equipment List, and Module Outline. Be sure to allow ample time to prepare your own training or lesson plan and gather all required materials and equipment.

**Air Quality Equipment
Annotated Instructor's Guide**

Module 03204-07

MODULE OVERVIEW

This module covers common accessories used to control air quality, including dehumidifiers, humidifiers, and filters. It also covers energy conservation equipment.

PREREQUISITES

Prior to training with this module, it is recommended that the trainee shall have successfully completed *Core Curriculum*; *HVAC Level One*; and *HVAC Level Two*, Modules 03201-07 through 03203-07.

OBJECTIVES

Upon completion of this module, the trainee will be able to do the following:

1. Explain why it is important to control humidity in a building.
2. Recognize the various kinds of humidifiers used with HVAC systems and explain why each is used.
3. Demonstrate how to install and service the humidifiers used in HVAC systems.
4. Recognize the kinds of air filters used with HVAC systems and explain why each is used.
5. Demonstrate how to install and service the filters used in HVAC systems.
6. Use a manometer or differential pressure gauge to measure the friction loss of an air filter.
7. Identify accessories commonly used with air conditioning systems to improve indoor air quality and reduce energy cost, and explain the function of each, including:
   - Humidity control devices
   - Air filtration devices
   - Energy conservation devices
8. Demonstrate or describe how to clean an electronic air cleaner.

PERFORMANCE TASKS

Under the supervision of the instructor, the trainee should be able to do the following:

1. Demonstrate how to inspect, clean, and replace humidifiers.
2. Inspect disposable/permanent air filters for mechanical damage and cleanliness.
3. Clean permanent-type air filters.
4. Measure the differential pressure drop across an air filter with a manometer.

MATERIALS AND EQUIPMENT LIST

Overhead projector and screen
Transparencies
Blank acetate sheets
Transparency pens
Whiteboard/chalkboard
Markers/chalk
Pencils and scratch paper
Humidifiers
Disposable air filters
Electronic air cleaner
Various types of air filters

Tools for removing and cleaning air filters
Manometer
Operating air filtration system
Manufacturers' literature on energy and heat recovery ventilators
Ultraviolet light purification system
Carbon monoxide and carbon dioxide monitors
Copies of the Quick Quiz*
Module Examinations**
Performance Profile Sheets**

* Located in the back of this module.
**Located in the Test Booklet.

SAFETY CONSIDERATIONS

Ensure that the trainees are equipped with appropriate personal protective equipment and know how to use it properly. The module requires that trainees work with air filters and testing equipment. Ensure all trainees are briefed on appropriate safety procedures.

ADDITIONAL RESOURCES

This module is intended to present thorough resources for task training. The following reference works are suggested for both instructors and motivated trainees interested in further study. These are optional materials for continued education rather than for task training.

Air Conditioning Systems, Principles, Equipment, and Service. 2000. Prentice Hall.
Refrigeration and Air Conditioning: An Introduction to HVAC. 2003. Prentice Hall.

TEACHING TIME FOR THIS MODULE

An outline for use in developing your lesson plan is presented below. Note that each Roman numeral in the outline equates to one session of instruction. Each session has a suggested time period of 2½ hours. This includes 10 minutes at the beginning of each session for administrative tasks and one 10-minute break during the session. Approximately 5 hours are suggested to cover *Air Quality Equipment*. You will need to adjust the time required for hands-on activity and testing based on your class size and resources. Because laboratories often correspond to Performance Tasks, the proficiency of the trainees may be noted during these exercises for Performance Testing purposes.

Topic **Planned Time**

Session I. Introduction, Humidity Control, and Indoor Air Quality
    A. Introduction
    B. Process and Comfort Air Conditioning
    C. Humidity Control
    D. Laboratory
        Trainees practice inspecting, cleaning, and replacing humidifiers. This laboratory corresponds to Performance Task 1.
    E. Mechanical Air Filters
    F. Laboratory
        Trainees practice inspecting disposable/permanent air filters. This laboratory corresponds to Performance Task 2.
    G. Laboratory
        Trainees practice cleaning permanent air filters. This laboratory corresponds to Performance Task 3.

Session II. Indoor Air Quality II, Review, and Testing
    A. Laboratory
        Trainees practice measuring the differential pressure drop across an air filter with a manometer. This laboratory corresponds to Performance Task 4.
    B. Air Conditioning Energy Conservation Equipment
    C. Ultraviolet Light Air Purification Systems
    D. Carbon Monoxide and Carbon Dioxide Monitors
    E. Review

F. Module Examination
   1. Trainees must score 70% or higher to receive recognition from NCCER.
   2. Record the testing results on Craft Training Report Form 200, and submit the results to the Training Program Sponsor.
G. Performance Testing
   1. Trainees must perform each task to the satisfaction of the instructor to receive recognition from NCCER. If applicable, proficiency noted during laboratory exercises can be used to satisfy the Performance Testing requirements.
   2. Record the testing results on Craft Training Report Form 200, and submit the results to the Training Program Sponsor.

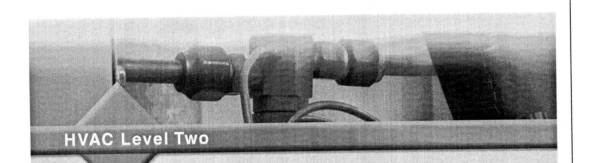

HVAC Level Two

03204-07

Air Quality Equipment

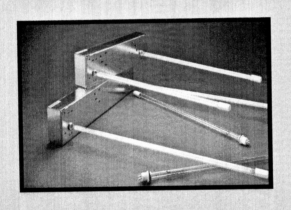

Assign reading of Module 03204-07.

03204-07
Air Quality Equipment

Topics to be presented in this module include:

1.0.0	Introduction	4.2
2.0.0	Process and Comfort Air Conditioning	4.2
3.0.0	Humidity Control	4.5
4.0.0	Indoor Air Quality	4.10
5.0.0	Air Conditioning Energy Conservation Equipment	4.14
6.0.0	Ultraviolet Light Air Purification Systems	4.21
7.0.0	Carbon Monoxide and Carbon Dioxide Monitors	4.22

Overview

There are many accessories that can be added to an HVAC system in order to improve the comfort level and the quality of the air in a building. Other accessories can improve the operating efficiency of the system, resulting in lower cost to the user. The quality of the air we breathe in our homes and elsewhere is affected by dust, molds, and other contaminants. Some accessories are designed to eliminate these contaminants. Other accessories, such as humidifiers, are designed to overcome problems that reduce the indoor comfort level. Because these accessories have become very commonplace, it is essential that HVAC specialists know how to properly install and service them.

Instructor's Notes:

Objectives

When you have completed this module, you will be able to do the following:

1. Explain why it is important to control humidity in a building.
2. Recognize the various kinds of humidifiers used with HVAC systems and explain why each is used.
3. Demonstrate how to install and service the humidifiers used in HVAC systems.
4. Recognize the kinds of air filters used with HVAC systems and explain why each is used.
5. Demonstrate how to install and service the filters used in HVAC systems.
6. Use a manometer or differential pressure gauge to measure the friction loss of an air filter.
7. Identify accessories commonly used with air conditioning systems to improve indoor air quality and reduce energy cost, and explain the function of each, including:
   - Humidity control devices
   - Air filtration devices
   - Energy conservation devices
8. Demonstrate or describe how to clean an electronic air cleaner.

Trade Terms

Dehumidifier
Economizer
Energy recovery ventilator (ERV)
Enthalpy
Evaporation
Free cooling
Heat recovery ventilator (HRV)
Humidifier
Mechanical cooling
Micron
Outgassing

Required Trainee Materials

1. Pencil and paper
2. Appropriate personal protective equipment

Prerequisites

Before you begin this module, it is recommended that you successfully complete *Core Curriculum*; *HVAC Level One*; and *HVAC Level Two*, Modules 03201-07 through 03203-07.

This course map shows all of the modules in the second level of the HVAC curriculum. The suggested training order begins at the bottom and proceeds up. Skill levels increase as you advance on the course map. The local Training Program Sponsor may adjust the training order.

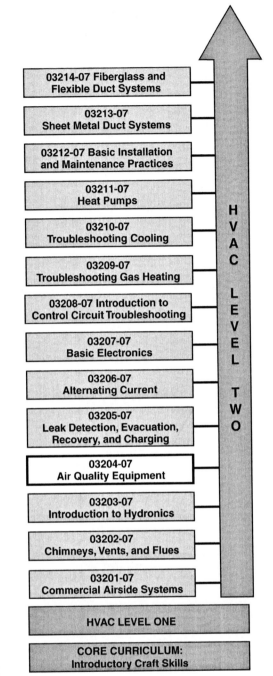

Ensure that you have everything required to teach the course. Check the Materials and Equipment list at the front of this module.

See the general Teaching Tip at the end of this module.

Explain that terms shown in bold are defined in the Glossary at the back of this module.

Show Transparency 1, Objectives, and Transparency 2, Performance Tasks. Review the goals of the module, and explain what will be expected of the trainee.

Review the modules covered in Level Two and explain how this module fits in.

Introduce the topic of air quality control.

Discuss the differences between air conditioning for manufacturing processes and for human comfort.

Identify the functions of comfort air conditioning.

Describe the comfort zone.

See the Teaching Tip for Sections 2.0.0–7.0.0 at the end of this module.

1.0.0 ♦ INTRODUCTION

Air conditioning is the process of treating indoor air to control its temperature, humidity, cleanliness, and distribution. It must do this in both summer and winter. Air conditioning provides comfort for humans and support of controlled manufacturing processes and materials.

Control of the temperature involves automatic control of the heating and cooling system(s) to maintain the best temperature range in any weather. Control of humidity usually involves adding moisture to the conditioned air in the winter, or dehumidifying it in the summer to remove moisture.

Methods used to control the cleanliness of air, such as with filtering devices, are usually the same for all seasons. The methods and devices used to control the temperature and air distribution systems of heating and cooling units are described in detail in several other modules.

This module deals with the HVAC accessories used to control the humidity, cleanliness, and overall indoor quality of air.

Indoor Air Quality

In the early and mid 1900s, building ventilation standards called for approximately 15 cfm of outside air for each building occupant, primarily to dilute and remove body odors. As a result of the 1973 oil embargo, however, national energy conservation measures called for a reduction in the amount of outdoor air provided for ventilation to 5 cfm per occupant. In many cases, these reduced outdoor air ventilation rates were inadequate to maintain the health and comfort of building occupants. Inadequate ventilation, which can occur if HVAC systems do not effectively distribute air to people in the building, is thought to be an important factor in a condition that came to be known as Sick Building Syndrome. In an effort to achieve acceptable indoor air quality while minimizing energy consumption, the American Society of Heating, Refrigerating and Air-Conditioning Engineers (ASHRAE) revised its ventilation standard to provide a minimum of 15 cfm of outdoor air per person (20 cfm/person in office spaces). Up to 60 cfm/person may be required in some spaces (such as smoking lounges) depending on the activities that normally occur in that space.

Source: U.S. Environmental Protection Agency

2.0.0 ♦ PROCESS AND COMFORT AIR CONDITIONING

Air conditioning in support of a manufacturing process or refrigeration of materials is based on the nature of the process or material, and will vary from system to system. On the other hand, air conditioning in support of human comfort is well known and basic requirements rarely vary from system to system. Control of the air for human comfort is sometimes called comfort air conditioning.

Comfort air conditioning performs the following functions:

- Heating and cooling
- Humidification and dehumidification
- Ventilation
- Filtration
- Air circulation

2.1.0 Comfort Air Conditioning

Normally, a comfort air conditioning system must maintain the temperature and humidity of the conditioned rooms within an acceptable range called the comfort zone. *Figure 1* shows the comfort zone plotted on a psychrometric chart.

As shown in *Figure 1*, the comfortable dry-bulb temperatures range from about 68°F to 78°F. The comfortable relative humidity range is from about 30 to 60 percent. The lowest temperatures and humidities typically apply to winter, while the highest temperatures and humidities apply to summer.

2.2.0 Maintaining Body Comfort

The temperature of the inner human body is normally 98.6°F. In order to maintain this temperature, the body must reject any excess heat. In a

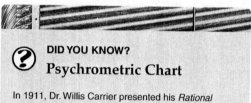

DID YOU KNOW?
Psychrometric Chart

In 1911, Dr. Willis Carrier presented his *Rational Psychrometric Formulas* to the American Society of Mechanical Engineers. This formula led to the development of the psychrometric chart. A psychrometric chart gives a graphical representation of the interrelationships that exist for all properties of air.

4.2   HVAC ♦ LEVEL TWO

Instructor's Notes:

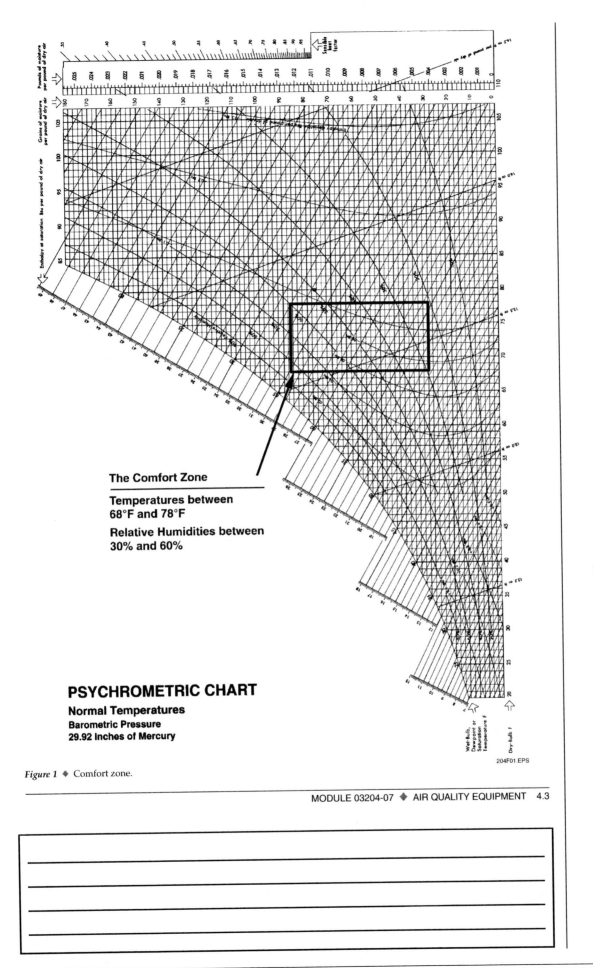

Figure 1 ♦ Comfort zone.

Show Transparency 3 (Figure 2). Explain how people transfer body heat to the surrounding air. Identify the conditions that must be controlled to maintain comfort.

Explain conduction and how it affects human comfort.

Explain how heat is transferred through convection. Discuss the importance of airflow.

Discuss radiation and surface temperatures.

conditioned room, the body always generates more heat than is needed to maintain its internal temperature. This excess heat is constantly being transferred to the room air. You will recall that heat is transferred from one place to another by three methods: conduction, convection, and radiation. Excess body heat is also transferred via these three methods. In addition, **evaporation** in the form of perspiration is another way that the body rejects heat. *Figure 2* shows the four ways that bodies transfer heat. It also shows the properties of air and conditions in a building that must be controlled in order to provide comfort. As shown, these include the room air and surface temperatures, air motion, and relative humidity.

2.2.1 Conduction

Conduction is the transfer of heat from a higher-temperature area to a lower-temperature area via direct contact. It can occur through a substance in contact with the higher- and lower-temperature materials, or directly from one substance to another. Heat generated by a person is conducted to objects the person touches and to the air surrounding them. Air temperature controls conduction; therefore, an acceptable temperature must

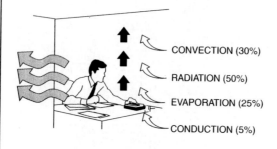

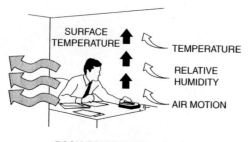

Figure 2 ♦ Transfer of body heat and ways to control it.

be maintained in a conditioned space. The cooler the air surrounding a person, the faster heat leaves the body. When the room temperature is too low, occupants will complain about feeling cold. When the temperature is too warm, they will complain about feeling hot.

2.2.2 Convection

Convection is heat transfer by the movement of a fluid (gas or liquid) from one place to another. In a room, convection air currents carry heat to the body or away from it. As body heat is conducted into the air close to a person, the air next to the person becomes warmer than the air which is farther away. Since warm air is lighter, or less dense, than cool air, the warm air rises and is replaced by cooler air in a continuing process.

The velocity of airflow in a room is important to comfort. If the room air moves too slowly, the temperatures within the room become uneven and the occupants feel stuffy. If it moves too fast, the occupants may complain about draftiness and feeling cold. In a properly conditioned room, the air motion is typically between 15 and 45 feet per minute (fpm). The natural movement of convection air currents in a room is usually too slow to maintain comfort. This is one reason why forced-air systems are used. The blower in a forced-air system causes the air to move more rapidly than it would by natural means.

2.2.3 Radiation

Radiation is the movement of heat in the form of invisible rays or waves, similar to light. Radiant heat travels from a warmer object to a colder object without heating up the area in between. Radiant heat flows from a body to the cooler surfaces around it. If a surface is warmer than the body, radiant heat will flow from the surface to the body.

Even when the room air temperature is fine, the temperature of surrounding surfaces can cause discomfort because of radiant heat flow. The surface temperatures of ceilings, walls, floors, and windows is determined mainly by how well insulated they are. The closer these temperatures are to the room air temperature, the more comfortable the room will be. This reduces the discomfort caused by radiant heat loss from a body in the winter. It also reduces the discomfort caused by radiant heat gain in the summer. Typically, ideal room comfort exists when all room surfaces are between 70°F and 80°F. To eliminate the radiation effect, the surface temperature must be within 1° of the room air temperature.

2.2.4 Evaporation

Evaporation is the condition in which the heat absorbed by a liquid causes it to change into a vapor. Evaporation transfers heat from bodies to the surrounding air. Moisture, in the form of perspiration, is given off through the pores in the skin. As body heat causes this moisture to evaporate, it turns into water vapor. Evaporation from our bodies goes on constantly, whether perspiration is visible or not. Actually, when perspiration is visible on the skin, the body is producing more heat than it can reject at the normal rate. Evaporation is the most important factor for keeping the body cool. Two things affect evaporation: air motion and relative humidity.

If there were no air movement in a room, the layer of air close to the body would rapidly absorb all the water vapor it could hold, causing evaporation of moisture from the body to stop. Normally, convection currents constantly move the air that has absorbed evaporated moisture away from the body, replacing it with drier air. This drier air allows the evaporation process to continue. The faster the air movement in a room, the more rapid the evaporation of moisture from a body. This causes the body to feel a greater sensation of coolness.

The relative humidity (RH) of room air greatly affects evaporation of moisture from the body. Relative humidity is a measure of how much moisture is in the air, expressed as a percentage. The lower the percentage, the drier the air and the greater the air's ability to absorb more moisture. For each pound of water evaporated from the skin, about 1,000 Btus of heat are removed from the body. The relative humidity of the air around a body affects the rate of evaporation. The lower the relative humidity, the faster evaporation occurs. The higher the relative humidity, the slower evaporation occurs.

3.0.0 ♦ HUMIDITY CONTROL

Proper humidity control improves comfort and health conditions in all seasons. Also, it often reduces the cost of equipment operation. The recommended RH levels are about 30 percent in winter and 50 percent in summer. RH levels that are too low or too high can cause problems.

Low RH levels can cause:

- Dry, itchy skin
- Static electricity shocks
- Clothing static cling
- Sinus problems
- A chilly feeling
- Sickly pets and plants
- Loose furniture joints

High RH levels can cause:

- Condensation on windows and inside exterior walls in cold weather
- Moist environments in warm weather that can lead to:
  - Development of bacteria, viruses, fungi, and mite infestations
  - The warping of wood
  - Reduction of personal comfort

3.1.0 Humidification

In the fall or winter, heated buildings tend to be dry, making the occupants feel uncomfortably cool or cold. This is because the drier indoor air accelerates the body's process for rejecting excess heat by evaporation. Most of the moisture within a building comes in from the outside by infiltration or ventilation. Because of its cold temperature in the winter, outside air brought into a building contains less moisture. For example, if the outdoor air is 20°F and has 50 percent RH, its specific humidity is only about 7.5 grains/lb (*Figure 3*). The specific humidity is low because at 20°F, the air can hold very little water vapor before it becomes saturated. When this same air enters a building heated to 72°F, its moisture content is still only 7.5 grains/lb, but its RH drops to 11 percent. This makes the air much drier. *Table 1* shows some examples of winter outdoor air temperatures and RH related to the corresponding indoor RH for the same air. It also shows the recommended indoor RH levels that should be maintained to keep the building occupants comfortable for the same conditions. To achieve the recommended RH levels in the winter, a **humidifier** is almost always installed in air conditioning systems used in cold climates. A humidifier is a device used to add and control humidity.

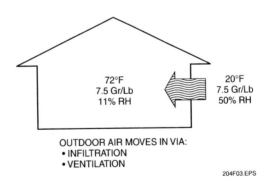

Figure 3 ♦ Comparison of outside and inside air with the same specific humidity.

Discuss the role of evaporation in keeping a body cool. Explain how relative humidity affects evaporation.

Explain the importance of maintaining proper relative humidity. Identify problems caused by low relative humidity.

Show Transparency 4 (Figure 3). Compare indoor and outdoor air with the same relative humidity.

Show Transparency 5 (Table 1). Discuss the recommended indoor relative humidity in winter.

Explain that humidifiers are typically installed in air conditioning systems in cold climates.

Measure the relative humidity in your classroom.

See the Teaching Tip for Section 3.0.0 at the end of this module.

MODULE 03204-07 ♦ AIR QUALITY EQUIPMENT   4.5

Explain how high humidity makes higher temperatures uncomfortable. Explain that dehumidifiers remove moisture from the air.

Identify the four basic types of humidifiers.

Describe a wetted-element humidifier.

Provide several types of humidifiers for trainees to examine.

Humidity in a Home
Compare the relative humidity level within the average home during the winter to the humidity level of a desert.

Table 1 Recommended Indoor RH in Winter

Outdoor Temperature (°F)	Outdoor RH (%)	Indoor RH (%)	Recommend (Safe) Indoor RH (%)
–10	30 TO 70	2	20
0	30 TO 70	5	25
10	30 TO 70	7	30
20	30 TO 70	11	35
30	30 TO 70	15	40

Based on an indoor temperature of 72°F

204T01.EPS

3.2.0 Dehumidification

In the summer when the RH is above 60 percent, people complain about feeling hot and sticky, regardless of the room temperature. This is because the moisture-laden room air slows down the body's process for rejecting excess heat by evaporation. To reduce the high humidity levels that occur in the summer, the conditioned air is normally dehumidified by the operation of the system cooling unit. For buildings that have only a heating system, the air can be dehumidified on a room-by-room basis using individual **dehumidifiers**. A dehumidifier is a device used to remove moisture from the air.

3.3.0 Humidifiers

All humidifiers are designed to introduce water vapor into the conditioned environment. A consistent supply of water is required, and often a drain connection as well. There are four basic types of humidifiers:

- Wetted-element
- Atomizing
- Infrared
- Steam

Heat plays a critical part in the addition of humidity into the conditioned environment or system airstream. While some types rely on an external heat source to function, others incorporate the needed heat into the humidification apparatus.

3.3.1 Wetted-Element Humidifiers

Wetted-element humidifiers are evaporative units commonly used in forced-air systems. They use a porous media that has been moistened with water. Air passes over the media by the action of either an air pressure differential or a self-contained fan. The air current picks up moisture from the media and carries it into the duct system.

Humidifier Settings
Although 40 to 60 percent relative humidity is considered the optimum comfort zone, humidifiers are designed to operate in the 35 percent relative humidity range. This is because most homes cannot tolerate wintertime humidity levels in the 40 to 60 percent range for any length of time without causing water damage from condensation.

Dehumidification
To increase dehumidification in the cooling mode, it is acceptable to lower the evaporator blower speed to produce a slightly colder coil. The colder coil and lower volume of air moving across it allow more moisture to be removed. Many manufacturers offer a kit that includes a humidistat and relay that automatically puts the cooling unit in dehumidification mode when the humidity is high.

Instructor's Notes:

Wetted-element humidifiers can be mounted in many ways, depending on the type and manufacturer. Always follow the manufacturer's instructions for the type of humidifier being installed.

There are three common types of wetted-element humidifiers: rotating drum, bypass, and fan-powered.

- *Rotating drum* – A rotating drum humidifier consists of a drum covered with a screen or sponge pad. The drum is motor-driven and rotates very slowly through a pan of water. The bottom part of the pad is immersed in the water and, as the pad rotates, the entire pad becomes wet. Air is drawn across the wet pad via a bypass duct because of the air pressure difference between the supply and return sides of the system. The water level in the pan is controlled by a float valve. A humidistat controls the operation of the drum. Typically, the humidistat is located near the thermostat or in the return air duct. If mounted in the return duct, it must be out of the path of any radiant heat. The humidifier can be wired so the drum operates only when the furnace is on and the humidistat senses a drop in humidity.
- *Bypass* – The bypass humidifier (*Figure 4*) uses a porous wetted element that is mounted vertically. Water applied to the top of the element moves evenly across the entire top surface. The porous material allows the water to migrate slowly from the top to the bottom to maintain a uniform wet surface. A pan with a drain is placed under the element to catch any water that does not evaporate. This humidifier is normally mounted on the supply air plenum or duct, with a bypass pipe connected to the return air plenum or duct. It may also be mounted on the return duct with a bypass duct connected to the supply. Warm air is drawn through the wetted element by the difference in air pressure between the supply and return sides of the system. The moving air picks up moisture by evaporation, then returns to the duct system via the bypass pipe. Water flow applied to the wet element is controlled by a humidistat, which allows water to flow only when it senses a drop in humidity.
- *Fan-powered* – The fan-powered humidifier mounts on the furnace plenum and operates basically the same as the bypass type. However, airflow through the wet element is caused by a separate fan built into the humidifier. Air bypass openings in the humidifier allow the fan to draw warm air directly from the plenum and then push it through the wet element and back into the plenum. This eliminates the need for a bypass pipe.

Identify the three common types of wetted-element humidifiers and explain their functions and applications.

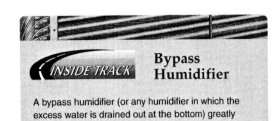

Bypass Humidifier

A bypass humidifier (or any humidifier in which the excess water is drained out at the bottom) greatly reduces the amount of mineral deposits that build up over time as the result of water evaporating.

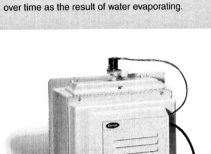

Figure 4 ♦ Bypass humidifier.

Wetted-element units rely solely on the airstream to evaporate moisture from the element. As a result, they are quite limited in their capacity. Moisture is evaporated much more quickly and in greater volume with a warm, dry airstream. Although wetted-element units can be used without the addition of heat, that is, with only room temperature air in circulation, very little moisture is generally added. As a result, their actual capacity is highly dependent upon heating cycle rates and discharge air temperatures. Water that is not picked up by the airstream simply remains on the element or is drained away.

The level of maintenance required for the wetted-element humidifier depends a great deal on the quality of the water supply. As moisture is evaporated from the element, minerals and other debris remain as deposits. As these deposits grow, the effectiveness of the wetted element to absorb and transfer water to the airstream is reduced. In some cases, the element can be cleaned, while in other units the element is simply replaced as needed. In climates where humidification needs are high, wetted elements may require several inspections and maintenance tasks performed each season to ensure proper performance.

Show Transparency 6 (Figure 5). Describe an atomizing humidifier.

Explain the operation of the three types of atomizing humidifiers. Discuss the advantages and disadvantages.

Describe infrared humidifiers and discuss their applications.

Describe steam humidifiers.

3.3.2 Atomizing Humidifiers

Atomizing humidifiers (*Figure 5*) convert water into small droplets for release into the airstream. They do this in one of three ways: using a spinning disc, a high-pressure spray nozzle, or ultrasonic frequencies.

- *Spinning-disc* humidifiers use a circular wheel or cone that rotates at a fairly high speed. Water is fed into the spinning disc and centrifugal force converts it into small droplets. This type of atomizing humidifier is commonly used in self-contained units.
- *High-pressure spray nozzle* humidifiers either blow water through a metered orifice into the duct airstream, or spray water onto an evaporative pad where it is absorbed into the airstream as vapor. They can be mounted on the plenum, on the side of the duct, or under the duct.
- *Ultrasonic* humidifiers contain a crystal known as a piezoelectric crystal. This crystal vibrates at a high frequency (above 16,000 Hertz) when an electric current is applied to it. Water dripping onto the vibrating crystal is atomized and injected into the airstream.

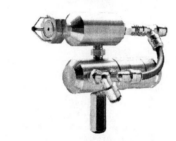

SPRAY NOZZLE

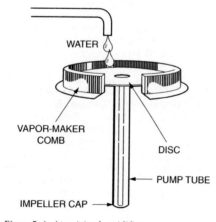

Figure 5 ♦ Atomizing humidifiers.

Like wetted-element units, atomizing humidifiers rely on heat in the airstream to pick up and absorb the moisture. However, these units introduce water in vapor form to the airstream in a consistent volume. Any water vapor that is not picked up by the airstream simply falls to the bottom of the duct, where it can create significant problems, including corrosion and rust, large mineral deposits, mold and mildew, or even structural damage. With these units, it is imperative that they operate only when there is a sufficient volume of dry, warm air in circulation to incorporate all vapor produced by the unit.

It is also important to note that minerals contained in the water supply remain in the water vapor produced by atomizing units. Once the water vapor is fully incorporated into the airstream, the minerals and other debris generally become a dry, white dust that precipitates out of the air supply both in the duct and the conditioned space. The presence of this excess dust often leads system users to assume poor air filtration or poor housekeeping is causing the problem.

3.3.3 Infrared Humidifiers

The infrared humidifier consists of a horizontal water reservoir and infrared lamps with reflectors. Infrared humidifiers do not use heat from the heating system to vaporize water. Instead, energy from the infrared lamps is reflected into the water, where the radiant heat evaporates the water rapidly into the airstream. Because infrared humidifiers do not use heat generated by the system, they can be installed almost anywhere in the ductwork.

Infrared humidification is most often used in computer and data processing facilities where precise control of temperature and humidity conditions are required. They are generally incorporated directly into the equipment by the manufacturer. Since water is evaporated directly from the pan, minerals are left behind which can accumulate quickly into rock-hard deposits if not addressed. Most units use a timed flushing cycle to reduce build-up, but eliminating these deposits altogether is unlikely. Proper filtration of the water source can also reduce deposits significantly. Water treatments that help prevent bonding of the minerals to the water pan surface are also useful.

3.3.4 Steam Humidifiers

Steam humidifiers offer the advantage of high capacity and a variety of approaches to steam generation. With the help of the additional heat contained in the steam, as well as the smaller size of the vapor particle when compared to atomizing units, moisture is more readily absorbed into the airstream. *Figure 6B* shows one type of self-contained unit that uses an electrode-type canister and a short steam transfer hose. The other device, (*Figure 6A*), is an example of a steam dispersion header placed inside the air duct. These units can be attached to a remote steam generator or used with a much larger source of steam, such as a boiler. Electrode-type steam units are often completely self-contained for use in laboratories, data processing facilities, and other areas requiring precise humidity control. Using a blower section and steam distribution cabinet or manifold, they operate independently of the heating or cooling system.

Electrode-type units pass electricity through water to create heat and generate steam. One significant advantage to electric steam humidifiers over many other types is their ability to modulate steam production to match environmental requirements. Although some types of humidifiers can use de-mineralized or distilled water to reduce mineral deposits, the electrode-type humidifier canister depends on some mineral content for sufficiently high conductivity in the water. Water conditions play a major role in the life of the generating canister. The vast majority of canisters are hermetically sealed and not serviceable. Once the electrodes become fouled with mineral deposits and efficiency is substantially reduced, the canister is replaced as a unit.

Another type of electric steam unit utilizes a simple electric heating element immersed in water. These units do not depend on water conductivity and are more flexible in the water quality provided as a result. This type, shown in *Figure 7*, is available for residential and light commercial duct-mounted applications, as well as larger commercial systems. Timed flushing cycles help reduce mineral build-up in the pan, and the expansion and contraction process of the heating element helps to keep minerals from bonding to the element itself. Most smaller units of this type employ simple ON-OFF control, while larger, more advanced units have the capability to modulate voltage to the element as a means of capacity control.

In spite of the precision control available from steam humidifiers, it is important to note that a sufficiently dry airstream or conditioned space is still required to prevent fall-out of excess moisture. Excessive amounts of steam injected into the airstream can cause water damage, rapid corrosion, and mold/mildew growth over time.

Explain the operations of different types of steam humidifiers.

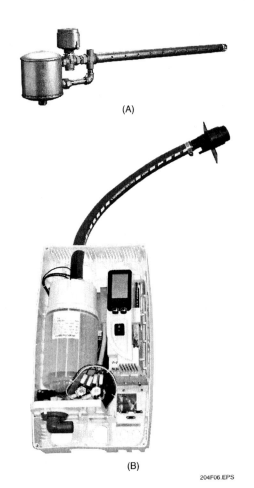

Figure 6 ♦ Steam humidifiers.

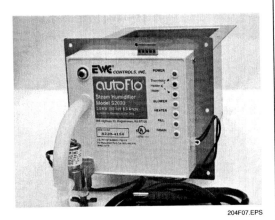

Figure 7 ♦ Steam humidifier with electric heating element.

Show Transparency 7 (Figure 8). Discuss humidifier sizing.

Explain the factors that define loose, average, and tight houses.

Show trainees how to inspect, clean, and replace humidifiers.

Have trainees practice cleaning, inspecting, and replacing humidifiers. Note the proficiency of each trainee. This laboratory corresponds to Performance Task 1.

Explain how energy conservation measures can affect indoor air quality.

See the Teaching Tip for Section 4.0.0 at the end of this module.

3.3.5 Humidifier Capacity

Humidifier capacity is typically rated in gallons of water per day. The required capacity depends on the volume of the building or area in square feet (ft^2). It also depends on the building's construction relative to its airtightness. *Figure 8* shows a typical graph used for the selection of residential humidifiers. Similar graphs and charts are available for commercial and industrial humidifiers.

Loose, average, and tight houses are defined as follows:

- *Loose* – The building has little insulation, no vapor barriers, and no storm doors or windows. In homes, this can also mean there is an undampered fireplace. The air exchange rate is about 1.5 changes per hour.
- *Average* – The building is insulated, has vapor barriers, and has loose storm doors or windows. In homes, this can also mean there is a dampered fireplace. The air exchange is about 1.0 change per hour.
- *Tight* – The building is well insulated, has vapor barriers, and tight storm doors or windows. In homes, this can also mean there is a dampered fireplace. The air exchange is about 0.5 change per hour.

4.0.0 ◆ INDOOR AIR QUALITY

As energy conservation became more of a concern, the immediate response was to tighten up the construction of homes and commercial buildings to retain the heated or cooled environment inside.

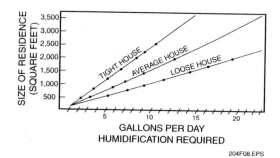

Figure 8 ◆ Humidifier sizing chart.

Energy was saved, but the tight construction trapped contaminants within the buildings, causing a variety of health problems. These included allergic reactions and respiratory problems.

Investigators found that the closed environment of some buildings and homes encouraged the growth of molds and spores. The stale air contributed to the spread of disease. Secondhand cigarette smoke became such an issue that smoking is now banned in most public buildings in the United States. Even some of the building materials, such as adhesives, carpets, and wall coverings, **outgassed** (emitted) substances that caused allergic reactions. Clearly, something had to be done to improve the quality of indoor air. The American Society of Heating, Refrigeration and Air Conditioning Engineers (ASHRAE) has established standards (*ASHRAE 62*) governing the design of HVAC systems to ensure that indoor air quality (IAQ) standards are met and maintained.

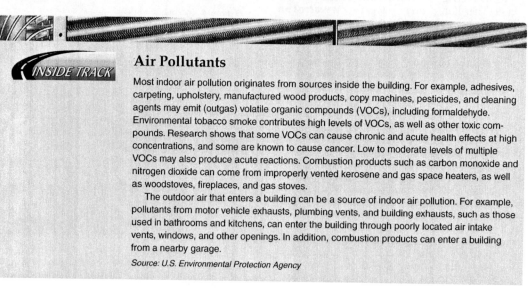

Air Pollutants

Most indoor air pollution originates from sources inside the building. For example, adhesives, carpeting, upholstery, manufactured wood products, copy machines, pesticides, and cleaning agents may emit (outgas) volatile organic compounds (VOCs), including formaldehyde. Environmental tobacco smoke contributes high levels of VOCs, as well as other toxic compounds. Research shows that some VOCs can cause chronic and acute health effects at high concentrations, and some are known to cause cancer. Low to moderate levels of multiple VOCs may also produce acute reactions. Combustion products such as carbon monoxide and nitrogen dioxide can come from improperly vented kerosene and gas space heaters, as well as woodstoves, fireplaces, and gas stoves.

The outdoor air that enters a building can be a source of indoor air pollution. For example, pollutants from motor vehicle exhausts, plumbing vents, and building exhausts, such as those used in bathrooms and kitchens, can enter the building through poorly located air intake vents, windows, and other openings. In addition, combustion products can enter a building from a nearby garage.

Source: U.S. Environmental Protection Agency

Properly designed and maintained HVAC systems play a key role in maintaining good IAQ. Humidifiers discussed earlier in this module work to maintain the correct moisture levels in the air. Air filters and cleaners remove unwanted particles from the air. **Energy recovery ventilators (ERVs)**, **heat recovery ventilators (HRVs)**, and **economizers** ensure that outdoor air is constantly being brought into the building as stale, contaminated air is being discharged from the building. These items of equipment, and others, are covered in the remainder of this module.

Figure 9 shows some of the common particles that contaminate air. As shown, these particles have diameters that range in size from less than 0.01 **micron** to more than 100 microns. A micron is a unit of length that is one-millionth of a meter, or about ½₅,₄₀₀ of an inch. The size of about 99 percent of the airborne particles is less than 1 micron. The remaining 1 percent consists of the larger, heavier particles such as dust, lint, and pollen. Several types of filtration devices are used to remove contaminants from the air, making it cleaner and healthier to breathe. Both mechanical filters and electronic air cleaners are in common use.

4.1.0 Mechanical Air Filters

There are many kinds of mechanical filters. Some of the more common ones include:

- Conventional disposable
- Extended-surface (mini-pleated) disposable
- Bag-type disposable
- Activated carbon disposable
- Electrostatic permanent
- Steel/aluminum mesh permanent
- High-efficiency HEPA

> **NOTE**
> It is extremely important that air filters be replaced in accordance with the manufacturers' instructions. It is also important to instruct building owners about filter replacement. Plugged filters can reduce airflow and will lead to inefficient system operation and possible equipment failure.

4.1.1 Conventional Disposable Filters

Conventional disposable filters (*Figure 10*) are available in bulk rolls of filtering material or in frames. They are typically made from fiberglass, hog-hair, or open-cell foam material. The filtering material is coated with a nondrying, nontoxic, adhesive coating that catches airborne particles. Even when the filter is coated with dust, this adhesive remains effective. The filter material gets progressively more dense as the air passes through it. For this reason, this type of filter must be placed in the airstream correctly. Normally, framed versions of these filters are marked with an arrow that points in the direction of airflow when installed in the duct system. Conventional filters do a good job of protecting the furnace and air distribution system from accumulating dust and dirt. However, they do not remove smaller particles, such as pollen, spores, and smoke.

Explain the importance of an HVAC system in maintaining indoor air quality.

Show Transparency 8 (Figure 9). Discuss common airborne particles.

Identify the common types of air filters.

Describe conventional disposable filters and identify their applications.

Provide common types of disposable air filters for trainees to examine.

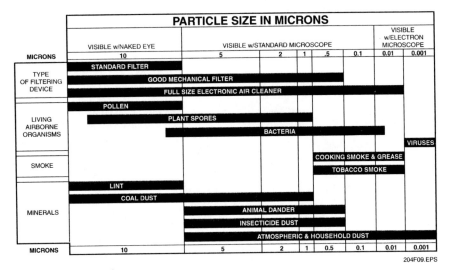

Figure 9 ♦ Sizes of airborne particles.

Describe common types of air filters and identify their applications. Include the following types: extended-surface disposable filters, bag-type disposable filters, activated carbon disposable filters, and electrostatic permanent filters.

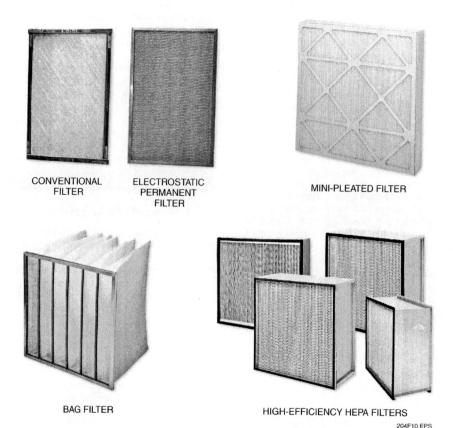

Figure 10 ◆ Mechanical filters.

4.1.2 Extended-Surface Disposable Filters

Extended-surface disposable filters are used when a higher level of cleanliness is required, such as in computer or electronic equipment rooms. They are also used when conditions prohibit the use of fiberglass or hog-hair filters. Extended-surface filters use a pleated, nonwoven, cotton-synthetic material reinforced with a metal backing. The filter material is installed in a rigid frame. Extended-surface filters typically have a usable life four times that of conventional disposable filters.

4.1.3 Bag-Type Disposable Filters

Bag-type disposable filters are typically used where tiny airborne particles cannot be tolerated, such as in hospital operating and recovery rooms, or pharmaceutical process rooms. These filters use an ultra-fine fiberglass material that is reinforced on the leaving-air side by nylon backing material within the bags.

4.1.4 Activated Carbon Disposable Filters

Activated carbon disposable filters provide the dual air cleaning functions of particle filtration and odor removal. The filter media and a honeycomb cell of odor-absorbing carbon are enclosed in a frame for installation in the system. Activated carbon acts like a porous sponge that absorbs odors.

4.1.5 Electrostatic Permanent Filters

Electrostatic filters clean the air using an electrostatic charge that is generated by the airflow as it passes over layered pairs of woven synthetic (plastic) filter material. Electrostatic action

Filter Replacement

Always replace an air filter with a similar or identical one. Different filters have different resistances to system overflow. If a filter with a high resistance is used to replace a filter with a much lower resistance, system performance may suffer and equipment damage or failure could result.

IAQ in Schools

According to the EPA, nearly 55 million U.S. citizens spend their days in primary and secondary schools. Studies have shown that 20 percent of the schools have reported unsatisfactory levels of indoor air quality. These statistics represent a challenge to implement programs that improve the air quality in our schools.

weakly attracts small, naturally charged particles, such as smoke particles, and causes them to collect and settle on the filter material. Electrostatic filters are cleaned by washing.

4.1.6 Steel/Aluminum Mesh Filters

Steel/aluminum mesh filters are permanent filters used in commercial and institutional buildings such as restaurants, hotels, and schools. The steel/aluminum filter mesh is cleaned by washing.

4.2.0 Electronic Air Cleaners

Electronic air cleaners trap airborne particles and odors. They can be standalone units or may be mounted in the air conditioning system. Electronic filters have a high-voltage, solid-state power supply. The high voltage produced by the power supply, which can range from 6,000 to 10,000 volts DC, is used to electrically charge (ionize) all particles in the air that pass through the filter. *Figure 11* shows a typical electronic air cleaner.

As shown, the filter portion consists of a prefilter, ionizer section, collector section, and charcoal filter section. As the air is moved through the filter, larger particles are trapped by the prefilter section. Smaller particles pass through the prefilter to the ionizing section. This section consists of a fine, tungsten-wire grid connected to the high-voltage power supply. An ionizing field created by the high voltage on the wire grid charges the particles in the airstream as they pass through the grid. These charged particles are then drawn into the collector section.

The collector section consists of a series of equally spaced, parallel collector plates. These plates are connected to the high-voltage power supply so that the even and the odd numbered plates are at a positive and negative DC voltage,

Describe steel/aluminum mesh filters.

Show Transparency 9 (Figure 11). Explain the operation of an electronic air cleaner.

Provide an electronic air cleaner for trainees to examine.

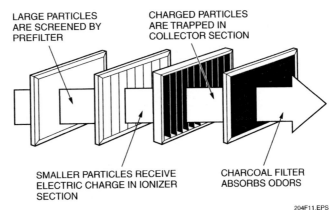

Figure 11 ♦ Electronic air cleaner.

Describe gas-phase air filtration.

Show Transparency 10 (Figure 12) and point out the placement of air filters and electronic air cleaners.

Show trainees how to inspect and clean air filters.

Have trainees practice inspecting and cleaning air filters. Note the proficiency of each trainee. This laboratory corresponds to Performance Tasks 2 and 3.

Show trainees how to measure the differential pressure drop across an air filter.

Have trainees practice measuring the differential pressure drop across an air filter with a manometer. Note the proficiency of each trainee. This laboratory corresponds to Performance Task 4.

Have the trainees review Sections 5.0.0–7.0.0.

Ensure that you have everything required for teaching this session.

respectively. As the ionized particles flow between the plates, they are attracted and held on oppositely charged collector plates. The air, cleaned of pollutants, then passes through the charcoal filter section where any odors are absorbed. From there it is passed on to the conditioned space. The pollutants remain held in the collector section until they are removed when the filter is cleaned. Electronic filter operation is sensitive to airflow. When operated at airflow rates above those recommended by the manufacturer, they can become quite inefficient. When the airflow is reduced below the manufacturer's recommended minimum, enough ozone can be generated to cause an annoying odor.

4.2.1 Gas-Phase Air Filtration

Gas-phase air filters are made to remove both particulate and odor contaminants. The system uses dry scrubbing, gas-phase filtration media. These types of filters are used mainly in commercial applications, such as office buildings, museums, airports, hospitals, and hotels. Gas-phase filtration systems are typically custom designed to meet the unique requirements for a specific application. This is because the control media used for a specific filter differs depending on the type of odor contaminant gas involved. The gas-phase media used removes the airborne gaseous contaminants from the air either by adsorption, where the gaseous molecules are captured and held to the media surface, or chemisorption, where the media reacts with gaseous molecules to change their chemical form to a nontoxic end product. Typically, the gas-phase filtration media is made in a dry, pellet bulk form used to fill filtration modules, or it is supplied as a coating on pleated fiber filters. Charcoal is a common filtration medium.

4.3.0 Filter and Electronic Air Cleaner Installation and Servicing

The location and installation of mechanical and electronic filtration devices should be as directed in the manufacturers' instructions. Typically, both types can be placed in similar locations within a duct system. *Figure 12* shows typical mounting locations.

Disposable filters should be replaced, and permanent ones cleaned, when they lose their efficiency or become so clogged that they produce a pressure drop that is too high. Visual inspection is one way to determine if a filter needs replacement. If it has turned black, the frame is bent or warped, or the filtering material is ripped or

Air Filter Installation

Most air filters have an airflow directional arrow printed or stamped on their frames. Always install the filter so that the tip of the arrow points in the direction of the airflow.

punctured, replace the filter. Filter cleanliness can be checked by placing a strong light on one side of the filter and looking through the filter from the other side to see how much light can be seen and how uniform the pattern is. Another method is to use a manometer or differential pressure gauge to determine if a filter has a pressure drop that is too high. The two ports on the manometer are connected to measure the airflow on the two opposite sides of the filter. Typically, if the pressure drop exceeds more than 25 percent of the pressure drop across the fan, it is too high. If possible, check the manufacturer's specifications to find out what the normal pressure drop of the filter should be.

Permanent mechanical filters are cleaned by vacuuming and/or washing with a mixture of mild detergent and warm water, followed by a rinse. Normally, the collector filter in electronic air cleaners can be cleaned in the same way. Be sure to disconnect the power before attempting any cleaning. Follow the manufacturer's instructions for cleaning the unit.

5.0.0 ♦ AIR CONDITIONING ENERGY-CONSERVATION EQUIPMENT

Many kinds of energy-conservation equipment are being incorporated into air conditioning systems. Three popular conservation methods are:

- Energy and heat recovery ventilators
- Economizers
- Zoned control

5.1.0 Energy and Heat Recovery Ventilators

Energy-efficient homes and buildings do a good job of keeping conditioned heated or cooled air in, but they also seal in air that has been recirculated within the building many times. This causes the air to become stale and contaminated with

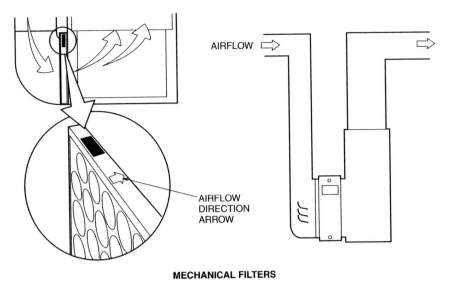

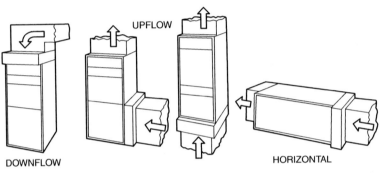

Figure 12 ♦ Typical filter and electronic air cleaner placement.

Identify the three popular energy conservation methods being incorporated into air conditioning systems.

Emphasize the importance of good ventilation. Discuss proper ventilation exchange rates.

Identify the two common types of ventilators.

airborne particles. Without proper ventilation, there is more opportunity for mold to grow, moisture can condense on windows, people may become sick more often, and a musty smell may be noticed. For this reason, more and more states now require mechanical ventilation in every new residential building. For the healthiest living environment, ASHRAE standards recommend that a building's indoor air be exchanged for fresh outdoor air at a rate of 0.35 air changes per hour. An alternate method recommended by ASHRAE calls for an exchange rate of 15 cfm per person, 20 cfm per bathroom, and 25 cfm per kitchen. Ventilators are one type of HVAC equipment that can be used to help solve poor indoor air quality problems within a building by bringing a controlled amount of outside air into the building. In addition to helping maintain good indoor air quality, ventilators also help to conserve energy.

There are two types of ventilators: energy recovery ventilators (ERVs) and heat recovery ventilators (HRVs). ERVs are used to supply fresh air and recover both heating and cooling energy year-round. ERVs are used in most localities in the United States. HRVs are used to supply fresh air and recover heat energy during the heating season. They typically are installed in homes in colder climates that have longer heating seasons, such as those in the northern part of the United States and Canada.

Show Transparency 11 (Figure 13). Explain how recovery ventilators work.

Compare and contrast energy and heat recovery ventilation and discuss their applications.

Provide manufacturers' literature on energy and heat recovery ventilators for trainees to examine.

Describe an economizer.

There are several manufacturers and designs for ERV and HRV equipment. The residential ERV and HRV units of one major manufacturer described here are similar both in construction and operation. Both have a heat exchanger central core and one or more blowers to push air through the unit. According to the U.S. Department of Energy, most models are capable of recovering about 60 to 80 percent of the energy from the exiting air and delivering the energy to the incoming fresh air. Typically, an ERV/HRV improves the indoor air by changing the air about every three hours.

Air from the living space is passed through the ERV or HRV and exhausted outside (*Figure 13*). At the same time, fresh air is brought in from the outside and sent through the unit. When the two airstreams pass through the heat exchanger core, most of the heat or cooling from the outgoing indoor air is transferred to the incoming fresh outdoor air. The core design allows this transfer of heat and cooling between the entering and leaving airstreams to occur without mixing the two airstreams. The result is a constant stream of fresh air being delivered to the living space.

The main difference between an ERV and HRV is the way the heat exchanger core works. In the HRV, only sensible heat is transferred. That's why HRVs are used mainly in colder climates. In the ERV, the core has the capability of transferring both sensible and latent heat, allowing it to transfer heat in the winter and remove moisture from the air during the summer cooling season. This makes the use of ERVs popular in humid climates, such as in the southeast. Upon installation of an ERV or HRV, balancing of the air distribution system is critical to make sure that the amounts of incoming and outgoing air are equal.

5.2.0 Economizers

An economizer is an accessory often found on commercial rooftop packaged heating/cooling units. The economizer mixes outdoor air with

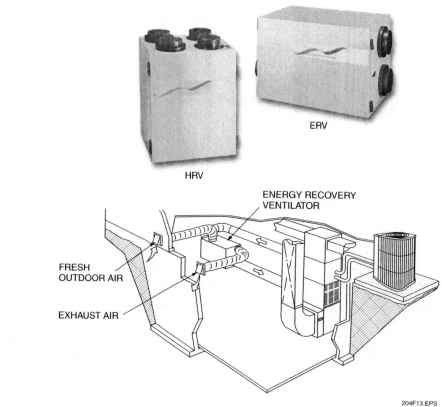

Figure 13 ◆ Recovery ventilators.

Discuss the benefits of an economizer.

Show Transparency 12 (Figure 14). Describe the operation of an economizer.

Inside Track

Economizer

Economizers save energy by pulling in outdoor air to reduce the cooling load. A typical economizer installation is shown here.

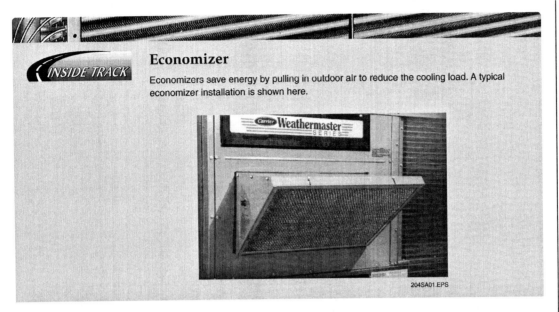

conditioned air in a proportion that depends on the outdoor air temperature and humidity. The controlled use of outdoor air reduces the amount of conditioned air needed, and thus reduces the operating cost of the system. In addition to using outdoor air for cooling, the economizer also controls building ventilation by introducing a minimum amount of outdoor air into the indoor environment.

The benefit of using an economizer is related mainly to the cooling mode of operation. Four conditions are used to control operation of an economizer: outside air, return air, mixed air temperature, and the required ventilation air.

Outdoor air conditions are sensed by either an outdoor air thermostat that senses the outdoor dry-bulb temperature or an **enthalpy** controller (sensor) that senses both the temperature and humidity. Enthalpy-based controls are typically used in high humidity areas. When the outdoor air sensor detects that the outdoor air is above its setpoint, cooling for the building is provided in the conventional way by the air conditioner compressor (**mechanical cooling**). When the outdoor air falls below the sensor setpoint, the economizer control system acts to cool the building by turning off the system compressor and using the indoor fan to bring outside air into the building through motor-actuated dampers. This mode of operation is called **free cooling**. The outdoor sensor setpoint is normally selected by the system designer and is based on both comfort and economy.

Figure 14 shows a basic economizer. As shown, the economizer consists of a mixed air thermostat (MAT) located in the conditioned space; damper motor and control circuit; and outdoor and return air dampers. During the free cooling mode, the MAT monitors the average air temperature on the face of the indoor coil and compares it to a predetermined setpoint. This thermostat is called a mixed air thermostat because it senses the temperature of the building air, which is a mixture of both outdoor and return air coming from the conditioned space. The MAT provides a voltage input to the damper motor and control unit.

The function of the damper motor and its control circuit is to control the position (open, closed, or somewhere in between) of the system outdoor air and return air dampers. It does this through linkage rods connected to the damper motor's shaft. The damper motor has two windings that determine the direction of shaft rotation. When energized, one of the windings will cause the shaft to turn clockwise and the other winding will cause it to turn counterclockwise. Only one winding can be energized at a time. The direction of rotation is determined by the voltage level applied to the damper motor from the MAT. This voltage level changes in response to changes in the condition of the indoor air sensed by the thermostat. The correct damper motor winding is energized by the action of a motor balancing circuit, which is part of the motor damper and control unit. This causes the damper motor to rotate and reposition the vanes of the outdoor air and

MODULE 03204-07 ♦ AIR QUALITY EQUIPMENT 4.17

Show Transparency 13 (Figure 15). Describe evaporative pre-coolers and explain their function.

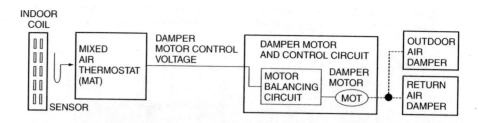

Mode of Operation	Outdoor Air Damper	Return Air Damper
Off	Closed	Wide open
Fan only and mechanical cooling	Opens to minimum position for ventilation	Modulates to complement outdoor air damper
Free cooling	Modulates to provide the proper mixed air temperature	Modulates to complement outdoor air damper
Heating	Opens to minimum position for ventilation	Modulates to complement outdoor air damper

204F14.EPS

Figure 14 ♦ Basic economizer unit.

return air dampers as needed. The damper motor continues driving the dampers until turned off by the motor balancing circuit. In this way, the pre-determined mixed air temperature level is maintained in the building.

The damper motor and control unit circuitry also controls the operation of the dampers to provide for ventilation during all operating modes of the air conditioning system. The damper positions for all modes of operation are summarized in *Figure 14*.

5.3.0 Evaporative Pre-Coolers

Evaporative pre-coolers are used to pre-cool air that is, in turn, used to cool equipment such as heat exchangers. A typical example of how an evaporative pre-cooler can be used is with a conventional air conditioner condenser coil (*Figure 15*). The condenser coil is used to dissipate heat to the outside air during the cooling cycle of the air conditioner. Without pre-cooling, the condenser coil uses outside air to dissipate the heat. If that heat is 120°F, then the efficiency of the heat transfer is greatly reduced. With evaporative pre-cooling, the 120°F air temperature is reduced to about 90°F, which greatly improves the efficiency of the heat transfer of the condenser coil. At lower temperatures across the condenser coil, the air conditioner efficiency improves, the energy used is reduced, and the useful life of the compressor is

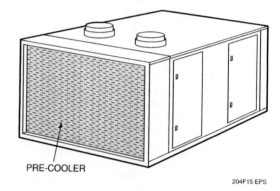

Figure 15 ♦ Evaporative pre-cooler mounted on an air conditioning unit.

increased. The air conditioner produces colder air at a lower energy cost.

As shown, the pre-cooler is placed over the air intake of the condenser unit. The pre-cooler unit has wet and dry sections and a pump and sump-type water distribution system with a metering valve and bleed-off line. Water is delivered to the evaporative cooler from its source by PVC Schedule 40 pipe. The wet section contains a rigid media over which the water is circulated. The dry section contains one or more blowers used to force air across the media pad in the wet section.

Direct evaporative cooling results from warm air being blown over the water-soaked media. In the wet section, the water is discharged upwards against a water deflector which redirects it back evenly over the full width of the media. The water absorbs the heat from the air, bringing the dry-bulb temperature closer to the wet-bulb temperature. It is the difference between the wet- and dry-bulb temperatures that determines the efficiency of the evaporative cooling system. The greater the wet-bulb/dry-bulb temperature differential, the greater the transfer of heat from the air to the water.

5.4.0 Zoned Control

Zoned control is the division of a building into a number of separately controlled spaces, or zones, where different heating or cooling temperatures can be maintained at the same time. This provides a way to overcome variations in cooling and heating loads that occur in different areas of the building at different times. Zoned control allows the occupants to set conditions in each zone independently. Energy can be saved when a zone is unoccupied by lowering or raising the thermostat setpoint temperature to control cooling or heating, respectively. When climatic conditions allow, cooling or heating in any zone can be turned off while comfort conditions are maintained in the rest of the zones.

Zones are set up by grouping one or more rooms in a building where there is little variation in heating or cooling needs, allowing them to be controlled by one thermostat. The selection of zones is usually based on either usage or exposure. When based on usage, the factors include:

- The zone is occupied only during the day or night. For example, bedrooms are only occupied at night. When unoccupied during the day, the thermostat can be set to save energy. Similarly, business offices are typically occupied only during the day, allowing the thermostat to be set to save energy during the night.
- The zone is occupied on an irregular basis (for example, recreational areas, and meeting and conference room areas).
- The zones are occupied at specific times of the day by the same group of people (for example, classroom, shop, cafeteria, and laboratory areas in a school).

When setting up zones by exposure, the factors involved include:

- Portions of the building have large areas of glass, or poorly shaded areas, with cooling loads caused mainly by solar heat gain through the glass.
- The building is occupied by a small number of people so the internal heat gain from people is small compared to the overall cooling requirements for the building.

There are several types of zoned systems in use. *Figure 16* shows a typical forced-air zoned system. As shown, it consists of four zones supplied from a central cooling/heating unit. Supply of the conditioned air to each zone is through a damper controlled by the zone thermostat.

Explain how the efficiency of an evaporative cooler is measured.

Describe zone control.

Discuss the factors for designing zones based on usage.

Discuss the factors for selecting zones based on exposure.

Show Transparency 14 (Figure 16). Describe a typical forced-air zoned system.

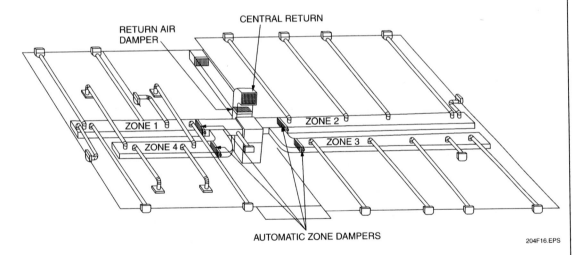

Figure 16 ♦ Typical forced-air zoned system.

Describe other types of zoned systems.

Show Transparency 15 (Figure 19). Describe residential zone control.

Some forced-air zoned systems use separate heating and cooling systems for each zone. Other types of zoned systems include:

- Hydronic (hot water) systems that use separate hot water loops to each zone controlled by zone valves or separate circulator pumps.
- Electric heat perimeter systems in which each area or room is typically set up as its own zone.
- Ductless, split systems in which each zone uses a separate indoor fan coil with its own thermostat. A single outdoor condensing unit is connected to two or more indoor coils with parallel refrigerant circuits controlled by solenoid valves.

5.4.1 Residential Zone Control

Zone control is not limited to large commercial installations. In fact, it has become common in residential applications as a means of improving energy efficiency and reducing system operating cost. In such applications, zone dampers like the one shown in *Figure 17* are used to control the flow of conditioned air to the zones. Each zone has a thermostat, which interfaces to a zone controller (*Figure 18*). The zone controller, in turn, sends a signal to the zone damper motor to open or close the damper. In a typical design, each zone is on a separate trunk line (*Figure 19*). The level of control varies from one system to another. Some systems are able to modulate the damper opening to reduce or increase the flow of conditioned air by incremental amounts.

Figure 17 ◆ Residential zone damper.

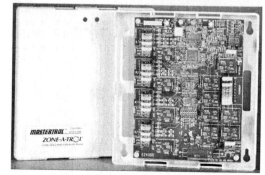

Figure 18 ◆ Residential zone controller.

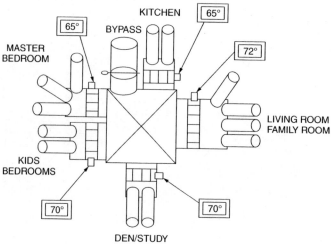

Figure 19 ◆ Example of a residential zone layout.

4.20 HVAC ◆ LEVEL TWO

Instructor's Notes:

Residential zones can be structured in a variety of ways. In a multi-story structure, each floor may be set up as a zone. In other cases, the zones may be divided into living areas, one or more sleeping areas, basement, and so forth. In multi-zone applications, it may be necessary to include a bypass damper (*Figure 20*) to relieve excess static pressure that occurs when several dampers are closed. The bypass damper dumps excess air pressure into an area such as a basement. Variable speed systems are also used to eliminate the problem of excess air pressure.

6.0.0 ◆ ULTRAVIOLET LIGHT AIR PURIFICATION SYSTEMS

Ultraviolet (UV) light air purification equipment can be used in HVAC air distribution systems to help prevent the growth of bacteria and other microorganisms that are known to cause indoor air problems and musty, mold-related odors. There are many manufacturers and designs of UV air purification equipment, but they all share the same basic principles of operation. C-band UV light (UVC) energy in the 240- to 280-nanometer wavelength range destroys microorganisms by penetrating their cell walls and damaging the protein structure of the cells and/or chemically altering the cell's DNA. Once this occurs, the organisms die or cannot reproduce. Germicidal effectiveness (killing power) is directly related to the UV dose applied, which is a function of time and intensity.

HVAC system air purification by UVC light is done in one of two ways: purification of a fixed object, or purification of the moving airstream. In fixed-object purification, the HVAC supply side evaporator/indoor coil and drain pan is continuously irradiated with light rays generated by stationary quartz UVC lamps, probes, etc. (emitters). Over time, the UVC rays destroy bacteria and viruses that are present on the fixed object. The time required to destroy microorganisms on fixed objects depends on a number of factors, including the distance at which the UVC emitter is mounted from the fixed object, the size and intensity or killing power of the UVC emitter, and the temperature of the air and UVC emitter.

In UV purification of the moving airstream, the air in a duct system is irradiated as it moves past a stationary UVC emitter. Achieving air purification using this method is much more difficult because of the short time (dwell time) that the air moving past the emitter is irradiated. Typically, the air moves past the UVC emitter at a speed of about 600 fpm or faster, spending only about 20 milliseconds in front of the probe/emitter. The intensity or killing power of the UVC emitter, how fast the UV ray intensity decreases with distance as the air moves away from the emitter, and how far into the airstream the UV rays penetrate determine the UV purification efficiency on the moving air. Because of the short dwell time that a portion of the moving air is subjected to the UV light rays, purification of the airstream normally requires the use of multiple UV light sources and reflectors that are capable of producing much stronger UV light rays than needed for a fixed object. For this reason, fixed object air purification systems are more widely used.

Figure 21 shows an example of a typical UVC air purification unit. It is designed to protect coils, drain pans, and humidifiers from mold and bacterial growth while killing some airborne microorganisms. It consists of a housing, power supply, and emitters. The components are incorporated into one assembly that is mounted outside the equipment at the cooling coil ductwork, with the emitters protruding into the center of the coil and airstream.

Explain how bypass dampers are used for residential zone control.

Describe ultraviolet light purification systems and explain their operation and applications.

Provide an ultraviolet light purification system for trainees to examine.

Figure 20 ◆ Bypass damper.

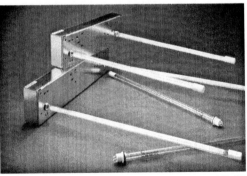

Figure 21 ◆ Example of a UVC air purification unit.

Discuss the hazards of carbon monoxide and carbon dioxide.

Describe carbon dioxide and identify dangerous levels. Explain that sensors are used to monitor levels of carbon dioxide in a building.

Provide carbon monoxide and carbon dioxide monitors for trainees to examine.

7.0.0 ◆ CARBON MONOXIDE AND CARBON DIOXIDE MONITORS

Carbon monoxide and carbon dioxide monitors are widely used HVAC accessories. The deadly effects of carbon monoxide (CO) are well known. Carbon monoxide gas results from incomplete combustion of carbon fuel. Carbon monoxide is colorless, odorless, and tasteless. A carbon monoxide (CO) detector is both a safety and IAQ monitor. Early detection of high CO is almost impossible without a CO detector. Stationary CO monitors (*Figure 22*) are made for use in automated systems. They are installed in strategic locations throughout a building and normally activate a contact closure and sound an alarm when a high level of CO is detected. Portable CO detectors are also available for use mainly when testing the operation of HVAC combustion equipment.

Carbon dioxide (CO_2) is also a colorless, odorless, tasteless gas derived from combustion and metabolic processes such as human breathing. The concentration level of CO_2 in a building is commonly used as one indicator of IAQ. Low concentrations of CO_2 produced by people are always present in buildings. Concentrations below 1,000 parts per million (ppm) generally indicate that the building's ventilation is adequate to deal with the routine byproducts of human occupancy. CO_2 levels above 1,000 ppm can indicate a ventilation problem. At higher building concentrations, some loss of the occupants' mental awareness is noted.

CO_2 sensors (*Figure 23*) are used to monitor CO_2 levels in a building. Some models can be used as a ventilation controller in demand-based ventilation control systems. When used as a ventilation controller, the CO_2 sensor determines the need for ventilation based on the CO_2 concentration and modulates the position of the building's dampers to maintain acceptable ventilation. If a space is unoccupied, the CO_2 controller will set the air intake volume at a minimum setting that allows established ventilation rates to be maintained while reducing over-ventilation.

In addition to CO and CO_2 monitors, there are many other types of gas monitors used to monitor the quality of air.

Figure 22 ◆ Carbon monoxide (CO) monitor.

Figure 23 ◆ Carbon dioxide (CO_2) monitor.

Instructor's Notes:

Review Questions

1. Human comfort can be maintained by controlling the room _____.
   a. air temperature and humidity
   b. convection
   c. radiation
   d. evaporation

2. Heat is transferred between humans and surrounding surfaces such as windows or walls by _____.
   a. conduction
   b. convection
   c. radiation
   d. evaporation

3. The relative humidity in a room affects the transfer of heat from a body mainly through _____.
   a. conduction
   b. evaporation
   c. radiation
   d. convection

4. Too much humidity can cause _____.
   a. dry, itchy skin
   b. static electricity shocks
   c. growth of bacteria
   d. sinus problems

5. Which humidifier uses heat from the system to vaporize the water?
   a. Ultrasonic
   b. Infrared
   c. Steam
   d. Fan-powered

6. Which humidifier is *not* an evaporative humidifier?
   a. Spinning disc
   b. Rotating disc
   c. Rotating drum
   d. Bypass

7. A 1,500 square foot tight house needs a humidifier that supplies about _____ gallons per day.
   a. 5
   b. 7
   c. 10
   d. 12

8. Which type of disposable filter would most likely be used in a hospital operating room?
   a. Extended surface
   b. Bag-type
   c. Conventional
   d. Electronic

9. A device that uses a heat exchanger to transfer heat from the system exhaust air to ventilation air entering a building is a(n) _____.
   a. economizer
   b. zone damper control
   c. heat recovery ventilator
   d. mixed air thermostat

10. When dividing a building into zones based on exposure, you must consider _____.
    a. areas in the building with greater amounts of glass
    b. if the zone is occupied at specific times of the day
    c. if the zone is occupied only during the day or night
    d. if the zone is occupied on an irregular basis

Have the trainees complete the Review Questions, and go over the answers prior to administering the Module Examination.

Summarize the major concepts presented in the module.

Administer the Module Examination. Record the results on Craft Training Report Form 200, and submit the results to the Training Program Sponsor.

Administer the Performance Test, and fill out Performance Profile Sheets for each trainee. If desired, trainee proficiency noted during laboratory sessions may be used to complete the Performance Test. Record the results on Craft Training Report Form 200, and submit the results to the Training Program Sponsor.

Summary

Air conditioning is the process of treating indoor air to control indoor air quality, including its temperature, humidity, cleanliness, and distribution. Air conditioning is done to provide comfort for humans and to support controlled manufacturing processes and materials.

Air conditioning for supporting a manufacturing process or refrigerating materials is based on the nature of the process or material and will vary from system to system. Air conditioning in support of human comfort is well-known, and basic requirements rarely vary from system to system. Control of the air for human comfort is sometimes called comfort air conditioning. It provides:

- Heating and cooling
- Humidification and dehumidification
- Ventilation
- Filtration
- Air circulation

Several accessories are used with the basic air conditioning system to enhance human health and comfort and/or conserve energy in the conditioned space. Common accessories include:

- Humidifiers
- Mechanical filters and electronic air cleaners
- Energy and heat recovery ventilators
- Economizer equipment
- Evaporative pre-coolers
- Zoned air conditioning systems
- Purification systems
- Gas monitors

Notes

Instructor's Notes:

Trade Terms Introduced in This Module

Dehumidifier: A device used to remove moisture from the air.

Economizer: An HVAC device that substitutes outdoor air for the cooled air produced by the air conditioning system, when outdoor air conditions permit. It also controls the amount of outdoor air used to ventilate a building.

Energy recovery ventilator (ERV): HVAC equipment used to supply fresh air and recover both heating and cooling energy year-round.

Enthalpy: The total heat content of a substance. In HVAC, the total heat content of the air and water vapor mixture as measured from a predetermined base or point.

Evaporation: The condition in which the heat absorbed by a liquid causes it to change into a vapor.

Free cooling: A mode of economizer operation. It is the cooling provided by outside air rather than the compressor.

Heat recovery ventilator (HRV): HVAC equipment that saves energy by using a heat exchanger to transfer heat from the heating system exhaust air to the cold ventilation air that is entering the building.

Humidifier: A device used to control humidity.

Mechanical cooling: A mode of economizer operation. It is the cooling provided in the conventional manner by the compressor.

Micron: One-millionth of a meter (about $1/25,400$ of an inch). It is also a precise measurement of pressure used with electronic vacuum measuring instruments and vacuum pumps. One inch of mercury equals 25,400 microns.

Outgassing: The slow release of a gas that was trapped or absorbed by a material.

Additional Resources

This module is intended to be a thorough resource for task training. The following reference works are suggested for further study. These are optional materials for continued education rather than for task training.

Air Conditioning Systems, Principles, Equipment, and Service. 2000. Prentice Hall.

Refrigeration and Air Conditioning: An Introduction to HVAC. 2003. Prentice Hall.

Figure Credits

Carrier Corporation, 204F01, 204F04, 204F09, 204F13, 204SA01

Axair Nortec, 204F05 (top), 204F06A

General Filters, Inc., www.generalaire.com, 204F06B

Topaz Publications, Inc., 204F07, 204F10, 204F11 (photo)

Zonefirst, 204F17-204F19

Jackson Systems, 204F20

Steril-Aire, Inc., 204F21

Brooks Equipment Co., Inc., 204F22

Digital Control Systems, Inc., 204F23

MODULE 03204-07 — TEACHING TIPS

The following are suggested activities or instructional methods to help you teach the material in this module.

General

When you call on someone to answer a question, the rest of the class relaxes or even tunes out because they expect that the question and answer will take place only between you and the trainee you called on. Instead, use this technique to involve more trainees in answering questions and to keep them on their toes.

1. Ask trainees to define a term or explain a concept.
2. After one trainee has answered, ask a trainee seated nearby if the answer is right. Then ask whether a trainee in the back of the room agrees.
3. Ask trainees to explain why they think an answer is right or wrong.
4. Use the session to clear up incorrect ideas and encourage trainees to learn from their mistakes.

Sections 2.0.0 through 7.0.0

Trade Terms Quiz

This Trade Terms Quiz will familiarize trainees with the terms and definitions that are commonly used when describing air quality equipment. You will need photocopies of the quiz provided in the following page. Trainees will need pencils. If you allow trainees to use the Trainee Guide, decrease the amount of time you give them to complete the quiz.

1. Make a photocopy of the quiz for each trainee.
2. Give trainees between 5 and 10 minutes to complete the quiz.
3. Go over the answers to the quiz.
4. Ask trainees if they have questions.

Answers to Quick Quiz

1. j
2. f
3. a
4. k
5. c
6. h
7. b
8. d
9. i
10. g
11. e

Quick Quiz **Trade Terms**

For each description listed, identify the term that the text best describes. Write the corresponding letter in the blank provided.

_____ 1. The size of 99 percent of airborne particles is less than one _____.

_____ 2. The most important factor in keeping a body cool is _____.

_____ 3. Devices designed to remove moisture from the air are called _____.

_____ 4. People can have allergic reactions to chemicals that are _____ from adhesives and carpets.

_____ 5. Units used to supply fresh air and recover heating and cooling energy are called _____.

_____ 6. Devices designed to introduce water vapor into the conditioned environment are called _____.

_____ 7. A(n) _____ mixes outdoor air with conditioned air in proportions that depend on outdoor air temperature and humidity.

_____ 8. In high humidity areas, _____ controllers are typically used.

_____ 9. Cooling provided by an air conditioner compressor is known as _____.

_____ 10. Units used to supply fresh air and recover heat energy during the heating season are called _____.

_____ 11. Cooling indoor air by using a fan to bring in outside air is known as _____.

    a. dehumidifiers
    b. economizer
    c. energy recovery ventilator
    d. enthalpy
    f. evaporation
    e. free cooling
    g. heat recovery ventilator
    h. humidifiers
    i. mechanical cooling
    j. micron
    k. outgassing

Section 3.0.0 *Humidification*

This exercise familiarizes trainees with humidity controls. Trainees will need pencils and paper. You will need to obtain a humidifier. Allow 20 to 30 minutes for this exercise.

1. Discuss humidity control. Measure the humidity and temperature in the classroom at the beginning of the session.
2. Run a humidifier in the classroom throughout the session. Measure the temperature and humidity at the end of the session.
3. Have trainees compare their comfort levels from the beginning of the session to the end of the session. Answer any questions they may have.

Section 4.0.0 *Indoor Air Quality*

This exercise familiarizes trainees with challenges to maintaining good indoor air quality. Trainees will need pencils and paper. You will need to assign homework. Allow 20 to 30 minutes for this exercise.

Several government and trade association websites have extensive information on indoor air quality, including the following, which trainees can use to research this assignment:

US EPA: www.epa.gov/iaq

OSHA: www.osha.gov/SLTC/indoorairquality/index.html

1. Discuss challenges to maintaining good indoor air quality. Assign homework to research one aspect of maintaining good indoor air quality.
2. Have trainees report their findings to the class.
3. After their presentations, answer any questions they may have.

MODULE 03204-07 — ANSWERS TO REVIEW QUESTIONS

Answer	Section
1. a	2.2.0
2. c	2.2.3
3. b	2.2.4
4. c	3.0.0
5. d	3.3.1
6. a	3.3.2
7. b	3.3.5; Figure 8
8. b	4.1.3
9. c	5.1.0
10. a	5.4.0

CONTREN® LEARNING SERIES — USER UPDATE

NCCER makes every effort to keep these textbooks up-to-date and free of technical errors. We appreciate your help in this process. If you have an idea for improving this textbook, or if you find an error, a typographical mistake, or an inaccuracy in NCCER's Contren® textbooks, please write us, using this form or a photocopy. Be sure to include the exact module number, page number, a detailed description, and the correction, if applicable. Your input will be brought to the attention of the Technical Review Committee. Thank you for your assistance.

Instructors – If you found that additional materials were necessary in order to teach this module effectively, please let us know so that we may include them in the Equipment/Materials list in the Annotated Instructor's Guide.

Write: Product Development and Revision
National Center for Construction Education and Research
3600 NW 43rd St, Bldg G, Gainesville, FL 32606

Fax: 352-334-0932

E-mail: curriculum@nccer.org

Craft _____ Module Name _____

Copyright Date _____ Module Number _____ Page Number(s) _____

Description

_____

_____

_____

_____

(Optional) Correction

_____

_____

_____

(Optional) Your Name and Address

_____

_____

_____

Module 03205-07

Leak Detection, Evacuation, Recovery, and Charging

NCCER STANDARDIZED CRAFT TRAINING PROGRAM

The National Center for Construction Education and Research (NCCER) provides a standardized national program of accredited craft training. Key features of the program include instructor certification, competency-based training, and performance testing. The program provides trainees, instructors, and companies with a standard form of recognition through a National Craft Training Registry. The program is described in full in the *Guidelines for Accreditation*, published by NCCER. For more information on standardized craft training, contact the NCCER by writing us at 3600 NW 43rd St., Bldg. G, Gainesville, FL 32606; calling 352-334-0911; or emailing info@nccer.org. More information may be found at our website, www.nccer.org.

HOW TO USE THIS ANNOTATED INSTRUCTOR'S GUIDE

Each page presents two sections of information. The larger section displays each page exactly as it appears in the Trainee Module. The narrow column ties suggested trainee and instructor actions to each page and provides icons (detailed below) to call your attention to material, safety, audiovisual, or testing requirements. The bottom of each page includes space for your notes.

The **Audiovisual** icon indicates an appropriate time to show a transparency or other audiovisual aid.

The **Classroom** icon prompts you to define a term, stress a point, ask trainees to explain a concept, or give examples.

The **Demonstration** icon directs you to show trainees how to perform tasks.

The **Examination** icon tells you to administer the written module examination.

The **Homework** icon is placed where you may wish to assign reading for the next class, assign a project, or advise trainees to prepare for an examination.

The **Laboratory** icon is used when trainees are to practice performing tasks.

The **Materials** icon is a reminder for you to gather materials needed for classes, labs, and testing.

The **Performance Testing** icon tells you to administer a performance test or a portion thereof.

The **Safety** icon is used to emphasize safety issues. It is often keyed to *Caution* and *Warning!* statements in the Trainee Module.

The **Teaching Tip** icon indicates additional guidance is available, such as how to conduct an exercise, get the most educational value from a field trip, or encourage class participation. Teaching Tips may expand on a feature (*Think About It*, *Did You Know?*) or provide *Quick Quizzes* or similar exercises. You will be referred to the Teaching Tips section at the back of the module if there is additional material.

The **Combination** icon indicates that the laboratory listed corresponds with a performance task. If desired, you can note the proficiency of the trainees during the laboratory, and use it to satisfy performance testing requirements.

PREPARATION

Before teaching this module, you should review the Objectives, Performance Tasks, Materials and Equipment List, and Module Outline. Be sure to allow ample time to prepare your own training or lesson plan and gather all required materials and equipment.

Leak Detection, Evacuation, Recovery, and Charging
Annotated Instructor's Guide

Module 03205-07

MODULE OVERVIEW

This module introduces the trainee to the leak detection, evacuation, recovery, and charging service procedures used to troubleshoot, repair, and/or maintain proper operation of the mechanical refrigeration systems.

PREREQUISITES

Prior to training with this module, it is recommended that the trainee shall have successfully completed *Core Curriculum*; *HVAC Level One*; and *HVAC Level Two*, Modules 03201-07 through 03204-07.

OBJECTIVES

Upon completion of this module, the trainee will be able to do the following:

1. Identify the common types of leak detectors and explain how each is used.
2. Perform leak detection tests using selected methods.
3. Identify the service equipment used for evacuating a system and explain why each item of equipment is used.
4. Perform system evacuation and dehydration.
5. Identify the service equipment used for recovering refrigerant from a system and for recycling the recovered refrigerant, and explain why each item of equipment is used.
6. Perform a refrigerant recovery.
7. Evacuate a system to a deep vacuum.
8. Identify the service equipment used for charging refrigerant into a system, and explain why each item of equipment is used.
9. Use nitrogen to purge a system.
10. Charge refrigerant into a system by the following methods:
    - Weight
    - Superheat
    - Subcooling
    - Charging pressure chart

PERFORMANCE TASKS

Under the supervision of the instructor, the trainee should be able to do the following:

1. Identify the common types of leak detectors and explain the advantages and disadvantages associated with each type.
2. Use selected electronic, ultrasonic, liquid (bubble), and ultraviolet/fluorescent leak detectors to leak test a pressurized operational system.
3. Under supervision, use a recovery and/or recovery/recycle unit to recover the refrigerant from a system.
4. Under supervision, use a mixture of nitrogen and a trace amount of HCFC-22 refrigerant to pressurize a refrigerant system in preparation for leak testing.
5. Under supervision, demonstrate and/or describe how to evacuate a system using the deep vacuum method.
6. Perform a vacuum leak test on an evacuated system.
7. Under supervision, demonstrate how to evacuate a system using the triple evacuation method.
8. Under supervision, demonstrate how to use dry nitrogen as the moisture-absorbing gas when triple evacuating a system.
9. Under supervision, demonstrate how to charge a system by weight.
10. Under supervision, demonstrate how to charge a system using the superheat method.
11. Under supervision, demonstrate how to charge a system using the subcooling method.
12. Under supervision, demonstrate how to charge a system using the charging pressure charts method.

MATERIALS AND EQUIPMENT LIST

Overhead projector and screen
Transparencies
Blank acetate sheets
Transparency pens
Whiteboard/chalkboard
Markers/chalk
Pencils and scratch paper
Appropriate personal protective equipment
Common leak detection devices
Various types of refrigerant cylinders
Cylinders of nitrogen
HCFC-22 refrigerant
Pressure regulator-relief valve
System pressure charging charts
Superheat charging calculator (optional)
Subcooling calculator (optional)
Videotape/DVD *Evacuation and Charging* (optional)
TV/VCR/DVD player (optional)

Operating air conditioning and/or refrigeration system(s)
Vacuum pump
Electronic vacuum gauges
Electronic charging scale
Leak detectors
   Halide torch
   Electronic
   Ultrasonic
   Ultraviolet/fluorescent
   Liquid (bubble)
Gauge manifold set
Certified recovery or recovery/recycle unit
Charging cylinder
Thermometers
Copies of Quick Quizzes*
Module Examinations**
Performance Profile Sheets**

* Located in the back of this module.
**Located in the Test Booklet.

SAFETY CONSIDERATIONS

Ensure that the trainees are equipped with appropriate personal protective equipment and know how to use it properly. This module may require trainees to visit job sites. Make sure that all trainees are briefed on site safety procedures. This module requires that trainees work with refrigerants. Ensure that all trainees are properly briefed on refrigerant and pressure systems safety procedures.

ADDITIONAL RESOURCES

This module is intended to present thorough resources for task training. The following reference work is suggested for both instructors and motivated trainees interested in further study. This is optional material for continued education rather than for task training.

Refrigerant Service Techniques, 1993. Syracuse, NY: Carrier Corporation.

TEACHING TIME FOR THIS MODULE

An outline for use in developing your lesson plan is presented below. Note that each Roman numeral in the outline equates to one session of instruction. Each session has a suggested time period of 2½ hours. This includes 10 minutes at the beginning of each session for administrative tasks and one 10-minute break during the session. Approximately 20 hours are suggested to cover *Leak Detection, Evacuation, Recovery, and Charging*. You will need to adjust the time required for hands-on activity and testing based on your class size and resources. Because laboratories often correspond to Performance Tasks, the proficiency of the trainees may be noted during these exercises for Performance Testing purposes.

Topic	Planned Time

Session I. Introduction and Leak Detection
    A. Introduction _____
    B. Detection Devices _____
    C. Laboratory _____

        Trainees practice identifying various leak detection devices. This laboratory corresponds to Performance Task 1.

    D. Leak Testing _____
    E. Laboratory _____

        Under your supervision, have trainees pressurize a system in preparation for leak testing, and then practice using selected leak detection devices to test the system. This laboratory corresponds to Performance Tasks 4 and 2.

Session II. Refrigerant Containment
    A. Refrigerant Containment _____
    B. Refrigerant Recovery _____
    C. Laboratory _____

        Trainees practice recovering refrigerant from a system. This laboratory corresponds to Performance Task 3.

Sessions III and IV. Evacuation
    A. Evacuation _____
    B. Service Equipment Used for Evacuation _____
    C. Methods of Evacuation _____
    D. Deep Vacuum Evacuation Method _____
    E. Laboratory _____

        Trainees practice demonstrating or describing how to evacuate a system using the deep vacuum method and performing a vacuum leak test on the system. This laboratory corresponds to Performance Tasks 5 and 6.

    F. Triple Evacuation Method _____
    G. Laboratory _____

        Trainees practice demonstrating or describing how to evacuate a system using the triple evacuation method with dry nitrogen as the moisture absorbing gas. This laboratory corresponds to Performance Tasks 7 and 8.

Session V. Charging I
    A. Servicing Equipment Used for Charging _____
    B. Charge Determination and Accuracy _____
    C. Charging by Weight _____
    D. Laboratory _____

        Trainees practice demonstrating or describing how to charge a system by weight. This laboratory corresponds to Performance Task 9.

Session VI. Charging II
    A. Charging by Superheat _____
    B. Laboratory _____

        Trainees practice demonstrating or describing how to charge a system by superheat. This laboratory corresponds to Performance Task 10.

    C. Charging by Subcooling _____
    D. Laboratory _____

        Trainees practice demonstrating or describing how to charge a system by subcooling. This laboratory corresponds to Performance Task 11.

Session VII. Charging III
 A. Charging Using Pressure Charts
 B. Laboratory

 Trainees practice demonstrating or describing how to charge a system by using pressure charts. This laboratory corresponds to Performance Task 12.
 C. Using Zeotrope Refrigerants

Session VIII. Review and Testing
 A. Review
 B. Module Examination
   1. Trainees must score 70% or higher to receive recognition from NCCER.
   2. Record the testing results on Craft Training Report Form 200, and submit the results to the Training Program Sponsor.
 C. Performance Testing
   1. Trainees must perform each task to the satisfaction of the instructor to receive recognition from NCCER. If applicable, proficiency noted during laboratory exercises can be used to satisfy the Performance Testing requirements.
   2. Record the testing results on Craft Training Report Form 200, and submit the results to the Training Program Sponsor.

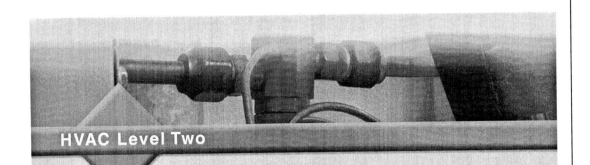

HVAC Level Two

03205-07

Leak Detection, Evacuation, Recovery, and Charging

Assign reading of Module 03205-07.

03205-07
Leak Detection, Evacuation, Recovery, and Charging

Topics to be presented in this module include:

1.0.0	Introduction	5.2
2.0.0	Leak Detection	5.2
3.0.0	Refrigerant Containment	5.7
4.0.0	Evacuation	5.9
5.0.0	Charging	5.18
6.0.0	Using Zeotrope Refrigerants	5.31

Overview

Few tasks performed by HVAC service technicians are more important than installing and maintaining the proper refrigerant charge. An improperly charged system will not perform efficiently and may even be damaged. Once a system is charged with the correct amount of refrigerant, any reduction in the charge is likely to be caused by a leak in the system. Any time that happens, it is up to the service tech to locate and repair the leak before adding refrigerant. This task requires specialized instruments and techniques.

If a system must be opened for repair, the refrigerant charge must first be transferred to approved containers using equipment made for this purpose. When the system is repaired and sealed, the system must be recharged, again using specialized equipment and methods. Any technician who installs or services HVAC systems must learn how to perform the critical tasks of leak detection, evacuation, recovery, and charging of the mechanical refrigeration system.

Instructor's Notes:

Objectives

When you have completed this module, you will be able to do the following:

1. Identify the common types of leak detectors and explain how each is used.
2. Perform leak detection tests using selected methods.
3. Identify the service equipment used for evacuating a system and explain why each item of equipment is used.
4. Perform system evacuation and dehydration.
5. Identify the service equipment used for recovering refrigerant from a system and for recycling the recovered refrigerant, and explain why each item of equipment is used.
6. Perform a refrigerant recovery.
7. Evacuate a system to a deep vacuum.
8. Identify the service equipment used for charging refrigerant into a system, and explain why each item of equipment is used.
9. Use nitrogen to purge a system.
10. Charge refrigerant into a system by the following methods:
    - Weight
    - Superheat
    - Subcooling
    - Charging pressure chart

Trade Terms

Fractionation
Reclamation
Recovery
Recycle
Temperature glide

Required Trainee Materials

1. Paper and paper
2. Appropriate personal protective equipment

Prerequisites

Before you begin this module, it is recommended that you successfully complete *Core Curriculum*; *HVAC Level One*; and *HVAC Level Two*, Modules 03201-07 through 03204-07.

This course map shows all of the modules in the second level of the HVAC curriculum. The suggested training order begins at the bottom and proceeds up. Skill levels increase as you advance on the course map. The local Training Program Sponsor may adjust the training order.

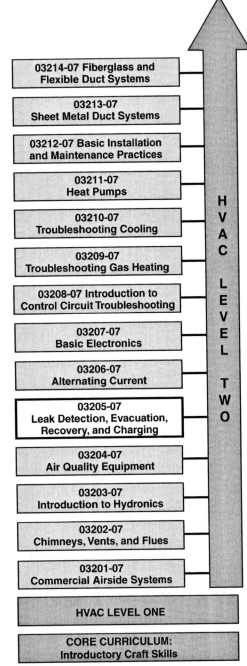

Ensure that you have everything required to teach the course. Check the Materials and Equipment list at the front of this module.

See the general Teaching Tip at the end of this module.

Explain that terms shown in bold are defined in the Glossary at the back of this module.

Show Transparency 1, Objectives, and Transparency 2, Performance Tasks. Review the goals of the module, and explain what will be expected of the trainee.

Review the modules covered in Level Two and explain how this module fits in.

List the basic service procedures performed on mechanical refrigeration systems.

Explain why leak testing is performed.

List common leak detection devices.

Describe electronic leak detectors and explain how they are used.

Discuss certification of refrigerant technicians.

See the Teaching Tips for Sections 1.0.0–6.0.0 and Section 2.0.0 at the end of this module.

Provide common leak detection devices for trainees to examine.

1.0.0 ♦ INTRODUCTION

This module covers the basic service procedures used to troubleshoot, repair and/or maintain correct operation of the mechanical refrigeration system:

- Leak detection
- Evacuation and dehydration
- **Recovery**
- Charging

These service procedures are performed when installing new systems and when servicing existing ones. Failure to perform any one of these procedures correctly can result in poor system operation, and may even cause system failure. They must also be performed in order to make sure that the venting requirements of the Clean Air Act are not violated.

2.0.0 ♦ LEAK DETECTION

Refrigerant leaks are one of the major causes of trouble in refrigeration systems. Leaks must be detected and repaired because they allow all or part of the system refrigerant to be lost, allow air and moisture to enter and contaminate the system, and cause our environment to become contaminated with ozone-depleting compounds. Leak testing is normally performed:

- When an operating system has low refrigerant charge
- After the assembly but before evacuating/dehydrating and charging a new system
- After making a repair on a closed refrigerant system
- When acid-moisture testing indicates moisture and/or acid in a system

2.1.0 Detection Devices

There are many leak detection devices ranging from very simple to complex. This section describes four common leak detectors:

- Electronic
- Ultrasonic
- Ultraviolet/fluorescent
- Liquid (bubble)

Technician Certification

As a result of the Clean Air Act, which governs the release of refrigerants into the atmosphere, the Environmental Protection Agency (EPA) requires the certification of all technicians who service air conditioning and refrigeration equipment. This includes anyone who performs installation, maintenance, or repair on such systems. Technicians must be certified in all equipment categories that they intend to service or install. For non-mobile or non-vehicular equipment these certifications are for Small Appliance (Type I), High-Pressure Appliance (Type II), Low-Pressure Appliance (Type III), or a Universal certificate (Type IV) that covers all types.

Certification is achieved by passing the Section 608 Technician Certification Program test for each category, as given by an EPA-approved certifying agency. The list of EPA-certified agencies includes the Air Conditioning Contractors of America (ACCA), the Air Conditioning & Refrigeration Institute (ARI), and a host of other agencies. A complete list of EPA-approved certification testing agencies can be found on the EPA website at www.epa.gov.

2.1.1 Electronic Detectors

Electronic leak detectors, commonly called sniffers, are accurate and easy to use (*Figure 1*). They typically detect leak rates of about ½ ounce per year. Electronic leak detectors should be operated and calibrated as directed by the manufacturer. Most of these detectors have an air filter that must be checked and replaced as needed. Usually, the leak detector sensor tip is placed next to each component or the piping in the system, and slowly moved at a rate of about one inch per second while searching for the leak. If possible, drafts should be reduced by shutting off fans and other devices that cause air movement. When a refrigerant leak is detected, the leak detector produces an audible signal, a flashing light, or both, depending on the detector in use.

NOTE

It is never acceptable to simply top-off the refrigerant charge if it is low. The mechanical refrigeration system is a sealed system that will retain its charge unless there is a refrigerant leak or the refrigerant has been allowed to escape. If a system has a low refrigerant charge, the cause of the problem must be located and corrected before refrigerant is added to the system.

Instructor's Notes:

WARNING!

Do not use electronic leak detectors in atmospheres that contain flammable or explosive vapors. The sensor operates at extremely high temperatures and may explode if it comes in contact with a combustible gas.

2.1.2 Ultrasonic Leak Detectors

Ultrasonic leak detectors (*Figure 2*) are accurate and easy to use. Ultrasonic sound frequencies are sound waves beyond the range of human hearing. Ultrasonic leak detectors detect the ultrasonic sounds that a refrigerant gas makes as it leaks from a pressurized system. As the gas leaks to the atmosphere, its flow becomes turbulent. This turbulent flow has a high content of ultrasonic waves that are sensed by the detector. These devices can also detect leaks in a system under vacuum because they hear the turbulence caused by the air as it is drawn from the outside into the system. Ultrasonic leak detectors include a detector unit and a headset. Ultrasonic leak detectors should be operated and calibrated as directed by the manufacturer. Usually, you put on the headset, turn on the instrument, and begin listening for leaks as you slowly move the detector sensor around the system. The leak will be audible in the headphones before it is indicated by the light on the detector.

Because the ultrasonic detector works from sound waves, it can be used to detect any type of refrigerant gas, including nitrogen. Ultrasonic detectors operate well in windy areas or areas with stray gases or fumes. One disadvantage of the ultrasonic detector is that sounds from other sources in the test area may affect its ability to detect a leak. Another disadvantage is that the ultrasonic detector may not be able to detect very small leaks.

2.1.3 Ultraviolet/Fluorescent Leak Detectors

This method involves injecting a small quantity of a fluorescent additive into the oil/refrigerant charge of an operating system. Areas of refrigerant leakage are observed as a yellow-green glow when viewed under the beam of a high-intensity ultraviolet (UV) lamp. This method of leak detection can be used to leak test CFC, HCFC, and HFC refrigerant systems as well as refrigerant blends using synthetic lubricants. One manufacturer's test kit used for this purpose includes a high-intensity ultraviolet lamp (*Figure 3*), formula capsules filled with a premeasured amount of special fluorescent solution, and miscellaneous self-sealing fittings.

INSIDE TRACK — Leak Detectors

Some electronic leak detectors are not capable of sensing some of the newer HFC refrigerants. Check with the equipment manufacturer to determine which refrigerants a particular device can detect.

Low-quality, inexpensive leak detectors may fail to detect refrigerant leaks in the field. Avoid these devices. Use only dependable, high-quality leak detectors when searching for refrigerant leaks.

Figure 1 ♦ Electronic leak detector.

Figure 2 ♦ Ultrasonic leak detector.

Identify the safety hazards of working with an electronic leak detection device around flammable or explosive vapors.

Discuss ultrasonic leak detectors. Identify their advantages and disadvantages.

Describe ultraviolet/fluorescent leak detectors and their applications.

Explain how ultraviolet/fluorescent leak detectors are used.

CAUTION

Before injecting any dye, fluid, or chemical into the oil/refrigerant charge of an operating system, check with the equipment manufacturer and make sure the substance used will not harm the system. Introducing a substance to the system that is not approved by the manufacturer may void the equipment warranty.

The fluorescent solution is injected into the system where it circulates and mixes with the refrigerant and compressor oil. The solution from the capsule can be injected into the system by any of the following methods:

- Using refrigerant from a drum to move the solution into the low side while throttling refrigerant flow to produce a safe mist
- Connecting the capsule between the high and low sides of the system and allowing the high pressure to move the solution while throttling refrigerant flow
- Adding the solution to the oil container and then pumping it into the system
- Pouring premeasured solution using a special injector device (*Figure 3*)

At spots where small amounts of refrigerant are leaking from the system, the fluorescent solution also leaks out. After enough time has been allowed for the solution to circulate and mix

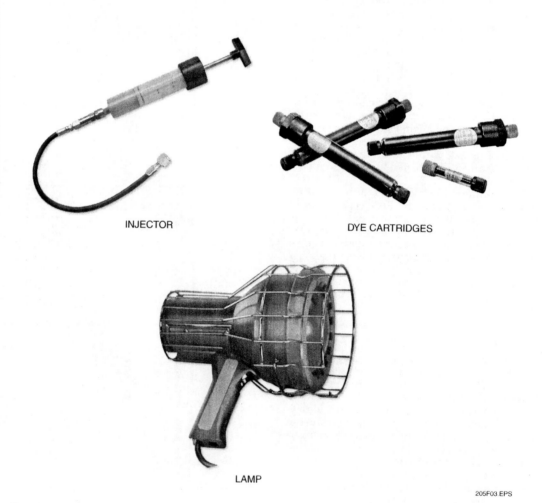

INJECTOR

DYE CARTRIDGES

LAMP

Figure 3 ♦ Ultraviolet/fluorescent leak detector kit.

thoroughly, the ultraviolet lamp is used to scan solder joints, fittings, couplings, and seals in the system to detect leaks. The source of any leak is shown by a bright, fluorescent yellow-green glow of the solution.

After any leaks have been repaired, the fluorescent solution can be removed from the exterior of the leak site with any general-purpose cleaner suitable for removing oil residue. The fluorescent solution injected into the system remains in the system even after any leaks have been repaired. This allows for future leak testing of the system on a regular basis. Even if the full refrigerant charge is lost due to a leak, the solution remains mixed with the compressor oil in the system. Unless the oil is drained, it stays in the system.

2.1.4 Liquid (Bubble) Detectors

Liquid detector or bubble solutions can be used to locate small leaks in a refrigerant system. Often, a liquid detector is used to find the exact source of an apparent leak that was found using one of the other types of leak detectors. The sudsy solution is brushed around the suspected area to find the leak. Escaping gas under pressure at the leak causes bubbles to form, providing a visual indication that pinpoints the leak. It may take several minutes for the system pressure to form a bubble. The main advantages of liquid detectors are their low cost and ease of use. A disadvantage is that larger leaks may blow through the solution, causing no bubbles to appear. Commercial bubble solutions are recommended because they are safer to use with metals. They also provide longer-lasting bubbles. Leak testing using solutions made from common laundry or kitchen detergents should be avoided. Many of these detergents contain chlorides which can corrode most of the metals used in refrigeration systems.

2.2.0 Leak Testing

Leak testing can be performed on a fully operational system, or on systems that have a partial refrigerant charge or no charge at all. Leak testing should be performed before charging a new system or after making a repair on an existing system. Leak testing can be one of the most challenging aspects of servicing HVAC systems, since leaks may be tiny and difficult to find, and piping may be difficult to access. However, you must be able to perform effective leak testing in order to comply with EPA and various environmental regulations. *Figure 4* shows a typical equipment hookup for a leak detection test.

2.2.1 Operational System

If the system is operational, a visual check might reveal the source of a suspected leak. Since oil is mixed with refrigerant inside the system, the presence of oil around tubing joints, fittings, and on coil surfaces can indicate a leak. Look closely at all mechanical fittings, since vibration can loosen them over time. Use your eyes and ears. Large leaks can sometimes be heard. If this check leads to a component or piping suspected to be leaking, confirm that a leak exists using a leak detector. If the sight and sound method fails to pinpoint the leak, test the entire system with a leak detector. After finding and marking all leaks, recover the system refrigerant before attempting to make any repairs. Recover down to 0 psig to prevent contamination of the recovered refrigerant with air.

2.2.2 Systems Without a Refrigerant Charge

The method recommended by the EPA for leak testing a refrigeration system uses a trace amount of HCFC-22 refrigerant mixed with dry nitrogen. This method is also used to test systems that normally use a refrigerant other than HCFC-22. First, pressurize the system with a small (trace) quantity of HCFC-22 to a pressure of about 10 psig. Then, add dry nitrogen as needed to further increase the system pressure to the desired test level. When pressurizing the system with nitrogen, be sure not to exceed the maximum test pressure limits stamped on the unit's nameplate or listed in the manufacturer's service literature for the unit. A safe maximum is about 125 psig.

Discuss liquid leak detectors and explain how they are used.

Show trainees how to identify common types of leak detection devices; discuss the advantages and disadvantages of each type.

Have trainees practice identifying common types of leak detection devices. Have them explain the advantages and disadvantages of each type. Note the proficiency of each trainee. This laboratory corresponds to Performance Task 1.

Show Transparency 3 (Figure 4). Describe the equipment hookup for leak detecting.

Explain that a preliminary inspection can often pinpoint the source of a leak.

Provide an overview of the procedures for leak testing a system without a refrigerant charge.

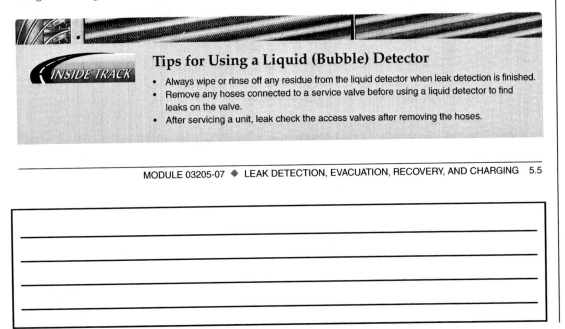

Tips for Using a Liquid (Bubble) Detector
- Always wipe or rinse off any residue from the liquid detector when leak detection is finished.
- Remove any hoses connected to a service valve before using a liquid detector to find leaks on the valve.
- After servicing a unit, leak check the access valves after removing the hoses.

Identify the safety precautions necessary when working with pressurized gas.

Provide an overview of the procedure for leak testing a system with a partial refrigerant charge.

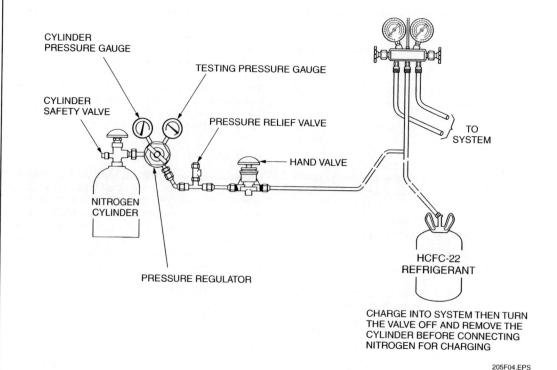

Figure 4 ♦ Equipment hookup for leak detecting.

The small amount of refrigerant used in this mixture is adequate to be detected with an electronic leak detector. The dry nitrogen provides the system pressure needed to perform the test. If you are using an ultrasonic leak detector, the system can be pressurized to the leak test level with nitrogen only. After locating and marking all leaks in the system, the mixture of trace refrigerant and nitrogen (or nitrogen alone) is vented to the atmosphere.

 WARNING!
Never use oxygen, compressed air, or flammable gas to pressurize a system. An explosion will result when oil and oxygen are mixed.

Nitrogen is a high-pressure gas. At full cylinder pressure (about 2,000 psig), nitrogen can rupture a refrigerant cylinder and/or the refrigeration system. Use a pressure regulator on the nitrogen cylinder to limit the pressure. Also, use an overpressure relief valve between the cylinder pressure regulator and the system. The relief valve should be adjusted to open at about 2 psig above the system test pressure, but never more than 150 psig.

When charging the system with both a refrigerant and nitrogen, always put the refrigerant in first. Always turn the valve off and remove the refrigerant cylinder before connecting the nitrogen cylinder.

2.2.3 Systems With a Partial Charge

When a system with a partial refrigerant charge has insufficient pressure to support the leak test, the partial charge must first be recovered. The system is then pressurized with an R-22 refrigerant and nitrogen mixture the same as would be done with an empty system.

 NOTE
It is a violation of EPA regulations to add nitrogen to an existing refrigerant charge in a system for the purposes of leak detection. The release of any such uncontrolled mixture to the atmosphere after the leak test is over is judged to be a release of pure CFCs or HCFCs.

2.3.0 Disassembling Brazed Joints for Leak Repair

When repairing a leak, it may be necessary to disassemble brazed piping. The procedures used to disassemble a brazed joint are basically the same as those used when assembling a brazed joint.

When disassembling a brazed joint, the flux is first brushed around the joint area. It takes as much heat to take a joint apart as it does to braze it. Watch the condition of the flux to see when the filler metal has melted enough to allow the tubing to be pulled from a fitting. After the joint is taken apart, clean the disassembled parts with warm water to remove the flux.

> **CAUTION**
>
> When using a torch to disassemble a brazed joint, the system pressure must be 0 psig. Even the slightest pressure inside the system can cause refrigerant and oil trapped in the piping to spray out as the joint is disassembled.
>
> To prevent pressure from building up as heat is applied, make a pressure vent line by attaching an open-ended service hose to a service port on the unit. Another option is to cut the rear of the joint with a tubing cutter.

3.0.0 ◆ REFRIGERANT CONTAINMENT

Since the passage of the Clean Air Act Amendments in 1990, it has been illegal to release refrigerants into the atmosphere. Most refrigerants being handled in any service procedure must be recovered, **recycled**, reclaimed, or destroyed. The EPA has specific meanings for the terms recovery, recycle, and **reclamation**:

- *Recovery* – The removal and temporary storage of refrigerant in containers approved for that purpose. Recovery does not provide for any cleaning or filtration of the refrigerant.
- *Recycle* – The circulation of recovered refrigerant through filtering devices that remove moisture, acid, and other contaminants. This does not mean that the recycled refrigerant meets the purity standards for new refrigerants.
- *Reclamation* – The remanufacture of used refrigerant to bring it up to the standards required of new refrigerant. Reclamation is not a field service procedure. It is a complicated process done only at reprocessing or manufacturing facilities.

Refrigerant must be recovered from the different types of equipment (appliances) to levels established by the EPA. These levels are based on the amount of evacuation that must be achieved in a system during recovery. *Table 1* shows required evacuation levels for stationary equipment. When recovering refrigerant, except from small appliances, a recovery unit must be used that will evacuate the system to the level shown in the table. Small appliances are those in which the refrigerant is sealed within the unit at the factory and the amount of charge is five pounds or less. An example of a small appliance is a room air conditioner. For small appliances, the system is considered completely recovered when 90 percent of the refrigerant is removed if the appliance has a running compressor, or 80 percent of the refrigerant is removed if the unit has a non-operating compressor. For practical purposes, small appliances can be considered recovered to an acceptable level when evacuated to 4 in. Hg.

There are some exceptions to these EPA recovery evacuation levels. If evacuation to the specified levels is not achievable because of leaks in the system, or if recovery to these levels would contaminate the refrigerant being recovered, you must perform the following:

- Isolate the leaking components from the rest of the system, if possible.
- Evacuate non-leaking components to the required levels, if they are to be opened.
- Evacuate leaking components to the lowest level attainable without substantially contaminating the refrigerant. This level cannot exceed 0 psig.

3.2.0 Refrigerant Recovery and Recovery/Recycle Units

Recovery of refrigerant from a system to the evacuation levels specified by the EPA requires the use of a certified refrigerant recovery unit or recovery/recycle unit. Because the recovery and/or recycling process must be accomplished without releasing any refrigerant into the air, or contaminating the refrigerant itself, only authorized technicians should perform these tasks on certified units.

3.2.1 Recovery Unit

Generally speaking, the greater the vacuum pulled by the recovery unit, the higher the probability that a high percentage of the refrigerant is

Describe the procedure for disassembling brazed piping.

Show trainees how to use selected leak detectors to test a pressurized operational system.

Under your supervision, have trainees practice using selected leak detectors to test a pressurized operational system. Note the proficiency of each trainee. This laboratory corresponds to Performance Task 2.

Have the trainees review Sections 3.0.0–3.2.3.

Ensure that you have everything required for teaching this session.

Define recovery, recycle, and reclamation.

Show Transparency 4 (Table 1). Discuss the requirements for various appliances.

Discuss exceptions to EPA evacuation levels and additional procedures required.

Explain that refrigerant recovery must be performed by an authorized technician with certified equipment.

Discuss the function and operation of a recovery unit.

Identify the conditions required to perform recovery.

Provide a typical recovery unit for trainees to examine.

Discuss the function and operation of a recycle unit.

Explain that special service procedures are required when using a recovery or recycle unit to process a different refrigerant than last processed.

Table 1 Required Levels of Evacuation for Stationary Appliances Except for Small Appliances

Type of Appliance	Inches of Mercury Vacuum* using Recovery/Recycle Unit Manufactured:	
	Before November 15, 1993	After November 15, 1993
HCFC-22 appliance normally containing less than 200 pounds of refrigerant	0	0
HCFC-22 appliance normally containing more than 200 pounds of refrigerant	4	10
High-pressure appliance normally containing less than 200 pounds of refrigerant (includes equipment that uses CFC-12, CFC-500, CFC-502, and CFC-114, HFC-134a, or HFC-410A refrigerant)	4	10
High-pressure appliance normally containing more than 200 pounds of refrigerant (includes equipment that uses CFC-12, CFC-500, CFC-502, CFC-114, or HFC-410A refrigerant)	4	15
Very high-pressure appliance (includes equipment that uses CFC-13, CFC-503, or HFC-23 refrigerant)	0	0
Low-pressure appliance (CFC-11, CFC-113, or HCFC-123 refrigerant)	25	25mm Hg absolute

*Note: Based on Standard Atmospheric Pressure of 29.9 in. Hg.

recovered. Recovery units are not vacuum pumps and do not provide that function. If the dehydration/evacuation of a system is required, a vacuum pump must be used.

Most recovery units are capable of recovering refrigerant from a system in either the vapor or liquid state. Make sure that the unit used is capable of recovering the type of refrigerant used in the system.

Test the recovered refrigerant for the presence of acid or moisture, to verify its purity, before recharging it back into the system. When the refrigerant being recovered is highly contaminated, or when liquid refrigerant is being recovered, an external filter-drier should be installed in the utility hose line of the gauge manifold set connected to the recovery unit. Be sure to orient the filter-drier for correct flow direction.

Because there is a wide difference in the capabilities of the various recovery or recovery/recycle units, recovery is performed by following the manufacturer's instructions for the unit being used.

Figure 5 shows a typical recovery unit connected to a system. Currently, recovered refrigerant can only be reused in the same system from which it was removed, or in another system belonging to the same owner.

Recovery must be performed under the following conditions:

- Before a system is opened to make repairs
- Before pressurizing a system for leak testing
- Before disposing of any system or component
- When necessary to remove excess refrigerant from an overcharged system

3.2.2 Recycle Unit

Recycling of recovered refrigerant is normally done at the job site by a certified technician using a certified recycle unit. Make sure that the unit is capable of processing the type of refrigerant you plan to recycle. Follow the manufacturer's instructions for the recycle unit being used.

A recycle unit dehydrates and cleans the recovered refrigerant so that it can be returned to the system. After recycling, the refrigerant is cleaner, but its purity is not the same as that of new refrigerant. The dehydrating and purifying capability of recycle units varies by model. Most units circulate the refrigerant through a filtration/drying process to achieve the desired quality. Acid/moisture testing of recycled refrigerant should be performed to verify the quality of the refrigerant before putting it back into the system. Currently, recycled refrigerant can only be reused in the same system from which it was recovered, or another system owned by the same customer.

3.2.3 Recovering or Recycling Different Refrigerants

Before using either a recovery unit or recycle unit to process a different refrigerant than was last processed, the recovery or recycle unit compressor oil should be drained (if applicable) and replaced with new oil. All filter-driers should be replaced and the recovery or recycle unit must be evacuated. Always use a recovery cylinder dedicated for use with the type of refrigerant to be recovered or recycled. Make sure to perform any

Figure 5 ◆ Recovery unit connected to a refrigerant system.

required service specified in the manufacturer's instructions for the unit being used.

A recovery machine with high temperature and pressure capabilities is needed to recover R-410A gas. The recovery machine must be able to pull the deep vacuum that is needed to remove all the refrigerant. The portable recovery machine shown in *Figure 6* is one such machine.

4.0.0 ◆ EVACUATION

Refrigeration systems are intended to contain only refrigerant and oil. Anything else in a closed system is considered a contaminant. Contaminants such as air and moisture are major causes of compressor failure.

Evacuation is important because it is the only way that air or moisture in a system can be removed. Air is a noncondensible that can accumulate in the condenser, taking up space needed for condensing the refrigerant. This results in an increase in the condensing temperature that makes the system work harder. It also promotes the creation of acids by chemical reaction with the oil and refrigerant mixture.

Under the heat of compression, moisture will react with the refrigerant to form hydrochloric and hydrofluoric acids. These acids can cause corrosion of metals and breakdown of the insulation on the motor windings.

Show trainees how to use a recycle and/or recovery unit to recover the refrigerant from a system.

Under your supervision, have trainees practice using a recycle and/or recovery unit to recover the refrigerant from a system. Note the proficiency of each trainee. This laboratory corresponds to Performance Task 3.

Have the trainees review Sections 4.0.0–4.2.2.

Ensure that you have everything required for teaching this session.

Explain that evacuation is the only way to effectively remove air or moisture from a system.

Tips for Refrigerant Recovery

These tips for trouble-free refrigerant recovery were compiled by Bacharach Inc., a manufacturer of refrigerant recovery and recycling equipment.

- Always identify the refrigerant you are recovering. This will minimize cross contamination and help you plan for the equipment needed while recovering.
- Pay attention to your hoses. Use the shortest hoses possible for the job. Long hoses increase the recovery time. Remove all restrictions in the hoses. Hoses with ball valves on the ends are better than hoses that are self-sealing. Remove Schrader core valves when possible from service ports.
- Do not run your recovery machine with a long extension cord or a cord of too light a gauge wire.
- Always recover refrigerant with an in-line filter-drier.
- Always pump liquid out of the system first, then recover the remaining vapors. This will speed up recovery rates significantly.
- With large amounts of refrigerant, use the liquid push-pull recovery method. This is three times faster than recovering liquid directly. Refer to the liquid push-pull instructions in your equipment's manual.
- When possible, recover from both the high and low side service port on the system. This will speed up your recovery rate.
- Climatic conditions affect recovery rates. Recovery rates can be adversely affected by either extreme hot or cold temperatures.
- Always purchase the right recovery machine for the job. A recovery machine with a lower capacity than required by the job will wear out prematurely.
- When recovering R-410 gas, you need a recovery machine with high pressure and temperature capabilities, able to pull the deep vacuum necessary for complete refrigerant removal.

Explain when evacuation is necessary.

Explain that evacuation requires the use of a vacuum pump and indicator.

Discuss the function and operation of a vacuum pump.

Show Transparency 5 (Figure 8). Explain that by lowering the pressure in the system, the vacuum pump allows the moisture to vaporize at a much lower temperature and makes it easier to remove.

Explain that a good vacuum pump must be able to evacuate a system down to 500 microns.

Describe the operation of vacuum pumps.

Show the video *Evacuation and Charging*.

Provide a vacuum pump for trainees to examine.

Moisture in the refrigerant can also cause oil sludge, which reduces the lubrication properties of the oil in the compressor. Moisture can freeze at the expansion device, slowing down or blocking the flow of refrigerant. The presence of moisture in a system can be determined using an acid/moisture test kit.

Generally, evacuation of a system is performed:

- After assembly and pressure checking, but prior to charging a system assembled in the field
- After an existing system is opened to the atmosphere as a result of parts replacement or other service procedures
- When an acid/moisture test shows the system is contaminated

4.1.0 Service Equipment Used for Evacuation

Evacuation of a system requires the use of a good vacuum pump and an accurate vacuum indicator. These devices are connected through a gauge manifold set to the system.

4.1.1 Vacuum Pump

The vacuum pump (*Figure 7*) is used to remove the air and moisture trapped in a system. The vacuum pump works to create a pressure differential between the system and the pump. This causes air and moisture vapor trapped in the system at a higher pressure to move into the lower pressure (vacuum) area created in the vacuum pump. The air and moisture vapor removed from the system are further processed through the vacuum pump and discharged into the atmosphere. When the vacuum pump lowers the pressure in the system enough, as determined by the ambient temperature of the system, liquid moisture trapped in the system will boil and change into vapor. Like free air, this water vapor is then pulled out of the system, processed through the pump, and exhausted to the atmosphere. *Figure 8* shows a chart of the boiling points of water at various levels of vacuum. As shown, the more the atmospheric pressure is lowered, the lower the boiling point of water.

The use of a quality, high-level vacuum pump is a must to properly evacuate and dehydrate a system. Typically, a good pump is capable of evacuating a system down to 500 microns (29.90 in. Hg vacuum). Microns are a precise measurement of pressure. One inch of mercury equals 25,400 microns. Most vacuum pumps use a direct-drive, two-stage, rotary-type vane pump driven by the attached motor. Most have an oil level sight gauge, oil port, and an oil drain. A pump with a gas ballast between the pump's first and second stages should be used. This feature permits relatively dry air from the atmosphere to enter the second stage of the pump, where it combines with the moist vapors passing through the pump from the refrigerant system. This helps to prevent the moisture from condensing into a liquid and mixing with the vacuum pump oil in the pump crankcase. The pump gas ballast valve should be opened during the early stages of a pumpdown to introduce the air into the second stage of the pump. The valve can be opened or closed at any time the pump is running. When first starting the pump, the valve should be closed. If it is left open, the pump oil may be discharged during the first revolution.

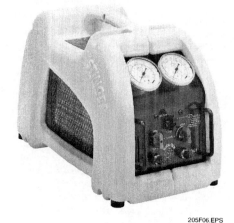

Figure 6 ♦ Portable recovery unit.

Figure 7 ♦ Vacuum pump.

How quickly a vacuum pump can evacuate a system is an important service consideration. The more air the pump is capable of moving, measured in cubic feet per minute (cfm), the faster it can reach an acceptable vacuum level. The cfm requirements for a vacuum pump vary from system to system. *Table 2* lists suggested minimum amounts.

NOTE
The speed of evacuation can be increased by removing the Schrader cores from the service valves and using larger hoses.

Vacuum pumps need periodic maintenance. Since the oil in the vacuum pump becomes contaminated through normal use, it should be changed after every 10 hours of pump operation. Always change the oil immediately after pumping down a wet or contaminated refrigerant system. Follow the manufacturer's instructions for servicing the vacuum pump.

4.1.2 Electronic Vacuum Gauge

An electronic vacuum or micron gauge is used to measure high vacuums. An accurate vacuum gauge must be used to measure the 500-micron range vacuum levels that must be achieved to properly evacuate and dehydrate refrigeration systems.

Even though the compound gauge on a gauge manifold set is capable of measuring a vacuum, it should not be used because the scale calibration is not accurate enough to read the specific vacuum levels needed in evacuation.

Electronic vacuum gauges are designed for use with high-vacuum pumps and can often read as low as 1.0 micron. A sensing tube is mounted at

Show Transparency 6 (Table 2). Explain minimum pump sizes for various applications.

Describe maintenance procedures for a vacuum pump.

Discuss the operation of an electronic vacuum gauge.

Provide an electronic vacuum gauge for trainees to examine.

Temperature in °C	Temperature in °F	Inches of Vacuum	Microns*	Pressure (lbs/sq in)
100°	212°	0.00	759,968	14.696
96°	205°	4.92	535,000	12.279
90°	194°	9.23	525,526	10.162
80°	176°	15.94	355,092	6.866
70°	158°	20.72	233,680	4.519
60°	140°	24.04	149,352	2.888
50°	122°	26.28	92,456	1.788
40°	104°	27.75	55,118	1.066
30°	86°	28.67	31,750	0.614
26.6°	80°	28.92	25,400	0.491
24.4°	76°	29.02	22,860	0.442
22.2°	72°	29.12	20,320	0.393
20.6°	69°	29.22	17,780	0.344
17.8°	64°	29.32	15,240	0.295
15°	59°	29.42	12,700	0.246
10.7°	53°	29.52	10,160	0.196
7.2°	45°	29.62	7,620	0.147
0°	32°	29.74	4,572	0.088
−6.1°	21°	29.82	2,540	0.049
−14.4°	6°	29.87	1,270	0.0245
−31.1°	−24°	29.91	254	0.0049
−37.2°	−35°	29.915	127	0.00245
−51.1°	−60°	29.919	25.4	0.00049
−56.6°	−70°	29.9195	12.7	0.00024
−67.8°	−90°	29.9199	2.54	0.000049

*Remaining pressure in system in microns
1.000 inch = 25,400 microns = 2.540 cm = 25.40 mm
0.100 inch = 2,540 microns = 0.254 cm = 2.54 mm
0.039 inch = 1,000 microns = 0.100 cm = 1.00 mm

205F08.EPS

Figure 8 ♦ Boiling temperatures of water at various levels of vacuum.

Note that the location of the sensor affects the reading of a vacuum gauge.

Discuss the function and operation of a gauge manifold set.

Explain that the two most common methods of evacuation are the deep vacuum method and the triple vacuum method.

Provide a gauge manifold set for trainees to examine.

Table 2 Recommended Minimum Vacuum Pump Sizes for Various Applications

System Size	Minimum Vacuum Pump Size
Up to 10 tons Domestic refrigeration	6 cfm
Up to 30 tons Small residential A/C systems	6 to 12 cfm
Up to 50 tons Rooftop A/C systems	10 to 15 cfm
Up to 70 tons Large commercial systems	12 to 20 cfm

some point in the vacuum line to provide a calibrated output. This output, in microns, is displayed on an analog meter scale, a digital display, or a light-emitting diode (LED) sequence display. *Figure 9* shows typical digital meter vacuum gauges.

The lower the micron reading displayed on the gauge, the closer the vacuum conditions are to a perfect vacuum (0 micron). For the LED-type gauge, the LED indicators, each representing a specific vacuum level in microns, turn off sequentially from the highest micron level to the lowest as the system vacuum pressure goes deeper. The location of the sensor tube affects the vacuum reading. The closer it is located to the vacuum source, the lower the reading, which may give an inaccurate reading. When measuring the vacuum in a system, isolate the vacuum pump with a vacuum valve and allow the pressure in the system to balance before taking a reading. Always operate the vacuum gauge as directed in the manufacturer's instructions. The vacuum gauge should be hooked with a tee to the suction hose. If the micron level rises after the vacuum pump valve and the evacuation gauge manifold set are closed, there is a leak in the system or in the gauge manifold set.

4.1.3 Gauge Manifold Set and Service Vacuum Hoses

Either a two-valve gauge manifold set with evacuation-quality service hoses or a high-capacity evacuation gauge manifold set can be used for evacuation. Typically, an evacuation gauge manifold set has four valves, enlarged internal passages, and ⅜" hose connections to speed up the evacuation process (*Figure 10*). The use of four valves enables the gauge manifold set to be connected to all the equipment needed to perform both the evacuation and charging of a system, without disconnecting and reconnecting the service hoses.

4.2.0 Methods of Evacuation

The deep vacuum and triple vacuum methods of system evacuation are the two most frequently used. However, some manufacturers may have specific requirements for evacuation. The deep vacuum method is typically used after a repair that required the system charge to be recovered and the system opened. Use of the triple evacuation method is recommended when a system has been especially wet. The amount of moisture in a system can be determined by using an acid/moisture test on the system refrigerant before or during refrigerant recovery.

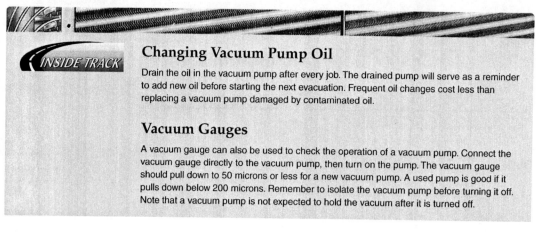

Changing Vacuum Pump Oil

Drain the oil in the vacuum pump after every job. The drained pump will serve as a reminder to add new oil before starting the next evacuation. Frequent oil changes cost less than replacing a vacuum pump damaged by contaminated oil.

Vacuum Gauges

A vacuum gauge can also be used to check the operation of a vacuum pump. Connect the vacuum gauge directly to the vacuum pump, then turn on the pump. The vacuum gauge should pull down to 50 microns or less for a new vacuum pump. A used pump is good if it pulls down below 200 microns. Remember to isolate the vacuum pump before turning it off. Note that a vacuum pump is not expected to hold the vacuum after it is turned off.

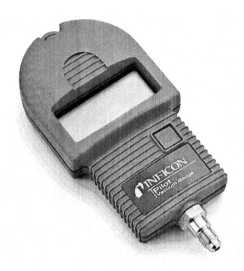

Figure 9 ♦ Vacuum gauges.

4.2.1 Deep Vacuum Evacuation Method

The deep vacuum method of evacuation relies on the use of the vacuum pump alone to remove moisture from the system. A deep vacuum is any vacuum of 500 microns or less (29.90 in. Hg vacuum or more). When a deep vacuum is established in a closed system, air and other noncondensibles in the system are reduced to a negligible level. As the pressure is reduced, the boiling point of water is also reduced. As long as the ambient temperature surrounding the system is higher than the boiling point of the internal moisture, the moisture will be boiled off and expelled.

If a deep vacuum is rapidly achieved with an oversized vacuum pump, water in the refrigerant lines may turn into ice. The deep vacuum evacuation method is performed as follows:

Step 1 Connect the vacuum pump, vacuum gauge, and gauge manifold set to the service valves on the system (or component) being evacuated, as shown in *Figure 11*.

Figure 10 ♦ Evacuation gauge manifold set.

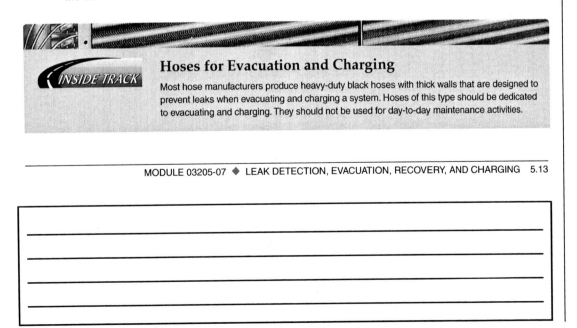

Hoses for Evacuation and Charging

Most hose manufacturers produce heavy-duty black hoses with thick walls that are designed to prevent leaks when evacuating and charging a system. Hoses of this type should be dedicated to evacuating and charging. They should not be used for day-to-day maintenance activities.

Provide an overview of the deep vacuum evacuation method.

Show Transparency 7 (Figure 11). Review the procedure for performing a deep vacuum evacuation.

Show trainees how to evacuate a system using the deep vacuum method and perform a vacuum leak test on the evacuated system.

Under your supervision, have trainees practice evacuating a system using the deep vacuum method and performing a vacuum leak test on the evacuated system. Note the proficiency of each trainee. This laboratory corresponds to Performance Tasks 5 and 6.

Step 2   Turn on the vacuum pump, open the system service valves, and evacuate the system to 500 microns. During the early stages of evacuation, stop the pump at least once and monitor the vacuum gauge to see if a rapid loss of vacuum occurs. Make sure to isolate the vacuum pump from the system with a valve to prevent vacuum loss through the pump. If there is a loss of vacuum, check the system and/or vacuum pump and vacuum gauge hookup for a leak. Repair any leaks, then proceed with evacuating the system.

Step 3   Once the 500-micron vacuum level is reached, close the valves to the gauge manifold set, and use the vacuum shutoff valve to isolate the vacuum pump from the system. Turn off the vacuum pump.

Step 4   To ensure that the system is adequately evacuated, the final equilibrium pressure of the system must be checked after the system has been pumped down below 500 microns, but before it is charged with refrigerant. Vacuum leak-test the system by watching the reading on the vacuum gauge to note any change in the level of vacuum in the system.

- A constant reading on the indicator of 500 microns indicates a leaktight, dry system.
- If the indicator shows a pressure rise but levels off between 1,000 and 2,000 microns, this indicates that the system is leaktight, but moisture or liquid refrigerant is in the system.
- If the gauge shows a pressure rise and the pressure continues to rise without leveling off, a leak exists in the system, gauges, or the connecting tubing.

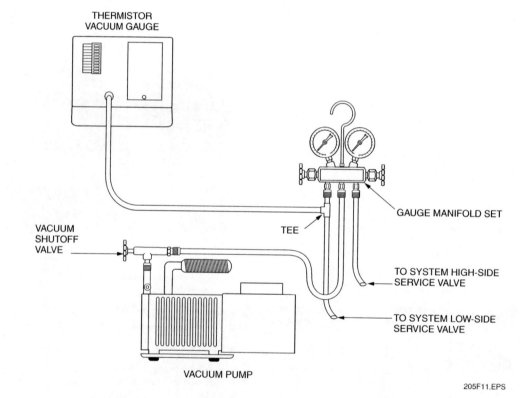

Figure 11 ♦ Service equipment connected for deep evacuation method.

Instructor's Notes:

4.2.2 Triple Evacuation Method

Triple evacuation is recommended by most manufacturers for wet or contaminated systems. The triple evacuation method evacuates the system three times to a vacuum of about 1,000 microns (29.88 in. Hg vacuum). After the first and second evacuations, dry nitrogen is charged into the system to absorb moisture during the time period between each evacuation. The method given in this section uses dry nitrogen because it absorbs moisture well, is readily available, and compared to refrigerant, is inexpensive. Unlike refrigerant, nitrogen does not have to be recovered from the system between each evacuation. It can be released to the atmosphere.

CAUTION
Be careful not to let any air into the system when changing over from the nitrogen purge to the vacuum pump between evacuations.

Nitrogen is a high-pressure gas. At full cylinder pressure (about 2,000 psig), nitrogen can rupture the refrigeration system. Be certain to use a pressure regulator on the nitrogen cylinder to limit the pressure. Also, make sure to use an overpressure relief valve between the cylinder pressure regulator and the system. The relief valve should be adjusted to open at about 2 psig above the system test pressure, but never more than 150 psig. A typical nitrogen cylinder hookup used for the triple evacuation procedure is shown in *Figure 12*.

WARNING!
Never use oxygen, compressed air, or flammable gas to pressurize a system. An explosion will result when oil and oxygen are mixed.

Figure 12 shows the hookup of the service equipment for the triple evacuation method. *Figure 13* shows the sequence of the triple evacuation process. The process can take about three hours when performed as recommended. This time can vary depending on how deep a vacuum is drawn in each step, how large the system is, and vacuum pump capacity. The system is evacuated to 1,000 microns (29.88 in. Hg) during the three evacuations. The vacuum pump continues to run at this level or lower for at least 15 minutes during each evacuation. Evacuating to this level will cause liquid water in the system to boil anywhere with a temperature of 35°F or higher. After the first evacuation, the vacuum pump is turned off and the system is vacuum leak tested as previously described for the deep evacuation method.

Between the first and second evacuations, the system is pressurized with dry nitrogen to 10 psig or higher and allowed to sit for up to an hour so that the nitrogen can absorb the moisture. A pressure of 10 psig provides more than enough nitrogen to absorb moisture in the system.

An optional triple evacuation method is to pull the system down to about 2,500 microns (29.72 in. Hg vacuum) for the first and second evacuations. For the last evacuation, a deep vacuum of 500 microns (29.90 in. Hg vacuum) is drawn and the system is vacuum leak tested in the same manner as previously described for the deep vacuum method.

Using nitrogen to pressurize systems for triple evacuation requires that certain precautions be followed.

Smoking Vacuum Pumps
During the initial pulldown of a vacuum, the vacuum pump can be very noisy and may emit smoke. As the vacuum deepens, the pump should operate more quietly and stop smoking. If the noise and smoke do not decrease, there may be a vacuum leak in the system.

Measuring Microns
Most vacuum micron gauges do not measure accurately above 2,500 microns. When pulling a system down to levels above 2,500 microns, use an appropriate, accurate gauge capable of reading these higher values.

Provide an overview of the triple evacuation method.

Show Transparencies 8 and 9 (Figures 12 and 13). Review the procedure for performing a triple evacuation.

Discuss the safety precautions necessary for working with pressurized gas.

Show trainees how to evacuate a system using the triple evacuation method with dry nitrogen as the moisture-absorbing gas.

Under your supervision, have trainees practice evacuating a system using the triple evacuation method with dry nitrogen as the moisture-absorbing gas. Note the proficiency of each trainee. This laboratory corresponds to Performance Tasks 7 and 8.

Have the trainees review Section 5.0.0–5.3.2.

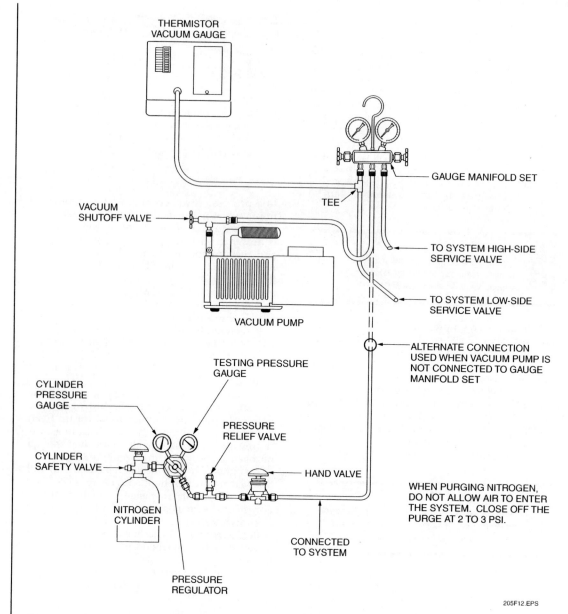

Figure 12 ♦ Service equipment connected for triple evacuation method.

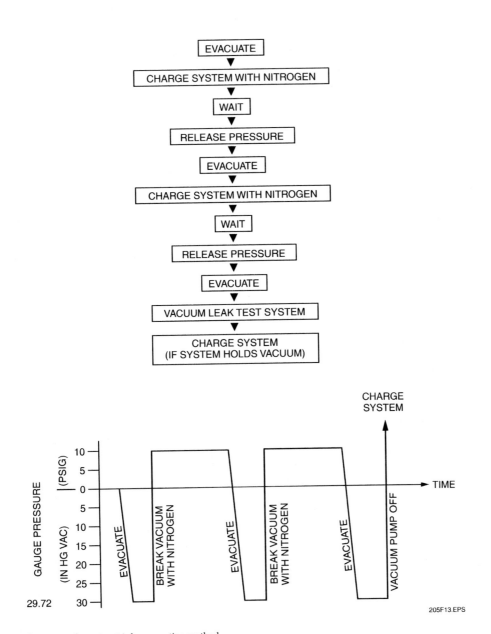

Figure 13 ♦ Sequence of events – triple evacuation method.

Ensure that you have everything required for teaching this session.

List the common methods of system charging.

List the types of service equipment used when charging a system.

Show Transparency 10 (Figure 14). Discuss the various types of refrigerant cylinders.

Provide various types of refrigerant cylinders for trainees to examine.

5.0.0 ♦ CHARGING

A system must be charged with refrigerant after it has been repaired, leak tested, and evacuated, in order to return it to service. Also, new equipment may need to be charged before starting it up. Some models come factory-charged, but may need adjustment in the field before being placed into service. This section describes the service equipment and methods commonly used to charge systems, including:

- Liquid charging by weight
- Vapor charging by weight
- Charging by superheat
- Charging by subcooling
- Charging using pressure charts

5.1.0 Service Equipment Used for Charging

This section will cover the service equipment used for charging. Any of several different methods can be used to charge a system with refrigerant. The equipment used depends on the charging method selected.

5.1.1 Refrigerant Cylinders

Refrigerant may be added to a system in either a vapor (gas) or liquid state (*Figure 14*). If using refrigerant contained in a disposable cylinder to charge the system, the cylinder must be set in the upright position in order to charge with vapor. To charge with liquid, the cylinder must be turned upside down. Refillable cylinders used with

RECOVERY (REFILLABLE) CYLINDER

UPRIGHT FOR VAPOR   UPSIDE DOWN FOR LIQUID

DISPOSABLE CYLINDER

205F14.EPS

Figure 14 ♦ Selecting vapor and liquid state of refrigerant in cylinders.

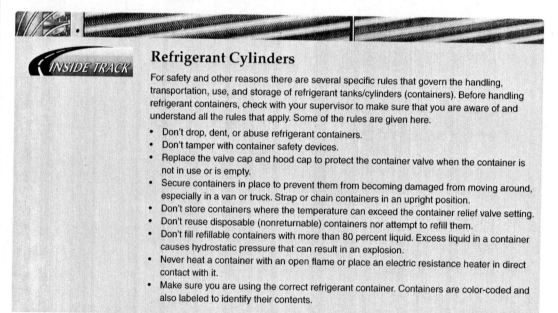

Refrigerant Cylinders

For safety and other reasons there are several specific rules that govern the handling, transportation, use, and storage of refrigerant tanks/cylinders (containers). Before handling refrigerant containers, check with your supervisor to make sure that you are aware of and understand all the rules that apply. Some of the rules are given here.

- Don't drop, dent, or abuse refrigerant containers.
- Don't tamper with container safety devices.
- Replace the valve cap and hood cap to protect the container valve when the container is not in use or is empty.
- Secure containers in place to prevent them from becoming damaged from moving around, especially in a van or truck. Strap or chain containers in an upright position.
- Don't store containers where the temperature can exceed the container relief valve setting.
- Don't reuse disposable (nonreturnable) containers nor attempt to refill them.
- Don't fill refillable containers with more than 80 percent liquid. Excess liquid in a container causes hydrostatic pressure that can result in an explosion.
- Never heat a container with an open flame or place an electric resistance heater in direct contact with it.
- Make sure you are using the correct refrigerant container. Containers are color-coded and also labeled to identify their contents.

Instructor's Notes:

recovery or recycle units remain upright for charging with either liquid or vapor refrigerant. Recovery cylinders have valves that are used to select the desired vapor or liquid state.

5.1.2 Charging Scales

Installing a full refrigerant charge into a system is best done by weight. This requires the use of an accurate refrigerant charging scale. Charging scales (*Figure 15*) can be used to charge both small and large systems. Selection of a charging scale is based mainly on the type and size of the system being charged. Since proper system operation depends on having the correct charge, the scale's most important feature is its accuracy. The accuracy should be matched to the size of the system being charged. A rule of thumb is that the accuracy should be within one percent of the required total system charge. To prevent damaging the scale, ensure that the scale weighing platform is strong enough to handle the largest size cylinder you intend to use.

Simple charging scales, electronic charging scales, and fully programmable automatic electronic charging scales are in common use. Depending on the model and its intended use, charging scales are calibrated to weigh refrigerant in pounds, ounces, kilograms, and/or grams. Typically, they display the cylinder/refrigerant weight using a liquid crystal display (LCD) or equivalent. Programmable models can control the flow of refrigerant and can be set to automatically dispense a preset amount of refrigerant and turn off when that amount is reached. Most stop charging if the cylinder empties before the full system charge is reached. As with other precise service instruments, the charging scale should always be operated and calibrated as directed in the manufacturer's instructions. Do not use bathroom or produce scales for refrigerant charging because they are not accurate enough. The system must be in a vacuum in order to weigh in the required charge.

5.1.3 Charging Cylinder

A charging cylinder is an accurate measuring device that can be used to charge either liquid or vapor refrigerant into a system (*Figure 16*). The use of a particular charging cylinder is based on the type and size of the system being charged. To maintain the accuracy of the charge, the cylinder used should be large enough to hold the total charge needed for the system. Typically, charging cylinders come in three capacities: 2½ pounds, 5 pounds, and 10 pounds.

Figure 15 ◆ Electronic refrigerant charging scale.

Figure 16 ◆ Refrigerant charging cylinder.

Some charging cylinders contain an electric warming element that heats the refrigerant and builds up pressure in the cylinder. This helps push the refrigerant out of the cylinder, particularly in cold weather.

The charging cylinder contains calibrated shrouds for three or more different refrigerants. It has a sight glass to show the amount of refrigerant in the cylinder. On top of the charging cylinder is a gauge to measure the refrigerant pressure and a hand valve. On the bottom is another hand valve. The top and bottom hand valves are used

Discuss the types of charging scales.

Emphasize that charging scales must be accurate to within one percent of the required total system charge.

Discuss various types of charging cylinders.

Provide charging scales and charging cylinders for trainees to examine.

Explain that the manufacturer's instructions must be followed to set up the recovery/recycle unit for the type of charging being performed.

Discuss the function of the system sight glass on an example system.

to charge a system with vapor or liquid refrigerant, respectively. The bottom hand valve is also used to fill the cylinder.

To reduce the time needed for charging, cylinders have heaters to overcome the equalization of pressure between the system and charging cylinder. Charging cylinders normally have an automatic pressure relief valve for safety. This valve automatically resets when the safe pressure is restored.

The charging cylinder is usually evacuated and filled through the bottom hand valve with liquid refrigerant supplied from a bulk tank. A direct measurement of weight is obtained by rotating the shroud to align the sight glass with the calibration that matches the gauge pressure reading. The weight measurement is shown on the shroud at the level of refrigerant in the sight glass.

When charging the system, the top or bottom hand valve is used to control the flow of the desired vapor (top valve) or liquid refrigerant (bottom valve) into the system. At the same time, the level of the liquid refrigerant in the sight glass is watched as the charge is transferred. When the total charge is in the system, the cylinder valve is closed.

To be sure of accurate system charging, the charging cylinder should always be operated and calibrated as directed in the manufacturer's instructions.

Note that if you are filling a charging cylinder and it is necessary to remove vapor refrigerant to allow more liquid to enter, that refrigerant must be recovered.

5.1.4 Recovery/Recycle Unit

When charging with a recovery/recycle unit, follow the manufacturer's instructions for either vapor or liquid charging. Depending on the unit, one or more of the other methods described in this section may also have to be used in order to achieve a properly charged system.

5.1.5 System Sight Glass

When a system is equipped with a sight glass (*Figure 17*) in the liquid line close to the inlet of the metering device, it can be used to charge the system if the metering device is a thermostatic expansion valve (TXV, also abbreviated as TEV).

When the system is properly charged, only a clear stream of liquid refrigerant passes through the sight glass, even under fairly heavy loads. Under extreme loads outside of normal operating

Figure 17 ◆ Moisture-liquid indicator (sight glass).

conditions, the sight glass may continue to display a fast-moving stream of bubbles until load conditions are reduced. With the help of refraction provided by the glass, a clear glass can be distinguished from an empty glass through a magnifying effect—a full sight glass will appear magnified inside. Under normal operating conditions, a bubbling or flashing sight glass often indicates a shortage of refrigerant, representing a mixed vapor-liquid condition.

 NOTE
With 400-series refrigerants, an occasional bubble may appear, even though the charge is normal.

For proper operation and full capacity, only a solid stream of liquid should enter the TXV. However, the sight glass is only one tool used to assist in charging a unit with a TXV as a metering device. A sight glass with a clear liquid stream merely demonstrates that clear liquid is reaching the TXV. The system could very well be overcharged, or the TXV may be feeding very little refrigerant under the present load. The sight glass should never be used alone as a charging tool. To ensure proper charging with TXV-equipped systems, it is imperative that other factors be evaluated, including subcooling. On initial charging, once a clear sight glass is obtained, more charge is likely needed to ensure sufficient refrigerant is available for peak load conditions.

There are occasions when the sight glass may display a very lazy stream of liquid. It will be evident that it is not completely full, but the stream of liquid will move along very slowly, and generally without bubbles. This can occur when loads are very light and the TXV is modulating its position very close to its seat. The resulting limited flow slows the movement of refrigerant throughout the system, and a vapor pocket forms in the top of the glass. When the TXV opens further to increase flow, the sight glass will suddenly clear with a solid stream of liquid. This condition, described as a low flow sight glass, is not an indication of an undercharge, but merely an indication that the TXV is nearly closed.

As noted above, charging with a sight glass is only valid with systems that are equipped with a TXV. Sight glasses are of little value in systems equipped with fixed restrictors. While charging to a clear sight glass, condensing temperatures should be held near 130°F. To raise the condensing temperature during the charging process, plastic or other suitable material can be used to block the condenser coil. As the charge is added, the material can be removed.

On older systems operating less efficiently than today's 13-SEER units, and larger commercial systems which may operate at lower SEER values, technicians could add 25°F to 30°F to the ambient temperature to determine the target condensing temperature. However, this practice is not recommended on newer systems, and is unnecessary when proper charging techniques employing subcooling measurements and a review of all system operating parameters are used. The 130°F target condensing temperature is well above a normal ambient temperature + 25 to 30°F, and thus the need for material to block the condenser temporarily to simulate higher ambient temperatures.

This procedure should charge the TXV system correctly and provide the right level of subcooling necessary for cooling comfort. You should always use other charging methods to verify the correct sight glass charge.

The use of a sight glass with a moisture indicator in a non-TXV system can also have value. The moisture indicator provides the technician with some insight as to the condition of the refrigerant. If the refrigerate has excess moisture it will need to be dehydrated, which may include changing the refrigerant, oil, and driers in addition to system evacuation. Do not trust a moisture indicator. Use an acid-moisture test kit to verify that moisture contamination exists in the system before beginning a lengthy refrigerant dehydration process. Occasionally, a moisture indicator will give a false moisture reading. The color of the moisture indicator varies among sight glass manufacturers.

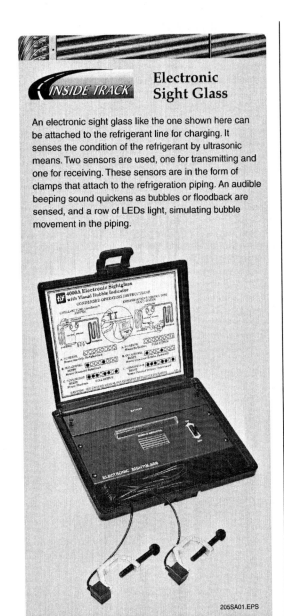

INSIDE TRACK

Electronic Sight Glass

An electronic sight glass like the one shown here can be attached to the refrigerant line for charging. It senses the condition of the refrigerant by ultrasonic means. Two sensors are used, one for transmitting and one for receiving. These sensors are in the form of clamps that attach to the refrigeration piping. An audible beeping sound quickens as bubbles or floodback are sensed, and a row of LEDs light, simulating bubble movement in the piping.

5.2.0 Charge Determination and Accuracy

Regardless of system size, the operating charge of a system determines how efficiently and economically the system runs. An overcharged system can lead to high temperatures and pressures with

Explain that the required charge is normally found on the unit nameplate. Discuss how to determine the total charge for built-up systems.

Explain that charging by weight involves either liquid charging or vapor charging. Point out that liquid charging is the faster method.

Show Transparency 11 (Figure 18). Explain the procedure for liquid charging by weight.

the possibility of component failure. An undercharged system leads to insufficient cooling, and on units with hermetic compressors, may lead to compressor motor shutdown or failure. Many of the problems that occur in refrigeration and air conditioning systems are the direct result of overcharging or undercharging. Both are serious errors that must be avoided. Whether charging old or new systems, or adjusting the charge on a system, the full or partial charge must be introduced in the right way and in the right quantity.

For split systems and packaged equipment, the unit nameplate usually shows the amount of refrigerant charge required. This is also listed in the manufacturer's service literature for the equipment. The amount of charge needed for new units is determined by what charge, if any, it has when shipped from the factory. Always consult the manufacturer's literature for the unit or component to get the specific charge information. Be sure to take into account any additional refrigerant charge that needs to be added to the total system charge in order to compensate for accessories, such as filter-driers, receivers, etc.

Nameplates on split systems normally give the charge weight for the system based on the use of a standard line set length. Always be sure to consult the manufacturer's service literature to find out the length of the standard line set. If the system being serviced uses a line set that exceeds the standard length, find the amount of additional refrigerant charge that must be added to compensate for each foot of increased length.

For built-up systems, where each component is bought separately and is installed in the field, the total charge weight must be determined by adding the capacities of the individual components and accessories, plus the capacity of the connecting piping. The capacity of each component can be found in the manufacturer's literature for the component, or it may be marked on the component. One rule of thumb to estimate the charge needed for a built-up system calls for two pounds of refrigerant per each ton of system capacity. This is a ballpark estimate only. During the actual charging procedure, the charge put into the system must be precisely measured. Charging a built-up system is usually accomplished by charging the system with an adequate quantity of liquid and/or vapor refrigerant to allow the system to be started and run without tripping the safety controls. Then, a precise charge is obtained by adjusting the amount of refrigerant in the system up or down to obtain a proper superheat or subcooling temperature, or discharge pressure, depending on the type of system and the metering device.

5.3.0 Charging by Weight

Charging by weight is a precise method for charging a system. It is used if a complete charge is to be added and the weight of the charge is known. As previously described, the total weight can be found either from the unit nameplate or the manufacturer's literature. Be sure to include in the total charge weight any additional volume needed to compensate for added accessories or long line set lengths. Both liquid and vapor refrigerant can be charged into a system by weight.

> **CAUTION**
> In many systems using a water loop, such as a geothermal system, vapor-charge to a saturated vapor temperature before adding liquid. Additionally, flowing or draining all water loops is recommended.

5.3.1 Liquid Charging by Weight

Liquid charging goes much faster than vapor charging. This is because the density of a liquid is much greater than that of a vapor. The result is that the same size charging hoses can deliver many times more liquid charge per minute than if charging with a vapor. Because it is faster, liquid charging is the first choice to charge an empty system. When conditions are right, the entire charge may be introduced into the system in the liquid state. At other times, the flow of liquid refrigerant may slow down to a trickle or stop before the entire charge can be weighed in. When this happens, liquid charging is stopped and the remainder of the required total charge is introduced into the system using the vapor charging by weight method.

Figure 18 shows the hookup of the service equipment used for liquid charging by weight. Liquid charging is done by adding the liquid refrigerant from the refrigerant cylinder (or charging cylinder) into the high-pressure side of the system through the high-side service valve. To prevent compressor damage, the compressor must be turned off, but the air handler should be turned on. Also, liquid refrigerant must never be added into the low-pressure side of the system or into the compressor discharge service port. If a system is equipped with a liquid charging valve between the condenser and metering device, it is better to charge the system through the liquid charging valve. Note that liquid charging valves tend to be used on systems with capacities over 20 tons.

5.22 HVAC ◆ LEVEL TWO

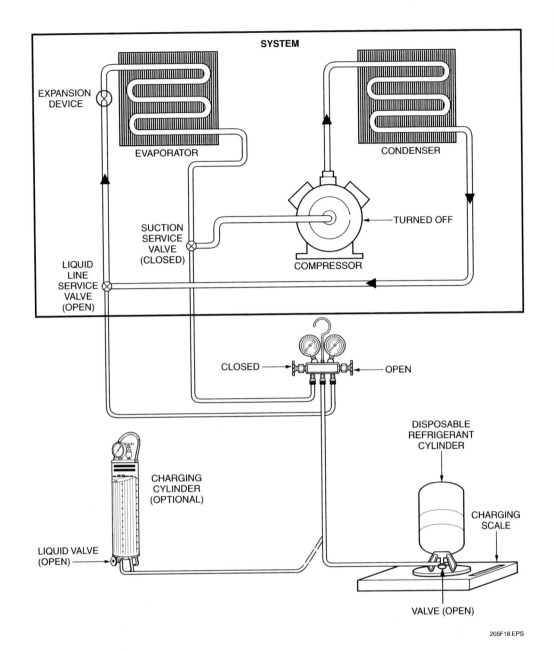

Figure 18 ♦ Service equipment connected for charging liquid refrigerant by weight.

Discuss the importance of ensuring that water remains flowing while refrigerant is added to the system.

Show Transparency 12 (Figure 19). Explain the procedure for vapor charging by weight.

Show trainees how to charge a system by weight.

Under your supervision, have trainees practice charging a system by weight. Note the proficiency of each trainee. This laboratory corresponds to Performance Task 9.

See the Teaching Tip for Section 5.3.2 at the end of this module.

Have the trainees review Section 5.4.0–5.5.0.

Ensure that you have everything required for teaching this session.

Discuss the hazards and precautions necessary when warming cylinders.

CAUTION

On water-cooled systems or chillers, it is imperative to ensure that water remains flowing while liquid refrigerant is added to the system. Failure to ensure water flow may result in freezing of the stagnant water in the condenser or evaporator coil and could lead to serious system damage.

5.3.2 Vapor Charging by Weight

Vapor charging by weight is normally performed in order to finish charging a system where the whole charge could not be introduced in the liquid state. Vapor charging by weight can also be used to charge an empty system, but is seldom used for this purpose. The correct method used to vapor charge a system with a partial charge depends on how much of the total charge weight is already in the system. If more than 50 percent is in the system, the liquid line service valve is closed, the compressor is turned on, and the refrigerant vapor is charged into the low-side of the system through the opened suction service valve (*Figure 19*). The system is vapor-charged in this manner until the required total charge by weight has entered the system.

If less than 50 percent of the total charge is in the system, the compressor is turned off, the low-side and high-side service valves are opened, and the refrigerant vapor is charged into both the low and high sides of the system through the opened service valves. When more than 50 percent of the total charge has entered the system, the high-side service valve can be closed, allowing the remainder of the refrigerant vapor to be charged through the low-side of the system with the compressor turned on. This method would also be used to charge an empty system using the vapor charging by weight method.

As vapor is removed from the refrigerant cylinder, the system and container pressures tend to equalize, and the flow of vapor may slow down or even stop. When this occurs, the cylinder can be warmed by placing it in warm water (90°F to 100°F). This will help to continue the flow of refrigerant by creating a pressure differential between the cylinder and the rest of the system.

WARNING!

When warming the container, observe these precautions:

- Never heat a refrigerant cylinder above 100°F.
- Never apply a direct flame to a refrigerant cylinder.
- Never place a refrigerant cylinder on a hot plate or in direct contact with any other electric heater.

5.4.0 Charging by Superheat

Superheat is the heat added to a refrigerant after it has all been changed into a vapor. By knowing the amount of superheat in the suction line, you can tell if the system is properly charged.

Some manufacturers recommend the superheat method for checking or adjusting the charge of a system already in operation. Maintaining correct superheat is critical, because if liquid refrigerant returns to the compressor, it can cause damage and possible failure.

Charging for superheat is performed on systems with a fixed metering device such as a capillary tube or fixed orifice (*Figure 20*). This method uses superheat and suction line temperature tables (*Figure 21*) to first find the proper superheat level, then the required suction line temperature. These values are based on the temperatures and pressures measured in the operating system. The tables are attached to the equipment or contained in the manufacturer's service literature.

To charge by superheat, proceed as follows:

Step 1 Operate the system for at least 15 minutes. The filter, condenser, and evaporator coil must be clean. Make sure good airflow is established throughout the system ductwork. Measure the following:

- Suction (vapor) line pressure
- Suction line temperature
- Outdoor dry-bulb temperature entering the condenser
- Indoor wet-bulb temperature entering the evaporator
- Indoor dry-bulb temperature entering the evaporator

Instructor's Notes:

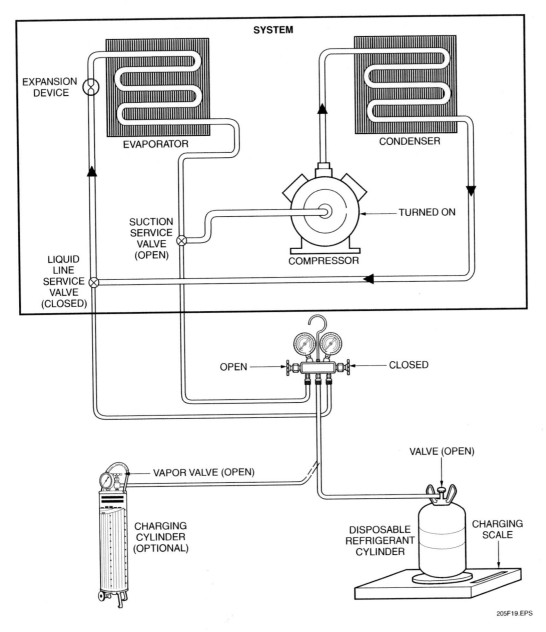

Figure 19 ♦ Service equipment connected for charging vapor refrigerant by weight.

Show Transparencies 13–15 (Figures 20 and 21). Explain the procedure for charging by superheat.

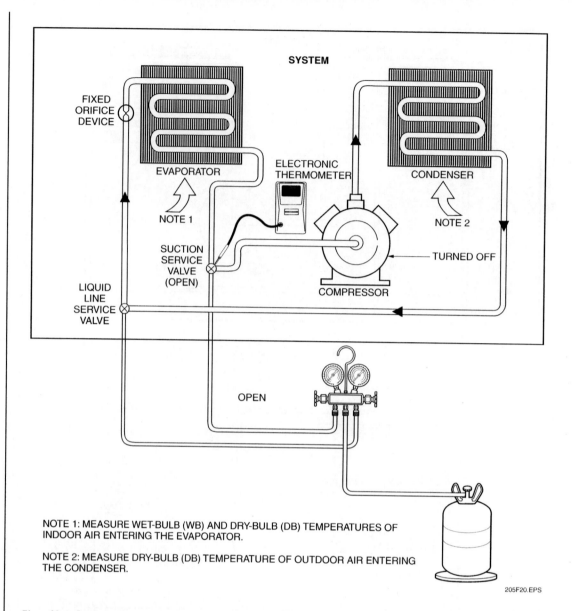

Figure 20 ♦ Service equipment connected for charging by superheat.

SUPERHEAT CHARGING CHART
(SUPERHEAT ENTERING SUCTION SERVICE VALVE)

INDOOR AIR CONDITIONS ENTERING EVAPORATOR			OUTDOOR AIR DRY BULB TEMPERATURE (°F) ENTERING CONDENSER COIL									
WB	DB	RH %	65	70	75	80	90	(95)	100	105	110	110
61	65	80	16	12	8	6						
	70	60	18	14	10	6		Charging at these conditions can result in damage to the compressor. Charge to 5 degrees superheat and check superheat when conditions are more favorable.				
	75	45	20	16	12	8	6					
	80	33	21	17	14	10	6					
	85	23	23	19	16	12	7					
63	70	68	21	17	13	10	8	6				
	75	52	23	19	16	12	9	6				
	80	39	24	20	17	14	10	7				
	85	29	25	21	18	15	12	8	6			
	90	20	26	22	19	16	13	9	7			
65	70	77	24	20	17	13	10	8	6			
	75	59	25	22	19	15	12	10	7			
	80	45	27	24	21	18	14	11	8	6		
	85	33	28	25	22	19	15	12	9	6		
	90	25	29	26	23	20	16	13	10	7		
67	70	86	27	24	21	17	14	11	8	6		
	75	66	28	26	22	18	15	13	10	8	5	
	80	50	30	27	24	21	18	15	12	10	7	
	85	39	31	28	25	22	19	16	13	11	8	
	90	30	32	29	26	23	20	17	14	12	9	6
69	70	95	30	27	25	21	18	15	12	9	7	6
	75	75	31	28	25	22	19	16	13	10	8	6
	80	58	32	29	27	24	21	19	16	14	11	8
	85	45	34	31	28	26	23	20	17	15	12	9
	90	35	35	32	30	27	24	21	19	16	13	11
(71)	75	82	34	31	29	26	23	21	18	15	12	9
	80	65	36	32	30	28	24	22	20	18	15	13
	85	51	37	34	31	29	26	24	21	19	16	14
	(90)	39	38	35	32	30	28	(25)	22	20	17	16
	95	30	39	35	33	30	29	25	23	21	18	17
73	75	92	37	34	32	29	27	24	22	20	17	15
	80	72	37	35	33	30	28	26	24	22	19	17
	85	58	38	36	34	31	29	27	25	23	20	18
	90	45	38	36	34	31	30	28	25	23	21	19
	95	35	38	36	35	32	31	28	26	24	21	19

NOTES: Superheat measurements should be taken at condensing unit service valves.
Charge system within 2°F of superheat indicated.
Recommended minimum superheat is 5°F.

White area in the chart is the optimum window for charging.

DB = dry-bulb temperature (°F).
WB = wet-bulb temperature (°F).
RH = approximate % of indoor relative humidity.

Find the required superheat at the intersection of the measured outdoor dry-bulb (DB) air temperature and the indoor wet-bulb (WB) and dry-bulb (DB) air temperature.

Example: The intersection of a 95°F DB outdoor temperature and a 71°F WB/90°F DB indoor temperature yields 25°F as the required superheat.

205F21A.EPS

Figure 21 ♦ Table used for charging for proper superheat (fixed orifice) with R-22 (sheet 1).

Show Transparency 16 (Figure 22). Explain how to use a superheat charging calculator.

Show trainees how to charge a system by superheat.

Under your supervision, have trainees practice charging a system by superheat. Note the proficiency of each trainee. This laboratory corresponds to Performance Task 10.

Provide a superheat calculator for trainees to examine.

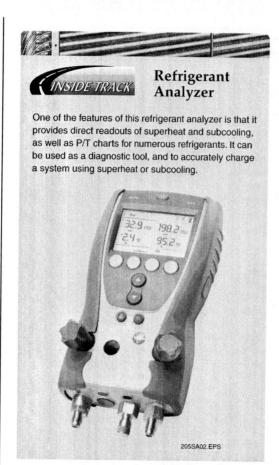

Inside Track — Refrigerant Analyzer

One of the features of this refrigerant analyzer is that it provides direct readouts of superheat and subcooling, as well as P/T charts for numerous refrigerants. It can be used as a diagnostic tool, and to accurately charge a system using superheat or subcooling.

205SA02.EPS

Step 2 Using the Superheat Charging Chart in *Figure 21, Sheet 1*, find the required superheat at the intersection point of the measured outdoor dry-bulb air temperature and the indoor wet-bulb temperature.

Step 3 Using the Required Suction-Line Temperature Chart in *Figure 21, Sheet 2*, find the required suction line temperature at the intersection of the superheat temperature and the measured suction line pressure.

Step 4 Compare the actual suction line temperature measured in the system to the required suction line temperature found in the chart to determine if an adjustment in the system refrigerant charge is needed. A tolerance of ±2°F is allowed before any adjustment is required. This may vary by manufacturer.

Step 5 If the measured superheat level is too high, add vapor refrigerant through the low-side service port with the compressor running until the superheat drops to the correct level.

Step 6 If the measured superheat level is too low, remove vapor refrigerant from the unit using a recovery unit and proper recovery procedures until the superheat rises to the correct level.

Step 7 If the air temperature entering the condenser coil changes, or the suction line pressure changes, repeat the procedure and charge to the new suction line temperature indicated by the chart.

Slide rule-type superheat charging calculators used to charge systems by the superheat method are available from equipment manufacturers. *Figure 22* shows the superheat calculator available from one such equipment manufacturer. As shown, complete instructions for use are printed on the calculator.

REQUIRED SUCTION-LINE TEMPERATURE (F) (ENTERING SUCTION SERVICE VALVE)

SUPERHEAT TEMP (F)	SUCTION PRESSURE AT SERVICE PORT (PSIG)								
	61.5	64.2	67.1	70.0	73.0	76.0	79.2	82.4	85.7
0	35	37	39	41	43	45	47	49	51
2	37	39	41	43	45	47	49	51	53
4	39	41	43	45	47	49	51	53	55
6	41	43	45	47	49	51	53	55	57
8	43	45	47	49	51	53	55	57	59
10	45	47	49	51	53	55	57	59	61
12	47	49	51	53	55	57	59	61	63
14	49	51	53	55	57	59	61	63	65
16	51	53	55	57	59	61	63	65	67
18	53	55	57	59	61	63	65	67	69
20	55	57	59	61	63	65	67	69	71
22	57	59	61	63	65	67	69	71	73
24	59	61	63	65	67	69	71	73	75
26	61	63	65	67	69	71	73	75	77
28	63	65	67	69	71	73	75	77	79
30	65	67	69	71	73	75	77	79	81
32	67	69	71	73	75	77	79	81	83
34	69	71	73	75	77	79	81	83	85
36	71	73	75	77	79	81	83	85	87
38	73	75	77	79	81	83	85	87	89
40	75	77	79	81	83	85	87	89	91

Find the required suction line temperature at the intersection of the superheat temperature (°F) and the measured suction line pressure (psig).

Example: The intersection of 22°F superheat and a suction line pressure of 70 PSIG yields a required suction temperature of 63°F.

205F21B.EPS

Figure 21 ♦ Table used for charging for proper superheat (fixed orifice) with R-22 (sheet 2).

Instructor's Notes:

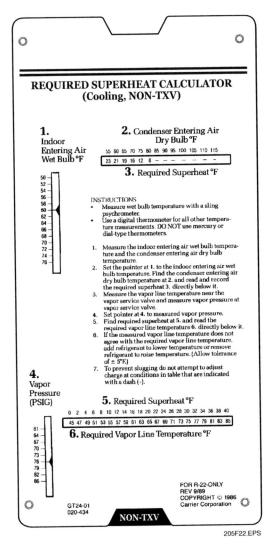

Figure 22 ♦ Superheat charging calculator.

measures the temperature of the subcooled refrigerant in the liquid line as a means of determining if the proper amount of liquid refrigerant is being applied to the TXV metering device.

If the liquid line temperature is incorrect, it can be corrected by adjusting the amount of refrigerant charge in the system.

The subcooling method is based on the existing pressure and temperature of the liquid line in an operating system. To charge by subcooling, use the following procedure.

Step 1 Determine the correct subcooling temperature specified by the equipment manufacturer. This can be found on the unit nameplate or in the manufacturer's service literature for the unit being serviced.

Step 2 Operate the system until it is stabilized, then measure the system liquid line temperature and pressure.

Step 3 Use the measured liquid line pressure and a standard pressure/temperature chart to find the saturated temperature of the refrigerant in the liquid line (*Figure 23*).

Step 4 Calculate the subcooling in the liquid line by subtracting the actual temperature measured in the liquid line from the saturated temperature. Compare the manufacturer's value for subcooling and the actual subcooling temperature that was calculated using the measured and saturated liquid line temperatures. If the measured subcooling temperature is within ±3°F of the required temperature, no adjustment in refrigerant charge is necessary.

Step 5 If the measured subcooling is too high, use a recovery unit to remove refrigerant vapor from the system. This will raise the temperature in the liquid line and lower the subcooling to the required level.

Step 6 If the measured temperature is too low, add refrigerant to the system.

Step 7 Check for correct superheat.

Slide rule-type subcooling charging calculators used to charge systems by the subcooling method are available from some of the major equipment manufacturers. *Figure 24* shows the subcooling calculator available from one such manufacturer. Complete instructions for use are printed on the calculator.

5.5.0 Charging by Subcooling

Devices like the thermostatic expansion valve (TXV) maintain a constant superheat over a wide range of load and charge conditions. Because of this, the superheat method of charging cannot be used with systems that contain TXVs or similar devices. In this situation, some manufacturers recommend measuring subcooling in the liquid line to obtain a correct charge.

Subcooling is the temperature removed from a liquid refrigerant after all the refrigerant has condensed into a liquid. The subcooling method

Show Transparency 17 (Figure 23). Explain the procedure for charging by subcooling.

Show Transparency 18 (Figure 24). Explain how to use a subcooling charging calculator.

Show trainees how to charge a system by subcooling.

Under your supervision, have trainees practice charging a system by subcooling. Note the proficiency of each trainee. This laboratory corresponds to Performance Task 11.

Provide a subcooling calculator for trainees to examine.

Have the trainees review Section 5.6.0–6.0.0.

Ensure that you have everything required for teaching this session.

Show Transparency 19 (Figure 25). Explain the procedure for charging using pressure charts.

Show trainees how to charge a system by using pressure charts.

Under your supervision, have trainees practice charging a system by using pressure charts. Note the proficiency of each trainee. This laboratory corresponds to Performance Task 12.

°F	R-22	R-134A	R-410A	°F	R-22	R-134A	R-410A
46	77.6	41.1	132.2	100	195.9	124.3	316.4
48	80.7	43.3	137.2	102	201.8	128.5	325.6
50	84.0	45.5	142.2	104	207.7	132.9	334.9
52	87.3	47.7	147.4	106	213.8	137.3	344.4
54	90.8	50.1	152.8	108	220.0	142.8	354.2
56	94.3	52.3	158.2	110	226.4	146.5	364.1
68	97.9	55.0	163.8	112	232.8	151.3	374.2
60	101.6	57.5	169.6	114	239.4	156.1	384.6
62	105.4	60.1	175.4	116	246.1	161.1	395.2
64	109.3	62.7	181.5	118	252.9	166.1	405.9
66	113.2	65.5	187.6	120	259.9	171.3	416.9
68	117.3	68.3	193.9	122	267.0	176.6	428.2
70	121.4	71.2	200.4	124	274.3	182.0	439.6
72	125.7	74.2	207.0	126	281.6	187.5	451.3
74	130.0	77.2	213.7	128	289.1	193.1	463.2
76	134.5	80.3	220.6	130	296.8	198.9	475.4
78	139.0	83.5	227.7	132	304.6	204.7	487.8
80	143.6	86.8	234.9	134	312.5	210.7	500.5
82	148.4	90.2	242.3	136	320.6	216.8	513.4
84	153.2	93.6	249.8	138	328.9	223.0	526.6
86	158.2	97.1	257.5	140	337.3	229.4	540.1
88	163.2	100.7	265.4	142	345.8	235.8	553.9
90	168.4	104.4	273.5	144	354.5	242.4	567.9
92	173.7	108.2	281.7	146	363.3	249.2	582.3
94	179.1	112.1	290.1	148	372.3	256.0	596.9
96	184.6	116.1	298.7	150	381.5	263.0	611.9
98	190.2	120.1	307.5				

SATURATION TEMPERATURE (FROM TP CHART)
— LIQUID LINE TEMPERATURE (MEASURED)

SUBCOOLING VALUE

EXAMPLE:
1. MEASURED LIQUID LINE PRESSURE (R-22) = 274 PSIG
2. SATURATED TEMPERATURE FROM CHART = 124°F
3. MEASURED LIQUID LINE TEMPERATURE = 114°F
4. SUBCOOLING = 124°F – 114°F = 10°F

205F23.EPS

Figure 23 ◆ Charging for proper subcooling.

5.6.0 Charging Using Pressure Charts

Charging pressure charts are used with units where the system charge weight is unknown. Pressure charts (*Figure 25*) are normally attached to the equipment or contained in the manufacturer's service literature. To charge using pressure charts, proceed as follows:

Step 1 Charge the system with an estimated amount of refrigerant, then operate the system until stabilized. Measure the following:
- Suction and discharge pressures
- Outdoor dry-bulb air temperature entering the condenser

Step 2 Using the pressure chart, find the required discharge line pressure at the intersection of the measured suction line pressure and condenser entering air temperature.

Step 3 If the measured discharge pressure matches the required discharge pressure, the refrigerant charge is considered correct and no adjustment in charge is necessary.

Step 4 If the measured discharge pressure is too high, remove excess refrigerant using a recovery unit and proper recovery procedures to lower the discharge pressure.

Step 5 If the measured discharge pressure is too low, add refrigerant vapor to raise the discharge pressure.

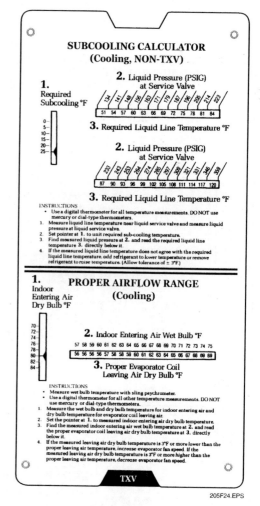

Figure 24 ◆ Subcooling calculator.

5.30 HVAC ◆ LEVEL TWO

Instructor's Notes:

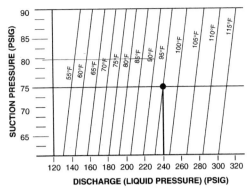

Find the required discharge line pressure at the intersection of the measured suction line pressure and condenser entering air temperature.

Example: the intersection of 75 psig suction line pressure and 95°F condenser entering air yields a required discharge line pressure of 240 PSIG.

Figure 25 ♦ Pressure chart.

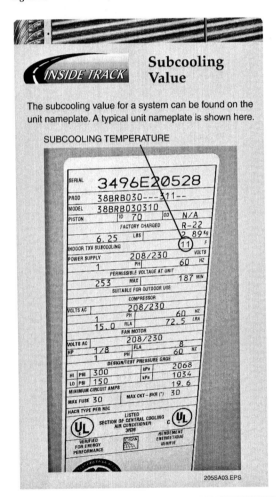

6.0.0 ♦ USING ZEOTROPE REFRIGERANTS

Refrigerants including zeotrope (near-azeotrope) blended refrigerants were discussed in detail in *HVAC Level One*. Some important points about zeotrope refrigerants relevant to the content of this module are briefly described here.

A zeotrope is a blend of two or more refrigerants (components). Currently, most zeotropes are a blend of three refrigerants. In a zeotrope refrigerant, unlike other refrigerants, the components retain their individual characteristics and evaporate or condense over a range of temperatures at a given pressure. This causes a zeotrope to have a different liquid saturation temperature than vapor saturation temperature for a given pressure. When working with zeotropes, you must understand the meaning of two terms: **fractionation** and **temperature glide**.

- *Fractionation* – When each of the refrigerants used in a blended refrigerant retains its own chemical properties, causing each of them to leak at a different rate if released into the atmosphere. The result is that the precise proportions of the refrigerants used in the blend are altered, thereby changing the properties of the blend to something other than that specified by the manufacturer.
- *Temperature glide* – A range of temperatures in which a zeotrope refrigerant will evaporate and condense for a given pressure.

After a leak in a system using a zeotrope refrigerant is repaired, do not charge the system with the refrigerant recovered from the system or use a partial charge of new refrigerant to top off the system. This is because fractionation has altered the chemical properties of the recovered refrigerant. Recharge the system with new refrigerant after the leak is repaired and the system has been evacuated. Charging with new refrigerant guarantees that the correct system operating temperatures and pressures can be achieved. It also protects the

Describe zeotropes.

Explain that zeotropes experience both fractionation and temperature glide.

Emphasize the importance of using new refrigerants for charging.

See the Teaching Tip for Section 6.0.0 at the end of this module.

Discuss special procedures used with zeotropes to ensure a proper charge.

Show Transparency 20 (Figure 26). Explain how the calculations are performed when using superheat or subcooling to charge a system with zeotrope refrigerants.

Have trainees perform sample calculations. Answer any questions they may have.

system from possible damage resulting from the use of a refrigerant that does not have the specific properties the manufacturer intended for use in the system.

When charging a system that uses a zeotrope refrigerant, the system should be charged using liquid refrigerant. This is necessary to avoid fractionation and to be sure that the proper refrigerant blend composition is charged into the system. If it is necessary to charge an operating system with refrigerant vapor, the refrigerant must be removed from the cylinder as a liquid, then fed from the cylinder to the low side of the system through a metering device to make sure that all of the liquid refrigerant is converted to vapor before entering the system.

When using the superheat or subcooling method of charging on a system that uses a zeotrope refrigerant, the calculation of superheat or subcooling is done in a slightly different manner than with other refrigerants. This is because of the temperature glide. The pressure-temperature relationship for a zeotrope refrigerant has two temperatures for a given pressure: the saturated vapor temperature and the saturated liquid temperature. When calculating superheat, be sure to use the saturated vapor temperature in the calculation. When calculating subcooling, be sure to use the saturated liquid temperature in the subcooling calculation. Using the wrong temperature will cause a superheat or subcooling calculation error and possible damage to the unit.

$$\begin{array}{r} \text{Actual suction line temperature} \\ - \text{Suction line saturated vapor temperature} \\ \hline = \text{Superheat} \end{array}$$

$$\begin{array}{r} \text{Liquid line saturated liquid temperature} \\ - \text{Actual liquid line temperature} \\ \hline = \text{Subcooling} \end{array}$$

Figure 26 shows an example of the superheat and subcooling calculations.

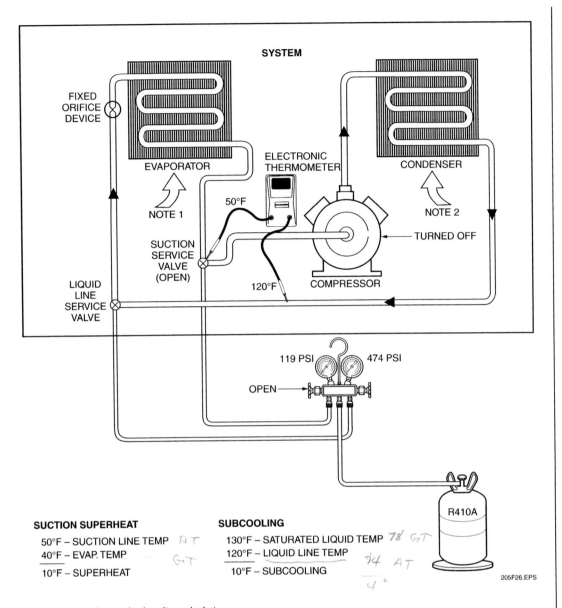

Figure 26 ◆ Superheat and subcooling calculations.

Have trainees complete the Review Questions, and go over the answers prior to administering the Module Examination.

Review Questions

1. Ultrasonic leak detection can best be used to check systems _____.
   a. under pressure or in a vacuum
   b. using liquid and vapor methods
   c. using high-side and low-side methods
   d. in highly noisy areas

2. After the application of a commercial bubble solution, a small leak may take _____ to form a bubble.
   a. several minutes
   b. 45 minutes
   c. 1 hour
   d. 4 hours

3. Because of the Clean Air Act Amendments, all CFC, HCFC, and HFC refrigerants must be _____.
   a. vented
   b. recovered
   c. charged in the liquid state
   d. charged in the vapor state

4. Recovery of a refrigerant means that the refrigerant is _____.
   a. removed from a system and stored in approved containers
   b. circulated through filtering devices to remove contaminants
   c. removed from a system and cleaned
   d. removed from a system and remanufactured to new refrigerant quality

5. Reclamation of a refrigerant means that it is _____.
   a. processed through a recycle unit
   b. processed so that it meets the standards for new refrigerant
   c. tested with a moisture/acid test kit
   d. mixed with new refrigerant

6. Recycling of a refrigerant means that it is _____.
   a. treated so that it is as dry and clean as new refrigerant
   b. run through a cleaning and decontamination filtering device
   c. mixed with new refrigerant
   d. recovered from a system more than once

7. A certified recovery unit used to recover refrigerant from a system containing 12 pounds of HCFC-22 refrigerant must be capable of evacuating the system to _____ in. Hg vacuum.
   a. 0
   b. 4
   c. 10
   d. 15

8. Contaminants that may be found in a poorly maintained system are _____.
   a. moisture
   b. air
   c. both moisture and air
   d. alcohol

9. Before using a recovery unit to process a different refrigerant than the last one used, you should _____.
   a. perform an acid/moisture test on the refrigerant
   b. install an external filter drier
   c. remove the vapors of old refrigerant
   d. drain and replace the compressor oil

10. The purpose of evacuating a system using a vacuum pump is to lower the system pressure so any moisture can _____.
    a. form
    b. condense
    c. boil off or vaporize
    d. be drained

11. If a system is evacuated to a level of 25,400 microns, the liquid water in the system will boil at _____ and above.
    a. 32°F
    b. 76°F
    c. 80°F
    d. 86°F

12. The most accurate way to measure a vacuum is with a _____.
    a. vacuum gauge
    b. gauge manifold set
    c. compound gauge
    d. wet-bulb indicator

Instructor's Notes:

Review Questions

13. The deep vacuum method of evacuation requires that the system be evacuated to a level of _____ microns.
    a. 500
    b. 1,000
    c. 2,000
    d. 50,000

14. During a vacuum test, the vacuum gauge indicates a continuous pressure rise. This usually means _____.
    a. a leak-tight, dry system
    b. a wet system
    c. acid has formed in the system
    d. there is a leak in the system

15. When charging refrigerant by weight, you can charge with _____.
    a. liquid refrigerant
    b. vapor refrigerant
    c. both liquid and vapor refrigerant
    d. nitrogen

16. When charging by superheat, the outdoor and indoor temperatures are used to find the _____.
    a. superheat in the suction line
    b. suction line pressure
    c. liquid line temperature
    d. required discharge line temperature

17. The subcooling charging method gives an indication of the amount of _____.
    a. pressure in the suction line
    b. refrigerant liquid at the expansion device
    c. refrigerant vapor just before the expansion device
    d. pressure in the liquid line

18. When charging using pressure charts, what is the required discharge line pressure if the suction pressure is 80 psig and the condenser entering air is 90°F? (Hint: Use the pressure chart in *Figure 25*.)
    a. 200 psig
    b. 220 psig
    c. 230 psig
    d. 240 psig

19. When each of the refrigerants used in a blended refrigerant leak at a different rate if released into the atmosphere, changing the precise proportions of the refrigerants used in the blend, the process is called _____.
    a. azeotrope reaction
    b. fractionation
    c. superheat
    d. subcooling

20. A range of temperatures in which a zeotrope refrigerant will evaporate and condense for a given pressure is called _____.
    a. fractionation
    b. subcooling
    c. temperature glide
    d. superheat

Summarize the major concepts presented in the module.

Administer the Module Examination. Record the results on Craft Training Report Form 200, and submit the results to the Training Program Sponsor.

Administer the Performance Test, and fill out Performance Profile Sheets for each trainee. If desired, trainee proficiency noted during laboratory sessions may be used to complete the Performance Test. Record the results on Craft Training Report Form 200, and submit the results to the Training Program Sponsor.

Summary

The four basic service procedures used to troubleshoot, repair, and/or maintain correct operation of a mechanical refrigeration system are leak detection, evacuation and dehydration, recovery, and charging.

These procedures are performed both when installing new systems and when servicing existing ones. Failure to perform any one of these procedures correctly can result in poor system operation and may even cause system failure. They must also be performed in order to make sure that the venting requirements of the Clean Air Act are not violated.

If the refrigerant charge is low, a leak is usually the cause. The leak can be located using electronic leak detecting devices or a leak detecting solution. Once located, the leak must be repaired, which generally involves opening the refrigerant piping system. Before the system is opened, the refrigerant charge must be recovered using approved equipment. Contaminants will enter an open system, so the system must be evacuated to remove moisture before he refrigerant charge is returned to the system.

Notes

Trade Terms Introduced in This Module

Fractionation: A term related to refrigerant blends; refers to the process by which each of the refrigerants in the blend leaks at a different rate if released into the atmosphere.

Reclamation: The remanufacture of used refrigerant to bring it up to the standards required of new refrigerant. Reclamation is not a field service procedure. It is a complicated process done only at reprocessing or manufacturing facilities.

Recovery: The removal and temporary storage of refrigerant in containers approved for that purpose. Recovery does not provide for any cleaning or filtration of the refrigerant.

Recycle: To circulate recovered refrigerant through filtering devices that remove moisture, acid, and other contaminants. This does not mean that it meets the purity standards for new refrigerants.

Temperature glide: A range of temperatures in which a zeotrope refrigerant will evaporate and condense for a given pressure.

Additional Resources and References

Additional Resources

This module is intended to be a thorough resource for task training. The following reference work is suggested for further study. This is optional material for continued education rather than for task training.

Refrigerant Service Techniques, 1993. Syracuse, NY: Carrier Corporation.

Figure Credits

INFICON, Inc., 205F01, 205F02, 205F09 (left), 205F15

Spectronics Corporation, 205F03

Topaz Publications, Inc., 205F05, 205SA03, 205F14 (photo)

Bacharach, Inc., 205F06

Robinair, a business unit of SPX Corporation, 205F07, 205F10, 205F16

Extech Instruments, 205F08 (photo)

Instrutech, Inc., 205F09 (right)

Emerson Climate Technologies, 205F17

TIF, a business unit of SPX Corporation, 205SA01

Carrier Corporation, 205F22, 205F24

Testo, Inc., 205SA02

MODULE 03205-07 — TEACHING TIPS

The following are suggested activities or instructional methods to help you teach the material in this module.

General

When you call on someone to answer a question, the rest of the class relaxes or even tunes out because they expect that the question and answer will take place only between you and the trainee you called on. Instead, use this technique to involve more trainees in answering questions and to keep them on their toes.

1. Ask trainees to define a term or explain a concept.
2. After one trainee has answered, ask a trainee seated nearby if the answer is right. Then ask whether a trainee in the back of the room agrees.
3. Ask trainees to explain why they think an answer is right or wrong.
4. Use the session to clear up incorrect ideas and encourage trainees to learn from their mistakes.

Sections 1.0.0 through 6.0.0

Quick Quiz

This Quick Quiz will familiarize trainees with the terms and definitions that are commonly used in leak detection, evacuation, recovery, and charging. You will need photocopies of the quiz provided in the following page. Trainees will need pencils. If you allow trainees to use the Trainee Guide, decrease the amount of time you give them to complete the quiz.

1. Make a photocopy of the quiz for each trainee.
2. Give trainees between 5 and 10 minutes to complete the quiz.
3. Go over the answers to the quiz.
4. Ask trainees if they have questions.

Answers to Quick Quiz

1. c
2. d
3. b
4. e
5. f
6. a
7. g

Quick Quiz *Leak Detection, Evacuation, Recovery, and Charging*

For each description listed, identify the term that the text best describes. Write the corresponding letter in the blank provided.

_____ 1. Removal and temporary storage of refrigerants is known as _____.

_____ 2. To _____ a recovered refrigerant is to filter it through a device that removes moisture, acid, and other contaminants.

_____ 3. The process where refrigerants are remanufactured to bring them up to required purity standards is known as _____.

_____ 4. Some manufacturers recommend the _____ method for checking or adjusting the charge of a system already in operation.

_____ 5. Systems that contain TXVs or similar devices must be charged using the _____ method.

_____ 6. Do not recharge a system using zeotrope refrigerants using recovered refrigerants because _____ has altered the properties of the recovered refrigerant.

_____ 7. The calculation of superheat for zeotrope refrigerants is slightly different due to _____.

   a. Fractionation
   b. Reclamation
   c. Recovery
   d. Recycle
   e. Superheat
   d. Subcooling
   e. Temperature glide

Section 2.0.0 *Identifying Leak Detection Equipment*

This exercise will familiarize trainees with various types of leak detection equipment. Trainees will need appropriate personal protective equipment, pencils, and paper. You will need to arrange workstations with a different type of leak detector at each station. Allow 30 to 45 minutes for this exercise. This exercise can correspond to Performance Tasks 1 and 2.

1. Describe various type of leak detection equipment. Discuss their advantages and disadvantages.
2. Have trainees rotate between the various workstations and identify the type of leak detector at each station. Have them note the advantages and disadvantages of each type.
3. If desired, have each trainee use each leak detector to leak test a pressurized system.
4. After all trainees have visited each station, have one trainee identify the type of leak detector at their station and identify its advantages and disadvantages. Answer any questions trainees may have.

Section 5.3.2 *Equipment Connections for Charging Vapor Refrigerant By Weight*

This Quick Quiz will familiarize trainees with equipment connections for charging vapor refrigerant by weight. You will need photocopies of the quiz provided on the following page. Trainees will need pencils. If you allow trainees to use the Trainee Module, decrease the amount of time you give them to complete the quiz.

1. Make a photocopy of the quiz for each trainee.
2. Give trainees between 5 and 10 minutes to complete the quiz.
3. Go over the answers to the quiz.
4. Ask trainees if they have questions.

Answers to Quick Quiz

1. closed
2. open
3. open
4. closed
5. open
6. open

Quick Quiz — Equipment Connections for Charging Vapor Refrigerant By Weight

Locate the valves listed below in the service equipment set up. Determine which valves are open and which are closed. For each valve shown, write open or closed in the corresponding blank provided.

1. Liquid line service valve
2. Suction service valve
3. Gauge manifold 1
4. Gauge manifold 2
5. Vapor valve
6. Cylinder valve

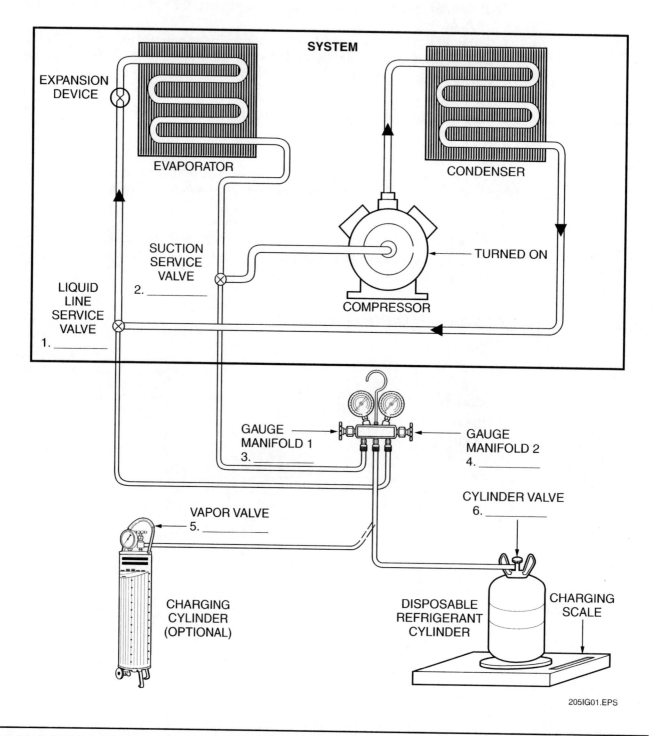

Section 6.0.0 *Using Zeotrope Refrigerants*

This exercise will familiarize trainees with using zeotrope refrigerants. Trainees will need pencils and paper. You will need to arrange for a manufacturer's representative to give a presentation on zeotrope refrigerants. Allow 20 to 30 minutes for this exercise.

Alternatively, have trainees research zeotrope refrigerants online and present the information to the class.

1. Tell trainees that a guest speaker will be presenting information on zeotrope refrigerants.
2. Have trainees brainstorm questions for the guest speaker before the speaker arrives.
3. Introduce the speaker. Ask the presenter to speak about different procedures used when charging systems with zeotrope refrigerants.
4. Have the trainees take notes and write down questions during the presentation.
5. After the presentation, answer any questions trainees may have.

MODULE 03205-07 — ANSWERS TO REVIEW QUESTIONS

Answer	Section
1. a	2.1.2
2. a	2.1.4
3. b	3.0.0
4. a	3.0.0
5. b	3.0.0
6. b	3.0.0
7. a	3.0.0; Table 1
8. c	4.0.0
9. d	3.2.3
10. c	4.1.1
11. c	4.1.1; Figure 8
12. a	4.1.2
13. a	4.2.1
14. d	4.2.1
15. c	5.3.0
16. a	5.4.0
17. b	5.5.0
18. c	5.6.0; Figure 25
19. b	6.0.0
20. c	6.0.0

CONTREN® LEARNING SERIES — USER UPDATE

NCCER makes every effort to keep these textbooks up-to-date and free of technical errors. We appreciate your help in this process. If you have an idea for improving this textbook, or if you find an error, a typographical mistake, or an inaccuracy in NCCER's Contren® textbooks, please write us, using this form or a photocopy. Be sure to include the exact module number, page number, a detailed description, and the correction, if applicable. Your input will be brought to the attention of the Technical Review Committee. Thank you for your assistance.

Instructors – If you found that additional materials were necessary in order to teach this module effectively, please let us know so that we may include them in the Equipment/Materials list in the Annotated Instructor's Guide.

Write: Product Development and Revision
National Center for Construction Education and Research
3600 NW 43rd St, Bldg G, Gainesville, FL 32606

Fax: 352-334-0932

E-mail: curriculum@nccer.org

Craft _____ Module Name _____

Copyright Date _____ Module Number _____ Page Number(s) _____

Description
_____
_____
_____
_____

(Optional) Correction
_____
_____
_____

(Optional) Your Name and Address
_____
_____
_____

Module 03206-07

Alternating Current

NCCER STANDARDIZED CRAFT TRAINING PROGRAM

The National Center for Construction Education and Research (NCCER) provides a standardized national program of accredited craft training. Key features of the program include instructor certification, competency-based training, and performance testing. The program provides trainees, instructors, and companies with a standard form of recognition through a National Craft Training Registry. The program is described in full in the *Guidelines for Accreditation*, published by NCCER. For more information on standardized craft training, contact the NCCER by writing us at 3600 NW 43rd St., Bldg. G, Gainesville, FL 32606; calling 352-334-0911; or emailing info@nccer.org. More information may be found at our website, www.nccer.org.

HOW TO USE THIS ANNOTATED INSTRUCTOR'S GUIDE

Each page presents two sections of information. The larger section displays each page exactly as it appears in the Trainee Module. The narrow column ties suggested trainee and instructor actions to each page and provides icons (detailed below) to call your attention to material, safety, audiovisual, or testing requirements. The bottom of each page includes space for your notes.

 The **Audiovisual** icon indicates an appropriate time to show a transparency or other audiovisual aid.

 The **Classroom** icon prompts you to define a term, stress a point, ask trainees to explain a concept, or give examples.

 The **Demonstration** icon directs you to show trainees how to perform tasks.

 The **Examination** icon tells you to administer the written module examination.

 The **Homework** icon is placed where you may wish to assign reading for the next class, assign a project, or advise trainees to prepare for an examination.

 The **Laboratory** icon is used when trainees are to practice performing tasks.

 The **Materials** icon is a reminder for you to gather materials needed for classes, labs, and testing.

 The **Performance Testing** icon tells you to administer a performance test or a portion thereof.

 The **Safety** icon is used to emphasize safety issues. It is often keyed to *Caution* and *Warning!* statements in the Trainee Module.

 The **Teaching Tip** icon indicates additional guidance is available, such as how to conduct an exercise, get the most educational value from a field trip, or encourage class participation. Teaching Tips may expand on a feature (*Think About It, Did You Know?*) or provide *Quick Quizzes* or similar exercises. You will be referred to the Teaching Tips section at the back of the module if there is additional material.

 The **Combination** icon indicates that the laboratory listed corresponds with a performance task. If desired, you can note the proficiency of the trainees during the laboratory, and use it to satisfy performance testing requirements.

PREPARATION

Before teaching this module, you should review the Objectives, Performance Tasks, Materials and Equipment List, and Module Outline. Be sure to allow ample time to prepare your own training or lesson plan and gather all required materials and equipment.

Alternating Current
Annotated Instructor's Guide

Module 03206-07

MODULE OVERVIEW

This module introduces the trainee to the production, transmission, and uses of alternating current in the HVAC field.

PREREQUISITES

Prior to training with this module, it is recommended that the trainee shall have successfully completed *Core Curriculum*; *HVAC Level One*; and *HVAC Level Two*, Modules 03201-07 through 03205-07.

OBJECTIVES

Upon completion of this module, the trainee will be able to do the following:

1. Describe the operation of various types of transformers.
2. Explain how alternating current is developed and draw a sine wave.
3. Identify single-phase and three-phase wiring arrangements.
4. Explain how phase shift occurs in inductors and capacitors.
5. Describe the types of capacitors and their applications.
6. Explain the operation of single-phase and three-phase induction motors.
7. Identify the various types of single-phase motors and their applications.
8. State and demonstrate the safety precautions that must be followed when working with electrical equipment.
9. Test AC components, including capacitors, transformers, and motors.

PERFORMANCE TASKS

Under the supervision of the instructor, the trainee should be able to do the following:

1. Identify the components used in a given AC circuit and explain their functions.
2. Identify types of single-phase and three-phase power distribution systems from electrical circuit diagrams.
3. Following applicable safety practices, test AC components, including transformers, capacitors, and motor windings.
4. Identify various types of AC motors from schematic diagrams.

MATERIALS AND EQUIPMENT LIST

Overhead projector and screen
Transparencies
Blank acetate sheets
Transparency pens
Whiteboard/chalkboard
Markers/chalk
Pencils and scratch paper
Appropriate personal protective equipment

Iron cores
Wire
Batteries
Various types of transformers
AC and DC power generation kits (optional)
Properly operating heating/cooling unit
Service entrance panel with circuit breakers
Fused disconnect box

continued

One each of the following for each team of two or three trainees:
- Split-phase induction motor
- Permanent split capacitor motor
- Capacitor-start motor
- Capacitor-start, capacitor-run motor
- Shaded-pole motor
- Tap wound motor
- Three-phase motor
- Various capacitors
- Inductor (coil)
- Various transformers
- Assortment of inoperative motors
- Oscilloscope
- Ohmmeter
- Megohmmeter (megger)
- Clamp-on ammeter
- In-line ammeter
- Capacitor analyzer
- Voltmeter or multimeter
- Copies of Quick Quiz*
- Module Examinations**
- Performance Profile Sheet**

* Located in the back of this module.
**Located in the Test Booklet.

SAFETY CONSIDERATIONS

Ensure that the trainees are equipped with appropriate personal protective equipment and know how to use it properly. This module requires that trainees work with electrical circuits. Ensure that all trainees are properly briefed on electrical safety procedures.

ADDITIONAL RESOURCES

This module is intended to present thorough resources for task training. The following reference works are suggested for both instructors and motivated trainees interested in further study. These are optional materials for continued education rather than for task training.

ARI Refrigeration and Air Conditioning; An Introduction to HVAC/R, 4th Edition, Prentice Hall.

General Training—Electricity (GTE), 1993. Syracuse, NY: Carrier Corporation.

HVAC Servicing Procedures, 1995. Syracuse, NY: Carrier Corporation.

Pocket Guide to Electrical Installations Under NEC 2002, Volumes I and II, 2001. Quincy, MA: National Fire Protection Association.

TEACHING TIME FOR THIS MODULE

An outline for use in developing your lesson plan is presented below. Note that each Roman numeral in the outline equates to one session of instruction. Each session has a suggested time period of 2½ hours. This includes 10 minutes at the beginning of each session for administrative tasks and one 10-minute break during the session. Approximately 7½ hours are suggested to cover *Alternating Current*. You will need to adjust the time required for hands-on activity and testing based on your class size and resources. Because laboratories often correspond to Performance Tasks, the proficiency of the trainees may be noted during these exercises for Performance Testing purposes.

Topic   **Planned Time**

Session I. Introduction, Transformers, Power Generation, and Using AC Power

    A. Introduction _____

    B. Transformers _____

    C. Power Generation _____

    D. Laboratory _____

        Trainees practice identifying types of power distribution systems from electrical circuit diagrams. This laboratory corresponds to Performance Task 2.

    E. Using AC Power _____

Session II. Induction Motors and Testing AC Components
    A. Induction Motors

    B. Laboratory

        Have trainees practice identifying various types of AC motors from schematic diagrams. This laboratory corresponds to Performance Task 4.

    C. Laboratory

        Have trainees practice identifying AC components and explaining their functions. This laboratory corresponds to Performance Task 1.

    D. Testing AC Components

Session III. Safety, AC Voltage on Circuit Diagrams, Review and Testing
    A. Safety

    B. Laboratory

        Following applicable safety practices, have trainees practice testing various AC components. This laboratory corresponds to Performance Task 3.

    C. Review

    D. Module Examination

        1. Trainees must score 70% or higher to receive recognition from NCCER.

        2. Record the testing results on Craft Training Report Form 200, and submit the results to the Training Program Sponsor.

    E. Performance Testing

        1. Trainees must perform each task to the satisfaction of the instructor to receive recognition from NCCER. If applicable, proficiency noted during laboratory exercises can be used to satisfy the Performance Testing requirements.

        2. Record the testing results on Craft Training Report Form 200, and submit the results to the Training Program Sponsor.

HVAC Level Two

03206-07

Alternating Current

Assign reading of Module 03206-07.

03206-07
Alternating Current

Topics to be presented in this module include:

1.0.0	Introduction	.6.2
2.0.0	Transformers	.6.2
3.0.0	Power Generation	.6.5
4.0.0	Using AC Power	.6.16
5.0.0	Induction Motors	.6.18
6.0.0	Testing AC Components	.6.24
7.0.0	Safety	.6.29
8.0.0	AC Voltage on Circuit Diagrams	.6.30

Overview

The electrical power used to operate HVAC equipment is alternating current, or AC power. Compressor motors, fan motors, and transformers depend on the constantly changing nature of AC to function. Because of its constantly changing nature, the principles you learned earlier about calculating voltage, current, resistance, and power cannot be directly applied to AC inductive circuits.

Instructor's Notes:

Objectives

When you have completed this module, you will be able to do the following:

1. Describe the operation of various types of transformers.
2. Explain how alternating current is developed and draw a sine wave.
3. Identify single-phase and three-phase wiring arrangements.
4. Explain how phase shift occurs in inductors and capacitors.
5. Describe the types of capacitors and their applications.
6. Explain the operation of single-phase and three-phase induction motors.
7. Identify the various types of single-phase motors and their applications.
8. State and demonstrate the safety precautions that must be followed when working with electrical equipment.
9. Test AC components, including capacitors, transformers, and motors.

Trade Terms

Alternator
Armature
Capacitor
Commutator
Dielectric
Effective voltage
Frequency
Fusible link
Hertz (Hz)
Induction

Induction motor
Inertia
Isolation transformer
Megohmmeter (megger)
Microfarad
Root-mean-square (rms) voltage

Rotor
Run capacitor
Run winding
Sinusoidal (sine) wave
Start winding
Stator
Synchronous speed
Torque
Turns ratio

Required Trainee Materials

1. Pencil and paper
2. Appropriate personal protective equipment

Prerequisites

Before you begin this module, it is recommended that you successfully complete *Core Curriculum*; *HVAC Level One*; and *HVAC Level Two*, Modules 03201-07 through 03205-07.

This course map shows all of the modules in the second level of the HVAC curriculum. The suggested training order begins at the bottom and proceeds up. Skill levels increase as you advance on the course map. The local Training Program Sponsor may adjust the training order.

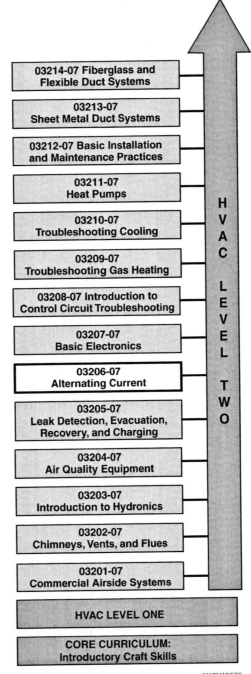

Ensure that you have everything required to teach the course. Check the Materials and Equipment list at the front of this module.

See the general Teaching Tip at the end of this module.

Explain that terms shown in bold are defined in the Glossary at the back of this module.

Show Transparency 1, Objectives, and Transparency 2, Performance Tasks. Review the goals of the module, and explain what will be expected of the trainee.

Review the modules covered in Level Two and explain how this module fits in.

1.0.0 ♦ INTRODUCTION

The public utilities that provide us with electricity produce AC voltage. DC voltage, when needed in AC-powered systems, is obtained by converting the available AC to DC. Batteries are also a common source of DC voltage.

Alternating current is defined as current that flows first in one direction, then in the opposite direction. The direction of current flow reverses at established intervals.

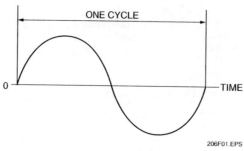

Figure 1 ♦ Alternating current sine wave.

The number of cycles that occur each second is known as the **frequency**; the standard frequency for AC systems in the United States is 60 **Hertz (Hz)**, or 60 cycles per second. A single cycle of AC voltage looks like the **sinusoidal (sine) wave** shown in *Figure 1*.

2.0.0 ♦ TRANSFORMERS

The transformer is the key component in an AC-powered system. A transformer generally consists of two or more coils of wire wound around a laminated iron core. One winding is connected to the power source and is called the primary winding. A separate winding is placed around the opposite side of the iron core and is called the secondary winding (*Figure 2*).

There is no physical electrical connection between the primary and secondary windings; electrical energy is transferred from the primary winding to the secondary winding by **induction**.

Transformers have no moving parts and therefore require very little maintenance. In operation, the primary circuit draws power from the source

Current Wars

After inventing the electric light bulb in 1879, Thomas Edison began work on a system for delivering electricity to homes and businesses. His system relied on direct current (DC)—electric current that always flows in one direction. However, DC transmission over long distances proved impractical. Transmitting direct current at the low voltages useful for lighting or motor operation required the use of thick, expensive copper wire. In fact, the service areas of Edison's DC generating stations were limited to about a square mile and mainly served the downtown areas of large cities. While Edison was pioneering DC distribution, electrical engineers in Europe were experimenting with alternating current (AC), which reverses direction at regular intervals.

The American businessman George Westinghouse saw the value in AC. High-voltage AC power could be distributed over longer distances using thinner, less expensive copper wires. Experts theorized that special devices (transformers) could step the voltage level up and down. Increasing the voltage level would allow it to be distributed across a wider region, while reducing the voltage level would enable the current to be used safely in homes and shops. Westinghouse hired a young electrical engineer named William Stanley, Jr., who developed the first effective transformer and demonstrated the first AC lighting system. Around the same time, the inventor Nikola Tesla filed patents for other devices run by AC. After Westinghouse bought these patents, a full-scale industrial war, known as the current wars, erupted. At stake was whether Edison's direct current or Westinghouse's alternating current would electrify America. Edison claimed that alternating current was extremely dangerous and called for outlawing the high voltages transmitted by AC. Westinghouse countered that transformers safely reduced AC voltages before they entered buildings.

Within ten years, the value of the alternating current system had been convincingly demonstrated. AC proved to be more practical and economical. Eventually, even Thomas Edison was forced to admit he had been wrong, and General Electric, the company he founded, began building and installing high-voltage AC transmission systems.

6.2 HVAC ♦ LEVEL TWO

Instructor's Notes:

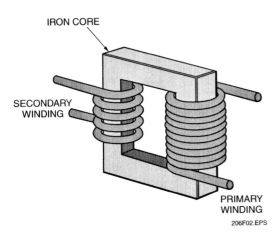

Figure 2 ◆ Basic components of a transformer.

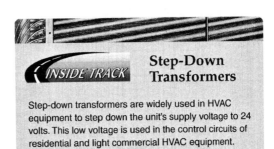

Step-Down Transformers

Step-down transformers are widely used in HVAC equipment to step down the unit's supply voltage to 24 volts. This low voltage is used in the control circuits of residential and light commercial HVAC equipment.

and the secondary circuit delivers the power to the load device. The power transferred from the primary winding to the secondary winding depends on the current flowing in the primary circuit. The amount of AC voltage induced in the secondary winding is directly related to the number of turns of wire in the secondary side as compared to the primary.

The output voltage of a transformer can be calculated by comparing the number of turns of wire in the primary winding to the number of turns in the secondary winding. If the number of turns in the secondary winding is twice that in the primary winding, its output voltage is twice that of the primary, and it is a step-up transformer. If the secondary winding has half the number of turns of the primary, then its output voltage is half that of the primary voltage, and it is a step-down transformer. This relationship is stated by the formula:

$$\frac{E_p}{E_s} = \frac{N_p}{N_s}$$

Where:
 E = voltage
 N = number of turns
 P = primary winding
 S = secondary winding

The relationship between the primary and secondary windings is called the **turns ratio**. It can be used to calculate unknown values in both step-up and step-down transformers. Some transformers contain more than one secondary winding, as shown in *Figure 3*. They are designed so that more than one voltage can be obtained from the secondary. Others have multi-tap primary windings, allowing them to be used with different levels of input voltage, as shown in *Figure 4*.

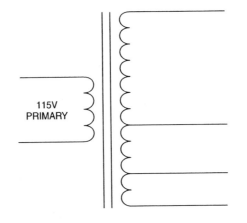

MULTIPLE-TAP SECONDARY WINDING

Figure 3 ◆ Multiple-tap secondary winding.

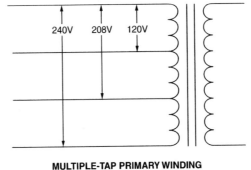

MULTIPLE-TAP PRIMARY WINDING

Figure 4 ◆ Multiple-tap primary winding.

Discuss the operation of transformers and explain how to calculate the output voltage of a transformer.

Show Transparencies 5 and 6 (Figures 3 and 4). Discuss multi-tap transformer windings.

Provide various types of transformers for trainees to examine.

Explain that the value of current changes inversely with voltage.

Discuss the operation and applications of isolation transformers.

Phasing of Transformers

It is sometimes necessary to connect two transformers in parallel to achieve an increased VA rating, in order to carry a specific load. When doing so, it is necessary that the two transformers be identical. Also, the secondary windings of the two transformers must be connected such that their output voltages are in phase; otherwise, the transformers can be damaged or destroyed. Some transformers are color-coded or numbered to indicate phasing. Others are not. A procedure for determining if the transformers are properly phased is described here.

Step 1 Connect the primary leads of the two transformers in parallel.

Step 2 Connect the secondary lead from one of the transformers to either secondary lead of the second transformer.

Step 3 Turn on the power, then use a multimeter to measure the AC voltage between the remaining two open secondary leads.

Step 4 If the meter reads zero volts, the connection is correct. Turn off the power and connect the two secondary leads.

Step 5 If the meter reads twice the rated voltage, turn off the power, then reverse the secondary leads.

Repeat the test. The meter should now read zero volts, indicating proper phasing.

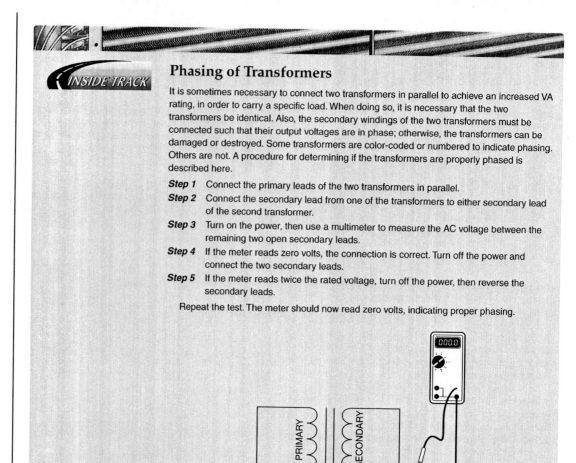

The value of current changes inversely with voltage. For example, if the voltage across a step-up transformer is doubled, the current is halved. Sometimes, a replaceable fuse, **fusible link**, or manually resettable external circuit breaker is used to protect the secondary winding from excessive current.

2.1.0 Isolation Transformers

Transformers with a one-to-one ratio are designed to deliver an output voltage equal to the input voltage. This special type of transformer is intended either to provide safety to a user or to minimize interference in certain electronic equipment. These transformers are known as **isolation transformers**.

Instructor's Notes:

2.2.0 Three-Phase Transformers

The types of transformers discussed previously are designed to work with single-phase voltage. With larger systems, such as those used in commercial buildings and factories, three-phase power is usually required. This requires special transformers.

Figure 5 shows two types of three-phase transformers. In these transformers, three legs of AC voltage are received on the primary windings, and equal voltage is induced into each of the secondary windings.

Three-phase power distribution arrangements and output voltages are discussed in detail later in this module.

2.3.0 Transformer Selection

Transformers are rated or sized according to the amount of power the secondary winding or circuit can handle. This power capability is expressed in VA (voltage times amperage). For example, if the secondary voltage of a transformer is 24V, and the amperage capacity of the circuit is 2A, the VA rating of that particular transformer is 48VA (24V × 2A = 48VA). Thus, the operating capability of that transformer is 48VA, which means that it should not be replaced by a transformer with a smaller VA rating. It can, however, be replaced by one with a larger VA rating.

Careful consideration should be given to the selection of transformers that are used to power a low-voltage control system. Inductive devices such as contactors, relays, solenoid valves, and motors require more power on startup than during steady operation. Thus, the transformer must be able to handle a startup current surge.

3.0.0 ◆ POWER GENERATION

Most electricity is produced by mechanically driven generators. The energy is generated by passing coils of wire through a magnetic field.

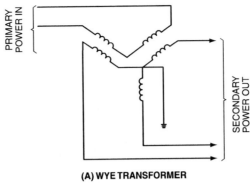
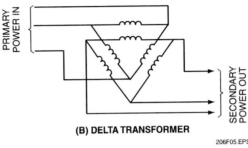

Figure 5 ◆ Three-phase transformers.

Generators can be designed to produce DC or AC.

The principles of the direct current, mechanically driven generator are illustrated in *Figure 6*. A simple horseshoe magnet forms a magnetic field as magnetic lines of force travel from one pole to the other. A loop of wire supported on a shaft is rotated within this field. As the loop turns, an electric current is generated and flows to terminals (segments) on the **commutator**. Two stationary contacts, or brushes, touch these commutator segments as the loop of wire rotates.

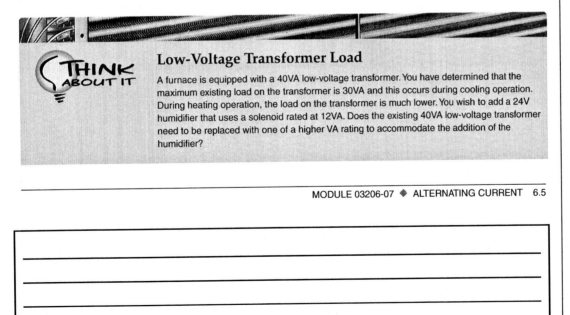

MODULE 03206-07 ◆ ALTERNATING CURRENT 6.5

Show Transparency 9 (Figure 7). Describe the operation of AC generators.

Show Transparency 10 (Figure 8). Explain how the position of the armature generates a sine wave.

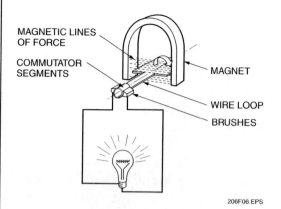

Figure 6 ♦ DC generation.

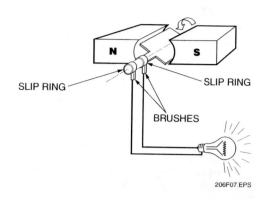

Figure 7 ♦ AC generation.

Current generated in the loop is carried by these brushes to an external circuit, where it is used as electrical energy.

The commutator is in two halves. When the loop of wire rotates one-half turn, each brush makes contact with the other half of the commutator. This is necessary to keep the current flowing in the same direction in the circuit; thus direct current is generated.

An alternating current generator, frequently called an **alternator**, operates on much the same principle as the DC generator (*Figure 7*). The rotation of the conductor through magnetic lines of force generates current in the conductor. Instead of being connected to segments on a commutator, the conductor ends are each connected to slip rings. The slip rings are fastened to the shaft and rotate with it. Stationary brushes are in contact with the slip rings. Current is carried through the brushes to the load.

During one revolution of the conductor, each side passes through the magnetic lines of force, first in one direction and then in the other. Therefore, current flows first in one direction and then the other, thus producing alternating current.

3.1.0 Sine Wave Generation

Figure 8 shows how the sine wave is generated. Voltage is induced as long as the conductor (**armature**) is moving through the magnetic field. The amount of voltage induced depends on the strength of the magnetic field and the angle and speed at which the conductor cuts the lines of force.

In *Figure 8*, the armature has been divided into a dark half and a light half for clarity. (No load is shown in this illustration.) In part A, the armature loop is moving parallel to the lines of force. The armature is not cutting through any lines of force, so minimum voltage is induced.

As the armature rotates toward the position shown in part B, it cuts more and more lines of force per second, as an increasingly strong voltage is induced. When it reaches the position shown in part B, the maximum voltage is attained because during its rotation, the armature has then cut through the maximum number of lines of force per second.

As the rotation proceeds toward the position shown in part C, the number of lines of force cut per second is reduced. The induced voltage decreases from its maximum value to zero voltage when the armature again reaches this position.

At this point in time (part C), the armature has rotated halfway around. The first half of the curve in *Figure 8* shows the varying strength of the voltage during this half turn. Part A, or the beginning position of the armature, can be labeled 0 degrees. When the armature has reached the C position, it has turned through 180 angular degrees. The voltage generated during that time is one alternation.

As the armature continues moving, it rotates through another 180 degrees, back to its starting point and again one alternation of voltage is generated. Two alternations are equal to 360 degrees of rotation, or one complete rotation.

In this basic generator scheme, with one north pole and one south pole affecting the armature, one complete rotation will always be required in order to generate one pair of alternations.

Figure 8 also shows the appearance of the AC voltage produced by this kind of generator. The curve shown is a sine wave. One sine wave corresponds to two alternations. With the two-pole generator just discussed, one sine wave is

6.6   HVAC ♦ LEVEL TWO

Instructor's Notes:

Explain the difference between peak voltage and effective (rms) voltage.

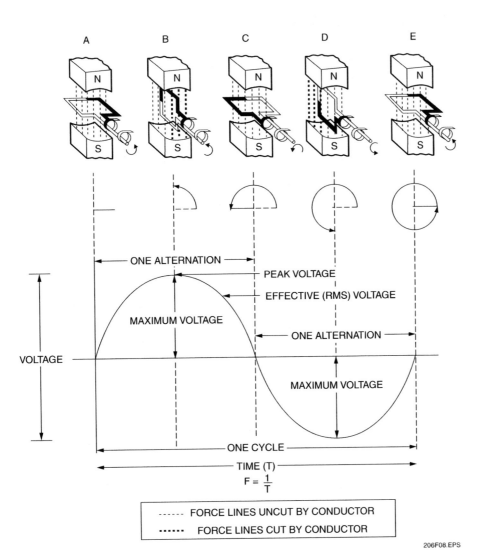

Figure 8 ♦ Sine wave generation.

generated by a complete 360-degree rotation of an armature.

The direction of the voltage induced in this generator (known as the polarity of the voltage) is reversed every alternation (180 degrees). This current flows through the load first in one direction, then in the other direction. The top half of the sine wave curve represents flow in one direction. The bottom half of the curve, lying under the horizontal axis, represents the flow of voltage in the opposite direction.

Because the amplitude of the voltage waveform varies between zero and peak, the amount of voltage available to do work—the **effective voltage**—is less than the peak voltage. Effective voltage, also known as **root-mean-square (rms) voltage**, is 0.707 times the peak voltage. Therefore, a peak of 169.71 volts produces a working voltage of 120 volts.

$$169.71 \times 0.707 = 120$$

If you know the effective voltage and want to determine the peak voltage, multiply the effective voltage by 1.414, which is the square root of 2.

Show Transparencies 11 and 12 (Figures 9 and 10). Discuss the relationship between armature speed and frequency.

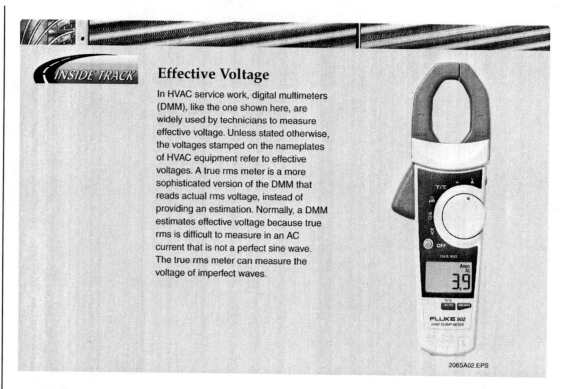

Effective Voltage

In HVAC service work, digital multimeters (DMM), like the one shown here, are widely used by technicians to measure effective voltage. Unless stated otherwise, the voltages stamped on the nameplates of HVAC equipment refer to effective voltages. A true rms meter is a more sophisticated version of the DMM that reads actual rms voltage, instead of providing an estimation. Normally, a DMM estimates effective voltage because true rms is difficult to measure in an AC current that is not a perfect sine wave. The true rms meter can measure the voltage of imperfect waves.

3.2.0 Frequency

The speed at which an armature rotates affects the frequency. If the speed of the armature increases, so does the frequency. If a single loop rotates in a two-pole field (produced by one north pole and one south pole), the current flows once in each direction to complete one pair of alternations, or one cycle, as the armature turns through 360 degrees (*Figure 9*). If there is one rotation per second, the frequency will be 1Hz. If the armature makes two complete rotations in one second, the frequency will be 2Hz (*Figure 10*). For three rotations per second, the frequency will be 3Hz, etc.

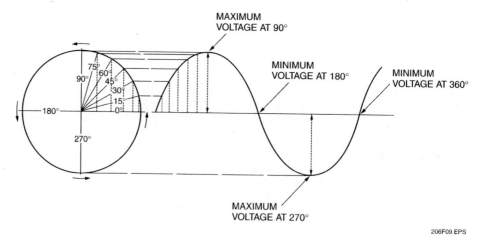

Figure 9 ◆ Sine wave plot.

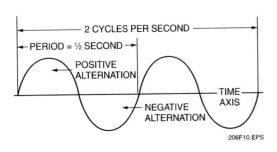

Figure 10 ♦ Voltage frequency of 2Hz.

The more pairs of poles are added to a generator, the higher the frequency will become, given the same armature speed. Generators that operate at low speeds need more pairs of poles than high-speed generators to provide voltage at the same frequency.

3.3.0 Single-Phase Power

Power generated at the power station is transmitted as a very high AC voltage, then stepped down to usable levels using transformers (*Figure 11*). The voltage received at the service entrance panel of a residence is usually about 240 volts.

Discuss frequency of commercial power. See the answer to the "Think About It" at the end of this module.

Show Transparency 13 (Figure 11). Discuss power distribution.

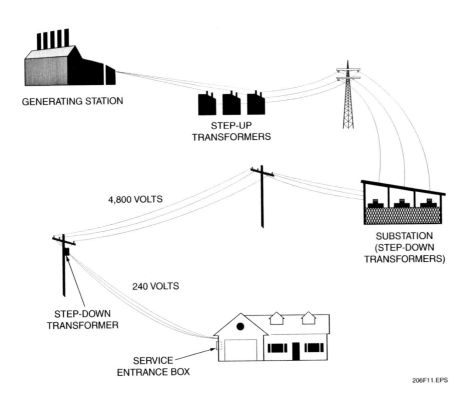

Figure 11 ♦ AC power distribution.

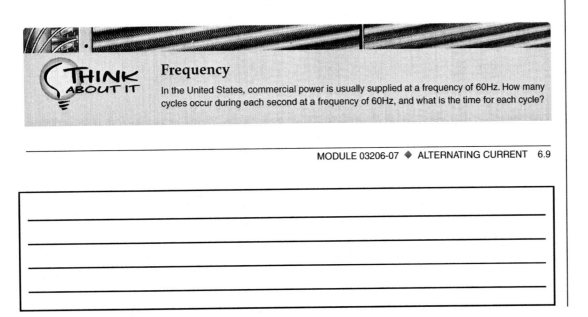

Frequency

In the United States, commercial power is usually supplied at a frequency of 60Hz. How many cycles occur during each second at a frequency of 60Hz, and what is the time for each cycle?

Show Transparency 14 (Figure 12). Explain that an Edison hookup is commonly used to deliver power to a residence.

Show Transparencies 15–17 (Figures 13–15). Describe a service entrance panel and the 240V and 120V circuits.

Have trainees locate the service entrance panel and related circuits at their home or apartment.

An Edison hookup (*Figure 12*), is a common wiring arrangement for a power transformer that delivers power to a residence. The neutral line is connected to the building ground, which may be a copper pipe that runs underground and/or a copper rod that has been driven into the ground.

The two hot legs (L) are commonly termed L1 and L2. The secondary winding of the transformer is center tapped and this leg is called the neutral (N). If the voltage is measured between L1 and L2, it would be found to be 240, whereas the voltage between either L1 or L2 and N would be 120. The neutral is electrically grounded at the power transformer. Other grounds may exist at the electric meter and the main circuit panel base, depending upon local codes.

After being stepped down, the power enters the structure through the service entrance (*Figure 13*). The service entrance is connected to a main fuse box or circuit breaker box. The circuits are broken down into a series of branch circuits within the main entrance box.

The 240V branch circuits (*Figure 14*) serve major appliances. The branch circuits are fused in both legs with fuses rated for the current draw of the appliance. The wire connecting these circuits to the appliance must be heavy enough to carry the load as determined by the size of the fuse or circuit breaker.

The 120V branch circuits (*Figure 15*) are also broken down into fused connections. Each branch of the 120V circuit is fused or protected in the hot leg only. The 120V branches are taken from each side of the 240V legs in order to balance the loads as closely as possible. Local electrical codes govern the area served by each branch.

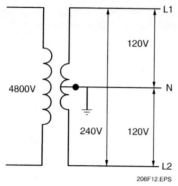

Figure 12 ♦ Edison hookup.

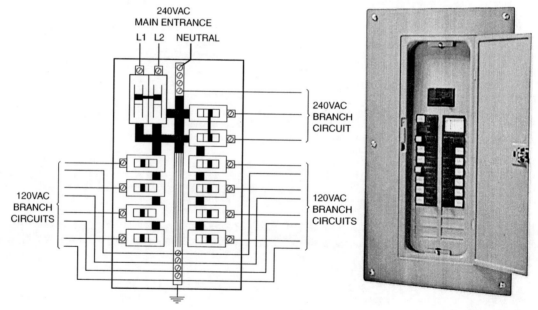

Figure 13 ♦ Service entrance panel.

6.10 HVAC ♦ LEVEL TWO

Instructor's Notes:

Why Do Power Companies Generate and Distribute AC Power Instead of DC Power?

The transformer is the key. Power plants generate and distribute AC power because it permits the use of transformers, which makes power delivery more economical. Transformers used at generation plants step the AC voltage up, which decreases the current. Decreased current allows smaller-sized wires to be used for the power transmission lines. Smaller wire is less expensive and easier to support over the long distances that the power must travel from the generation plant to remotely located substations. At the substations, transformers are again used to step AC voltages back down to a level suitable for distribution to homes and businesses.

There is no such thing as a DC transformer. This means DC power would have to be transmitted at low voltages and high currents over very large-sized wires, making the process very uneconomical. When DC is required for special applications, the AC voltage may be converted to DC voltage by using rectifiers, which make the change electrically, or by using AC motor-DC generator sets, which make the change mechanically.

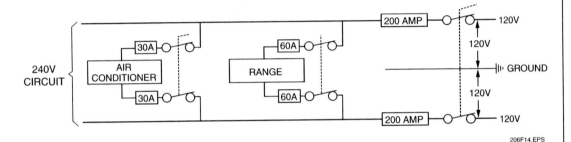

Figure 14 ♦ 240V branch circuits.

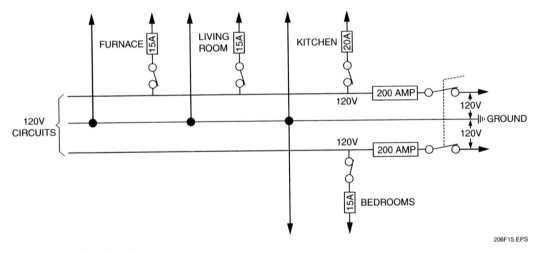

Figure 15 ♦ 120V branch circuits.

MODULE 03206-07 ♦ ALTERNATING CURRENT 6.11

Show Transparency 18 (Figure 16). Describe a typical air conditioner branch circuit.

Explain that the *National Electrical Code®* has specific requirements for proper locations of disconnect switches.

Provide a fused disconnect box for trainees to examine.

Explain that commercial HVAC equipment typically requires three-phase power.

Figure 16 shows how the voltage from a 240V branch circuit might be distributed to an air conditioning compressor and its control circuit.

The *National Electrical Code®* requires that a power disconnect be installed on or within sight of the unit. The disconnect has either a manual switch that can be locked out, or a removable fuse plug that disconnects power when pulled out of the unit.

3.4.0 Three-Phase Power

Single-phase power is adequate to supply residences and small commercial businesses. In commercial and industrial uses, the power demand is greater, especially where large electric motors are used. Motors larger than one horsepower are usually three-phase motors.

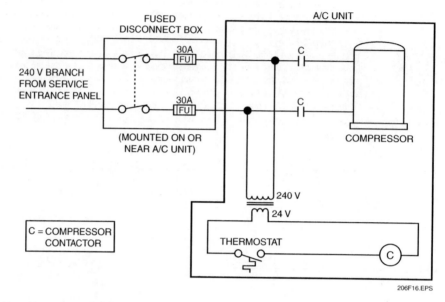

Figure 16 ♦ Air conditioner branch circuits.

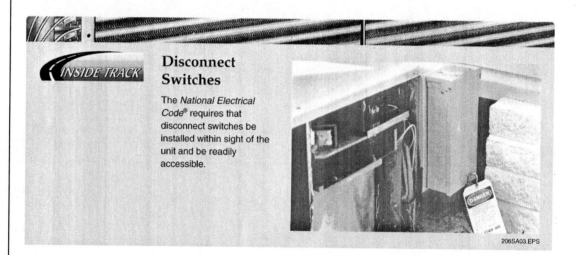

Disconnect Switches

The *National Electrical Code®* requires that disconnect switches be installed within sight of the unit and be readily accessible.

6.12 HVAC ♦ LEVEL TWO

The discussion of power generation has focused on a single conductor rotating in a magnetic field. If three rotating conductors are placed 120 degrees apart, three equal voltages are generated. As shown in *Figure 17*, they occur 120 degrees apart in time; in other words, they are 120 degrees out-of-phase with one another.

There are several ways in which three-phase power sources can be connected, depending on the voltage(s) and the amount of current required. The four-wire closed delta arrangement (*Figure 18*) uses three transformers to provide 120V single-phase, 240V three-phase, and 208V single-phase power (the 208V leg should never be loaded in a delta configuration). The four-wire open delta (*Figure 19*) uses two single-phase transformers to produce 240V three-phase power, and 120V, 240V, and 208V single-phase power. This connection method is not as stable as the closed delta; thus it is more difficult to keep the current in the three legs balanced.

In a three-phase system, a current imbalance in one leg can cause overheating in the other two legs. Three-phase systems therefore need to be checked periodically to make sure they are balanced. If the current is out of balance by more than 10 percent or the voltage is out of balance by more than 2 percent, the imbalance must be corrected. Sometimes the problem is at the source and must be corrected by the power company. This is a common occurrence following a power outage.

A four-wire wye system (*Figure 20*) may be used to supply power for industrial use where high voltage is required to run large machines.

Show Transparency 19 (Figure 17). Explain how three-phase power is generated.

Show Transparencies 20 and 21 (Figures 18 and 19). Discuss open and closed delta systems.

Show Transparency 22 (Figure 20). Discuss wye systems.

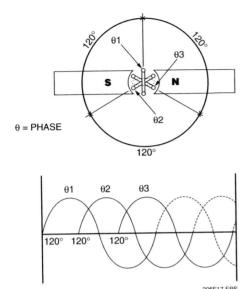

Figure 17 ◆ Three-phase voltage.

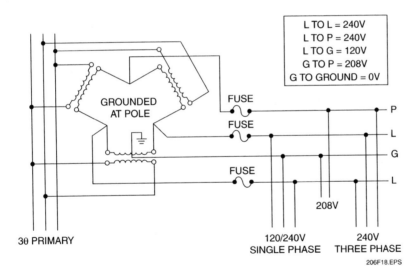

Figure 18 ◆ Four-wire closed delta.

Show Transparency 23 (Figure 21). Explain how three-phase power is supplied to a compressor circuit.

Show trainees how to identify single-phase and three-phase power distribution systems from electrical circuit diagrams.

Provide a selection of circuit diagrams and have trainees identify single-phase and three-phase power circuits from the diagrams. Note the proficiency of each trainee. This laboratory corresponds to Performance Task 2.

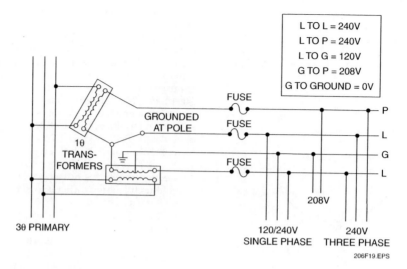

Figure 19 ♦ Four-wire open delta.

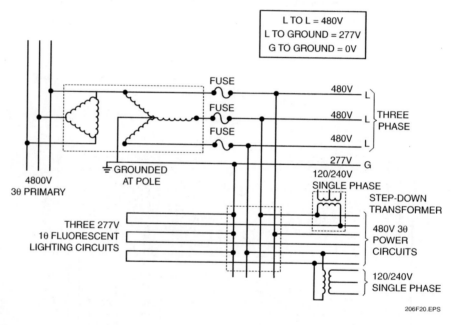

Figure 20 ♦ Four-wire wye.

This arrangement produces 480V three-phase power as well as the 277V single-phase power required for fluorescent lighting. *Figure 21* shows a wye-connected source supplying power to a three-phase compressor circuit. In air conditioning systems, single-phase compressor motors are used only up to a cooling capacity of about 60,000 Btuh. Three-phase compressor motors are used for large-capacity systems.

6.14 HVAC ♦ LEVEL TWO

Instructor's Notes:

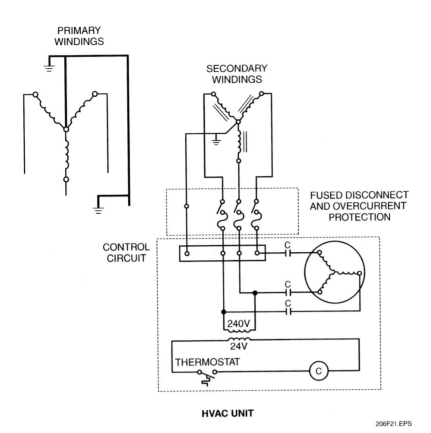

Figure 21 ♦ Three-phase compressor circuit.

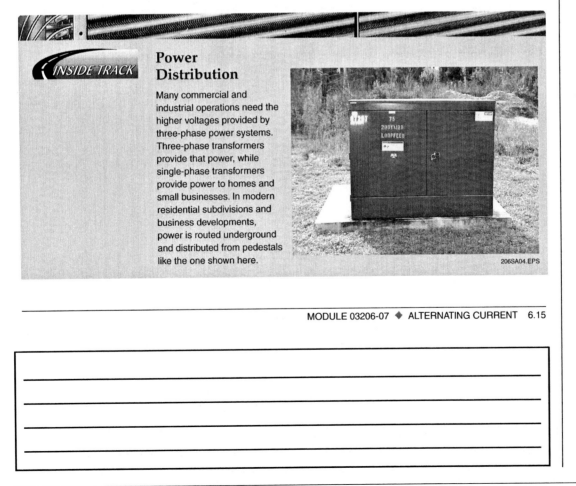

Power Distribution

Many commercial and industrial operations need the higher voltages provided by three-phase power systems. Three-phase transformers provide that power, while single-phase transformers provide power to homes and small businesses. In modern residential subdivisions and business developments, power is routed underground and distributed from pedestals like the one shown here.

Discuss the relationship between voltage and current imbalance in three-phase systems.

Describe the effects of current imbalance.

Show Transparency 24 (Figure 22). Explain that the current is in phase with the voltage in a resistive circuit.

3.5.0 Voltage and Current Imbalance in Three-Phase Systems

Voltage imbalance is very important when working with three-phase equipment. A small imbalance in the phase-to-phase voltages can result in a much greater current imbalance. With a current imbalance, the heat generated in motor windings and other inductive loads will be increased. Both current and heat can cause nuisance overload trips and may cause motor/equipment failures. For this reason, the voltage imbalance between any two legs of the voltage applied to a three-phase motor or system should not exceed 2 percent. If a voltage imbalance of more than 2 percent exists at the input to the equipment, the problem in the building or utility power distribution system should be corrected before operating the equipment.

Current imbalance between any two legs of a three-phase system should not exceed 10 percent. A current imbalance may occur without a voltage imbalance. This can happen when an electrical terminal, contact, etc. becomes loose or corroded, causing a high resistance in the leg. Since current follows the path of least resistance, the current in the other two legs will increase, causing more heat to be generated in the devices supplied by those legs.

Procedures for determining the voltage and current imbalance in a three-phase system are covered in detail in *Introduction to Control Circuit Troubleshooting*.

4.0.0 ◆ USING AC POWER

One important characteristic that distinguishes AC from DC is that AC acts differently in different kinds of circuits. HVAC components (such as motors) take advantage of this characteristic.

4.1.0 Resistive Circuits

Heating elements in an electric furnace are examples of resistive loads. In a resistive circuit, the current waveform occurs in phase with the voltage waveform (*Figure 22*). The amount of power consumed by the load is determined by the formula: P (power) = E (voltage) × I (current).

All electrical loads generate some heat. Electric heaters, stove burners, and clothes dryer elements take advantage of this fact. One unit of measure for heat is the British thermal unit (Btu).

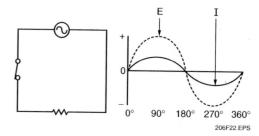

Figure 22 ◆ Resistive circuit.

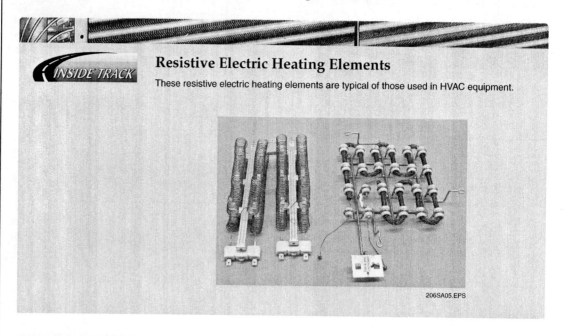

Resistive Electric Heating Elements
These resistive electric heating elements are typical of those used in HVAC equipment.

One watt represents 3.414 Btus per hour. A heater rated at 1,000 watts produces 3,414 Btus per hour. It is important to understand this principle because the heating load of a building is usually stated in Btus, while the capacity of a heater may be stated in watts.

4.2.0 Inductive Circuits

When alternating current flows through a coil of wire such as a motor winding, the magnetic field produced by one turn of the wire induces a voltage in adjacent turns. The voltage induced in this manner is opposite in polarity to the applied voltage, and therefore opposes current flow. For that reason, the current through a coil lags the voltage. In a purely inductive circuit, the current waveform will lag the voltage waveform by 90 degrees (*Figure 23*).

4.3.0 Capacitors

A **capacitor** (*Figure 24*) is an electrical storage device that charges and discharges as the applied voltage changes. It consists of two metal plates separated by an insulating material known as a **dielectric**. Its capacity, which is measured in **microfarads**, is determined by the size of the plates, the distance between the plates, and the type of dielectric material used.

Current will not flow through a dielectric material. When an AC voltage is applied across the plates of the capacitor, electrons will flow from one plate, through the load, and collect on the other plate, creating a charge with the same polarity as that of the input waveform. When the input waveform changes direction, the capacitor discharges (that is, the charge built up on one plate flows rapidly to the other plate), causing current flow through the load.

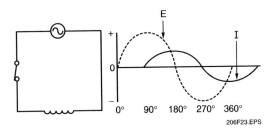

Figure 23 ◆ Inductive circuit.

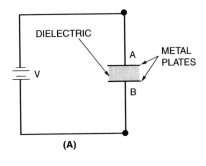

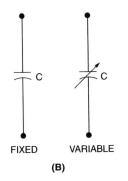

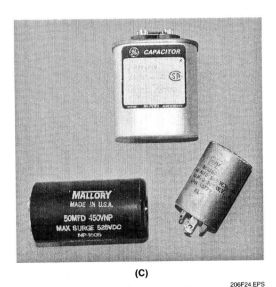

Figure 24 ◆ Capacitor operation.

Show Transparency 27 (Figure 25). Explain that the voltage waveform lags the current waveform by 90 degrees in a capacitive circuit.

Have the trainees review Sections 5.0.0–6.7.0.

Ensure that you have everything required for teaching this session.

List types of induction motors commonly used in HVAC systems.

Identify the components of a single-phase motor.

Because the voltage across the capacitor is created by current flowing from one plate to the other, the voltage across the capacitor lags the current. In a purely capacitive circuit, voltage lags current by 90 degrees. The phase shift is the opposite of that in an inductor, where voltage leads current. *Figure 25* shows the phase relationships in resistive, inductive, and capacitive circuits.

There are two common types of capacitors. In the oil-filled capacitor, paper soaked with an insulating fluid acts as the dielectric. This capacitor has large plates and a large amount of dielectric oil in order to dissipate heat. It can thus remain in the circuit all the time. The electrolytic capacitor is smaller and contains less dielectric material. It will overheat and be damaged if it is left in the circuit.

When electrolytic capacitors are used, they must be switched out of the circuit as soon as the circuit is started.

Capacitors and inductors and the phase shift they provide play an important role in helping to start single-phase **induction motors** and make them operate more efficiently.

5.0.0 ♦ INDUCTION MOTORS

AC induction motors are the primary load devices in HVAC equipment. They range in size from fractions of a horsepower to hundreds of horsepower. They drive compressors, blowers, condenser fans, ventilating fans, induced-draft fans, and humidifiers, among other things. Electromagnetism and induction are the keys to the operation of an AC motor.

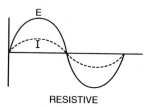
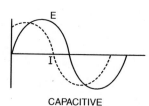
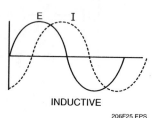

Figure 25 ♦ Phase relationships.

5.1.0 Single-Phase Motors

The main components of a single-phase motor are the **rotor** and the **stator** (*Figure 26*). The stator is fixed and the rotor turns. The motor accomplishes work by converting electrical energy into

Inside Track — Inductive Loads
The motors used in HVAC equipment are inductive loads.

ELI the ICE Man

Remembering the phrase "ELI the ICE man" is an easy way to remember the phase relationships that always exist between voltage and current in inductive and capacitive circuits. An inductive circuit is a circuit where there is more inductive reactance than capacitive reactance. The L in ELI indicates inductance. The E (voltage) is stated before the I (current) in ELI, meaning that the voltage leads the current in an inductive circuit.

A capacitive circuit is a circuit in which there is more capacitive reactance than inductive reactance. This is indicated by the C in ICE. The I (current) is stated before the E (voltage) in ICE, meaning that the current leads the voltage in a capacitive circuit.

Show Transparency 28 (Figure 27). Discuss the basic operation of a single-phase motor.

Explain that single-phase motors use various devices to overcome starting problems.

Show Transparency 29 (Figure 28). Discuss the operation and applications of a split-phase motor.

Provide single-phase and split-phase motors for trainees to examine.

where the stator field is exerting neither push nor pull, the motor will not restart when power is reapplied. Various ways of dealing with these problems are discussed next.

5.1.1 Split-Phase Motors

In the split-phase motor (*Figure 28*), an additional winding (**start winding**) is added to the stator. The start winding contains many turns of very fine wire; therefore the current buildup is slower than that of the **run winding**, which has fewer turns of wire. The phase difference between the two windings creates a torque that starts the rotor turning. A centrifugal switch opens to remove the winding from the circuit once the motor has started.

The rotor of this motor consists of conductive bars set in an iron core and connected at the ends by a copper ring. This type of rotor is called a squirrel cage rotor because the arrangement of the copper bars resembles a cage. Split-phase motors are used in pumps, oil burners, and other applications requiring ⅓ horsepower or less.

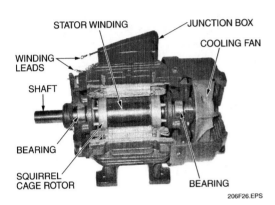

Figure 26 ◆ Basic parts of a motor.

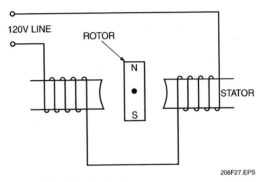

Figure 27 ◆ AC motor basics.

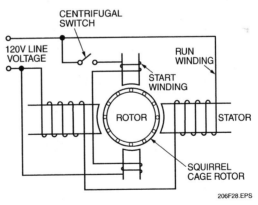

Figure 28 ◆ Split-phase AC motor.

mechanical energy that is delivered by attaching a mechanism such as a fan to the rotor shaft.

To operate the motor, voltage is applied to the stator winding (*Figure 27*), which contains many turns of wire. The magnetic field created by the current through the stator winding attracts or repels the rotor, causing it to turn. Because the polarity of the AC voltage applied to the stator is constantly changing, the relationship of the north and south magnetic poles of the stator and rotor constantly changes. This creates a rotating magnetic field that continuously pulls and pushes the rotor and keeps it turning.

The problem with single-phase motors is getting them started. First, it takes extra energy (**torque**) to overcome **inertia** and start the rotor turning. Second, if the rotor stops at a position

Take apart a split-phase motor to show trainees the windings.

Discuss start mechanisms for split-phase motors.

Show Transparency 30 (Figure 30). Discuss the operation and applications of permanent split capacitor motors.

Provide permanent split capacitor motors for trainees to examine.

Figure 29A shows a centrifugally actuated start winding switch that is closed when the motor is at rest. The start winding switch is opened and closed by the movement of an actuator disk against a contact lever. The actuator disk is moved back and forth on the rotor shaft by a centrifugal weight assembly. When the motor is at rest, springs retract the weights and cause the actuator disk to move toward the bearing at the end of the shaft. This pushes on the contact lever, closing the switch contacts. After the motor starts and reaches about 75 percent of its rated speed, the weights swing out against the spring tension and retract the contact disk from the contact lever. This opens the switch contacts, removing the starting winding from the circuit. If the switch does not open after starting, the motor will operate at a reduced speed until it overheats the starting winding and activates the thermal relay. *Figure 29B* shows the starting winding and the running winding.

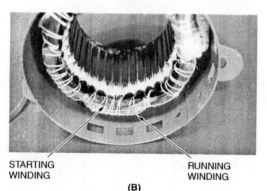

Figure 29 ◆ Split-phase induction motor.

INSIDE TRACK — Capacitors

Capacitors are widely used with motors in HVAC systems to facilitate motor starting and enhance operation.

5.1.2 Permanent Split Capacitor Motors

In this type of motor (*Figure 30*), the **run capacitor** provides a phase shift between the run and start windings that helps start the motor, then remains in the circuit to improve running efficiency. These motors are often used to drive blowers for applications that require ⅛ to ¾ horsepower.

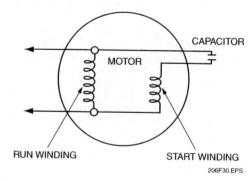

Figure 30 ◆ PSC motor

5.1.3 Capacitor-Start Motors

Capacitor-start motors range in size from fractional horsepower to as high as 10 horsepower. Their high starting torque makes them suitable for powering fans and some refrigeration compressors. The capacitor, which is wired in series with the start winding, provides high starting torque with relatively low starting current. It is an electrolytic capacitor, and must therefore be switched out of the circuit as soon as the motor reaches ⅔ to ¾ of its rated speed. Two methods are commonly used to do this. In the type shown in *Figure 31*, a centrifugal switch is used.

In the more common type shown in *Figure 32*, a starting relay is used. When operating, all electric motors have some electrical voltage generating capacity resulting from induced voltages generated in the motor stator windings by the motion of the rotating rotor. The voltage generated is known as back electromotive force (back EMF). It is important to know that the back EMF generated in a motor normally has a much higher potential than the line voltage applied to the motor. For example, a motor being driven by a 230V line source can generate back EMF voltages of over 400 volts.

Show Transparencies 31 and 32 (Figures 31 and 32). Discuss the operation and applications of capacitor-start motors. Explain that these motors may incorporate either a centrifugal switch or a start relay to control the start capacitor.

Provide capacitor-start motors for trainees to examine.

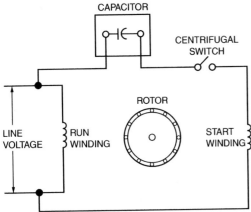

Figure 31 ◆ Capacitor-start motor with centrifugal switch.

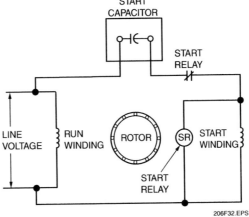

Figure 32 ◆ Capacitor-start motor with start relay.

Current Relays

Current-sensing relays are used to monitor electrical currents. The conductors are fed through the "donut," which senses the current flow. When the current flow reaches a preset level, the relay trips.

MODULE 03206-07 ◆ ALTERNATING CURRENT 6.21

Show Transparency 33 (Figure 33). Discuss the operation and applications of capacitor-start, capacitor-run motors.

Show Transparency 34 (Figure 34). Discuss the operation and applications of shaded-pole motors.

Provide capacitor-start, capacitor-run and shaded-pole motors for trainees to examine.

Explain that the speed of a motor is determined by the number of windings and the applied voltage.

Motor Run and Start Capacitors

The run capacitors used in HVAC motor circuits are of the oil-filled type and remain in the circuit at all times. The start capacitors typically used to start single-phase compressors remain in the circuit for a fraction of a second to help the motor start. A special start relay is commonly used whenever a start capacitor is installed in a system. The start relay functions to quickly remove the start capacitor from the system before it overheats. These relays are position-sensitive and must be placed with a certain side up.

It is very important that capacitors and start relays be sized correctly. If you must install these components to help a compressor start, you must use the exact capacitor and start relay specified by the compressor manufacturer.

In a capacitor-start motor, the back EMF voltage, not line voltage, is used to energize and de-energize the start relay. By design, the start relay will energize at some predetermined back EMF voltage level that is always higher than the line voltage.

When the motor is first turned on, the start relay is de-energized and the start capacitor and start winding are connected in the circuit through the normally closed contacts of the start relay. As the motor picks up speed, the back EMF in the start winding builds up and eventually causes the start relay to energize. This opens the relay contacts and removes the capacitor and start winding from the circuit. As long as the motor runs at or above this speed, the back EMF generated in the start winding remains applied across the parallel-connected start relay coil. This keeps the relay energized and the start capacitor and winding out of the circuit.

5.1.4 Capacitor-Start, Capacitor-Run Motor

The capacitor-start, capacitor-run motor (*Figure 33*) is used to drive refrigerant compressors. It combines high starting torque with smooth, quiet, efficient operation. Like the capacitor-start motor, the start capacitor is switched out of the circuit when the motor comes up to speed. The motor then runs as a permanent split capacitor motor.

5.1.5 Shaded-Pole Motor

In the shaded-pole motor (*Figure 34*), the rotor pivots within two pairs of stator windings. A groove in the stator pole separates a small portion of the stator from the remainder. A metal band (shading coil) is placed around the smaller portion. This band causes a slight phase shift, which is enough to provide torque to start the motor. Shaded-pole motors are used to drive small fans and pumps. This type of motor will turn in the direction of the shading coil, which in this case is clockwise.

5.1.6 Multi-Speed Motors

The speed of a motor, which is measured in revolutions per minute (rpm), is determined by the number of stator windings and the frequency of

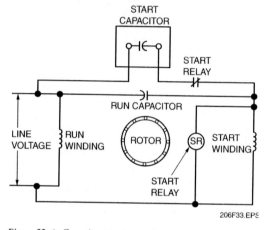

Figure 33 ♦ Capacitor-start, capacitor-run motor.

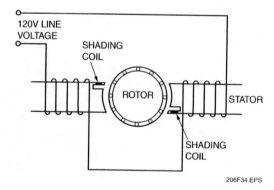

Figure 34 ♦ Shaded-pole motor.

Inside Track

Capacitor-Start, Capacitor-Run Motors

These motors run quietly and smoothly and have the high starting torque often needed for compressors. Their construction is similar to that of a split-phase motor; they use a centrifugal switch to remove the start capacitor from the starting winding, while leaving the run capacitor connected. These motors have a higher power factor than ordinary split-phase motors.

the applied voltage. The maximum speed at which a motor can run is known as its **synchronous speed**. Adding a load to a motor causes some slippage, or inefficiency, in the operation of the motor. Therefore, motors are generally rated at 95 to 97 percent of their synchronous speed.

The speed of some motors can be changed using taps on the stator winding (*Figure 35*). Instead of using the entire stator coil, a portion is used—the larger the portion, the lower the speed. This creates more impedance, resulting in lower current, which reduces the power available. This method allows a single motor to be used in different applications. In cases where the same air handler is used for both heating and cooling, the control circuit will automatically select low speed for heating or high speed for cooling.

5.2.0 Three-Phase Motors

Three-phase motors are used primarily in large commercial systems. They require three-phase voltage, which is not readily available in residential areas. Three-phase motors offer several important advantages over single-phase motors:

- They have a higher starting torque and require no special starting equipment such as capacitors, start relays, or start windings. Because the stator windings are 120 degrees out of phase, at least one of the windings is always applying torque to the rotor.

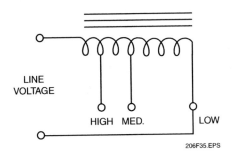

Figure 35 ♦ Speed taps.

Explain the difference between operating speed and synchronous speed.

Show Transparency 35 (Figure 35). Explain that changing speed taps on the motor winding can alter a motor's speed.

List the advantages of three-phase motors.

Provide three-phase motors for trainees to examine.

Show Transparency 36 (Figure 36). Explain that the winding arrangement of the motor depends on the application.

Provide a selection of circuit diagrams. Show trainees how to identify the types of AC motors from the diagrams.

Have trainees practice identifying AC motors from the diagrams. Note the proficiency of each trainee. This laboratory corresponds to Performance Task 4.

Provide a selection of AC components, including transformers, capacitors, and motors. Show trainees how to identify the components and explain their functions.

Have trainees practice identifying components used in AC circuits and explain their functions. Note the proficiency of each trainee. This laboratory corresponds to Performance Task 1.

List the types of equipment used to test AC circuits.

Discuss capacitor analyzers and their applications.

Describe applications of wattmeters.

- The direction of rotation can be easily reversed by reversing any two of the three supply voltage connections.
- They run very smoothly.
- There is less running pulsation because at least one stator winding is always applying torque to the rotor.

Three-phase motor stators are connected in either a delta or wye configuration, depending on the application (*Figure 36*). A delta-connected motor provides more current, and therefore more starting torque, but also consumes more power than the wye-connected motor. The motor's voltage rating is also a factor in selecting the three-phase hookup.

6.0.0 ◆ TESTING AC COMPONENTS

Most components of AC circuits can be tested with the standard electrical test meters—the voltmeter, ammeter, and ohmmeter. A multimeter, or VOM, will handle most voltage and resistance readings and a clamp-on ammeter will suffice for current readings. In addition to these instruments, a capacitor analyzer, wattmeter, and **megohmmeter (megger)** will sometimes be needed.

6.1.0 Capacitor Analyzer

A capacitor analyzer (*Figure 37*) is a special-purpose instrument used to test capacitors. This analyzer will check capacitors for current leakage, insulation breakdown, capacity, shorts, and opens. An example of its use is to determine the value of a capacitor when you are unable to read the value. This sometimes happens when replacing defective start or run capacitors. Many digital multimeters have a capacitor test feature.

6.2.0 Wattmeter

Rather than measure voltage and current and then calculate power, a wattmeter can be connected into a circuit to measure power directly. Not only does a wattmeter simplify power measurements, it also has two other advantages.

First, voltage and current in an AC circuit are not always in phase; current sometimes either leads or lags the voltage. When this happens, multiplying the voltage times the current yields apparent power, not true power. However, the wattmeter takes this into account and always indicates true power.

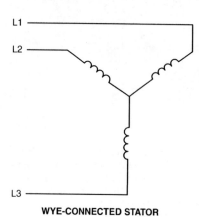

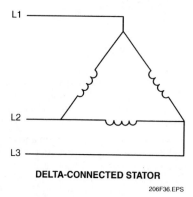

Figure 36 ◆ Three-phase motors.

Figure 37 ◆ Capacitor analyzer.

Second, voltmeters and ammeters consume power. The amount consumed depends on the levels of the voltage and current in the circuit, and cannot be accurately predicted. Therefore, very accurate power measurements cannot be made by measuring voltage and current and then calculating power. Some wattmeters compensate for internal power losses so that only the power dissipated in the circuit is measured. If the wattmeter does not compensate for these losses, the power that is dissipated is sometimes marked on the meter or else can be easily determined so that a very accurate measurement can be made. Typically, the accuracy of a wattmeter is within ±1 percent.

The basic wattmeter consists of two stationary coils connected in a series and one movable coil (*Figure 38*). The movable coil, wound with many turns of fine wire, has a high resistance. The stationary current and voltage coils, wound with a few turns of a larger wire, have a low resistance. The interaction of the magnetic fields around the different coils will cause the movable coil and its pointer to rotate in proportion to the voltage across the load and the current through the load. Thus, the meter indicates E times I, or power.

6.3.0 Megohmmeter (Megger)

Normally, an ohmmeter is not used to measure extremely high resistances, such as those involving conductor insulations, insulation between motor or transformer windings, and so on. To adequately test such high resistances, it is necessary to use a much higher potential than is furnished by the battery of an ohmmeter. Test voltages ranging from 50V to 5,000V can be supplied by the megohmmeter, or megger. Meggers measure resistance in megohms (equal to one million ohms). There are three types of meggers: hand, battery, and electric.

The megger has two coils (*Figure 39*). Coil A is in series with resistor R_2 across the output of the generator. This coil is wound so that it causes the pointer to move toward the high-resistance end of the scale when the generator is operating. Coil B is in series with R_1 and the unknown resistance (R_x) to be measured. This coil is wound so that it causes the pointer to move toward the low- or zero-resistance end of the scale when the generator is operating.

When an extremely high resistance appears across the input terminals of the megger, the current through coil A is greater, causing the pointer to deflect toward infinity. Conversely, when a relatively low resistance appears across the input

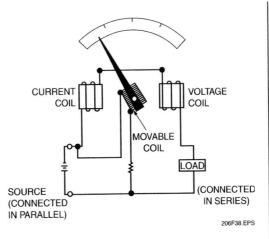

Figure 38 ◆ Wattmeter schematic.

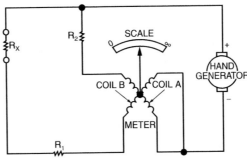

Figure 39 ◆ Megger schematic.

terminals, the current through coil B is greater, and causes the pointer to deflect toward zero.

In this type of megger, hand generators are used to produce the test voltage. To avoid excessive test voltages, most hand meggers are equipped with friction clutches. When the generator is cranked faster than its rated speed, the clutch slips and the generator speed and output voltage are maintained at their rated values.

Newer meggers use the same operational principles. Instead of having a scaled meter movement, these meters give the value of resistance in a digital readout. The digital readout makes reading the measurement much easier and helps eliminate errors.

Explain that the generator voltage is present on the test leads of a megger. Emphasize the importance of not touching the test leads while the megger is being used.

Discuss safety precautions associated with the use of meggers.

Explain that it is often necessary to know the conditions of an electrical circuit over a period of time.

 WARNING!
When a megger is used, the generator voltage is present on the test leads. This voltage could be hazardous to you or the equipment you are testing. Never touch the test leads while the tester is being used. Isolate the item you are testing from the circuit before using the megger.

6.3.1 Safety Precautions

When you use a megger, you could be injured or cause damage to the equipment that you are working on if the following minimum safety precautions are not observed:

- Use meggers on high-resistance measurements only (such as insulation measurements or to check two separate conductors in a cable).
- Never touch the test leads while the handle is being cranked.
- De-energize the circuit and verify that it is off before connecting the meter.
- Disconnect the item being checked from other circuitry, if possible, before using the meter.

6.4.0 Recording Instruments

It is often necessary to know the conditions that exist in an electrical circuit over a period of time to determine such things as peak loads, voltage fluctuations, etc.

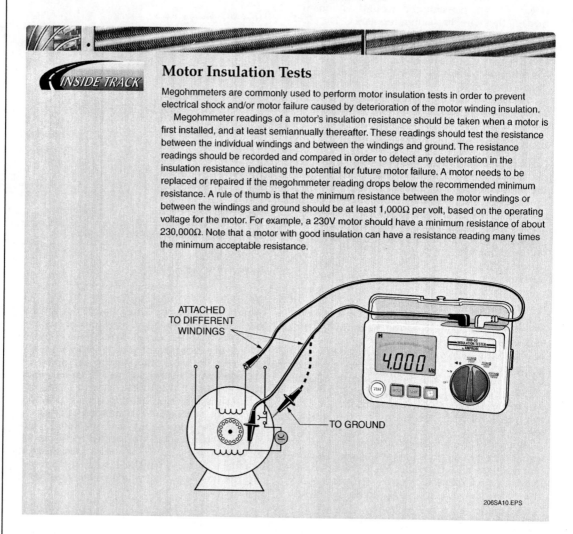

Motor Insulation Tests

Megohmmeters are commonly used to perform motor insulation tests in order to prevent electrical shock and/or motor failure caused by deterioration of the motor winding insulation.

Megohmmeter readings of a motor's insulation resistance should be taken when a motor is first installed, and at least semiannually thereafter. These readings should test the resistance between the individual windings and between the windings and ground. The resistance readings should be recorded and compared in order to detect any deterioration in the insulation resistance indicating the potential for future motor failure. A motor needs to be replaced or repaired if the megohmmeter reading drops below the recommended minimum resistance. A rule of thumb is that the minimum resistance between the motor windings or between the windings and ground should be at least 1,000Ω per volt, based on the operating voltage for the motor. For example, a 230V motor should have a minimum resistance of about 230,000Ω. Note that a motor with good insulation can have a resistance reading many times the minimum acceptable resistance.

6.26 HVAC ♦ LEVEL TWO

It may be neither practical nor economical to assign a worker to watch an indicating instrument and record its readings. An automatic recording instrument can be connected to take continuous readings, and the record can be collected for review and analysis.

The term recording instrument describes many instruments that make a permanent record of measured quantities over a period of time. Recording instruments can be divided into three general groups:

- Instruments that record electrical quantities, including potential difference, current, power, resistance, and frequency
- Instruments that record nonelectrical quantities by electrical means (such as a temperature recorder that uses a potentiometer system to record thermocouple output)
- Instruments that record nonelectrical quantities by mechanical means (such as a temperature recorder that uses a bimetallic element to move a pen across an advancing strip of paper)

Recording instruments are basically the same as the indicating meters already discussed, but they have recording mechanisms attached to them. They are generally made of the same parts, use the same electrical mechanisms, and are connected in the same way. The only basic difference is the permanent record.

Strip chart recorders are the most widely used recording instruments for electrical measurement. Their name comes from the fact that the record is made on a strip of paper, usually four to six inches wide and perhaps up to 60' long.

Strip chart recorders offer several advantages in electrical measurement. The long charts allow the recording to cover a considerable length of time with little attention. Also, strip chart recorders can be operated at a relatively high speed if very detailed records are needed.

6.5.0 Checking Inductive Loads

The most common inductive load is the stator winding of a motor. Stator windings typically have a very low resistance; one or two ohms is fairly common. The resistances will vary from manufacturer to manufacturer and from one type of motor to another. The larger the motor, the smaller the resistance. The start winding of a single-phase motor will have a higher resistance than the run winding; perhaps three to four times higher. The winding resistance is not provided on the motor nameplate, so the resistance values themselves may not be of much use unless you are familiar with the particular motor.

An open winding will be readily apparent (*Figure 40*), as will a severe short. A partial short may only be recognizable if you are familiar with the motor or if there is an unusual difference between the resistances of the start and run windings.

A three-phase motor is easier to check; the resistances of the three windings should be the same.

When resistance and continuity checks are made, one end of the target component or series of components must be disconnected from the circuit. Otherwise, the ohmmeter circuit current might read the resistance of a parallel circuit. In the upper section of *Figure 40*, for example, the meter would read the resistance of the parallel path because it offers much less resistance than the open coil. That is why the target coil is disconnected from the circuit.

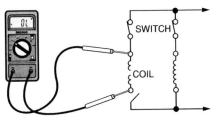

INFINITE READING = COIL OPEN

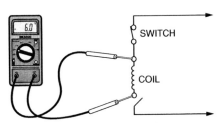

NORMAL READING = COIL GOOD

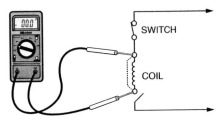

BELOW NORMAL READING = COIL SHORTED

Figure 40 ♦ Inductive load check.

Show Transparency 40 (Figure 41). Emphasize the importance of discharging capacitors before testing them. Show trainees how to discharge a capacitor.

Explain that capacitors contain a dangerous residual charge and must be handled with caution.

Explain how to test a capacitor.

Show trainees how to test a capacitor.

Explain that some older capacitors may contain PCBs and must be handled with caution and disposed of properly.

Show Transparency 41 (Figure 42). Describe how to check the continuity of a fuse. List the resistance values for various conditions.

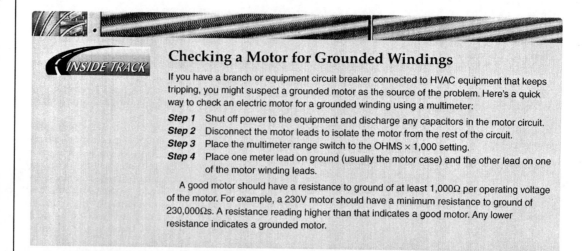

Checking a Motor for Grounded Windings

If you have a branch or equipment circuit breaker connected to HVAC equipment that keeps tripping, you might suspect a grounded motor as the source of the problem. Here's a quick way to check an electric motor for a grounded winding using a multimeter:

Step 1   Shut off power to the equipment and discharge any capacitors in the motor circuit.
Step 2   Disconnect the motor leads to isolate the motor from the rest of the circuit.
Step 3   Place the multimeter range switch to the OHMS × 1,000 setting.
Step 4   Place one meter lead on ground (usually the motor case) and the other lead on one of the motor winding leads.

A good motor should have a resistance to ground of at least 1,000Ω per operating voltage of the motor. For example, a 230V motor should have a minimum resistance to ground of 230,000Ωs. A resistance reading higher than that indicates a good motor. Any lower resistance indicates a grounded motor.

6.6.0 Checking Capacitors

A capacitor tester is needed to effectively test a capacitor. A capacitor tester can be used to determine the capacitor value, and can also be used to check for leakage. Before connecting the capacitor tester, turn off the power and bleed off any capacitor charge using a capacitor discharging tool like the one shown in *Figure 41*. It is a good idea to do this even if the capacitor is equipped with a bleeder resistor, in case the bleeder is defective. Don't use a screwdriver placed directly across both terminals of the capacitor to discharge a capacitor, as this can damage the capacitor.

 WARNING!
Never place your fingers across the capacitor terminals; the residual charge on the capacitor can be dangerous.

When you connect the probes to the capacitor, the meter will read the actual capacitor value, which can then be compared to the value marked on the capacitor, or on the schematic. A typical capacitor tester will also check for shorts and opens and will indicate if the capacitor is leaky.

 WARNING!
Some older capacitors may contain PCB (polychlorinated biphenyl), which was at one time widely used as a dielectric in capacitors and power distribution transformers. PCB is known to cause cancer in humans. If you have a defective oil-filled capacitor or transformer, do not attempt to pry it open or puncture it. If it is leaking, don't touch the oil or breathe the fumes. Dispose of it in accordance with applicable local or national codes.

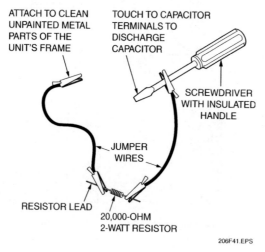

Figure 41 ♦ Capacitor discharging tool.

6.7.0 Checking Fuses

The best way to test a fuse is by measuring continuity (*Figure 42*). To check fuses, always open the unit disconnect switch, then remove the fuses using an insulated fuse puller. Test the fuses for continuity using an analog or digital multimeter (VOM/DMM).

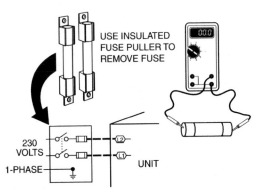

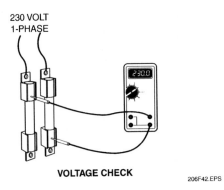

Figure 42 ♦ Fuse checks.

fuse. If voltage is measured on the load side of the fuse, the fuse is good; if not, the fuse is bad. Repeat this procedure so that all fuses are measured with one test lead on the load side and the other test lead on the line side of a different fuse. This method tests one fuse at a time.

7.0.0 ♦ SAFETY

Alternating current can be a deadly force if carelessly handled. One tenth of an ampere of alternating current flowing through a vital organ can prove fatal; therefore, safety precautions must be observed whenever working around or with electricity. The amount and duration of the current flow, the parts of the body involved, and the frequency of the current determine the extent of body damage. Damage is greatest when the current flow is through or near nerve centers and vital organs.

People differ in their resistance to electric shock. Consequently, an amount of current that may cause only a painful shock to one person might be fatal to another. *Table 1* presents the effect of 60Hz current flowing through the body from hand-to-hand or hand-to-foot. The table shows that at approximately one milliampere or mA (0.001A), shock is perceptible. At approximately 10mA, the shock would be sufficient to prevent voluntary control of the muscles; at approximately 100mA (0.1A), the shock is fatal if it lasts more than one second.

High-frequency currents (200Hz and above) have a tendency to flow along the surface of the skin, usually causing severe burns. The current may not penetrate the body. In addition to the possibility of burns and death, involuntary movements as a result of electrical shock can cause other types of serious injuries resulting from falls or contact with rotating machinery or hot surfaces.

Two conditions must be present for an electric current to flow through the body and cause electric shock. First, the body or some part of the body must form part of a closed circuit. Second, there must be a voltage somewhere in the closed circuit. To prevent electric shock, you must make certain that your body never forms part of a closed circuit. Your body must also be well insulated from the ground.

Practically all electric shocks are due to human error, rather than equipment failure. Nearly all deaths due to electrical shock are due to the worker's failure to observe safety precautions, failure to repair equipment for electrical defects, or failure to remedy all defects found by tests and inspections.

If a short exists (zero ohms) across the fuse, it is usually good. If an open exists (infinite resistance) across the fuse, it is blown. A blown fuse is usually caused by some abnormal overload condition, such as a short circuit within the equipment or an overloaded motor.

> **WARNING!**
> Replacing a blown fuse without locating and correcting the cause can result in injury to personnel and damage to the equipment.

Fuses can also be tested with the circuit energized. Set the meter to AC voltage on a range that is higher than the highest voltage expected. Turn on the power and place one of the meter leads on the input (line) side of the fuse. Touch the other test lead to the load side of another

Explain the dangers of replacing a blown fuse without correcting the cause.

Note that fuses can also be checked with the circuit energized, but this method is less preferable than the continuity check.

Have the trainees review Sections 7.0.0–8.0.0.

Ensure that you have everything required for teaching this session.

Show Transparency 42 (Table 1). Discuss the effects of current on the human body.

List the safety precautions associated with working on or near electric circuits.

Show trainees how to follow applicable safety procedures and test AC components, including transformers, capacitors, and motor windings.

Under your supervision, have trainees practice testing AC components. Note the proficiency of each trainee. This laboratory corresponds to Performance Task 3.

Show Transparency 43 (Figure 43). Discuss the ways in which voltages are depicted on circuit diagrams.

Table 1 Current Effects on the Human Body

Current Value	Typical Effects
Less than 1 milliamp	No sensation.
1 to 20 milliamps	Sensation of shock, possibly painful. May lose some muscular control between 10 and 20 milliamps.
20 to 50 milliamps	Painful shock, severe muscular contractions, breathing difficulties.
50 to 200 milliamps	Up to 100 milliamps, same symptoms as above, only more severe. Between 100 and 200 milliamps ventricular fibrillation may occur. This typically results in almost immediate death unless special medical equipment and treatment are available.
Over 200 milliamps	Severe burns and muscular contractions. The chest muscles contract and stop the heart for the duration of the shock, resulting in death.

The following are recommended precautions:

- Never cut off the ground prong of a grounded plug or use one that has been cut off.
- Never touch any electrical wire without ensuring that it is not a live wire.
- Never switch an appliance on or off while standing in or touching a wet surface or area.
- Always turn off power at the main disconnect before working on an electrical circuit or device. Lock and tag the power switch.
- Always unplug an electrical appliance before working on it.
- Replace all worn power cords.
- Unplug cords by pulling on the plug; do not pull the cord.
- Notify the power company or utility whenever a power line is touching the ground.
- If it becomes necessary to work on live electrical wiring, try to use one hand only. If you are shocked while using only one hand, current will probably flow through the hand and arm, then through the feet to the ground. If a shock is conducted through both hands and arms, the electrical path would be through the heart and lungs and would be more likely to be fatal.
- Use protective equipment such as rubber gloves and insulated boots.
- Use tools with dielectric insulation.
- Remove metal jewelry such as rings and watches.

The *National Electrical Code*® (*NEC*®), when used together with the electrical code for your local area, provides the minimum requirements for the installation of electrical systems. Always use the latest edition of the *NEC*® as your on-the-job reference. It specifies the minimum provisions necessary for protecting people and property from electrical hazards. In some areas, different editions of the *NEC*® may be in use, so be sure to use the edition specified by your employer.

8.0.0 ◆ AC VOLTAGE ON CIRCUIT DIAGRAMS

The schematic diagrams you will see in your work will generally be divided into high-voltage sections and low-voltage sections (see *Figure 43*). The high-voltage section will contain the line voltage distribution circuits and the primary loads. In an air conditioning system, these would be the compressor motor, fan motors, and resistance heaters. The low-voltage section, which contains the control devices such as the thermostat and control relays, will often operate at 24 volts.

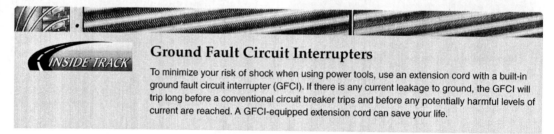

Ground Fault Circuit Interrupters

To minimize your risk of shock when using power tools, use an extension cord with a built-in ground fault circuit interrupter (GFCI). If there is any current leakage to ground, the GFCI will trip long before a conventional circuit breaker trips and before any potentially harmful levels of current are reached. A GFCI-equipped extension cord can save your life.

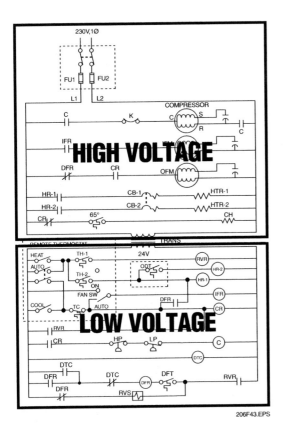

Figure 43 ♦ High-voltage and low-voltage circuits.

The low voltage is obtained by using a control transformer to step down the line voltage. In some large systems using three-phase line voltages of 240V and higher, a control voltage of 120V may be used for some of the control devices.

As shown in *Figure 44*, single-phase line voltage is often represented on ladder diagrams as two vertical lines labeled L1 and L2, representing the two 240V lines from the secondary of the pole transformer.

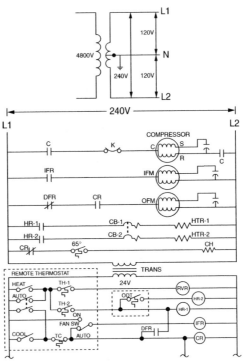

C	Compressor contactor	LPS	Low-pressure switch
CB	Circuit Breaker	ODT	Outdoor thermostat
CH	Crankcase heater	OFM	Outdoor fan motor
CR	Control relay	RC	Run capacitor
DFR	Defrost relay	RVS	Reversing valve solenoid
HPS	High-pressure switch	T	Thermistor
HR	Heater relay	TC	Cooling thermostat
IFM	Indoor fan motor	TH	Heating thermostat
IFR	Indoor fan relay	TRAN	Transformer
K	Thermal overload	-------	Mechanical connection

Figure 44 ♦ Ladder diagram.

Have trainees complete the Review Questions, and go over the answers prior to administering the Module Examination.

Review Questions

1. A transformer has 200 windings in its primary, which receives 120 volts AC. The secondary has 400 windings. The voltage across the secondary is _____ volts.
   a. 60
   b. 120
   c. 240
   d. 400

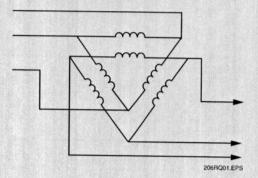

Figure 1

2. The schematic diagram shown in *Figure 1* represents a(n) _____.
   a. delta-connected three-phase transformer
   b. wye-connected three-phase transformer
   c. three-phase motor
   d. autotransformer

3. The effective (rms) voltage of a sine wave with a peak voltage of 200V is _____.
   a. 100
   b. 120.5
   c. 141.4
   d. 200

4. The voltage at the 90-degree point of a sine wave with a maximum voltage of 10V is _____.
   a. –10V
   b. 0V
   c. 5V
   d. +10V

5. The voltage at the 270-degree point of a sine wave with a maximum voltage of 10V is _____.
   a. –10V
   b. 0V
   c. 5V
   d. +10V

6. In an inductive circuit _____.
   a. current leads voltage
   b. voltage leads current
   c. voltage and current are in phase
   d. voltage lags current by 90 degrees

7. A run capacitor in a single-phase motor can remain in the circuit after the motor starts because _____.
   a. single-phase motors don't draw much current
   b. it has a paper dielectric and foil plates
   c. single-phase motors don't generate much heat
   d. it has large plates and dielectric oil to dissipate heat

8. The stator winding of a single-phase motor _____ when voltage is applied.
   a. rotates
   b. remains stationary
   c. moves up and down like a piston
   d. remains stationary for a few seconds, then begins rotating

9. The start winding of a split-phase motor _____.
   a. has fewer turns of wire than the run winding
   b. is always left in the circuit after the motor has started
   c. has more turns of wire than the run winding
   d. is 120 degrees out of phase with the run winding

10. A _____ single-phase motor is most likely to be used to drive a refrigeration compressor.
    a. permanent split capacitor
    b. capacitor-start, capacitor-run
    c. shaded-pole
    d. split phase

6.32 HVAC ♦ LEVEL TWO

Review Questions

11. You are likely to find shaded-pole motors in _____.
    a. large blower units
    b. compressors
    c. small fans
    d. centrifugal chillers

12. Which of the following applies to three-phase motors?
    a. They must be delta-connected.
    b. They require starting devices.
    c. They generate less starting torque than single-phase motors.
    d. They may be delta-connected or wye-connected.

13. A megger is used to measure _____.
    a. capacitance
    b. inductance
    c. voltage
    d. resistance

14. When testing the windings of a single-phase induction motor, the resistance of the start winding will be _____.
    a. about the same as that of the run winding
    b. about half that of the run winding
    c. three to four times greater than that of the run winding
    d. about 500 ohms

15. The majority of electrical shocks are caused by _____.
    a. lightning
    b. incorrect use of GFCIs
    c. workers' failure to observe safety precautions
    d. defective test equipment

Summarize the major concepts presented in the module.

Administer the Module Examination. Record the results on Craft Training Report Form 200, and submit the results to the Training Program Sponsor.

Administer the Performance Test, and fill out Performance Profile Sheets for each trainee. If desired, trainee proficiency noted during laboratory sessions may be used to complete the Performance Test. Record the results on Craft Training Report Form 200, and submit the results to the Training Program Sponsor.

Summary

The HVAC equipment you encounter will be powered by AC voltage. Some of the internal circuits may use DC, which will be obtained by rectifying the AC. Single-phase AC is used to power most homes and small commercial operations. Where more power is needed, three-phase power is available from the local utility.

AC induction motors drive the compressors and fans used in HVAC equipment. There are several types of single-phase motors used in HVAC equipment. They are selected for their starting torque and running characteristics, which are determined by the arrangement of stator windings and the use of capacitors to provide phase shift. Three-phase motors are used where higher torque is required.

In addition to basic test instruments, troubleshooting of AC circuits may require a capacitor, wattmeter, megger, and sometimes a chart recorder.

The importance of safety in the lab and on the job cannot be overstated. Learn and follow the established practices for the safety of yourself and your co-workers.

Notes

Instructor's Notes:

Trade Terms Introduced in This Module

Alternator: A device that generates alternating current by means of conductors rotated in a magnetic field.

Armature: The rotating component of a generator.

Capacitor: An electrical storage device containing two metal plates separated by an insulating (dielectric) material.

Commutator: The movable contact surface on an electric generator or motor.

Dielectric: A material that strongly resists the passage of current.

Effective voltage: See *root-mean-square (rms) voltage*.

Frequency: The number of complete cycles of an alternating current, sound wave, or vibrating object that occur in a period of time.

Fusible link: A circuit protective device that melts, opening the circuit, when the current is excessive.

Hertz (Hz): The unit of measure for the frequency of alternating current. One Hertz equals one cycle per second.

Induction: To generate a current in a conductor by placing it in a moving magnetic field.

Induction motor: An AC motor.

Inertia: The tendency of a body in motion to remain in motion and a body at rest to remain at rest.

Isolation transformer: A transformer with a one-to-one turns ratio. It is used for personnel safety and to prevent electrical interference.

Megohmmeter (megger): A test instrument used to test high-resistance circuits.

Microfarad: One-millionth of a farad. Used to rate capacitors.

Root-mean-square (rms) voltage: The value of AC voltage that will produce as much power when connected across a load as an equivalent amount of DC voltage. Also known as effective voltage.

Rotor: The rotating component of an induction motor.

Run capacitor: A capacitor that remains in the motor circuit while the motor is running to improve running efficiency.

Run winding: The stator winding of a motor that draws current during the entire running cycle of the motor.

Sinusoidal (sine) wave: The waveform created by an AC generator.

Start winding: The stator winding of a motor that is used to provide starting torque.

Stator: The stationary windings of a motor.

Synchronous speed: The maximum rated speed of a motor.

Torque: The force that must be generated to turn a motor.

Turns ratio: The ratio between the number of turns in the primary and secondary windings of a transformer.

Additional Resources and References

Additional Resources

This module is intended to be a thorough resource for task training. The following reference works are suggested for further study. These are optional materials for continued education rather than for task training.

ARI Refrigeration and Air Conditioning; An Introduction to HVAC/R, 4th Edition, Prentice Hall.

General Training—Electricity (GTE), 1993. Syracuse, NY: Carrier Corporation.

HVAC Servicing Procedures, 1995. Syracuse, NY: Carrier Corporation.

Pocket Guide to Electrical Installations Under NEC 2002, Volumes I and II, 2001. Quincy, MA: National Fire Protection Association.

Figure Credits

Fluke Corporation, reproduced with permission, 206SA02

Square D/Schneider Electric, 206F13 (photo)

Carrier Corporation, 206SA03, 206F43, 206F44

Topaz Publications, Inc., 206SA04, 206SA05, 206F24 (photo), 206SA06, 206F26, 206F29, 206SA07, 206SA09

CR Magnetics, 206SA08

Supco, Inc., 206F37

MODULE 03206-07 — TEACHING TIPS

The following are suggested activities or instructional methods to help you teach the material in this module.

General

When you call on someone to answer a question, the rest of the class relaxes or even tunes out because they expect that the question and answer will take place only between you and the trainee you called on. Instead, use this technique to involve more trainees in answering questions and to keep them on their toes.

1. Ask trainees to define a term or explain a concept.
2. After one trainee has answered, ask a trainee seated nearby if the answer is right. Then ask whether a trainee in the back of the room agrees.
3. Ask trainees to explain why they think an answer is right or wrong.
4. Use the session to clear up incorrect ideas and encourage trainees to learn from their mistakes.

Sections 2.0.0 through 5.0.0

Quick Quiz

This Quick Quiz will familiarize trainees with the terms and definitions that are commonly used when dealing with alternating current. You will need photocopies of the quiz provided on the following page. Trainees will need pencils. If you allow trainees to use the Trainee Guide, decrease the amount of time you give them to complete the quiz.

1. Make a photocopy of the quiz for each trainee.
2. Give trainees between 5 and 10 minutes to complete the quiz.
3. Go over the answers to the quiz.
4. Ask trainees if they have questions.

Answers to Quick Quiz

1. f
2. m
3. g
4. d
5. i
6. a
7. o
8. e
9. c
10. j
11. n
12. k
13. h
14. l
15. b

Quick Quiz — *Alternating Current*

For each description listed, identify the term that the text best describes. Write the corresponding letter in the blank provided.

_____ 1. Sixty cycles per second is the standard _____ used in the United States.

_____ 2. A single cycle of AC voltage looks like a(n) _____.

_____ 3. In a transformer, electrical energy is transferred from the primary winding to the secondary winding by _____.

_____ 4. The relationship between the primary and secondary windings is called the _____.

_____ 5. A(n) _____ is intended to provide safety to the user or minimize interference in certain electronic equipment.

_____ 6. An alternating current generator is frequently called a(n) _____.

_____ 7. The maximum speed at which a motor can run is called the _____.

_____ 8. The amount of voltage available to do work is called the _____.

_____ 9. A(n) _____ is an electrical storage device that charges and discharges as the applied voltage changes.

_____ 10. The capacity of a capacitor is measured in _____.

_____ 11. In a motor, voltage is applied to the _____ windings.

_____ 12. The rotating magnetic field in a motor continuously pulls and pushes the _____ and keeps it running.

_____ 13. It takes extra energy to overcome _____ and start the motor turning.

_____ 14. In a permanent split capacitor motor, the _____ provides a phase shift between the run and start windings.

_____ 15. Alternating current is generated by a(n) _____ rotating through a magnetic field.

a. alternator
b. armature
c. capacitor
d. turns ratio
e. effective voltage
f. frequency
g. induction
h. inertia
i. isolation transformer
j. microfarads
k. rotor
l. run capacitor
m. sine wave
n. stator
o. synchronous speed

Section 2.0.0 *Transformers*

This exercise will familiarize trainees with transformers. You will need to supply trainees with the basic components to build a simple transformer. Trainees will need appropriate personal protective equipment, pencils, and paper. Allow 30 to 45 minutes for this exercise.

1. Describe transformers and explain how they operate.
2. Supply trainees with an iron core, wire, and a battery. Have them construct a simple transformer and test it using a multimeter.
3. If desired, have each trainees calculate the expected voltage and verify it with a multimeter.
4. Answer any questions trainees may have.

Section 2.3.0 *Think About It – Low-Voltage Transformer Load*

The maximum load on the existing transformer is 30V during cooling operation. Since the 12VA of additional load is for a humidifier that only operates during the heating mode, the transformer is adequate since it was predetermined that the load on the transformer during heating operation was less than 30VA. If, however, the additional 12VA load was added as a cooling-related component, the total load in cooling would exceed 40VA and a transformer with a higher VA rating would have to be installed.

Section 3.0.0 *Power Generation*

This exercise will familiarize trainees with power generation. Trainees will need appropriate personal protective equipment, pencils, and paper. You will need to arrange for a tour of a local power plant. Allow 30 to 45 minutes for this exercise.

Alternatively, obtain simple power generation kits. Demonstrate generation of AC and DC power generation.

1. Discuss power generation with trainees.
2. Brainstorm questions with trainees.
3. Take a tour of a local power plant. Discuss power generation, transmission, safety controls, and grid management.
4. Have the trainees take notes and write down any additional questions during the tour.
5. After the tour, answer any questions trainees may have.

Section 3.2.0 *Think About It – Frequency*

A frequency of 60 Hz means that only sixty cycles occur during each second, with each cycle lasting $\frac{1}{60}$ of a second.

Section 6.0.0   *Identification and Testing AC Components*

This exercise will give trainees practice in identifying and testing AC components. Trainees will need appropriate personal protective equipment, pencils, and paper. You will need to obtain various types of working and non-working AC components and test equipment. Allow 30 to 45 minutes for this exercise. This exercise corresponds to Performance Tasks 1 and 3.

Obtain old or broken equipment or appliances from local plants, a disposal facility, an equipment recycler, or the local scrap yard.

1. Describe various components and explain how to test them. Set up workstations with one piece of equipment and testing equipment. Demonstrate various testing procedures and safety precautions.
2. Have trainees circulate to each of the workstations and test each piece of equipment. Have them note if the component is working or non-working.
3. Once all trainees have visited each station, have one trainee stand and report the test results of the equipment at their station. Answer any questions they may have.

MODULE 03206-07 — ANSWERS TO REVIEW QUESTIONS

Answer	Section
1. c*	2.0.0
2. a	2.2.0; Figure 5
3. c	3.1.0
4. b	3.1.0
5. a	3.1.0
6. b	4.2.0
7. d	4.3.0
8. b	5.1.0
9. c	5.1.1
10. b	5.1.4
11. c	5.1.5
12. d	5.2.0
13. d	6.3.0
14. c	6.5.0
15. c	7.0.0

*The secondary has twice as many windings as the primary, so the voltage is doubled: $2 \times 120 = 240$.

CONTREN® LEARNING SERIES — USER UPDATE

NCCER makes every effort to keep these textbooks up-to-date and free of technical errors. We appreciate your help in this process. If you have an idea for improving this textbook, or if you find an error, a typographical mistake, or an inaccuracy in NCCER's Contren® textbooks, please write us, using this form or a photocopy. Be sure to include the exact module number, page number, a detailed description, and the correction, if applicable. Your input will be brought to the attention of the Technical Review Committee. Thank you for your assistance.

Instructors – If you found that additional materials were necessary in order to teach this module effectively, please let us know so that we may include them in the Equipment/Materials list in the Annotated Instructor's Guide.

Write: Product Development and Revision
National Center for Construction Education and Research
3600 NW 43rd St, Bldg G, Gainesville, FL 32606

Fax: 352-334-0932

E-mail: curriculum@nccer.org

Craft _____ Module Name _____

Copyright Date _____ Module Number _____ Page Number(s) _____

Description
_____
_____
_____
_____

(Optional) Correction
_____
_____
_____

(Optional) Your Name and Address
_____
_____
_____

Module 03207-07

Basic Electronics

NCCER STANDARDIZED CRAFT TRAINING PROGRAM

The National Center for Construction Education and Research (NCCER) provides a standardized national program of accredited craft training. Key features of the program include instructor certification, competency-based training, and performance testing. The program provides trainees, instructors, and companies with a standard form of recognition through a National Craft Training Registry. The program is described in full in the *Guidelines for Accreditation*, published by NCCER. For more information on standardized craft training, contact the NCCER by writing us at 3600 NW 43rd St., Bldg. G, Gainesville, FL 32606; calling 352-334-0911; or emailing info@nccer.org. More information may be found at our website, www.nccer.org.

HOW TO USE THIS ANNOTATED INSTRUCTOR'S GUIDE

Each page presents two sections of information. The larger section displays each page exactly as it appears in the Trainee Module. The narrow column ties suggested trainee and instructor actions to each page and provides icons (detailed below) to call your attention to material, safety, audiovisual, or testing requirements. The bottom of each page includes space for your notes.

The **Audiovisual** icon indicates an appropriate time to show a transparency or other audiovisual aid.

The **Classroom** icon prompts you to define a term, stress a point, ask trainees to explain a concept, or give examples.

The **Demonstration** icon directs you to show trainees how to perform tasks.

The **Examination** icon tells you to administer the written module examination.

The **Homework** icon is placed where you may wish to assign reading for the next class, assign a project, or advise trainees to prepare for an examination.

The **Laboratory** icon is used when trainees are to practice performing tasks.

The **Materials** icon is a reminder for you to gather materials needed for classes, labs, and testing.

The **Performance Testing** icon tells you to administer a performance test or a portion thereof.

The **Safety** icon is used to emphasize safety issues. It is often keyed to *Caution* and *Warning!* statements in the Trainee Module.

The **Teaching Tip** icon indicates additional guidance is available, such as how to conduct an exercise, get the most educational value from a field trip, or encourage class participation. Teaching Tips may expand on a feature (*Think About It, Did You Know?*) or provide *Quick Quizzes* or similar exercises. You will be referred to the Teaching Tips section at the back of the module if there is additional material.

The **Combination** icon indicates that the laboratory listed corresponds with a performance task. If desired, you can note the proficiency of the trainees during the laboratory, and use it to satisfy performance testing requirements.

PREPARATION

Before teaching this module, you should review the Objectives, Performance Tasks, Materials and Equipment List, and Module Outline. Be sure to allow ample time to prepare your own training or lesson plan and gather all required materials and equipment.

Basic Electronics
Annotated Instructor's Guide

Module 03207-07

MODULE OVERVIEW

This module introduces the trainee to electronic components and circuits used in HVAC systems.

PREREQUISITES

Prior to training with this module, it is recommended that the trainee shall have successfully completed *Core Curriculum*; *HVAC Level One*; and *HVAC Level Two*, Modules 03201-07 through 03206-07.

OBJECTIVES

Upon completion of this module, the trainee will be able to do the following:

1. Explain the basic theory of electronics and semiconductors.
2. Explain how various semiconductor devices such as diodes, LEDs, and photo diodes work, and how they are used in power and control circuits.
3. Identify different types of resistors and explain how their resistance values can be determined.
4. Describe the operation and function of thermistors and cad cells.
5. Test semiconductor components.
6. Identify the connectors on a personal computer.

PERFORMANCE TASKS

Under the supervision of the instructor, the trainee should be able to do the following:

1. Identify various semiconductor components.
2. Test a cad-cell flame detector.
3. Test thermistors.

MATERIALS AND EQUIPMENT LIST

Overhead projector and screen
Transparencies
Blank acetate sheets
Transparency pens
Whiteboard/chalkboard
Markers/chalk
Pencils and scratch paper
Index cards
Appropriate personal protective equipment
Ohmmeter
Voltmeter
Selection of wire-wound and carbon composition resistors
Printed circuitboard
TV/VCR/DVD (optional)
Video *Understanding Electronic Controls* (optional)

Selection of electronic components, including:
  Diodes
  LEDs
  Photo diodes
  Thermistors
  Thyristors
  Integrated circuit chips
  Cad cell flame detector
  Silicon-controlled rectifiers
  Diacs
  Triacs
Cooling system or furnace with built-in diagnostic capability
Personal computer system and examples of storage media
Copies of Quick Quiz*
Module Examinations**
Performance Profile Sheets**

* Located in the back of this module.
**Located in the Test Booklet.

SAFETY CONSIDERATIONS

Ensure that the trainees are equipped with appropriate personal protective equipment and know how to use it properly. This module requires that trainees work with electrical circuits. Ensure that all trainees are properly briefed on electrical safety procedures.

ADDITIONAL RESOURCES

This module is intended to present thorough resources for task training. The following reference works are suggested for both instructors and motivated trainees interested in further study. These are optional materials for continued education rather than for task training.

Electronics Fundamentals: Circuits, Devices, and Applications, 2007. Thomas L. Floyd. Prentice Hall.
Electric Circuit Fundamentals, 2004. Thomas L. Floyd. Prentice Hall.

TEACHING TIME FOR THIS MODULE

An outline for use in developing your lesson plan is presented below. Note that each Roman numeral in the outline equates to one session of instruction. Each session has a suggested time period of 2½ hours. This includes 10 minutes at the beginning of each session for administrative tasks and one 10-minute break during the session. Approximately 5 hours are suggested to cover *Basic Electronics*. You will need to adjust the time required for hands-on activity and testing based on your class size and resources. Because laboratories often correspond to Performance Tasks, the proficiency of the trainees may be noted during these exercises for Performance Testing purposes.

Topic   **Planned Time**

Session I. Introduction to Electronics
- A. Introduction
- B. Theory of Electronics
- C. Semiconductor Fundamentals
- D. Electronic Components and Circuits
- E. Laboratory

    Trainees practice identifying various types of semiconductor components. This laboratory corresponds to Performance Task 1.

- F. Laboratory

    Trainees practice testing thermistors and flame-cell detectors. This laboratory corresponds to Performance Tasks 2 and 3.

- G. Printed Circuit Boards

Session II. Computers, Review, and Testing
- A. Introduction to Computers
- B. Review
- C. Module Examination
    1. Trainees must score 70% or higher to receive recognition from NCCER.
    2. Record the testing results on Craft Training Report Form 200, and submit the results to the Training Program Sponsor.
- D. Performance Testing
    1. Trainees must perform each task to the satisfaction of the instructor to receive recognition from NCCER. If applicable, proficiency noted during laboratory exercises can be used to satisfy the Performance Testing requirements.
    2. Record the testing results on Craft Training Report Form 200, and submit the results to the Training Program Sponsor.

THE AIR SYSTEM

While the air system may seem like just a collection of ductwork, it is actually a complex entity that requires very specialized knowledge and careful design. Some HVAC technicians specialize in air distribution systems.

In cold climates, the dry winter air can make building occupants uncomfortable.

Humidifiers like this one are added to furnaces to inject moisture into the air.

High-efficiency pleated air filters cost more than the common "horsehair" filter, but the benefits in cleaner air far outweigh the cost difference.

A good HVAC technician will learn how to use specialized devices to test and balance an air system so that occupants in all parts of the building are equally comfortable.

An electronic air cleaner provides the highest degree of air purity.

HVAC SYSTEMS AND THEIR COMPONENTS

There are many kinds of heating and cooling systems. The HVAC technician must be familiar with all types.

TXVs are often used on commercial cooling systems. If the sensing bulb is not properly installed and insulated, system operation will be erratic.

Many types of accessories are used to protect the expensive compressors used in commercial systems. The filter-drier keeps moisture and debris from entering the compressor.

Trucks, ships, and food warehouses use low-temperature refrigeration systems that will keep food frozen on the way to its destination.

There are many types of heating systems. One of the most popular is the high-efficiency gas-fired condensing furnace.

Some HVAC technicians specialize in commercial systems where they learn to service high-capacity cooling devices such as this centrifugal chiller.

Whether it's a ¼-HP rotary compressor or a 25-HP semi-hermetic compressor, the compressor is the most costly part of any cooling system. The bigger the compressor, the more costly it will be.

HVAC CONTROLS

Most of the problems encountered by an HVAC technician will be in the control system. At its most basic level, the control system consists of a thermostat and a few switching devices. More complex systems have computer-based control systems with numerous safety devices to protect the occupants and the equipment.

Modern systems are controlled by microprocessor-based electronic circuits that simplify troubleshooting and provide control features not possible with older control systems.

The thermostat is the primary control on all HVAC systems. The electro-mechanical thermostat is still in use, but is rapidly being replaced by the more versatile and precise programmable electronic thermostat.

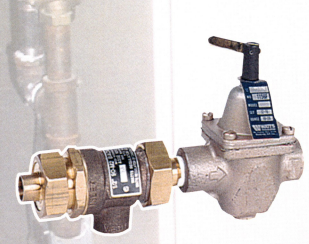

Knowledge of valves is important for anyone working on hydronic heating and cooling systems.

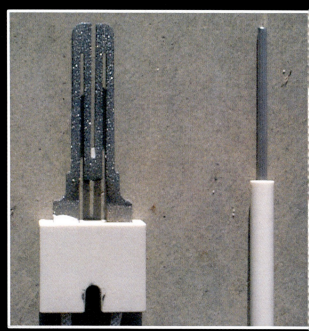

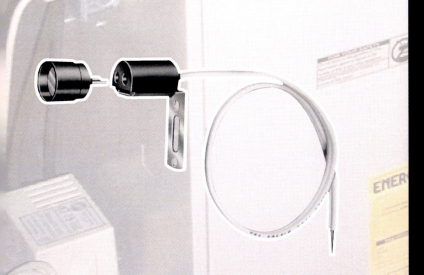

Oil-fired systems have their own special controls, including this cad cell that senses whether a burner flame is present before releasing oil into the combustion chamber.

Safety controls are of utmost importance in a furnace. The gas igniter has evolved from a standing pilot flame to the much safer hot surface igniter and glow coil.

HVAC

TEST INSTRUMENTS

In order to install, test, and troubleshoot air distribution, heating/cooling, and control subsystems, the HVAC service technician must know how and when to use a large variety of test instruments.

The gauge manifold and the multimeter are unquestionably the HVAC technician's most important test instruments.

Manometers are used to measure the air pressure in ductwork, as well as gas pressure in a burner manifold.

While thermometers can measure dry-bulb temperature, the sling psychrometer can also be used to measure wet-bulb temperature.

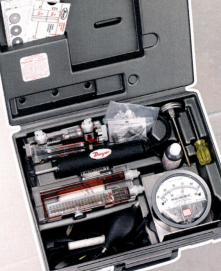

Oil furnaces require specialized instruments such as this combustion efficiency test kit.

A velometer, such as this rotating-vane type is used to check air velocity in duct systems.

The ability to measure temperature quickly and accurately is essential to the HVAC technician.

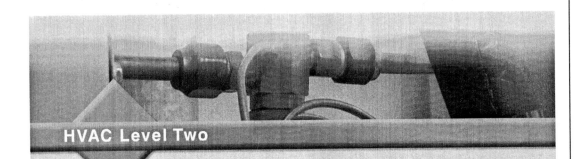

HVAC Level Two

03207-07
Basic Electronics

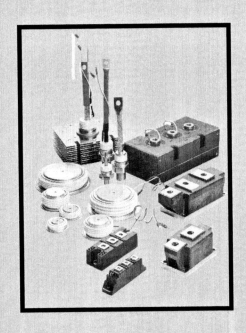

Assign reading of Module 03207-07.

03207-07
Basic Electronics

Topics to be presented in this module include:

1.0.0	Introduction	7.2
2.0.0	Theory of Electronics	7.2
3.0.0	Semiconductor Fundamentals	7.3
4.0.0	Electronic Components and Circuits	7.4
5.0.0	Printed Circuit Boards	7.17
6.0.0	Introduction to Computers	7.19

Overview

Most HVAC products hitting the market today use electronic control devices. Although you may encounter electromechanical control devices, they are outdated. For example, thermistors are now commonly used as temperature sensing devices in place of bimetallic switches. Most zone/room thermostats marketed today are programmable electronic devices. Heating and cooling systems now have "black box" control modules or printed circuit boards that do all their thinking and provide information on the status of the system. Microprocessor-based controls are now found in everything from residential furnaces and air conditioning systems to the systems used to heat and cool industrial complexes.

Anyone doing installation or service work these days is likely to encounter these electronic devices. In addition, service technicians must know how to program a thermostat and how to troubleshoot a heating or cooling system in which all the controls are solid-state electronic devices.

Instructor's Notes:

Objectives

When you have completed this module, you will be able to do the following:

1. Explain the basic theory of electronics and semiconductors.
2. Explain how various semiconductor devices such as diodes, LEDs, and photo diodes work, and how they are used in power and control circuits.
3. Identify different types of resistors and explain how their resistance values can be determined.
4. Describe the operation and function of thermistors and cad cells.
5. Test semiconductor components.
6. Identify the connectors on a personal computer.

Trade Terms

Anode
Basic input/output system (BIOS)
Bridge rectifier
Cathode
Centrifugal force
Chip
Diac
Electromechanical components
Electronics
Free electrons
Full-wave rectifier
Half-wave rectifier
Integrated circuit
Light-emitting diode (LED)
Microminiaturization
Microprocessor
Monochrome
Photo diode
Pixel
Rectification
Semiconductor
Silicon-controlled rectifier (SCR)
Triac
Valence electrons

Required Trainee Materials

1. Pencil and paper
2. Appropriate personal protective equipment

Prerequisites

Before you begin this module, it is recommended that you successfully complete *Core Curriculum*; *HVAC Level One*; and *HVAC Level Two*, Modules 03201-07 through 03206-07.

This course map shows all of the modules in the second level of the HVAC curriculum. The suggested training order begins at the bottom and proceeds up. Skill levels increase as you advance on the course map. The local Training Program Sponsor may adjust the training order.

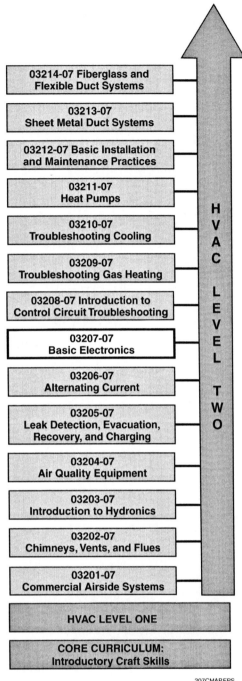

Ensure that you have everything required to teach the course. Check the Materials and Equipment list at the front of this module.

See the general Teaching Tip at the end of this module.

Explain that terms shown in bold are defined in the Glossary at the back of this module.

Show Transparency 1, Objectives, and Transparency 2, Performance Tasks. Review the goals of the module, and explain what will be expected of the trainee.

Review the modules covered in Level Two and explain how this module fits in.

Discuss the advantages of electronic circuits.

Show Transparency 3 (Figure 1). Describe the structure of an atom.

Show Transparency 4 (Figure 2). Discuss the electrical and centrifugal forces in an atom.

Discuss the advantages of electronic circuits. Explain that atoms are normally electrically neutral.

See the Teaching Tip for Sections 1.0.0–6.0.0 at the end of this module.

1.0.0 ◆ INTRODUCTION

The science of **electronics** plays a large role in the control of HVAC systems as it does in many other aspects of our lives. Most of the switching and sensing functions performed by the **electromechanical components** you studied in earlier lessons can now be done with electronic circuits and devices. You will encounter electromechanical devices for some time to come. During your career, however, you can expect to see electronic controls completely replace controls with moving parts.

Electronic circuits have some major advantages. For one thing, they are very small; thousands of circuits can fit on an **integrated circuit** or **chip** no larger than the end of your thumb. For contrast: computers built in the 1950s needed rooms full of equipment; by the 1990s, more processing power than those early models possessed would fit easily in the palm of your hand. **Microminiaturization** is a term that was coined in the computer age; the circuits used in modern computers are so tiny, they can only be seen with a powerful microscope.

Electronic circuits can do a lot more than conventional circuits. They can, for example, process a lot of information about the status of the system and the conditioned space, and use the information to precisely control system operation. This capability results in improved comfort control and operating efficiency. Electronic circuits are also easier to service and less likely to fail than conventional circuits.

2.0.0 ◆ THEORY OF ELECTRONICS

If you could view the flow of electrons through a high-powered microscope, at first glance you might think you were studying astronomy rather than electricity. The atom consists of a central nucleus composed of protons and neutrons, surrounded by orbiting electrons, as shown in *Figure 1*. The nucleus is relatively large when compared with the orbiting electrons, just as our sun is large in comparison to its orbiting planets.

In an atom, the orbiting electrons are held in place by the electric force between the electron and the nucleus. The law of electric charges states that opposite charges attract and like charges repel. The positively charged protons in the nucleus, therefore, attract the negatively charged electrons. If this force of attraction were the only one in effect, the electrons would be pulled closer and closer to the nucleus and eventually be absorbed into it. However, this force of attraction is balanced by the **centrifugal force** that results

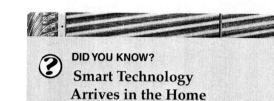

DID YOU KNOW?
Smart Technology Arrives in the Home

The increased use of electronic programmable climate control systems has allowed homeowners the same precise control of air flow and temperature that is common in high-tech commercial applications.

With advances in computer technology, wireless devices, and high-speed web connections, zone control can be automated and synchronized with up-to-the-minute weather information. Air flow, heating, cooling, and electromechanical controls are integrated to create optimal energy efficiency for application in residences.

from the motion of the electrons as they orbit around the nucleus (*Figure 2*). The law of centrifugal force states that a spinning object will pull away from its center point. The faster an object spins, the greater the centrifugal force becomes.

Since the protons and electrons of an atom are equal in number, and equal and opposite in charge, they neutralize each other electrically. Thus, each atom is normally electrically neutral—that is, it exhibits neither a positive nor a negative charge. However, under certain conditions, an atom can become unbalanced by losing or gaining electrons. If an atom loses a negatively

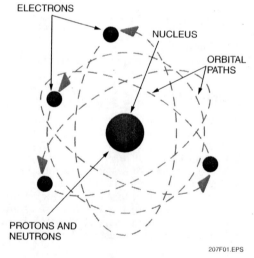

Figure 1 ◆ Structure of an atom.

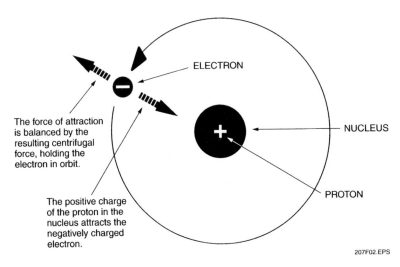

Figure 2 ♦ Electron in orbit around the nucleus.

Describe what happens when an atom gains or loses an electron.

Explain that semiconductors are used to control the current flow in solid-state electronics.

Show Transparency 5 (Figure 3). Discuss the functions and atomic structure of conductors.

charged electron, the atom will exhibit a positive charge and is then referred to as a positive ion. Similarly, an atom that gains an additional negatively charged electron becomes negatively charged itself and is then called a negative ion. In either case, an unbalanced condition is created in the atom, causing the formerly neutral atom to become charged. When one atom is charged and there is an unlike charge in another nearby atom, electrons can flow between the two. This flow of electrons is an electrical current.

3.0.0 ♦ SEMICONDUCTOR FUNDAMENTALS

Semiconductors are the basis for what is known as solid-state electronics. Solid-state electronics is in turn the basis for all modern microminiature electronics such as the tiny integrated circuit and **microprocessor** chips used in computers.

The ability to control the amount of conductivity in semiconductors makes them ideal for use in integrated circuits. In order to understand how semiconductors work, it is first necessary to review the principles of conductors and insulators.

3.1.0 Conductors

Conductors readily carry electrical current. Good electrical conductors are usually also good heat conductors. Conductors are generally made from materials such as metals that have comparatively large, heavy atoms.

In each atom, there is a specific number of electrons that can be contained in each orbit, or shell. The outer shell of an atom is the valence shell, and the electrons contained in the valence shell are known as **valence electrons**.

Conductors are materials that have only one or two valence electrons in their atoms, as shown in *Figure 3*. These electrons can be easily knocked out of their orbits and are therefore known as **free electrons**. An atom that has only one valence electron makes the best conductor because the electron is loosely held in orbit and is easily released to create current flow.

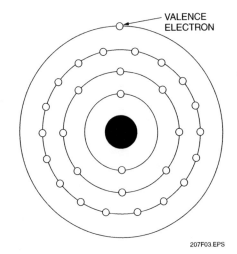

Figure 3 ♦ Atom of a conductor (copper).

List common conductors.

Show Transparency 6 (Figure 4). Discuss the functions and atomic structure of insulators. List common insulators.

Show Transparency 7 (Figure 5). Discuss the functions and atomic structure of semiconductors. List common semiconductors.

Compare P-type and N-type semiconductors. Note that the arrangement of these materials determines the device formed.

Gold and silver are excellent conductors, but they are too expensive to use on a large scale. However, in special applications requiring high conductivity, contacts may be plated with gold or silver. You would be most likely to find such conductors in precision devices where small currents are common and a high degree of accuracy is essential.

Copper is the most widely used conductor because it has excellent conductivity, while being much less expensive than precious metals such as gold and silver. Copper is used as the conductor in most types of wire and provides the printed current path on printed circuit boards. Aluminum is also used as a conductor, but it is not as good as copper. Aluminum may be prohibited in some applications such as household wiring because of its tendency to overheat.

3.2.0 Insulators

As you already know, insulators are materials that resist (and sometimes totally prevent) the passage of electrical current. Rubber, glass, and some plastics are common insulators. The atoms of insulating materials are characterized by having more than four valence electrons in their atomic structures. *Figure 4* shows the structure of an insulator atom. Note that it has eight valence electrons; this is the maximum number of electrons for the third shell of an atom. Therefore, this atom has no free electrons and will not easily pass electric current.

3.3.0 Semiconductors

Semiconductors (*Figure 5*) are materials that are neither good conductors nor good insulators. The materials used as semiconductors, such as germanium and silicon, have more free electrons than an insulator, but fewer than a conductor. Silicon is more commonly used because it withstands heat better.

The factor that makes semiconductors valuable in electronic circuits is that their conductivity can be readily controlled. Semiconductors can be made to have positive or negative characteristics by adding certain impurities through a process known as doping.

When a substance with five valence electrons (such as indium or gallium) is added to the semiconductor material, the semiconductor material will no longer be electrically neutral. Instead, it will take on a positive charge and be known as a P-type material.

When substances like arsenic or antimony, which have three valence electrons, are added to the semiconductor material, the material takes on a negative charge and is known as an N-type material.

4.0.0 ♦ ELECTRONIC COMPONENTS AND CIRCUITS

All solid-state (semiconductor) devices are made from a combination of P-type and N-type materials. The type of device formed is determined by how the P-type and N-type materials are

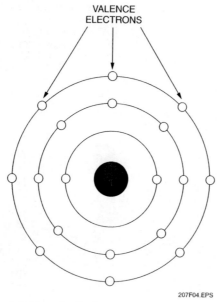

Figure 4 ♦ Atom of an insulator.

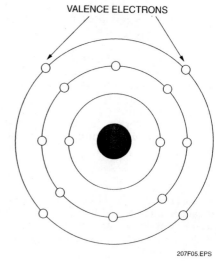

Figure 5 ♦ Atom of a semiconductor.

connected or joined, the number of layers of material, and the thickness of various layers. For instance, a diode is often called a PN junction because it is made by joining a piece of P-type material and a piece of N-type material, as shown in *Figure 6*. The contact surface is the PN junction.

Diodes allow current to flow in one direction, but not in the other. This unidirectional current capability is the distinguishing feature of the diode. The activity occurring at the PN junction is responsible for the unidirectional property of the diode. Diodes are discussed at length in the following section.

4.1.0 Diodes

Modern HVAC systems rely heavily on electronic controls, which use low-level DC voltages. Some HVAC systems also use special controls powered by DC motors when very precise control is required. The electricity furnished by the power company is AC; it must be converted to DC to be suitable for most electronic circuits. The process of converting AC to DC is known as **rectification**.

Diodes are used extensively to convert AC to DC. A diode conducts current only when the voltage at its **anode** is positive with respect to the voltage at its **cathode** (*Figure 7*). At that time, it is said to be forward biased. When the voltage at the anode is negative with respect to the cathode (reverse bias), current will not flow unless the voltage is so high that it overwhelms the diode. Most circuits using diodes are designed so that the diode will not conduct current unless the anode is positive with respect to the cathode.

There are several ways in which diodes are marked to indicate the cathode and the anode (*Figure 8*). Note that in some cases, diodes are marked with the schematic symbol, or there may be a band at one end to indicate the cathode. Other types of diodes use the shape of the diode housing to indicate the cathode end; that is, the cathode end is either beveled or enlarged to ensure proper identification. When in doubt, the polarity of a diode may be determined with an ohmmeter as shown in *Figure 9*.

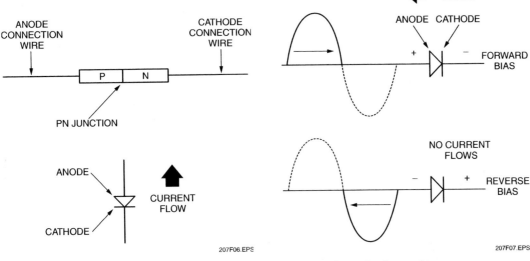

Figure 6 ◆ Material structure of a diode.

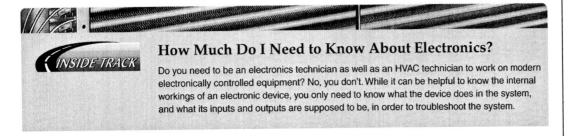

Figure 7 ◆ Forward and reverse bias.

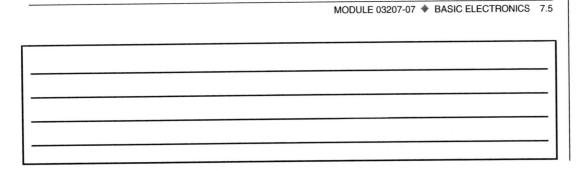

How Much Do I Need to Know About Electronics?

Do you need to be an electronics technician as well as an HVAC technician to work on modern electronically controlled equipment? No, you don't. While it can be helpful to know the internal workings of an electronic device, you only need to know what the device does in the system, and what its inputs and outputs are supposed to be, in order to troubleshoot the system.

Show Transparency 11 (Figure 9). Explain how to check the polarity of a diode with an ohmmeter.

Show trainees how to test a diode using an ohmmeter.

Diodes can also be tested with an ohmmeter. The leads of the ohmmeter are placed on the anode and cathode of the diode. The meter selector should be placed at the lowest ohms scale.

The diode will only conduct an electric current in one direction. If the ohmmeter shows a low resistance reading in both directions, the diode is faulty. A good diode will block current flow in one direction and not in the other. Therefore, if the diode indicates flow in both directions or no flow in both directions (leads or meter polarity reversed), it is defective. LEDs can be checked in the same manner as regular diodes.

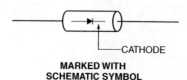

MARKED WITH SCHEMATIC SYMBOL

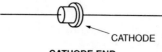

CATHODE END PHYSICALLY LARGER

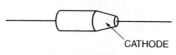

CATHODE END BEVELED

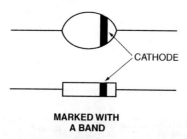

MARKED WITH A BAND

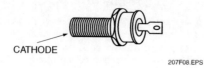

Figure 8 ♦ Diode component identification.

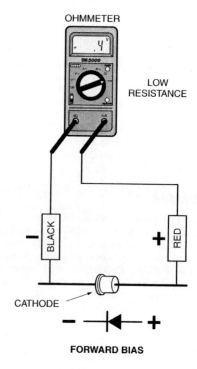

FORWARD BIAS

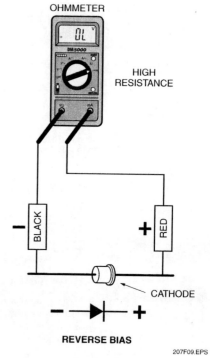

REVERSE BIAS

Figure 9 ♦ Testing a diode with an ohmmeter.

7.6 HVAC ♦ LEVEL TWO

4.1.1 Rectifiers

In a **half-wave rectifier** (*Figure 10*), the single diode conducts current only when the AC applied to its anode is on its positive half-cycle. The result is a pulsating DC voltage. A capacitor connected across the load can filter some of the AC component (ripple), but the voltage is not clean enough to operate electronic circuits.

Figure 11 shows a **full-wave rectifier** with a special center-tapped transformer. In this circuit, one of the diodes conducts on each half-cycle of the AC input. This produces a smoother pulsating DC voltage. Again, a filter capacitor can be used to eliminate almost all of the ripple.

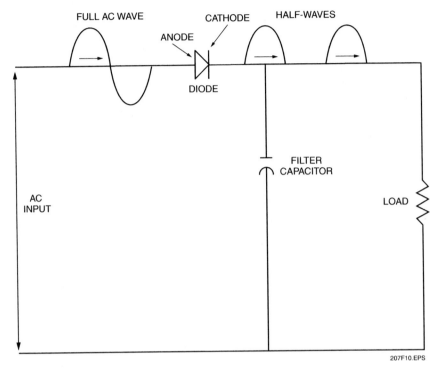

Figure 10 ♦ Half-wave rectifier.

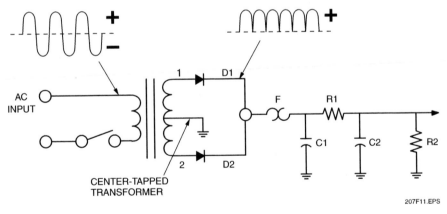

Figure 11 ♦ Full-wave rectifier.

Show Transparency 14 (Figure 12). Describe the operation of a bridge rectifier.

Show Transparency 15 (Figure 13). Describe the operation of a three-phase rectifier.

Explain that a light-emitting diode (LED) gives off visible light when it is energized.

Show Transparencies 16 and 17 (Figures 14 and 15). Discuss the functions and operation of LEDs.

Explain how a photo diode differs from an LED. Identify the symbols used for LEDs and photo diodes.

Describe a liquid crystal display (LCD).

Provide examples of light-emitting diodes for trainees to examine.

The **bridge rectifier** (*Figure 12*) contains four diodes, two of which conduct on each half-cycle. The bridge rectifier provides a smooth DC output, and is the type most commonly used in electronic circuits.

An advantage of the bridge rectifier is that it does not need a center-tapped transformer. A filter and voltage regulator added to the output of the rectifier provide the precise, stable DC voltage needed for electronic devices.

In a three-phase power system, the three-phase rectifier shown in *Figure 13* is used.

4.2.0 Light-Emitting Diode

A **light-emitting diode (LED)** is, as the name implies, a diode that will give off visible light when it is energized. In any forward-biased diode, some energy is given off in the form of photons. In some types of diodes, the number of photons of light energy emitted is sufficient to create a very visible light source.

The process of giving off light by applying an electrical source of energy is called electroluminescence (*Figure 14*).

Note in *Figure 15* that the symbol for an LED is similar to that of a conventional diode except that an arrow is pointing away from the diode.

The **photo diode** is another solid-state device that is turned on by light. The schematic symbol for the photo diode is exactly like that of a standard LED except that the arrow is reversed, as shown in *Figure 15*. The photo diode must have light in order to operate. It acts like a conventional switch. That is, light turns the circuit on, and the absence of light opens the circuit.

The liquid crystal display (LCD) is another method used to display information in electronic systems. The LCD is a segmented display containing conductive material in a semi-liquid state.

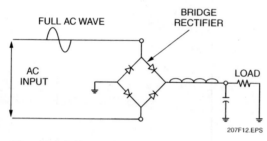

Figure 12 ◆ Bridge rectifier.

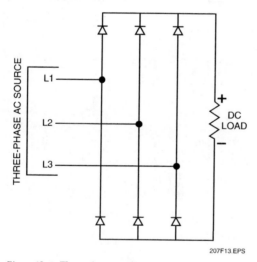

Figure 13 ◆ Three-phase rectifier.

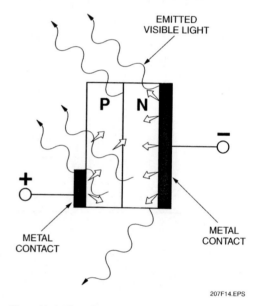

Figure 14 ◆ Electroluminescence in an LED.

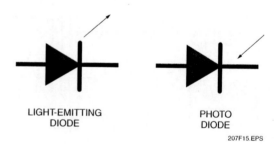

Figure 15 ◆ Schematic symbols for LEDs and photo diodes.

7.8 HVAC ◆ LEVEL TWO

Different combinations of segments are electrically excited to create the display of a number or letter. *Figure 16* shows how numbers are formed. The number 8 requires all seven segments to be excited, while other numbers use fewer segments. Letters are formed in a similar manner, except that diagonal segments are needed to represent letters such as X and M.

When used in a circuit, an LED is generally operated at about 20mA or less. For example, if an LED is to be connected to a 9VDC circuit, a current-limiting resistor must be connected in series with the LED. Ohm's law may be used to calculate the required resistance as follows:

$$R = \frac{E}{I}$$

$$R = \frac{9VDC}{0.020A}$$

$$R = 450\Omega$$

Therefore, a 450Ω resistor or a resistor of the closest standard size (without going under 450Ω) should be used to limit the current flow through the LED.

NOTE

LEDs use the same identifying marks as conventional diodes. Most manufacturers use a flat surface in the LED case near one of the leads. The lead closest to this flat surface is the cathode.

LEDs are used as pilot lights on electronic equipment and as numerical displays. Many programmable HVAC controls use LEDs to indicate when a process is in operation. LEDs are also used in the opto-isolation circuit of solid-state relays for both motor controls and HVAC control systems.

A light-sensing diode (photo diode) can be checked with an ohmmeter by varying the amount of light available to the sensor. To test a photo diode, set the ohmmeter at the lowest scale and connect the leads to the anode and cathode.

Figure 16 ♦ Seven-segment display.

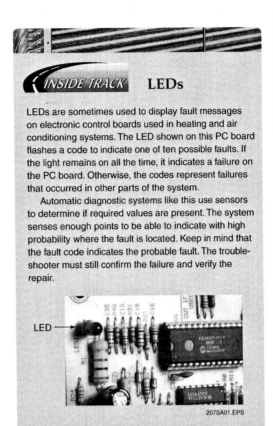

INSIDE TRACK — LEDs

LEDs are sometimes used to display fault messages on electronic control boards used in heating and air conditioning systems. The LED shown on this PC board flashes a code to indicate one of ten possible faults. If the light remains on all the time, it indicates a failure on the PC board. Otherwise, the codes represent failures that occurred in other parts of the system.

Automatic diagnostic systems like this use sensors to determine if required values are present. The system senses enough points to be able to indicate with high probability where the fault is located. Keep in mind that the fault code indicates the probable fault. The troubleshooter must still confirm the failure and verify the repair.

If the polarity of the leads is correct, the meter reading will fluctuate with the varying light input. If there is no instrument needle deflection, reverse the meter leads and check again. The photo diode should conduct current in only one direction.

4.3.0 Silicon-Controlled Rectifiers

Silicon-controlled rectifiers (SCRs), along with **diacs** and **triacs**, belong to a class of semiconductors known as thyristors. One characteristic of SCRs is that they act as an open circuit until a triggering current is applied to their gate. Once that happens, the SCR acts as a low-resistance current path from anode to cathode. It will continue to conduct, even if the gate signal is removed, until either the current is reduced below a certain level or the SCR is turned off. These devices are used for a variety of purposes, including the following:

- AC power controllers
- Emergency lighting circuits

Show Transparency 18 (Figure 16). Explain how numbers are formed on LCDs.

Explain that LEDs require the use of current-limiting resistors.

Show trainees how to calculate the required resistance using Ohm's law.

Explain how to test a photo diode using an ohmmeter.

Explain the functions and operation of silicon-controlled rectifiers (SCRs).

List applications where SCRs are used.

Compare diodes and SCRs.

Show Transparency 19 (Figure 17). Discuss the characteristics of SCRs.

Show Transparencies 20 and 21 (Figures 18 and 19). Describe the functions, operations, and applications of diacs and triacs.

Provide examples of SCRs, diacs, and triacs for trainees to examine.

- Lamp dimmers
- Motor speed controls
- Ignition systems

The SCR is similar to a diode except that it has three terminals. Like a common diode, current will only flow through the SCR in one direction. However, in addition to needing the correct voltage polarity at the anode and cathode, the SCR also requires a gate voltage of the same polarity as the voltage applied to the anode. Once fired, the SCR will remain on until the cathode-to-anode current falls below a value known as the holding current. Once the SCR is off, another positive gate voltage must be applied before it will start conducting again.

The SCR is made from four adjoining layers of semiconductor material in a PNPN arrangement, as shown in *Figure 17*. The SCR symbol is also shown. Note that it is the same as the diode symbol except for the addition of a gating lead.

There is also a light-activated version of the SCR known as an LASCR. Its symbol is the same as that of the regular SCR, with the addition of two diagonal arrows representing light, similar to the photo diode previously covered.

The ability of an SCR to turn on at different points in the conducting cycle can be used to vary the amount of power delivered to a load. This type of variable control is called phase control. With such control, the speed of an electric motor, the brilliance of a lamp, or the output of an electric resistance heating unit can be controlled.

4.4.0 Diacs

A diac can be thought of as an AC switch. Because it is bidirectional, current will flow through it on either half of the AC waveform. One major use of the diac is as a control for a triac. Although the diac is not gated, it will not conduct until the applied voltage exceeds its breakover voltage.

The basic construction and symbols for a diac are shown in *Figure 18*. Note that the diac has two symbols, either of which may be found on schematic diagrams.

4.5.0 Triacs

The triac can be viewed as two SCRs turned in opposite directions with a common gate (*Figure 19*). It could also be viewed as a diac with a gate terminal added. One important distinction, however, is that the voltage applied across the triac does not have to exceed a breakover voltage in order for conduction to begin.

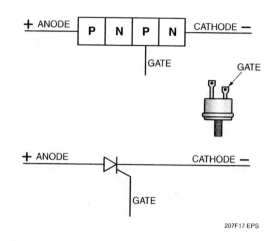

Figure 17 ◆ SCR characteristics and symbol.

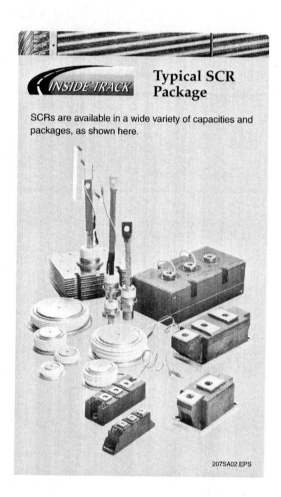

Typical SCR Package

SCRs are available in a wide variety of capacities and packages, as shown here.

Instructor's Notes:

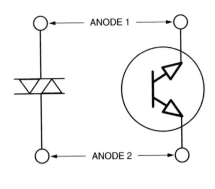

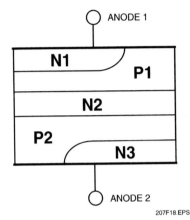

Figure 18 ♦ Basic diac construction and symbols.

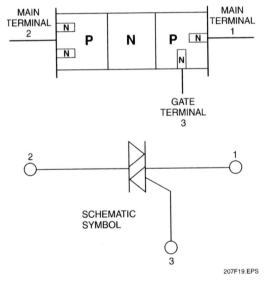

Figure 19 ♦ Basic triac construction and symbol.

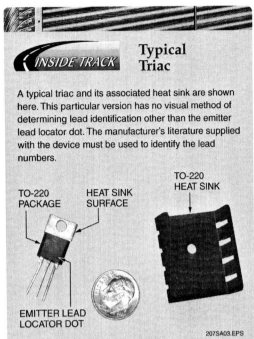

INSIDE TRACK

Typical Triac

A typical triac and its associated heat sink are shown here. This particular version has no visual method of determining lead identification other than the emitter lead locator dot. The manufacturer's literature supplied with the device must be used to identify the lead numbers.

Like SCRs, triacs are used in phase control applications to control the average power applied to loads. Examples include light dimmers and photocell light switches.

4.6.0 Resistors

The two most common types of resistors used in electronic circuits are the wire-wound and the carbon composition. A wire-wound resistor consists of a length of nickel wire wound on a ceramic tube and covered with porcelain. Low-resistance connecting wires are provided and the resistance value is usually printed on the side (*Figure 20*). Carbon composition resistors are constructed by molding mixtures of powdered carbon and insulating materials into a cylindrical shape. An outer sheath of insulating material provides mechanical and electrical protection; connecting wires are provided at each end. Carbon composition resistors are smaller and less expensive than the wire-wound type. However, the wire-wound type is the more rugged of the two, and is able to handle more power than the carbon type.

While most resistors have standard fixed values, variable or adjustable resistors are also used a great deal in electronics. The two most common symbols for a variable resistor are shown in *Figure 21*.

Show Transparency 22 (Figure 20). Discuss the two types of resistors commonly used in electronic circuits.

Show Transparency 23 (Figure 21). Identify the symbols used for variable resistors.

Provide various types of resistors for trainees to examine.

Discuss the functions and operation of variable resistors.

Show Transparencies 24 and 25 (Figures 22 and 23). Explain that resistors are identified using color codes.

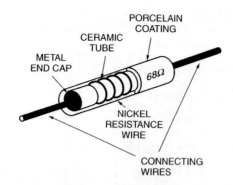

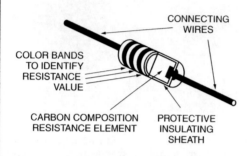

Figure 20 ◆ Common resistors.

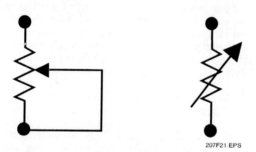

Figure 21 ◆ Symbols used for variable resistors.

A variable resistor consists of a coil of closely wound insulated resistance wire formed into a partial circle. The coil has a low-resistance terminal at each end; a third terminal is connected to a movable contact. The movable contact can be set to any point on a connecting track that extends over one (uninsulated) edge of the coil.

Resistor Applications

A variable resistor is commonly used as a heat anticipator in a room thermostat. The heat anticipator causes the heating thermostat to open just before the room reaches the thermostat set point. This prevents the room temperature from overshooting the set point and causing the room to become too warm.

Fixed resistors are used in thermostats as cooling compensators. A cooling compensator is used to make up for the lag between the call for cooling and the time the system actually begins cooling the space. It turns the thermostat on just before the room temperature reaches the setpoint.

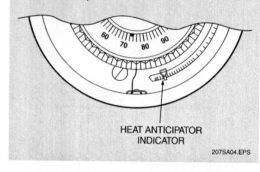

Using the adjustable contact, the resistance from either end terminal to the center terminal may be adjusted from zero to the maximum resistance.

4.6.1 Resistor Color Codes

Because carbon resistors are physically small (some are less than 1 cm in length), it is not practical to print the resistance value on the side. Instead, a color code in the form of color bands is used to identify the resistance value and tolerance. The color code is illustrated in *Figure 22*.

Starting from one end of the resistor, the first two bands identify the first and second digits of the resistance value, and the third band indicates the number of zeros. An exception to this is when the third band is either silver or gold, which indicates a 0.01 or 0.1 multiplier, respectively.

The fourth band is always either silver or gold. In this position, silver indicates a ±10 percent tolerance and gold indicates ±5 percent tolerance.

Where no fourth band is present, the resistor tolerance is ±20 percent.

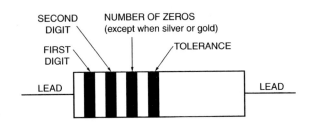

0	BLACK	7	VIOLET
1	BROWN	8	GREY
2	RED	9	WHITE
3	ORANGE	0.1	GOLD
4	YELLOW	0.01	SILVER
5	GREEN	±5%	GOLD - TOLERANCE
6	BLUE	±10%	SILVER - TOLERANCE

Figure 22 ♦ Resistor color codes.

Put this information to practical use by determining the range of values for the carbon resistor in *Figure 23*.

Brown = 1   Black = 0   Red = 2 zeros
Gold = ±5 percent

First digit of 1 + second digit of 0 + 2 zeros = 1,000Ω

Since this resistor has a value of 1,000Ω ±5 percent, the resistor can range in value from 950Ω to 1,050Ω.

4.7.0 Thermistors

Thermistors are temperature-sensitive semiconductor devices. Their resistance varies in a predictable way with variations in temperature. This allows them to be used in a variety of HVAC control applications. Thermistors have either a positive or a negative coefficient of resistance. If the resistance increases as temperature rises, it has a positive coefficient of resistance. If resistance increases as the temperature drops, it has a negative coefficient of resistance. Thermistors come in different sizes and shapes to suit a variety of applications.

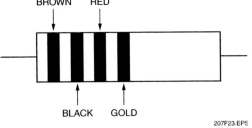

Figure 23 ♦ Sample color codes on a fixed resistor.

Show trainees how to identify resistor values and tolerance using the color codes.

Discuss the functions and operation of thermistors.

Provide thermistors for trainees to examine.

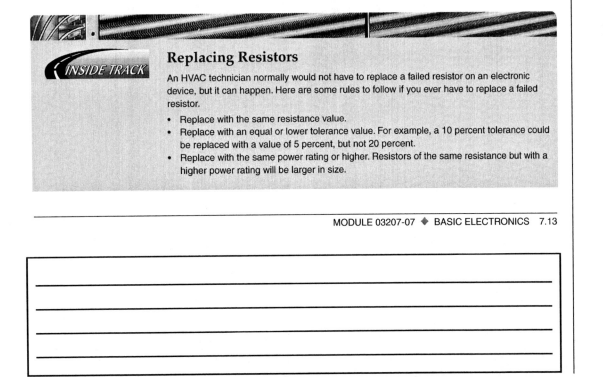

Replacing Resistors

An HVAC technician normally would not have to replace a failed resistor on an electronic device, but it can happen. Here are some rules to follow if you ever have to replace a failed resistor.

- Replace with the same resistance value.
- Replace with an equal or lower tolerance value. For example, a 10 percent tolerance could be replaced with a value of 5 percent, but not 20 percent.
- Replace with the same power rating or higher. Resistors of the same resistance but with a higher power rating will be larger in size.

Explain that thermistors are used to sense temperatures.

Show Transparency 26 (Figure 24). Explain that in addition to being used as temperature sensors, thermistors can be used in a bridge circuit to make vacuum measurements.

List common HVAC applications for thermistors.

Show trainees how to identify various semiconductor components.

See the Teaching Tip for Section 4.0.0 at the end of this module.

Have trainees practice identifying semiconductor components. Note the proficiency of each trainee. This laboratory corresponds to Performance Task 1.

Show Transparency 27 (Figure 25). Explain how to test a thermal-electric expansion valve using a voltmeter.

Thermistors are used to sense temperature changes. They are also used as motor protective devices. Some typical applications of thermistors are: electronic thermometers, room thermostats, duct sensors, electronic expansion valve sensors, and selected control circuits.

Vacuum measurement can also be obtained with thermistors. Two thermistors can be wired in a bridge circuit (*Figure 24*) so that any change in one thermistor will produce a reading on a current meter that is directly calibrated in microns. One thermistor is placed in the vacuum and the other thermistor is placed in the ambient air. With this configuration, the rate of heat loss can be directly converted into a vacuum reading.

Thermistors are used in temperature differential controls, such as the defrost control in some heat pumps. Two thermistors are used; one senses coil temperature and the other senses air temperature. They are wired in a circuit so that a 15°F to 25°F temperature differential will produce enough difference in resistance to turn a relay OFF or ON.

Another use of a thermistor is as a start-assist device for a compressor motor. A ceramic thermistor with a steep-slope positive temperature coefficient is wired in parallel with the run capacitor on a permanent split capacitor motor, increasing the starting torque by 200 to 300 percent.

Other uses of thermistors include sequence switching and current in-rush surge suppression.

4.7.1 Testing a Thermal-Electric Expansion Valve Sensor

When a thermistor is used as a sensing device for the operation of a thermal-electric (TE) expansion valve, both the expansion valve and the thermistor can be checked as follows. With the leads of the voltmeter placed across the terminals of the electric valve (*Figure 25*) and the system operating at or near peak load, the voltmeter reading should be within the range of 15V to 20V. With the system operating at lower loads, the reading should remain between 8V and 14V. The size of the expansion valve in relation to the capacity of the system will be directly related to the time that the valve registers the voltage limits. Therefore, the manufacturer's recommendations should be consulted. The readings should be observed for two or three minutes.

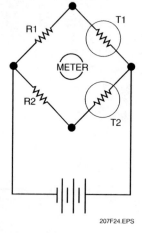

Figure 24 ♦ Bridge circuit.

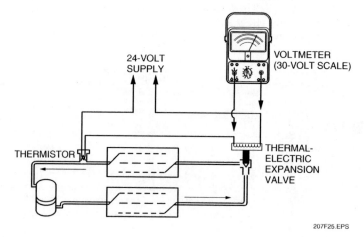

Figure 25 ♦ TE expansion valve.

4.7.2 Testing Motor Protection Thermistors

Thermistors used on the internal windings for overload protection of three-phase motors can be checked by taking a resistance reading through them. An ohmmeter should be used and connected as illustrated in *Figure 26*.

A short or ground in the thermistor will be indicated by a reading of zero or a value approaching zero. An open would be indicated by a reading of infinity. The procedure is as follows:

Step 1 Shut off the disconnect switch and allow the equipment to cool.

Step 2 Connect the ohmmeter leads as indicated in the illustration and record the resistance in ohms.

The values will change with a change in temperature, but at room temperature (75°F) the reading should be about 75Ω. Again, the resistance of the winding temperature sensors will vary by manufacturer and application, so unit specifications should be consulted before condemning or accepting the devices.

4.8.0 Cadmium Sulfide Detector

A cadmium sulfide flame detector (cad cell) is a device mounted in an oil burner that looks down the tube at the flame. Cadmium sulfide (*Figure 27*) is photoconductive; its electrical resistance is high in darkness but lower in the presence of visible light. The more intense the light, the lower the internal resistance. The size of the cadmium sulfide conductor determines the sensitivity of the cell to the light and the amount of current it can conduct. The resistance of the cell may exceed 100,000Ω in darkness but could drop to less than 1,500Ω in the presence of a flame. High internal resistance prevents current from flowing across the cell. This keeps the primary control from energizing, thus shutting off the flow of fuel.

The cadmium sulfide flame detector can also be checked with an ohmmeter (*Figure 28*). Set the ohmmeter to the highest or intermediate scale. In darkness, the resistance of the cell should measure in excess of 100,000Ω. In the presence of an oil burner flame or other comparable light, the cell resistance should drop to less than 1,500Ω. Note that the resistance, in the presence of light or flame, will vary between makes and models. The manufacturer's instructions should specify the limits.

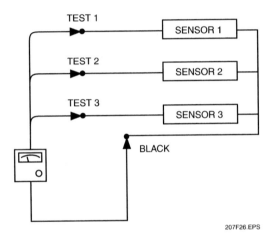

Figure 26 ♦ Overload sensor test.

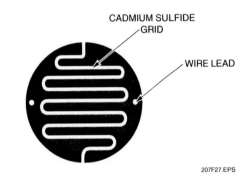

Figure 27 ♦ Cadmium sulfide flame detector.

Show Transparency 28 (Figure 26). Explain how to test a motor protection thermistor using an ohmmeter.

Show trainees how to test thermistors.

Have trainees practice testing thermistors. Note the proficiency of each trainee. This laboratory corresponds to Performance Task 3.

Show Transparencies 29 and 30 (Figures 27 and 28). Explain that cadmium sulfide detectors (cad cells) are used to determine the presence of a flame. Explain that cad cells can be tested using an ohmmeter.

Show trainees how to test a cad-cell flame detector.

Have trainees practice testing cad-cell flame detectors. Note the proficiency of each trainee. This laboratory corresponds to Performance Task 2.

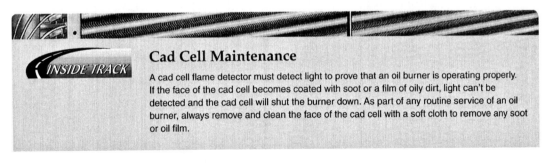

Cad Cell Maintenance

A cad cell flame detector must detect light to prove that an oil burner is operating properly. If the face of the cad cell becomes coated with soot or a film of oily dirt, light can't be detected and the cad cell will shut the burner down. As part of any routine service of an oil burner, always remove and clean the face of the cad cell with a soft cloth to remove any soot or oil film.

Show Transparency 31 (Figure 29). Discuss the operation and applications of electronically commutated motors.

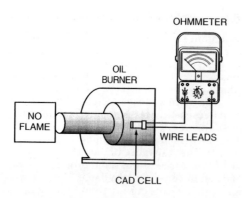
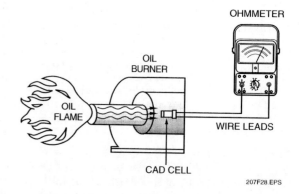

Figure 28 ♦ Flame detector testing.

4.9.0 Electronically Commutated Motors

Electronically commutated motors (ECMs) are direct-current (DC) motors used in variable-speed applications. DC motors are considered far better for applications in which continuously variable control is needed. Like AC motors, DC motors rely on the interaction of the magnetic fields between two electromagnets, one fixed and the other rotating. Because DC voltage does not fluctuate, it is necessary to create a rotating magnetic field using a commutator, as shown in *Figure 29*. In many DC motors, the commutator is connected to the DC voltage source by brushes that remain in constant contact with the commutator and transfer power to it. Because brushes need periodic replacement, a brushless DC motor is used with ECMs, so that the motor can be used in non-reparable components, such as hermetic compressors.

The ECM receives its power from the system AC power source, and must convert that power to DC using a rectifier that is built into the ECM control. The electronic circuits in the ECM control convert the DC voltage to a voltage that is effectively three-phase AC; that is, three out-of-phase voltage waveforms. This voltage is applied to the three stator windings of the motor.

One application for the ECM is a variable-speed compressor motor. Rather than just cycling on and off in response to the thermostat, the ECM adjusts the speed of the compressor in proportion to the amount of cooling needed. This approach improves cooling efficiency and reduces energy costs.

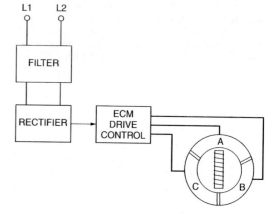

SIMPLIFIED ECM CIRCUIT

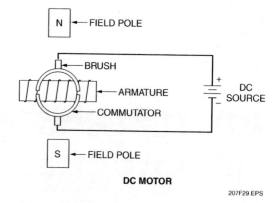

DC MOTOR

Figure 29 ♦ Simplified ECM circuit.

4.10.0 Variable Frequency Drives

A variable frequency drive (VFD) is a motor control system that provides variable control of AC motors electronically. Most AC motors operate at a single speed. In the past, if a load required different speeds, it was common to mechanically adjust the load output using pulleys or to employ a method of throttling the load.

VFDs adjust the speed of the motor by adjusting the voltage and frequency of the electrical power supplied to the motor. The changes are made in response to system load requirements. The VFD is a typical closed-loop system. Sensors determine the load requirements and feed that information to the VFD control. A typical VFD contains a converter, an inverter, a control unit, and a sensor (*Figure 30*).

The sensor, which could be a temperature-sensing device, determines the instantaneous load requirement and feeds that information to the control unit. The rectifier converts the 60Hz system power signal into a DC voltage. The inverter converts the DC voltage into an adjustable-frequency, adjustable-voltage AC voltage. A technique known as pulse width modulation (PWM) is commonly used for the inversion process. PWM produces a current waveform that closely matches the power line waveform. This reduces the likelihood of motor overheating.

The control unit controls the amplitude and frequency of the voltage applied to the motor in response to the demand level indicated by the sensor.

5.0.0 ◆ PRINTED CIRCUIT BOARDS

Most of the electronic circuits you encounter will be mounted on printed circuit (PC) boards. In many HVAC systems, all the control circuits—relays, capacitors, diodes, etc.—are located on a single PC board (see *Figure 31*). The components are mounted on the top of the board; their electrical leads are inserted through holes in the board and soldered to terminal points on the bottom of the board. There is very little, if any, wiring on the PC board. Instead, a copper foil is bonded to the bottom of the board. The desired circuit is imprinted on the foil by a machine, and the copper is then chemically etched away from the unprinted areas. The printed copper acts as the conductor between the components on the circuit. Instead of a wiring harness and plug to connect the circuit to the outside, an edge connector is often built into the board. The edge connector then is plugged into a connector mounted on the hardware.

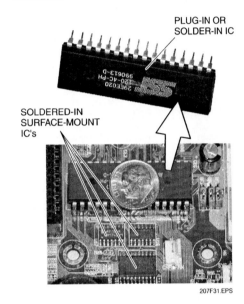

Figure 31 ◆ Printed circuit board.

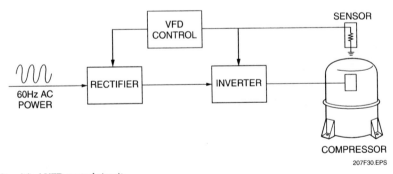

Figure 30 ◆ Simplified VFD control circuit.

Discuss the importance of determining the cause of a system failure before replacing the circuit board.

Discuss the functions and operation of an integrated circuit chip.

Explain microminiaturization.

Provide an integrated circuit chip for trainees to examine.

Some electronic circuits are packaged in sealed modules. This method is common with electronic control devices that perform a specific function and can be used in a number of different systems.

A very important feature that distinguishes packaged electronic controls from circuits built of discrete (separate) components is that the electronic circuit is treated as a black box; that is, if there is a control circuit failure, the entire board or module is replaced. In conventional circuits, on the other hand, you have to analyze the circuit and isolate the fault to the failed component; a bad relay, for example.

When an electronically controlled system is not working, it is tempting to just replace the control module or PC board without checking it. This will result in one of three outcomes, only one of which is desirable. The one good outcome is that it might fix the problem. A more likely outcome is that it won't. Electronic circuits are very reliable, and they have no moving parts, so it is not that common for an electronic control to fail. The worst possible outcome is that something external to the control caused it to fail, and will also cause its replacement to fail. It can be very embarrassing to explain to a customer why you charged them for a repair that didn't work.

Before replacing an electronic circuit, the troubleshooter must verify that the circuit has actually failed, and determine if an outside source caused the failure. This is done by first verifying that the printed circuit board or module is receiving the necessary supply voltages and control signals, and that they are at the proper levels. Once that is done, the outputs need to be verified. If the device is receiving the required inputs, but fails to produce the expected outputs, it can be assumed that the device has failed.

5.1.0 Integrated Circuit Chips

An integrated circuit chip (*Figure 32*) is a tiny wafer of semiconductor material containing microminiature electronic circuits designed to perform a specific function or functions. To get

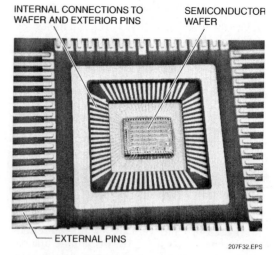

Figure 32 ◆ An integrated circuit (IC) chip.

a perspective on what microminiature means, think about a multi-function digital wristwatch. All the complex timekeeping, calendar, and display functions are contained on a single integrated circuit chip that you might have trouble finding if you looked inside the watch.

5.2.0 Microprocessors

Microminiaturization enables tiny devices smaller than the tip of your little finger to perform work that, in the early days of computers, used to take a roomful of electronic equipment to do. The semiconductor makes microminiaturization possible.

Semiconductors are materials in which the capacity to conduct electricity can be controlled by varying the voltage applied. In this case, we are talking about low-level DC voltages in the range of 5V to 15V. Heat, light, and pressure are also used to control current flow in semiconductors. Silicon and germanium are the two most widely used semiconductor materials.

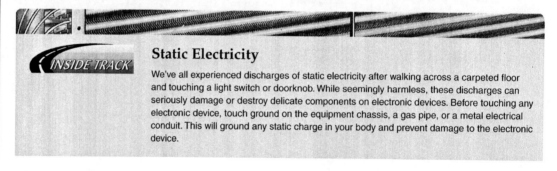

Static Electricity

We've all experienced discharges of static electricity after walking across a carpeted floor and touching a light switch or doorknob. While seemingly harmless, these discharges can seriously damage or destroy delicate components on electronic devices. Before touching any electronic device, touch ground on the equipment chassis, a gas pipe, or a metal electrical conduit. This will ground any static charge in your body and prevent damage to the electronic device.

Some integrated circuits can be programmed to perform complex tasks such as decision-making and mathematical calculations. These are known as microprocessors (*Figure 33*). They are the brains of the personal computer and are used to perform logical and analytical functions in many computerized HVAC systems. One of the most common applications is the programmable thermostat.

In microprocessor-controlled systems, integrated circuit chips and microprocessor chips are usually mounted with other components such as relays, resistors, and capacitors, on PC boards or encapsulated modules. The major advantage of a microprocessor-controlled system is its ability to provide very precise control. The microprocessor collects temperature, pressure, and humidity information from sensors located at strategic points in the system and the conditioned space. Information on the status of safety controls may also be supplied to the microprocessor. The microprocessor can evaluate the information and change system operation to meet changing conditions. Conventional controls are very limited in this sense; it would take many relays and hundreds of feet of wiring to accomplish even the most basic logic functions performed by a microprocessor.

5.3.0 Diagnostic Capability

Another important feature of microprocessor-controlled systems is their ability to recognize, isolate, and report faults. In small systems, the microprocessor receives sensor information such as temperatures and pressures and analyzes this information to locate a fault. The microprocessor is programmed to recognize patterns and relate those patterns to system components. Once the problem is isolated, the system will use a digital readout or a flashing light code to identify where the problem is located. Larger, more complex systems may have programmed tests that the technician can select to help isolate a malfunction.

Figure 33 ♦ A microprocessor chip.

5.4.0 Electrostatic Discharge Sensitivity

One important thing to remember about integrated circuits and microprocessors is that they can be damaged by static electricity. Be sure to ground yourself to something metal, or use a grounding wrist strap before handling them, and then don't touch the connector pins or wiring runs.

In later modules, we will discuss specific electronic control devices and circuits.

6.0.0 ♦ INTRODUCTION TO COMPUTERS

Modern commercial buildings use computers to manage building systems such as HVAC, lighting, and security. In such buildings, the performance of these systems is monitored at a central computer workstation, which may be located in another building, and perhaps even another city. In addition to monitoring system functions, the computer is used as the control point for the building systems. Here's an example: If you needed to change the thermostat setpoint for all the offices in a 20-story building, you would no

---

Troubleshooting an Electronic Furnace Control

A furnace contains an electronic control. Normal inputs to the control are 24 volts and Y, W, and G control signals from the room thermostat. Outputs from the control include 115 volts to operate the inducer motor and blower motors and 24 volts to operate the gas valve, blower relay, and cooling unit. On a no-heat service call, you find that 24 volts and all control signals from the room thermostat are present. A 24-volt output is available to the gas valve. The blower and inducer motor operate but there is no burner operation. Should you replace the electronic control in the furnace to correct the problem?

Define common terms related to computer systems.

doubt prefer to do it once, rather than go from office to office changing all the thermostats. Computer-controlled building systems make this possible by electronically linking all the thermostats to a central control point.

Residential systems are now employing some of the same technology used in commercial systems. Smart homes, once a futuristic concept, are now a reality.

As an HVAC service technician, you can expect to work with computer-controlled systems. For that reason, it is essential that you become familiar with the terms and equipment associated with computers. This section reviews computer fundamentals for those who may have a limited knowledge of computers, and the special terminology associated with them.

6.1.0 Special Terms

Many special terms and abbreviations are used in the computer world. Here are some terms you may encounter in reading or talking about computers:

- *Bandwidth* – The speed at which data travels in transmission lines. Bandwidth is stated in cycles per second, which represents the number of bits of data per second. A typical telephone modem has a bandwidth of 56,000 bits per second (56Kbps), while a high-speed connection such as video cable has a bandwidth of 1.5 million bits per second. A file transferred on the 1.5Mbps line would move about 27 times faster than the same data transferred over a 56K line. A speed of 1.5Mbps is required to receive high-quality video.
- *Basic input/output system* – The **basic input/output system (BIOS)** is the first set of instructions to run when a computer is started (booted) up. The BIOS is stored in Read-Only Memory (ROM) located on the system board.
- *Binary digit (bit)* – The smallest unit of information in a computer system.
- *Byte* – One character, such as a number. It consists of 8 bits. A kilobyte (KB) is 1,024 bytes; a megabyte is approximately one million bytes; a gigabyte is approximately one billion bytes; and a terabyte is approximately one trillion bytes. These terms are used to define storage capacity.
- *Bus* – The wiring pathway between the internal elements of a computer. Buses are defined in terms of their width, which means how many bits of data they can carry, which in turn affects the processing speed of the computer. Two common busses are the ISA and PCI. The PCI bus is the faster of the two, and is able to handle 32-bit and 64-bit data. The ISA bus, in contrast, handles 16-bit data.
- *Cache* – Cache is a type of memory in which data is stored, or stock-piled ahead of time, so it is available for use when needed. Having key instructions and information readily available in cache allows the computer to work faster because it does not have to go the hard drive or other device to search for it.
- *DIP switch* – One of a set of tiny switches, located on a circuit board. DIP switches are used to configure the processor to perform certain functions. They are often used to select options.
- *Digital subscriber line (DSL)* – DSL is a method of providing high-speed communication over telephone lines. It is one of several such methods.
- *Handshake* – The process by which two computers initially establish communication. During the handshake, the computers determine if a connection is possible, then establish the best mode for the transmission.
- *Integrated drive electronics (IDE)* – IDE is a high-speed interface protocol associated with hard drives.
- *Integrated services digital network (ISDN)* – ISDN is a high-speed telecommunications connection. It has a bandwidth of 64Kbps, as compared with DSL which transfers at 150Kbps and up.
- *Network interface card (NIC)* – A NIC is a special printed circuit card or adapter that enables a workstation to connect to a computer network.
- *Parallel I/O* – I/O stands for input/output. A parallel I/O is one in which multiple data bits being transferred from one device to another are sent simultaneously on separate wires. Printers are typically connected to parallel I/O connections on computers.
- *Partition* – A section of a hard drive that is treated as a separate drive by the operating system. Partitions allow a single hard drive to have multiple formats. A computer user might put application software on one partition, and data files on another, or use one partition for one operating system, and a second one for a different operating system.
- *Plug and play* – A special process in which the BIOS recognizes peripheral devices such as printers, scanners, and drives, and automatically configures the system to interface with them. Before plug and play, the operator had to specifically configure the computer to handle each device as it was added.

- *Random access memory (RAM)* – RAM is the main temporary storage for information in a computer. Information stored in RAM can be accessed and changed very quickly by the computer. RAM is volatile storage. This means that the contents are erased if the power is shut off. RAM consists of memory chips located on a memory board connected to the main circuit board (motherboard). The memory modules are known as SIMMs (single in-line memory modules), which have memory chips on one side of the board, or DIMMs (dual in-line memory modules), which have memory chips on both sides of the board. The amount of RAM a computer has determines what applications it can use and how many applications it can have running at once. There are two types of RAM: DRAM (D is for dynamic) is the most common. It must be refreshed often by the computer to retain the information stored in it. SRAM (S is for static) retains information without being refreshed. It is more expensive, however, so it is only used where necessary, such as in video and cache applications. SDRAM and Rambus DRAM (RDRAM) are newer, faster versions of DRAM.
- *Read-only memory (ROM)* – ROM chips are preprogrammed with instructions or information for the computer in which they are used. One important allocation of ROM in a PC is the storage of the BIOS, which contains the boot-up instructions for the PC.
- *Small computer system interface (SCSI)* – Pronounced "scuzzy." SCSI is an interface specification for connecting peripheral devices to a computer. It supports several high-speed devices through a single 50-pin or 68-pin cable.
- *Serial input/output (serial I/O)* – A method of transferring data between two devices one bit at a time. Modems and some printers are connected to a serial port.
- *T-1 line* – A high-speed communication line used for data transfer. It consists of 24 64Kbps channels, which can be used separately, combined into clusters, or combined into a single connection that will provide a 1.5Mbps data transfer rate.
- *T-3 line* – A very high-speed communication line consisting of 43 64Kbps channels, which can be combined into a single connection that will provide a 43Mbps transfer rate.
- *Virtual memory* – Hard disk space allocated to augment RAM. It is not as fast as RAM because of the disk access time, but there are some uses for which it is suitable. One of these is to serve as RAM when the PC is running multiple programs that require more RAM than the computer has available.

6.2.0 Mainframe Computers

Computers fall into two major classifications: mainframe computers and personal computers (PCs).

A mainframe computer (*Figure 34*) is intended for enterprise-wide applications where a large amount of processing is required. Large businesses and government entities use mainframe computers to handle their accounting and payrolls, keep track of inventory, and manage the flow of information and products. A single computer may be accessed by many people using personal computers or "dumb" terminals that consist of a monitor and keyboard. Clients of mainframe computers are linked to it in a network by cabling. External links are provided by telecommunications systems.

6.3.0 Personal Computers

Personal computers have become so commonplace that it is now unusual to meet someone who does not use one. People use them at home to play video games, obtain movies and music, correspond with friends and family, and do their shopping. At work, people use them to control their environments, manage their schedules,

Figure 34 ♦ Mainframe computer.

Explain that there are two major classifications of computers: mainframes and personal computers (PCs).

Explain that mainframe computers are used to control large amounts of data processing.

Discuss the applications of personal computers.

Explain that PCs are either Macintosh or IBM compatible.

Discuss the common components of a personal computer system.

Provide a sample PC system for trainees to examine, and have them identify its components.

access information, and send information and correspondence around the world.

There are two basic types of PCs: The Apple Macintosh and the IBM PC Compatible. The latter, although originally developed by IBM, has been cloned by many companies and is the standard for business computing and most home computers. The Macintosh (or Mac) was the first PC to use a mouse, and has typically been favored by graphic designers and desktop publishers.

A tower configuration PC with expansion slots to accommodate additional special-purpose PC boards, space for built-in storage drives, and connectors (ports) to hook up peripherals such as scanners, printers, and game devices is shown in *Figure 35*.

The components found in a typical system include:

- *Computer case* – The case contains the processing circuits and other devices. It will be described further when we take a look inside.

- *Diskette (floppy) drive* – The floppy drive supports a 3½" magnetic diskette that will store about 1.4 megabytes of data. It is therefore a convenient medium for storing and transferring text files outside the computer, but is not very useful for graphics files, which consume a lot more storage space than text. For many years, the floppy disk was a major means of transferring files from one computer to another, and a floppy disk drive was standard on every PC. With the advent of other read-write devices with much greater capacity (for example, CD-R), the use of floppies has diminished.

- *Compact Disc Read-Only Memory (CD-ROM)* – These disks are portable and can hold up to 650 megabytes of data. They look exactly like audio CDs. CD-ROMs are very inexpensive to reproduce, but need special equipment to manufacture. They are mainly used to mass-distribute software.

Figure 35 ♦ Multimedia computer (tower configuration).

DID YOU KNOW?
The First Computer

The UNIVAC®, which stands for universal automatic computer, is recognized as the first computer developed for commercial use. The original UNIVAC® was developed by the Remington Rand Corporation under a government contract, and was delivered to the U.S. Census Bureau in 1951. The UNIVAC® was designed to perform arithmetic computations—in short, to "crunch" numbers. By today's standards, it was huge. The computer and its peripheral devices required approximately 360 square feet of floor space, and the equipment weighed nearly 16,000 pounds. Ten large magnetic tape drives were required to store data and programs. Today, a single two-pound notebook computer can do more work, and do it faster, than the original eight-ton computer system.

Instructor's Notes:

- *CD-Recordable (CD-R)* – The CD-R is a special blank CD which can be written to only once, using a special CD-R drive. Once it has been written to, it can then be read in most CD-ROM drives. The low cost and flexibility make this format ideal for backups and archiving. The availability of inexpensive CD-R drives has made this an extremely popular medium.
- *CD-Rewritable (CD-RW)* – The CD-RW is another recordable CD format. These disks can be erased and written to repeatedly, but are less popular because they are more expensive and can only be read with special drives.
- *Digital versatile disk (DVD) drive* – The DVD is newer than the CD-ROM and has a much greater capacity to store information. It has four different storage modes, the lowest of which can store 4.7 gigabytes of data, which is about seven times the capacity of a CD-ROM. In its highest capacity mode (both sides, two layers per side), it can store more than 17 gigabytes.
- *Monitor* – The monitor is the display device. It receives information from the video card inside the CPU case.
- *Keyboard* – The keyboard allows the operator to enter alphabet characters and numbers. It also contains function and control keys that are used by computer programs to perform special functions. The use of function keys is less common in a mouse-driven system where interaction with the computer is done by clicking on graphic objects rather than entering keystrokes. However, the keys are still functional and can be used in place of the mouse for many tasks.
- *Speakers* – With the advent of multimedia, audio was introduced to the PC. Now, it is unusual to find a PC without a set of speakers. Speakers with a built-in amplifier provide the best quality sound. Powered speakers have their own power source; the sound level can be adjusted using a volume control on the main speaker, rather than doing it in the operating system. Speakers require an audio card or special audio circuits on the motherboard.
- *Mouse* – The mouse is so-called because of its shape and long tail of cord. Trackballs are another common input device (*Figure 36*).

6.3.1 Monitors

The monitor that appeared in the first PCs was a **monochrome** device capable of displaying only text and line drawings (simple plots, etc). Since then, the monitor has evolved through several

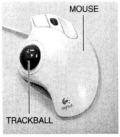

Figure 36 ♦ Computer input devices.

generations of devices capable of displaying color graphics, starting with the color graphic adapter (CGA) standard. CGA was followed by the enhanced graphic adapter (EGA) and video graphics array (VGA), which emerged in 1987. These were followed with super VGA (SVGA), XVGA, and Video Electronic Standards Association (VESA)-compliant monitors, which brought the screen image to new quality levels.

The quality of a monitor is measured in terms of its resolution and the number of colors it is capable of presenting. The screen image is made of up of **pixels** (short for "picture elements"), which are small dots that appear on the screen. Resolution is measured in terms of the number of pixels that appear, both horizontally and vertically. The more pixels, the higher the quality of the image.

Your computer must be set up to view the resolution of the subject image. This is done in Windows by going to the Display function on the Control Panel and resizing the window for a different resolution. Otherwise, the image on the screen will be undersized or oversized.

For example, if your monitor is set for 1280 × 1024 pixel resolution, and you are viewing a 800 × 600 pixel image, the image will be small in relation to the available viewing area. Changing the screen resolution will bring the image back to its normal size. Changing the resolution can be done from the display icon at the bottom of the screen on some Windows computers.

The number of colors (color depth) is another important measure of image quality. Early CGA monitors were capable of displaying 16 colors (4-bit color) with a resolution of 320 × 200 pixels. VGA brought that capability up to 256 colors (8-bit color) and 640 × 480 resolution, and VESA-compliant designs yielded 1280 × 1024 resolution with 16.8 million colors (24-bit color).

Describe the various types of monitors in common use.

Explain the relationship between the number of pixels displayed on the screen and the screen resolution.

Show Transparency 33 (Figure 37). Discuss the type of ports typically found on a PC.

Point out the common ports on a sample PC.

 **Touch Screens**

Touch screens are popular computer input devices that you may find in HVAC control systems. The touch screen takes the place of the keyboard and mouse by providing selections that the operator touches with a finger or stylus. The touch screen overlays the computer monitor. The personal digital assistant (PDA) is one common touch screen device.

Typically, an electric current flows through the touch sensor, which overlays the computer screen. Touching the screen at a particular point changes the current. A special controller and dedicated software translate the change into a signal that the computer is able to process. The change is comparable to clicking a mouse cursor at a particular point on the computer monitor.

Monitors range in size from 14" to 21", with 15" to 17" being commonplace in homes and businesses. The size refers to the distance from one corner of the screen to the other (diagonally); the actual display area will be smaller. Although monitors typically look like the ones depicted earlier, flat screen displays are available and show signs of coming into general use.

6.3.2 Connections

On most PCs, the connection devices are located at the rear of the unit (*Figure 37*). In the computer world, connectors are referred to as ports.

- *Parallel ports* – A PC will have one or two parallel ports, which are designed to mate with 25-pin male connectors. They are designed to interface with peripheral devices such as printers, scanners, and tape drives. The computer designates parallel ports as LPT (line print terminal). If the PC has two parallel ports, they will be designated LPT 1 and LPT 2.
- *Serial ports* – A serial port is usually a 9-pin connector, but some are 25-pin connectors. They can be distinguished from other ports because they are male connectors. The serial port is used to connect such components as a mouse or external modem to the PC. Serial ports are designated COM ports.
- *Monitor port* – The monitor port is a 15-pin connector. The cable that carries video from the computer to the monitor is plugged in here.
- *Keyboard port* – The keyboard is connected to the computer with a round 9-pin connector. The connector is keyed to prevent the cable from being connected incorrectly.

7.24 HVAC ♦ LEVEL TWO

Instructor's Notes:

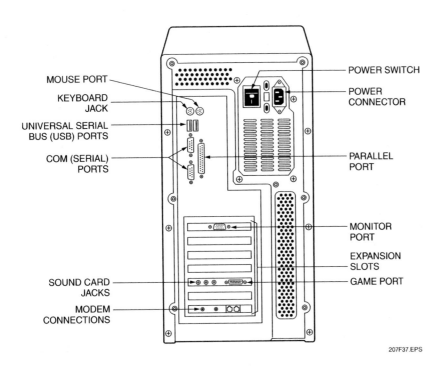

Figure 37 ♦ Rear view of a tower configuration PC.

Explain the difference between computer memory and non-volatile storage media.

- *Universal serial bus (USB) ports* – A USB port allows the user to connect a wide variety of peripheral devices to the PC. One of the problems with earlier PCs is that they did not have enough ports to connect all of the external peripherals a person might want to use. If you wanted to use a document scanner, for example, you might have to disconnect the printer to obtain a connection. On many PCs, the USB ports are located on the front of the computer to allow easy swapping of peripheral devices.
- *Game port* – The game port is a 15-pin connector used to connect a joystick or similar device used for games and simulations.
- *Phone jack* – This jack is used to connect the modem to the phone system. It is the same type of jack used in standard telephone circuits.
- *Mouse port* – Some versions of the mouse are connected with a 6-pin connector. Other versions are connected to a serial port.

6.4.0 Computer Storage Media

Since computer memory is erased when you shut the power off, computers need a way to keep information without power. This is called non-volatile storage. The computer can retrieve information from a non-volatile storage device and load it into RAM (reading) to work with it, then record information back to the device (writing). *Figure 38* shows some common storage media.

The most important form of non-volatile storage is the hard drive. Most computers contain some form of hard drive. This is a high-capacity, high-speed magnetic disk, mounted permanently inside the computer. The computer's operating system, applications, and other data are stored on the hard drive. The capacity of modern hard drives is measured in gigabytes, and is increasing rapidly. A computer can have more than one hard disk drive. Some computers do not contain internal hard drives, but store everything remotely on a network storage device.

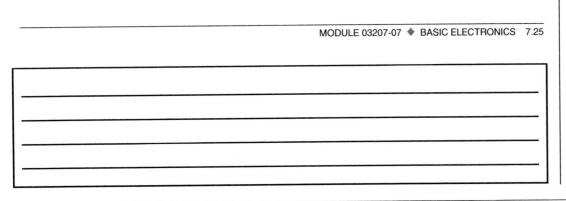

Discuss common types of removable media storage devices.

Provide examples of removable media for trainees to examine.

Figure 38 ◆ Data storage devices.

The efficiency of a hard drive is determined by two critical factors:

- The rotational speed of the disk, measured in revolutions per minute (rpm). The faster it rotates, the faster it can find and retrieve information.
- Access (seek) time, measured in milliseconds (ms). Hard drives typically have access times ranging from 5 to 15 milliseconds. The shorter the time, the faster (and more expensive) the drive.

Transferring information to other computers, or making backup copies of important data from the hard disk is usually done with removable media. Removable media are storage devices that can be removed and replaced while the computer is running. There are several kinds of removable media available.

The CD-ROM is popular for distributing software to many computers, because it holds a lot of data and is very inexpensive to mass-produce. The disadvantage of CD-ROM is that it is read-only. Information on a CD-ROM cannot be changed, and the equipment for producing CD-ROMs is very expensive. To get around this limitation, CD-R technology was developed. CD-R (Compact Disc-Recordable) disks have a special coating that can be changed by exposure to a laser beam. These disks can be written to by special CD-R drives. Once they have been written to, these disks will function like normal CD-ROMs in most drives. The process of writing information to a CD-R is called burning. The availability of inexpensive CD-R drives and media has made this an extremely popular format for archiving and storing large files. Some of these drives also incorporate CD-RW technology, which makes it possible to erase and rewrite information on the same disk. CD-RWs are not reliable for long-term storage due to their gradual breakdown. Because they are more expensive and cannot be read by normal CD-ROM drives, these disks are not as popular as CD-R disks.

DVD-ROM is similar to CD-ROM, but is a newer technology with a much larger storage capacity. Recordable and rewritable DVDs are also available.

Inside Track

Hard Drive Operation

If you are familiar with the phonograph, you have a rough idea of how a hard drive works. The storage device is a hard magnetic disk. The data is read onto the disk and extracted from the disk by arms that contain read-write heads. The heads are millionths of an inch from the disk, but do not actually touch it. The disk spins at a very high speed—5,400 rpm is considered slow; 10,000 rpm is a fast drive. As the disk spins, the arms move at a high speed to place or extract data as commanded by the operating system. The disk is divided into tracks and sectors by the operating system in order to simplify data access.

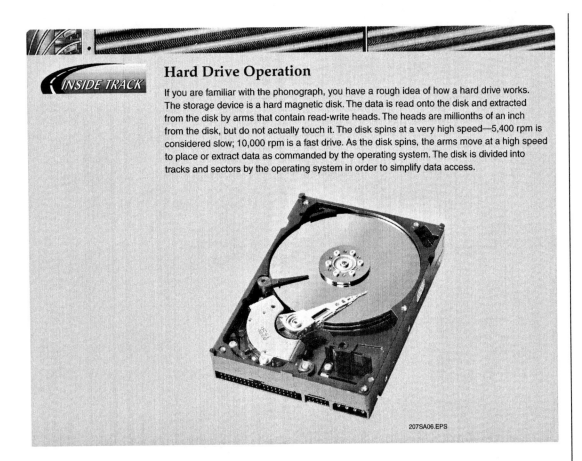

207SA06.EPS

The diskette, or floppy disk, is becoming less popular as the need to store large files increases. A standard 3.5" floppy disk has a capacity of about 1.4 megabytes. This is enough to hold a few documents or small programs, but not enough for multimedia use. It would take hundreds of floppy disks to hold as much information as a single CD-ROM. Floppy disks are slow, and not reliable for long-term storage. Their biggest advantage is that they are the cheapest storage medium that can be easily written to, as well as read.

Digital magnetic tape drives can store large amounts of information reliably. There are many different kinds of digital tape. Currently, a large-capacity tape can hold hundreds of gigabytes, and as technology improves, this capacity is constantly growing. Digital tape is typically used to store a large amount of data at once. The most important use of tapes is making backups of the information on a hard disk in case the hard disk fails. Compared to other storage media, tapes are slow.

There are many other types of removable storage media, including memory cards, removable hard drives, and USB flash disks. Each type has advantages and disadvantages, and it is not unusual to find more than one kind of storage device within the same system.

Have trainees complete the Review Questions, and go over the answers prior to administering the Module Examination.

Review Questions

1. An electrical charge occurs when an atom gains or loses _____.
   a. weight
   b. protons
   c. electrons
   d. protons

2. Which of the following is true of semiconductors?
   a. They are better conductors than insulators.
   b. They are better insulators than conductors.
   c. When mixed with impurities they become insulators.
   d. They are neither good insulators nor good conductors.

3. All semiconductor devices are _____.
   a. made from P-type and N-type material
   b. made from diodes
   c. forward-biased
   d. reverse-biased

4. For a diode to conduct current the _____.
   a. anode must be negative with respect to the cathode
   b. anode must be positive with respect to the cathode
   c. diode must be reverse-biased
   d. diode must be used in a half-wave rectifier

5. A bridge rectifier might be used in place of a full-wave rectifier because it _____.
   a. is more effective
   b. produces AC while the full-wave rectifier produces DC
   c. does not require a center-tapped transformer
   d. uses fewer diodes

6. Color-coding is normally used on _____ resistors.
   a. wire-wound
   b. carbon composition
   c. wire-wound and carbon composition
   d. variable

7. If a thermistor has a positive coefficient of resistance, its resistance will _____ as temperature increases.
   a. increase
   b. decrease
   c. remain the same
   d. decrease then return to its original value

8. A cadmium sulfide flame detector is sensitive to _____.
   a. light
   b. heat
   c. smoke
   d. vibration

9. If it is necessary to handle a PC board containing an integrated circuit chip, you should wear _____.
   a. a dust mask
   b. a hard hat
   c. a grounding wrist strap
   d. safety glasses

10. A computer's BIOS is normally located _____.
    a. on the hard drive
    b. in RAM
    c. on the back of the unit
    d. in ROM

7.28 HVAC ♦ LEVEL TWO

Instructor's Notes:

Summary

The science of electronics has made it possible to replace mechanical and electromechanical devices with electronic devices that take up less space, work much faster, and last longer. Electronic devices and circuits provide greater precision as well as a greater range of control. These capabilities add up to improved comfort control, greater operating efficiency, and easier servicing. Modular packaging of electronic controls, along with built-in diagnostic and testing capabilities, make it easier for service technicians to troubleshoot and repair HVAC systems.

Summarize the major concepts presented in the module.

Administer the Module Examination. Record the results on Craft Training Report Form 200, and submit the results to the Training Program Sponsor.

Administer the Performance Test, and fill out Performance Profile Sheets for each trainee. If desired, trainee proficiency noted during laboratory sessions may be used to complete the Performance Test. Record the results on Craft Training Report Form 200, and submit the results to the Training Program Sponsor.

Notes

Trade Terms Introduced in This Module

Anode: The positive terminal of a diode.

Basic input/output system (BIOS): The basic method by which a computer exchanges information.

Bridge rectifier: A rectifier circuit that uses four diodes, two of which conduct current on each half-cycle. Has the advantage of not needing a center-tapped transformer.

Cathode: The negative terminal of a diode.

Centrifugal force: The force that makes rotating objects tend to move away from the center of rotation.

Chip: A common term used to describe an integrated circuit.

Diac: A three-layer diode designed for use as a trigger in AC power control units.

Electromechanical components: Electrical devices that contain moving parts.

Electronics: The science that deals with the behavior and effects of electron movement in conductors, insulators, and semiconductors.

Free electrons: Valence electrons that can easily be knocked out of orbit.

Full-wave rectifier: A rectifier circuit that uses two diodes.

Half-wave rectifier: A rectifier circuit that uses a single diode.

Integrated circuit: A plug-in circuit containing microminiature electronic circuits. Sometimes called a chip.

Light-emitting diode (LED): A diode that gives off light when current flows through it.

Microminiaturization: The technology that allows the manufacture of microscopic electronic circuits.

Microprocessor: An integrated circuit chip designed to perform computing functions. The microprocessor is the heart of a personal computer.

Monochrome: Able to display a single color.

Photo diode: A diode that conducts current when exposed to light.

Pixel: An abbreviation for picture element. A pixel is a single dot in a graphic image on a computer screen.

Rectification: The conversion of AC into DC using diodes.

Semiconductor: A material that contains four valence electrons and is used in the manufacture of integrated circuits.

Silicon-controlled rectifier (SCR): A device that is used mainly to convert AC voltage into DC voltage. To do so, however, the gate of the SCR must be triggered before the device will conduct current.

Triac: A bi-directional triode thyristor that functions as an electrically controlled switch for AC loads.

Valence electrons: Electrons located in the outer orbit of an atom.

Additional Resources and References

Additional Resources

This module is intended to be a thorough resource for task training. The following reference works are suggested for further study. These are optional materials for continued education rather than for task training.

Electronics Fundamentals: Circuits, Devices, and Applications, 2007. Thomas L. Floyd. Prentice Hall.

Electric Circuit Fundamentals, 2004. Thomas L. Floyd. Prentice Hall.

Figure Credits

Carrier Corporation, 207SA01

Powerex, Inc., 207SA02

Topaz Publications, Inc., 207SA03, 207F31, 207F33, 207F35, 207F36, 207F38

DPA Components International, 207F32

Sun Microsystems Inc., 207F34

Dukane Communication Systems, 207SA05

Western Digital Corporation, 207SA06

MODULE 03207-07 — TEACHING TIPS

The following are suggested activities or instructional methods to help you teach the material in this module.

General

When you call on someone to answer a question, the rest of the class relaxes or even tunes out because they expect that the question and answer will take place only between you and the trainee you called on. Instead, use this technique to involve more trainees in answering questions and to keep them on their toes.

1. Ask trainees to define a term or explain a concept.
2. After one trainee has answered, ask a trainee seated nearby if the answer is right. Then ask whether a trainee in the back of the room agrees.
3. Ask trainees to explain why they think an answer is right or wrong.
4. Use the session to clear up incorrect ideas and encourage trainees to learn from their mistakes.

Sections 1.0.0 through 6.0.0

Quick Quiz

This Quick Quiz will familiarize trainees with the terms and definitions that are commonly used in basic electronics. You will need photocopies of the quiz provided in the following page. Trainees will need pencils. If you allow trainees to use the Trainee Guide, decrease the amount of time you give them to complete the quiz.

1. Make a photocopy of the quiz for each trainee.
2. Give trainees between 5 and 10 minutes to complete the quiz.
3. Go over the answers to the quiz.
4. Ask trainees if they have questions.

Answers to Quick Quiz

1. l
2. o
3. k
4. a
5. g
6. c
7. i
8. f
9. m
10. n
11. d
12. b
13. j
14. h
15. e

Quick Quiz   *Basic Electronics*

For each description listed, identify the term that the text best describes. Write the corresponding letter in the blank provided.

_____ 1. Materials called _____ are the basis for solid-state electronics and are used to manufacture integrated circuits.

_____ 2. Electrons contained in the outer shell of an atom are called _____.

_____ 3. The process of _____ converts AC to DC using diodes.

_____ 4. Most circuits using diodes are designed so that the diode will not conduct current unless the _____ is positive with respect to the cathode.

_____ 5. A(n) _____ produces pulsating DC voltage that is not clean enough to operate electronic circuits.

_____ 6. The _____ contains four diodes and provides a smooth DC output.

_____ 7. A solid-state device that is turned on by light is a(n) _____

_____ 8. A(n) _____ can be thought of as an AC switch.

_____ 9. A group of semiconductors known as thyristors include diacs, triacs, and _____.

_____ 10. Light dimmers use _____ to control the power applied to the load.

_____ 11. On some diodes, the diode housing is beveled or enlarged to indicate the _____ end.

_____ 12. The first set of instructions that is run when a computer is started is the _____.

_____ 13. The image that appears on a computer screen is made of of small dots or _____.

_____ 14. The first PC monitors were _____ and could only display text and line drawings.

_____ 15. Thousands of circuits can fit on an integrated circuit or _____ that is the size of your thumbnail.

a. anode
b. basic input/output system (BIOS)
c. bridge rectifier
d. cathode
e. chip
f. diac
g. half-wave rectifier
h. monochrome
i. photo diode
j. pixels
k. rectification
l. semiconductors
m. silicon-controlled rectifiers (SCRs)
n. triacs
o. valence electrons

Section 4.0.0 *Identification of Semiconductor Components*

This exercise will give trainees practice in identifying semiconductor components. Trainees will need appropriate personal protective equipment, pencils, and paper. You will need to obtain various types of semiconductor components. Allow 15 to 20 minutes for this exercise. This exercise corresponds to Performance Task 1.

Alternatively, have trainees create flash cards depicting the symbol for each type of component. As you hold up a component, have all trainees hold up their corresponding flash card.

1. Describe various components and explain how to identify them. Set up workstations with one component. If desired, demonstrate various testing procedures and safety precautions.
2. Have trainees circulate to each of the workstations and identify each component. If desired, have the test the components or write the symbol for that component.
3. Once all trainees have visited each station, have one trainee stand and identify the component at their station. Answer any questions they may have.

Section 5.3.0 *Think About It – Troubleshooting an Electronic Furnace Control*

To determine if an electronic control is faulty, check all input and output signals. In this example, all input and output signals appear to be in order. There is 24V power available to energize the gas valve, but there is no burner in operation. In this case, the problem appears to be external to the electronic control, so it should not be replaced.

MODULE 03207-07 — ANSWERS TO REVIEW QUESTIONS

Answer	Section
1. c	2.0.0
2. d	3.3.0
3. a	4.0.0
4. b	4.1.0
5. c	4.1.1
6. b	4.6.1
7. a	4.7.0
8. a	4.8.0
9. c	5.4.0
10. d	6.1.0

CONTREN® LEARNING SERIES — USER UPDATE

NCCER makes every effort to keep these textbooks up-to-date and free of technical errors. We appreciate your help in this process. If you have an idea for improving this textbook, or if you find an error, a typographical mistake, or an inaccuracy in NCCER's Contren® textbooks, please write us, using this form or a photocopy. Be sure to include the exact module number, page number, a detailed description, and the correction, if applicable. Your input will be brought to the attention of the Technical Review Committee. Thank you for your assistance.

Instructors – If you found that additional materials were necessary in order to teach this module effectively, please let us know so that we may include them in the Equipment/Materials list in the Annotated Instructor's Guide.

Write: Product Development and Revision
National Center for Construction Education and Research
3600 NW 43rd St, Bldg G, Gainesville, FL 32606

Fax: 352-334-0932

E-mail: curriculum@nccer.org

Craft _____ Module Name _____

Copyright Date _____ Module Number _____ Page Number(s) _____

Description
_____
_____
_____
_____

(Optional) Correction
_____
_____
_____

(Optional) Your Name and Address
_____
_____
_____

Module 03208-07

Introduction to Control Circuit Troubleshooting

NCCER STANDARDIZED CRAFT TRAINING PROGRAM

The National Center for Construction Education and Research (NCCER) provides a standardized national program of accredited craft training. Key features of the program include instructor certification, competency-based training, and performance testing. The program provides trainees, instructors, and companies with a standard form of recognition through a National Craft Training Registry. The program is described in full in the *Guidelines for Accreditation*, published by NCCER. For more information on standardized craft training, contact the NCCER by writing us at 3600 NW 43rd St., Bldg. G, Gainesville, FL 32606; calling 352-334-0911; or emailing info@nccer.org. More information may be found at our website, www.nccer.org.

HOW TO USE THIS ANNOTATED INSTRUCTOR'S GUIDE

Each page presents two sections of information. The larger section displays each page exactly as it appears in the Trainee Module. The narrow column ties suggested trainee and instructor actions to each page and provides icons (detailed below) to call your attention to material, safety, audiovisual, or testing requirements. The bottom of each page includes space for your notes.

The **Audiovisual** icon indicates an appropriate time to show a transparency or other audiovisual aid.

The **Classroom** icon prompts you to define a term, stress a point, ask trainees to explain a concept, or give examples.

The **Demonstration** icon directs you to show trainees how to perform tasks.

The **Examination** icon tells you to administer the written module examination.

The **Homework** icon is placed where you may wish to assign reading for the next class, assign a project, or advise trainees to prepare for an examination.

The **Laboratory** icon is used when trainees are to practice performing tasks.

The **Materials** icon is a reminder for you to gather materials needed for classes, labs, and testing.

The **Performance Testing** icon tells you to administer a performance test or a portion thereof.

The **Safety** icon is used to emphasize safety issues. It is often keyed to *Caution* and *Warning!* statements in the Trainee Module.

The **Teaching Tip** icon indicates additional guidance is available, such as how to conduct an exercise, get the most educational value from a field trip, or encourage class participation. Teaching Tips may expand on a feature (*Think About It, Did You Know?*) or provide *Quick Quizzes* or similar exercises. You will be referred to the Teaching Tips section at the back of the module if there is additional material.

The **Combination** icon indicates that the laboratory listed corresponds with a performance task. If desired, you can note the proficiency of the trainees during the laboratory, and use it to satisfy performance testing requirements.

PREPARATION

Before teaching this module, you should review the Objectives, Performance Tasks, Materials and Equipment List, and Module Outline. Be sure to allow ample time to prepare your own training or lesson plan and gather all required materials and equipment.

Introduction to Control Circuit Troubleshooting
Annotated Instructor's Guide

Module 03208-07

MODULE OVERVIEW

This module covers the various types of thermostats used in HVAC systems. It also covers hydronic, pneumatic, and digital controls and introduces the trainee to control circuit analysis and troubleshooting.

PREREQUISITES

Prior to training with this module, it is recommended that the trainee shall have successfully completed *Core Curriculum*; *HVAC Level One*; and *HVAC Level Two*, Modules 03201-07 through 03207-07.

OBJECTIVES

Upon completion of this module, the trainee will be able to do the following:

1. Explain the function of a thermostat in an HVAC system.
2. Describe different types of thermostats and explain how they are used.
3. Demonstrate the correct installation and adjustment of a thermostat.
4. Explain the basic principles applicable to all control systems.
5. Identify the various types of electromechanical, electronic, and pneumatic HVAC controls, and explain their function and operation.
6. Describe a systematic approach for electrical troubleshooting of HVAC equipment and components.
7. Recognize and use equipment manufacturer's troubleshooting aids to troubleshoot HVAC equipment.
8. Demonstrate how to isolate electrical problems to faulty power distribution, load, or control circuits.
9. Identify the service instruments needed to troubleshoot HVAC electrical equipment.
10. Make electrical troubleshooting checks and measurements on circuits and components common to all HVAC equipment.
11. Isolate and correct malfunctions in a cooling system control circuit.

PERFORMANCE TASKS

Under the supervision of the instructor, the trainee should be able to do the following:

1. Identify various types of thermostats and explain their operation and uses.
2. Install a conventional 24V bimetal thermostat and hook it up using the standard coding system for thermostat wiring.
3. Check and adjust a thermostat, including heat anticipator setting and indicator adjustment.
4. Program an electronic programmable thermostat.
5. Identify electrical, electronic, and pneumatic components and circuits, recognize their diagram symbols, and explain their functions.
6. Interpret control circuit diagrams.
7. Perform electrical tests and troubleshooting as follows:
   - Single- and three-phase input voltage measurements
   - Fuse and circuit breaker checks
   - Resistive and inductive load checks
   - Switch and contactor/relay checks
   - Control transformer checks
8. Perform electrical tests and troubleshooting of compressor and fan motors as follows:
   - Start and run capacitor checks
   - Start relay and start thermistor checks
   - Open, shorted, and grounded winding check

MATERIALS AND EQUIPMENT LIST

Overhead projector and screen
Transparencies
Blank acetate sheets
Transparency pens
Whiteboard/chalkboard
Markers/chalk
Pencils and scratch paper
Appropriate personal protective equipment
Length of wire with alligator clips
Thermostat wire
Wire stripper
Fuses
Videotapes/DVDs (optional):
  Electrical Components and Their Symbols
  Electrical Troubleshooting
  An Introduction to Hydronic Heating
TV/VCR/DVD player (optional)
Programmable electronic thermostat
Mercury bulb thermostat
Line voltage thermostat

Selection of 24V thermostats:
  Heating only
  Cooling only
  Heating-cooling
  Automatic changeover
  Heat pump thermostat with emergency heat control
Level
Clamp-on ammeter
Lockout devices and tags
Relays
Contactors
Motor starters
Multimeter
Circuit breakers
Operating heating/cooling system
Manufacturers' troubleshooting aids
Copies of Quick Quiz*
Module Examinations**
Performance Profile Sheets**

* Located in the back of this module.
**Located in the Test Booklet.

SAFETY CONSIDERATIONS

Ensure that the trainees are equipped with appropriate personal protective equipment and know how to use it properly. This module requires trainees to work with heating/cooling appliances. Make sure that all trainees are briefed on appropriate safety procedures. Emphasize electrical safety and lockout/tagout procedures.

ADDITIONAL RESOURCES

This module is intended to present thorough resources for task training. The following reference works are suggested for both instructors and motivated trainees interested in further study. These are optional materials for continued education rather than for task training.

Air Conditioning Systems, Principles, Equipment, and Service. Latest Edition. Joseph Moravek. Upper Saddle River, NJ: Prentice Hall.

HVAC Servicing Procedures, 1995. Syracuse, NY: Carrier Corporation.

Pocket Guide to Electrical Installations Under NEC 2002, Volumes I and II, 2001. Quincy, MA: National Fire Protection Association.

Refrigeration and Air Conditioning, An Introduction to HVAC/R, Fourth Edition. Larry Jeffus. Air Conditioning and Refrigeration Institute. Upper Saddle River, NJ: Prentice Hall.

TEACHING TIME FOR THIS MODULE

An outline for use in developing your lesson plan is presented below. Note that each Roman numeral in the outline equates to one session of instruction. Each session has a suggested time period of 2½ hours. This includes 10 minutes at the beginning of each session for administrative tasks and one 10-minute break during the session. Approximately 30 hours are suggested to cover *Introduction to Control Circuit Troubleshooting*. You will need to adjust the time required for hands-on activity and testing based on your class size and resources. Because laboratories often correspond to Performance Tasks, the proficiency of the trainees may be noted during these exercises for Performance Testing purposes.

Topic **Planned Time**

Session I. Introduction and Thermostats

    A. Introduction _____

    B. Thermostats _____

    C. Laboratory _____

    Trainees practice identifying various types of thermostats and explaining the operation and uses of each. This laboratory corresponds to Performance Task 1.

    D. Laboratory _____

    Under your supervision, have trainees install a conventional 24V bimetal thermostat using the standard coding for thermostat wiring and make any necessary checks and adjustments. This laboratory corresponds to Performance Tasks 2 and 3.

    E. Laboratory _____

    Have trainees program an electronic thermostat. This laboratory corresponds to Performance Task 4.

Session II. HVAC Control Systems

    A. HVAC Control Systems _____

    B. Laboratory _____

    Have trainees practice identifying and describing the function of various electrical, electronic, and pneumatic components and circuits. This laboratory corresponds to Performance Task 5.

Session III. Control Circuit Sequence of Operation

    A. Control Circuit Sequence of Operation _____

    B. Laboratory _____

    Have trainees practice interpreting control circuit diagrams for selected HVAC equipment. This laboratory corresponds to Performance Task 6.

Session IV. Organization and Safety

    A. Using an Organized Approach to Electrical Troubleshooting _____

    B. Safety _____

Session V. Troubleshooting I: Input Power, Load, and Control Circuits

    A. HVAC System Troubleshooting _____

    B. HVAC Equipment Input Power, Load, and Control Circuits _____

Sessions VI and VII. Troubleshooting II: Electrical System

    A. Electrical Troubleshooting Common to All HVAC Equipment _____

    B. Laboratory _____

    Under your supervision, have trainees perform various electrical tests and troubleshooting, including input voltage measurements and components testing. This laboratory corresponds to Performance Task 7.

Sessions VIII and IX. Troubleshooting III: Motors
  A. Motors and Motor Circuit Troubleshooting
  B. Laboratory

  Under your supervision, have trainees perform various electrical tests and troubleshooting of compressors and fan motors. This laboratory corresponds to Performance Task 8.

Session X. Hydronic and Pneumatic Systems
  A. Hydronic Systems
  B. Pneumatic Systems

Session XI. Digital Systems
  A. HVAC Digital Control Systems

Session XII. Review and Testing
  A. Review
  B. Module Examination
    1. Trainees must score 70% or higher to receive recognition from NCCER.
    2. Record the testing results on Craft Training Report Form 200, and submit the results to the Training Program Sponsor.
  C. Performance Testing
    1. Trainees must perform each task to the satisfaction of the instructor to receive recognition from NCCER. If applicable, proficiency noted during laboratory exercises can be used to satisfy the Performance Testing requirements.
    2. Record the testing results on Craft Training Report Form 200, and submit the results to the Training Program Sponsor.

HVAC Level Two

03208-07

Introduction to Control Circuit Troubleshooting

Assign reading of Module 03208-07.

03208-07
Introduction to Control Circuit Troubleshooting

Topics to be presented in this module include:

1.0.0	Introduction	8.2
2.0.0	Thermostats	8.2
3.0.0	HVAC Control Systems	8.13
4.0.0	Control Circuit Sequence of Operation	8.23
5.0.0	Using an Organized Approach to Electrical Troubleshooting	8.27
6.0.0	Safety	8.32
7.0.0	HVAC System Troubleshooting	8.33
8.0.0	HVAC Equipment Input Power, Load, and Control Circuits	8.33
9.0.0	Electrical Troubleshooting Common to all HVAC Equipment	8.37
10.0.0	Motors and Motor Circuit Troubleshooting	8.44
11.0.0	Hydronic Controls	8.50
12.0.0	Pneumatic Controls	8.53
13.0.0	HVAC Digital Control Systems	8.57

Overview

Most malfunctions that occur in heating and air conditioning equipment are caused by failures in the power distribution or control system. In order to find and correct these malfunctions, service technicians must be able to interpret the schematic and wiring diagrams and other service literature provided by the manufacturer. The successful troubleshooter is able to read these diagrams and determine the correct operating sequence. Once the correct operating sequence is known, the troubleshooter can determine where the system has failed. Without that knowledge, troubleshooting is no more than a guessing game.

Instructor's Notes:

Objectives

When you have completed this module, you will be able to do the following:

1. Explain the function of a thermostat in an HVAC system.
2. Describe different types of thermostats and explain how they are used.
3. Demonstrate the correct installation and adjustment of a thermostat.
4. Explain the basic principles applicable to all control systems.
5. Identify the various types of electromechanical, electronic, and pneumatic HVAC controls, and explain their function and operation.
6. Describe a systematic approach for electrical troubleshooting of HVAC equipment and components.
7. Recognize and use equipment manufacturer's troubleshooting aids to troubleshoot HVAC equipment.
8. Demonstrate how to isolate electrical problems to faulty power distribution, load, or control circuits.
9. Identify the service instruments needed to troubleshoot HVAC electrical equipment.
10. Make electrical troubleshooting checks and measurements on circuits and components common to all HVAC equipment.
11. Isolate and correct malfunctions in a cooling system control circuit.

Trade Terms

Actuator
Analog-to-digital converter
Automatic changeover thermostat
Bleed control
Cooling compensator
Deadband
Differential
Droop
Electric-pneumatic (E-P) relay
Fault isolation diagram
Invar®
Label diagram
Ladder diagram
Pneumatic-electric (P-E) relay
Sub-base
Thermostat base
Troubleshooting
Troubleshooting table
Wiring diagram

Required Trainee Materials

1. Pencil and paper
2. Appropriate personal protective equipment

Prerequisites

Before you begin this module, it is recommended that you successfully complete *Core Curriculum*; *HVAC Level One*; and *HVAC Level Two*, Modules 03201-07 through 03207-07.

This course map shows all of the modules in the second level of the HVAC curriculum. The suggested training order begins at the bottom and proceeds up. Skill levels increase as you advance on the course map. The local Training Program Sponsor may adjust the training order.

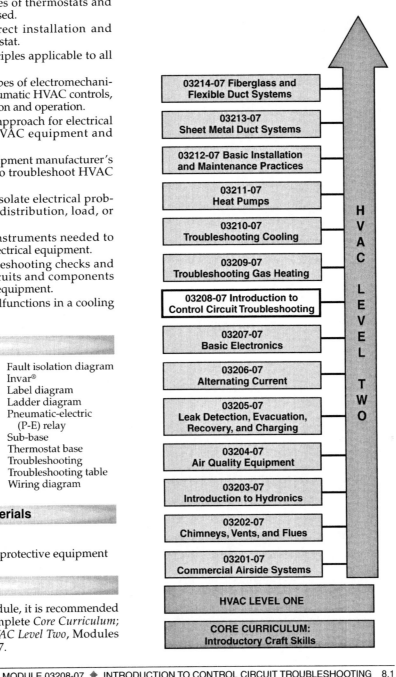

Ensure that you have everything required to teach the course. Check the Materials and Equipment list at the front of this module.

See the general Teaching Tip at the end of this module.

Explain that terms shown in bold are defined in the Glossary at the back of this module.

Show Transparency 1, Objectives, and Transparency 2, Performance Tasks. Review the goals of the module, and explain what will be expected of the trainee.

Review the modules covered in Level Two and explain how this module fits in.

List the types of control devices used in HVAC systems.

Note that the room thermostat is the primary control in an HVAC system.

Explain that residential systems typically use one thermostat, while commercial systems often have more than one zone of heating/cooling, each with its own thermostat.

Point out that thermostats either use electronic or bimetal sensing elements.

Show Transparency 3 (Figure 1). Discuss the construction and operation of bimetal sensing elements.

See the Teaching Tips for Section 2.0.0 at the end of this module.

Provide a mercury bulb thermostat for the trainees to examine.

1.0.0 ♦ INTRODUCTION

The first half of this module describes the operation of several common HVAC electrical control devices. You have been introduced to some of these devices in your HVAC Level One training. This module expands on this information and also covers several additional devices.

In an HVAC system, control devices such as relays, contactors, switches, and thermostats interact to control every aspect of system operation. Control devices are typically used in circuits that operate to stop and start HVAC system load devices such as motors, compressors, and heaters.

The second half of this module introduces the task of electrical **troubleshooting** with a focus placed on methods that are common to most types of HVAC equipment. Troubleshooting methods that are unique to the specific types of HVAC equipment, such as gas heating, cooling, and heat pumps are covered in later modules. Control devices used in hydronic and pneumatic control systems and digital control systems are introduced in this module.

2.0.0 ♦ THERMOSTATS

The room thermostat is the primary control in an HVAC system. It can be as simple as the single temperature-sensitive switch you learned about in HVAC Level One. It can also be a complex collection of sensing elements and switching devices that provide many levels of control. The thermostats referred to in this module are the control devices that are mounted on a wall in the conditioned space.

Most residences, and many small commercial businesses such as retail stores and shops, have a single thermostat. Large office buildings, shopping malls, and factories will be divided into cooling and heating zones, and will have several thermostats—one for each zone. An important fact to remember about zoned heating and cooling is that each zone is independent of the others. Therefore, each thermostat must control either a separate system or the airflow from a common system.

2.1.0 Principles of Operation

Programmable electronic thermostats using electronic sensing elements are becoming increasingly popular; however, thermostats with bimetal sensing elements are still being manufactured and are still used in line-voltage thermostats. Although not quite as precise as the electronic thermostat, bimetal devices are considerably less expensive and have proven to be effective and reliable. Most bimetal thermostats will maintain the space temperature within ±2° of the setpoint. A good electronic thermostat will maintain the temperature within ±1°.

A bimetal element (*Figure 1*) is composed of two different metals bonded together; one is usually copper or brass. The other, a special metal called **Invar**®, contains 36 percent nickel. When heated, the copper or brass has a more rapid expansion rate than the Invar®, and changes the shape of the element. The movement that occurs when the bimetal changes shape is used to open or close switch contacts in the thermostat.

While bimetal elements are constructed in various shapes, the spiral-wound element is the most compact in construction and the most widely used. In *Figure 1*, for example, a glass bulb containing mercury (a conductor) is attached to a coiled bimetal strip. When the bulb is tipped in one direction, the mercury makes an electrical connection between the contacts and the switch is closed. When the bulb is tipped in the opposite direction, the mercury moves to the other end of the bulb and the switch is opened. Thermostat switching action should take place rapidly to prevent arcing, which could cause damage to the switch contacts. A magnet is used to provide rapid action to help eliminate the arcing potential in some bimetal thermostats.

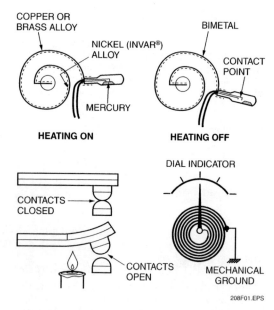

Figure 1 ♦ Bimetal sensing elements.

8.2 HVAC ♦ LEVEL TWO

Instructor's Notes:

WARNING!
Mercury is toxic. Even short-term exposure may result in damage to the lungs and central nervous system. Mercury is also an environmental hazard. Do not dispose of thermostats containing mercury bulbs in regular trash. Contact your local waste management or environmental authority for disposal/recycling instructions.

Most residential and small commercial thermostats are of the low-voltage (24V) type. With low-voltage control circuits, there is less risk of electrical shock and less chance of fire from short circuits. Low-voltage components are also less expensive and less likely to produce arcing, coil burnout, and contact failure. Some self-generating systems use a millivolt power supply that generates about 750mV to operate the thermostat circuit. These thermostats are very similar in construction and design to low-voltage thermostats; however, they are not interchangeable with 24V thermostats.

NOTE
Mercury-bulb thermostats are still in use but are being phased out. Digital non-programmable thermostats are replacing mercury-bulb types.

2.2.0 Heating-Only Thermostats

A wall-mounted heating-only thermostat typically contains a temperature-sensitive switch as shown in *Figure 2*. In this arrangement, the heating device (such as a furnace) will not come on unless the thermostatic switch is calling for heat.

When the temperature in the conditioned space reaches the thermostat setpoint, the thermostatic switch will open. Because of the residual heat in the heat exchangers and the continued rotation of the fan as it slowly comes to a stop, the temperature will overshoot the setpoint. To avoid the discomfort that this condition might cause, the thermostat contains an adjustable heat anticipator that causes the thermostat to open before the temperature in the space reaches the setpoint.

As discussed in an earlier module, the heat anticipator is a small resistance heater in series with the switch contacts. The anticipator heats the bimetal strip, causing the contacts to open early.

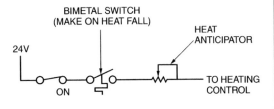
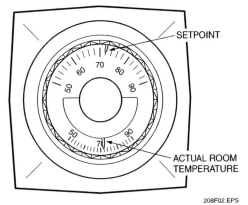

Figure 2 ◆ Heating-only thermostat.

2.3.0 Cooling-Only Thermostats

The cooling thermostat (*Figure 3*) is the opposite of a heating thermostat. When the bimetal coil heats up and unwinds, the mercury switch closes its contacts and starts the compressor. When the cooling thermostat is turned down to make the conditioned space cooler, it tips the mercury bulb so that the coil must cool more and wind tighter to turn the cooling system off.

Cooling thermostats contain a device called a **cooling compensator** to help improve indoor comfort. The cooling compensator (*Figure 4*) is a fixed resistance in parallel with the thermostatic switch. (Heating anticipators are adjustable.) No current flows through the compensator when cooling is on because it has a much higher resistance than the switch contacts. In this case, the contacts are essentially a short circuit. When the thermostat is open, however, a small current can flow through the compensator and the contactor coil.

Because of the size of the compensator, the current is not enough to energize the contactor. The heat created by the current flowing through the compensator makes the thermostatic switch contacts close sooner than they would without the

Discuss the safety hazards associated with exposure to mercury.

Point out that most residential and small commercial thermostats operate at low voltages (typically 24V).

Show Transparency 4 (Figure 2). Describe the operation of a heating-only thermostat.

Provide a heating-only thermostat for the trainees to examine.

Show Transparencies 5 and 6 (Figures 3 and 4). Describe the operation of a cooling-only thermostat.

Provide a cooling-only thermostat for the trainees to examine.

Show Transparencies 7 and 8 (Figures 5 and 6). Describe the operation of a heating-cooling thermostat. Note that this thermostat must be manually switched between heating and cooling.

Provide a heating-cooling thermostat for the trainees to examine.

Discuss the disadvantages of a heating-cooling thermostat.

compensator. In this way, the cooling compensator accounts for the lag between the call for cooling and the time when the system actually begins to cool the space.

2.4.0 Heating-Cooling Thermostats

When heating and cooling are combined for year-round comfort, it is impractical to use a separate thermostat for each mode. Therefore, the two are combined into one heating-cooling thermostat (*Figure 5*). When a mercury bulb design is used, a set of contacts is located at one end of the bulb for heating and the other end of the bulb for cooling (*Figure 6*). The cooling contacts close the control circuit on a rise in temperature and open the circuit on a drop in temperature. The bulb with both sets of contacts is attached to a single bimetal element.

Unless a switch is provided in the heating-cooling thermostat, the thermostat will continuously switch back and forth from heating to cooling. In effect, heating and cooling will combat each other for control. A switch provides a means to direct the control to cooling, while disconnecting the heating control circuit. Likewise, when the switch is moved to heating, the switch connects the heating components, while electrically isolating the cooling circuit. When the switch is in the center or OFF position, neither the heating nor cooling control circuits are energized.

2.5.0 Heating-Cooling Automatic Changeover Thermostats

The disadvantage of a heating-cooling thermostat is that the building occupant must determine

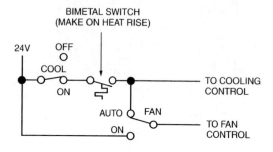

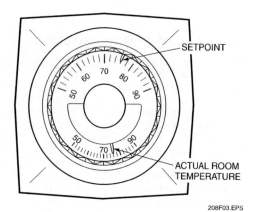

Figure 3 ♦ Cooling-only thermostat.

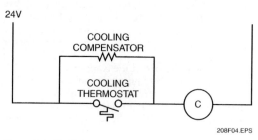

Figure 4 ♦ Cooling compensator.

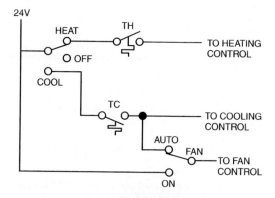

Figure 5 ♦ Heating-cooling thermostat.

Instructor's Notes:

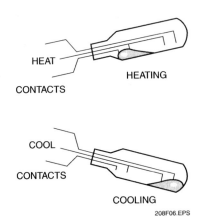

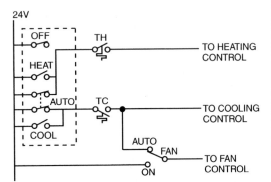

Figure 6 ♦ Heating-cooling contacts.

whether heating or cooling is needed at a particular time and set the thermostat switch accordingly. In some climates, that is very impractical; the need could change several times a day.

The **automatic changeover thermostat** automatically selects the mode, depending on the heating and cooling setpoints. The thermostat shown in *Figure 7* is essentially the same as that shown in *Figure 5*, with the exception that in *Figure 7* there is an AUTO position on the main control switch. The occupant can still select either heating or cooling. When the switch is in the AUTO position, however, the thermostat makes the selection. All that is necessary is for one of the thermostatic switches to close, indicating that the conditioned space is too warm or too cold.

A thermostat contains a built-in mechanical **differential**, which is the difference between the cut-in and cut-out points of a thermostat. The differential is normally 2°F. For example, if the heating setpoint is 70°F, the furnace will turn on at 70°F and run until the temperature is 72°F.

Automatic changeover thermostats also have a minimum interlock setting, commonly known as the **deadband**. The deadband is a built-in feature that prevents the heating and cooling setpoints from being any closer together than 3°F and fighting each other to control the space temperature. In other words, the deadband prevents cooling operation from cooling down the structure to the point that heating operation comes on to try to warm the structure back up.

2.6.0 Two-Stage Thermostats

A two-stage indoor thermostat is normally used to control heat pumps. During the cooling season, it functions as a conventional air condition-

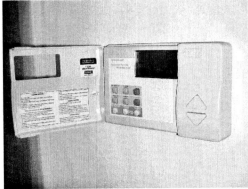

Figure 7 ♦ Automatic changeover thermostat.

ing thermostat. During the heating season, the first stage controls the compressor of the heat pump. The second stage, usually preset 1°F to 2°F below the first stage, allows the supplementary electric heat to be energized if the indoor temperature continues to drop while the heat pump is operating. A two-stage thermostat is also used to control both cooling and heating on systems using two-speed compressors.

Heat pump thermostats usually have an emergency or auxiliary heat switch (*Figure 8*). If the heat pump becomes inoperative, this switch locks out compressor operation and heats the area with supplementary electric heat until the problem can be corrected. An indicator light, usually red, is mounted on the thermostat. It will come on when the selector switch is in the emergency heat position. As soon as the unit has been repaired, return the switch to the normal operating position.

Because of the wide range of heating and cooling applications, multi-stage heating-cooling thermostats usually come in two pieces. The **subbase** contains the wiring terminals and control

Discuss programmable thermostats.

Show Transparency 11 (Figure 9). Point out the features of modern programmable thermostats compared to earlier motor-driven versions.

Provide a programmable thermostat for trainees to examine.

switches. The **thermostat base** is selected for the specific heating and cooling needs of the application. The product lines offered by thermostat manufacturers have a variety of sub-bases and bases that can be combined to meet numerous applications.

2.7.0 Programmable Thermostats

Programmable thermostats are self-contained controls with the timer, temperature sensor, and switching devices all located in the unit mounted on the wall. Early programmables looked very much like conventional thermostats (*Figure 9*). This thermostat contains a motor-driven time clock driving a wheel containing cams that can be set to raise and lower the temperature settings at desired intervals. It is a self-contained version of the arrangement described previously, in which two thermostats and a timer were used. This technology is now obsolete.

Modern electronic programmable thermostats (*Figure 10*) use microprocessors and integrated circuits to provide a wide variety of control and energy-saving features. Their control panels use touch-screen technology and their indicators are digital readouts rather than analog needle positioning. Different thermostats offer different features; the more sophisticated (and expensive) the thermostat, the more features it offers. The following are some of the features available on electronic thermostats:

- *Better temperature control* – Temperatures are sensed electronically, allowing temperatures to be held within a much tighter range than possible with bimetal thermostats. Some versions have built-in humidity control.
- *Installation versatility* – Wireless versions of electronic thermostats are available for applications where it is difficult or impossible to pull thermostat cable. A built-in transmitter sends information to a remote receiver connected to the equipment being controlled.

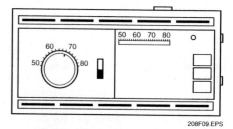

Figure 9 ◆ Motor-driven programmable thermostat.

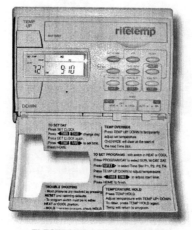

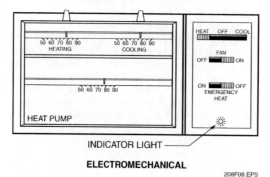

Figure 8 ◆ Heat pump thermostat.

Figure 10 ◆ Electronic programmable thermostat.

8.6   HVAC ◆ LEVEL TWO

- *Override control* – This feature allows the occupant to override the program when desired. For example, you might override the night setback on Monday night so the thermostat isn't lowered before the football game is over.
- *Multiple programs* – This feature allows the occupant to design and select different schedules for different conditions. For example, you could program a special schedule for a vacation away from home.
- *Battery backup* – This feature prevents program loss in the event of a power failure.
- *Staggered startup for multi-unit systems* – This feature avoids excessive current drain. This is an important feature in office buildings, shopping malls, and hotels.
- *Maintenance tracking* – This feature indicates when maintenance is to be performed (for example, when to replace filters).

The savings available from programmable thermostats are significant. For example, it is estimated that a setback of 10°F for both daytime and nighttime can result in a 20 percent energy savings. A 5°F setback will yield a 10 percent energy savings.

2.8.0 Line Voltage Thermostats

Just about all the thermostats you encounter will operate at low voltages (for example, 24V). There are also thermostats that operate at line voltages (for example, 240V). They are commonly used in controlling electric baseboard heat. Line voltage thermostats may be controlled by a bimetal sensing element or a hydraulic sensing bulb. The latter controls the thermostat by means of pressure. *Figure 11* shows a thermostat that is actuated by pressure on a bellows. The sensing bulb contains refrigerant which increases in pressure as the temperature increases. The increasing pressure acts to expand the bellows, thus pushing the switch toward the closed position.

Line voltage thermostats often suffer from a mechanical condition known as thermal offset, commonly called **droop**. Droop causes the thermostat control point to drift away from the selected setpoint. Because it is caused by heat, droop most often occurs in line voltage thermostats, which are subject to heat from high current. It may also occur in 24V thermostats that use anticipators.

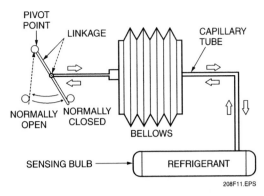

Figure 11 ◆ Remote bulb thermostat.

2.9.0 Thermostat Installation

Even the best, most sensitive thermostat cannot perform correctly if it is poorly installed. Selecting the proper location for the thermostat is the first step in any installation procedure.

2.9.1 Installation Guidelines

The thermostat should be installed in the space in which it will be called upon to control the temperature and other conditioning factors. The thermostat should be installed on a solid inside wall that is free from vibration that could affect operation by making the thermostat contacts chatter. For the same reason, it should not be located on a wall near slamming doors or near stairways. The following practices should be observed when installing a thermostat:

- The installer must be a trained, experienced technician.
- The manufacturer's instructions must be carefully read prior to installing. Failure to follow them could lead to product damage or a hazardous condition.
- The rating should be checked in the instructions and on the unit to make sure the thermostat is suitable for the particular application.

Thermistors

Modern electronic thermostats often use thermistors to sense temperature. A thermistor is a resistor that varies its resistance with temperature. The resistance sensed at a specific temperature is fed to the microprocessor-controlled circuits within the room thermostat. The microprocessor then provides very precise outputs based on that temperature input.

Review guidelines for thermostat location and installation.

Discuss why it is important for trainees to ground themselves before handling sensitive electronic devices.

Explain how to replace an existing thermostat.

Show Transparency 13 (Figure 12). Describe the procedure for mounting a thermostat.

Show Transparency 14 (Figure 13). Explain that mercury bulb thermostats must be carefully leveled to ensure proper operation.

Show Transparency 15 (Figure 14). Describe the wiring connections for a typical thermostat.

Emphasize the importance of following all applicable codes when wiring any electrical device.

- When the installation is complete, the thermostat must be operationally checked and adjusted as indicated in the installation instructions.

CAUTION

Thermostats containing solid-state devices are sensitive to static electricity when not mounted on the wall; therefore, you must discharge body static electricity before handling the instrument. This can be accomplished by touching a metal doorknob or similar hardware. Touch only the front cover when holding the device.

When unpacking the new thermostat and wallplate or sub-base, handle them with care. Rough handling can damage the thermostat. Save all instructional information, screws, and literature for later reference and use.

Locate the thermostat and wallplate or sub-base about five feet above the floor in an area with good air circulation at room temperature. Avoid locations that create the following conditions:

- Drafts or dead air spots
- Hot or cold air from ducts or diffusers
- Radiant heat from direct sunlight or hidden heat from appliances
- Concealed supply ducts and chimneys (to avoid radiant heat)
- Unheated areas behind the thermostat, such as an outside wall or garage

To replace an existing heating thermostat, first turn furnace power off. This de-energizes the 24V transformer in the furnace. Loosen the screws on the existing thermostat base and lift the thermostat away from the wall and wallplate or sub-base. Where applicable, remove the existing wallplate or sub-base from the wall. Disconnect and label each wire with the letter or number on the wiring terminal as each wire is removed, in order to avoid miswiring upon installation of the new thermostat.

Install the new wallplate or sub-base (*Figure 12*). Mount the wallplate or sub-base directly onto the wall with the screws enclosed in the package. If the wallplate or sub-base must be mounted on a vertical outlet box, use the proper adapter ring or cover plate. Mercury bulb thermostats must be carefully leveled (*Figure 13*). If they are not, the thermostat will not operate properly. The wallplate or sub-base of an electronic thermostat usually does not require leveling except for appearance.

2.9.2 Thermostat Wiring

The thermostat is connected through multi-conductor thermostat wire to a terminal strip or junction box in the air conditioning unit. A standard coding method is used in the HVAC industry to designate wiring terminals and wire colors. *Figure 14* shows the coding method and illustrates how the terminals of the heating-cooling thermostat discussed earlier would be designated. The terminal designation arrangement is fairly standard among thermostat and equipment manufacturers. It is not safe to assume, however, that the person who installed an existing system followed the color scheme in wiring the control circuits. There are additional codes for more complex thermostats; for example, the letter O designates orange and is connected to the reversing valve control in a heat pump (B is sometimes used, however). When there are multiple stages of heating or cooling, those terminals are designated with the appropriate letter plus a number.

Thermostat wiring should be done in accordance with national and local electrical codes. All wiring connections should be tight. To avoid damaging the control wire conductor, always use

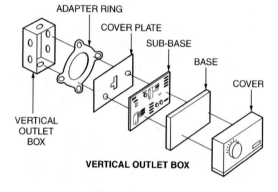
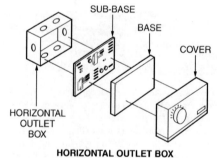

Figure 12 ♦ Thermostat mounting.

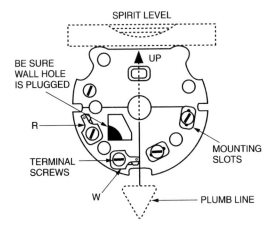

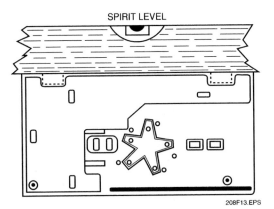

Figure 13 ♦ Thermostat leveling.

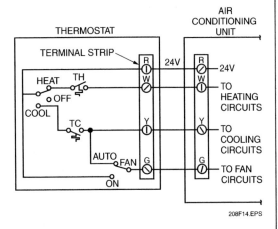

Figure 14 ♦ Thermostat wiring.

Show Transparency 16 (Figure 15). Discuss the methods used to determine current draw.

a stripping tool designed to strip small-gauge wire. Color-coded wiring should be used where possible for easy reference should system troubleshooting be required at a later date.

Wires should not be spliced, but if splicing is absolutely necessary, soldered splices are recommended. If wires are stapled to prevent movement, make sure that the staple does not go through the wire insulation. Seal the wire opening in the wall space behind the thermostat so it is not affected by drafts within the wall stud space.

2.9.3 Checking Current Draw

The next step is to determine the current draw. This may be done by locating the current draw of the primary control in the heating unit, or the heat anticipator setting on the existing thermostat (*Figure 15*). The current draw is usually printed

Thermostat Wiring

Thermostat wires come in a variety of conductor configurations. Simple two-conductor wire can be used in a heating-only installation. Thermostat wires with multiple conductors are readily available from any HVAC parts distributor for use in more complex installations. Normally, 18-gauge thermostat wire is adequate for most installations. However, long runs of thermostat wire can produce a voltage drop that might prevent equipment operation. If you have a run of thermostat wire that seems excessively long, use 16-gauge or heavier wire to reduce voltage drop.

Show Transparency 17 (Figure 16). Explain that the current draw may be marked on the gas valve.

Stress that the power must be off before making any adjustment to the current draw.

Show Transparency 18 (Figure 17). Explain how to check the current draw with an ammeter.

Show Transparency 19 (Figure 18). Explain that some thermostats include an adjustable current draw feature.

Explain how a heat anticipator operates and point out that it should be set at the amperage indicated on the primary control.

Updated Thermostat Wiring

In addition to the traditional terminals such as R, Y, G, W, and O found on room thermostats, many modern electronic thermostats contain a C (24-volt common) terminal. These modern thermostats require power for various internal circuits and by connecting the thermostat to R and C, 24-volt power is always available to the device.

on the furnace nameplate and/or a primary control such as the gas valve (*Figure 16*), the relay, or the oil burner control. It may also be found in the manufacturer's installation and service literature.

WARNING!
To prevent electrical shock or equipment damage, make sure the power is off before connecting the wiring for the current draw adjustment.

If the current draw or heat anticipator setting cannot be found, shut off the power and connect a clamp-on ammeter of the appropriate range (0 to 2 amperes) around a wire connected between the R and W terminals of the existing wallplate or sub-base (*Figure 17*). Since the current is so low, you will have to wrap 11 passes of wire (10 turns) around the jaws of the clamp-on ammeter to get a reading. Divide this reading by 10 to get the actual current draw of the circuit. Then operate the system for one minute, take the reading, and shut off the power.

Some thermostats have an adjustable current draw feature to allow proper operation regardless of system current draw. An example is shown in *Figure 18*. To use this feature, connect the white wire to the W terminal. After the current has been determined, wire the red wire to R-Lo if the current draw is greater than 0.15A and less than 0.60A. If the current draw is greater than 0.60A and less than 1.2A, connect the red wire to the R-Hi terminal. This should help guarantee accurate temperature control.

2.9.4 Adjusting Heat Anticipators

The heat anticipator is adjustable and should be set at the amperage indicated on the primary control or the amperage value specified in the equip-

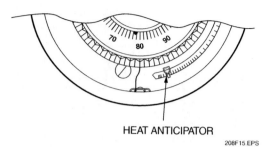

Figure 15 ♦ Thermostat heat anticipator.

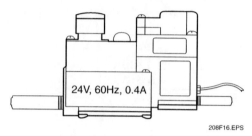

Figure 16 ♦ Gas valve electric ratings.

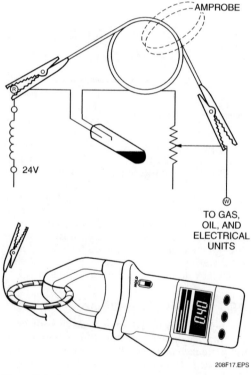

Figure 17 ♦ Amperage check.

8.10   HVAC ♦ LEVEL TWO

Instructor's Notes:

ment manufacturer's literature. Small variations from the required setting can be made by the service technician to improve performance on individual jobs.

Changing the setting of the anticipator changes the resistance of the wire resistor. This shortens or lengthens the heating cycle. Some heat anticipators have arrows to indicate the heating cycle adjustment.

Some heat anticipators have a fixed resistance, while others are equipped with an adjustable dial to change the resistance. The adjustable heat anticipator has a slide wire adjustment with the pointer scale marked in tenths of an ampere. This is used to set the anticipator to agree with the control amperage draw of the particular furnace. Furnaces are provided with an information sticker near the burner which states the amperage drawn by the control circuit of that particular furnace. This is the amperage at which the thermostat heat anticipator should be set under ideal conditions.

For example, if the amperage draw of a control circuit is shown as 0.45A, the installer should adjust the anticipator setting to 0.45A on the scale. The heat anticipator adjustment determines the length of the thermostat call for heat cycle by artificially heating the bimetal coil. As more heat is directed at the bimetal coil, a shorter heating cycle will occur. Conversely, as less heat is directed at the bimetal coil, a longer heating cycle will occur until the thermostat satisfies the call for heat.

When the control circuit amperage is high, less of the heater wire is needed; when the control amperage is low, more of the heater wire is needed. The control circuit amperage draw should be measured for each heating system as previously described.

2.9.5 Cycle Rate

Electronic programmable thermostats may require a change in cycle rate for correct equipment operation. The heating cycle rate can be found in the furnace manufacturer's literature. The heating cycle rate can be adjusted by following the procedure as outlined in the thermostat manufacturer's instructions.

2.9.6 Final Check

The final step is to check the heating and/or cooling system when the thermostat is installed. With the thermostat in the heating mode, turn the power on, place the system switch at HEAT, and leave the fan switch in the AUTO position. Turn

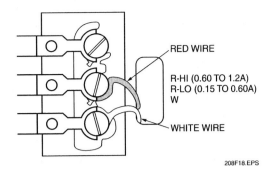

Figure 18 ♦ Connection points.

the setpoint dial to at least 5°F above the room temperature. The burner should come on within 15 seconds. The fan will start after a short delay. Then turn the setpoint dial to 5°F below the room temperature. The main burner should shut off within 15 seconds, but the blower may continue to run for one to two minutes. Next, set the thermostat to the cooling mode.

CAUTION
To avoid compressor damage, do not check cooling operation unless the outside temperature is at least 50°F and the crankcase heater (if so equipped) has been on for at least 24 hours.

If the outside temperature is at least 50°F, return power to the unit, set the thermostat to the COOL position, and the fan switch to AUTO. Leave the setpoint of the thermostat at 5°F below the room temperature. The cooling system should come on either immediately, or after any start delay, if the unit is so equipped. The indoor fan should come on immediately.

After the cooling system has come on, set the thermostat to at least 5°F above the room temperature. The cooling equipment should turn off within 15 seconds. Place the system switch to OFF. Move the setpoint dial to various positions. The system should not respond for heating or cooling.

NOTE
Many electronic programmable room thermostats have a built-in compressor start delay feature. Be aware of this possibility when checking room thermostat operation in the cooling mode.

Explain how to adjust a heat anticipator.

Note that some thermostats may require cycle rate adjustments.

Review the procedure used to check for proper thermostat operation.

Point out the special requirements for testing a system in the cooling mode.

Provide an overview of the procedure used to check and adjust the calibration of a thermostat.

Demonstrate how to install a conventional 24V bimetal thermostat using the standard coding for thermostat wiring, including making any necessary checks and adjustments.

Under your supervision, have the trainees install a conventional 24V bimetal thermostat using the standard coding for thermostat wiring, including making any necessary checks and adjustments. Note the proficiency of each trainee. This laboratory corresponds to Performance Tasks 2 and 3.

Have the trainees review Sections 3.0.0–3.7.4.

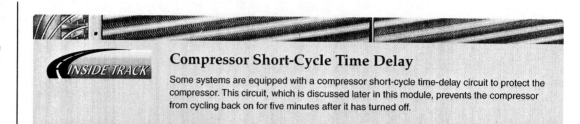

Compressor Short-Cycle Time Delay

Some systems are equipped with a compressor short-cycle time-delay circuit to protect the compressor. This circuit, which is discussed later in this module, prevents the compressor from cycling back on for five minutes after it has turned off.

2.9.7 Adjusting the Thermostat

Thermostats are calibrated or preset at the factory for accurate temperature response and will not normally need recalibration. If a thermostat seems out of adjustment, the first thing to do is to check for accurate leveling. If that doesn't solve the problem, check and/or adjust the calibration as follows:

Step 1 Move the temperature setting to the lowest setting. Set the system switch to HEAT and wait about ten minutes.

Step 2 Remove the thermostat cover and move the thermostat temperature selector lever toward a higher temperature setting until the mercury switch just makes contact. This can be done by observing the mercury droplet in the bulb.

Step 3 If the thermostat pointer and the setting lever read about the same at the instant the switch makes contact, no recalibration is needed. If recalibration is necessary, follow the manufacturer's instructions. A typical recalibration procedure is as follows:

a. With the system switch on HEAT, move the setting lever several degrees above room temperature.

b. Insert the end of an Allen (hex) wrench in the socket at the top center of the main bimetal coil. Use an open-end wrench to hold the hex nut under the coil.

c. Hold the setting lever so it will not move, and tur the Allen wrench clockwise until the mercury switch breaks contact. Remove the wrench.

d. Select a low temperature setting. Wait at least five minutes until the thermostat loses any heat gained from your hands and its own operation.

e. Slowly move the setting lever up the scale until it reads the same as the thermometer.

f. Reinsert the Allen wrench. Holding the setting lever so it won't move, carefully turn the Allen wrench counterclockwise until the mercury just makes contact (turn it no further than necessary).

g. Recheck the calibration. Note that the calibration process may have to be repeated if the calibration is still off.

h. When you are satisfied with the calibration, replace the cover and set the thermostat lever switches for the desired operation.

NOTE

An advantage of electronic programmable room thermostats is that they do not need to be recalibrated.

Thermostat Calibration

When calibrating a room thermostat, you may have to repeat the steps several times because the heat from your hands and your breath can affect the thermostat operation. Avoid breathing on the thermostat and keep handling to a minimum to achieve a more accurate calibration.

Instructor's Notes:

3.0.0 ♦ HVAC CONTROL SYSTEMS

Most HVAC control systems are designed to automatically maintain the desired heating, cooling, and ventilation conditions set into the system. The controls for a small system such as a window air conditioner are very simple—a couple of control switches, a thermostat, and a couple of relays. As the system gets larger and provides more features, the controls become more complicated. For example, add gas heating to a packaged cooling unit, and the size and complexity of the control circuits will more than double. Make it a heat pump instead, and the control complexity will triple.

The use of electronic controls is now widespread across all HVAC product lines. It is rare to find a modern unit that does not contain some kind of electronic control. Manufacturers put these controls in their equipment to increase reliability and energy efficiency. Electronic controls also allow engineers to design more versatility into the products than was ever possible with electromechanical controls.

Large commercial systems may use pneumatic and electronic controls in conjunction with conventional electrical controls. These systems may have thirty or forty control devices, whereas a window air conditioner has just a handful.

The good news is that there are only a few different kinds of control devices. Once you learn to recognize them and understand the role each plays, it won't matter how many are used to control a particular system.

All automatic control systems have the following basic characteristics in common:

- The sensing element (thermistor, thermostat, pressurestat, or humidistat) measures changes in temperature, pressure, and humidity.
- The control mechanism translates the changes into energy that can be used by devices such as motors and valves.
- The connecting wiring, pneumatic piping, and mechanical linkages transmit the energy to the motor, valve, or other devices that act at the point where the change is needed.
- The device then uses the energy to achieve some change. For example, motors operate compressors, fans, or dampers. Valves control the flow of gas to burners or cooling coils and permit the flow of air in pneumatic systems. Valves also control the flow of liquids in chilled-water systems.
- The sensing elements in the control detect the change in conditions and signal the control mechanism.

Integrated Systems

Manufacturers now provide matched, integrated comfort systems to provide superior indoor comfort. All components of the system, including the electronic room thermostat, have electronic circuits within them that instantly recognize each other and work together in such a way as to provide optimum indoor comfort over a wide range of load conditions.

- The control stops the motor, closes the valves, or terminates the action of the component being used. As a result, the call for change is ended.

3.1.0 Relays, Contactors, and Starters

HVAC control system circuits contain relays, contactors, and motor starters. As you will see in the following sections, these devices are physically different in size and configuration, but their principles of operation are the same. Because of their extensive use in control and power distribution circuits, it is extremely important that you understand how these devices operate. Without this understanding, you will have difficulty in reading schematics and troubleshooting circuits. Refer to the *Appendix* at the back of this module for common schematic symbols.

3.1.1 Relays

A relay operates to stop or permit the flow of electricity. Sometimes, a relay is used to reroute the flow of electricity in a different direction. Relays can be hard-wired into a circuit, or plug-in relays (*Figure 19*) can be used. Plug-in relays make troubleshooting and replacement much easier. Instead of having to connect and disconnect wiring, a new relay can simply be snapped into place.

NOTE
In today's equipment, relays that were once stand-alone components are now mounted (soldered) to an electronic control. If you suspect a relay soldered on an electronic control is bad, the entire electronic control will have to be replaced.

Ensure that you have everything required for teaching this session.

Discuss the types of control devices used in HVAC systems.

List the basic components common to all control systems.

Emphasize the importance of understanding the operation of relays, contactors, and starters.

Explain the function of a relay.

Explain the operation of a relay.

Show Transparency 20 (Figure 21). Discuss the operation and applications of single-pole, single-throw (SPST) relays.

Show Transparency 21 (Figure 22). Discuss the operation and applications of single-pole, double-throw (SPDT) relays.

Provide various types of relays for trainees to examine.

Relays

The invention of the relay is credited to Samuel Morse, one of the inventors of the telegraph. Morse developed the relay to boost signal strength. An incoming signal activated an electromagnet, which closed a battery circuit, thereby transmitting the signal to the next relay.

Figure 19 ♦ Examples of plug-in relays.

The operation of a relay is sometimes difficult to grasp, because there seems to be a lot going on inside the relay's sealed enclosure that can't be seen. In its basic form, the relay consists of two parts: an electromagnetic coil and a set of contacts. When the coil is energized by the application of the proper voltage, it causes the position of the relay's contacts to change. Contacts are identified as being either normally closed or normally open. This refers to their position when the relay coil is de-energized. All relay contacts shown on schematic diagrams are shown with the relay in the de-energized position.

Figure 20 shows the open and closed contacts of a typical relay. It also shows the schematic symbol for normally closed and normally open relay contacts. Remember this is the position of the contacts when the relay coil is de-energized. As you can see, a normally closed set of contacts will allow electric current to flow through the contacts, while a normally open set of contacts does not allow the flow of electric current.

Figure 21(A) shows the schematic symbol for a simple relay consisting of the relay coil and a set of contacts. As shown, this relay is a normally closed relay. It is also classified as a single-pole, single-throw (SPST) relay because it only has one set of contacts and one current path. When the relay coil is de-energized as shown, current present at terminal 3 can travel through the closed contacts to terminal 4 for subsequent application to the remainder of control circuit. When voltage is applied across the coil of the relay via terminals 1 and 2, the coil energizes and the relay contacts open, preventing any current applied at terminal 3 from flowing through the contacts to terminal 4. Remember, when the coil is energized, the normally closed contacts open. The last thing that needs to be determined about a relay is its coil voltage. Most relay coils used in HVAC control circuits operate on 24V. If such a relay needs to be replaced, use an SPST normally closed relay with a 24V coil.

It is important to understand that no electrical connection exists between the coil of a relay and its contacts within the relay housing. Opening and closing of the relay contacts happens because of electromechanical linkage between the coil and contacts. *Figure 21(B)* shows a mechanical representation of the same relay just described. As shown, terminals 1 and 2 are still the coil connections, and terminals 3 and 4 are still the contact connections. When power is applied to the relay coil, an electromagnetic field is created by the coil. This electromagnetic field pulls the plunger up, causing the normally closed contacts to open. There are other ways to accomplish this in relays, but this diagram should help you understand what's going on inside the relay mechanically.

Typically, the power applied to the coil might be 24V, while 120V may be applied through the contacts. This is one common situation but there are a number of other coil/contact voltage combinations. There are some applications in which the same voltage is applied to both the coil and the contacts of a relay. In this case, the connection between the two is always made using terminals outside of the relay. Again, it is important to remember that there is no electrical connection between the coil and contacts of a relay within the housing of the relay.

Now that you understand how a basic SPST normally closed relay works, let's move on to other types of relays. The counterpart of the first relay is the SPST normally open relay. This relay is similar to the SPST normally closed relay, except that when the coil is de-energized, the contacts are normally open. When the relay coil is energized, the contacts close.

The next relay is the single-pole, double-throw (SPDT) relay (*Figure 22*). In the SPST relay described earlier, the power applied at terminal 3

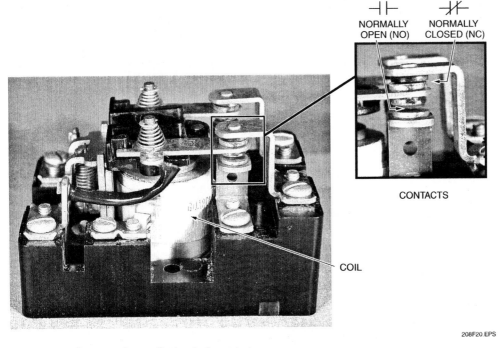

Figure 20 ◆ Normally open and normally closed relay contacts.

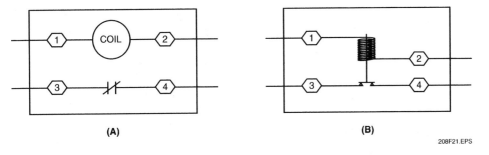

Figure 21 ◆ Single-pole, single-throw (SPST) relay. (A) Schematic symbol. (B) Electromechanical presentation.

was applied to terminal 4, or not applied to terminal 4, depending on whether the contacts where closed or open, respectively. With a double-throw relay, we can allow power to be directed between two different terminals. As shown, with the coil de-energized, power coming in on the common terminal 3 is applied through the normally closed set of contacts to terminal 5. If the coil is energized, the positions of the relay contact sets change, causing the normally open contacts to close and the normally closed contacts to open. This causes the power applied from terminal 3 to be redirected from terminal 5 to terminal 4 for subsequent application to a different branch of the unit's control circuits.

The next relay is the double-pole, double-throw (DPDT) relay. As shown in *Figure 23(A)*, another set of double-throw contacts is added to a basic SPDT relay to form this type of relay. When the coil is energized, it causes both sets of contacts to change position simultaneously. It is important to remember that there is no electrical connection between the first set of contacts and the second set of contacts, just as there is no connection between

Keep Transparency 22 (Figure 23) showing. Explain how additional poles can be added to a relay. Describe a ladder diagram.

Discuss common relay designations.

Show Transparency 23 (Figure 24). Explain that electronic relays use triacs and LEDs to control switching action.

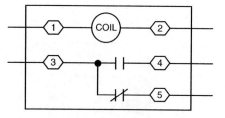

Figure 22 ◆ Single-pole, double-throw relay.

the coil circuitry and the contact circuitry. Again, if one or more relay terminals must be connected together, the connections are made at terminals external to the relay housing.

Now that you understand the concept of adding contact sets, we can add poles to our relay at will. You might find applications that require the use of a 3PDT (three-pole, double-throw) relay or even a 4PDT (four-pole, double-throw) relay. Regardless of the relay coil/contact configuration, each pole is electrically isolated from the coil and from all other poles. When the coil is energized, the position of all the poles change simultaneously.

Since each pole in a relay is electrically isolated from each other and from the coil, the poles from the same relay can be wired into different branches of the equipment's control circuits. Even though the coil and poles are physically housed in the same relay assembly, they are commonly shown in different areas of schematic and **ladder diagrams** used for troubleshooting. Schematic and ladder diagrams are covered later in this module.

Because the relay coil and contacts are often shown in different areas of a schematic or ladder diagram, we need a method of designating which relay is associated with which contact set. An example is shown in *Figure 23(B)*, where the coil is designated R1 (relay 1) and all related contact sets are also identified with numbers that begin with R1. As shown, the first pole has contacts R1-1 and R1-2 (relay 1/contact set 1, and relay 1/contact set 2). The second pole has contacts R1-3 and R1-4. This method of relay designation is also used for three- and four-pole relays. For example, if you look at a schematic or ladder diagram and you see a normally open contact set designated R1-3, you know that the coil of relay R1 controls the contacts. If you energize the R1 coil, the R1-3 contacts will close. Likewise, all the other sets of R1 contacts are also going to change position.

NOTE
You will see many different designations for coils. For example, you may find CR (control relay), CC (compressor contactor), or FR (fan relay). There are dozens of different designations and they vary widely from manufacturer to manufacturer. Always consult the legend provided with the unit diagram.

3.1.2 Electronic Solid-State Relays

Unlike the electromechanical relays just discussed, solid-state relays have no moving parts. They use electronic devices such as triacs (a special type of current gateway) and light-emitting diodes (LEDs) to control switching action.

In the example shown in *Figure 24*, the light created by current flow through the LED will

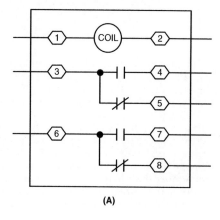
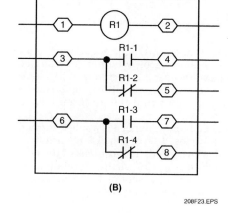

Figure 23 ◆ Double-pole double-throw (DPDT) relay.

8.16   HVAC ◆ LEVEL TWO

Instructor's Notes:

activate a photo sensitive device that triggers the triac. Solid-state relays are used in general control applications, as well as in motor starting circuits.

3.1.3 Contactors and Starters

Contactors (*Figure 25*) are a type of heavy-duty relay. They are used to start and stop high-current, nonmotor loads or in motor circuits where overload protection is provided separately. A contactor does not provide overload protection.

A contactor is a normally open, single-throw device with one or more poles. It operates in the same way as a relay and is often represented on a schematic in a similar way. When the coil is energized, the movable contacts are closed against the stationary contacts, thus completing the circuit. When power is removed from the coil, the contacts open, stopping the flow of electric current to the related load.

A motor starter is used to stop and start motors and provide overload protection. Standard starters include overload protection but they do not include a disconnection means or short circuit protection. Thermal overloads in the starter sense excessive current flowing to a motor and protect the motor from overload. If more current is flowing than the motor is designed to handle, the overload causes the motor to shut down. Starters usually have auxiliary contacts (additional poles). Combination starters are also available. They include a standard starter and a fused or nonfused switch or circuit breaker in the same enclosure.

3.2.0 Motor Speed Controls

Greater efficiency can be achieved by varying the speed of fan motors to adapt to changing heating and cooling loads. In this way, the equipment consumes only as much power as is needed to meet the demand. For example, HVAC equipment commonly uses variable-speed blowers. In addition, two-stage compressors are used in small systems, while larger systems use unloaders or multiple compressors to adapt to changing loads.

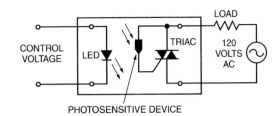

Figure 24 ♦ Solid-state relay.

Figure 25 ♦ Typical contactor.

Discuss the operation and applications of various contactors and motor starters.

Provide a variety of contactors and motor starters for trainees to examine.

Show the video *Electrical Components and Their Symbols*.

Explain that greater efficiency can be achieved by varying the speed of fan motors.

Electronic Solid-State Relays

Some advantages of solid-state relays are longer life, higher reliability, high-speed switching, and high resistance to shock and vibration. The absence of mechanical contacts eliminates contact bounce, arcing when contacts open, and hazards from explosives and flammable gases.

MODULE 03208-07 ♦ INTRODUCTION TO CONTROL CIRCUIT TROUBLESHOOTING 8.17

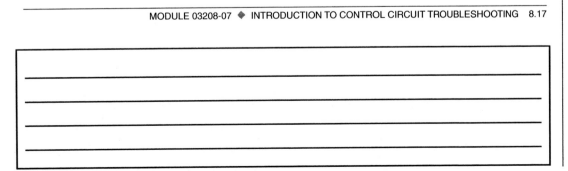

Review the operation of hot surface ignitors.

Show Transparencies 24–26 (Figures 26–28). Discuss the types of motor speed controls used in HVAC systems including autotransformers, triacs, potentiometers, and electronic controls.

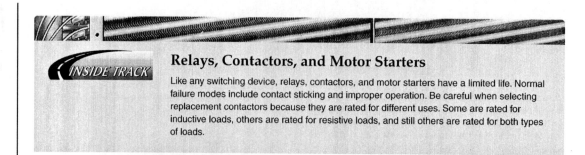

Relays, Contactors, and Motor Starters

Like any switching device, relays, contactors, and motor starters have a limited life. Normal failure modes include contact sticking and improper operation. Be careful when selecting replacement contactors because they are rated for different uses. Some are rated for inductive loads, others are rated for resistive loads, and still others are rated for both types of loads.

Some variable controls are connected or adjusted manually to vary the speed. For example, a furnace blower motor can be adjusted during installation to run at a speed that is optimum for the particular application.

A potentiometer can also be used to adjust motor speed. A potentiometer is a resistor with a wiper arm. The position of the wiper arm determines how much resistance is offered by the potentiometer, and therefore determines how much current flows in the circuit. *Figure 26* shows a potentiometer with two triacs.

The triac is the most common motor control device. A knob on the control device is used to adjust a potentiometer; the setting of the potentiometer determines the motor speed. This circuit is similar to that of a light dimmer.

In *Figure 27*, the potentiometer is controlled by a bellows that responds to temperature changes detected by the sensor. As the bellows moves, the wiper arm of the potentiometer also moves. This changes the amount of current flowing to the damper motor, and thus changes the position of the damper to compensate for the temperature change. This is an example of a continuously variable motor control.

Modern electronic motor controls use microprocessor chips to achieve continuous control. The furnace control system shown in the simplified diagram in *Figure 28* controls the combustion system as well as the blower speed. The microprocessor monitors information such as the length of the last heating cycle and how often the furnace is cycling on and off. It then optimizes the heating cycle by selecting the appropriate fan speed and adjusting the length of the low-fire and high-fire cycles to match the conditions in the space. It is also able to sense any changes within the duct system such as a dirty filter or a closed zone valve, and vary the blower speed to maintain the correct cfm in the system.

NOTE
Variable-speed, electronically commutated fan motors operate and behave differently than conventional fan motors. When troubleshooting systems using these types of motors, always follow the manufacturer's troubleshooting procedures.

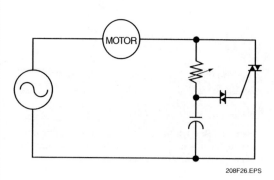

Figure 26 ♦ Motor speed control using TRIACs.

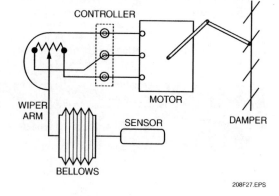

Figure 27 ♦ Potentiometer damper control.

Instructor's Notes:

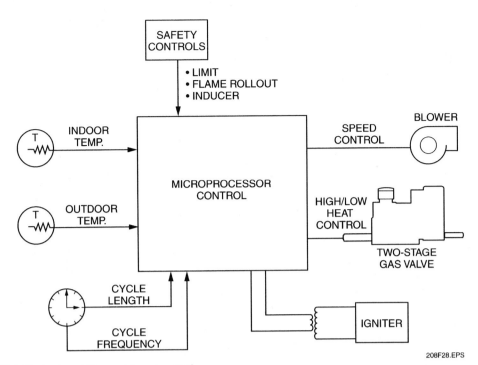

Show Transparency 27 (Figure 29). Discuss the operation and applications of lockout relays.

Explain that load devices must never be added in parallel with a lockout relay.

Figure 28 ♦ Electronic variable-speed furnace control.

3.3.0 Lockout Control Circuit

The purpose of the lockout relay in a control circuit is to prevent the automatic restart of the HVAC equipment. If the lockout relay has been activated, the system may be reset only by interrupting the power supply to the control circuit; for example, resetting the thermostat (in the case of an HVAC system), or turning the main power switch off and then on again.

In the circuit in *Figure 29*, the lockout relay coil, due to its high resistance, is not energized during normal operation. However, when any one of the safety controls opens the circuit to the compressor contactor coil, current flows through the lockout relay coil, causing it to become energized and to open its contacts. These contacts remain open, keeping the compressor contactor circuit open until the power is interrupted after the safety control has reset. Performance depends on the resistance of the lockout relay coil being much greater than the resistance of the compressor contactor coil.

If the lockout relay becomes defective, it should be replaced with an exact duplicate to maintain the proper resistance balance. This type of relay is sometimes called an impedance relay.

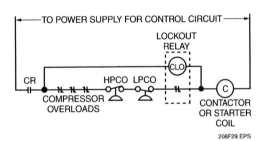

Figure 29 ♦ Lockout relay used in an HVAC control circuit.

It is permissible to add a control relay coil in parallel with the contactor coil when a system demands another control. The resistance of the contactor coil and the relay coil in parallel decreases the total resistance and does not affect the operation of the lockout relay.

 WARNING!
Never put additional lockout relays, lights, or other load devices in parallel with the lockout relay coil. Doing so might defeat the lockout and create a very hazardous situation.

MODULE 03208-07 ♦ INTRODUCTION TO CONTROL CIRCUIT TROUBLESHOOTING 8.19

Show Transparency 28 (Figure 30). Discuss the operation and applications of time-delay relays.

Explain that a compressor short-cycle timer prevents the compressor from restarting until the system pressures have equalized.

Identify control circuit safety switches used in HVAC systems.

Discuss the operation and application of pressure switches.

3.4.0 Time-Delay Relay

The purpose of a time-delay relay is to delay the normal operation of a compressor or motor for a predetermined length of time after the control system has been energized. The length of the delay depends on the time built into the relay coil, and may vary from a fraction of a second to several minutes. A common use for a time-delay relay is to delay the startup or shutdown of a furnace blower to improve heating efficiency.

Electrical systems containing several motors may also use time delays to start the motors one at a time in order to limit the inrush current. For example, the schematic drawing in *Figure 30* shows four compressor control circuits with time-delay relays for sequencing the starting of each motor. In this type of motor design, if all three stages of the thermostat are closed and electrical power is supplied to the units, compressor contactor coil 1 and time-delay relay 1 become energized to start compressor 1. After the specified time delay, the contacts of time-delay relay 1 close to energize compressor contactor coil 2, which starts compressor 2, and time-delay relay 2, which starts compressor 3, and so on, until all four compressors are in operation.

3.5.0 Compressor Short-Cycle Timer

Attempting to start a compressor against high head pressure can damage the compressor. When a refrigerant system shuts down, it should not be restarted until the pressures in the system have had time to equalize. Short-cycling can be caused by a momentary power interruption or by an occupant changing the thermostat setting.

A compressor short-cycle protection circuit contains a timing function that prevents the compressor contactor from re-energizing for a specified period of time after the compressor shuts off. Lockout periods are typically 2 to 5 minutes.

Short-cycle timers are available as self-contained modules that can be direct-wired into a unit. They are often sold as optional accessories, and are often included as a feature in electronic programmable room thermostats.

Electromechanical versions use a motor-driven cam to run a timing mechanism, while electronic versions use an electronic timer.

3.6.0 Control Circuit Safety Switches

Refrigerant system compressor control circuits normally include several different types of safety switches. These include pressure switches, freezestats, and outdoor thermostats.

3.6.1 Pressure Switches

Many systems use one or more pressure switches in the compressor control circuit. These are safety devices designed to protect the compressor. A pressure switch is normally closed and wired in series with the compressor contactor control circuit. *Figures 29* and *30* show examples of low-pressure cutout (LPCO) and high-pressure cutout (HPCO) switches used in such circuits.

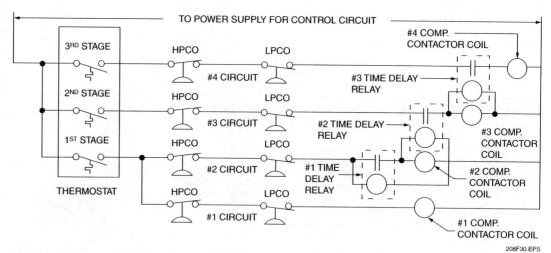

Figure 30 ◆ Four compressor control circuits with time-delay relays.

High-pressure switches are designed to open if the compressor head pressure is too high. Low-pressure switches are designed to open if the suction pressure is too low. Pressure switches use a bellows mechanism that presses against switch contacts that have preset open and close settings. When the system pressure begins to rise above or drop below the normal operating pressure, the related high-pressure or low-pressure switch will open, causing the compressor contactor to de-energize.

Some manufacturers use a type of low-pressure switch called a loss-of-charge switch that removes power to the equipment if the refrigerant charge is low. Loss-of-charge switches are often used in heat pumps.

3.6.2 Freezestat

Another type of safety switch is the freeze-up protection thermostat commonly called a freezestat. Its purpose is to prevent evaporator coil freeze-up. The freezestat switch is a normally closed bimetal switch that is usually attached to one of the endbells in the evaporator coil. It will open if the refrigerant temperature drops below a predetermined setpoint.

3.6.3 Outdoor Thermostats

Some equipment uses an outdoor thermostat to shut off the equipment when the ambient temperature drops below a predetermined setpoint (typically between 55°F and 65°F). This prevents equipment damage. Outdoor thermostats perform other functions, including controlling crankcase heaters and initiating and terminating defrost in heat pumps.

3.7.0 Furnace Controls

Common furnace controls include the fan control, limit control, thermocouple, and inducer proving switch. These controls are described in this section. Note that controls used with boilers are covered later in the hydronic controls section of this module.

3.7.1 Fan Control

A fan control (*Figure 31*) is a temperature-actuated control that, when heated, will close a set of contacts to start the indoor fan motor. The sensing element of the fan control is positioned inside one of the heat exchangers where the temperature is the highest.

The fan control is normally actuated by a bimetal element that opens or closes the contacts in response to a temperature change. The fan control may be set to bring the fan on at about 150°F and to stop the fan at about 100°F.

In operation, the burner or heating element provides heat to the heat exchangers for a few seconds to warm the heating chamber before the

Discuss the operation and application of freezestats and outdoor thermostats.

Identify the common types of furnace controls.

Show Transparency 29 (Figure 31). Explain the operation and function of fan controls.

Pressure Problems

High-pressure and low-pressure switches prevent system operation if the system pressure exceeds a preset range. Common causes of high head pressure are dirty condenser coils or a failed condenser fan motor. Common causes of low suction pressure include a loss of charge and low evaporator airflow.

Temperature-Sensing Devices

In modern equipment, thermistors have taken over the role of sensing temperature, a role previously handled by devices such as freezestats and outdoor thermostats. Here's why. Freezestats and outdoor thermostats only open or close at a preset temperature. This limits their versatility. Thermistors, on the other hand, provide an infinite number of temperature inputs to a system, allowing the system to respond in a variety of ways. For example, an outdoor temperature thermistor can simultaneously provide outdoor temperature information to a room thermostat and a defrost control board. The two devices use the temperature information in different ways. The room thermostat uses the information to display the outdoor temperature, while the heat pump control board uses the information to decide if supplemental heat is required.

Discuss the function and operation of furnace controls, including limit controls, thermocouples, and inducer switches.

Show trainees how to identify and describe the functions of various electrical, electronic, and pneumatic components and circuits.

Have the trainees identify and describe the function of various electrical, electronic, and pneumatic components and circuits. Note the proficiency of each trainee. This laboratory corresponds to Performance Task 5.

Have the trainees review Section 4.0.0.

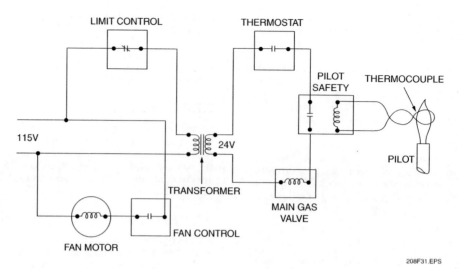

Figure 31 ♦ Complete 24V furnace control circuit.

fan is started. This prevents blowing cold air into the conditioned space. When the thermostat is satisfied, the main burner or heating element stops providing heat, but the fan continues to operate until the temperature in the heating chamber has been reduced. This removes excess heat from the heat exchangers and improves efficiency by using residual heat. In modern equipment, many manufacturers now use a timed on and off delay of the furnace blower motor instead of a temperature-actuated control.

3.7.2 Limit Control

A limit control is also a heat-actuated switch with a bimetal sensing element positioned near the heat exchangers. This is a safety control that is wired into the primary side of the transformer. If the temperature inside the heating chamber or plenum reaches approximately 200°F, the power will be shut off to the transformer, which also stops all power to the temperature control circuit. Depending on the application, limit switches may have an automatic or manual reset.

3.7.3 Thermocouple

A thermocouple is a device that uses dissimilar metals to control electron flow. The hot junction of the thermocouple is located in the pilot flame of a gas-fired furnace. When heat is applied to the welded junction, a small voltage is produced. This small voltage is measured in millivolts (mV) and is the power used to operate the pilot safety control. The output of a thermocouple is about 30mV. This simple device can cause many problems if the connections are not kept clean and tight or the pilot flame is inadequate to generate the correct voltage.

3.7.4 Inducer Fan-Proving Switch

Newer furnaces that use an inducer motor have either a centrifugal switch or a pressure-operated switch in the heating control circuit to monitor the operation of the inducer motor. If the inducer motor fails to operate or come up to speed, the switch prevents control voltage from being applied to the furnace gas valve.

The centrifugal switch has a set of contacts located near the inducer motor shaft. The centrifugal force created by the spinning shaft throws the contacts outward, closing a switch that allows control power to be applied to the furnace gas valve.

The disadvantage of a centrifugal switch is that it can stick in either the closed or open position. This can keep the unit from firing up, or in some cases, allow the unit to continue firing without a demand from the thermostat.

The pressure switch is the more common type of inducer switch. It has tubing connected to the housing, and in some cases, to the burner enclosure. When the inducer motor is operating, a pressure is created that causes the pressure switch to operate, allowing control power to be applied to the furnace gas valve.

Instructor's Notes:

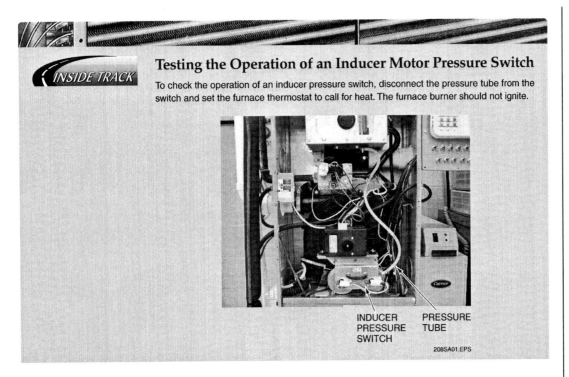

Testing the Operation of an Inducer Motor Pressure Switch

To check the operation of an inducer pressure switch, disconnect the pressure tube from the switch and set the furnace thermostat to call for heat. The furnace burner should not ignite.

Ensure that you have everything required for teaching this session.

Describe control circuits.

Show Transparency 30 (Figure 32). Trace the operation of a simplified control circuit.

4.0.0 ♦ CONTROL CIRCUIT SEQUENCE OF OPERATION

Control circuits can be relatively simple, yet they are sometimes difficult for the new technician to understand. The function of most control circuits is simply to control the start and stop of motors. To begin to understand a basic control circuit, you need to know when the motor should start and when it should stop. This is called the sequence of operation.

The sequence of operation determines the types of devices used in the control circuits of each HVAC system. When sitting down to design a control circuit, the first thing an engineer must know is the required sequence of operation—that is, what motor is supposed to be running and when. Then, a control circuit sequence can be designed.

An automatic air conditioning circuit is simple in nature, and is a good place to begin an analysis of control circuits. *Figure 32* shows a basic cooling system control circuit.

Power from the power plug is applied to the ON-OFF switch. When the switch is in the ON position, power is passed to the cooling thermostat (TC). When the thermostat calls for cooling and closes, power is passed simultaneously to both the compressor and fan motor, causing both

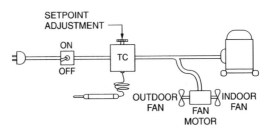
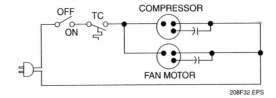

Figure 32 ♦ Basic cooling system control circuit.

motors to turn on and run. When the thermostat is satisfied and opens, power is no longer applied to the motors, causing both motors to turn off. This describes the sequence of operation for this circuit. As you can see, it is very simple, but you must know what is supposed to be happening at any given time.

Show Transparency 31 (Figure 33). Trace the operation of a cooling-only control circuit.

Inside Track

Split-System Condensing Units

This is an example of a typical outdoor (condensing) unit used in a split-system residential air conditioning system. The compressor and outdoor fan are located in this unit.

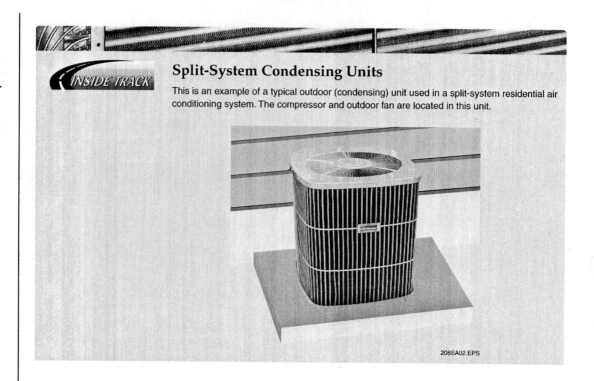

Figure 33 shows a control circuit that is more typical of a basic cooling-only system. You can see that some additional features have been added to this circuit. A transformer has been added because we want to use 24V for the control circuit. This makes the control circuit components and wiring less expensive and easier to install. An indoor fan motor (IFM) has also been added to the circuit. Because the control circuit is now operating at 24V, the control relays (C and IFR) are added to provide control for the motors.

Let's trace this circuit to determine its control sequence. As shown, line voltage is applied at terminals L1 and L2, making it present at the contactors (C), the indoor fan relay (IFR), and the primary of the system transformer at all times. With the transformer energized, the 24V output from its secondary is applied through one control circuit path to the VENT terminal of the FAN switch and by a second control circuit path to the ON-OFF switch. Remember that the power is applied to both control circuit paths at the same time. We now have two control circuit paths to analyze. Let's take them one at a time, starting with the simplest circuit, which is the one that controls the fan.

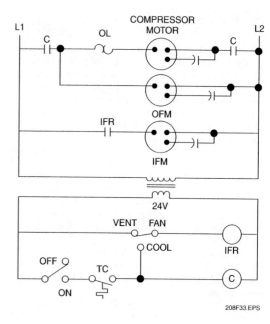

Figure 33 ♦ Typical cooling system control circuit.

8.24 HVAC ♦ LEVEL TWO

When the FAN switch is in the VENT position as shown, power is applied through the first control circuit path to the coil of the indoor fan relay (IFR), energizing the relay. This causes the related IFR contacts to close, applying power to the indoor fan motor (IFM), which energizes the fan motor. This mode of operation allows the occupant to use the fan for ventilation without operating the compressor for cooling. If the FAN switch is placed in the COOL position, the control circuit path to the indoor fan relay remains open, causing the indoor fan relay to be de-energized.

Now look at the second control circuit path. With the ON-OFF switch closed, power is applied to the cooling thermostat (TC). When the thermostat calls for cooling, the thermostat contacts close and apply power simultaneously to the COOL terminal of the FAN switch and to the coil of contactor C. With power applied to the contactor coil, it energizes and closes its contacts. This allows power to be applied through C to the compressor motor and the outdoor fan motor (OFM), turning them both on. Assuming the FAN switch is still in the VENT position as shown, no power is applied through the FAN switch COOL terminal to the indoor fan relay (IFR) through this path. However, the IFR coil is already energized through the VENT terminal of the FAN switch, allowing the indoor fan motor (IFM) to operate as needed to circulate the cooled air.

With the ON-OFF switch closed, the thermostat calling for cooling, and the FAN switch placed in the COOL position, both the contactor coil (C) and the indoor fan motor coil (IFR) are energized through the operation of the second control circuit only. As before, this causes the compressor, outdoor fan, and indoor fan motors to be turned on simultaneously as required for the cooling mode of operation.

No matter how complex the control circuit appears to be, you will find the basic control arrangement shown in *Figure 33* (or something very much like it) at the heart of the circuit. Any additional circuits will represent special features used to improve equipment safety or operating efficiency. *Figure 34* illustrates the point. This diagram is for a combined cooling and gas heating unit. The circuit looks different and there are several more components, but if you trace out the cooling control function, you will see that it is essentially the same as the simple circuit discussed earlier. The additional features are not that complicated. First of all, the heating controls have been added near the bottom of the diagram, along with a heating-cooling thermostat. The cooling control has more extras, such as a compressor short-cycle protection circuit, a crankcase heater, and a current-sensitive overload device. Take away these components and you have a circuit that's identical to the one in *Figure 33*.

The diagram appears different because instead of drawing L1 and L2 and the two sides of the transformer secondary as the verticals on a ladder, they are shown emanating from common terminals. This reflects the way the circuit is actually wired, and is a common method used by manufacturers to draw control circuits. Ladder diagrams are helpful, but they are not supplied by all manufacturers.

Some of the features of the circuit shown in *Figure 34* are as follows:

- The unit has a two-speed indoor fan that runs on high speed for cooling and low speed for heating. In heating, the indoor fan is controlled by the time-delay relay.
- Operation of the inducer fan must be proven before the gas valve is turned on. The inducer pressure switch (PS), located in the draft hood, will close when the induced-draft motor is running at the required speed. If the induced-draft fan stops, the stack pressure will drop and the switch will open, disabling the gas supply. The induced-draft motor (IDM) is energized by the induced-draft relay (IDR) as soon as the thermostat closes.
- The heating section has two additional safety devices in series with the gas valve. A flame rollout switch (RS) will open if the burner flame escapes from the burner box. This usually indicates insufficient air for combustion, or a leak in the burner box. The limit switch (LS) is a heat-sensitive switch that extends into the heat exchanger. If the heat is excessive, it will disable the gas valve. One cause of excessive heat is insufficient air flowing over the heat exchangers. This could be caused by a blower failure or a dirty filter.

Continue a discussion of a cooling-only circuit.

Show Transparency 32 (Figure 34). Trace the operation of a heating/cooling control circuit.

Show trainees how to interpret control circuit diagrams for selected HVAC equipment; identify the devices used and the sequence of operation for each system.

Have trainees interpret control circuit diagrams for selected HVAC equipment. Have them identify the devices used and the sequence of operation for each system. Note the proficiency of each trainee. This laboratory corresponds to Performance Task 6.

Have the trainees review Sections 5.0.0–6.4.0.

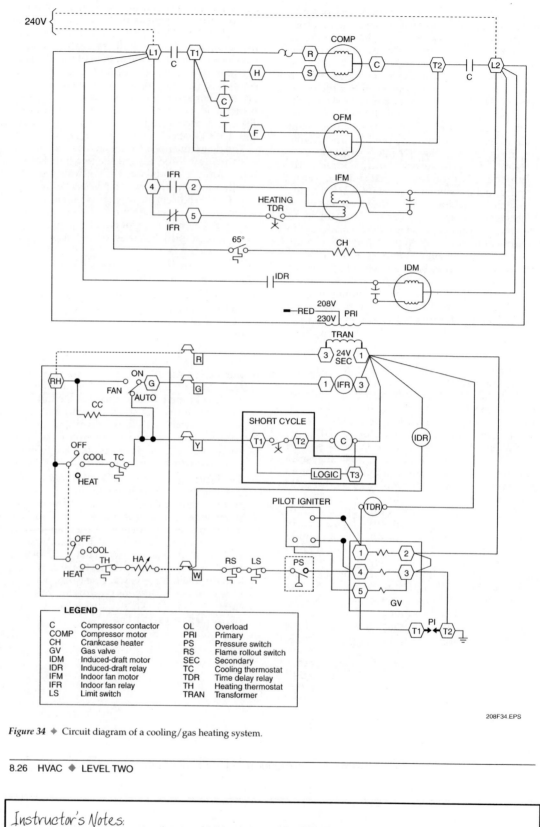

Figure 34 ♦ Circuit diagram of a cooling/gas heating system.

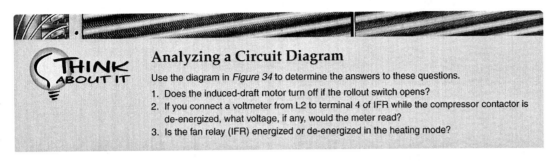

Analyzing a Circuit Diagram

Use the diagram in *Figure 34* to determine the answers to these questions.

1. Does the induced-draft motor turn off if the rollout switch opens?
2. If you connect a voltmeter from L2 to terminal 4 of IFR while the compressor contactor is de-energized, what voltage, if any, would the meter read?
3. Is the fan relay (IFR) energized or de-energized in the heating mode?

5.0.0 ♦ USING AN ORGANIZED APPROACH TO ELECTRICAL TROUBLESHOOTING

Troubleshooting can be defined as a procedure by which the technician locates the source of a problem, then makes the repairs and/or adjustments to correct the cause of a problem so that it will not recur. Troubleshooting can be divided into the five basic elements listed here. The vast majority of problems can usually be found quickly if a systematic approach is used. This includes the following:

- Customer interviews
- Physical examination of the system
- Basic system analysis
- Use of manufacturer's troubleshooting aids
- Fault isolation in equipment problem area

5.1.0 Customer Interviews

The troubleshooting procedure should begin with the technician learning all that can be learned about the customer's complaint by talking to the customer and the service dispatcher. Talking with the customer prior to working on the equipment is always recommended because it can provide valuable information on equipment operation that can aid in the troubleshooting process. It can also identify the source of a problem that is not related to the HVAC equipment, thereby eliminating unnecessary equipment maintenance.

The first evidence of trouble with the HVAC system is often a complaint from an individual who is too cold, too hot, or bothered by drafts. In many cases, the problem behind such a complaint may not be an equipment or control system malfunction, but a personal comfort problem. If the system appears to be operating correctly but individuals are complaining about comfort problems, the technician should check for one or more of the following conditions before assuming the HVAC equipment is malfunctioning.

- *Air distribution and circulation problems* – Persons outside of the immediate area that is controlled by a thermostat may feel too hot or cold. The thermostat senses only the temperature at its particular location. Temperature levels in all the other areas controlled by the same thermostat are subject to variation that can be caused by poor air distribution and/or room air circulation.
- *False heat loads* – Direct sunlight on the thermostat or heat from lamps, appliances, and pipes can cause overcooling of a zone. Direct sunlight or artificial sources of radiant heat can also cause heating discomfort.
- *Covered grilles and diffusers* – Occupants frequently cover part or all of a discharge grille, causing improper heating or cooling.
- *Occupant locations* – Occupants located adjacent to outside walls or windows may be subject to air infiltration or radiant cooling or heating from a wall.
- *Insufficient conditioned air supply* – May be caused by poor air distribution design or fan speed, dirty filters, or lack of the proper amount or size of return air outlets.
- *Overcrowding* – Overheating will result if more people or mechanical equipment occupy a conditioned area than the space was designed to hold.
- *System size* – Extreme weather conditions may exceed the capacity of the heating or cooling equipment.
- *Drafts* – Forced-air systems rely on the movement of air to deliver the desired conditioning. To many people, even a slight air motion may be uncomfortable. This problem may be alleviated by moving the occupant's work station or redirecting the airflow.
- *Stale air* – A stuffy or smoky atmosphere usually results from an insufficient fresh air supply, air that is too humid, or air that has inadequate exhaust. Stale air can also be caused by overcrowding or low air circulation.

Ensure that you have everything required for teaching this session.

Show Transparency 32 (Figure 34) again. Have trainees answer the questions in the "Think About It." See the answers at the end of this module.

List the tasks involved in a systematic approach to troubleshooting.

Emphasize that the customer interview often provides valuable information on equipment operation.

Discuss the types of situations that can cause a properly operating system to provide poor comfort control.

Describe how to physically examine a malfunctioning system.

Explain that the basic process of system analysis involves comparing the unit's current operation against its expected operation.

Identify the troubleshooting aids typically provided by the equipment manufacturer.

Show Transparencies 33–36 (Figures 35–38). Describe and explain how to use label diagrams, troubleshooting tables, fault isolation diagrams, and diagnostic equipment and tests.

Provide samples of manufacturers' troubleshooting aids for trainees to examine.

5.2.0 Physical Examination of the System

Many problems can be identified by simply using your senses to check the system. Conduct a preliminary, power-off visual system inspection followed, if possible, by a preliminary power-on system inspection using your senses.

- Look for evidence of leaks and physical damage.
- Look for dirt accumulation on filters and coils that can affect system operation.
- Listen for unusual sounds that could indicate a malfunction and possibly lead you to its source.
- Check for odors, especially the smell of overheating or gas.

 NOTE
Oil stains on coils or tubing may indicate a refrigerant leak.

5.3.0 Basic System Analysis

The proper diagnosis of a problem requires that you know what the unit should be doing when it is operating properly. If you are not familiar with how a particular unit should operate, you must first study the manufacturer's service literature to acquaint yourself with the equipment's modes and sequence of operation.

The second part of the diagnosis is to find out what the unit actually is doing, and what symptoms are exhibited by the improperly operating unit. Finding out what the unit is doing is accomplished both by carefully listening to the customer's complaints and by analyzing the operation of the unit yourself. As applicable, this means making electrical, temperature, pressure, and/or airflow measurements at key points in the system. The set of measured values can then be compared with a set of typical readings for a properly operating system as previously recorded on system operating logs or in the manufacturer's service literature. This process can often quickly pinpoint the system problem.

5.4.0 Use of Manufacturers' Troubleshooting Aids

To aid in the isolation of faults, many manufacturers provide troubleshooting information marked on the equipment or contained in the service instructions for a particular product. This information typically includes the following:

- Label diagrams
- Troubleshooting tables
- Fault isolation diagrams
- Diagnostic equipment and tests

5.4.1 Label Diagrams

Label diagrams (*Figure 35*) are usually placed in a convenient location inside the equipment, typically on the inside of a control circuit access panel. They normally show a component arrangement diagram, **wiring diagram**, legend, and notes pertaining to the equipment.

The component arrangement diagram shows where the components are physically located in the unit. It is useful because it helps you locate and identify the components shown on the wiring diagram. The legend identifies the meanings of the abbreviations used on the label diagram.

The wiring diagram, sometimes called a schematic, provides a picture of what the unit does electrically and shows the actual external and internal wiring of the unit. Many label diagrams also contain a ladder diagram (*Figure 36*). Ladder diagrams do not usually show wire color and physical connection information. This makes it more useful by focusing on the functional, not the physical, aspects of the equipment. Wiring and ladder diagrams are the primary troubleshooting aids for isolating electrical problems.

5.4.2 Troubleshooting Tables and Fault Isolation Diagrams

Troubleshooting tables and/or fault isolation diagrams are usually contained in the manufacturer's installation, startup, and service instructions for a particular product. As shown in *Figure 37*, troubleshooting tables are intended to guide you to a corrective action based on your observations of system operation. Fault isolation diagrams (*Figure 38*), also called troubleshooting trees, normally start with a failure symptom observation and take you through a logical decision-action process to isolate the failure.

5.4.3 Diagnostic Equipment and Tests

Many manufacturers incorporate electronically controlled or semi-automatic testing features in their equipment to help isolate malfunctions. Depending on the equipment, these built-in diagnostic devices can be simple or complex. Some units contain microprocessor controllers that can run a complete check of all system functions, then report back the results by means of a system

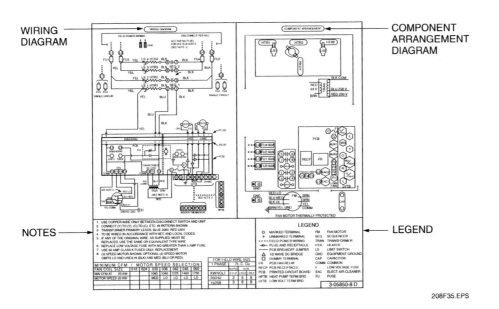

Explain that troubleshooting aids may only isolate the problem to a specific area, and further checks may have to be performed to isolate the faulty component.

Figure 35 ♦ Typical label diagram.

numeric display or flashing LED display. Normally, these built-in test functions isolate a fault to a functional problem area. For example, if a test indicates a compressor failure, this means that the failure has been isolated to the compressor and its related control circuits and wiring. The technician must perform additional troubleshooting within the problem area to find out exactly where the fault is located.

In addition to built-in diagnostic equipment, many manufacturers have developed a series of stand-alone electronic module testers that are available to troubleshoot the different electronic control modules commonly used in the manufacturer's product line. The module tester is usually plugged into the control module in the equipment under test.

Troubleshooting involves the testing of each module control circuit using a sequential troubleshooting process that is performed per the manufacturer's instructions provided with the module tester.

5.5.0 Fault Isolation in the Equipment Problem Area

Once troubleshooting aids have isolated a problem to a functional equipment area, it may be necessary to make additional measurements and use a step-by-step process of elimination to isolate the specific cause of the problem. This is usually the case for the more difficult problems you will encounter.

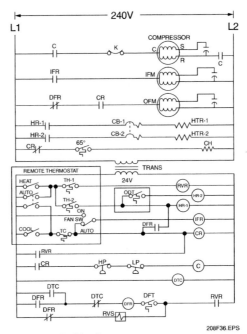

Figure 36 ♦ Ladder diagram.

Malfunction	Probable Cause	Corrective Action
Compressor motor and condenser motor will not start, but fan/coil unit (blower motor) operates normally	Check the thermostat system switch to ascertain that it is set to COOL.	Make necessary adjustments to settings.
	Check the thermostat to make sure that it is set below room temperature.	Make necessary adjustments.
	Check the thermostat to see if it is level. Most thermostats must be mounted level; any deviation will ruin their calibration.	Remove cover plate, place a spirit level on top of the thermostat base, loosen the mounting screws, and adjust the base until it is level; then tighten the mounting screws.
	Check all low-voltage connections for tightness.	Tighten.
	Make a low-voltage check with a voltmeter on the condensate float switch; the condensate may not be draining.	The float switch is normally found in the fan/coil unit. Repair or replace.
	Low air flow could be causing the trouble, so check the air filters.	Clean or replace.
	Make a low-voltage check of the antifrost control.	Replace if defective.
	Check all duct connections to the fan-coil unit.	Repair if necessary.
Compressor, condenser, and fan/coil unit motors will not start	Check the thermostat system switch setting to ascertain that COOL.	Adjust as necessary.

208F37.EPS

Figure 37 ◆ Typical troubleshooting diagram.

208F38.EPS

Figure 38 ◆ Typical fault isolation diagram.

Instructor's Notes:

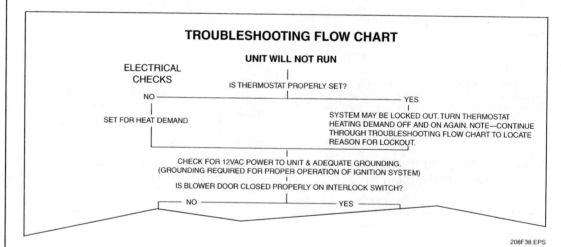

 **Wiring Diagram Exercise**

Using the schematic diagram of the packaged heat pump, answer the following questions:

1. Where would your voltmeter leads be placed to check for the correct incoming voltage to the unit?
2. You place voltmeter leads across terminals R and X of the low-voltage terminal strip and read 24V. What does this tell you?
3. During defrost, the defrost relay (DR) is energized. During a normal defrost, you place voltmeter leads across terminal 2 of the defrost relay and terminal X on the low-voltage terminal strip. What should the voltmeter read?
4. Assume a room thermostat is connected to the low-voltage terminal strip. Across which terminals on the low-voltage terminal strip would you have to place the voltmeter leads to confirm that a call for second-stage heat was coming from the room thermostat? What voltage would you expect to measure?
5. During defrost, the defrost relay (DR) is energized. Knowing this, what happens to the outdoor fan motor during defrost?
6. You wish to know if the demand defrost control is receiving the correct input power. Across which terminals of the demand defrost control would you check for power and what is the voltage you would expect to read?
7. The compressor in this unit occasionally won't start and trips the power supply circuit breaker. You wish to check the compressor's current draw during startup. What test instrument would you use to do this and where in the circuit would you attach or apply the instrument?
8. The blower control (BC) has three power leads and requires two inputs to operate properly. What are the two input signals and at what points are they applied to the control?

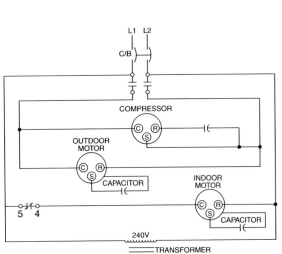
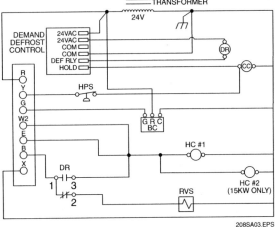

9. If the compressor fails in this heat pump, it is still possible to heat the home. How is this done? At what terminals on the low-voltage terminal strip would you expect to measure a signal that would make this happen?
10. Since this unit energizes the defrost relay during heat pump defrost operation, what does this fact tell you about the state of the reversing valve solenoid (RVS) during cooling operation? Is the RVS energized or de-energized during the cooling mode?

Discuss the safety precautions necessary when working with electrical equipment.

Discuss the OSHA lockout/tagout rule.

Explain that OSHA provides for minimum safety requirements in its lockout/tagout procedure, but more stringent local or company requirements take precedence.

See the Teaching Tip for Section 6.3.0 at the end of this module.

Explain how to prepare for a lockout/tagout.

6.0.0 ♦ SAFETY

This section covers safety practices and procedures.

6.1.0 Safety Practices

Practically all electric shocks are due to human error, rather than equipment failure. Nearly all deaths due to electric shock are due to the worker's failure to observe safety precautions, failure to repair equipment for electrical defects, or failure to remedy all defects found by tests and inspections.

The following is a list of recommended safety practices:

- Always turn off power at the main disconnect before working on an electrical circuit or device. Lock and tag the power switch.
- Use a voltmeter to verify that the power to the unit is actually off. Remember that even though the power may be switched off, there is still potential at the input side of the shutoff switch.
- Never switch equipment on or off while standing in or touching a wet surface or area.
- Notify the power company or utility whenever a power line is touching the ground.
- If it becomes necessary to work on live electrical wiring, try to use one hand only. If you are shocked while using only one hand, current will probably flow through the hand and arm, then down the side and through the feet to the ground. If a shock is conducted through both hands and arms, the electrical path would be through the heart and lungs and would be more likely to be fatal.
- Use protective equipment such as rubber gloves and insulated boots.
- Use tools with dielectric insulation.
- Remove metal jewelry such as rings and watches.

The *National Electrical Code® (NEC®)*, when used together with the electrical code for your local area, provides the minimum requirements for the installation of electrical systems. Always use the latest edition of the code as your on-the-job reference. It specifies the minimum provisions necessary for protecting people and property from electrical hazards. In some areas, different editions of the code may be in use, so be sure to use the edition specified by your employer.

6.2.0 OSHA Lockout/Tagout Rule

In addition to these general electrical safety rules, OSHA released *29 CFR 1926.417* (Lockout/Tagout) in December 1991. This rule covers the specific procedure to be followed for the "servicing and maintenance of machines and equipment in which the unexpected energization or startup of the machines or equipment, or releases of stored energy, could cause injury to employees. This standard establishes minimum performance requirements for the control of such hazardous energy."

The purpose of the OSHA procedure is to make sure that machinery is isolated from all potentially hazardous energy, and tagged and locked out before employees perform any servicing or maintenance activities where the unexpected energization, startup, or release of stored energy could cause injury.

 WARNING!
The OSHA procedure provides only the minimum requirements for the lockout/tagout procedure. Consult the lockout/tagout procedure for your company and the local area in which you are working. Remember, your life could depend on this procedure. It is critical that you use the correct procedure for your site.

6.3.0 Lockout/Tagout Procedure

To prepare for a lockout/tagout, make a survey to locate and identify all isolating devices (see *Figure 39*). You need to be certain which switch(es), valve(s), or other energy-isolating devices apply to the equipment to be locked and tagged. More than one energy source (electrical, mechanical, or others) may be involved.

Figure 39 ♦ Lock out and tag HVAC equipment.

Instructor's Notes:

The following procedure outlines the general steps for lockout/tagout (note that this procedure is provided only as an example). Each employer will designate who is qualified to use the procedure.

Step 1 Notify all affected employees of the lockout/tagout and why it is necessary. The authorized employee will know the type of energy that the machine or equipment uses and the associated hazards.

Step 2 If the machine or equipment is operating, shut it down using the normal procedure. For example, press the stop button or open the toggle switch.

Step 3 Operate the switch, valve, or other energy-isolating device(s) so that the equipment is isolated from its energy source(s). Stored energy must be dissipated or restrained by repositioning, blocking, or bleeding down. Examples of stored energy are springs, elevated machine members, rotating flywheels, hydraulic systems, and air, gas, steam, or water pressure.

Step 4 Lock and tag the energy isolating devices with assigned individual lock(s) and tag(s).

Step 5 After making sure that no personnel are exposed, operate the start button or other normal operating controls to make certain the equipment will not operate. This will confirm that the energy sources have been disconnected.

The equipment is now locked and tagged.

All equipment must be locked and tagged to protect against accidental operation when such operation could cause injury to personnel. Never try to operate any switch, valve, or energy isolating device when it is locked and tagged.

CAUTION
Return operating control(s) to their neutral or OFF position after the test.

6.4.0 Restoring Machines or Equipment

After service and/or maintenance is complete, the equipment is ready for normal operation. Use the following procedure to restore the machines or equipment to their normal operating condition.

Step 1 Check the area around the machines or equipment to make sure that all personnel are at a safe distance.

Step 2 Remove all tools from the machine or equipment and reinstall any guards.

Step 3 Again making sure that all personnel are in the clear, remove all lockout/tagout devices.

Step 4 Operate the energy isolating devices to restore energy to the machine or equipment.

7.0.0 ♦ HVAC SYSTEM TROUBLESHOOTING

Troubleshooting HVAC systems covers a wide range of electrical and mechanical problems. Obviously, not all problems fit easily into these two categories. For example, a loose or corroded wire may cause a compressor in a cooling system to cycle on and off intermittently. Although the problem is electrical, it looks like a mechanical refrigeration problem.

Another example is a compressor that fails from shorted windings because of the acids formed in a poorly evacuated system. This is an example of a mechanical refrigeration problem that looks like an electrical problem.

The methods used to troubleshoot mechanical problems in cooling, heating, and other HVAC equipment tend to be unique to the type of equipment being serviced. For this reason, no further information about troubleshooting mechanical problems is given in this module. Information about troubleshooting HVAC mechanical problems is described in *Troubleshooting Cooling*.

The methods used to troubleshoot electrical problems and components are very similar regardless of the type of HVAC equipment being serviced. The remainder of this module describes those procedures that are common to troubleshooting electrical problems in most types of HVAC equipment.

8.0.0 ♦ HVAC EQUIPMENT INPUT POWER, LOAD, AND CONTROL CIRCUITS

Troubleshooting electrical problems in HVAC equipment may appear complex. However, the process can be simplified if the unit's electrical components are divided into smaller functional circuit areas based on the operation they perform in the equipment.

Review the procedures for lockout/tagout.

Show trainees how to perform a typical lockout/tagout procedure.

Point out that all equipment should be returned to its OFF position after confirming that it will not operate.

Review the procedures for restoring equipment to operation.

Show trainees how to restore equipment to operation following service or maintenance procedures.

Have the trainees review Sections 7.0.0–8.2.0.

Ensure that you have everything required for teaching this session.

Explain that troubleshooting HVAC systems involves isolating both mechanical and electrical problems. Point out that this module focuses on the electrical problems typically encountered in HVAC systems.

Explain that complex electrical problems can be broken down into smaller functional circuit areas.

Show Transparency 37 (Figure 40). Explain that HVAC equipment can normally be divided into three functional circuit areas: input power, loads, and controls.

Discuss input power distribution circuits.

Show Transparency 38 (Figure 41). Explain that the input power and loads operate at line voltage, while the control circuits typically operate at low voltage.

Discuss control circuits.

Most HVAC equipment can be divided into three functional circuit areas (*Figure 40*):

- Input power distribution circuits
- Load circuits
- Control circuits

Input power distribution circuits serve as the power source for the entire unit. They operate at either single-phase or three-phase line voltages, and act to distribute the input power to the various loads in the unit. Power circuits usually consist of the field-installed power wiring from the main electrical service to a disconnect switch located near the unit, and from the disconnect switch to the unit. The input power and distribution circuits include protective devices such as fuses and/or circuit breakers.

Loads are devices that consume power to do work. Compressor motors, fan motors, heater elements, and the primary winding of transformers are all loads normally found in cooling and heating units. Because the input power distribution circuits and the load circuits are both energized and operate at the input voltage level, they are often called the high-voltage circuits (*Figure 41*).

Control circuits provide a link between loads and the input power. Control circuits start, stop, or otherwise control the operation of a load. They usually contain one or more control devices such as relays, switches, and thermostats that work to apply or remove power from the loads. The more complex the system, the more control devices it will have. When a load such as a compressor motor is not working, you have to determine whether the problem is in the load itself, or in the circuits controlling the load.

As discussed earlier, control circuits generally operate at 24V. This low voltage is obtained by using a control transformer to step down the line voltage. Because most control circuits operate at 24V, they are often called the low-voltage circuits. In some larger systems using three-phase line voltages of 240V and higher, a control voltage of 120V or higher may be used for some of the control devices. For this reason, you must always measure the control circuit voltage. Never assume the control circuit voltage is a low voltage.

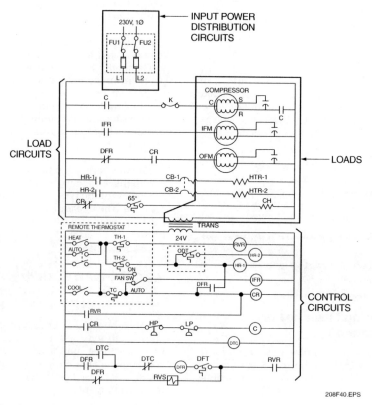

Figure 40 ◆ HVAC equipment functional circuit areas.

8.34 HVAC ◆ LEVEL TWO

Instructor's Notes:

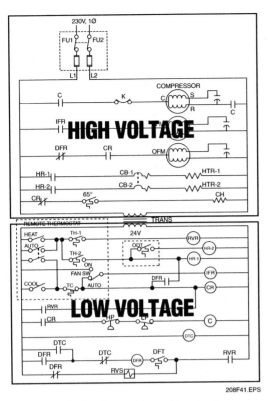

Figure 41 ◆ High-voltage and low-voltage circuits.

8.1.0 Isolating to a Faulty Circuit via the Process of Elimination

Isolating a faulty functional circuit (input power distribution circuit, load circuit, or control circuit) can be relatively easy based on an analysis of the equipment operation and a process of elimination. Also, talking to the customer prior to working on the equipment is always recommended because it can provide valuable information that can aid in the troubleshooting process.

For example, suppose you answer a service call on the heat pump system shown in *Figure 40*. The customer complains that the unit is running but is blowing warm air instead of cool air. Your preliminary check of system operation reveals that the compressor is not running when the thermostat is calling for cooling.

You begin fault isolation through the process of elimination. Because the indoor fan motor runs, you can immediately eliminate the input power distribution circuits as the source of the problem. Next, find out if the compressor will run in the reverse cycle heating mode by setting

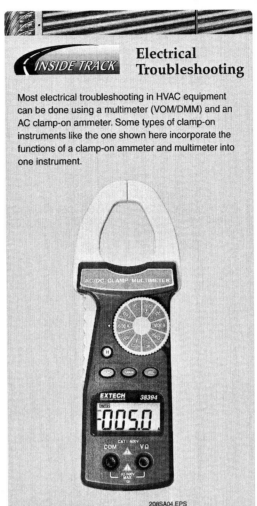

Inside Track: Electrical Troubleshooting

Most electrical troubleshooting in HVAC equipment can be done using a multimeter (VOM/DMM) and an AC clamp-on ammeter. Some types of clamp-on instruments like the one shown here incorporate the functions of a clamp-on ammeter and multimeter into one instrument.

the thermostat to call for heating. (If it is too hot, thermostat TH-1 can be jumpered to simulate a call for heating.) For the purpose of this example, assume the compressor runs. Now you can eliminate the compressor load circuits and everything in the compressor contactor energizing path, including the control relay (CR).

This isolates the problem to the only devices left in the control circuit that are unique to the cooling mode. By studying the schematic (*Figure 40*), you identify these components to be the cooling thermostat (TC) and the related COOL control switch. The thermostat can be eliminated as the cause of the problem if the unit works when the control switch is set to the AUTO mode.

Explain how to use the process of elimination to isolate a problem in a control circuit.

Show Transparency 39 (Figure 42). Explain how system measurements can be used to isolate a fault to a specific circuit component.

Show the video *Electrical Troubleshooting*.

8.2.0 Isolating to a Faulty Circuit Component

Once the source of an electrical problem has been isolated to the malfunctioning load or control circuit, the next step is to make a series of voltage measurements across the components in the malfunctioning circuit to find the faulty component. As shown in *Figure 42*, the measurements can start from the line or control voltage side of the circuit and move toward the load device, such as a motor or a relay coil. Measurements are made until either no voltage is observed, or until the voltage has been measured across all the components in the circuit. Note that when there are many devices in the circuit under test, the measurements can be made by starting at the midpoint in the circuit (divide by two), then working towards either the source of line or control voltage or the load device, depending on whether voltage was or was not measured at the midpoint. As a result of taking these voltage measurements, one of the following situations should exist.

At some point within the circuit under test, no voltage will be indicated on the meter. This pinpoints an open component or set of switch contacts between the last measurement point and the previous measurement point. *Figure 42(A)* shows an example of this situation. In this case, the contacts of the low-pressure switch are open, preventing the compressor contactor coil (C) from energizing.

If the open is caused by a set of contactor or relay contacts, you must find out if the related contactor or relay coil is not being energized or is bad. *Figure 42(B)* shows an example of this situation. In this case, the contacts (CR) in the control circuit are open, preventing the compressor contactor (C) from energizing. These contacts close when the control relay coil (CR) is energized. Before assuming that the problem is caused by the open contacts (CR), you must troubleshoot the control circuit containing the related relay coil (CR) to find out if the coil is energized or de-energized.

If the coil is de-energized, you must further troubleshoot its control circuit to find out why. For example, if the thermostat cooling switch contacts are open, the relay coil (CR) will not be energized.

If voltage is measured at the contactor coil, motor, or other load device, and the device is not working, the load device is most likely at fault. You should turn off the power to the unit, then disconnect the device from the circuit and test it to confirm that it is defective. *Figure 42(C)* shows

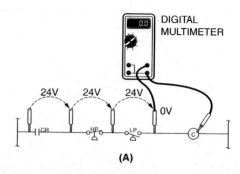

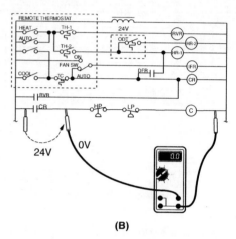

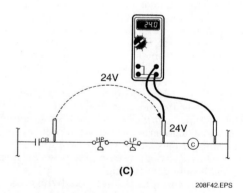

Figure 42 ◆ Isolating to a faulty circuit component.
(A) Open low-pressure switch contacts.
(B) Open relay contacts – check related coil control circuit.
(C) Load device (contactor coil) is bad.

an example of this situation in which 24V power is applied to the compressor contactor (C), but it is not energized. In this case, the contactor is probably bad.

9.0.0 ◆ ELECTRICAL TROUBLESHOOTING COMMON TO ALL HVAC EQUIPMENT

This section covers the electrical troubleshooting procedures common to all HVAC equipment. Specific procedures for motors, hydronic controls, and pneumatic controls are covered later in this module.

9.1.0 Input Voltage Measurements

All HVAC equipment is designed to operate within a specific range of system voltages including a safety factor, typically ±10 percent. This safety factor is added to compensate for temporary supply voltage fluctuations that might occur. Continuous operation of HVAC equipment outside the intended range of voltages can damage the equipment.

9.1.1 Effects of High and Low Voltage

Too high or too low an operating voltage can cause overheating and possible failure of motors and other devices. The power supply voltage applied to the equipment should be measured and checked against the supply voltage indicated on the unit nameplate. See *Figure 43*.

Measuring Input Voltage

When the contactor closes in a cooling unit and applies power to loads such as the compressor and outdoor fan motor, the voltage level may drop about 3 percent from the measured open circuit voltage (contactor open). This is acceptable as long as the voltage does not drop below the manufacturer's stated minimum voltage.

If low voltage exists at the equipment, the voltage should be measured at the electrical service entrance to make sure that a voltage drop does not exist in the branch circuit or feeder that supplies the HVAC equipment. If a voltage drop exists, it may be necessary to install larger wires between the service and the equipment.

Operating voltages applied to motors in the equipment must be maintained within limits from the voltage value given on the motor nameplate. The voltage tolerances used for most HVAC motors are as follows:

- *Single-voltage rated motors* – The input supply voltage should be within ±10 percent of the nameplate voltage. For example, a motor with a nameplate single voltage rating of 230V should have an input voltage that ranges between 207V (–10 percent of 230V) and 253V (+10 percent of 230V).

Have the trainees review Sections 9.0.0–9.6.3.

Ensure that you have everything required for teaching this session.

Point out that electrical troubleshooting procedures involve various system measurements.

Discuss the effects of inadequate or excess operating voltages.

Show Transparency 40 (Figure 43). Explain the procedure for checking the system operating voltage.

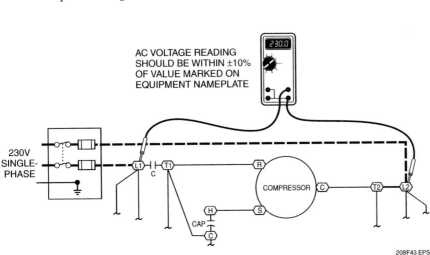

Figure 43 ◆ Single-phase input voltage checks.

MODULE 03208-07 ◆ INTRODUCTION TO CONTROL CIRCUIT TROUBLESHOOTING 8.37

Show Transparency 41 (Figure 44). Explain how to calculate the voltage and current imbalance in a three-phase system.

Provide example problems for the trainees to solve.

- *Dual-voltage rated motors* – The input supply voltage should be within ±10 percent of the nameplate voltage. For example, a motor with a nameplate dual voltage rating of 208V/230V should have an input voltage that ranges between 187V (–10 percent of 208V) and 253V (+10 percent of 230V).

9.1.2 Voltage Phase Imbalance

Voltage imbalance becomes very important when working with three-phase equipment. A small imbalance in phase-to-phase voltage can result in a much greater current imbalance. This current imbalance increases the heat generated in the motor windings. Both current and heat can cause nuisance overload trips and may cause motor failure. For this reason, the voltage imbalance between any two legs of the supply voltage applied to a three-phase system or motor should not exceed 2 percent. If a voltage imbalance of more than 2 percent exists at the input to the HVAC equipment, correct the problem in the building or utility power distribution system before operating the equipment. *Figure 44* shows how the amount of voltage imbalance is determined in a three-phase system using the formula:

$$\text{percent imbalance} = \frac{\text{maximum deviation from avg.}}{\text{average voltage}} \times 100$$

The current imbalance in any one leg of a three-phase system should not exceed 10 percent. A current imbalance may occur without a voltage imbalance. This can occur when an electrical terminal, contact, etc. becomes loose or corroded, causing a high resistance in the leg. Since current follows the path of least resistance, the current in the other two legs will increase, causing more heat to be generated in the devices supplied by those legs. The current imbalance in a three-phase system is determined in the same way as voltage imbalance, but average current is substituted for average voltage.

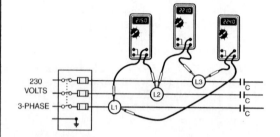
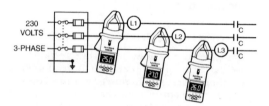

Figure 44 ♦ Three-phase voltage and current checks. (A) Calculating voltage imbalance. (B) Calculating current imbalance.

9.2.0 Fuse/Circuit Breaker Checks

Fuses and/or circuit breakers are normally the first components checked when a unit is totally inoperative.

NOTE
The system must be operating when voltage and current imbalance tests are performed.

9.2.1 Fuse Checks

The best way to test a fuse is by measuring continuity (*Figure 45*). To check fuses, always open the unit disconnect switch, then remove the fuses using an insulated fuse puller. Test the fuses for continuity using an analog or digital multimeter (VOM/DMM). If a short (zero ohms) exists across the fuse, it is usually good. If an open (infinite resistance) exists across the fuse, it is bad. A blown fuse is usually caused by some abnormal overload condition, such as a short circuit within the equipment or an overloaded motor. Replacing a blown fuse without locating and correcting the cause can result in damage to the equipment.

Fuses can also be tested with the circuit energized. This is accomplished by shutting off power to the unit, then disconnecting the wires from the load side of each fuse. This eliminates the possibility of current being fed back to the meter through a short circuit within the unit. Set the multimeter to measure AC voltage on a range that is higher than the highest voltage expected. Turn on the power and place one of the multimeter test leads on the line side of a fuse. Touch the other test lead to the load side of another fuse. If voltage is measured on the load side of a fuse, the fuse is good; if not, the fuse is bad. Repeat this procedure so that all fuses are measured with one test lead on the load side and the other test lead on the input side of a different fuse. This method tests one fuse at a time. If the measurement was performed with both test leads on the load side of the fuses, and the multimeter showed no reading, you would know that a fuse was blown, but not which one.

9.2.2 Circuit Breaker Checks

At the power distribution panel, set the circuit breaker to OFF. If required, remove the panel that covers the circuit breaker to expose the body of the breaker and the wires connected to its terminals. Set the multimeter to measure AC voltage on a range that is higher than the highest voltage expected.

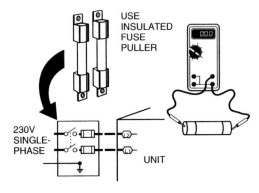
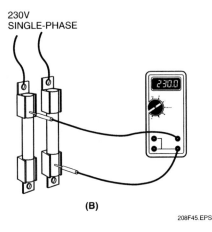

Figure 45 ◆ Fuse checks. (A) Continuity check. (B) Voltage check.

See *Figure 46*. Measure the voltage applied to the circuit breaker input terminals:

- A to neutral or ground (single-pole breaker)
- A to B (two-pole breaker)
- A to B, B to C, and C to A (three-pole breaker)

Make sure that the breaker is closed by first setting it to the OFF position, then setting it to the ON position. Measure the voltage at the circuit breaker output terminals:

- A1 to neutral or ground (single-pole breaker)
- A1 to B1 (two-pole breaker)
- A1 to B1, B1 to C1, and C1 to A1 (three-pole breaker)

Point out that when a system is completely inoperative, circuit breakers and fuses are usually checked first.

Show Transparency 42 (Figure 45). Describe how to test fuses using continuity and voltage checks.

Show Transparency 43 (Figure 46). Describe how to test circuit breakers using voltage and current checks.

Show trainees how to test circuit breakers.

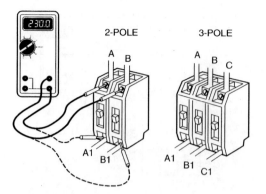

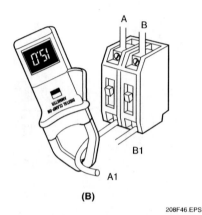

Figure 46 ♦ Circuit breaker checks. (A) Voltage check. (B) Current check.

The measured input and output voltages should be the same. If the voltage is significantly lower than that measured at the input to the circuit breaker, visually inspect the circuit breaker for loose wires and terminals or signs of overheating. If none are found, the circuit breaker should be replaced.

If the circuit breaker shows signs of overheating, or trips when voltage is applied to the equipment, reset it, then check the current flow through the breaker using an AC clamp-on ammeter. Set up the AC clamp-on ammeter to measure AC current on a range that is higher than the highest current expected. Check the ampere rating marked on the breaker. It is usually stamped on the breaker lever or body. One wire at a time, measure the current flow in the wires connected to the circuit breaker output terminals:

- A1 (single-pole breaker)
- A1, B1 (two-pole breaker)
- A1, B1, and C1 (three-pole breaker)

If the circuit breaker trips at a current below its rating or is not tripping at a higher current, the circuit breaker should be replaced. Be sure that the breaker is not being tripped because of high ambient temperature.

Inside Track: Circuit Breakers

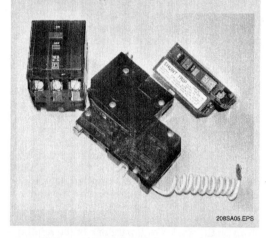

Circuit breakers are available in a wide variety of sizes and types to suit various applications. Some types of circuit breakers are listed as heating, air conditioning, and refrigeration (HACR) circuit breakers. HACR breakers have a built-in time delay that allows a higher-than-rated current to momentarily flow in the circuit. This compensates for the large starting current drawn by such loads. The equipment being protected must also be marked by the manufacturer as suitable for protection by this type of breaker.

NOTE

Some service entrance panels will only accept replacement circuit breakers of the same brand as the panel manufacturer. Other service entrance panels will accept circuit breakers made by other manufacturers.

9.3.0 Resistive and Inductive Load Checks

Electric crankcase heaters and electric heater elements are examples of resistive loads found in HVAC equipment. Inductive loads include contactor, relay, and motor starter coils. They also include control transformers, solenoid valves, and some gas valves. Compressor motors, fan motors, and other motors are also inductive loads. Because motors require special troubleshooting methods, they are covered as a separate category later on in this module.

Once a resistive or inductive load has been identified as the probable cause of an electrical problem, it should be tested to confirm that it is good or bad. The best way to test a resistive or inductive load is by measuring the resistance across the terminals of the device. Before measuring resistance, make sure to electrically isolate the component being measured by disconnecting at least one lead of the component from the circuit. This is important in order to achieve an accurate resistance reading. Otherwise, the meter will read the resistance of other components that are connected in parallel with the component to be measured. As shown in *Figure 47*, a reading of zero ohms indicates a shorted load, while a reading of infinite resistance indicates an open load. In either case, the device should be replaced. When the multimeter indicates a measurable resistance, it usually indicates that the device is good. If a low resistance is measured on a contactor, relay, or starter coil, place one meter probe to ground or to the unit frame. Touch the other probe to each coil terminal. If a resistance is measured from either terminal to ground, replace the device.

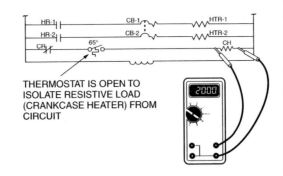
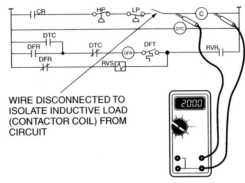

METER READS
MEASURABLE RESISTANCE = GOOD LOAD
ZERO RESISTANCE = SHORTED LOAD
INFINITE RESISTANCE = OPEN LOAD

Figure 47 ♦ Resistive and inductive load resistance checks.

List the resistive and inductive loads found in HVAC systems.

Show Transparency 44 (Figure 47). Describe how to check the resistance of resistive and inductive loads.

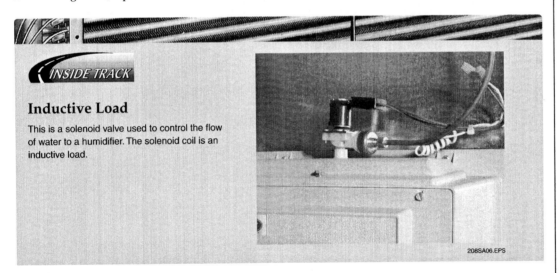

Inside Track

Inductive Load

This is a solenoid valve used to control the flow of water to a humidifier. The solenoid coil is an inductive load.

MODULE 03208-07 ♦ INTRODUCTION TO CONTROL CIRCUIT TROUBLESHOOTING 8.41

Measuring Resistance

The actual resistance value measured for resistive and inductive loads can vary widely depending on the type of device. Ideally, the exact resistance value for the device can be found in the manufacturer's service literature. Another way to judge the resistance reading is by comparing the resistance of the device being tested with that of a similar device that is known to be good.

Measuring Thermistor Resistance

Thermistors are widely used in modern HVAC equipment. To check the resistance of a thermistor, many manufacturers provide a resistance versus temperature chart. If you know the temperature of the thermistor, you can compare its actual resistance to the value on the chart. Immersing the thermistor in ice water allows the resistance to be measured against a known temperature (32°F). On the curve shown, this thermistor should have a resistance of about 33,000 ohms at 32°F.

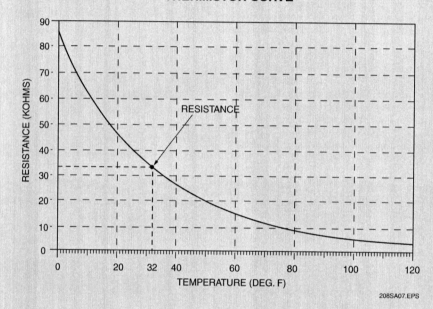

THERMISTOR CURVE

9.4.0 Switch and Contactor/Relay Contact Checks

Once a switch or contactor/relay contact has been identified as the probable cause of an electrical problem, the contacts should be tested to confirm their position. Switches and contactor/relay contacts can be tested by making a continuity measurement to determine whether the contacts are open or closed (*Figure 48*). If the switch contacts are open, the multimeter indicates an infinite resistance reading. If the switch contacts are closed, the multimeter indicates a short (zero ohms).

NOTE
The power must be turned off to make the continuity check. As such, the check can only confirm the status of normally open and normally closed switch contacts. It does not reflect the status of the contacts when the system is powered up.

9.5.0 Control Transformer Checks

Control transformers are usually checked by measuring the voltages across the secondary and primary windings (*Figure 49*). Typically, the secondary winding is measured first. The multimeter should be set to measure AC voltage on a range that is higher than the control voltage expected. If the voltage measured across the secondary winding is within ±10 percent of the required voltage, the transformer is good. If no voltage is measured at the secondary winding, the voltage across the primary winding must be measured. If so equipped, also check the secondary fuse to see if it is blown.

If the voltage measured at the primary winding is within ±10 percent of the required voltage, the transformer most likely is bad. This can be confirmed by performing a continuity check of the transformer primary and secondary windings.

If no voltage or low voltage is measured across the primary winding, the power supply voltage to the equipment should be checked. If the power supply voltage is OK, troubleshoot the circuit wiring between the power supply and control transformer primary winding.

NOTE
Many transformers are protected by a fuse, fusible link, or circuit breaker located in the secondary side. If no or low voltage is present, check that the fuse is not blown or the circuit breaker tripped.

ZERO OHMS = CLOSED CONTACTS

INFINITE RESISTANCE = OPEN CONTACTS

Figure 48 ♦ Relay contact checks.

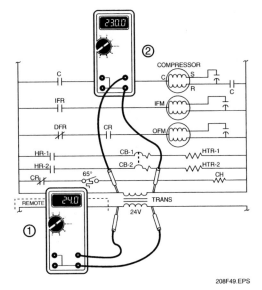

Figure 49 ♦ Control transformer checks.

9.6.0 Thermostat Checks

Troubleshooting procedures for a typical non-electronic heating/cooling thermostat are covered here. They can easily be checked by using an insulated jumper wire connected across the thermostat's R (24V) terminal and fan (G), heat (W), or cool (Y) terminal, as applicable.

NOTE
Because of the diversity and complexity of electronic thermostat design, manufacturers normally provide troubleshooting aids in their service literature. Always follow the manufacturer's troubleshooting instructions.

Show Transparency 45 (Figure 48). Explain that contactors and relays can be tested using a continuity check.

Show Transparency 46 (Figure 49). Explain that transformers can be tested by measuring the voltage across the windings.

Explain that thermostats can be tested by using a jumper wire to energize each mode of operation.

Show Transparency 47 (Figure 50). Explain the procedure for checking the fan switch operation.

Explain the procedure for checking the cooling and heating operation.

Show trainees how to perform various electrical tests and troubleshooting, including input voltage measurements and component testing.

Under your supervision, have the trainees perform various electrical tests and troubleshooting, including input voltage measurements and component testing. Note the proficiency of each trainee. This laboratory corresponds to Performance Task 7.

Have the trainees review Sections 10.0.0–10.6.2.

Ensure that you have everything required for teaching this session.

Review the types of motors commonly used in HVAC systems.

9.6.1 Fan Switch Operation Checks

With the thermostat FAN switch set to the ON position, the indoor fan motor should be running. If not, connect the jumper wire across the thermostat's R and G terminals (*Figure 50*). If the indoor fan runs with the jumper in place, this proves that 24V control voltage is being applied to the thermostat; therefore, the thermostat FAN switch or related fan wiring within the thermostat is bad.

9.6.2 Cooling Operation Checks

With the thermostat FAN switch set to AUTO, the HEAT/OFF/COOL switch set to COOL, and the thermostat set to call for cooling, the compressor, outdoor fan motor, and indoor fan motor should all be running. If not, temporarily connect the jumper wire across the thermostat's R and Y terminals. If the compressor and outdoor fan motors run, the thermostat's HEAT/OFF/COOL switch, cooling thermostat (TC), or related wiring is bad. If the indoor fan motor fails to run, connect the jumper across the thermostat R and G terminals. If the indoor fan runs, the thermostat FAN switch is bad.

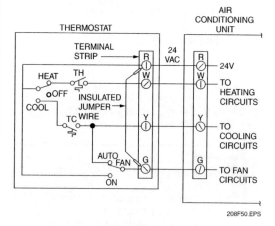

Figure 50 ◆ Troubleshooting fan switch function of a heating/cooling thermostat.

9.6.3 Heating Operation Checks

With the thermostat HEAT/OFF/COOL switch set to HEAT, and the thermostat set to call for heat, the furnace burner should be ignited. If not, connect the jumper wire across the thermostat's R and W terminals. If the furnace burner ignites, the thermostat HEAT/OFF/COOL switch, heating thermostat (TH), or related wiring is at fault.

10.0.0 ◆ MOTORS AND MOTOR CIRCUIT TROUBLESHOOTING

The operation and uses of motors in HVAC equipment has been studied previously, but is reviewed briefly here. Six types of single-phase motors are commonly used in HVAC equipment: shaded-pole; split-phase; permanent split capacitor; capacitor start; capacitor start, capacitor run; and electronically commutated motors (ECM). Permanent split capacitor (PSC) and capacitor start, capacitor run (CSR) motors are most often used in single-phase hermetic compressors because of their good running characteristics and high efficiency. Indoor and outdoor fan motors and blower motors are usually single- or multi-speed PSC motors. Use of electronically commutated motors (ECMs) is also on the increase in fan and blower motor applications. Shaded-pole motors are typically used in low-torque applications such as small direct-drive fan and blower motors.

Both PSC and CSR motors (*Figure 51*) have at least three external terminals leading to two internal windings. The main or run winding (R) contains relatively few turns of heavy wire. The start winding (S) contains a greater number of turns of lighter wire. The point where the two windings meet internally is called the common (C). The arrangement of the motor windings used in both the PSC and CSR motors are the same. The configuration of the motor as a PSC or CSR motor is determined by the run and/or start circuit components used with the motor. The PSC motor has a run capacitor permanently connected across the run and start windings. The CSR uses an extra capacitor called a start capacitor to aid in starting. As shown, the start capacitor and the contacts of a start relay (SR) are connected in parallel with the run capacitor. When the motor is turned on and reaches about 75 percent of full speed, these contacts open and remove the start capacitor from the circuit. The start relay method of removing the start capacitor from the circuit is commonly used with hermetic and semi-hermetic compressor motors. A start relay can be used with all CSR motors; however, in non-compressor applications, a centrifugal

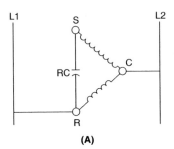
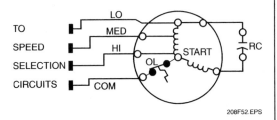

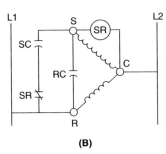

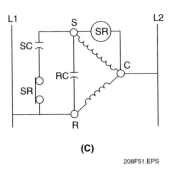

Figure 52 ♦ PSC multi-speed motor.

Figure 51 ♦ PSC and CSR motors. (A) PSC. (B) CSR with start relay. (C) CSR with centrifugal switch.

switch is frequently used to disconnect the start capacitor from the circuit when the motor comes up to speed. Start capacitor failures are often the cause of compressor and other motor problems.

Multi-speed PSC motors (*Figure 52*) used to drive fans and/or blowers in HVAC equipment are capable of operating at two or more speeds. The motor's speed can be changed by switching the motor leads, terminal taps, or by the use of speed control switches or relays. In many heating/cooling units, the motor speed is selected automatically by the control circuits, as determined by the mode of operation. Normally, slower fan speeds are used with heating modes of operation, and higher speeds are used with cooling modes of operation. There are many types of multi-speed motors. As shown, the speed is changed by connecting the line voltage either to the low-speed tap (LO), medium-speed tap (MED), or high-speed tap (HI) of the motor. The specific taps used are selected when installing the unit. A control relay with contacts located in the motor load circuit is normally used to prevent more than one motor winding speed tap from being energized at the same time, a condition that would destroy the motor.

Because multi-speed motors use tapped windings, series-connected winding sections, and/or other wiring configurations that enable operation at different speeds, they may fail in such a way that the motor will not run at one or more speeds, but runs at other speeds. When troubleshooting multi-speed motors, it is important to eliminate the speed selection circuits external to the motor as the cause of the problem before condemning the motor itself.

Variable-speed, electronically commutated fan motors operate and behave differently than conventional fan motors. To determine if these types of motors are defective, it is often necessary to measure low-level input voltage signals. Often the voltages to be measured are DC voltages. When troubleshooting systems that use these types of motors, always use the correct measuring instrument and always follow the specific instructions found in the manufacturer's troubleshooting procedures.

Three-phase motors (*Figure 53*) are generally used when high starting torque is needed or when the motor requirements are greater than 1hp. All have at least three internal windings, with each winding having an equal resistance and the same number of wire turns. Six- and nine-lead, three-phase motors are also found in large applications where part winding start is necessary to reduce the initial inrush current at motor startup. Three-phase motors have good starting and running characteristics and high efficiency. Three-phase motors require no external starting relays or capacitors.

Show Transparency 48 (Figure 51). Discuss PSC and CSR motors.

Show Transparency 49 (Figure 52). Discuss multi-speed motors.

Show Transparency 50 (Figure 53). Discuss three-phase motors.

Discuss safety precautions necessary when testing motors.

Show Transparency 51 (Figure 54). Explain that capacitors can be tested using a multimeter or capacitor analyzer.

Explain the function of the starter relays and how they tend to fail.

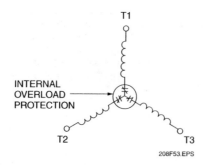

Figure 53 ♦ Three-lead, wye-connected, single-voltage, three-phase motor.

10.1.0 Precautions for Motor Testing

In addition to the standard safety precautions that must be taken when working on electrical equipment, the following precautions must be adhered to when troubleshooting compressors and other motors.

WARNING!

If damaged, the terminals of hermetic and semi-hermetic compressor motors have been known to blow out when disturbed in a pressurized system. To avoid injury, do not disconnect or connect wiring at the compressor terminals. When testing compressors, do not place test probes on the compressor terminals. Instead, use terminal points downstream from the compressor. To be safe, measurements and connecting/disconnecting wiring should only be done at the compressor terminals with the power off.

The capacitors used in motor circuits can hold a high-voltage charge after the system power is turned off. Always discharge capacitors before touching them.

10.2.0 Start and Run Capacitor Checks

The start and/or run circuits on single-phase motors use capacitors. Capacitors affect the wattage, amperage draw, torque, speed, and efficiency of a motor. Run capacitors are connected in the motor circuit at all times; therefore, they are referred to as continuous-duty capacitors. They are usually larger in physical size, but have lower capacitance ratings than start capacitors. Because run capacitors are in the circuit at all times, they are typically filled with a dielectric fluid that acts to dissipate heat. A shorted capacitor may provide a visual indication of its failure.

The pop-out hole at the top of a start capacitor may appear bulged or blown. A run capacitor may be bulged and/or leaking. If a capacitor is found to be defective, always replace it with one specified by the manufacturer.

If you must know the exact capacitance value of a capacitor, use a capacitor tester (*Figure 54*). Follow the tester manufacturer's instructions to perform the test. Typically, the measured microfarad value for a start capacitor should be ±20 percent of the value shown on the capacitor label. If the measured value is outside the range of ±20 percent, replace the capacitor. For a run capacitor, the measured value should be ±10 percent of the value shown on the capacitor. If the measured value is outside the range of ±10 percent, replace the capacitor.

10.3.0 Start Relay Checks

The start relay is used to remove the start capacitor from the motor starting circuit when the motor reaches about 75 to 80 percent of its operating speed. Start relays are made that can be actuated by either current or voltage. The start relays used with HVAC equipment motors are normally voltage-actuated relays (potential relays); therefore, the remainder of this discussion will cover the testing of a voltage-actuated start relay.

Start relays tend to fail with their contacts closed. This results in the start capacitor remaining in the start circuit, causing the motor's start winding to overheat and fail. It may also result in failure of the start capacitor. When it is necessary to replace a start relay, an identical replacement must be used. Substitution of a relay with a different pickup voltage can cause damage to the

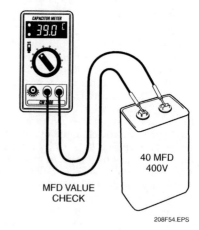

Figure 54 ♦ Capacitor checks.

start capacitor or motor start winding. Also, the replacement relay must be positioned and wired exactly as the original.

Start relays can be tested by measuring the motor start winding current with a clamp-on ammeter. The use of an analog clamp-on ammeter is recommended because it is easier to observe the current reading. If using a digital clamp-on ammeter, one with a MIN/MAX current capability must be used. Testing begins by first finding the full load amps (FLA) rating for the motor as marked on the compressor or motor nameplate. The clamp-on ammeter is then set up to measure AC current on a range scale that is higher than the motor FLA.

With the power turned off, the clamp-on ammeter jaws are placed around the wire that connects the motor start capacitor to the start relay contacts, as shown in *Figure 55*. Power is turned on to the unit while watching the clamp-on ammeter indication to observe the current flow in the start capacitor circuit as the motor starts.

When the start relay is operating properly, the clamp-on ammeter current indication should momentarily indicate current flow, then fall back to zero as the motor comes up to speed. This shows that the start relay is good because its contacts have opened.

If current continues to be read on the clamp-on ammeter after the motor is up to speed, the relay contacts are stuck closed. This means the relay is bad and should be replaced. The start capacitor and/or motor start winding can be damaged when the start relay contacts are stuck closed.

If no current is shown on the clamp-on ammeter, the relay contacts may be stuck open, the related start capacitor may have failed open, or the related wiring may be open. In this instance, the relay contacts can be checked for continuity with the unit power turned off. Contacts that are stuck open will measure infinite resistance.

10.4.0 Start Thermistor Checks

Start thermistors can be used to provide additional starting torque for PSC compressors. The start thermistor is a temperature-sensitive device whose electrical resistance changes as a result of a change in temperature. Positive temperature coefficient (PTC) thermistors increase their resistance with an increase in temperature. PTC thermistors are commonly used in the start circuits of PSC motors. *Figure 56* shows a PSC compressor motor with a PTC start thermistor. As shown, the PTC thermistor is placed across the run capacitor.

At room temperature, the PTC thermistor resistance is very low, about 25Ω to 50Ω. When the compressor is turned on, the application of voltage provides an initial surge of high current through the start winding, because the low resistance of the PTC thermistor is effectively bypassing (shorting) the run capacitor. This surge results in increased motor starting torque. The temperature increase created by the high current causes the PTC thermistor resistance to increase very rapidly to several thousand ohms, blocking current flow and effectively removing the thermistor from across the run capacitor. The motor then runs as a normal PSC motor. A small leakage

Show Transparency 52 (Figure 55). Describe how to test start relays using a clamp-on ammeter.

Show Transparency 53 (Figure 56). Describe how to test a start thermistor by making a resistance measurement.

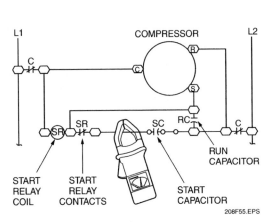

Figure 55 ♦ Start relay check.

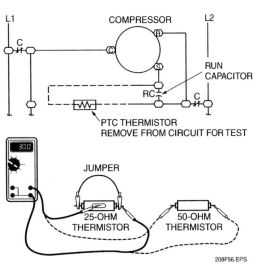

Figure 56 ♦ Start thermistor check.

Show Transparency 54 (Figure 57). Explain that unmarked motor terminals can be identified using a series of resistance checks.

Explain that a multimeter is used to check for open or shorted motor windings.

current through the thermistor keeps the thermistor heated and its resistance high while the motor is running. Circuit operation remains this way until the motor is turned off. After a cool-down period, the thermistor's resistance will once again be the low value needed to start the motor.

The thermistor is tested with the equipment power off, the capacitors discharged, and the thermistor under test isolated from the remainder of the circuit. Testing the thermistor is done by making a resistance measurement of the thermistor. Before attempting to measure the resistance, you should wait at least 10 minutes to allow the thermistor to cool to ambient temperature. The cold resistance of any PTC thermistor should be about 100 to 180 percent of the thermistor ohm rating. For example, a thermistor rated at 25Ω should have a cold resistance of 25Ω to 45Ω. If the PTC thermistor resistance is much lower or more than 200 percent higher than its rating, the thermistor should be replaced.

> **NOTE**
> If the resistance of a start thermistor is acceptable and the compressor still is not starting consistently, a start capacitor and start relay may have to be installed to ensure consistent compressor starting.

10.5.0 Identifying Unmarked Terminals of a PSC/CSR Motor

Sometimes the terminals on a single-phase motor are not marked, or are hard to identify. The terminals can be identified by using a multimeter to measure the resistance of the motor windings. First, the multimeter is used to find the two terminals across which the highest resistance is measured (*Figure 57*). These are the run (R) and start (S) terminals. The remaining terminal is the common (C) terminal. Next, put one lead of the multimeter on the common (C) terminal, and find which of the remaining terminals gives the highest resistance reading. This is the start winding (S) terminal. The remaining terminal is the run winding (R) terminal.

10.6.0 Open, Shorted, or Grounded Winding Checks

Motors are tested for open, shorted, and/or grounded windings with the equipment power off, the capacitors discharged, and the motor leads disconnected from all the related components, including the run and start capacitors and the start relay.

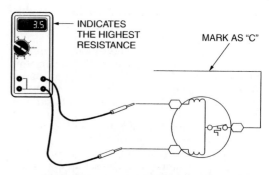

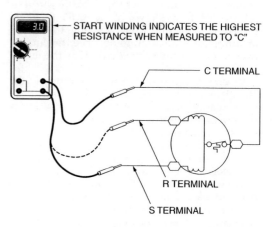

Figure 57 ◆ Identifying unmarked terminals of a PSC/CSR motor.

Electronically commutated motors often have the electronic speed control circuits built into the motor, making it difficult or impossible to measure the motor windings. When troubleshooting these types of motors, always follow the specific troubleshooting instructions in the manufacturer's service literature.

10.6.1 Open or Shorted Winding Checks

Testing for shorted or open windings is done by measuring the resistance of the windings with a multimeter. Be sure to use a multimeter that has an accurate low range (R × 1Ω) scale because the resistance of some undamaged windings can be as low as ½Ω. Perform the test with the multimeter set to measure resistance on the R × 1Ω scale. Make sure that the meter is zeroed. One lead of the multimeter is connected to one of the motor leads, as shown in *Figure 58*. Touch the other lead to the remaining motor leads, one lead at a time, and observe the meter indication. If the multimeter reads a measurable resistance, the windings are probably good. If the multimeter reads zero resistance at one or more leads, the motor has a shorted winding; if it reads an infinite resistance, it has an open winding.

When checking a motor with an internal motor protection device, always make sure that the motor has had adequate time to cool off so that the protective device has time to reset. (This may take an hour or more.)

10.6.2 Grounded Winding Check

Testing a motor for grounded windings is done by measuring the resistance of the windings with a multimeter. Perform the test with the multimeter set up to measure resistance on the R × 10,000 scale. Connect one lead of the multimeter to a good ground connection, such as the motor or compressor frame or compressor discharge/suction line. Poor electrical contact because of a coat of paint, layer of dirt, or corrosion can cause an inaccurate measurement and hide a grounded winding. The other meter lead is then touched to each of the motor leads, one lead at a time, while watching the meter indication (*Figure 59*). An infinite or high resistance should be measured from each lead to ground. If a high resistance reading is indicated, it should not be less than 1,000Ω per volt. For example, on a 230V motor, the resistance should not be less than 230,000Ωs (230V × 1,000Ω/V = 230,000Ω). This indicates that the motor winding is not grounded.

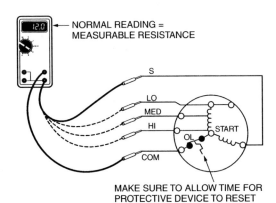

Figure 58 ◆ Motor open or shorted winding check.

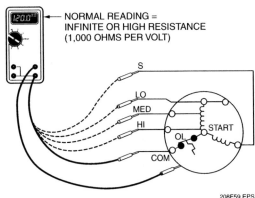

Figure 59 ◆ Grounded winding check.

Show Transparencies 55 and 56 (Figures 58 and 59). Describe the procedures used to check for open, shorted, or grounded motor windings.

Checking for Open or Shorted Motor Windings

Some rules of thumb commonly used to judge the condition of windings are as follows:

- If testing a single-phase PSC/CSR compressor motor, the resistance of the start winding is typically three to five times that of the run winding. For non-compressor motors, the resistances can vary widely depending on the design of the motor.
- When testing a single-phase multi-speed motor run winding, the highest resistance is normally measured between the common lead and the low (LO) speed lead, and the lowest resistance between the common lead and the high (HI) speed lead. The resistance measured between the common lead and the medium speed lead (MED) should be somewhere between that measured for the LO and HI leads.
- If testing a three-phase motor, the motor windings are usually judged to be good if the resistance measured across each winding is nearly identical to the other two windings.

Discuss hazards and safety precautions necessary when performing resistance tests on damaged motor windings.

Show trainees how to perform various electrical tests and troubleshooting of compressors and fan motors.

Under your supervision, have the trainees perform various electrical tests and troubleshooting of compressors and fan motors. Note the proficiency of each trainee This laboratory corresponds to Performance Task 8.

Have the trainees review Sections 11.0.0–12.3.0.

Ensure that you have everything required for teaching this session.

Show Transparency 57 (Figure 60). Discuss the basic operation of a hydronic system. List the control components used in this type of system.

Discuss the function and operation of an aquastat.

If you measure a low or zero resistance, or a measurable resistance that is less than 1,000Ω per volt, this usually indicates that the motor winding is grounded. However, if testing a hermetic or semi-hermetic compressor motor, note that erroneous resistance readings to ground can be measured if liquid refrigerant is present in the compressor shell. In this instance, it is recommended that the refrigerant be recovered from the system, then the compressor be retested before condemning it.

> **WARNING!**
> Never measure the resistance of compressor motor windings at the compressor terminals, especially if the compressor motor is suspected of being damaged. Certain types of compressor motor electrical failures can damage the compressor terminals. Touching damaged compressor terminals, especially if the compressor is under pressure, can release that pressure, causing personal injury. If testing at the compressor terminal is required, first recover all refrigerant from the system. The preferred method of measuring compressor motor winding resistance is through the leads connected to the compressor terminals.

11.0.0 ◆ HYDRONIC CONTROLS

In many parts of the United States, especially New England, the Middle Atlantic States, and the Upper Midwest, hot water or hydronic heat is very popular. With this type of heat, hot water (or in some cases, steam) is circulated through pipes to radiators in different parts of the building. Boilers are pressurized vessels and have unique safety and installation requirements. The nature of hydronic heat dictates specialized components and controls not found in other areas of HVAC.

A simple hydronic system, such as the one shown in *Figure 60*, contains a hot water boiler where the water is heated, a circulating pump to move the heated water throughout the system, a tank to absorb water as it expands when heated, a device called an aquastat that controls the water temperature, and a relief valve to bleed excess pressure so that boiler pressures do not reach explosive levels. The water may be heated by an oil or gas burner. These burners and many of the controls associated with them are similar to the burner controls on gas or oil-fired warm air furnaces.

11.1.0 Aquastat

The aquastat performs several important functions. In its simplest form, it is nothing more than a limit switch set to prevent water from exceeding a certain temperature. An aquastat, such as the one shown in *Figure 61(A)*, is usually mounted on the boiler and contains a temperature-sensing element that is inserted in a well in the boiler jacket. With such an aquastat, the water in the boiler will stay cold until there is a call for heat. The burner will then ignite and start to warm the water. The burner will operate until the room thermostat is satisfied or the water in the boiler reaches the high limit temperature setting.

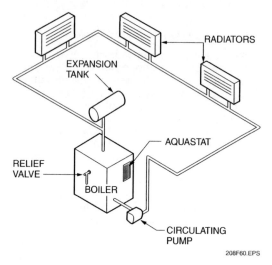

Figure 60 ◆ Simple hydronic system.

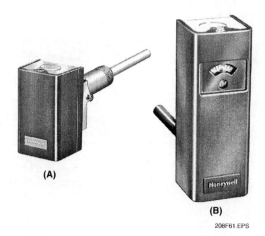

Figure 61 ◆ Aquastat controls.

8.50 HVAC ◆ LEVEL TWO

Instructor's Notes:

A more complex aquastat, such as the one shown in *Figure 61(B)*, provides additional functions. In addition to a high limit setting, it contains a low limit setting that prevents the water in the boiler from getting too cold. This feature is necessary if domestic hot water is provided by a heat exchanger (called a tankless coil) that is inserted in the boiler jacket. This type of aquastat also contains a circulator control that prevents circulator operation until the water is warm enough to provide heat from the radiators.

Other specialized aquastats, especially those used on oil-fired boilers, have the oil burner primary control built into the aquastat. *Figure 62* shows how an aquastat with circulator and high and low limits would operate on an oil-fired boiler. If the water temperature is too low, contacts R-B remain closed, allowing the water to heat up but with no circulation. As the water heats up, R-W closes and R-B opens. If there is a call for heat, power from the switching relay feeds through the closed high limit switch to keep the burner running and through R-W to power the circulator pump.

11.2.0 Reset Controller

To achieve maximum comfort from a hydronic heating system, some systems incorporate an aquastat device called a reset controller. This device monitors the outdoor temperature and adjusts the boiler water temperature for maximum comfort and energy savings. It allows the boiler water temperature to rise on cold days to supply more heat but limits the boiler water temperature to lower levels on milder days when less heat is required. In addition to a temperature sensor in the boiler jacket, the reset controller contains an outside temperature sensor (*Figure 63*) that is typically mounted on the north side of the structure away from direct sunlight.

11.3.0 Low Water Cutoff

Loss of water in a boiler can have catastrophic results. The boiler may produce large amounts of steam, overheat, and build up explosive pressures. To prevent this, a low water cutoff device may be installed in the boiler. The simplest devices are nothing more than a float that activates a switch if the water level drops. The switch shuts off burner operation to prevent overheating and/or activates an alarm or warning light. Electronic versions (*Figure 64*) have a probe that is inserted in the boiler to monitor the water level. Poor water quality caused by contamination will affect the operation of electronic low-water cutoffs. These devices also shut off burner operation and/or sound an alarm when the water level drops too far.

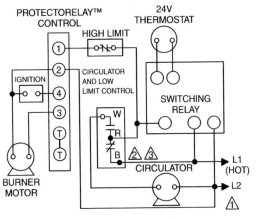

Figure 62 ♦ Aquastat used for low limit and circulator control in an oil-fired hydronic system.

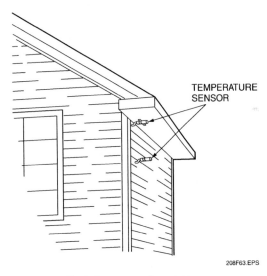

Figure 63 ♦ Typical reset controller outside temperature sensor mounting.

Explain that the circulator pump in a hydronic system is similar to the blower in a forced air system.

Discuss the function and operation of a zone valve.

Figure 64 ◆ Low water cutoff control.

Figure 65 ◆ Circulator pump.

11.4.0 Circulator Pump

The circulator pump in a hydronic system performs a job similar to the blower in a forced air furnace. The major difference is that the pump moves water instead of air. Circulating pumps can be mounted on the floor or directly in the return water line. *Figure 65* shows a typical circulator pump. The pump motor is direct coupled to the pump's centrifugal impeller through a shaft seal in the impeller housing. Circulator operation can be controlled by the aquastat or by the zone valves (if so equipped). Most residential hydronic systems have one circulator pump. In some zoned installations, each zone has its own pump. Some older circulator pumps require periodic lubrication of the motor and pump, while newer models are sealed units with little or no maintenance required.

11.5.0 Zone Valves

Zoned control is easily accomplished with hydronic heating systems due to the simplicity of the zone valve. This device is simply an electrically operated valve. It is placed in the supply line to the zone and opens when the thermostat in the zone calls for heat. *Figure 66* shows a typical zone valve.

In residential applications, zone valves are powered by 24VAC. In commercial applications, higher voltages may be used. Most valves are motorized so that on a call for heating from the

Figure 66 ◆ Zone valve.

zone, the valve opens. Some valves are equipped with a switch that closes as the valve opens. This switch closure starts the circulator pump so that water can flow to the zone that is calling for heat. There are a number of ways to control burner and circulator operation on a gas or oil-fired boiler. Much of it depends on the aquastat used but the installer does have some control over the sequence of operation.

12.0.0 ♦ PNEUMATIC CONTROLS

Pneumatic control systems use compressed air to supply energy for the operation of valves, motors, relays, and other pneumatic control components. They are used primarily in commercial air conditioning systems, although they may sometimes be used in residential systems. Pneumatic control circuits consist of air piping, valves, orifices, and similar mechanical devices.

Thermostats control an air line; the air in the line can, in turn, operate the pneumatic motors, which in turn operate dampers, valves, and switches.

The following advantages are available through the use of pneumatic control systems, especially in commercial and industrial applications:

- Pneumatic equipment is adaptable to modulating operation, but two-position or positive operation can also be provided.
- A great variety of control sequences and combinations are available while using relatively simple pieces of equipment.
- Pneumatic equipment is relatively free of operating difficulties.
- It is very suitable for controlling explosion hazards.
- Installation and maintenance costs may be less than for electrical controls, particularly if building codes require the use of electrical conduit.

12.1.0 Basic Components

Pneumatic control systems (*Figure 67*) consist of the following components:

- A source of clean, dry, compressed air
- Air lines called mains that deliver air from the supply to the controlling devices
- Controlling devices, such as thermostats, humidistats, relays, and switches or controllers
- Branch circuits or branch lines that deliver air from the controlling devices to the controlled devices
- Operators or **actuators**, such as valves and motors, which are the controlled components of the system

The air source is an electrically driven compressor that is connected to the storage tank in which the pressure is maintained between fixed limits (usually 20 to 30 psi). Air leaving the tank is filtered to remove oil and dust. In many installations, a small refrigeration unit is included to condense out any entrained moisture. Pressure-

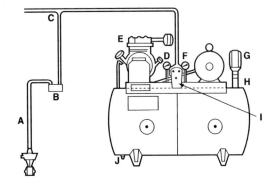

A. TO ACTUATOR
B. THERMOSTAT
C. TO OTHER PARTS OF SYSTEM
D. HIGH-PRESSURE GAUGE
E. COMPRESSOR UNIT
F. LOW-PRESSURE GAUGE
G. PRESSURE SWITCH
H. RELIEF VALVE
I. PRESSURE REGULATOR AND FILTER
J. DRAIN

208F67.EPS

Figure 67 ♦ Pneumatic system.

reducing valves control the air pressure to the main that feeds the controller (thermostat, etc.).

The controller regulates the positioning of the controlled device. It does this by taking air from the supply main at a constant pressure and delivering it through the branch line to the controlled component at a pressure that is varied according to the change in the measured condition.

For example, a change in the temperature of the conditioned space causes the thermostat to change the air pressure in the branch line. This change in pressure then causes the controlled component to move toward the open or closed position, depending on whether the room temperature has increased or decreased. When the valve of the controlled device moves toward the open position, more heat is added to the space; when the valve moves toward the closed position, less heat is added. A similar procedure applies to the cooling position.

There are four types of controllers or thermostats:

- A direct-acting thermostat increases the pressure of its branch line when the air temperature increases in the conditioned space.
- A reverse-acting thermostat decreases the pressure of its branch line when the air temperature increases in the conditioned space.

Show Transparencies 61 and 62 (Figures 68 and 69). Discuss the types of controllers used in pneumatic systems.

Show Transparencies 63–65 (Figures 70–72). Discuss the types of actuators used in pneumatic systems.

- A graduate thermometer gradually changes its branch line pressure when the air temperature changes in the space. This type of thermometer can maintain any pressure from 0 psi to 15 psi.
- A positive-acting thermostat abruptly changes its branch line pressure when the room temperature changes. In this instance, the branch line pressure is either 0 psi or 15 psi. The thermostat does not maintain pressures between these values. It is only a two-position control, either fully open or fully closed. Humidity and pressure controllers operate in a similar manner.

A bleed thermostat control (*Figure 68*) has a bimetal element which reacts to temperature and positions the vane near or away from the air nozzle. Thus, the pressure in the branch is relative to how much air is bled off. **Bleed controls**, if used directly, do not provide a wide range of control. Therefore, they are frequently coupled to a relay that is separately furnished with air for activating purposes; the bleed thermostat simply controls the relay action. Bleed controls maintain a constant drain on the compressed air source.

Non-bleed thermostat controllers (*Figure 69*) use air only when the branch line pressure is being increased. The air pressure is regulated by system valves. The valves eliminate the constant bleeding of air that is present in the bleed thermostat. The exhaust and air main valves are controlled by the action of the bellows resulting from changes in space temperature. The exhaust or bleeding action is relatively small and occurs only on a pressure increase.

Relays are installed between a controller and a controlled device. They are used to perform a function that cannot be accomplished by the controller. A diverting relay, for example, can supply branch air in one position and exhaust it in another position. It can also supply air to either of two branches without exhausting the other branch, or it can shut off the branch air without exhausting the branch.

The controlled devices (actuators) are mostly pneumatic damper motors and valves. The motor moves according to changes in the branch line pressure. A pneumatic motor contains one of three mechanisms: a bellows (*Figure 70*); a diaphragm; or a cylinder and piston. The damper linkage or valve stem is connected to the bellows and when the branch line pressure decreases, the bellows expand due to the internal spring pressure. As the branch line pressure increases, the bellows contract as a result of the increase in air pressure. The expansion and contraction of the bellows, piston, or diaphragm moves the linkage so that the damper opens and closes.

A normally open damper (*Figure 71*) is installed so that it moves toward the open position as the air pressure in the damper motor decreases.

A normally closed damper (*Figure 72*) is a damper that is installed so that it moves toward the closed position when the air pressure in the damper motor decreases.

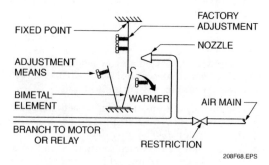

Figure 68 ◆ Bleed-type thermostat.

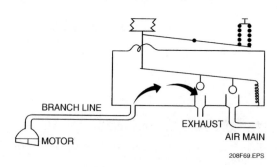

Figure 69 ◆ Non-bleed thermostat.

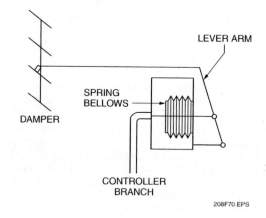

Figure 70 ◆ Pneumatic actuator.

8.54　HVAC ◆ LEVEL TWO

Instructor's Notes:

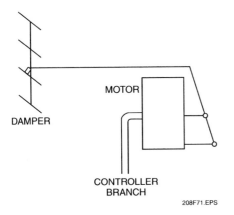

Figure 71 ♦ Normally open damper.

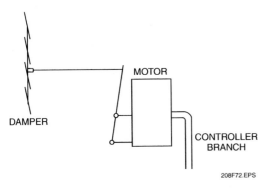

Figure 72 ♦ Normally closed damper.

As the branch line pressure changes, the movement of the bellows actuates the lever arm or valve stem. The spring exerts an opposing force so that a balanced, controlled position can be stabilized. The motor arm can be linked to a variety of functions.

There will always be some crossover between the air devices and the electrical system. The devices most widely used for this process are **pneumatic-electric (P-E) relays** and **electric-pneumatic (E-P) relays**.

P-E relays are simply pressure switches in which a pneumatic signal causes an electrical change. P-Es are used to turn on electrical devices such as pumps and electric heaters. All P-Es have switch differentials (the pressure change that causes the switch to make and break).

E-P relays are three-way solenoid valves (see *Figure 73*). They are generally used as interlocking devices where a circuit is enabled under certain conditions which are signaled electrically.

When the E-P is de-energized, the common and normally open ports are connected, and the normally closed port is blocked. When the E-P is energized, the common port is connected with the normally closed port, and the normally open port is blocked.

12.2.0 Pneumatic Control System

The components of a typical modulating air conditioning system are shown in *Figure 73*. The system controls both the room temperature and the temperature of the air as it enters the room. It also provides adjustment of the quantity of outside air used for ventilation. This system may be used in any structure that has a demand for good air circulation to all areas.

Solid-state electronics have made possible the accurate modulation of temperature, humidity, and/or air volume. The modulator is usually controlled by a thermostat, a remote bulb and bellows or diaphragm, and/or pressurestats or humidistats. Most modulating systems have continuously running blowers.

These controls are usually not used for residential and small commercial applications. They are more often used for large commercial or industrial buildings. They are required for large structures having complex air handling needs. Control of the volume, temperature, and relative humidity of air supplied to the conditioned space of large buildings must be regulated to suit the space load and varying occupant requirements. The regulating system is generally composed of proportioning thermostats, supply and return air controllers, damper and valve motors, plus other controls to start and stop fans and to position air dampers.

Modulating controls contain potentiometers that provide electrical signals to actuators when the temperature, humidity, air volume, or air pressure being controlled deviates from the setpoint. The strength of the electrical signal is directly proportional to the amount of deviation. In order to make a correction, the controller signals the actuator to assume a new position to correct the deviation. Modulating controls operate on low voltages, 24 volts AC or DC.

Modulating systems have been developed to fit the machine capacity very closely to the heating or cooling loads.

For example, in a cooling application, two or more compressors that are connected in parallel may be used. Each compressor is operated by a motor control. During operation, if the load increases and the temperature starts to rise, one compressor will begin to run. If the temperature

Explain that some hydronic controls are both pneumatic and electric.

Discuss the applications of these devices.

Show Transparency 66 (Figure 73). Explain that modulating systems are typically used in commercial applications.

See the Teaching Tip for Section 12.2.0 at the end of this module.

Show Transparency 67 (Figure 74). Discuss the function and operation of airflow (sail) switches.

Have the trainees review Sections 13.0.0–13.3.0.

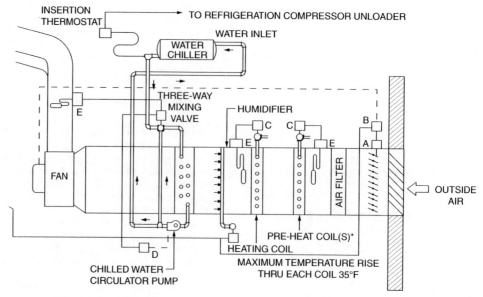

A. TWO POSITION DAMPER MOTOR
B. ELECTRIC-PNEUMATIC RELAY
C. MODULATING STEAM OR WATER VALVES
D. PNEUMATIC-ELECTRIC RELAY
E. MODULATING TEMPERATURE CONTROLLERS

Figure 73 ◆ Modulating system with E-P relay.

keeps rising, the second compressor will start to operate. Additional compressors may cut in until enough capacity is obtained. The control contains a special switching device that rotates the service of the various compressors so that each compressor will be used about the same amount of time. Some modulating refrigeration systems may use a multiple cylinder compressor, with each cylinder equipped with an unloader device. Variable-speed motors are also used to provide a modulated cooling capacity.

12.3.0 Airflow Control

An airflow switch (*Figure 74*), also known as a sail switch, is often installed in ductwork as a safety device. It is used in electric heating systems to guarantee that air is circulating through the air distribution system before the heating elements are turned on. It is also used in cooling systems to verify that there is air flowing across the condenser and evaporator before the compressor is turned on and to activate an electronic air cleaner. Duct airflow can also be sensed by using a pressure switch connected so that it senses air pressure inside the duct.

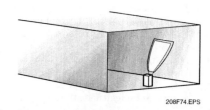

Figure 74 ◆ Sail switch.

8.56 HVAC ◆ LEVEL TWO

13.0.0 ◆ HVAC DIGITAL CONTROL SYSTEMS

The increasing size of modern buildings and building complexes, and the difficulty of monitoring and controlling the operation of up to several hundred elements of the heating, ventilating, and air conditioning systems has led to a revolution in the way manufacturers design HVAC controls. With this revolution in controls, occupants are enjoying a very high level of comfort and control. The microprocessor has led to this revolution in today's HVAC control systems.

The microprocessor is a high-speed instruction executer. You give it a list of things to do (instructions) and it simply executes or carries out those instructions over and over.

A goal of today's HVAC control designs is to build an integrated building system that is able to collect information and process that information to achieve maximum building comfort, energy usage, indoor air quality, and building safety and security.

As *Figure 75* illustrates, controlling the HVAC equipment is only one aspect of today's control systems, but it is an important one. When correctly integrated, it will prolong equipment life, increase customer satisfaction, aid in the repair of the equipment, and sometimes even communicate a problem before it can damage the equipment.

13.1.0 Direct Digital Control

HVAC, lighting, and other building systems controlled by traditional pneumatics, timers, switches, and thermostats may appear to function well. But slow response, calibration shifts, mechanical wear, and the inability to coordinate operation with other systems results in wasted energy and an inconsistent level of comfort.

Direct digital control (DDC), on the other hand, can integrate all building systems through a central computer. It can perform complex control sequences automatically, improving the

Ensure that you have everything required for teaching this session.

Show Transparency 68 (Figure 75). Discuss the components and advantages of centralized building management systems.

Define direct digital control (DDC).

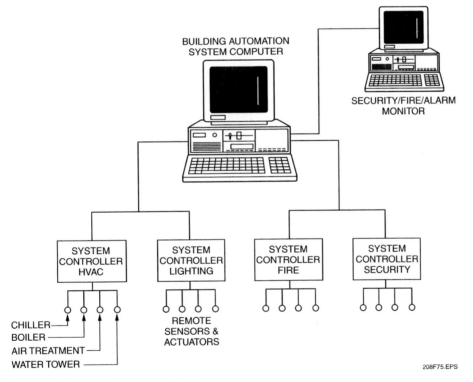

Figure 75 ◆ Centralized building management system.

MODULE 03208-07 ◆ INTRODUCTION TO CONTROL CIRCUIT TROUBLESHOOTING 8.57

Show Transparency 69 (Figure 76). Discuss the function and operation of the system controller module in a DDC system.

Show Transparencies 70 and 71 (Figures 77 and 78). Discuss control signals. Point out that analog signals are typically used to express a change in value, while digital signals express a binary state (on/off).

building environment and lowering costs at the same time. The greatest strength of DDC is its ability to communicate with the real world and to understand and sort the signals it receives. This communication helps it to reject obviously erroneous signals that older controllers were unable to detect.

Great care goes into the design of the controller input/output (I/O) function of the DDC system because it is like a nerve system to the controller (*Figure 76*). It receives all of the signals from the system and directs them to the microprocessor to be used based on the instructions stored within its memory.

13.2.0 Controlling Devices

All control systems require some means of communication between devices. This is accomplished by two basic methods: digital signals and analog signals (*Figure 77*). In the HVAC industry, these signals are called points.

The first category of control devices includes external digital devices. These are things such as relays, switches, lights, and other devices that can be operated in either a full-on or full-off (binary) condition.

The second category includes external analog devices. The types of devices in this group include thermistors, photocells, and DC motor controls. The problem of interfacing to an analog device is somewhat more complicated. In this case, we are attempting to take a signal with an infinite number of values (analog) and convert it to a form that can be represented and manipulated by two-state (binary) devices (digital). Most real-world processes are continuously changing (*Figure 78*).

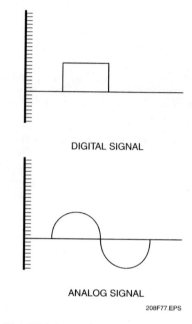

Figure 77 ♦ Digital vs. analog signal.

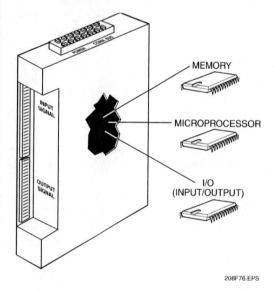

Figure 76 ♦ System controller module.

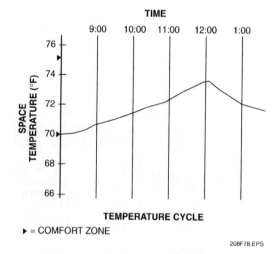

Figure 78 ♦ Changes are small and occur gradually.

Physical quantities such as pressure, temperature, liquid levels, and fluid flow tend to change value rather gradually. Changes of this type produce a large number of discrete values before ever reaching a final state. Converting from analog form to digital form requires the use of a circuit called an **analog-to-digital converter**.

The analog and digital signals can be further divided into inputs and outputs. Analog In (AI) signals are from sensors such as temperature, pressure, and humidity. Analog Out (AO) signals are analog commands, such as reset of system setpoints. Digital In (DI) signals are contact closures or openings, showing status or alarm conditions in a two-position mode. Digital Out (DO) signals are two-position commands like start/stop or open/close states.

13.3.0 Example of a Digital Control System

The key to a successful control system is the integration of all the sensors and unit controllers. Let's look at the integration of a typical constant volume HVAC system. A constant air volume system is one in which the volume of supply air remains constant and the temperature of the air is varied to achieve the desired comfort conditions. It is a single-zone system. If there is more than one zone, each must have its own dedicated unit. The constant air volume system contrasts with a variable air volume system in which the supply air temperature is held constant and the volume of air changes to meet the changing demand. In a variable air volume system, a single unit can serve several zones.

A schematic of a constant-volume system is shown in *Figure 79*. The control system is made up of several control loops: economizer control of mixed air; heating-cooling sequencing; and humidification-dehumidification sequencing.

When the unit fan is energized, as sensed by a static pressure sensor in the supply duct, the damper control system becomes activated. A mixed-air sensor maintains the mixed-air temperature by modulating the outdoor air, return

Describe an analog-to-digital converter.

Show Transparency 72 (Figure 79). Describe the operation of a typical constant-volume system with DDC control.

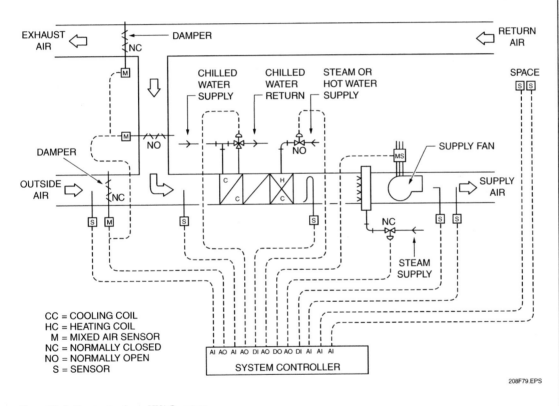

Figure 79 ♦ Constant-volume HVAC system.

Keep Transparency 72 (Figure 79) showing. Continue a discussion of constant-volume systems with DDC control.

See the Teaching Tip for Section 13.3.0 at the end of this module.

air, and exhaust dampers. When the outdoor air temperature exceeds the setting of the outdoor air sensor, the outdoor and exhaust air dampers return to the minimum open position, as programmed, to provide ventilation. The return air damper takes the corresponding open position. In large buildings, it is often required that some outside ventilation air be provided at all times. Therefore, the damper has a minimum open position.

A space temperature sensor, also through the controller, maintains the space temperature by modulating the heating coil valve in sequence with the chilled water coil valve. A space humidity sensor, also through the controller, maintains space humidity. Upon a drop in space relative humidity, the humidifier steam valve modulates toward the open position, subject to a duct-mounted high-limit humidity sensor. With a rise in space relative humidity, the humidifier steam valve modulates to the closed position, followed by the opening of the chilled water coil valve to provide dehumidification. During the dehumidification cycle, the space temperature sensor modulates the heating coil valve to maintain space temperature conditions.

A low-temperature controller, with its capillary located on the discharge side of the heating coil, will de-energize the unit fan, close the outdoor and exhaust dampers, and open the return damper if the discharge air temperature drops below its setting. Whenever the unit fan is de-energized, as sensed by the supply duct static pressure sensor, the damper control system will be de-energized, closing the outdoor and exhaust dampers, along with the humidifier steam control valve.

Review Questions

1. Low-voltage (24V) control circuits are used in HVAC systems because _____.
   a. step-down transformers produce only 24V output
   b. 24V power is safer and cheaper
   c. 24V power is readily available from the electric company
   d. it is the same voltage used to operate the compressor and fan motors

2. For what purpose is a heat anticipator used in heating thermostats?
   a. It turns on the heat just before the temperature reaches the thermostat setpoint.
   b. It keeps the system warm in cold weather.
   c. It turns on a light to let the occupants know when the furnace is about to come on.
   d. It opens the thermostat just before the heat in the conditioned space reaches the thermostat setpoint.

3. Which of the following is *not* true regarding a cooling compensator?
   a. It compensates for the lag between the thermostat call for cooling and the time when the system actually begins cooling the conditioned space.
   b. It is a variable resistor.
   c. It causes the cooling thermostat to close sooner than it otherwise would.
   d. It is usually placed in parallel with the cooling thermostat switch.

4. The differential in a thermostat is _____.
   a. the difference between the cut-in and cut-out points of the thermostat
   b. the difference between the cooling and heating setpoints
   c. normally at least 12°F
   d. the difference between the settings of the heat anticipator and the cooling compensator

5. Which of the following is *not* a common feature on a programmable electronic thermostat?
   a. Battery backup to prevent loss of programs
   b. Motor-driven heat anticipator
   c. Program override
   d. Digital readouts

6. A good location for a thermostat is _____.
   a. anywhere that it will receive direct sunlight
   b. about five feet above the floor on an outside wall
   c. about five feet above the floor on an inside wall
   d. in a corner where there is minimum air circulation

7. In a standard thermostat wiring scheme, the R terminal would be connected to _____.
   a. the fan control
   b. 24V power
   c. the reversing valve
   d. the cooling circuits

8. The correct current draw setting for a thermostat heat anticipator may be obtained from _____.
   a. the thermostat label
   b. the wallplate
   c. measurements taken across the thermostat R and G terminals
   d. the furnace literature

9. The purpose of a motor starter is to _____.
   a. jump-start the compressor if the power fails
   b. vary the speed of a motor as the conditions change
   c. check the motor current to make sure the motor is running
   d. control power application to the motor and provide motor overload protection

Have trainees complete the Review Questions, and go over the answers prior to administering the Module Examination.

Review Questions

10. The purpose of a lockout relay is to _____.
    a. make sure no one turns on the power when you want it off
    b. make sure the compressor doesn't turn on until the fans are running
    c. protect the compressor motor from start-up surges
    d. prevent automatic restart of HVAC equipment when a safety control resets

11. The purpose of a compressor short-cycle timer is to _____.
    a. prevent the compressor from restarting before system pressures can equalize
    b. make sure the compressor runs for at least five minutes once it cycles on
    c. keep track of how often the compressor turns on
    d. run the compressor every five minutes to make sure it doesn't freeze up

12. A limit control is provided in a furnace to _____.
    a. make sure the furnace puts out as much heat as possible
    b. shut off the fan
    c. prevent the furnace from overheating
    d. sense the presence of carbon monoxide

13. In the electrical control circuit of a combined heating/cooling system, the outdoor fan motor _____.
    a. usually has a continuous-on mode for ventilation
    b. runs at low speed in the heating mode
    c. is on whenever the compressor is on
    d. is off whenever the compressor is off

14. Listening to the customer's complaint does *not* provide _____.
    a. valuable information about the nature of an equipment problem
    b. information on what the unit should do
    c. information that may eliminate the HVAC equipment as the cause of the problem
    d. all of the information needed to make an accurate problem diagnosis

15. A troubleshooting aid normally given on a label diagram is the _____.
    a. troubleshooting tree
    b. troubleshooting table
    c. wiring diagram
    d. fault isolation diagram

16. When troubleshooting an electrical problem by the process of elimination, which statement is *not* true?
    a. If the problem is only with heating in a combined heating/cooling unit, the cooling components can be eliminated.
    b. If a component operates properly, it is probably not the problem.
    c. Areas of the diagram that show optional equipment that is not installed can be eliminated.
    d. If an outdoor fan motor runs but a related compressor does not, the compressor control circuit can be eliminated.

17. When troubleshooting an energized compressor load circuit, the multimeter indicates voltage on both sides of a set of switch contacts. This indicates that the switch contacts are _____.
    a. open
    b. bad
    c. closed
    d. designated as normally closed (NC) contacts

18. When troubleshooting in an energized fan motor load circuit, the multimeter indicates voltage on one side of a set of relay contacts but not on the other. This indicates that the _____.
    a. relay contacts are shorted
    b. relay contacts are closed
    c. relay contacts are open
    d. related relay coil is energized

Instructor's Notes:

Review Questions

19. The power supply voltage measured across the output terminals of a two-pole circuit breaker is lower than was measured across the input terminals of the same breaker. The problem can be _____.
    a. loose wires and/or corroded circuit breaker terminals
    b. the breaker has been subjected to extremely cold temperatures
    c. there is no problem
    d. the breaker is set to OFF

20. When testing a relay coil, the multimeter indicates a measurable resistance. This means that the coil is _____.
    a. shorted
    b. probably good
    c. open
    d. partially shorted

21. When troubleshooting the operation of a compressor start relay circuit by measuring the start winding current draw with a clamp-on ammeter, no current is measured on the clamp-on ammeter when the compressor is started. This means that the _____.
    a. start relay contacts are open
    b. start relay contacts are closed
    c. start relay coil is open
    d. start capacitor is shorted

22. On a single-phase PSC compressor, the highest resistance should be measured between _____.
    a. the common and start terminals
    b. the run and start terminals
    c. the common and run terminals
    d. depends on the winding characteristics of the PSC compressor being measured

23. When checking a 440V motor for grounded windings, the minimum resistance that should be measured between any lead and ground is _____.
    a. $1,000\Omega$
    b. $230,000\Omega$
    c. $440,000\Omega$
    d. infinity

24. A pneumatic-electric relay is one that _____.
    a. uses air pressure to open and close electrical relay contacts
    b. uses a solenoid to close a pneumatic valve
    c. supplies power to the air compressor
    d. uses hydraulic fluid pressure to open and close relay contacts

25. The form in which temperature information is likely to be received by a computer-controlled system is _____.
    a. binary
    b. analog
    c. digital
    d. audio

Summarize the major concepts presented in the module.

Administer the Module Examination. Record the results on Craft Training Report Form 200, and submit the results to the Training Program Sponsor.

Administer the Performance Test, and fill out Performance Profile Sheets for each trainee. If desired, trainee proficiency noted during laboratory sessions may be used to complete the Performance Test. Record the results on Craft Training Report Form 200, and submit the results to the Training Program Sponsor.

Summary

HVAC control systems are made up of a variety of electrical, electronic, and pneumatic controls that control all functions of an HVAC system.

The ability to analyze HVAC control systems is a critical skill for the service technician because a large percentage of the problems that occur in HVAC systems are control circuit faults. No matter how complex the control system, it consists of individual components that are combined to form control functions such as heating, cooling, fan control, and defrost. If you know how the components work, you can figure out how they fit together and how the system functions as a whole. Once you have these skills, you can troubleshoot any system.

Effective troubleshooting is a process by which the HVAC technician listens to a customer's complaint, performs an independent analysis of a problem, and then initiates and performs a systematic step-by-step approach to troubleshooting that results in the correction of the problem. The HVAC technician must understand the purpose and principles of operation of each component in the equipment being serviced. You must be able to tell whether or not a given device is functioning properly and to recognize the symptoms arising from the improper operation of any part of the equipment.

Notes

Instructor's Notes:

Trade Terms Introduced in This Module

Actuator: The portion of a regulating valve that converts one type of energy, such as pneumatic pressure, into mechanical energy (for example, opening or closing a valve).

Analog-to-digital converter: A device designed to convert analog signals such as temperature and humidity to a digital form that can be processed by logic circuits.

Automatic changeover thermostat: A thermostat that automatically selects heating or cooling.

Bleed control: A valve with a tiny opening that permits a small amount of fluid to pass.

Cooling compensator: A fixed resistor installed in a thermostat to act as a cooling anticipator.

Deadband: A temperature band, usually 3°F, that separates heating and cooling in an automatic changeover thermostat.

Differential: The difference between the cut-in and cut-out points of a thermostat.

Droop: A mechanical condition caused by heat that affects the accuracy of a bimetal thermostat.

Electric-pneumatic (E-P) relay: A three-way pneumatic solenoid valve.

Fault isolation diagram: A troubleshooting aid usually contained in the manufacturer's installation, startup, and service instructions for a particular product. Fault isolation diagrams are also called troubleshooting trees. Normally, fault isolation diagrams begin with a failure symptom then guide the technician through a logical decision-action process to isolate the cause of the failure.

Invar®: An alloy of steel containing 36 percent nickel. It is one of the two metals in a bimetal device.

Label diagram: A troubleshooting aid usually placed in a convenient location inside the equipment. It normally depicts a wiring diagram, a component arrangement diagram, a legend, and notes pertaining to the equipment.

Ladder diagram: A troubleshooting aid that depicts a simplified schematic diagram of the equipment. The load lines are arranged like the rungs of a ladder between vertical lines representing the voltage source. Normally, all the wire color and physical connection information is eliminated from the diagram to make it easier to use by focusing on the functional, not the physical, aspects of the equipment.

Pneumatic-electric (P-E) relay: A pressure switch in which a pneumatic signal causes an electrical change.

Sub-base: The portion of a two-part thermostat that contains the wiring terminals and control switches.

Thermostat base: The portion of a two-part thermostat that contains the heating and cooling thermostats.

Troubleshooting: A procedure by which the technician locates the source of a problem, then makes the repairs and/or adjustments to correct the cause of a problem so that it will not recur.

Troubleshooting table: A troubleshooting aid usually contained in the manufacturer's installation, startup, and service instructions for a particular product. Troubleshooting tables are intended to guide the technician to a corrective action based on observations of system operation.

Wiring diagram: A troubleshooting aid, sometimes called a schematic, that provides a picture of what the unit does electrically and shows the actual external and internal wiring of the unit.

Appendix

Schematic Symbols

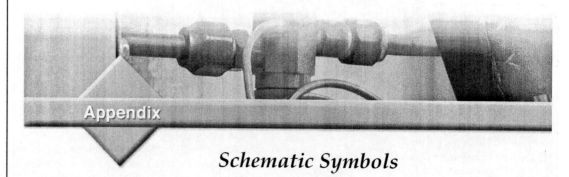

(1) Make on rise
(2) Make on fall

COILS			
AUTOMATIC TRANSFORMER	REACTORS		ADJUSTABLE
	IRON CORE	AIR CORE	SHOWN WITH IRON CORE

DIODE RECTIFIERS		MOTORS & COMPRESSORS		
HALF-WAVE	FULL-WAVE	THREE-PHASE	SINGLE-PHASE	
			TWO LEADS	CAPACITOR START/RUN

RESISTORS	
FIXED	VARIABLE

THERMOCOUPLE	LAMPS	BATTERY	GROUND		CAPACITOR	
			ELECTRICAL	MECHANICAL	FIXED	ADJUSTABLE
						X-SIDE NEAR GROUND

Additional Resources and References

Additional Resources

This module is intended to be a thorough resource for task training. The following reference works are suggested for further study. These are optional materials for continued education rather than for task training.

Air Conditioning Systems, Principles, Equipment, and Service. Latest Edition. Joseph Moravek. Upper Saddle River, NJ: Prentice Hall.

HVAC Servicing Procedures, 1995. Syracuse, NY: Carrier Corporation.

Pocket Guide to Electrical Installations Under NEC 2002, Volumes I and II, 2001. Quincy, MA: National Fire Protection Association.

Refrigeration and Air Conditioning, An Introduction to HVAC/R, Fourth Edition. Larry Jeffus. Air Conditioning and Refrigeration Institute. Upper Saddle River, NJ: Prentice Hall.

Figure Credits

Topaz Publications, Inc., 208F05 (photo), 208F07 (photo), 208F08 (photo), 208F19, 208F20, 208F25, 208SA01, 208SA05, 208SA06

Courtesy of Honeywell International Inc., 208F10, 208F61, 208F62

Tyco Electronics, 208F24

Carrier Corporation, 208SA02, 208F34–208F36, 208F39–208F41

Extech Instruments, 208SA04

Courtesy ITT, 208F64

Taco, Inc., 208F65, 208F66

MODULE 03208-07 — TEACHING TIPS

The following are suggested activities or instructional methods to help you teach the material in this module.

General When you call on someone to answer a question, the rest of the class relaxes or even tunes out because they expect that the question and answer will take place only between you and the trainee you called on. Instead, use this technique to involve more trainees in answering questions and to keep them on their toes.

1. Ask trainees to define a term or explain a concept.
2. After one trainee has answered, ask a trainee seated nearby if the answer is right. Then ask whether a trainee in the back of the room agrees.
3. Ask trainees to explain why they think an answer is right or wrong.
4. Use the session to clear up incorrect ideas and encourage trainees to learn from their mistakes.

Section 2.0.0 *Quick Quiz*

This Quick Quiz will familiarize trainees with the terms and definitions that are commonly used when discussing thermostats. You will need photocopies of the quiz provided in the following page. Trainees will need pencils. If you allow trainees to use the Trainee Guide, decrease the amount of time you give them to complete the quiz.

1. Make a photocopy of the quiz for each trainee.
2. Give trainees between 5 and 10 minutes to complete the quiz.
3. Go over the answers to the quiz.
4. Ask trainees if they have questions.

Answers to Quick Quiz

1. c
2. h
3. a
4. f
5. j
6. e
7. i
8. g
9. d
10. b

Quick Quiz   *Thermostats*

For each description listed, identify the term that the text best describes. Write the corresponding letter in the blank provided.

_____ 1. Cooling thermostats contain a device called a _____ to help improve indoor comfort.

_____ 2. _____ are self-contained controls with the times, temperature sensor, and switching devices all located in the unit mounted on the wall.

_____ 3. _____ automatically selects the mode, depending on the heating and cooling setpoints.

_____ 4. The two metals in a thermostat sensing element are typically copper or brass and _____.

_____ 5. _____ are normally used to control heat pumps.

_____ 6. On a thermostat, the _____ is typically 2°F.

_____ 7. On a two-stage indoor thermostat, the heating and cooling thermostats are on the _____.

_____ 8. _____ can operate on 240V and are typically used to control electric baseboard heat.

_____ 9. Automatic changeover thermostats also have a minimum interlock setting, commonly known as a _____.

_____ 10. Most _____ will maintain the space temperature within 3°F of the setpoint.

     a. automatic changeover thermostats
     b. bimetal thermostats
     c. cooling compensator
     d. deadband
     e. differential
     f. Invar®
     g. line voltage thermostats
     h. programmable thermostats
     i. thermostat base
     j. two-stage indoor thermostats

Section 2.0.0 *Identifying Thermostats*

This exercise will familiarize trainees with identifying various types of thermostats. Trainees will need pencils and paper. You will need various types of thermostats. Allow 20 to 30 minutes for this exercise. This exercise can correspond to Performance Task 1.

1. Set up several workstations with a different type of thermostat at each one.
2. Have trainees rotate between the workstations and identify the type of thermostat. If desired, have the trainees note a fact about the thermostat; for example, its applications.
3. When all trainees have circulated to all of the workstations, have one trainee identify the thermostat at his or her workstation.
4. Answer any questions trainee may have.

Section 4.0.0 *Think About It – Analyzing a Circuit Diagram*

1. No; as long as the heating thermostat TH is closed, there is 24V available to the induced-draft relay.
2. 240V; the condition of the compressor contactor has no bearing. Connecting the voltmeter to these two points is the same as connecting across the power source.
3. De-energized; the indoor fan relay can only be energized for cooling or ventilation.

Section 5.5.0   *Think About It – Wiring Diagram Exercise*

1. Place the meter leads across L1 and L3 of the compressor contactor (CC) with the contactor de-energized (no call for heating or cooling). This will display the input voltage available to the unit.
2. The low-voltage transformer is energized and the correct voltage is available to power components in the control circuit.
3. Since the defrost relay is energized and contacts 1 and 2 of the defrost relay are open, the voltmeter should indicate zero volts.
4. Since terminal W2 is the universal terminal designation for second-stage heating operation, voltmeter leads placed across the low-voltage terminals W2 and X should read 24V. Since this is a heat pump, the first-stage signal comes through terminal Y to start the compressor.
5. Contacts 4 and 5 of the defrost relay (DR) are open, stopping the outdoor fan for the duration of the defrost cycle. This aids in the removal of frost from the outdoor coil.
6. Terminals 24VAC and COM in the demand defrost control are connected directly across the 24V transformer. Placing voltmeter leads on these points should produce a reading of 24V.
7. A clamp-on ammeter can measure the current in any wire in the circuit. In this case, clamp the jaws of the ammeter around the lead going to the compressor common C terminal, turn the power on, and observe the current draw of the compressor.
8. One input is 24V power applied at the BC control R terminal. The other input is a 24V call for fan signal from the room thermostat applied at the G terminal. Note that the C terminal is not an input point but a connection to the common side of the 24V transformer.
9. The room thermostat can be put in the emergency heat mode. This locks out compressor operation and allows electric resistance heaters, energized through heater contactors HC #1 and HC #2, to heat the home. The 24V signal to energize the heaters in the emergency heat mode would be measured across terminals E and X on the low-voltage terminal strip.
10. Since the defrost relay energizes during defrost and opens the contacts between points 1 and 2 to de-energize the reversing valve solenoid, it must be assumed that the RVS is energized in heating. By de-energizing the solenoid during defrost, the unit temporarily reverts back to the cooling mode. In the actual cooling mode (not defrost), the RVS would also be de-energized.

Section 6.3.0 *Lockout/Tagout*

This exercise will familiarize trainees with electrical safety. Trainees will need pencils and paper. You will need to obtain a safety video or arrange for an OSHA or other safety professional to give a presentation on lockout/tagout or other electrical safety topics. Allow 20 to 30 minutes for this exercise.

Many organizations, such as Associated General Contractors, Petroleum Equipment Institute, university safety offices, local safety councils, or local OSHA offices, may have safety videos that can be borrowed free of charge for presentation to the class. Obtain one of the following or another safety training video:

Lockout/Tagout Of Energy Sources Training Video, 27 mins. Seton. $149.00
LOTO for Authorized & Affected Personnel Video, 17 mins. National Safety Compliance, http://www.osha-safety-training.net/loto/loto.html. $99.00

Alternatively, online training is available for free at the following locations:

OSHA Lockout Tagout Interactive Training:
http://www.osha.gov/dts/osta/lototraining/about.htm
Seton Compliance Resource Center
http://www.setonresourcecenter.com/safety/loto/
University of South Carolina Lockout/Tagout Online Training:
http://ehs.sc.edu/modules/Lockout%20Tagout/l'oto_intro.htm

1. Tell trainees that a guest speaker will be presenting information on electrical safety.
2. Have trainees brainstorm questions for the guest speaker before the speaker arrives.
3. Introduce the speaker or video. Ask the presenter to speak about electrical safety. Have them include personal protective equipment, lockout/tagout procedures, and other safety concerns.
4. Have the trainees take notes and write down questions during the presentation.
5. After the presentation, ask the speaker to spend some time to answer questions. Have the trainees ask their questions.

Section 12.2.0 *Pneumatic Control Systems*

This exercise will familiarize trainees with pneumatic control systems. Trainees will need pencils and paper. You will need to arrange for a manufacturer's representative to give a presentation on commercial pneumatic control systems. Allow 20 to 30 minutes for this exercise.

Alternatively, take trainees to a commercial site that uses a pneumatic control system and point out the various components of the system.

1. Tell trainees that a guest speaker will be presenting information on pneumatic control systems. Have trainees brainstorm questions for the guest speaker before the speaker arrives.
2. Have the trainees take notes and write down questions during the presentation.
3. After the presentation, answer any questions trainees may have.

Section 13.3.0 *Digital Control Systems*

This exercise will familiarize trainees with digital control systems. Trainees will need pencils and paper. You will need to arrange for a manufacturer's representative to give a presentation on commercial digital control systems. Allow 20 to 30 minutes for this exercise.

Alternatively, take trainees to a commercial site that uses a digital control system and point out the various components of the system.

1. Tell trainees that a guest speaker will be presenting information on digital control systems. Have trainees brainstorm questions for the guest speaker before the speaker arrives.
2. Have the trainees take notes and write down questions during the presentation.
3. After the presentation, answer any questions trainees may have.

MODULE 03208-07 — ANSWERS TO REVIEW QUESTIONS

Answer	Section
1. b	2.1.0
2. d	2.2.0
3. b	2.3.0
4. a	2.5.0
5. b	2.7.0
6. c	2.9.1
7. b	2.9.2
8. d	2.9.3
9. d	3.1.3
10. d	3.3.0
11. a	3.5.0
12. c	3.7.2
13. c	4.0.0
14. d	5.1.0
15. c	5.4.1
16. d	8.1.0
17. c	8.2.0
18. c	8.2.0
19. a	9.2.2
20. b	9.3.0
21. a	10.3.0
22. b	10.5.0
23. c	10.6.2
24. a	12.1.0
25. b	13.2.0

CONTREN® LEARNING SERIES — USER UPDATE

NCCER makes every effort to keep these textbooks up-to-date and free of technical errors. We appreciate your help in this process. If you have an idea for improving this textbook, or if you find an error, a typographical mistake, or an inaccuracy in NCCER's Contren® textbooks, please write us, using this form or a photocopy. Be sure to include the exact module number, page number, a detailed description, and the correction, if applicable. Your input will be brought to the attention of the Technical Review Committee. Thank you for your assistance.

Instructors – If you found that additional materials were necessary in order to teach this module effectively, please let us know so that we may include them in the Equipment/Materials list in the Annotated Instructor's Guide.

Write: Product Development and Revision
National Center for Construction Education and Research
3600 NW 43rd St, Bldg G, Gainesville, FL 32606

Fax: 352-334-0932

E-mail: curriculum@nccer.org

Craft _____ Module Name _____

Copyright Date _____ Module Number _____ Page Number(s) _____

Description
_____
_____
_____
_____

(Optional) Correction
_____
_____
_____

(Optional) Your Name and Address
_____
_____
_____

Module 03209-07

Troubleshooting Gas Heating

NCCER STANDARDIZED CRAFT TRAINING PROGRAM

The National Center for Construction Education and Research (NCCER) provides a standardized national program of accredited craft training. Key features of the program include instructor certification, competency-based training, and performance testing. The program provides trainees, instructors, and companies with a standard form of recognition through a National Craft Training Registry. The program is described in full in the *Guidelines for Accreditation*, published by NCCER. For more information on standardized craft training, contact the NCCER by writing us at 3600 NW 43rd St., Bldg. G, Gainesville, FL 32606; calling 352-334-0911; or emailing info@nccer.org. More information may be found at our website, www.nccer.org.

HOW TO USE THIS ANNOTATED INSTRUCTOR'S GUIDE

Each page presents two sections of information. The larger section displays each page exactly as it appears in the Trainee Module. The narrow column ties suggested trainee and instructor actions to each page and provides icons (detailed below) to call your attention to material, safety, audiovisual, or testing requirements. The bottom of each page includes space for your notes.

The **Audiovisual** icon indicates an appropriate time to show a transparency or other audiovisual aid.

The **Classroom** icon prompts you to define a term, stress a point, ask trainees to explain a concept, or give examples.

The **Demonstration** icon directs you to show trainees how to perform tasks.

The **Examination** icon tells you to administer the written module examination.

The **Homework** icon is placed where you may wish to assign reading for the next class, assign a project, or advise trainees to prepare for an examination.

The **Laboratory** icon is used when trainees are to practice performing tasks.

The **Materials** icon is a reminder for you to gather materials needed for classes, labs, and testing.

The **Performance Testing** icon tells you to administer a performance test or a portion thereof.

The **Safety** icon is used to emphasize safety issues. It is often keyed to *Caution* and *Warning!* statements in the Trainee Module.

The **Teaching Tip** icon indicates additional guidance is available, such as how to conduct an exercise, get the most educational value from a field trip, or encourage class participation. Teaching Tips may expand on a feature (*Think About It, Did You Know?*) or provide *Quick Quizzes* or similar exercises. You will be referred to the Teaching Tips section at the back of the module if there is additional material.

The **Combination** icon indicates that the laboratory listed corresponds with a performance task. If desired, you can note the proficiency of the trainees during the laboratory, and use it to satisfy performance testing requirements.

PREPARATION

Before teaching this module, you should review the Objectives, Performance Tasks, Materials and Equipment List, and Module Outline. Be sure to allow ample time to prepare your own training or lesson plan and gather all required materials and equipment.

Troubleshooting Gas Heating
Annotated Instructor's Guide

Module 03209-07

MODULE OVERVIEW

This module introduces the trainee to the procedures for recognizing, analyzing, and repairing malfunctions in gas heating equipment.

PREREQUISITES

Prior to training with this module, it is recommended that the trainee shall have successfully completed *Core Curriculum*; *HVAC Level One*; and *HVAC Level Two*, Modules 03201-07 through 03208-07.

OBJECTIVES

Upon completion of this module, the trainee will be able to do the following:

1. Describe the basic operating sequence for gas heating equipment.
2. Interpret control circuit diagrams for gas heating systems.
3. Describe the operation of various types of burner ignition methods.
4. Identify the tools and instruments used when troubleshooting gas heating systems.
5. Demonstrate using the tools and instruments required for troubleshooting gas heating systems.
6. Isolate and correct malfunctions in gas heating systems.

PERFORMANCE TASKS

Under the supervision of the instructor, the trainee should be able to do the following:

1. Develop a checklist for troubleshooting a gas heating appliance.
2. Select the tools and instruments needed to troubleshoot a gas heating appliance in a given situation.
3. Analyze control circuit diagram(s) for a selected gas heating appliance.
4. Isolate and correct malfunctions in a gas heating appliance.
   - Control circuits
   - Combustion system
   - Safety controls
   - Air system

MATERIALS AND EQUIPMENT LIST

Overhead projector and screen
Transparencies
Blank acetate sheets
Transparency pens
Whiteboard/chalkboard
Markers/chalk
Pencils and scratch paper
Appropriate personal protective equipment
Insulated jumper wires
Microammeter
Millivoltmeter

Multimeter
Operating natural-draft and induced-draft gas heating systems
Selection of pre-faulted components
Manufacturers' literature on gas-fired furnaces
Hot surface ignitors
Flame sensors
Copies of Quick Quiz*
Module Examinations**
Performance Profile Sheets**

* Located in the back of this module.
**Located in the Test Booklet.

SAFETY CONSIDERATIONS

Ensure that the trainees are equipped with appropriate personal protective equipment and know how to use it properly. This module requires trainees to work with gas heating appliances. Make sure that all trainees are briefed on appropriate safety procedures. Emphasize electrical safety.

ADDITIONAL RESOURCES

This module is intended to present thorough resources for task training. The following reference work is suggested for both instructors and motivated trainees interested in further study. This is optional material for continued education rather than for task training.

Refrigeration and Air Conditioning, An Introduction to HVAC/R, Fourth Edition. Larry Jeffus. Air Conditioning and Refrigeration Institute. Upper Saddle River, NJ: Prentice Hall.

TEACHING TIME FOR THIS MODULE

An outline for use in developing your lesson plan is presented below. Note that each Roman numeral in the outline equates to one session of instruction. Each session has a suggested time period of 2½ hours. This includes 10 minutes at the beginning of each session for administrative tasks and one 10-minute break during the session. Approximately 12½ hours are suggested to cover *Troubleshooting Gas Heating*. You will need to adjust the time required for hands-on activity and testing based on your class size and resources. Because laboratories often correspond to Performance Tasks, the proficiency of the trainees may be noted during these exercises for Performance Testing purposes.

Topic   **Planned Time**

Sessions I and II. Introduction and Control Circuits

   A. Introduction   _____

   B. Control Circuits   _____

   C. Laboratory   _____

      Trainees practice developing a checklist for troubleshooting gas heating systems. This laboratory corresponds to Performance Task 1.

   D. Laboratory   _____

      Trainees practice identifying the tools and instruments needed to troubleshoot a gas heating appliance. This laboratory corresponds to Performance Task 2.

   E. Laboratory   _____

      Use pre-faulted components, jumpers, or other means to insert safety and control circuit malfunctions into a gas furnace. Provide trainees with the wiring diagram for the unit and have them isolate and correct the fault under your supervision. This laboratory corresponds to Performance Tasks 3 and 4.

Session III. Combustion Systems

   A. Combustion Systems   _____

   B. Laboratory   _____

      Use pre-faulted components, jumpers, or other means to insert combustion system malfunctions into a gas furnace. Provide trainees with the wiring diagram for the unit and have them isolate and correct the fault under your supervision. This laboratory corresponds to Performance Tasks 3 and 4.

Session IV. Air Systems

   A. Air Systems   _____

   B. Laboratory   _____

      Use pre-faulted components, jumpers, or other means to insert air system malfunctions into a gas furnace. Provide trainees with the wiring diagram for the unit and have them isolate and correct the fault under your supervision. This laboratory corresponds to Performance Tasks 3 and 4.

Session V. Review and Testing
   A. Review
   B. Module Examination
      1. Trainees must score 70% or higher to receive recognition from NCCER.
      2. Record the testing results on Craft Training Report Form 200, and submit the results to the Training Program Sponsor.
   C. Performance Testing
      1. Trainees must perform each task to the satisfaction of the instructor to receive recognition from NCCER. If applicable, proficiency noted during laboratory exercises can be used to satisfy the Performance Testing requirements.
      2. Record the testing results on Craft Training Report Form 200, and submit the results to the Training Program Sponsor.

HVAC Level Two

03209-07
Troubleshooting Gas Heating

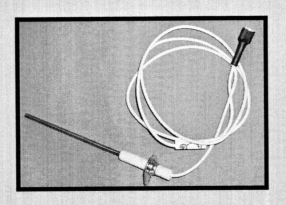

Assign reading of Module 03209-07.

03209-07
Troubleshooting Gas Heating

Topics to be presented in this module include:

1.0.0	Introduction	.9.2
2.0.0	Control Circuits	.9.2
3.0.0	Combustion Systems	.9.15
4.0.0	Air System	.9.21

Overview

Gas furnaces and other gas heating devices contain a variety of controls. Most of these controls are safety devices that either shut the system down in the event of a problem or prevent it from operating if certain conditions are not met in a specified sequence. The service technician must be able to determine how the control devices interact to control the system. The safety controls are not the only things that can go wrong in a gas appliance. The service technician must also learn to recognize combustion-related problems and test the combustion system using specialized test instruments.

Instructor's Notes:

Objectives

When you have completed this module, you will be able to do the following:

1. Describe the basic operating sequence for gas heating equipment.
2. Interpret control circuit diagrams for gas heating systems.
3. Describe the operation of various types of burner ignition methods.
4. Identify the tools and instruments used when troubleshooting gas heating systems.
5. Demonstrate using the tools and instruments required for troubleshooting gas heating systems.
6. Isolate and correct malfunctions in gas heating systems.

Trade Terms

Flame rectification
Modulating gas valve
Reverberator
Submerged flame
Two-stage gas valve
Vane switch

Required Trainee Materials

1. Pencil and paper
2. Appropriate personal protective equipment

Prerequisites

Before you begin this module, it is recommended that you successfully complete *Core Curriculum*; *HVAC Level One*; and *HVAC Level Two*, Modules 03201-07 through 03208-07.

This course map shows all of the modules in the second level of the HVAC curriculum. The suggested training order begins at the bottom and proceeds up. Skill levels increase as you advance on the course map. The local Training Program Sponsor may adjust the training order.

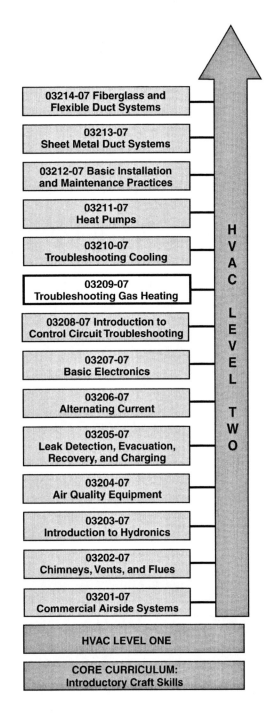

Ensure that you have everything required to teach the course. Check the Materials and Equipment list at the front of this module.

See the general Teaching Tip at the end of this module.

Explain that terms shown in bold are defined in the Glossary at the back of this module.

Show Transparency 1, Objectives, and Transparency 2, Performance Tasks. Review the goals of the module, and explain what will be expected of the trainee.

Review the modules covered in Level Two and explain how this module fits in.

Identify the skills needed to troubleshoot gas heating systems.

Explain that control circuits vary widely.

Discuss the functions of electronic controls and explain how they can be tested by testing their inputs.

Show Transparencies 3 and 4 (Figures 1 and 2). Explain control circuits and wiring diagrams for a typical packaged unit with natural-draft gas heat.

See the Teaching Tips for Sections 1.0.0–4.0.0 and Section 2.0.0 at the end of this module.

1.0.0 ♦ INTRODUCTION

In order to troubleshoot a gas heating system, the technician must consider three separate but interrelated subsystems: controls, air or water flow, and gas combustion. The ability to read circuit diagrams, use electrical test equipment, recognize and correct combustion-related problems, and measure and adjust airflow all play a role in troubleshooting gas-fired heating equipment.

2.0.0 ♦ CONTROL CIRCUITS

The control circuits for gas-fired systems come in many varieties. It is rare to find two furnace models that are electrically the same, even if they are made by the same company. Although packaged electronic controls have allowed manufacturers to standardize control circuits, you will find some variations from model to model, if only in the programming of the microprocessor.

This module includes some examples of circuit diagrams from different furnace manufacturers, along with a description of their operating sequences. By studying these circuits, you will reinforce and expand your knowledge of furnace controls. This will help you analyze the control circuit of an unfamiliar product in order to locate the cause of a malfunction.

In addition to gas furnaces and boilers, gas heating may also be provided by packaged heating and cooling systems. One major difference is that packaged units are installed outdoors and therefore need no vent piping. The control circuits of packaged units and gas furnaces are about the same.

The use of microprocessor-based electronic controls is now widespread across all HVAC product lines. Although older technology exists in millions of legacy systems, it is rare to find a modern heating unit that does not contain some kind of electronic control. Manufacturers put these controls in their equipment to increase reliability and energy efficiency. Electronic controls also allow engineers to design more versatility into the products than was ever possible with electro-mechanical controls. Electronic control boards are checked for correct operation by making sure that the board is receiving the correct inputs such as voltage signals from the room thermostat and inputs from temperature sensors such as limit switches. If all the inputs are correct, the control should provide appropriate outputs. For example, if the control is receiving a room thermostat input to W on the control, and all temperature inputs are within an acceptable range, the control will provide output signals to energize the ignitor, gas valve, and blower motor. If the correct input and output signals are present and the furnace does not run, the problem is not with the control. If the correct inputs are present but no outputs are present, the control may be defective. Input and output signals are typically measured with a multimeter. Troubleshooting of systems with microprocessor-based controls is covered in a later module.

2.1.0 Natural-Draft Packaged Units

Government-mandated energy efficiency requirements for gas furnaces cannot easily be met with furnaces using natural-draft venting technology. This has made natural-draft furnaces obsolete. However, natural-draft systems will be in use for some time to come, as they were still being installed in the 1990s.

Figure 1 represents the control circuits and sequence of operation for a packaged air conditioning unit with natural-draft gas heat. *Figure 2* is the wiring diagram for this unit. Because we are dealing with gas heating, only portions of the cooling system are shown.

INSIDE TRACK

Indoor Comfort

What is indoor comfort? To many consumers, comfort means hot air coming from the registers in winter and cold air coming from them in summer. True indoor comfort is more than just hot or cold. Today's equipment manufacturers are taking advantage of new technologies that allow them to provide improved humidity control, air quality, and energy efficiency that were only dreamed of a few years ago. Those new technologies include the widespread use of electronic controls, capacity control using staged gas valves and new compressor technology, and the use of variable-speed blower motors. You need to be prepared to deal with these new technologies when servicing modern gas heating systems.

Instructor's Notes:

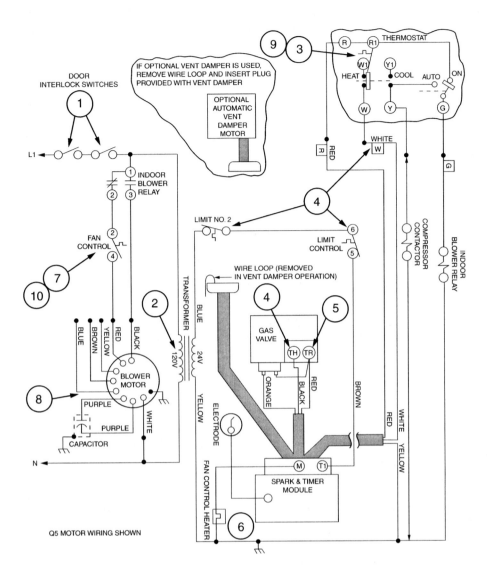

1. Line potential feeds through the door interlocks (if used). Access panels must be in place to energize machine.

2. Transformer provides 24V control circuit.

3. On a heating demand, the thermostat heating contacts close.

4. The 24V control circuit is complete to the pilot valve through limit controls, W leg of the thermostat, and internal ignitor circuits.

5. After the pilot flame has been proven, the main gas valve is energized. Main burners are ignited by the pilot flame.

6. As the main gas valve is energized, the fan control heater is also activated after a brief delay.

7. After a brief delay, the heater provides sufficient heat to close the fan control contacts.

8. This energizes the blower motor on the selected speed.

9. As heating demand is satisfied, the thermostat heating contacts open. This de-energizes the ignition control, gas valve, and fan control heater.

10. The blower motor continues running until the furnace temperature drops below fan control set point.

Figure 1 ◆ Natural-draft gas heating system.

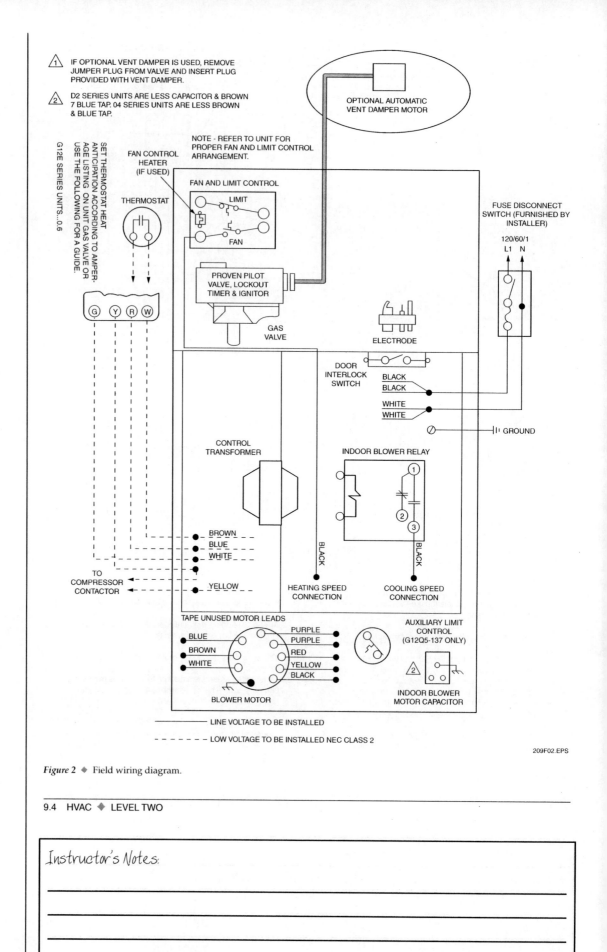

Figure 2 ♦ Field wiring diagram.

This type of unit is sometimes called a year-around air conditioner (YAC). It may also be known as a packaged gas-electric unit (gas heat, electric cooling). A packaged electric-electric unit would be an air conditioning package with electric resistance heaters. The basic operating sequence of any natural-draft unit is as follows:

1. The thermostat calls for heat.
2. The system verifies (proves) that the pilot is lighted.
3. Assuming all safety controls are closed, the gas valve is turned on and the gas ignites.
4. The blower comes on after the heat exchanger heats and closes the fan switch.
5. The thermostat opens. The blower continues to operate until the heat exchanger cools, opening the fan switch.

The heating side of the unit shown in *Figure 1* has a couple of variations from the basic sequence:

- It uses a spark ignitor rather than a standing pilot.
- There is a blower time delay at the beginning and end of the cycle.

2.2.0 Induced-Draft Packaged Units

Induced-draft furnaces are considerably more efficient than natural-draft furnaces; that is, more of the heat produced by the furnace is delivered to the indoor space. Unlike natural-draft furnaces, which rely on natural convection, induced-draft technology uses fans to draw combustion products through more restrictive and efficient heat exchangers to solve the problem. To prove that airflow is adequate through the heat exchanger before the burners can be fired, an airflow proving device such as a pressure switch is used. Adequate airflow will close switch contacts in the pressure switch, allowing the ignition sequence to proceed.

Figure 3 represents the control circuits for another packaged gas-electric unit with an induced-draft blower. The basic operating sequence of induced-draft equipment is slightly different than that of natural-draft equipment, primarily because of the addition of the inducer fan.

1. The thermostat calls for heat.
   - The combustion blower motor (induced-draft fan motor) is energized.
   - When the induced-draft fan motor is up to design rpm, the pressure it creates will close a pressure switch or airflow switch in the vent hood to enable the next step in the sequence.
2. The system verifies (proves) that the pilot is lighted.
3. Assuming all safety controls are closed, the gas valve is turned on and the gas ignites.
4. The blower comes on.
5. The furnace runs until the thermostat opens.

The control circuit shown in *Figure 3* has an induced-draft fan motor (referred to as the combustion fan motor in the operating sequence). Follow along on the diagram as you read through the operating sequence. As you progress, you will see that the operating sequence for this unit is somewhat different from the basic sequence just discussed. The differences include the following:

- In addition to the vane switch used to prove combustion airflow, a second vane switch must prove conditioned airflow before ignition can take place.
- It uses a spark ignitor rather than a standing pilot.
- It has a **two-stage gas valve**, and therefore uses a two-speed combustion fan motor.
- It has a blower-off time delay.

Discuss the operating sequence of a typical packaged unit with natural-draft gas heat.

Identify the differences between induced-draft and natural-draft furnaces. Explain that the key difference between the units is the need to prove adequate airflow.

Show Transparency 5 (Figure 3). Discuss the operating sequence of a typical packaged unit with induced-draft gas heat.

Discuss the differences in operating sequences between the two types of units.

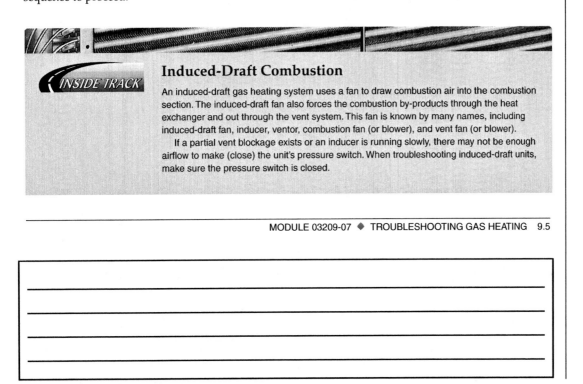

Induced-Draft Combustion

An induced-draft gas heating system uses a fan to draw combustion air into the combustion section. The induced-draft fan also forces the combustion by-products through the heat exchanger and out through the vent system. This fan is known by many names, including induced-draft fan, inducer, ventor, combustion fan (or blower), and vent fan (or blower).

If a partial vent blockage exists or an inducer is running slowly, there may not be enough airflow to make (close) the unit's pressure switch. When troubleshooting induced-draft units, make sure the pressure switch is closed.

MODULE 03209-07 ♦ TROUBLESHOOTING GAS HEATING 9.5

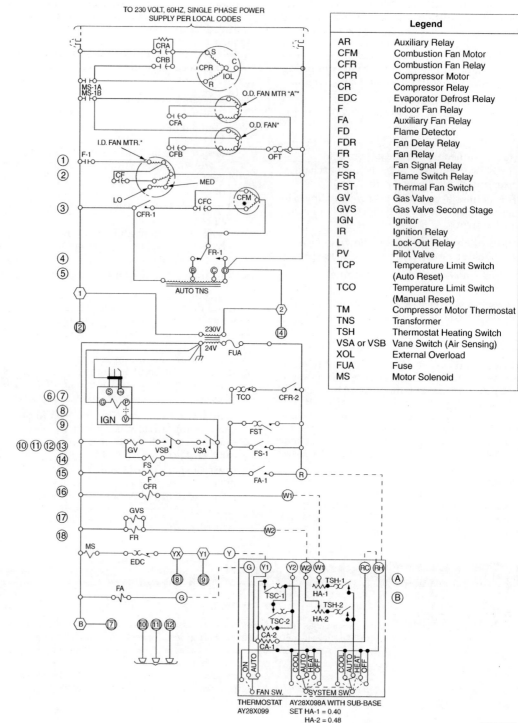

Figure 3 ◆ Induced-draft gas heat/electric cool unit.

1. The thermostat heating switch TSH-1 (A) closes, completing a 24V circuit through terminal W1, energizing the combustion fan relay coil CFR (16).
2. CFR contact CFR-1 (3) closes, energizing the combustion fan motor in low speed through the autotransformer (5), providing combustion airflow through the heat exchanger. Combustion airflow then closes vane switch VSB (11).
3. Contact CFR-2 (7) closes, completing a circuit to the electronic ignitor IGN (8) through the temperature limit switch TCO (6).
4. When power is applied, the ignitor control contact (8) closes, completing a circuit to the fan signal relay FS (14).
5. The fan signal relay contact FS-1 (13) closes, completing a circuit to the indoor fan relay coil F (15), energizing the indoor fan motor (2) through contact F-1 (1).
6. When indoor airflow is established, it closes vane switch VSA (12). The gas valve GV (10) circuit is then completed through the vane switch VSB (11) that was closed previously by combustion airflow.

NOTE
At this point, if flame is not established within ten seconds, the ignition sequence may be re-initiated by manually breaking and resetting either the high- or low-voltage circuit.

7. If additional heating is required, the thermostat heating switch TSH-2 (B) closes, completing a 24V circuit through terminal W2, energizing the second stage gas valve GVS (17) and the fan relay FR (18). The second stage gas valve (GVS) opens, increasing gas flow, while fan relay contact FR-1 (4) switches, bypassing the autotransformer. This allows the combustion fan motor cfm to run at high speed and increases combustion airflow.
8. The heating cycle is terminated when the thermostat TSH-1 (A) opens, breaking the circuit to the combustion fan relay CFR (16), opening contacts CFR-1 (3) and CFR-2 (7).
9. The thermal fan switch FST (9) allows the indoor fan relay F to be energized after the burners have been shut off, until the lower limit is reached during the cool-down period.

2.3.0 Induced-Draft Gas Furnace

A common induced-draft heating system is an 80 percent annual fuel utilization efficiency (AFUE) gas furnace (*Figure 4*). The sequence of operation is similar to the packaged unit operation. However, in this particular unit, the ignition is accomplished with a hot-surface ignitor (HSI), burner flame is monitored by a flame sensor, and the sequencing/monitoring of the furnace is accomplished by a sealed, solid-state controller containing micro-miniature relays.

The furnace shown in *Figure 4* is similar to one represented by the wiring diagram in *Figure 5*.

Show Transparency 5 (Figure 3) again. Review the operating sequence for induced-draft furnaces. Review the legend for the in-duced-draft packaged unit. Point out some components on the diagram and have trainees locate the definitions in the legend.

Describe an induced-draft gas furnace.

Show Transparency 6 (Figure 5). Review the components and wiring diagram for an induced-draft gas furnace.

See the Teaching Tip for Section 2.3.0 at the end of this module.

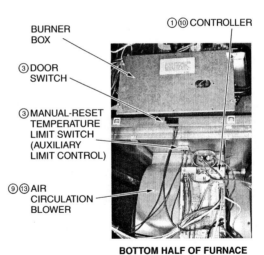

Figure 4 ♦ Typical 80 percent AFUE induced-gas gas furnace.

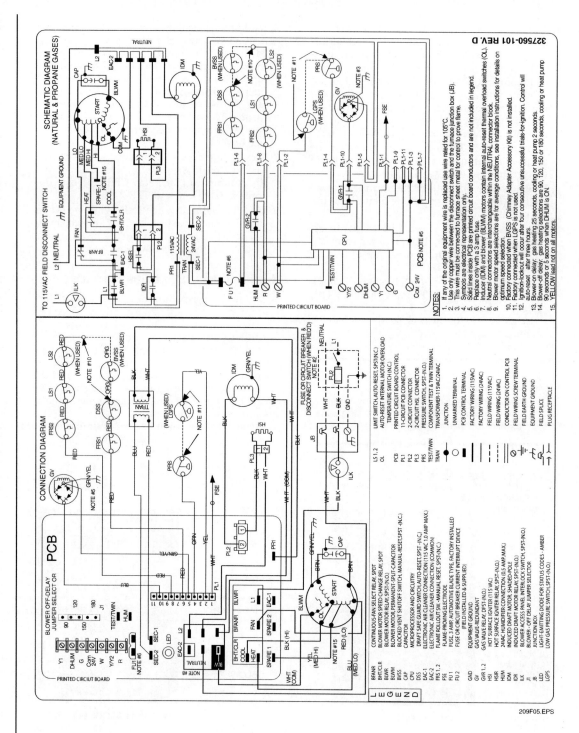

Figure 5 ♦ Typical 80 percent AFUE gas furnace wiring diagram.

1. A 24-volt call-for-heat signal from the room thermostat is applied across the R and W terminals of the control board. The electronic control board (CPU) performs a self-check and verifies that the pressure switch (PRS) contacts are open.
2. If the pressure switch contacts are open, the inducer motor (IDM) can then start. The inducer will run for 15 seconds to purge the heat exchanger. This causes the pressure switch contacts to close, proving that the inducer is running and that there is adequate airflow through the heat exchanger.
3. At the end of the purge period, the hot surface ignitor (HSI) is energized. After the ignitor warms up, the gas valve relay (GVR) contacts close, applying power to the gas valve (GV), to energize it. The ignitor will light the gas that is now flowing to the burners. The flame sensor (FSE) has two seconds to prove that an adequate burner flame is present. The ignitor is de-energized as soon as an adequate flame is proven. Modulating gas valves can provide very precise levels of indoor comfort. They are usually combined with a variable-speed blower motor that matches airflow to burner output. Typically the gas valve can modulate between 40 percent to 100 percent burner capacity in 5 percent increments. A microprocessor-based furnace control tells the gas valve at what level to modulate the burner flame based on various inputs such as indoor and outdoor temperature and the length of previous burner cycles.
4. Failure to sense an adequate burner flame will cause the control board to de-energize the gas valve and attempt (up to three times) to re-establish a flame. If these attempts fail, the control board will lock out any more attempts to re-establish a flame. Interrupting power will reset the control.
5. Once the burner flame is proven, the blower motor (BLWM) is energized 25 seconds after the gas valve opens. The blower motor operates on HEAT speed. The motor speed for heating is field-selectable.
6. If the room thermostat is satisfied, the 24-volt R to W signal is removed. This de-energizes the gas valve, causing the burner flames to extinguish. The inducer motor will run for a short period to purge any combustion products from the heat exchanger. The blower motor will continue to run for up to 180 seconds. The blower off-delay time is field-selectable.

> **NOTE**
> The limit switch prevents the furnace from overheating, a condition that is dangerous. Over the life of a properly operating and maintained furnace, the limit switch may never open. The most common cause of a furnace cycling on and off on the limit switch is a clogged air filter.

If the flame fails to be proven or if the power is interrupted during operation, the controller will normally attempt to recycle the ignition sequence a predetermined number of times before locking out. Some controllers are equipped with a watchdog timer that after lockout, under certain fault conditions, will cause the controller to retry the ignition sequence after a certain time period until the furnace operates or until a fault is corrected. This function is for unattended operation, when damage may occur if a structure is unheated. If a flame rollout limit switch has been tripped, the cause of the flame rollout must be corrected and the switch must be manually reset before the gas valve will function. If the furnace shuts down repeatedly due to a high-temperature overlimit switch activation, and the supply air filters, ducts, and blower are clean, check the air-temperature rise of the furnace. If it is normal, the switch may be defective. If it is not normal, too many registers or air terminals may have been closed, or the blower speed may require adjustment. *Appendix A* is a typical induced-draft gas furnace troubleshooting flowchart.

Door Interlock Switch

Door interlock switches have traditionally been used to disconnect power when the door of a cabinet containing electrical circuits is opened. The purpose is to reduce the hazard of electrical shock. On gas furnaces, the door interlock serves a different purpose. It prevents furnace operation if the blower door is removed. Keeping the door in place during furnace operation prevents harmful products of combustion from being drawn into the duct system by the blower. During troubleshooting procedures, it is acceptable to temporarily tape the door interlock switch closed. However, the tape must be removed when the troubleshooting procedure is complete.

Describe the operating sequence for an induced-draft gas furnace. Identify the differences between the induced-draft gas furnace and the packaged unit.

Explain how to troubleshoot an induced-draft gas furnace. Refer to *Appendix A*.

Show trainees how to develop a checklist for troubleshooting gas heating appliances.

Have trainees practice developing a troubleshooting checklist for gas heating appliances. Note the proficiency of each trainee. This laboratory corresponds to Performance Task 1.

Review the differences between an induced-draft mid-efficiency furnace and a high-efficiency condensing furnace.

Emphasize the importance of checking the heat exchangers of a condensing furnace.

Explain that the room thermostat should not be set lower than 60°F to avoid condensation that can damage the heat exchangers.

Show Transparency 7 (Figure 6). Explain the operation of a gas-fired boiler and compare it to a forced-air furnace.

Have trainees study the wiring diagram for a unit, identify a fault, and state the symptoms that fault would cause. Duplicate the fault on the equipment to verify the symptoms.

INSIDE TRACK

90 Percent Plus AFUE Condensing Gas Furnace

A condensing furnace has both a primary and secondary heat exchanger. Because the secondary heat exchanger extracts most of the remaining heat from the vent gases, condensation of some of the combustion products occurs in the secondary heat exchanger. This acidic condensate is drained from the secondary heat exchanger and routed through a condensate drain trap to an appropriate drain, either directly or by use of a condensate pump. Some local codes require that a replaceable neutralizer cartridge be installed in the condensate drain line before it is routed to a drain. The typical sequence of operation, along with the combustion devices, controller, and monitoring devices, are essentially the same as described for the 80 percent AFUE gas furnace.

Carbon monoxide (CO) emitting from cracked or defective heat exchangers in furnaces that use vent-mounted inducer blowers may be difficult to detect by instrumentation. This is because the heat exchangers operate under negative pressure (vacuum) created by the inducer blower. Therefore, it is important to always conduct a thorough visual inspection of the heat exchangers for cracks or other defects when performing any repair service or annual maintenance on these types of furnaces. Most manufacturers do not recommend operating condensing furnaces with return air temperatures below 55°F for prolonged periods. This is because condensation may occur in the primary heat exchanger, causing it to corrode and fail prematurely. Instruct the owners of such furnaces to set the room temperature thermostat no lower than 60°F if the building will be unoccupied for an extended period.

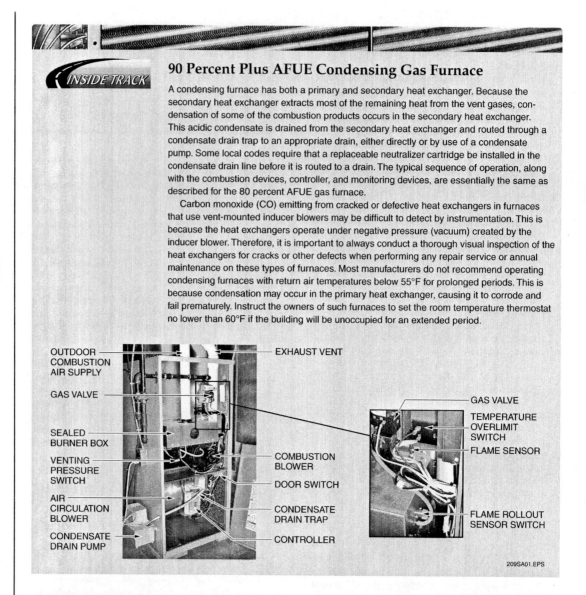

2.4.0 Boiler Operation

The operating sequence for a gas-fired boiler is different from that of a forced-air furnace. The following is the operating sequence for a small packaged gas-fired boiler. The boiler control circuit is shown in *Figure 6*.

1. A call-for-heat signal from the room thermostat is applied to the two "T" terminals on the aquastat, energizing the 1K relay.

2. 1K relay contacts 1K1 and 1K2 close. Assuming the high limit switch is closed, 120-volt power is applied to the circulator pump and the draft inducer.

3. The draft inducer comes up to speed. If there is no vent blockage, the pressure switch closes. This applies 24-volt power to the intermittent pilot control through the flame rollout switch.

4. The intermittent pilot control opens the pilot gas valve, allowing the spark generated by the pilot control to light the pilot.

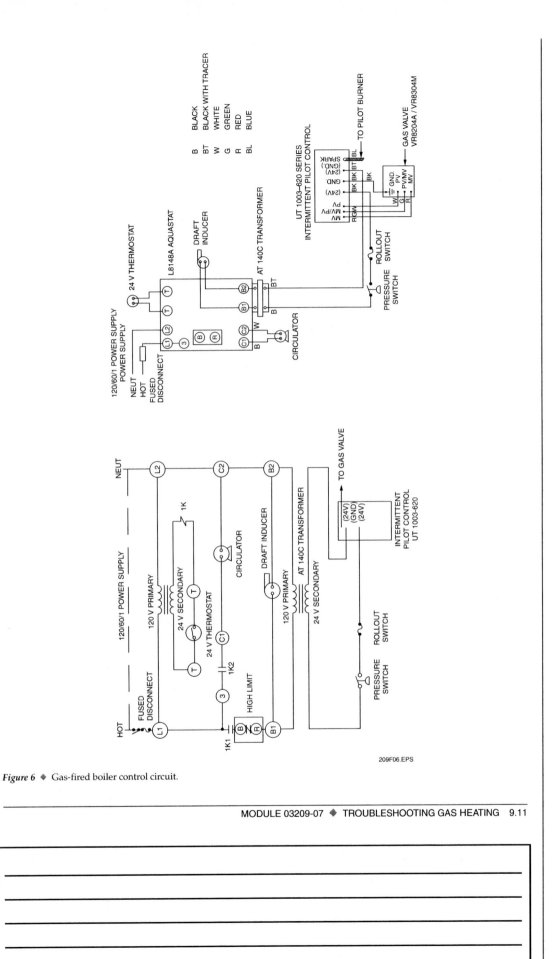

Figure 6 ◆ Gas-fired boiler control circuit.

Discuss the operation and applications of low-intensity and infrared gas-fired units.

Explain the difference between vented and non-vented units.

Explain the operation of a high-intensity infrared gas-fired furnace. Explain that the use of high-intensity infrared heaters is limited to buildings with high infiltration rates or exhaust fans.

Provide manufacturers' literature on infrared gas-fired furnaces for trainees to examine.

5. Once an adequate pilot flame is proven to the pilot control, the spark ceases.
6. The main gas valve now opens and the pilot ignites the main burners.
7. The main burner heats water within the boiler. If the water temperature exceeds the high limit setpoint, burner operation will cease, but the circulator will continue to run until the room thermostat is satisfied. When the water temperature drops below the setpoint, burner operation will resume, assuming the room thermostat is still calling for heat.
8. If the vent system becomes blocked, the pressure switch will open, removing power from the pilot control and stopping burner operation. The draft inducer stays powered as long as the room thermostat continues to call for heat and the limit switch contacts remain closed. If the vent blockage clears, burner operation resumes.
9. When the room thermostat is satisfied, the 1K relay de-energizes, opening the 1K1 and 1K2 contacts to stop burner, circulator pump, and draft inducer operation.

2.5.0 Low- and High-Intensity Infrared Gas-Fired Heaters

Low- and high-intensity infrared heaters are used in buildings with a high volume of air exchange due to numerous external door operations, or in high-bay buildings where forced-air heating is impractical. The basic purpose of infrared heaters is to heat objects, not air. This makes them ideal for commercial applications in factories, foundries, large workshops, vehicle repair garages, warehouses, aircraft hangars, and sports arenas. A few residential versions are available for limited use in exterior garages or workshops.

2.5.1 Low-Intensity Infrared Gas-Fired Heaters

Figure 7A shows a typical gas-fired, tubular, low-intensity infrared heater, also called an indirect radiant heater. They are mounted at a relatively high level in a building and are available in straight-tube and U-tube configurations. A gas flame is directed from a burner assembly through the steel tube, heating the tube. The hot outer surface of the tube produces infrared radiant energy. The energy is focused by a stainless steel reflector onto the floor, occupants, and other surfaces or objects under the heater. The effect is similar to that caused by the sun's infrared rays. The surfaces, occupants, and objects under the heater are warmed, creating a comfortable environment without producing drafts. Tubular heaters are available in either vented or non-vented versions. The vented versions are exhausted (vacuum vented), if more than one is connected to a vent, by a single induced-draft blower located at the vent exit to the outdoors. Single-vented or non-vented units may be exhausted (power vented) by a blower in the burner assembly at the input to the tube.

In a non-vented unit, the combustion products are discharged into the heated space for additional heat. These heaters are permitted in areas with a high outdoor air-infiltration rate, or when exhaust fans change the building air a set number of times over a specific period. The burner assembly operates the same as the burner assembly in an induced draft gas furnace.

The safety devices used in low-intensity infrared gas-fired heaters are similar to those used in induced-draft gas furnaces. The standing pilot devices used, as well as various electrical and electronic devices, including glow coils, are also similar to induced-draft gas furnace equipment.

Another version of a gas-fired low-intensity infrared heater is a vertical panel heater (*Figure 7B*). The unit shown can be individually vented or non-vented. The vent is a natural-draft vent, and ignition is achieved by standing pilot, so no power is required to operate the unit.

2.5.2 High-Intensity Infrared Gas-Fired Heaters

These non-vented units (*Figure 8*), also called direct-fired radiant burners, are used at high levels off the floor as spot or space heaters. The units operate by burning gas within two-layer porous ceramic grids (**submerged flame**). The grids attain a red heat level of about 1,600°F and produce an intense infrared radiant energy. In some cases, the grids can also be closely covered by an optional wire screen, called a **reverberator**, that increases the infrared radiation when heated.

High-intensity infrared gas-fired units must be mounted at a specified angle to allow the combustion products to flow upward off each grid and into the room. Reflectors around the perimeter of the grids focus the radiant energy to the desired location. These non-vented units are permitted for use in areas with a high outdoor air infiltration rate or when exhaust fans change the building air a set number of times over a specific period. The gas burner, located behind the grids in a plenum, operates in the same way as the burner in a natural-draft gas furnace.

The safety devices used in high-intensity infrared gas-fired heaters are similar to those used in induced-draft gas furnaces. The standing pilot devices used, as well as various electrical and electronic devices, including glow coils, are also similar to induced-draft gas furnace equipment.

Discuss gas furnace controllers.

Gas Furnace Controllers

Newer, solid-state gas furnace controllers are usually equipped with a diagnostic aid in the form of multiple or single light-emitting diodes (LEDs). Controllers with a single LED only light the LED when the ignition sequence is interrupted by a fault. The LED usually flashes in a code corresponding to the type of failure detected. For example, two flashes might mean the pressure switch is stuck closed, while three flashes could mean the pressure switch is stuck open. Controllers with multiple LEDs usually light the LEDs in a sequence corresponding to the ignition sequence of the controller during operation. If the ignition sequence fails, the last LED lighted is usually the last valid action completed. Most controllers with LEDs bear labels stating the meaning of the LEDs or listing the flash codes. To isolate the failure, voltages at the designated (or next) component in the sequence can be monitored as the ignition sequence is recycled. Consult the manufacturer's literature to verify the actual ignition sequence and components involved.

Older solid-state gas furnace controllers may not have diagnostics, or may have only limited diagnostics. The older controller shown has an LED that flashes to indicate an external failure has occurred. If lighted continuously, an internal controller failure is indicated. To determine an ignition-sequence failure, the furnace must be observed during the ignition sequence to visually check where the sequence is halting. Then, voltages representing inputs from, or outputs to, the components associated with the sequence failure can usually be monitored at test points on a connection block. This is done to isolate the malfunctioning component while the controller is recycled. With these types of controllers, consult the manufacturer's literature for the ignition sequence and test point information.

The operation of gas furnace controllers, along with other types of controllers, is explained in more detail in a later module.

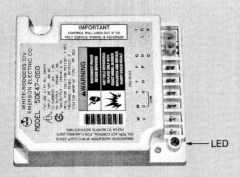

LED ON = FAILED CONTROLLER
LED FLASHING = EXTERNAL COMPONENT FAILURE
CONTROLLER WITH MINIMAL DIAGNOSTICS

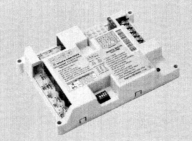

DIAGNOSTIC INDICATOR FLASH CODES
1. System lockouts (Retries of lockouts exceeded)
2. Pressure switch stuck closed
3. Pressure switch stuck open
4. Open high-temperature limit switch
5. Flame rollout sensed
6. 115 VAC power reversed/improper ground
7. Low flame sense signal

Continuous flash – flame sensed 0.5 sec without gas valve
Continuous on – internal control failure

Explain how infrared heater controls and related black-bulb sensors are used to control infrared radiant heating systems.

Discuss how infrared heater controls can be programmed in the same manner as programmable thermostats.

Review safety procedures and equipment that will be used in the laboratory session.

Identify the tools and instruments needed to troubleshoot a gas heating appliance.

Have trainees practice identifying the tools and instruments needed to troubleshoot a gas heating appliance. Note the proficiency of each trainee. This laboratory corresponds to Performance Task 2.

Figure 8 ◆ High-intensity infrared heater.

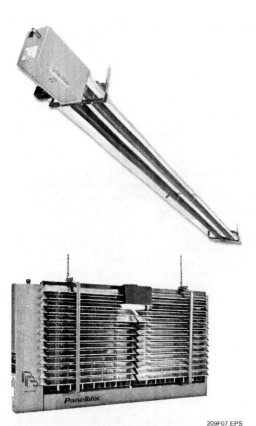

Figure 7 ◆ Low-intensity (A) infrared tubular heater and (B) vertical panel heater.

2.5.3 Infrared Heater Controls and Sensors

Multiple commercial infrared heaters are usually controlled from a central multi-zone control panel (*Figure 9*). The panels typically include a digitally programmed time switch. This switch allows switching instructions to be programmed for intervals of as little as one minute for each zone. Summer/winter changeovers as well as daylight saving time changes normally can be performed without disrupting operation. Battery backup options prevent programming loss in the event of a power failure. Each zone uses a black-bulb radiant sensor located in the heated area, coupled to a black-bulb thermostat mounted in the control panel. The thermostats shown in *Figure 9* have a mode switch, two setting dials for day and night temperature settings, a power-on indicator, and a zone-running indicator. The mode switch provides zone off, constant night, constant day, automatic, and time switch override functions. The

BLACK BULB RADIANT SENSOR

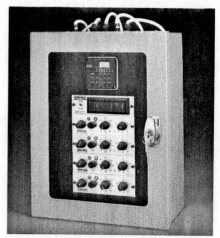

INFRARED HEATER CONTROL

Figure 9 ◆ Infrared heater control and sensor.

9.14 HVAC ◆ LEVEL TWO

Instructor's Notes:

black-bulb radiant sensor shown in *Figure 9* is in the form of a hemisphere of very thermally conductive material with a blackened surface. Encapsulated within the hemisphere is a highly temperature-sensitive element.

3.0.0 ◆ COMBUSTION SYSTEMS

Troubleshooting the combustion system in gas heating equipment consists primarily of observing the flame and the mechanical operation of the unit. Based on these observations, it is possible to identify the source of a problem. The troubleshooting chart in *Appendix B* identifies some of the common problems that occur in gas heating systems, along with their probable causes.

If a gas heating system is not working at all, the ignition system and gas valve may be malfunctioning. In older furnaces with standing pilots, thermocouples are used to control the main gas valve. Modern furnaces use **flame rectification** sensors in conjunction with intermittent pilot spark ignitors or direct ignition devices such as hot surface ignitors (*Figure 10*). Hot surface ignitors will wear out or crack after several years of use. When replacing a hot surface ignitor, consider replacing it with a more durable and reliable nitride ignitor.

3.1.0 Troubleshooting Thermocouples

A thermocouple generates only enough voltage to hold open the safety valve disc once it is manually held open by depressing the pilot valve. Thermocouples are typically rated at 30 millivolts, but operate at 18 to 25 millivolts under

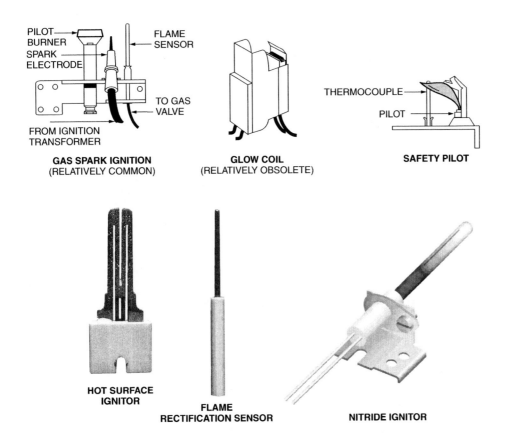

Figure 10 ◆ Gas furnace ignition devices.

Review the operation of a thermocouple as it is used to control the supply of gas to the gas valve.

Show Transparencies 9 and 10 (Figures 11 and 12). Explain how to test a thermocouple with and without a load.

Discuss visual clues that can help in determining whether a thermocouple is defective.

Identify various ways that a thermocouple can fail.

Soot Buildup

If you notice a lot of soot on the secondary heat exchanger of a condensing furnace, there may be a problem with the gas pressure. Perform a gas pressure check.

load. The voltage produces current through a holding coil that creates enough of a magnetic field to hold the disc open. It takes time to heat up the thermocouple, and that is why the pilot valve must be held open manually for about 60 seconds. If the pilot flame goes out after the warmup period, it probably means the thermocouple is not generating enough voltage to hold the safety valve open. This can be verified by checking to see if the thermocouple is producing sufficient voltage. If the thermocouple tests okay, the problem is most likely in the gas valve. On some gas valves, the holding coil can be replaced. In other cases, the entire gas valve assembly has to be replaced. *Figure 11* shows the method for testing a thermocouple.

When tested without a load, as shown in *Figure 11*, the thermocouple should generate 20 to 30 millivolts DC. An adapter (*Figure 12*) can be used to test the thermocouple under actual operating (loaded) conditions. In a test under load, the voltage will be about half that for an unloaded test.

Here are some other troubleshooting hints to consider when checking a thermocouple:

- The flame must be directed toward the tip of the thermocouple in order for a voltage to be generated.
- A flickering pilot flame, or one that lifts off the thermocouple, generally indicates too much draft or leakage in the air system. Lifting can also indicate a gas pressure problem.
- A floating pilot flame may indicate inadequate draft.
- A yellow pilot flame, caused by a partially blocked orifice in the pilot assembly, will not produce enough heat to generate an adequate voltage. Take apart the pilot assembly and clean the orifice to restore the flame to its proper blue color.

There are several ways a thermocouple can fail:

- The tip of the thermocouple may be burned from extended use.

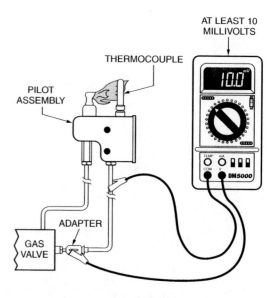

1. Disconnect the thermocouple from the gas valve.
2. Connect the thermocouple lead to the multimeter as shown.
3. Take the multimeter reading with the pilot flame on the thermocouple.

Figure 11 ♦ Thermocouple test (no load).

1. Connect the multimeter to the thermocouple lead and adapter as shown.
2. Take the multimeter reading with the pilot flame on the thermocouple.

Figure 12 ♦ Thermocouple test under load.

9.16 HVAC ♦ LEVEL TWO

Instructor's Notes:

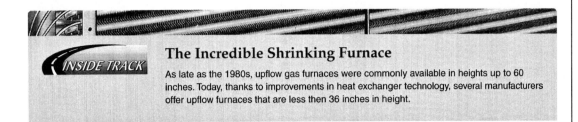

The Incredible Shrinking Furnace

As late as the 1980s, upflow gas furnaces were commonly available in heights up to 60 inches. Today, thanks to improvements in heat exchanger technology, several manufacturers offer upflow furnaces that are less then 36 inches in height.

- The sensing element can be misaligned so that the pilot flame does not make adequate contact with it.
- The lead can be damaged or kinked.
- If the connecting nut was tightened too much, the insulator may have collapsed and shorted-out the thermocouple. If it is not tight enough, moisture may have entered and corroded the connection. Finger-tighten the nut, then turn it one-quarter turn with a wrench.
- A dirty contact can block current flow. Do not touch the contact, because oil from your fingers can cause the same problem.

3.2.0 Troubleshooting Pilot Reignition and Direct Ignition Devices

Some systems use spark ignitors or glow coil ignitors to light the pilot when the thermostat calls for heat. Pilotless systems use either a hot surface ignitor (HSI), which is made of ceramic material that glows when a current flows through it, or a direct spark ignitor. Both ignite the gas at the burner.

The spark ignition device will have a built-in sensing element that sends a tiny current to the gas valve to prove combustion whenever the burners are lighted. When a hot surface ignitor is used, the flame-sensing rod is not part of the ignitor, but will be located somewhere on the burner box or manifold and positioned in the burner flames.

Regardless of the type of ignitor used, if the flame sensor is not producing an output, the gas supply will be shut off.

When troubleshooting spark ignitors or HSIs in an electronically controlled system, keep in mind that the system will lock out the ignition circuit if it fails to establish and maintain a flame after so many tries or so much elapsed time. Usually, either the control voltage or the primary power must be reset before the system will operate again. Some systems have a watchdog timer that will automatically recycle the operational sequence every hour or so, indefinitely.

> **NOTE**
> Most ignition devices require that the equipment they are installed in be connected to a good earth ground for proper operation.

3.2.1 Spark Ignitors

> **WARNING!**
> Be very careful when troubleshooting spark ignitors because they are operated by a very high voltage pulse (10,000 to 20,000 volts) from an ignition transformer. Do not touch the spark ignitor with your fingers or with a screwdriver. Do not attempt to disconnect the ignitor without turning off the power.

You should be able to see and hear whether the spark ignitor is working. If you can hear the spark, but the gas is not being ignited, the ceramic insulator may be cracked. If there is no spark at all, the spark gap may need to be readjusted according to the manufacturer's instructions. This is a very common problem. If the gap is correct but there is still no spark or the spark is weak or intermittent, the problem is most likely in the ignition module or the high-voltage wire to the ignitor. If there is a spark but no pilot flame, the pilot orifice may be plugged. During the summer, small insects may build webs or cocoons inside the orifice.

3.2.2 Glow Coil Ignitors

A glow coil ignitor uses a coil of platinum wire to ignite the pilot gas. They are available for use with standing and intermittent pilots. The glow coil is in series with the furnace safety controls in the 24V control circuit (*Figure 13*). In some designs, a series resistance drops about half the voltage, so the voltage drop across the glow coil is 11.5 to 12 volts. In other designs, the 24V transformer is center-tapped.

Describe the various methods used to light the pilot. Discuss considerations for troubleshooting spark ignitors or hot surface ignitors (HSI).

Explain how to identify a defective spark ignitor.

Remind trainees that spark ignitors generate a very high voltage pulse.

Show Transparency 11 (Figure 13). Review the operation of a glow coil ignitor. Explain that while these ignitors may be found on older equipment, they have largely been replaced by hot surface ignitors.

Review the operation of hot surface ignitors.

Discuss the electrical hazards and safety precautions involved when testing an HSI.

Provide HSIs for trainees to examine. Point out that they are very fragile and can be easily cracked or broken.

Show Transparency 12 (Figure 14). Describe the procedures for testing an HSI. Explain that multiple failures of the HSI in a given system may indicate that the line voltage is too high.

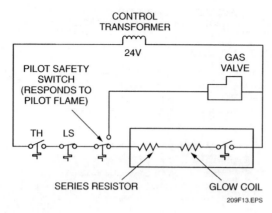

Figure 13 ♦ Glow coil circuit.

The advantage of the series resistance is that it can be adjusted in the field by increasing or decreasing the amount of resistance wire to obtain the right voltage across the glow coil. This is sometimes necessary because the voltage across the glow coil must be within half a volt in order to operate properly. If the voltage is too low, the coil will not ignite the pilot gas. If it is too high, it will burn out the coil.

The glow coil has its own thermal trip device that opens the circuit after the coil has been on for about 15 seconds. Without this protective device, the glow coil would burn out. The glow coil employs an obsolete technology that has been replaced with spark ignitors or HSI in most furnaces. However, some gas-fired infrared radiant heaters are available with glow coil ignition.

3.2.3 Hot Surface Ignitors

Hot surface ignitors (HSIs), sometimes called glow bars, are made of carborundum (silicon carbide). Carborundum is resistive to current flow. When electrical power is applied through the material, it heats up rapidly, becoming very hot and bright red. The carborundum element of the ignitor is very brittle and can be easily cracked or broken if handled carelessly or bumped during equipment cleaning. Contrary to popular belief, any oils transferred to the element during handling will not harm the element. If a bright white line is visible anywhere on an operating HSI element, the element is cracked, but hasn't completely failed. In this case, the HSI must be replaced. Ignitors made of nitride material are now available. They are more durable and reliable than silicon carbide ignitors and may be used to replace silicon carbide ignitors.

If an HSI fails to glow at the proper time during an ignition sequence and the required preceding actions (ventor running, pressure switch closed, etc.) have occurred, then one or more of the following has occurred:

- The element is open because it is cracked, broken, burned out, or the HSI wiring is defective.
- No power is being supplied to the element.

WARNING!
Testing an HSI involves measurement of 120VAC power. Use caution to avoid contacting the HSI element, any live wires, or bare electrical contacts when making the measurements. Make sure the 120VAC power to the equipment is turned off before disconnecting and connecting wiring or bare contacts.

Observe the HSI element without removing it from the burner section. If the HSI element is not visibly cracked, broken, or burned out, monitor the 120VAC power applied to the HSI during the ignition sequence at its inline connector inside the equipment (*Figure 14*). Cycle the equipment through its ignition sequence while monitoring the HSI voltage. If the proper voltage is detected, the HSI has failed and must be replaced. If multiple failures of the same HSI have occurred, make sure that the voltage does not exceed 125VAC. High operating voltages can significantly reduce the life of an HSI.

If no power is monitored at the inline connector, the power source (inline fuse, controller, or separate relay) for the HSI has failed and must be replaced. In this case, make sure the HSI leads are not damaged or shorted to any metal part of the furnace before reapplying power to the HSI.

NOTE
Even though the HSI may be glowing, it may not have enough strength to ignite the gas. Technicians have experienced problems with ignitors which have not failed completely, but their resistance values decrease over time and use, and they fail to ignite the burners. That is, they don't fail to open—they just experience a sufficient reduction in resistance so that they can't get hot enough to get the job done.

9.18 HVAC ♦ LEVEL TWO

Instructor's Notes:

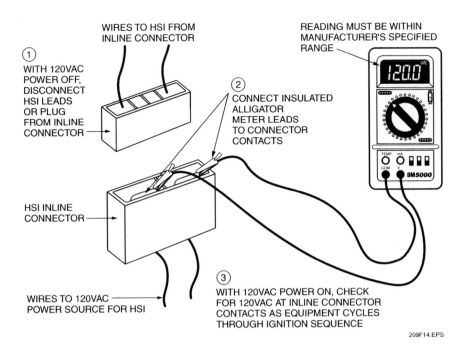

Figure 14 ♦ Checking HSI source voltage.

Discuss the operation of a flame sensor.

Show Transparency 13 (Figure 16). Describe the procedures for testing a flame sensor.

Provide flame sensors for trainees to examine.

3.3.0 Troubleshooting the Flame Sensor

In direct ignition systems, the flame-sensing rod is not an integral part of the ignitor. It is installed in the path of the burner flame (*Figure 15*). It uses flame rectification to produce a tiny DC current in the microampere range. This current must be sensed by the gas-fired equipment controller, or the main gas valve will be turned off.

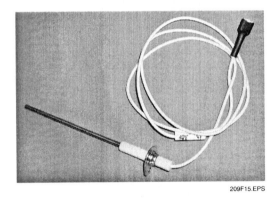

Figure 15 ♦ Flame sensor.

Figure 16 shows the method for testing a flame sensor. A multimeter capable of reading DC current in the low microampere range (less than 10 microamps) is connected in series between the flame sensor electrode and the flame-proving circuit. Then the power is turned on and the thermostat is set to call for heat. After the burners ignite, the current reading should be within the range specified by the manufacturer. The normal range is between 0.5 and 4.5 microamperes.

If the current reading is absent or too low, it could be caused by one of several problems:

- The flame-sensing rod is incorrectly positioned.
- The ignition module is not properly grounded.
- The sensor electrode is not making good electrical contact.
- The sensing rod has become oxidized. In this case, clean the rod with fine sandpaper.
- The controller or flame sensor insulation/wiring is defective.
- Improper 120V polarity.

Special equipment is available for testing spark generators, flame sensors, gas valves, and other ignition devices. One manufacturer provides optional accessories for a digital multimeter.

MODULE 03209-07 ♦ TROUBLESHOOTING GAS HEATING 9.19

Discuss the combined HSI and flame sensor.

Show Transparency 14 (Figure 17). Explain how to test a combined HSI and flame sensor.

Explain that the testing of flame sensors is done with the line voltage applied to the device.

Show trainees the correct method for testing the following: thermocouples, pilot recognition and direct ignition devices, spark ignitors, hot surface ignitors, flame sensors, and combination HSI and flame sensor devices.

Use pre-faulted components, jumpers, or other means to insert combustion malfunctions into a gas furnace or packaged air conditioner with gas heat. Provide trainees with the wiring diagram for the unit and have them isolate and correct the fault under your direct supervision. Note the proficiency of each trainee. This laboratory corresponds to Performance Tasks 3 and 4.

Have the trainees review Section 4.0.0.

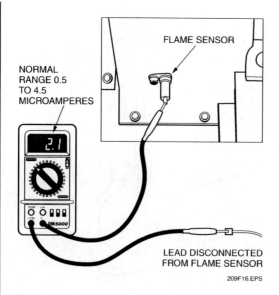

Figure 16 ◆ Flame sensor test.

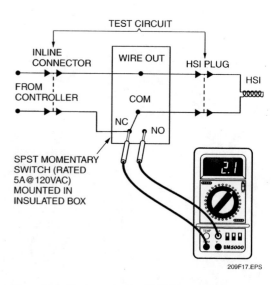

Figure 17 ◆ Testing a combined HSI/flame sensor.

These accessories permit voltage and current measurements in the very low ranges necessary for testing ignition devices.

A furnace manufacturer might provide a furnace tester for use with specific furnace models. The tester can be plugged into the furnace. Switches on the tester can then be manipulated to simulate various inputs.

3.3.1 Testing a Combined HSI/Flame Sensor

If desired, the flame signal can be checked using a test circuit as shown in *Figure 17*. The meter used must have a DC microammeter range of about 10 microamps.

WARNING!
Testing a combined HSI/flame sensor involves measurement of circuits operating with 120VAC power. Use caution to avoid contacting the HSI element, any live wires, or bare electrical contacts when making the measurements. Make sure the 120VAC power to the equipment is turned off before disconnecting and connecting wiring or bare contacts.

CAUTION
Do not push the momentary contact switch when testing a combined HSI/flame sensor until after the ignitor function de-energizes; otherwise, the meter will be damaged.

After turning off all primary 120VAC power, connect the test circuit at a convenient point (inline connector) between the controller and the HSI. Depending on the gas equipment manufacturer, the test circuit may require various inline connector plugs to interface to the equipment. Reapply power and cycle the system through an ignition sequence. When the gas ignites and after the ignitor function de-energizes, push and hold the momentary contact switch to read the DC flame signal on the meter. If no reading occurs, the switch may be in the wrong lead from the controller. In that case, turn off the primary 120VAC power, reverse the test leads to the controller, and repeat the process.

Check that the flame signal is within the prescribed limits for the controller manufacturer. If the flame signal is out of limit, or the flame goes out before a reliable reading can be obtained, replace the HSI and retest. If the problem reoccurs, replace the controller and retest. If the problem is solved with replacement of the controller, reinstall the original HSI and retest.

Instructor's Notes:

> ### Inside Track
> **Combined HSI/Flame Sensor**
>
> Instead of a separate HSI and flame sensor, many gas-fired systems use a controller that switches the connections to the HSI so that it functions as the ignitor for a prescribed time, then as a flame sensor using flame rectification. The controller accomplishes this internally by removing the AC heating current from the HSI after the ignition sequence, opening one of the HSI leads, and substituting a high-resistance circuit and an AC signal in the other lead to the HSI element. The element then functions the same as a separate flame sensor.
>
> As the HSI ages, a protective oxide coating slowly forms on the surface of the element. This coating is normal; however, it may reduce the strength of the flame signal to the controller. If the gas ignites, but immediately or occasionally goes out (nuisance trips) before the thermostat is satisfied, a weakened flame-sensing signal may be the cause. Always check the continuity of the burners and their flame carryover tube to ground, as well as the controller ground. If the grounds are satisfactory, visually check the HSI element for contamination. If not visibly contaminated, check that the HSI element is positioned so that ¾" to 1" of the element is continuously immersed in flame.
>
>

4.0.0 ◆ AIR SYSTEM

In addition to component failure, there are three general kinds of problems that can occur in the supply and return air system:

- *Obstructed airflow* – The most likely cause of an airflow problem is dirt. A dirty filter, dirty heat exchangers, or a dirty blower wheel will reduce airflow and prevent the appliance from providing adequate heating. Insufficient airflow can cause nuisance trips of the temperature limit switch because heat is allowed to accumulate around the heat exchangers.
- *Ductwork* – Problems with ductwork size are likely to occur only if the appliance has been replaced with one having significantly greater or smaller capacity, without resizing the ductwork. Such systems will often be noisy, especially if the ductwork is too small. If internally insulated ductwork is used, it is possible for the insulation to come loose and block airflow.
- *Blower motor* – If the appliance had been operating properly at some time in the past and the blower wheel has been periodically cleaned, it is unlikely that the blower size or speed is the source of a problem. If a blower has been speeded up to compensate for an undersized unit or ductwork, the system will often be noisy.

> **NOTE**
> Electronically-controlled, variable-speed motors are being used more and more as furnace blower motors. When checking airflow in a furnace that uses a variable-speed blower motor, always ensure that the blower is operating at its correct speed before checking airflow. During different operating modes, variable-speed blower motors will operate at different speeds similar to conventional multi-speed motors. One major difference is that the variable-speed motors come up to speed slowly and reduce speed slowly before stopping.

Use *Appendix B* to explore various troubleshooting scenarios.

Show trainees how to use a wiring diagram to isolate and correct air system faults on a gas heating appliance.

Use pre-faulted components, jumpers, or other means to insert air system malfunctions into a gas furnace or packaged air conditioner with gas heat. Provide trainees with the wiring diagram for the unit and have them isolate and correct the fault under your direct supervision. Note the proficiency of each trainee. This laboratory corresponds to Performance Tasks 3 and 4.

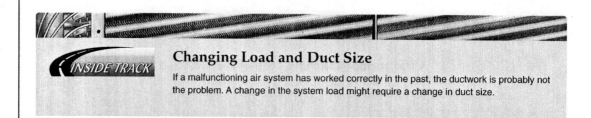

Changing Load and Duct Size

If a malfunctioning air system has worked correctly in the past, the ductwork is probably not the problem. A change in the system load might require a change in duct size.

The troubleshooting chart in *Appendix B* identifies symptoms and likely causes of air system problems.

Heating and cooling generally require different volumes of air. If a unit with a single-speed blower is used for both heating and cooling, it may be inadequate in one mode or the other. For example, a high-speed blower that is suitable for cooling may prevent the heat exchangers from getting warm enough to do their job. In such cases, it may be necessary to replace the blower and its controls with a two-speed system that automatically selects the correct blower speed. Another option, applicable to single-speed belt-driven blowers, is changing the pulley size or the drive belt size.

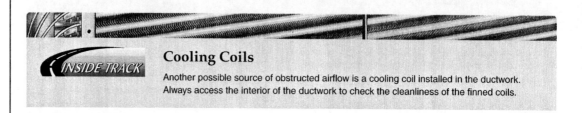

Cooling Coils

Another possible source of obstructed airflow is a cooling coil installed in the ductwork. Always access the interior of the ductwork to check the cleanliness of the finned coils.

Instructor's Notes:

Review Questions

1. Which of the following is *not* found in a natural-draft furnace?
   a. Blower
   b. Inducer fan
   c. Standing pilot
   d. Gas valve

2. Natural-draft furnaces are no longer available for new installations because they are _____.
   a. too expensive to manufacture
   b. not energy-efficient
   c. unsafe
   d. not compliant with the new electrical codes

3. The ignition device used in the system depicted in *Figure 1* is a _____.
   a. hot surface ignitor
   b. glow coil
   c. spark ignitor
   d. standing pilot

4. In the unit depicted in *Figure 3*, the method used to prove combustion airflow is _____.
   a. pressure switch
   b. shaft rotation sensor
   c. none; it is assumed to be running if the thermostat is closed
   d. vane switch

5. Refer to *Figure 3*. The unit shown in the ladder diagram provides a single stage of gas heating.
   a. True
   b. False

6. The basic purpose of infrared heaters is to heat _____.
   a. objects
   b. air
   c. water
   d. coolant

7. High-intensity infrared gas-fired heaters are also called _____.
   a. vented units
   b. indirect radiant heaters
   c. direct-fired radiant burners
   d. tubular heaters

8. Optional wire screens used to cover the grids of high-intensity infrared gas-fired heaters are called _____.
   a. u-tubes
   b. radiators
   c. vents
   d. reverberators

9. What voltage would you expect to read if you measured the output of a thermocouple that is disconnected from the gas valve?
   a. 10 to 15 millivolts
   b. 20 to 30 millivolts
   c. 12 volts
   d. 24 volts

10. Pilotless heating systems use either direct spark ignitors or _____.
    a. indirect spark ignitors
    b. hot surface ignitors
    c. thermocouple ignitors
    d. flame ignitors

11. Problems that can occur with a spark ignitor include all of the following *except* _____.
    a. incorrect spark gap
    b. faulty cold junction
    c. ignition module failure
    d. cracked ceramic insulator

12. The method used to obtain the 12 volts required for operation of a glow coil is _____.
    a. center-tapped control transformer
    b. series resistance
    c. center-tapped control transformer and series resistance
    d. 12VDC power supply

13. When testing a flame-sensing rod, the reading should be _____.
    a. 12 volts
    b. 30 ohms
    c. 0.5 to 4.5 milliamps
    d. 0.5 to 4.5 microamps

Review Questions

14. When checking the flame signal while testing a combined HSI/flame sensor, the meter used must have a DC microammeter range of about _____ microamps.
    a. 5
    b. 10
    c. 15
    d. 20

15. A dirty air filter in a gas furnace is likely to cause _____.
    a. nuisance trips of the temperature limit switch
    b. thermostat failure
    c. combustion noise
    d. power failure

Summary

Troubleshooting gas heating appliances requires skill in reading and interpreting electrical diagrams as well as the ability to evaluate and test the ignition, combustion, and air distribution systems. The HVAC technician must have good observation skills and a working knowledge of the various gas heating systems and accessories available. Technicians must also possess a thorough understanding of the safety precautions to be taken while working with gas heating equipment.

Although all gas heating controls have elements in common, there are wide variations in the ways in which controls are implemented. Packaged gas heating units include natural-draft and induced-draft units. Gas furnace types include induced-draft furnaces as well as low- and high-intensity gas-fired heaters. It is important to be able to analyze the various types of control circuits used in these devices. The HVAC technician must understand the operation of various types of control devices, including ignitors, gas valves, and flame sensors.

Summarize the major concepts presented in the module.

Administer the Module Examination. Record the results on Craft Training Report Form 200, and submit the results to the Training Program Sponsor.

Administer the Performance Test, and fill out Performance Profile Sheets for each trainee. If desired, trainee proficiency noted during laboratory sessions may be used to complete the Performance Test. Record the results on Craft Training Report Form 200, and submit the results to the Training Program Sponsor.

Trade Terms Introduced in This Module

Flame rectification: A process in which exposure of a sensing rod to a flame produces a tiny current that can be sensed and used to control the gas valve.

Modulating gas valve: A gas valve that provides precise control of burner capacity by modulating the flow of gas to the burners in small increments.

Reverberator: A wire screen used to cover the ceramic grids in a gas-fired heater that operates by submerged flame.

Submerged flame: A method of gas-fired heater operation that uses burning gas within two-layer porous ceramic grids.

Two-stage gas valve: A gas valve body containing two separate valves that supply fuel gas to the burners based on the amount of heat demand.

Vane switch: A switch that is actuated by a vane that reacts to the flow of water or air.

Appendix A

Typical Induced-Draft Gas Furnace Troubleshooting Flow Chart

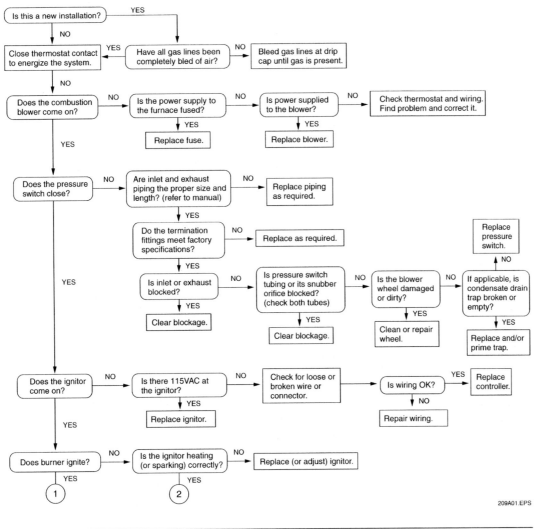

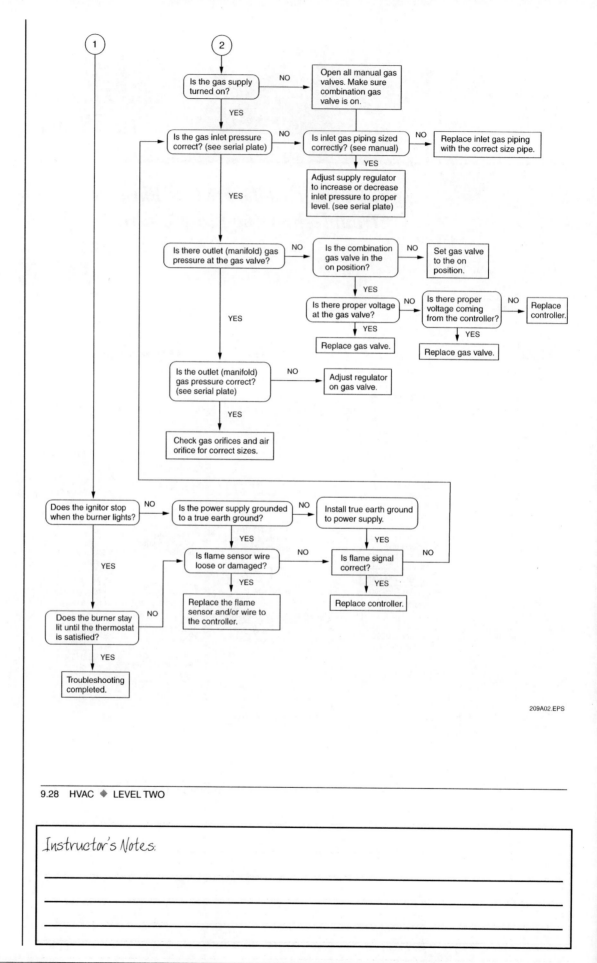

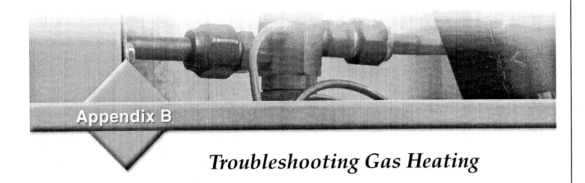

Appendix B

Troubleshooting Gas Heating

GENERALLY THE CAUSE Make these checks first.	OCCASIONALLY THE CAUSE Make these checks only if the first checks failed to locate the trouble.	RARELY THE CAUSE Make these checks only if other checks failed to locate the trouble.
PROBLEM: No heat – Burner fails to start		
Power failure Blown fuses or tripped circuit breaker Open disconnect switch or blown fuse Thermostat set too low Thermostat switch not in proper position Pilot Manual fuel valve Pilot valve	Control transformer Limit controls (manual reset) Pilot safety control Gas valve Thermocouple Fuel lines Gas meter	Faulty wiring Loose terminals Low voltage Thermostat faulty Improper burner adjustment Plugged burner orifices Gas pressure Thermostat not level (mercury bulb thermostat only)
PROBLEM: No heat – Inducer starts but burner fails to ignite		

Combustion air switch
Inducer motor
Primary or safety controls
Gas valve
Flame detector or electrode
Improper burner adjustment
Manual fuel valve
Pilot valve
	Ignition devices
	Regulator
	Plugged vent or air supply
	  (induced draft only)
	Plugged pressure switch line
	  at inducer (induced draft only)
		Low voltage
		Fuel lines
		Gas pressure
		Plugged burner orifice
		Condensate trap broken or empty
		  (condensing furnace only)

PROBLEM: Insufficient or no heat – Burner starts and fires, then switches off when safety switch trips		
Safety controls Flame detector or electrode Improper burner adjustment Dirty air filter Dirty or plugged heat exchanger	Combustion air switch Inducer motor	Faulty wiring Loose terminals Fuel lines Gas pressure
PROBLEM: Insufficient or no heat – Burner cycle too short		
Limit controls Thermostat not level (mercury bulb thermostat only) Thermostat location Blower belt broken or slipping Dirty filters Thermostat heat anticipation incorrect	Fan control Blower bearings Blower motor speed Low air volume Dirty blower wheel	Faulty wiring Loose terminals Low voltage Blower wheel Ductwork small or restricted Restrictions
PROBLEM: Insufficient or no heat – Burner runs continuously		
Blower belt slipping Dirty blower wheel Dirty filters Gas pressure	Orifice size Low air volume Improper burner adjustment Plugged burner ports Input too low Undersized furnace	Ductwork small or restricted Displaced or damaged baffles Insufficient insulation Excessive infiltration

Instructor's Notes:

GENERALLY THE CAUSE Make these checks first.	OCCASIONALLY THE CAUSE Make these checks only if the first checks failed to locate the trouble.	RARELY THE CAUSE Make these checks only if other checks failed to locate the trouble.
PROBLEM: Too much heat – Burner cycle too long		
Thermostat heat anticipation incorrect Thermostat out of calibration	Thermostat not level Thermostat location	
PROBLEM: Combustion noise		
Pilot Gas valve Improper burner adjustment Plugged burner orifice Air shutter adjustment (if so equipped)	Inducer motor Primary or safety controls Ignition transformer Flashback to venturi Gas pressure Pressure regulator	Faulty wiring Loose terminals Regulator Dirty or plugged heat exchanger Vent or flue Displaced or damaged baffles Blocked heat exchanger Input too high
PROBLEM: Mechanical noise		
Blower bearings Blower belt cracked or slipping	Inducer motor Blower motor Cabinet	Low voltage Control transformer Control relay or contactor Blower wheel Cracked or ruptured heat exchanger Displaced or damaged baffles Fuel lines
PROBLEM: Air noise		
Blower Cabinet Ductwork small or restricted Air leaks in ductwork	Other air system restrictions	Blower wheel Dirty filters
PROBLEM: Odor		
Vent or flue Flue gas spillage Fuel leaks Cracked or ruptured heat exchanger Blocked heat exchanger	Regulator Low air volume Improper burner adjustment Displaced or damaged baffles Humidifier stagnant water Water or moisture	Faulty wiring Loose terminals Control transformer Dirty filters Dirty or plugged heat exchanger Input too high Input too low Outdoor odors
PROBLEM: Cost of operation		
Combustion motor Blower motor Dirty filters Improper burner adjustment Equipment size	Low air volume Insufficient insulation Excessive infiltration	Low voltage Fan control Orifice Blower belt slipping Dirty blower wheel Ductwork small or restricted Vent or flue Input too high

Additional Resources and References

Additional Resources

This module is intended to be a thorough resource for task training. The following reference work is suggested for further study. This is optional material for continued education rather than for task training.

Refrigeration and Air Conditioning, An Introduction to HVAC/R, Fourth Edition. Larry Jeffus. Air Conditioning and Refrigeration Institute. Upper Saddle River, NJ: Prentice Hall.

Figure Credits

Trane, 209F03

Topaz Publications, Inc., 209F04, 209SA01, 209F10 (bottom left and center), 209F15

Carrier Corporation, 209F05, 209F06, 209SA03

Ambi-Rad USA, 209F07 (top), 209F09

Panelbloc, Inc., 209F07 (bottom)

White-Rodgers/Emerson Climate Technologies, 209SA02, 209F10 (bottom right)

Enerco Technical Products, Inc., 209F08

MODULE 03209-07 — TEACHING TIPS

The following are suggested activities or instructional methods to help you teach the material in this module.

General

When you call on someone to answer a question, the rest of the class relaxes or even tunes out because they expect that the question and answer will take place only between you and the trainee you called on. Instead, use this technique to involve more trainees in answering questions and to keep them on their toes.

1. Ask trainees to define a term or explain a concept.
2. After one trainee has answered, ask a trainee seated nearby if the answer is right. Then ask whether a trainee in the back of the room agrees.
3. Ask trainees to explain why they think an answer is right or wrong.
4. Use the session to clear up incorrect ideas and encourage trainees to learn from their mistakes.

Sections 1.0.0 through 4.0.0

Quick Quiz

This Quick Quiz will familiarize trainees with the terms and definitions that are commonly used when troubleshooting gas heating. You will need photocopies of the quiz provided on the following page. Trainees will need pencils. If you allow trainees to use the Trainee Guide, decrease the amount of time you give them to complete the quiz.

1. Make a photocopy of the quiz for each trainee.
2. Give trainees between 5 and 10 minutes to complete the quiz.
3. Go over the answers to the quiz.
4. Ask trainees if they have questions.

Answers to Quick Quiz

1. c
2. b
3. e
4. a
5. f
6. d

Quick Quiz *Troubleshooting Gas Heating*

For each description listed, identify the term that the text best describes. Write the corresponding letter in the blank provided.

_____ 1. In some cases, grids are closely covered by an optional wire screen, called a(n) _____.

_____ 2. Very precise levels of indoor comfort can be obtained by using _____.

_____ 3. An induced-draft packaged unit has a _____ and therefore uses a two-speed combustion motor.

_____ 4. Modern furnaces use _____ sensors in conjunction with intermittent pilot spark ignitors or direct ignition devices.

_____ 5. Before ignition can take place in an induced-draft unit, two _____ must be activated.

_____ 6. Furnaces in which the gas is burned in a two-layer porous ceramic grid are known as _____ units.

a. flame rectification
b. modulating gas valves
c. reverberator
d. submerged flame
e. two-stage gas valve
f. vane switches

Section 2.0.0 *Troubleshooting Gas-Fired Heating Equipment*

This exercise will familiarize trainees with troubleshooting various types of gas-fired heating equipment. Trainees will need pencils and paper. You will need to arrange for a manufacturer's representative to give a presentation on gas-fired heating equipment. Allow 20 to 30 minutes for this exercise.

1. Tell trainees that a guest speaker will be presenting information on gas-fired heating equipment. Have trainees brainstorm questions for the guest speaker before the speaker arrives.
2. Introduce the speaker. Ask the presenter to speak about troubleshooting various types of gas-fire heating systems including induced-draft, natural-draft and mid- and high-efficiency. Have them discuss various troubleshooting procedures.
3. Have the trainees take notes and write down questions during the presentation.
4. After the presentation, answer any questions trainees may have.

Section 2.0.0 *Control Circuits*

This exercise will familiarize trainees with circuit diagrams for gas-fired heating equipment. Trainees will need pencils and paper. You will need to obtain the wiring diagrams for several types of units. Allow 20 to 30 minutes for this exercise. This exercise corresponds to Performance Task 3.

1. Divide the class into several groups.
2. Have each group study a portion of the circuit diagram, and then have them teach the operating sequence for that portion to the rest of the class. Alternatively, assign complete diagrams to each group. For example, Group 1 could teach the natural-draft furnace and Group 2 the induced-draft furnace.
3. Have trainees teach the rest of the class what they learned by studying the wiring diagram. Answer any questions trainees may have.

Section 2.3.0 *Identifying Parts of a Gas-Fired Furnace*

This exercise will familiarize trainees with identifying various parts of a gas-fired furnace. Trainees will need pencils and paper. You will need an overhead projector, acetate sheets, and transparency markers. Allow 20 to 30 minutes for this exercise. This exercise corresponds to Performance Task 3.

1. Divide the class in half. On one sheet write Team One on one half and Team Two on the other half.
2. Call out the components in *Figure 4*. Have trainees locate the component on the circuit diagram in *Figure 5*.
3. To ensure complete participation, each team member can answer no more than two times (adjust according to the number of participants and items to be identified). Each correct answer gets a check on the acetate sheet. An incorrect answer means the other team has the opportunity to answer.
4. Keep score to encourage class participation. At the end of the game, answer any questions.

MODULE 03209-07 — ANSWERS TO REVIEW QUESTIONS

Answer	Section
1. b	2.1.0
2. b	2.1.0
3. c	2.1.0
4. d	2.2.0
5. b	2.2.0
6. a	2.5.0
7. c	2.5.2
8. d	2.5.2
9. b	3.1.0
10. b	3.2.0
11. b	3.2.1
12. c	3.2.2
13. d	3.3.0
14. b	3.3.1
15. a	4.0.0; Appendix B

CONTREN® LEARNING SERIES — USER UPDATE

NCCER makes every effort to keep these textbooks up-to-date and free of technical errors. We appreciate your help in this process. If you have an idea for improving this textbook, or if you find an error, a typographical mistake, or an inaccuracy in NCCER's Contren® textbooks, please write us, using this form or a photocopy. Be sure to include the exact module number, page number, a detailed description, and the correction, if applicable. Your input will be brought to the attention of the Technical Review Committee. Thank you for your assistance.

Instructors – If you found that additional materials were necessary in order to teach this module effectively, please let us know so that we may include them in the Equipment/Materials list in the Annotated Instructor's Guide.

Write: Product Development and Revision
National Center for Construction Education and Research
3600 NW 43rd St, Bldg G, Gainesville, FL 32606

Fax: 352-334-0932

E-mail: curriculum@nccer.org

Craft _____ Module Name _____

Copyright Date _____ Module Number _____ Page Number(s) _____

Description
_____
_____
_____
_____
_____

(Optional) Correction
_____
_____
_____

(Optional) Your Name and Address
_____
_____
_____

Module 03210-07

Troubleshooting Cooling

NCCER STANDARDIZED CRAFT TRAINING PROGRAM

The National Center for Construction Education and Research (NCCER) provides a standardized national program of accredited craft training. Key features of the program include instructor certification, competency-based training, and performance testing. The program provides trainees, instructors, and companies with a standard form of recognition through a National Craft Training Registry. The program is described in full in the *Guidelines for Accreditation*, published by NCCER. For more information on standardized craft training, contact the NCCER by writing us at 3600 NW 43rd St., Bldg. G, Gainesville, FL 32606; calling 352-334-0911; or emailing info@nccer.org. More information may be found at our website, www.nccer.org.

HOW TO USE THIS ANNOTATED INSTRUCTOR'S GUIDE

Each page presents two sections of information. The larger section displays each page exactly as it appears in the Trainee Module. The narrow column ties suggested trainee and instructor actions to each page and provides icons (detailed below) to call your attention to material, safety, audiovisual, or testing requirements. The bottom of each page includes space for your notes.

The **Audiovisual** icon indicates an appropriate time to show a transparency or other audiovisual aid.

The **Classroom** icon prompts you to define a term, stress a point, ask trainees to explain a concept, or give examples.

The **Demonstration** icon directs you to show trainees how to perform tasks.

The **Examination** icon tells you to administer the written module examination.

The **Homework** icon is placed where you may wish to assign reading for the next class, assign a project, or advise trainees to prepare for an examination.

The **Laboratory** icon is used when trainees are to practice performing tasks.

The **Materials** icon is a reminder for you to gather materials needed for classes, labs, and testing.

The **Performance Testing** icon tells you to administer a performance test or a portion thereof.

The **Safety** icon is used to emphasize safety issues. It is often keyed to *Caution* and *Warning!* statements in the Trainee Module.

The **Teaching Tip** icon indicates additional guidance is available, such as how to conduct an exercise, get the most educational value from a field trip, or encourage class participation. Teaching Tips may expand on a feature (*Think About It, Did You Know?*) or provide *Quick Quizzes* or similar exercises. You will be referred to the Teaching Tips section at the back of the module if there is additional material.

The **Combination** icon indicates that the laboratory listed corresponds with a performance task. If desired, you can note the proficiency of the trainees during the laboratory, and use it to satisfy performance testing requirements.

PREPARATION

Before teaching this module, you should review the Objectives, Performance Tasks, Materials and Equipment List, and Module Outline. Be sure to allow ample time to prepare your own training or lesson plan and gather all required materials and equipment.

Troubleshooting Cooling
Annotated Instructor's Guide

Module 03210-07

MODULE OVERVIEW

This module covers the troubleshooting methods used with cooling systems.

PREREQUISITES

Prior to training with this module, it is recommended that the trainee shall have successfully completed *Core Curriculum*; *HVAC Level One*; and *HVAC Level Two*, Modules 03201-07 through 03209-07.

OBJECTIVES

Upon completion of this module, the trainee will be able to do the following:

1. Describe a systematic approach for troubleshooting cooling systems and components.
2. Isolate problems to electrical and/or mechanical functions in cooling systems.
3. Recognize and use equipment manufacturer's troubleshooting aids to troubleshoot cooling systems.
4. Identify and use the service instruments needed to troubleshoot cooling systems.
5. Successfully troubleshoot selected problems in cooling equipment.
6. State the safety precautions associated with cooling troubleshooting.

PERFORMANCE TASKS

Under the supervision of the instructor, the trainee should be able to do the following:

1. Develop a checklist for troubleshooting cooling systems.
2. Select the tools and instruments needed to troubleshoot a cooling system in a given situation.
3. Analyze control circuit diagram(s) for a selected cooling system.
4. Isolate and correct malfunctions in a cooling appliance:
   - Electrical problems
   - Compressor electrical failures
   - System-related compressor problems
   - Refrigerant overcharge and undercharge
   - Evaporator and condenser problems
   - Metering device problems
   - Refrigerant lines and accessories
   - Noncondensibles and contamination

MATERIALS AND EQUIPMENT LIST

Overhead projector and screen
Transparencies
Blank acetate sheets
Transparency pens
Whiteboard/chalkboard
Markers/chalk
Pencils and scratch paper
Appropriate personal protective equipment
Dry-erase markers
Laminated copy of refrigeration cycle
Service literature for selected demonstration equipment
Cylinder of refrigerant of the type used in the demonstration system
Cylinder of nitrogen with regulators
Filter-driers
Oil test kit
Sealed tube-type acid/moisture test kit (Carrier TOTALTEST® or equivalent)
Samples of contaminated refrigerant oil
Refrigerant temperature-pressure charts*
Operating air conditioning and/or refrigeration system(s)
Boost start capacitors
Assortment of inoperative components as needed to simulate exercise troubleshooting problems
Lockout/tagout locks and tags
Multimeters (VOMs/DMMs)
Clamp-on ammeter
Compressor analyzers
Capacitor testers
Gauge manifold sets
Thermometers
Sling psychrometer
Manufacturer's standalone electronic module tester
Recovery/recycle unit and recovery cylinder
Fuse pullers
Inspection mirrors
Leak detectors
Mechanic's hand tool set
Copies of Quick Quiz**
Module Examinations***
Performance Profile Sheets***

* If the trainees do not already have temperature-pressure charts, you should be able to obtain these charts in card form from a local HVAC/R distributor.
** Located in the back of this module.
*** Located in the Test Booklet.

SAFETY CONSIDERATIONS

Ensure that the trainees are equipped with appropriate personal protective equipment and know how to use it properly. This module requires trainees to work with cooling appliances. Make sure that all trainees are briefed on appropriate safety procedures. Emphasize electrical safety.

ADDITIONAL RESOURCES

This module is intended to present thorough resources for task training. The following reference works are suggested for both instructors and motivated trainees interested in further study. These are optional materials for continued education rather than for task training.

Air Conditioning Systems, Principles, Equipment, and Service, 2001, Joseph Moravek. Upper Saddle River, NJ: Prentice Hall.

General Training Air Conditioning 2, GTAC, 1996. Syracuse, NY: Carrier Corporation.

HVAC Servicing Procedures, 1995. Syracuse, NY: Carrier Corporation.

TEACHING TIME FOR THIS MODULE

An outline for use in developing your lesson plan is presented below. Note that each Roman numeral in the outline equates to one session of instruction. Each session has a suggested time period of 2½ hours. This includes 10 minutes at the beginning of each session for administrative tasks and one 10-minute break during the session. Approximately 20 hours are suggested to cover *Troubleshooting Cooling*. You will need to adjust the time required for hands-on activity and testing based on your class size and resources. Because laboratories often correspond to Performance Tasks, the proficiency of the trainees may be noted during these exercises for Performance Testing purposes.

Topic   **Planned Time**

Session I. Introduction to Troubleshooting Cooling
    A. Introduction   _____
    B. Operation of the Mechanical Refrigeration (Cooling) System   _____
    C. Electrical Control of Mechanical Cooling Operation   _____
    D. Troubleshooting Approach   _____
    E. Laboratory   _____
       Trainees practice identifying the tools and instruments needed to troubleshoot a cooling appliance. This laboratory corresponds to Performance Task 2.

Sessions II and III. Troubleshooting I
    A. Electrical Troubleshooting   _____
    B. Mechanical Refrigeration Cycle Troubleshooting   _____

Sessions IV and V. Troubleshooting II
    A. Low Charge or Overcharge of Refrigerant   _____
    B. Evaporator and Condenser Airflow Problems   _____
    C. Compressor Problems and Causes   _____
    D. Metering Device Troubleshooting   _____

Sessions VI and VII. Troubleshooting III
    A. Troubleshooting Refrigeration Lines and Accessories   _____
    B. Noncondensibles and Contamination in a System   _____
    C. Condensate Water Disposal Problems   _____
    D. Laboratory   _____
       Trainees practice developing a checklist for troubleshooting cooling systems. This laboratory corresponds to Performance Task 1.
    E. Laboratory   _____
       Use pre-faulted components, jumpers, or other means to insert system malfunctions into a cooling system. Provide trainees with the wiring diagram for the unit and have them isolate and correct the fault(s). This laboratory corresponds to Performance Tasks 3 and 4.

Session VIII. Review and Testing
    A. Review   _____
    B. Module Examination   _____
       1. Trainees must score 70% or higher to receive recognition from NCCER.
       2. Record the testing results on Craft Training Report Form 200, and submit the results to the Training Program Sponsor.
    C. Performance Testing   _____
       1. Trainees must perform each task to the satisfaction of the instructor to receive recognition from NCCER. If applicable, proficiency noted during laboratory exercises can be used to satisfy the Performance Testing requirements.
       2. Record the testing results on Craft Training Report Form 200, and submit the results to the Training Program Sponsor.

03210-07
Troubleshooting Cooling

Assign reading of Module 03210-07.

03210-07
Troubleshooting Cooling

Topics to be presented in this module include:

1.0.0	Introduction	10.2
2.0.0	Operation of the Mechanical Refrigeration (Cooling) System	10.2
3.0.0	Electrical Control of Mechanical Cooling System Operation	10.7
4.0.0	Troubleshooting Approach	10.10
5.0.0	Electrical Troubleshooting	10.11
6.0.0	Mechanical Refrigeration Cycle Troubleshooting	10.20
7.0.0	Low Charge or Overcharge of Refrigerant	10.24
8.0.0	Evaporator and Condenser Airflow Problems	10.25
9.0.0	Compressor Problems and Causes	10.26
10.0.0	Metering Device Troubleshooting	10.29
11.0.0	Troubleshooting Refrigerant Lines and Accessories	10.32
12.0.0	Noncondensibles and Contamination in a System	10.34
13.0.0	Condensate Water Disposal Problems	10.36

Overview

It is often difficult to determine whether a failure in a comfort cooling system is caused by a control malfunction or a problem in the mechanical refrigeration cycle that is causing a control device to trip. The mechanical refrigeration system is a closed-loop system, so it is not easy to tell where in that system a problem has occurred. The best way to troubleshoot the mechanical refrigeration system is to know what the temperature and pressure values at key points in the system are supposed to be when the system is operating correctly. This provides the basis for identifying a pattern of symptoms that can be related to a particular cause. An experienced service technician learns to recognize these patterns.

Instructor's Notes:

Objectives

When you have completed this module, you will be able to do the following:

1. Describe a systematic approach for troubleshooting cooling systems and components.
2. Isolate problems to electrical and/or mechanical functions in cooling systems.
3. Recognize and use equipment manufacturer's troubleshooting aids to troubleshoot cooling systems.
4. Identify and use the service instruments needed to troubleshoot cooling systems.
5. Successfully troubleshoot selected problems in cooling equipment.
6. State the safety precautions associated with cooling troubleshooting.

Trade Terms

Burnout
Capacitance boost
Flooded starts
Flooding
Hygroscopic
Noncondensible gas
Seized (stuck) compressor
Short cycling
Slugging

Required Trainee Materials

1. Pencil and paper
2. Appropriate personal protective equipment

Prerequisites

Before you begin this module, it is recommended that you successfully complete *Core Curriculum*; *HVAC Level One*; and *HVAC Level Two*, Modules 03201-07 through 03209-07.

This course map shows all of the modules in the second level of the HVAC curriculum. The suggested training order begins at the bottom and proceeds up. Skill levels increase as you advance on the course map. The local Training Program Sponsor may adjust the training order.

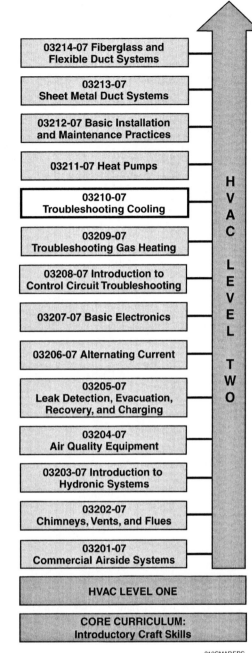

Ensure that you have everything required to teach the course. Check the Materials and Equipment list at the front of this module.

See the general Teaching Tip at the end of this module.

Explain that terms shown in bold are defined in the Glossary at the back of this module.

Show Transparency 1, Objectives, and Transparency 2, Performance Tasks. Review the goals of the module, and explain what will be expected of the trainee.

Review the modules covered in Level Two and explain how this module fits in.

Explain that efficient troubleshooting requires a systematic approach. The use of customer comments, manufacturer's wiring diagrams, and specifications will streamline the troubleshooting process.

Show Transparency 3 (Figure 1). Review the components in a basic mechanical refrigeration system. Review the functions of the lines in a basic refrigeration system.

See the Teaching Tip for Sections 1.0.0–4.0.0 at the end of this module.

1.0.0 ♦ INTRODUCTION

Troubleshooting cooling systems involves close examination of the equipment to detect telltale signs of what may be wrong. There are standard procedures to follow that make troubleshooting cooling systems easier than the trial-and-error method. By using manufacturers' wiring diagrams, specifications, and troubleshooting aids, the cause of almost any malfunction can be quickly isolated and resolved.

2.0.0 ♦ OPERATION OF THE MECHANICAL REFRIGERATION (COOLING) SYSTEM

An understanding of the basic refrigeration cycle is critical to the task of troubleshooting cooling systems. There are many types of systems used to provide cooling for personal comfort, food preservation, and industrial processes. Each of these systems uses a mechanical refrigeration system to produce the cooling. The operation of the mechanical refrigeration system is the same for all vapor compression systems. Varying from system to system are the type of refrigerant, the size and style of the components, and the installed locations of the basic components and the lines.

The basic refrigeration cycle is covered in the following sections, and an air conditioning system in more detail later in this module.

2.1.0 Basic Mechanical Refrigeration System

A basic refrigeration system (see *Figure 1*) includes the following components:

- *Evaporator* – A heat exchanger in which the heat from the area or item being cooled is absorbed and transferred to the refrigerant.
- *Compressor* – Creates the pressure differential needed to make the refrigeration cycle work. The compressor is referred to as the heart of the system.
- *Condenser* – A heat exchanger in which the heat absorbed by the refrigerant is transferred to the cooler outdoor air, or another cooler substance.
- *Expansion device* – Controls the flow of refrigerant to the evaporator. It provides a pressure drop that lowers the boiling point of the refrigerant just before it enters the evaporator. It is also called the metering device. In *Figure 1*, the metering device is a thermostatic expansion valve, commonly called a TXV.

INSIDE TRACK
Troubleshooting Cooling Systems

The accurate diagnosis and repair of a cooling system requires that the technician properly select and safely use test equipment and instruments.

210SA01.EPS

Also shown in *Figure 1* is the piping, called lines, used to connect the basic components in order to provide the path for refrigerant flow. Together, the components and lines form a closed refrigeration system. The arrows show the direction of flow for the refrigerant through the closed system. The lines are as follows:

- *Suction line* – The tubing that carries the low-temperature refrigerant gas from the evaporator to the compressor.
- *Hot gas line* – The tubing that carries hot refrigerant gas from the compressor to the condenser. This is also called the discharge line.
- *Liquid line* – The tubing that carries liquid refrigerant, formed in the condenser, to the expansion device.

Upon a call for cooling, the evaporator receives a low-temperature, low-pressure liquid/vapor refrigerant mix from the expansion device. The evaporator is mainly a series of tubing coils that exposes the cooler refrigerant to the warmer air

Instructor's Notes:

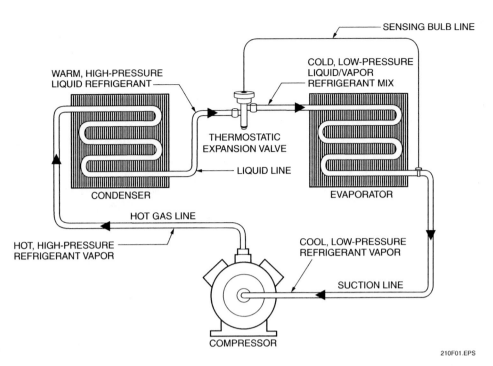

Figure 1 ♦ Basic refrigeration cycle.

Review the operation of a basic refrigeration system.

Identify new refrigerants commonly used instead of R-22.

List the requirements for persons using new refrigerants.

passing over them. Heat from the warmer air is transferred through the evaporator tubing into the cooler refrigerant, causing it to boil or vaporize. Remember that even though it has just boiled, it is still not considered hot because refrigerants boil at such low temperatures. So it is a cool, low-temperature, low-pressure refrigerant vapor that travels through the suction line to the compressor.

The compressor receives this vapor and compresses it. It then becomes a hot, high-temperature, high-pressure vapor. This travels to the condenser via the hot gas line, commonly called the discharge line.

Like the evaporator, the condenser is a series of tubing coils through which the refrigerant flows. As cooler air moves across the condenser tubing, the hot refrigerant vapor gives up superheat and cools. As it continues to give up heat to the outside air, it cools to the condensation point, where it begins to change from a vapor into a liquid. As more cooling takes place (subcooling) all of the refrigerant becomes liquid. This warm, high-temperature, high-pressure liquid travels through the liquid line to the input of the expansion device.

The expansion device regulates the flow of refrigerant into the evaporator. It also decreases the refrigerant pressure and temperature. By using a built-in restriction, such as a small hole or orifice, the expansion device converts the high-temperature, high-pressure refrigerant from the condenser into the low-temperature, low-pressure refrigerant needed to absorb heat in the evaporator.

2.2.0 New Refrigerants Used in Air Conditioners

Due to environmental concerns, refrigerants containing chlorine such as HCFC-22 (R-22) are being phased out. R-410a is an HFC refrigerant that does not contain any chlorine and appears to be the refrigerant of choice of the HVAC industry to replace R-22. Several major equipment manufacturers now offer products that use R-410a refrigerant. Products using R-410a present new requirements for persons installing, troubleshooting, and repairing them, including the following:

- The need for special tools and test equipment
- Preventing moisture from entering the refrigeration system
- The use of special replacement parts

Identify the special tools and test equipment needed when working with R-410a.

Discuss prevention of moisture contamination when working with refrigerants.

Discuss the hazards and safety precautions needed when handling certain lubricating oils.

INSIDE TRACK

Evaporator/Condenser Operation

Shown here are examples of the changes that occur to a refrigerant as it flows through a condenser coil and through an evaporator coil. The temperature loss or gain values that are shown at various points within the condenser and evaporator are typical for an R-22 air conditioning system operating at a given ambient temperature and load conditions. In the condenser coil, the hot refrigerant vapor applied at the input is converted into a warm liquid refrigerant output. This happens because the heat contained in the hot refrigerant flowing through the condenser is transferred to the cooler air being passed over the condenser coils. In the evaporator, the cold, liquid/vapor refrigerant applied at the input is converted to a cool refrigerant vapor output. This happens because the refrigerant flowing through the evaporator absorbs heat from the warmer air being passed over the evaporator coils.

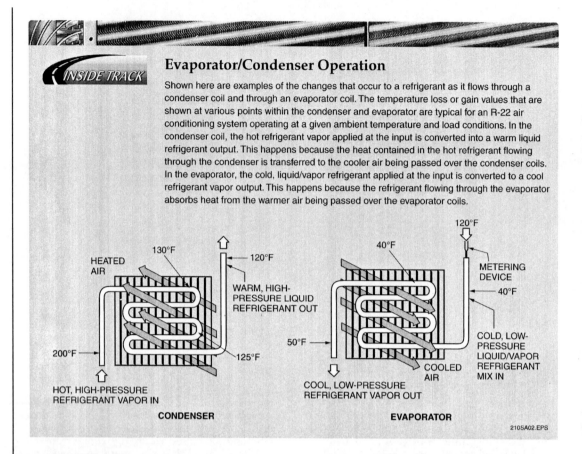

2.2.1 Special Tools and Test Equipment

The operating pressures of R-410a are much higher than those of a comparable R-22 system. For that reason, a special gauge manifold with gauges that read higher pressures and hoses rated for those higher pressures are required. For safety reasons, a gauge manifold and hoses designed for R-22 systems cannot be used with R-410a systems. Refrigerant recovery machines must be rated for the higher pressures of R-410a. The use of a recovery machine not rated for these higher pressures can damage the machine and pose a safety hazard. Since R-410a does not contain chlorine, some leak detectors cannot detect this refrigerant.

CAUTION

When working with R-410a refrigerant, always remember to use a leak detector rated for HFC refrigerants.

2.2.2 Preventing Moisture Contamination

Moisture is the enemy of all refrigeration systems, including those that contain HFC refrigerants. The compressor lubricating oil in R-410a compressors is very **hygroscopic**; that is, it readily absorbs moisture. To prevent moisture from contaminating an R-410a system, any opened system must be sealed immediately to prevent moisture from entering. Special R-410a filter-driers are required. It is also critical to fully evacuate the system to 500 microns before recharging.

WARNING!

The lubricating oil for R-410a compressors can irritate exposed skin and can damage certain types of roofing materials. Follow all manufacturer recommendations for handling this type of oil, and be sure to wear proper personal protective equipment (PPE).

2.2.3 Replacement Parts

Replacement parts for R-410a refrigeration system components are designed to handle the higher pressures of that refrigerant. It may be tempting to substitute similar-looking refrigeration system parts designed for an R-22 system. Do not under any circumstances use parts not rated for an R-410a system. The higher R-410a pressures may rupture the unauthorized parts, causing personal injury.

WARNING!
Never substitute other refrigeration system components in an R-410a system. Always use components rated for R-410a, and follow all manufacturer recommendations for proper installation.

2.3.0 Refrigeration Cycle for a Typical Air Conditioning System

Figure 2 shows a basic air conditioning system with the components divided into two sections based on pressure. The high-pressure side includes all the components in which the pressure of the refrigerant is at or above the condensing pressure. This is often referred to as head pressure, discharge pressure, or high-side pressure. The low-pressure side includes all the components in which the pressure is at or below the evaporating pressure. This is often called the suction pressure or low-side pressure. The dividing point between the sections cuts through the compressor and the expansion device.

The following describes a typical air conditioner that can use either HCFC-22 (R-22) or HFC-410a (R-410a). These are the two most commonly used refrigerants in today's air conditioning systems. This example demonstrates the temperature/pressure relationships that are typical of those that exist at key points in the system during operation. For our example, assume an indoor air temperature of 75°F and an outdoor air temperature of 95°F. The temperatures and pressures used in the example are representative values only. They will vary due to equipment and load conditions. Load conditions vary from system to system and are based largely on climatic conditions. For example, in the deserts of the U.S., the sensible heat load is very high, while the latent (humidity) load is low. In very humid areas such as along the Gulf of Mexico, the sensible and latent heat loads are both high. The numbers given in the description below correspond to the numbers shown in *Figure 2*. Follow along on the figure as the system is described.

1. A mixture (75 percent liquid, 25 percent vapor) is supplied from the expansion device to the evaporator. This mixture is at a pressure of 69/118 psig, which corresponds to the 40°F boiling point of both refrigerants. (See chart in *Figure 2*.) The 40°F boiling point used here is typical of the temperatures normally used for evaporators in air conditioning systems.

2. Because the refrigerant flowing through the evaporator is cooler (at 40°F) than the warmer inside room air (at 75°F) passing over the evaporator, it absorbs heat, causing the liquid refrigerant to boil and turn into a vapor. After traveling about 90 percent of the way through the evaporator tubing, all of the refrigerant has boiled into a vapor known as the saturated vapor.

3. During the remaining 10 percent of travel through the evaporator, the saturated vapor continues to absorb heat from the warmer air, thus raising its temperature to 50°F. In other words, the saturated vapor is superheated 10°F (50°F – 40°F = 10°F). This superheated vapor flows through the suction line and is drawn into the low-pressure side of the compressor. The cooled inside room air is re-circulated by the evaporator fan back into the room at a temperature of about 55°F.

4. The superheated vapor applied at the suction input of the compressor picks up an additional 3° to 5° of superheat because the vapor in the suction line absorbs more heat from the warmer surrounding air as it travels from the evaporator to the compressor.

5. After compression, the highly superheated gas from the compressor flows through the hot gas line to the condenser. This hot gas may be close to 200°F at 297/475 psig, and the hot gas line has gained about 70°F superheat (200°F – 130°F = 70°F). Typically, discharge superheat should range between 50°F and 150°F. This superheat must be removed before the refrigerant vapor can be condensed into a liquid. The 200°F refrigerant in the hot gas line easily gives up some of its superheat to the surrounding 95°F air. The hot gas line is not normally insulated, and the tubing is a good conductor of heat.

Discuss replacement parts for R-410a refrigerant systems.

Emphasize the importance of using properly rated replacement parts.

Show Transparency 4 (Figure 2). Explain the division between system high-side pressure and low-side pressure. Trace the refrigerant flow through the system, pointing out temperatures and pressures at critical locations.

Have the trainees trace the refrigerant flow on a laminated copy of the refrigeration cycle with a dry-erase marker. This helps to reinforce what is happening at critical points in the system.

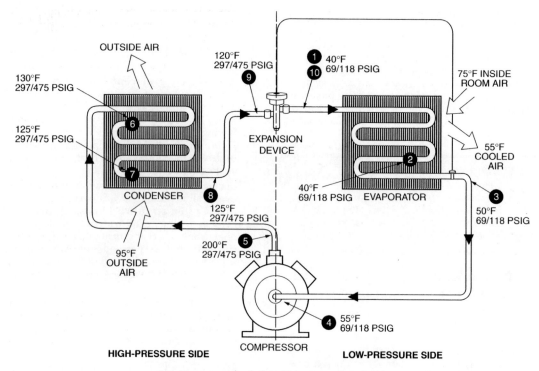

Figure 2 ♦ Typical air conditioning cycle for HCFC-22 (R-22) and HFC-410a (R-410a) refrigerants.

6. Because the refrigerant in the condenser is still hotter than the warmer outside air passing over the condenser, it easily gives up the remaining superheat. This drops its temperature to 130°F. As heat continues to be transferred from the vapor to the cooler outside air, the vapor begins to cool and condense into a liquid. After the refrigerant has traveled about three quarters of the way through the condenser, all of the refrigerant has condensed into a liquid.

The 130°F condensing temperature is set by the condenser design. A standard condenser is designed to have a condensing temperature about 35°F higher than the surrounding air. In this case, 95°F outside air is used to absorb the heat, so 95°F + 35°F = 130°F condensing temperature.

7. During the remaining one quarter of travel through the condenser, the liquid refrigerant continues to drop in temperature (subcooling). This lowers its temperature to about 125°F. In other words, the liquid refrigerant is subcooled 15°F (130°F – 115°F = 15°F).

8. Subcooled liquid refrigerant from the condenser flows through the liquid line to the expansion device. Because the liquid line is often uninsulated, the subcooling can increase or decrease due to length, location, and elevation change. In this case, 5° of subcooling occurred.

9. The expansion device controls the flow of liquid refrigerant to the evaporator. Subcooled liquid from the condenser enters at the high temperature of 120°F and high pressure of 297/475 psig. It leaves the expansion device at the low temperature of 40°F and low pressure of 69/118 psig, thereby lowering the boiling point of the liquid refrigerant supplied to the evaporator. In the expansion device, the subcooled liquid refrigerant at 297/475 psig is passed through a small opening or orifice. This changes the pressure of the liquid refrigerant from 297/475 psig to 69/118 psig, causing some of it to flash into vapor. This flash gas cools the remaining liquid to produce a mixture of about 75 percent liquid and 25 percent vapor. The pressure of this mixture is 69/118 psig, which corresponds to the 40°F boiling point temperature needed for correct evaporator operation. This low-temperature, low-pressure mixture from the expansion device then travels to the evaporator.

10. The refrigerant has now completed the refrigeration cycle and is ready to start the cycle again.

The pressures and temperatures just shown are typical for a standard efficiency air conditioner using a reciprocating compressor. Today's high-efficiency air conditioners are likely to have pressures and temperatures that are different from the typical example shown. For more information about the temperatures and pressures typical of a specific product, consult the equipment manufacturer's service literature.

3.0.0 ♦ ELECTRICAL CONTROL OF MECHANICAL COOLING SYSTEM OPERATION

Figure 3 shows an electrical control circuit for a residential system that might be used with a basic cooling-only system like the one just described. As shown, it has some features that are common to most cooling circuits:

- The control devices (thermostat, compressor contactor, and fan relay) operate off of 24VAC.
- The indoor fan (evaporator fan) has a separate control. By setting the FAN switch to ON, the occupant can use the fan for ventilation without operating the compressor. When the POWER switch is set to ON and the FAN switch is in the AUTO position, the fan relay (IFR) will energize whenever the cooling thermostat (TC) closes.
- The outdoor fan (condenser fan) motor (OFM) runs whenever the compressor is on. Follow this sequence: the cooling thermostat (TC) closes, completing the 24V path to the coil of the compressor contactor (C), causing it to energize. Its normally open contacts (C contacts) in the upper part of the circuit close, completing the current path to the compressor and outdoor fan motor.
- The compressor motor will have an internal automatic-reset overload to provide protection against damaging high temperatures.

No matter how complex a control circuit appears, you will find that it is always the same basic control arrangement shown in *Figure 3*, or something very similar. Everything else will be related to special features to improve equipment safety or operating efficiency, such as in *Figure 4*.

Describe an induced-draft furnace.

Show Transparency 5 (Figure 3). Explain the operation of common control devices. Point out that even the most complex circuit will have the same basic control arrangement.

See the Teaching Tips for Section 3.0.0 at the end of this module.

Show Transparency 6 (Figure 4). Compare the operating sequence for a common cooling unit to a typical heating operation.

R-22 Versus R-410a Temperatures and Pressures

R-410a is the refrigerant chosen by the HVAC industry to replace R-22 because it has similar characteristics. In operating systems, similar temperatures will be present. The major difference is in the operating pressures. R-410a operates at pressures much higher than R-22 for comparable temperatures.

Pressure-Temperature Chart for R-410A and R-22

°F	R-410A	R-22	°F	R-410A	R-22	°F	R-410A	R-22	°F	R-410A	R-22
−40	10.8	0.6	10	62.2	32.8	60	169.6	101.6	110	364.1	226.4
−39	11.5	1.0	11	63.7	33.8	61	172.5	103.5	111	369.1	229.6
−38	12.1	1.4	12	65.2	34.8	62	175.4	105.4	112	374.2	232.8
−37	12.8	4.8	13	66.8	35.8	62	178.4	107.3	113	379.4	236.1
−36	13.5	2.2	14	68.3	36.8	64	181.5	109.3	114	384.6	239.4
−35	14.2	2.6	15	69.9	37.8	65	184.5	111.2	115	389.9	242.8
−34	14.9	3.1	16	71.5	38.8	66	187.6	113.2	116	395.2	246.1
−33	15.6	3.5	17	73.2	39.9	67	190.7	115.3	117	400.5	249.5
−32	16.3	4.0	18	74.9	40.9	68	193.9	117.3	118	405.9	253.0
−31	17.1	4.5	19	76.6	42.0	69	197.1	119.4	119	411.4	256.5
−30	17.8	4.9	20	78.3	43.1	70	200.4	121.4	120	416.9	260.0
−29	18.6	5.4	21	80.0	44.2	71	203.6	123.5	121	422.5	263.5
−28	19.4	5.9	22	81.8	45.3	72	207.0	125.7	122	428.2	267.1
−27	20.2	6.4	23	83.6	46.5	73	210.3	127.8	123	433.9	270.7
−26	21.1	6.9	24	85.4	47.6	74	213.7	130.0	124	439.6	274.3
−25	21.9	7.4	25	87.2	48.8	75	217.1	132.2	125	445.4	278.0
−24	22.7	8.0	26	89.1	50.0	76	220.6	134.5	126	451.3	
−23	23.6	8.5	27	91.0	51.2	77	224.1	136.7	127	457.2	
−22	24.5	9.1	28	92.9	52.4	78	227.7	139.0	128		
−21	25.4	9.6	29	94.9	53.7	79	231.1	141.3	129		
−20	26.3			96.8	55.0	80	234.9	143.6	130		
−19	27.2				56.2	81	238.6	146.0	13		
−18						82	242.3	148.4			
−17							246.0	150.8			

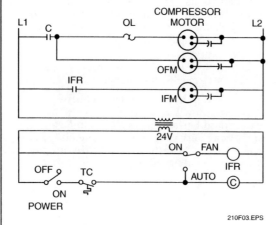

Figure 3 ♦ Cooling system control circuit.

Figure 4 is one manufacturer's diagram for a combined cooling and gas heating unit. Other manufacturers have similar diagrams. The circuit looks different and there are several more components needed for heating operation, but if you trace out the cooling control function, you will see that it is essentially the same. The differences are as follows:

- The control (ON/OFF) switch, fan control, and heating and cooling thermostats have all been incorporated into a combined heating/cooling thermostat unit.
- The cooling control circuit has more devices, such as a compressor short-cycle protection circuit and high- and low-pressure safety switches.
- A PTC start thermistor (ST) has been added across the compressor run capacitor.

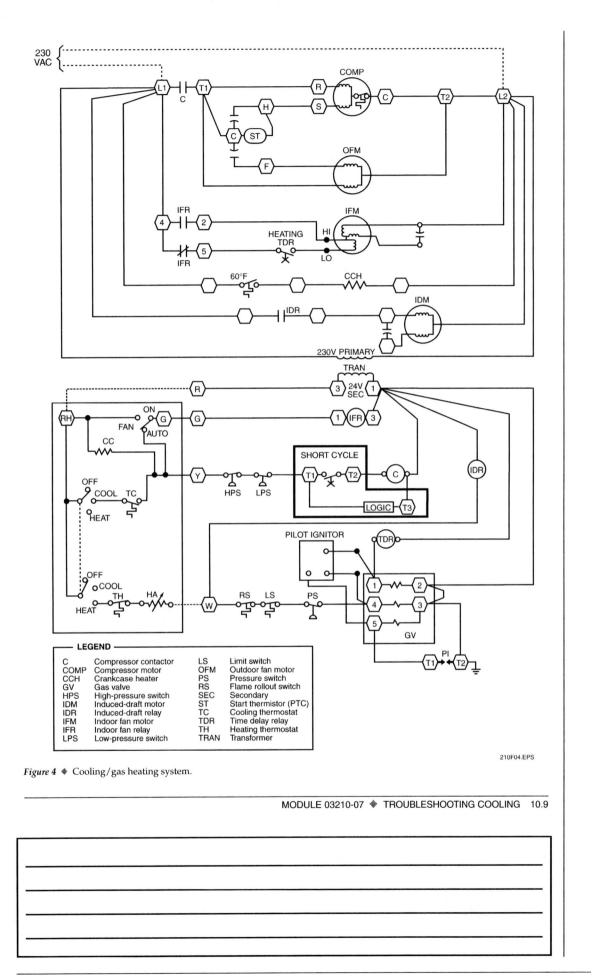

Figure 4 ♦ Cooling/gas heating system.

Keep Transparency 6 (Figure 4) showing. Identify the features specific to a cooling system. Explain the operating sequence for a common cooling unit.

Refer to Appendix A and Appendix B for information regarding a typical three-phase cooling unit.

Have the trainees trace the energized circuits on a laminated copy of the wiring schematic with a dry-erase marker. This helps to reinforce the operating sequence. Alternately, project Figure 4 onto a dry-erase board and have them trace the energized circuits.

Explain that troubleshooting problems are divided into two categories: electrical and mechanical.

Emphasize the importance of a logical and structured approach to troubleshooting.

See the Teaching Tip for Section 4.0.0 at the end of this module.

Show the trainees how to select tools and instruments for troubleshooting.

- A thermostat-controlled crankcase heater (CCH) circuit has been added.

Note that L1 and L2 are depicted as emanating from common terminals. This reflects the way the circuit is actually wired, and it is a common method used by manufacturers to draw control circuits. Some of the features of the circuit shown in *Figure 4* relative to cooling are as follows:

- The unit has a two-speed indoor fan motor (IFM) that runs on high speed for cooling and low speed for heating. In heating, the indoor fan is controlled by the time delay relay (TDR).
- High- and low-pressure safety cutout switches (HPS and LPS) are added in series with the compressor contactor coil to protect the compressor against operation when an abnormal high- or low-pressure problem exists.
- A short-cycle protection module has been added in series in the compressor contactor control circuit. This prevents the compressor from being restarted immediately after it has been turned off (**short cycling**). This protects the compressor from restarting against a high pressure differential. Short cycling can result when there is a momentary interruption of power to the compressor for any reason. If the compressor is running and the power is interrupted, the short-cycle module must first time out before power will be reapplied to the compressor. Timers typically have delays ranging from 30 seconds to 5 minutes. The specific time delay used depends on the application.

Even with the additions to the basic cooling system control circuits, the operating sequence of this unit in the cooling mode remains quite simple:

- In preparation for cooling, the unit must be energized. The FAN switch is set to AUTO and the function switch is set to COOL.
- The thermostat dial is set below room temperature, causing the cooling thermostat to call for cooling.
- The indoor fan starts and the outdoor fan and compressor start after a delay (about 3 to 5 seconds).
- When the thermostat setting is satisfied, the fans and compressor stop.
- If the thermostat calls for cooling immediately after the compressor has stopped, the compressor will not restart until the short-cycle module times out.

An example of a typical three-phase cooling system and its sequence of operation are shown in *Appendix A*. This example can be studied for additional practice in analyzing the operation of cooling systems. *Appendix B* contains additional troubleshooting information.

NOTE

Variable-speed, electronically commutated fan motors operate and behave differently than conventional fan motors. When troubleshooting systems using these types of motors, always follow the manufacturer's troubleshooting procedures.

4.0.0 ♦ TROUBLESHOOTING APPROACH

In this module, troubleshooting problems are divided into two categories: electrical problems and mechanical refrigeration system problems. This does not mean that all malfunctions fit easily into these two categories; often there are problems in which both areas are affected.

Whether troubleshooting an electrical or mechanical problem, the use of a logical, structured approach is essential. You must know how to diagnose specific control devices and system components. A good set of service tools and the ability to read electrical wiring diagrams is a must, otherwise you are only guessing at the cause of a problem.

Similarities Among Equipment

Regardless of the manufacturer, most HVAC equipment contain similar components and have similar sequences of operation. For example, if you know the sequence of operation for a packaged rooftop air conditioner made by one manufacturer, there is a good chance that similar packaged air conditioners made by other manufacturers will have similar sequences of operations.

10.10   HVAC ♦ LEVEL TWO

Instructor's Notes:

5.0.0 ♦ ELECTRICAL TROUBLESHOOTING

Looking at the electrical circuits of cooling equipment might make electrical troubleshooting seem complicated. However, using a systematic, logical approach simplifies the process. Before attempting any troubleshooting procedure, you should learn the operating sequence of the unit. Find out what is supposed to happen and when. Always refer to the manufacturer's label diagrams and product literature. For basic equipment, almost all electrical troubleshooting can be accomplished by the process of elimination. This means that when any load device is working normally, that device and all of its switches can be eliminated as a possible source of trouble. For example, if a particular system has four major load devices, and three are operating normally, then the problem is, by the process of elimination, isolated to the fourth load and its related control circuit.

5.1.0 Troubleshooting Electronic Modules/Boards

The use of electronic controls is now widespread across all HVAC product lines. It is rare to find a modern cooling unit that does not contain some kind of electronic control. Manufacturers put these controls in their equipment to increase reliability and increase energy efficiency. Electronic controls also allow engineers to deign more versatility into the products than was ever possible with electro-mechanical controls.

When troubleshooting a system that uses electronic controls, do not arbitrarily replace the board if there is a problem with the equipment. Use a systematic, logical approach to first make sure the module or board has failed. In some equipment, this means using a built-in diagnostic system to indicate that a board has failed. If the equipment does not have a built-in diagnostic capability, make sure you isolate the board as the cause of a problem by using the troubleshooting information in the manufacturer's service literature.

Electronic control boards are checked for correct operation by making sure that the board is receiving the correct inputs such as voltage signals from the room thermostat and inputs from temperature and pressure sensors. If all the inputs are correct, the control should provide appropriate outputs. For example, if the control is receiving room thermostat inputs to Y and G on the control, and all pressure and temperature inputs are within an acceptable range, the control will provide output signals to energize the compressor and outdoor fan contactor, and the indoor fan relay. If the correct input and output signals are present and the equipment does not run, the problem is not with the control. If the correct inputs are present but no outputs are present, the control may be defective. Input and output signals are typically measured with a multimeter.

Manufacturers often recommend the use of a module/board tester designed to troubleshoot specific electronic controls used in the manufacturer's product line. If troubleshooting using module/board testers, be sure to follow the instructions provided with the tester.

5.2.0 Troubleshooting Functional Circuits and Components

The electrical troubleshooting information in this module is limited to the electrical devices particular to cooling systems. Troubleshooting most of the electrical problems and components in HVAC equipment is performed in the same way, regardless of the type of equipment. Because of this, fault isolating and troubleshooting data common to all HVAC equipment, including cooling equipment, is described in *Introduction to Control*

Have trainees select the tools and instruments needed to troubleshoot a cooling system in a given situation. Note the proficiency of each trainee. This laboratory corresponds to Performance Task 2.

Have the trainees review Sections 5.0.0–6.4.0.

Ensure that you have everything required for teaching this session.

Discuss the importance of reviewing the manufacturer's literature prior to performing any troubleshooting procedures.

Describe the process for troubleshooting electronic controls.

Provide manufacturers' literature on troubleshooting cooling systems for trainees to examine.

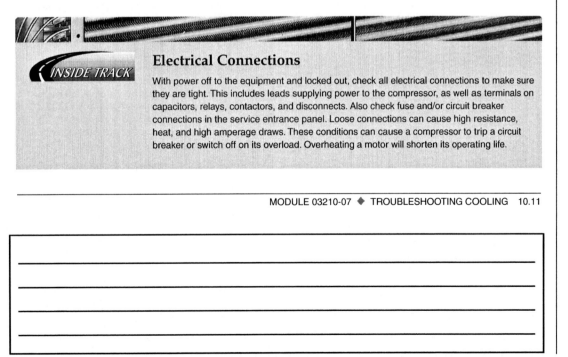

Electrical Connections

With power off to the equipment and locked out, check all electrical connections to make sure they are tight. This includes leads supplying power to the compressor, as well as terminals on capacitors, relays, contactors, and disconnects. Also check fuse and/or circuit breaker connections in the service entrance panel. Loose connections can cause high resistance, heat, and high amperage draws. These conditions can cause a compressor to trip a circuit breaker or switch off on its overload. Overheating a motor will shorten its operating life.

Review troubleshooting procedures covered in previous modules that apply to cooling systems.

Show Transparency 7 (Figure 5). Review the electrical troubleshooting flow chart. Explain that when a load is operating normally, that load and all of its switches can be eliminated as a possible source of trouble.

Inside Track

Inputs and Outputs

HVAC equipment manufacturers' product literature will contain information stating the correct inputs and outputs that can be measured at an electronic control. In the schematic of the electronic control shown, inputs are provided by thermistors and a low-pressure switch. Outputs include a signal to energize an outdoor fan relay (ODFR), a liquid line solenoid relay (LLSR), and a relay (K4) on the control board.

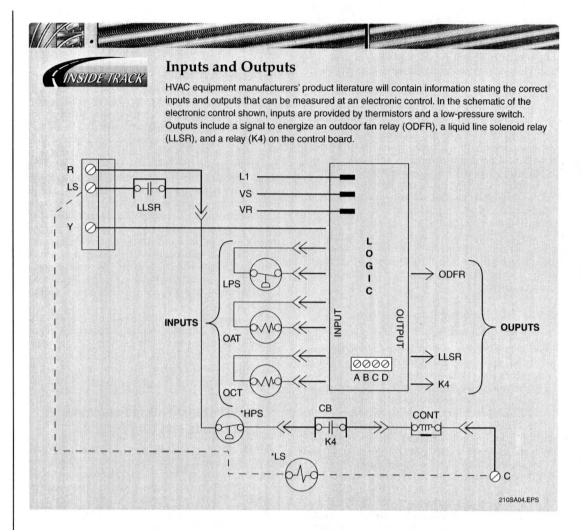

Circuit Troubleshooting. For quick reference, the troubleshooting data in that module which applies to cooling systems is listed here:

- Isolating to a faulty circuit via the process of elimination
- Isolating to a faulty component
- Single-phase and three-phase input voltage measurements
- Three-phase voltage and current imbalance measurements
- Fuse and circuit breaker checks
- Resistive and inductive load checks
- Switch and relay/contactor contact checks
- Control transformer checks
- Start and run capacitor checks
- Start relay checks
- Start thermistor checks
- Identifying unmarked terminals of a PSC/CSR motor
- Compressor/motor open, shorted, or grounded winding checks

Figure 5 shows an example of a flow chart that provides troubleshooting guidance to fault isolate a problem where the compressor will not run. It also serves to show that many of the measurements and checks just listed must be performed while accomplishing this particular troubleshooting task.

10.12 HVAC ◆ LEVEL TWO

Instructor's Notes:

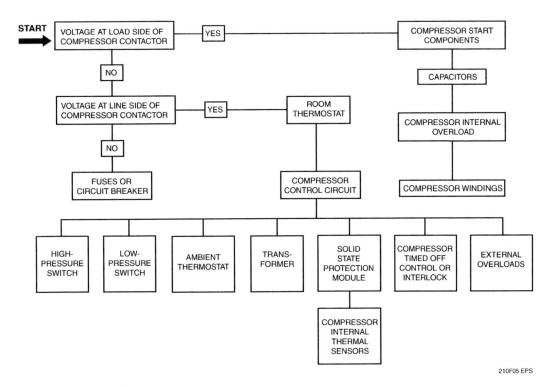

Figure 5 ♦ Electrical troubleshooting flow chart.

Emphasize the importance of visually checking the thermostat. Explain how to perform visual checks on the thermostat.

Show Transparency 8 (Figure 6). Describe the circuitry for a typical electro-mechanical thermostat.

5.3.0 Troubleshooting Thermostats

Troubleshooting thermostats involves visual checks, electrical checks for electro-mechanical thermostats, and electronic troubleshooting for electronic thermostats. When troubleshooting and servicing electronic thermostats, be sure to avoid damaging their microprocessors and integrated circuits.

5.3.1 Visual Checks

You must check the thermostat to ensure that it is the correct model for the system. Visually check for the following conditions that can impact thermostat or system operation:

- The thermostat should be mounted on an inside wall in a location that is free of vibration. This helps prevent contact chattering (also known as bounce) and the intermittent equipment operation that results.
- The thermostat should not be subjected to drafts or dead air spots, hot or cold air from ducts or diffusers, or radiant heat from direct sunlight, lamps, or hidden heat from appliances.
- Check to make sure the thermostat function switch is set for cooling.
- Check that the set point is below room temperature.
- Make sure the thermostat is level.
- Make sure the mercury bulb or bimetal element is not sticking.

5.3.2 Electrical Checks for Electro-Mechanical Thermostats

Thermostats used in residential and some commercial structures include electro-mechanical (non-electronic) cooling-only, heating-only, and heating/cooling automatic changeover types. *Figure 6* shows a residential heating/cooling thermostat typical of those in common use.

Regardless of the type, the approach to troubleshooting a thermostat for a cooling system problem is basically the same. If a unit is not running at all, first check the thermostat function switch to see if it is set for cooling, and that the thermostat set point is below room temperature. Next, check that 24V power is being applied to the thermostat by setting the FAN switch to ON.

Explain the procedures for checking electro-mechanical thermostats.

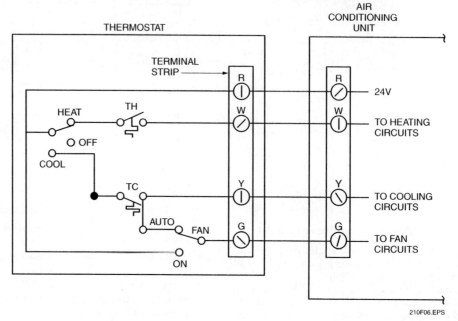

Figure 6 ♦ Electro-mechanical heating/cooling thermostat.

If the system blower starts, 24V power is being applied to the thermostat. If the system blower does not start, the thermostat should be checked for missing control voltage or an open circuit.

To check a thermostat while the power is on, proceed as follows:

Step 1 Identify the terminals and color codes of the connecting wires as they relate to the wiring diagram. Make sure that all wires in the sub-base are connected properly and that the connections are tight. *Figure 6* shows the standard coding method used for residential heating/cooling thermostats. More complex thermostats have additional codes. For example, when there are multiple stages of cooling or heating, those terminals are designated with the appropriate letter plus a number, such as Y1 and Y2.

Step 2 Measure the voltage between the R (24V) and C common (24V return) terminals on the furnace or air handler to check that the 24V control power is being applied to the input of the thermostat. If no control voltage is measured, troubleshoot the control transformer circuit. If the voltage is lower than about 21V, it may affect the operation of control relays and other control components. Troubleshoot the control transformer circuit for the cause of low voltage.

> ## Modern Room Thermostats
>
> In addition to the traditional terminals such as R, Y, G, and W found on room thermostats, many modern electronic thermostats contain a C (24-volt) terminal. These modern thermostats require power for various internal circuits. By connecting the thermostat to R and C, 24-volt power is always available to the device.

Step 3 Remove the thermostat from the sub-base, connect a jumper across the sub-base terminals listed here and observe equipment operation:
- R (24V) to G (fan) – Indoor fan should run.
- R (24V) to Y (cool) – Compressor and outdoor fan should run.

If the compressors and/or fans operate when the related sub-base terminals are jumpered, the sub-base and all wiring to that point is good. Reinstall the thermostat. If the loads do not operate with the thermostat in place, the problem is in the thermostat.

5.3.3 Troubleshooting Electronic Programmable Thermostats

Electronic programmable thermostats (*Figure 7*) are commonly used in commercial applications and are also becoming more popular for use in residential applications. These thermostats use microprocessors and integrated circuits to provide a wide variety of control and energy-saving features. Their control panels often use touchscreen technology and their indicators are digital readouts. Different thermostats offer different features; the more sophisticated the thermostat, the more features it offers. Because of their diversity and complexity, manufacturers of electronic programmable thermostats generally provide aids in their service literature to use when troubleshooting a particular thermostat. As an HVAC technician, you should always use this data.

> **CAUTION**
>
> Microprocessors and integrated circuits used in electronic thermostats are sensitive to, and can be damaged by, static electricity. Before handling the thermostat or touching its components, always discharge your body static electricity by grounding yourself to a metal object. It is also a good practice to never touch the connector pins or terminals of the microprocessor and/or integrated circuit. Before troubleshooting an electronic thermostat, refer to the manufacturer's service literature to make yourself aware of any other special precautions about handling the thermostat.

5.4.0 Troubleshooting Compressor Failures

If not corrected, many cooling system problems will result in compressor failure. If the compressor fails and the cause has not been found and eliminated, a replacement compressor will probably also fail. *Figure 8* summarizes the possible causes of the symptoms exhibited by the compressor for some common cooling system electrical problems. It also shows mechanical refrigeration system problems that can exist in a cooling system.

The measurements and checks used to electrically test compressor motors are provided in *Introduction to Control Circuit Troubleshooting*. The remainder of this section describes several compressor problems that are frequently encountered when troubleshooting.

Figure 7 ♦ Electronic programmable thermostat with built-in diagnostic features.

Discuss the common causes of compressor failure.

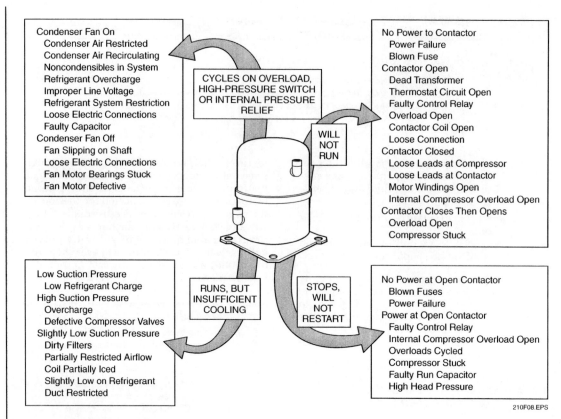

Figure 8 ♦ Compressor troubleshooting.

5.4.1 Seized Compressor

A **seized (stuck) compressor** hums but will not start, and draws locked rotor current for several seconds. Normally, locked rotor current, which can be four to six times the normal running current, lasts for a fraction of a second when the motor first starts. A seized compressor can be caused by a mechanical or electrical problem within the compressor or by an external electrical problem.

Before condemning a seized compressor, check for the following mechanical refrigeration system or electrical system conditions:

- Compressor contactor not making good contact on all poles
- Defective (open) start relay
- Start or run capacitor open or weak
- Unequalized system pressures (especially in units using PSC compressor motors)
- Low supply voltage

If the supply voltage is within ±10 percent of the motor's nameplate rating, but a single-phase PSC motor fails to start, a start kit may have to be installed to increase the motor starting torque to correct the problem.

Seized or Stuck Compressor

A seized or stuck compressor does not necessarily mean that the compressor has failed. For example, a compressor may be stuck due to external and correctable electrical problems such as low voltage or defective or missing compressor start circuit components. Correcting external electrical problems can turn a seized or stuck compressor into an operational compressor.

10.16 HVAC ♦ LEVEL TWO

Show Transparency 10 (Figure 9). Explain how a capacitance boost can start a seized compressor.

Reinforce safety practices for troubleshooting equipment with power applied to it.

Inside Track

Compressors

The reciprocating, scroll, and rotary compressors shown here are typical of those used in many split-system and packaged unit air conditioning systems.

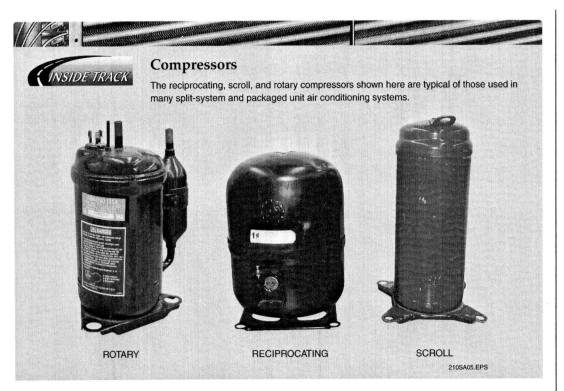

ROTARY RECIPROCATING SCROLL

If electrical tests prove a PSC compressor has no open, shorted, or grounded windings, and if all external factors check out to be good, you can attempt to start a stuck PSC compressor with a **capacitance boost** (*Figure 9*). Capacitance boosting involves momentarily connecting a start capacitor (200 to 300 MFD range) across the compressor's existing run capacitor when power to the compressor is turned on. Since the start capacitor is placed in parallel with the run capacitor, the values of the capacitors are combined, providing increased capacitance and increased torque to start the PSC motor. Once started, the compressor should continue to run at full speed with the start capacitor removed from the circuit. It is important that the start capacitor be removed because leaving it in the circuit for more than a few seconds will cause it to overheat and burn up. Allow the compressor to run for about ten minutes.

WARNING!
To avoid the possibility of electric shock, do not handle live wires. Switch the power off and lock it out before connecting the wires. Also, bridge the start capacitor to ensure that it is discharged before connecting wires to it.

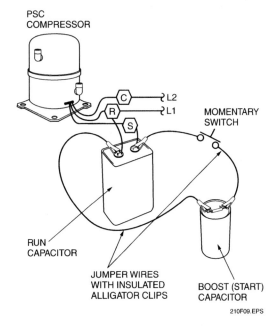

Figure 9 ♦ Starting a stuck PSC compressor with a capacitance boost.

MODULE 03210-07 ♦ TROUBLESHOOTING COOLING   10.17

Discuss the use of start kits for use with compressors.

Emphasize the importance of using the correct start kit for a compressor.

Check the Compressor Motor for a Grounded Condition

A motor can be grounded by a problem as simple as a motor winding that is touching the case. In this case the resistance reading will be very low. Resistance readings can be several thousand ohms and still be considered shorted. A good rule of thumb is that the resistance to ground should be greater than 1,000Ω per operating volt. For example, a motor that operates on 240V should have a minimum resistance of 240,000Ω to ground. Both refrigerant and oil contamination can create low resistance-to-ground readings. To improve these readings, change the refrigerant, oil, and driers.

Try restarting the compressor again without the capacitance boost. If the compressor starts the first time and starts on subsequent attempts, a permanently installed start capacitor and start relay are probably not needed. If the compressor will not start without a capacitance boost, permanently install a correctly sized start capacitor and start relay. Start kits designed for use with most compressors are available from HVAC distributors.

CAUTION
Always use the exact start kit specified for the compressor. Failure to use the appropriate start kit may cause damage to the compressor motor.

Never use a megger or attempt to start the compressor in a deep vacuum.

Troubleshooting Rotary Compressors

Rotary compressors sometimes become oil-locked. This is a condition in which the rotor will not turn due to the tight clearances in the compressor. Oil lock can occur in new compressor motors. To fix an oil-locked compressor, try to start the motor several times. If this does not work, install a soft- or hard-start kit. If the kit does not fix the problem, the compressor must be replaced.

Another problem experienced by rotary compressors is that of stuck vanes. When a vane sticks, there is no pumping action and the motor draws low amperage. The rotor is turning, but the vanes are not moving in and out to create compression pockets. Before you replace the compressor, try removing the suction line from the compressor shell. Use a heat gun to warm the tubing and suction area on the compressor. The heat may cause the stuck vane to release. Before reconnecting the suction line, start the compressor and place the palm of your hand over the suction fitting. If the vane is no longer stuck, the compressor will pull on the palm of your hand. This will indicate that the vane is now working and creating suction pressure.

A leaking vane will cause pressure readings that are lower than normal on the high side and higher than normal on the suction side.

A blocked accumulator will create a very high suction pressure. Install a pressure tap between the accumulator and the compressor. Compare the reading at the tap to the inlet suction pressure reading. Any pressure drop in excess of a few psi is considered excessive, and the accumulator will need to be changed. Do not operate a rotary compressor without an accumulator. The suction line is connected directly to the compression chamber. Any liquid refrigerant in the suction line can cause immediate damage to the compressor.

Never use a rotary compressor in a pump-down system. Pulling the compressor into a vacuum will cause the oil to wash away from bearing surfaces, leaving tight-fitting clearances without lubrication. This will damage the compressor.

Capacitors

Use a capacitance checker to determine the capacitance of run and start capacitors. The needle deflection of an analog or digital ohmmeter will not indicate the correct capacitance. A capacitor should operate within 10 percent of its microfarad rating.

If the compressor has been changed, a different capacitor rating may be required. This is sometimes overlooked when compressors are replaced. Every time a compressor or motor is changed, the capacitor should be changed with it. The terms of many warranties require installing a new capacitor with a compressor/motor replacement.

Three-phase motors can sometimes be unstuck by temporarily reversing or interchanging any two of the leads. After the motor has started, always turn off the power, then reconnect the compressor leads as shown on the unit wiring diagram.

5.4.2 Low or No Capacity (Compressor Runs But Does Not Pump)

If the system is properly charged, and the suction pressure is excessively high and/or the discharge pressure is excessively low, the system compressor most likely has an internal problem that is the cause for low system capacity. Reciprocating compressors depend upon their valves and rings to provide a seal between the high- and low-pressure sides of the system and the compressor. If either is damaged, correct operating suction and/or discharge pressures will never be developed.

A current measurement made with a clamp-on ammeter (*Figure 10*) can be used to help confirm that a compressor has an internal problem. If abnormally high suction and abnormally low discharge pressures are accompanied by a considerably lower than normal current draw, the compressor needs replacement.

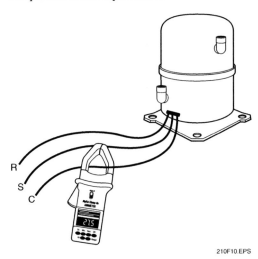

Figure 10 ♦ Compressor low capacity check.

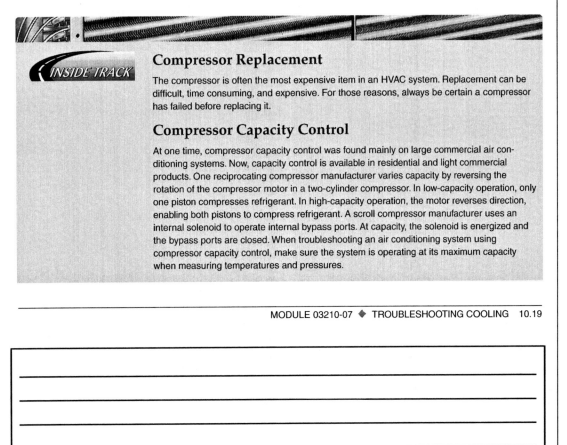

Compressor Replacement

The compressor is often the most expensive item in an HVAC system. Replacement can be difficult, time consuming, and expensive. For those reasons, always be certain a compressor has failed before replacing it.

Compressor Capacity Control

At one time, compressor capacity control was found mainly on large commercial air conditioning systems. Now, capacity control is available in residential and light commercial products. One reciprocating compressor manufacturer varies capacity by reversing the rotation of the compressor motor in a two-cylinder compressor. In low-capacity operation, only one piston compresses refrigerant. In high-capacity operation, the motor reverses direction, enabling both pistons to compress refrigerant. A scroll compressor manufacturer uses an internal solenoid to operate internal bypass ports. At capacity, the solenoid is energized and the bypass ports are closed. When troubleshooting an air conditioning system using compressor capacity control, make sure the system is operating at its maximum capacity when measuring temperatures and pressures.

Explain that suction and discharge pressures will rise and fall together with a properly running compressor.

Explain that the compressor is suspect if suction pressure rises while discharge pressure drops, or suction pressure drops while discharge pressure rises.

Show Transparency 11 (Figure 10). Explain how to check the electrical system of a compressor.

Show trainees how to perform electrical troubleshooting on equipment with preset faults.

Provide flame sensors for trainees to examine.

Review the systematic procedures for troubleshooting mechanical refrigeration systems.

Review the procedure for performing a preliminary inspection.

Stress the importance of following safe work practices when troubleshooting energized electrical components.

 **Troubleshooting Scroll Compressors**

Basic troubleshooting for scroll compressors is similar to that of other compressors. Expect a pressure differential between the suction and discharge sides of the compressor. Scroll compressors generally are installed in high-efficiency condensers. The discharge pressure will be lower than normal and suction pressure will be higher than that found on low-efficiency systems with other types of compressors.

Like reciprocating compressors, scroll compressors often require start-assist components, especially in the low-voltage starting conditions that often occur in the heat of summer. Because scroll compressors run backwards momentarily at shutdown, it is important to use short-cycle timers, either those provided by most digital thermostats or separate delay-on-break time-delay relays. Short-cycle timers provide the scroll compressor with the time required for pressures to equalize and rotation to stop completely, allowing the compressor to start without load and rotate in the proper direction.

If the compressor is restarted too quickly, rotation may continue in the opposite direction. Unlike reciprocating compressors, scroll compressors cannot pump properly when reversed. If rotating in the wrong direction, they will pump poorly and overheat. It is important to note that three-phase models must be properly phased to ensure proper rotation.

6.0.0 ◆ MECHANICAL REFRIGERATION CYCLE TROUBLESHOOTING

When troubleshooting the mechanical refrigeration cycle, it is sometimes difficult to determine the cause of a problem. Many times, what seems to be the problem is only a symptom. Also, many problems produce the same or similar symptoms. Isolating the problem requires a logical, systematic approach, which generally involves the following:

- Preliminary inspection
- Analyzing system operating conditions
- Using manufacturers' troubleshooting aids
- Troubleshooting system components

6.1.0 Preliminary Inspection

Preliminary inspection of the system is performed by using your senses of sight, sound, touch, and smell to identify possible problems with the system or compressor. The preliminary inspection is performed as follows:

 WARNING!
Be sure all electrical power to the equipment is turned off. Open, lock, and tag disconnects. Watch out for pressurized or hot components. Follow all safety instructions labeled on the equipment and given in the manufacturer's service manual for the equipment.

Step 1 Turn off the equipment. Lockout and tag equipment so it cannot be energized.

Step 2 Look for an evaporator or condenser mounted above the compressor that might dump liquid refrigerant into the compressor.

Step 3 Look for the following piping problems:

- Oversized or undersized refrigerant lines or excessive elbows or fittings
- Long or uninsulated suction line which might develop excessive superheat
- Liquid line running through an unconditioned space (hot or cold) which might affect subcooling
- Buried lines which might cause refrigerant to condense
- Extremely long liquid line which might hold an excessive amount of refrigerant

Step 4 At the evaporator and condenser, check for the following:

- Fin collars corroded
- Fins or coils dirty or damaged
- Supply plenum dirty
- Filters dirty or missing
- Fan belts at improper tension
- Blowers and fans dirty
- Incorrect blower speed
- Evaporator shows signs of freezing up

Instructor's Notes:

Step 5 As applicable, at the compressor:
- Check that the service valves are fully open.
- Check that the hold-down bolts are loosened or unloosened per manufacturer's instructions.
- Check that the crankcase heater is working.
- On open or semi-hermetic compressors, inspect cylinder heads to see if they are scorched or blistered from excessive heat.
- Inspect for rust streaks indicating condensation from cold return gas.
- Check that the oil level is at the proper height in the sightglass.

6.2.0 Analyzing System Operating Conditions

In order to tell if the system or compressor has a problem, the actual conditions that exist in the system must be known. If the compressor is operable, system operation should be monitored and the critical parameters measured. This is necessary so that the actual conditions can be compared against a set of normal system operating parameters in order to determine if there is a problem, and where it is.

6.2.1 System Operation Checks

Check the system operating conditions and record the values for system parameters as described here.

WARNING!
Watch out for rotating, pressurized, or hot components. Follow all safety instructions labeled on the equipment and given in the manufacturer's service manual for the equipment.

Step 1 Connect a gauge manifold set to the system gauge ports. Install thermometers on the suction line at the compressor input, on the hot gas discharge line, and on the liquid line at the input to the expansion device.

Step 2 Turn on the system and adjust the thermostat to call for cooling.

Step 3 Start the system and listen for abnormal sounds.
- Check for excessive vibration of the compressor, piping, motors, and fans.
- Note any compressor knocks or rattles, which may indicate liquid refrigerant is being drawn into the cylinder(s). If this condition continues for more than a few seconds, shut down the system and look for the cause of excess liquid return.

Step 4 Visually inspect the operating compressor for abnormalities. Check the compressor oil sightglass (if equipped). Heavy foaming at the sightglass should clear 5 to 10 minutes after startup. If not, there may be excessive refrigerant in the oil.

WARNING!
Danger exists if the compressor terminals are damaged and the system is pressurized. Disturbing the terminals to take measurements could cause them to blow out, causing injury. When making voltage, current, or continuity checks on a hermetic or semi-hermetic compressor in a pressurized system, always take measurements at terminal boards and test points away from the compressor.

Step 5 Measure and record the input voltage and current at the compressor contactor.
- The measured voltage should be within ±10 percent of the motor nameplate value.
- In a three-phase motor, the voltage imbalance between any two phases should not exceed 2 percent. Any current imbalance between phases should not exceed 10 percent.

Step 6 Check that all fans, motors, and pumps are operational and moving the proper amounts of air or water.

Step 7 As applicable, measure and record, or calculate, the following operating parameters:
- Indoor bulb and wet-bulb temperatures
- Outdoor dry-bulb temperature
- Suction pressure
- Saturated suction temperature
- Suction line temperature
- Superheat
- Discharge pressure
- Saturated discharge temperature
- Discharge line temperature
- Subcooling
- Liquid line temperature entering the expansion device

Show trainees how to perform a preliminary inspection.

Explain how to make basic system operating checks.

Reinforce safety practices for troubleshooting equipment with power supplied to it. Emphasize the danger of rotating parts.

Emphasize that compressor voltage checks should only be made at terminal strips and test points away from the compressor terminals. Point out that under pressure, damaged terminals could blow out.

Demonstrate how to make basic system operating checks.

Show Transparency 12 (Table 1). Explain how system conditions can be compared to normal system operating parameters to identify problems.

Show Transparency 13 (Figure 11). Explain how to compare the system operating conditions with the manufacturer's data to locate a malfunction in a cooling system.

- Oil pressure (semi-hermetic compressors)
- Compressor temperature at the bottom of the cylinder heads (semi-hermetic compressors)
- Compressor temperature at the top and bottom of the motor barrel (semi-hermetic compressors)
- Crankcase or shell temperature
- Compression ratio

6.2.2 Analyzing System Conditions

Once the actual conditions that exist in a system are known, they can be compared against a set of normal system operating parameters to determine if there is a problem. *Table 1* shows an example of the types of system parameters that should be recorded for the purpose of comparison. For this example, the data shown reflect the normal operation parameters for the system described in *Figure 2* at a 95°F outdoor temperature. It's important to remember that readings vary widely among systems because of equipment application, ambient conditions, and type of refrigerant used. For the specific equipment you are servicing, refer to any existing operating logs or the manufacturer's service manual to find typical readings. Better yet, if you service the same kind of systems for various customers, or periodically service the same system for one customer, it is recommended that you capture the data listed in *Table 1* when the system is operating properly. This data will prove to be a valuable reference source for later use should it become necessary to troubleshoot the system.

6.3.0 Manufacturer's Troubleshooting Aids

Based on the symptoms revealed by your analysis of system operation, consult the manufacturer's troubleshooting aids to identify the likely causes of the problem. Most refrigeration cycle problems result in abnormal system pressures. *Figure 11* summarizes the common system problems that relate to high and low head pressures and low suction pressures. Note that a low head pressure in combination with a high suction pressure and low compressor current normally indicates that the compressor has leaky valves or worn piston rings and should be replaced.

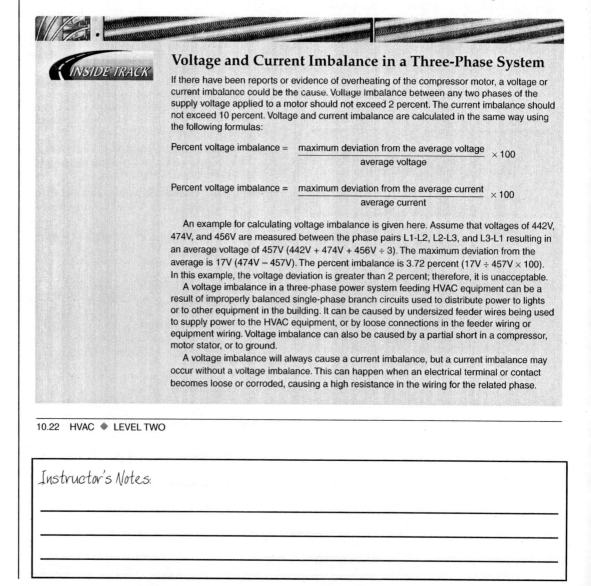

INSIDE TRACK

Voltage and Current Imbalance in a Three-Phase System

If there have been reports or evidence of overheating of the compressor motor, a voltage or current imbalance could be the cause. Voltage imbalance between any two phases of the supply voltage applied to a motor should not exceed 2 percent. The current imbalance should not exceed 10 percent. Voltage and current imbalance are calculated in the same way using the following formulas:

$$\text{Percent voltage imbalance} = \frac{\text{maximum deviation from the average voltage}}{\text{average voltage}} \times 100$$

$$\text{Percent voltage imbalance} = \frac{\text{maximum deviation from the average current}}{\text{average current}} \times 100$$

An example for calculating voltage imbalance is given here. Assume that voltages of 442V, 474V, and 456V are measured between the phase pairs L1-L2, L2-L3, and L3-L1 resulting in an average voltage of 457V (442V + 474V + 456V ÷ 3). The maximum deviation from the average is 17V (474V − 457V). The percent imbalance is 3.72 percent (17V ÷ 457V × 100). In this example, the voltage deviation is greater than 2 percent; therefore, it is unacceptable.

A voltage imbalance in a three-phase power system feeding HVAC equipment can be a result of improperly balanced single-phase branch circuits used to distribute power to lights or to other equipment in the building. It can be caused by undersized feeder wires being used to supply power to the HVAC equipment, or by loose connections in the feeder wiring or equipment wiring. Voltage imbalance can also be caused by a partial short in a compressor, motor stator, or to ground.

A voltage imbalance will always cause a current imbalance, but a current imbalance may occur without a voltage imbalance. This can happen when an electrical terminal or contact becomes loose or corroded, causing a high resistance in the wiring for the related phase.

Table 1 Typical Temperature-Pressure Data for the Example Shown in Figure 2

System Parameter	Typical Operational HCFC-22 Air Conditioning System	Typical Operational HFC-410a Air Conditioning System	Air Conditioning System Under Test
Suction Pressure (psig)	69 psig	118 psig	
Suction Line Temperature (°F)	52°F	52°F	
Saturated Suction Temperature (°F)	40°F	40°F	
Superheat (°F) – TXV	12°F	12°F	
Discharge Pressure (psig)	280 psig	460 psig	
Discharge Line Temperature (°F)	200°F	200°F	
Saturated Discharge Temperature (°F)	130°F	130°F	
Liquid Temperature Leaving Condenser (°F)	125°F	125°F	
Liquid Line Temperature Entering Expansion Device (°F)	120°F	120°F	
Subcooling (°F)	10°F	10°F	
Air Temperature Entering Evaporator (DB°F)	75°F	75°F	
Air Temperature Leaving Evaporator (DB°F)	55°F	55°F	
Air Temperature Drop Across Evaporator (DB°F)	20°F	20°F	
Outdoor Temperature Rise Across Condenser (DB°F)	20°F	20°F	
Compressor Compression Ratio (Absolute Discharge Pressure/Absolute Suction Pressure)	3.76 to 1	3.72 to 1	
Compressor Oil Pressure (psig)	86 psig	86 psig	

Explain how the process of elimination is used to isolate complicated system problems.

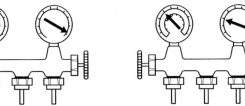

HIGH HEAD PRESSURE

1. Air or noncondensibles in the system
2. Obstructions in the condenser coil fins, such as dirt, etc.
3. Overcharge of refrigerant
4. Recirculation of condenser air
5. Higher-than-ambient temperature air entering condenser
6. Wrong rotation of condenser fan blade

LOW HEAD PRESSURE

1. Low refrigerant charge
2. Defective compressor valves
3. Low ambient temperature

LOW SUCTION PRESSURE

1. Loose or broken evaporator blower belt
2. Defective or overloaded blower belt
3. Obstructed or dirty evaporators
4. Dirty air filters
5. Low refrigerant charge
6. Dirty or faulty expansion valve (leaking around push rod)
7. Recirculation of evaporator air (compare return air temperature with conditioned space temperature)
8. Restriction in refrigerant system
9. Restricted or undersized duct work
10. Wrong rotation of evaporator blower

Figure 11 ♦ Causes of abnormal refrigerant pressure.

6.4.0 Troubleshooting System Components

The information given in manufacturers' troubleshooting aids usually isolates the problem to the specific failed component, especially when the problem is related to the evaporator, condenser, and the air or water flow circuits. However, for more difficult problems, the troubleshooting aid will only give potential cause(s) for a particular symptom. For example, the cause may be listed as a loss of charge, restriction in the refrigeration system, or air or noncondensibles in the system. For this type of problem, more system analysis using a process of elimination is usually needed to isolate the

Show trainees how to perform troubleshooting on mechanical refrigeration equipment with preset faults.

Have the trainees review Sections 7.0.0–10.3.0.

Ensure that you have everything required for teaching this session.

Identify symptoms associated with low and no refrigerant charge.

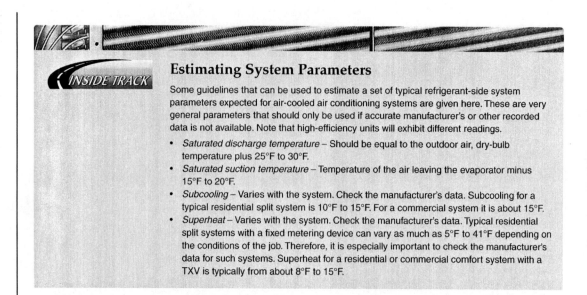

Estimating System Parameters

Some guidelines that can be used to estimate a set of typical refrigerant-side system parameters expected for air-cooled air conditioning systems are given here. These are very general parameters that should only be used if accurate manufacturer's or other recorded data is not available. Note that high-efficiency units will exhibit different readings.

- *Saturated discharge temperature* – Should be equal to the outdoor air, dry-bulb temperature plus 25°F to 30°F.
- *Saturated suction temperature* – Temperature of the air leaving the evaporator minus 15°F to 20°F.
- *Subcooling* – Varies with the system. Check the manufacturer's data. Subcooling for a typical residential split system is 10°F to 15°F. For a commercial system it is about 15°F.
- *Superheat* – Varies with the system. Check the manufacturer's data. Typical residential split systems with a fixed metering device can vary as much as 5°F to 41°F depending on the conditions of the job. Therefore, it is especially important to check the manufacturer's data for such systems. Superheat for a residential or commercial comfort system with a TXV is typically from about 8°F to 15°F.

problem. The remainder of this module provides guidelines to use when troubleshooting these problems.

7.0.0 ◆ LOW CHARGE OR OVERCHARGE OF REFRIGERANT

The refrigerant charge in a system must be neither too low nor too high. Too much or too little refrigerant causes various system problems.

7.1.0 Low Refrigerant Charge

Loss of refrigerant is a common problem encountered when servicing cooling systems. The customer usually complains of steadily deteriorating cooling performance. Low refrigerant results from leaks that occur as a result of poor installation practices, physical damage, or a factory defect. Suction and discharge pressures will be lower than normal with high superheat. If the system has a sightglass, it will show bubbles. Systems with an automatic expansion valve may not

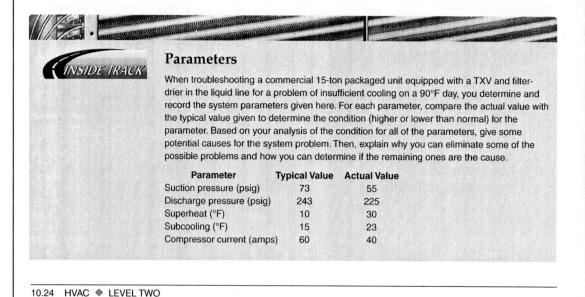

Parameters

When troubleshooting a commercial 15-ton packaged unit equipped with a TXV and filter-drier in the liquid line for a problem of insufficient cooling on a 90°F day, you determine and record the system parameters given here. For each parameter, compare the actual value with the typical value given to determine the condition (higher or lower than normal) for the parameter. Based on your analysis of the condition for all of the parameters, give some potential causes for the system problem. Then, explain why you can eliminate some of the possible problems and how you can determine if the remaining ones are the cause.

Parameter	Typical Value	Actual Value
Suction pressure (psig)	73	55
Discharge pressure (psig)	243	225
Superheat (°F)	10	30
Subcooling (°F)	15	23
Compressor current (amps)	60	40

have low suction pressure because the valve may be operating wide open, maintaining the setpoint for evaporator pressure. In extreme cases, condenser subcooling will be very low because there is little refrigerant to subcool.

Once you determine the cause of a system problem is a low charge, the first thing to do is to find and repair any leaks that may exist in the system. The methods used for leak detection, and the subsequent evacuation/dehydration and charging of the system, are covered in *Leak Detection, Evacuation, Recovery, and Charging*. Never simply add refrigerant to a system without repairing a leak:

- The release of refrigerant into the atmosphere damages the environment.
- Failure to repair a leak may violate EPA laws. Remember that a calculated annual leak rate of 15 percent is the threshold of repair for comfort cooling chillers and all other equipment with a charge over 50 pounds per circuit. Industrial and commercial refrigeration systems require leak repair if they leak at a rate greater than 35 percent.
- Repeated service calls to recharge a leaking system, and the failure to repair the problem, could result in customer frustration and dissatisfaction.

7.2.0 Overcharge of Refrigerant

The symptoms of refrigerant overcharge are similar from system to system. The discharge pressure is high, and the suction pressure may be high in systems equipped with a thermal expansion valve, capillary tube, or fixed-orifice metering device. The superheat will be low. A system with an automatic expansion valve will not have a high suction pressure because it maintains a constant suction pressure. If the discharge pressure is high enough in a capillary system, the capillary tube may allow enough refrigerant to pass so that it will cause **flooding** of the compressor. Once you determine that excess refrigerant charge is the cause of a system problem, recover the excess refrigerant as needed to achieve a correct system charge.

8.0.0 ◆ EVAPORATOR AND CONDENSER AIRFLOW PROBLEMS

Both the evaporator and the condenser may experience airflow problems. These must be corrected for the system to operate properly.

8.1.0 Evaporator Airflow Problems

To perform properly, a blower in a cooling system must move about 350 to 450 cubic feet per minute of air per ton of cooling across the evaporator coil. Too much or too little air can cause comfort problems. Too much air across the evaporator coil usually results in poor humidity control. Decreasing the blower speed usually will correct this problem. A dirty evaporator coil will cause low suction pressure and low superheat.

The problem of too little airflow is more common. Low airflow is generally indicated in refrigerant circuit performance by low suction pressures and, especially with fixed metering devices, a low superheat condition. The airflow

Discuss the consequences of topping off a refrigerant charge without identifying the source of the leak.

Discuss some of the causes of evaporator airflow problems.

Discuss the effects of refrigerant undercharge. See the answer to the "Think About It" at the end of this module.

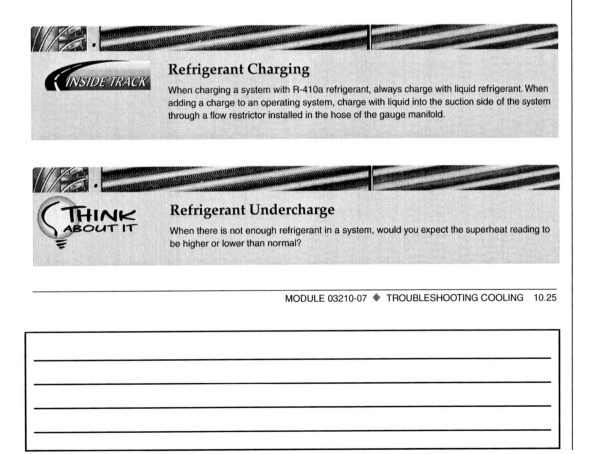

Refrigerant Charging

When charging a system with R-410a refrigerant, always charge with liquid refrigerant. When adding a charge to an operating system, charge with liquid into the suction side of the system through a flow restrictor installed in the hose of the gauge manifold.

Refrigerant Undercharge

When there is not enough refrigerant in a system, would you expect the superheat reading to be higher or lower than normal?

Discuss some of the causes of condenser airflow problems.

Explain the difference between flooding and slugging. Discuss some of the causes of liquid slugging in a system.

Point out that slugging and flooding can ruin even brand new compressors. Something as simple as not connecting the crankcase heater in a new installation can result in compressor failure in as little as a couple of days, depending on the conditions.

Discuss the effects of a dirty condenser coil. See the answer to the "Think About It" at the end of this module.

can be increased by increasing blower speed, providing that none of the following conditions are the real cause of the reduced airflow:

- The system air filter, blower wheel, or evaporator coil is dirty.
- The blower is rotating in the wrong direction.
- The belts on belt-driven blowers are loose or worn or the belt tension is incorrect.
- There is loose insulation in the ductwork.

Electronically controlled, variable-speed motors are being used more and more in evaporator blowers. When checking airflow in a system using a variable-speed blower-motor, always ensure that the blower is operating at its maximum speed before checking airflow. During many normal operating modes, variable-speed blower motors do not operate at their maximum speed.

8.2.0 Condenser Airflow Problems

To operate properly, the condenser coil must be able to reject heat absorbed by the refrigerant in the evaporator. When this does not happen, the system pressures and temperatures build up and, if severe enough, will cause protective devices to shut the compressor off. In the worst case, this can be caused by a failed condenser fan or related fan circuit components. Other possible causes for poor condenser airflow are as follows:

- The condenser fan is rotating in the wrong direction.
- The fan blade is not located properly in the fan orifice.
- The wrong fan blade has been used, or the pitch of a replacement blade is incorrect.
- Hot condenser discharge is being circulated to the coil instead of being carried away. This is usually caused by the unit being installed too close to buildings, or under decks or overhangs.

Dirty Condenser Coil

A system with a dirty condenser will have higher than normal discharge and suction pressure readings. Why?

- Higher-than-ambient air temperature is entering the condenser coil from other warm air sources, such as exhaust vents, dryer vents, and adjacent condensing units.
- Grass clippings or other airborne debris are present on the surface of the coil, reducing airflow.
- Airflow is blocked by shrubs and plants.
- The replacement motor has the wrong rpm.

9.0.0 ♦ COMPRESSOR PROBLEMS AND CAUSES

Compressor problems include **slugging**, flooding, **flooded starts**, and overheating. Damaged valves and rings may also cause problems with compressors.

9.1.0 Slugging

Slugging occurs when a compressor tries to compress liquid refrigerant, oil, or both, instead of superheated gas. If slugging occurs, it will occur at startup or during a rapid change in system operating conditions. It can sometimes be detected by a periodic knocking noise at the compressor. Common causes of slugging include the following:

- The system has an overcharge of refrigerant.
- The TXV is oversized or damaged, or the sensing bulb is loose.
- The system has an overcharge of oil.
- The crankcase heater is open.
- Condensed refrigerant is present in any cold part of the system, such as the evaporator, during the off cycle. Buried refrigerant lines or lines passing through cold spots can allow the refrigerant to condense back into a liquid at shutdown.

INSIDE TRACK — Slugging

Burying refrigerant lines may sound like a good idea, but it is not. Since it is usually cooler underground, refrigerant in buried lines will condense during shutdown and cause slugging when the compressor starts. For this reason, most HVAC equipment manufacturers prohibit or severely restrict the use of buried refrigerant lines with their equipment.

Inside Track

Evaporator Coil

This evaporator coil is typical of those used in the indoor unit (furnace or fan coil unit) of a split air conditioning system.

- Slugs of oil are trapped in the suction line because the suction gas does not have enough velocity to return the oil to the compressor. Normally, the oil and refrigerant mix. The oil is circulated through the system in very small drops as it is being swept along by the velocity of the refrigerant vapor. If it gets trapped in the system piping and returns all at once, it can cause slugging. This condition tends to be found in the suction line of built-up systems. It also occurs in systems that use compressors with unloaders, especially when the compressor runs unloaded for long periods of time.
- Poor piping design allows liquid refrigerant to return to the compressor during the off cycle.

9.2.0 Flooding

Flooding is caused by the continuous return of liquid refrigerant or liquid droplets in the suction vapor to the compressor during operation. Flooding dilutes the oil, resulting in crankcase foaming and overheating of bearing surfaces. If bad enough, it can damage the pistons, rings, and valves because liquid refrigerant acts like a solvent and washes the oil off the bearing surfaces. Flooding is most often caused by the uncontrolled flow of refrigerant that results from one or more of the following conditions:

- There is an oversized thermostatic expansion valve.
- The thermostatic expansion valve sensing element is broken, mislocated, making poor contact, or improperly insulated.

Explain that flooding is a continuous return of liquid to the compressor.

Discuss some of the causes of flooding in a system.

Explain that flooded starts are caused by oil in the compressor crankcase and explain how to minimize the problem.

Explain how overheating can lead to compressor failure.

Explain how to check and service the valves and rings on a compressor.

Under your supervision, have the trainees modify a system's refrigerant charge and obstruct the condenser and evaporator airflow. Note the changes to system operating parameters.

- The superheat setting is too low.
- There is low load on the evaporator caused by low airflow. Reduced airflow often causes frosting of the coil, which adds to the problem. Possible causes of restricted airflow include dirty filters, air restriction, dirty fan wheels, and loose or broken drive belts.
- There is an overcharge of refrigerant in systems that use fixed-orifice metering devices. Since fixed metering devices do not react to load change, an overcharge of refrigerant can raise the head pressure, which can increase the flow rate to a point at which there is more flow than available heat transfer.
- Poor piping design can contribute at startup.

9.3.0 Flooded Starts

Flooded starts are caused by the oil in the compressor crankcase absorbing refrigerant. This is more relevant during long compressor shutdowns at low temperature. Refrigerant will migrate to and condense in the cold compressor shell, where it mixes with the oil. Oil will absorb refrigerant under most conditions. The amount absorbed depends on the temperature of the oil and the pressure in the crankcase. On startup, the refrigerant-rich oil mixture is pumped through the oil pump of the compressor, resulting in marginal lubrication of the bearings. As the crankcase pressure drops after startup, the refrigerant will flash from a liquid to a gas, causing foaming. This foaming can restrict the oil passages and cause the oil pressure to build. It can also cause hydraulic slugging. The problem of flooded starts can be minimized by making sure that the system has a proper refrigerant charge and the correct amount of oil in the crankcase. Make sure the crankcase heater is working, or add one if necessary. Crankcase heaters are used to raise the temperature of the oil during shutdown and prevent the condensation of refrigerant. Piping should also be designed to prevent liquid from accumulating and flooding the compressor on startup.

9.4.0 Overheating

Compressors normally generate heat and are designed to handle this normal heat. When the compressor is overheated, the cause must be determined. High superheat, lubrication problems, condenser problems, or electrical problems are all possible causes. Temperatures between 275°F and 300°F in the hot gas discharge line cause oil and refrigerant to break down, with the potential for compressor failure.

9.5.0 Valves and Rings

In a reciprocating compressor, valves and rings provide a seal between the high-pressure and low-pressure sides. If they are damaged, the compressor must be replaced or the valves repaired (open or serviceable hermetic). Suction and discharge pressure checks can be used to test for this condition. Bad valves or rings may exist if the suction pressure will not pull down or discharge pressure will not build up when the system is properly charged and under normal load. As described earlier in this module, another check can be made by measuring the running current with an ammeter. If, under loaded conditions, the running current is considerably lower than normal and the suction/discharge pressures are abnormal, faulty valves or rings should be suspected. If the compressor has unloaders, make sure they are not activated.

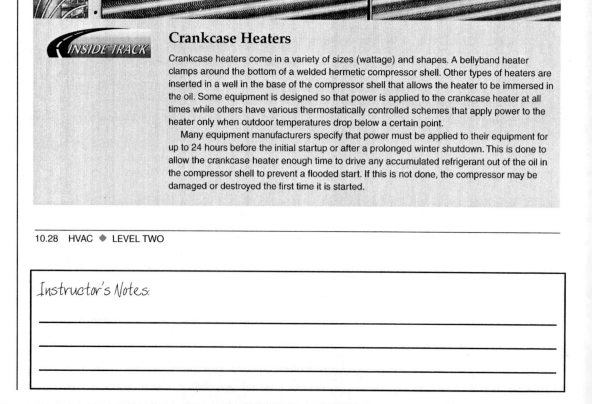

Crankcase Heaters

Crankcase heaters come in a variety of sizes (wattage) and shapes. A bellyband heater clamps around the bottom of a welded hermetic compressor shell. Other types of heaters are inserted in a well in the base of the compressor shell that allows the heater to be immersed in the oil. Some equipment is designed so that power is applied to the crankcase heater at all times while others have various thermostatically controlled schemes that apply power to the heater only when outdoor temperatures drop below a certain point.

Many equipment manufacturers specify that power must be applied to their equipment for up to 24 hours before the initial startup or after a prolonged winter shutdown. This is done to allow the crankcase heater enough time to drive any accumulated refrigerant out of the oil in the compressor shell to prevent a flooded start. If this is not done, the compressor may be damaged or destroyed the first time it is started.

Instructor's Notes:

Troubleshooting Overheating Problems

If a system is undercharged, it will overheat and cut off on its overload protection device.

If a compressor overheats due to lubrication problems, check for the following conditions, which will be indicated by high discharge temperatures and possible overload tripping:

- Low suction pressure
- High condensing pressures
- High compression ratios

Low suction pressure is normally the result of an undercharge condition, an incorrect pressure switch setting, a drop in suction line pressure, light load operating conditions, or a restricted evaporator coil.

High condensing pressures can be caused by inadequate air or water flow through the condenser, an undersized condenser coil, a refrigerant overcharge, or the presence of noncondensibles in the system.

High compression ratios are caused by a combination of low suction pressure and high condensing pressure.

Scroll Compressor Problems

If the valves or rings in a reciprocating compressor go bad, it will result in an inefficient compressor. Typical indicators are low discharge pressure combined with high suction pressure, which is the opposite of normal operation. If the same combination of symptoms occurs in a system with a scroll compressor, which has no valves at all, it could indicate there is a crack in the orbiting scroll plate.

Explain that the troubleshooting of thermostatic expansion valve (TXV) problems should always begin with a measurement of superheat.

Identify problems often associated with TXVs.

Explain that flooding can be caused by a TXV stuck in the open position. Identify other causes of flooding.

Discuss compressor current draw. See the answer to the "Think About It" at the end of this module.

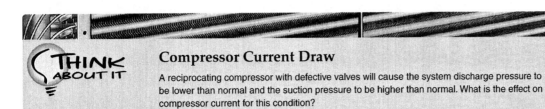

Compressor Current Draw

A reciprocating compressor with defective valves will cause the system discharge pressure to be lower than normal and the suction pressure to be higher than normal. What is the effect on compressor current for this condition?

10.0.0 ◆ METERING DEVICE TROUBLESHOOTING

Faulty metering devices may cause a variety of system problems. Metering devices include thermostatic expansion valves (TXVs), fixed-orifice devices, and capillary tubes.

10.1.0 Thermostatic Expansion Valves

TXVs are often blamed for system problems caused by any number of other faults. Troubleshooting TXV problems should always begin with a measurement of superheat. If too little refrigerant is fed to the evaporator, the superheat is high. If too much refrigerant is fed to the evaporator, the superheat is low. Although these symptoms may be thought to be caused by improper TXV control, more often the problem is elsewhere. Problems often blamed on TXVs are:

- Overfeeding (flooding)
- Underfeeding (starving)
- Erratic operation (hunting)

Flooding can be caused by the valve sticking in an open position. It may also be caused by the following valve or system problems:

- Improper superheat on the valve
- Improper location of the sensing bulb on the suction line
- Loose or poor thermal contact of the sensing bulb

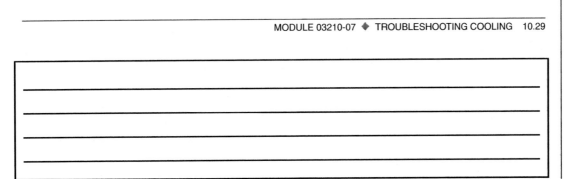

Show Transparency 14 (Figure 12). Explain TXV hunting.

Identify factors that will cause excessive hunting.

Identify installation errors that will cause system problems, and discuss how to correct them.

- A very light system load
- Excess oil in the system
- Wrong type of valve for the system refrigerant

Starving can be caused by the valve sticking closed, or by a partial loss of charge in the power element. The coil can be starved for the following reasons:

- Shortage of refrigerant
- Improper superheat setting or incorrect TXV selection (size)
- Plugged drier
- Plugged refrigerant distributor
- Improper valve location
- Plugged equalizer line

Hunting describes the changes in refrigerant flow as the TXV adapts to changing conditions. By their design, all valves hunt to some degree, since a valve tries to control the superheat of the gas leaving the evaporator by controlling the flow of liquid refrigerant entering the evaporator.

Hunting can only be verified by making several suction line temperature or pressure measurements over a period of time. If there is a repetitive pattern, such as that shown in *Figure 12*, hunting is occurring. Excessive hunting can be caused by sticking or binding of the TXV internal parts. It can also occur if the system load on the TXV drops below 30 percent. Other than faults in the valve itself, the following can cause excessive hunting:

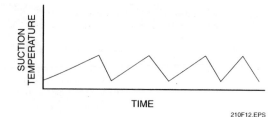

Figure 12 ♦ Suction temperature variations that indicate hunting.

- Oversized expansion valve
- Very light load
- Long refrigerant line
- Rapid changes in condensing pressure or temperature
- Rapid load changes
- Intermittent flashing in the liquid line

Some system problems caused by TXVs are the result of poor installation. Make sure the sensing bulb is securely fastened on the top of a clean, straight section of the suction line, close to the evaporator outlet. Also, check the insulation on the bulb to make sure outside air temperatures are not affecting it.

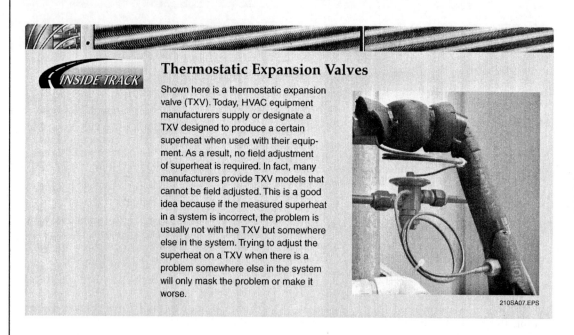

Thermostatic Expansion Valves

Shown here is a thermostatic expansion valve (TXV). Today, HVAC equipment manufacturers supply or designate a TXV designed to produce a certain superheat when used with their equipment. As a result, no field adjustment of superheat is required. In fact, many manufacturers provide TXV models that cannot be field adjusted. This is a good idea because if the measured superheat in a system is incorrect, the problem is usually not with the TXV but somewhere else in the system. Trying to adjust the superheat on a TXV when there is a problem somewhere else in the system will only mask the problem or make it worse.

TXV Sensing Bulbs

If the sensing bulb for a TXV is disconnected from the suction line, how will the TXV respond?

10.2.0 Distributors

A distributor is a device installed downstream of the metering device (usually a TXV) and is designed to ensure that refrigerant leaving the TXV gets evenly distributed to the various circuits in the evaporator coil. Distributors (*Figure 13*) do not function well at very low loads. Low-load conditions may be caused by oversizing. If a distributor load falls below 50 percent, the liquid-vapor mixture entering the evaporator may separate. The liquid will go to the bottom circuits and the vapor will go to the top circuits. Excessive superheat and liquid floodback to the compressor will result. A way to check for this problem is to look for high temperatures at the top of the evaporator and low temperatures at the bottom.

To correct this type of problem, it may be necessary to replace the nozzle with the next smaller size or arrange the controls to shut off an evaporator section when the load is below 50 percent.

10.3.0 Fixed-Orifice Devices and Capillary Tubes

Two problems occur with fixed-orifice devices and capillary tubes: they are the wrong size for the application, or they become partially or completely restricted. Normally, the sizes of both devices are factory-selected for use with a system. If a capillary tube is too long or the diameter too small, the evaporator will be starved for refrigerant, and excess liquid will build up in the condenser. The effect initially will be high head pressure and inadequate cooling. If the unit runs for awhile, the liquid trapped in the condenser will cool and the pressure will drop to normal. Use of an undersized fixed-orifice device will have the same effect. If a capillary tube is too short or the diameter too great, excess liquid will be fed to the evaporator. The same is true with an oversized fixed-orifice device. This could cause slugging at the compressor.

For any given pressure difference across a capillary tube, it can pass more weight of liquid

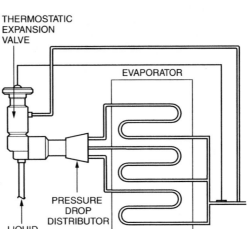

Figure 13 ◆ Multi-circuit evaporator fed by a distributor.

refrigerant than weight of gas because the liquid has a much greater density. This makes a capillary tube very sensitive to excess oil, excess subcooling, or flashing. Excess oil in circulation displaces some refrigerant in the capillary tube and reduces the actual weight of refrigerant flowing through the tube. This can starve the evaporator. Excess flashing due to a shortage of gas, liquid line restriction, or a capillary tube contacting a hot surface, will reduce the amount of liquid the tube can pass and also result in a starved evaporator coil.

Excess subcooling can reduce flashing in the capillary tube to such a point that too much refrigerant flows through the tube, causing flooding. Because of the risk to equipment reliability and the need to meet higher energy efficiency standards, one major equipment manufacturer has abandoned the use of fixed-orifice metering devices on the indoor coils of residential split-system products and has switched entirely to TXVs.

Ensure that you have everything required for teaching this session.

Show Transparency 16 (Figure 14). Explain how to confirm a line restriction.

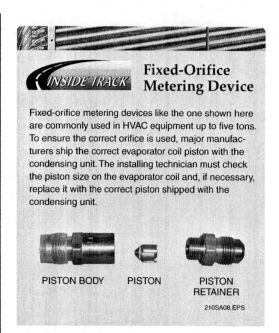

Fixed-Orifice Metering Device

Fixed-orifice metering devices like the one shown here are commonly used in HVAC equipment up to five tons. To ensure the correct orifice is used, major manufacturers ship the correct evaporator coil piston with the condensing unit. The installing technician must check the piston size on the evaporator coil and, if necessary, replace it with the correct piston shipped with the condensing unit.

PISTON BODY   PISTON   PISTON RETAINER

11.0.0 ♦ TROUBLESHOOTING REFRIGERANT LINES AND ACCESSORIES

A restriction in a refrigerant liquid line or related accessory (*Figure 14*) reduces or prevents the system refrigerant from getting to (or through) the evaporator. This results in low system capacity.

The cause of problems in a system with a restriction is often diagnosed as the system being low on refrigerant charge. This is because the symptoms of low suction and discharge pressures and high superheat are similar. The subcooling temperature can be used to help determine if the problem is a restriction rather than a low charge. Systems with a restriction in the liquid line usually show normal or increased subcooling. This is because the mass flow of refrigerant in the system slows down. Because of the restriction, most of the system liquid refrigerant remains in the condenser and/or receiver and subcools. The liquid refrigerant stays in the condenser longer, resulting in greater subcooling because it has a longer time to be cooled.

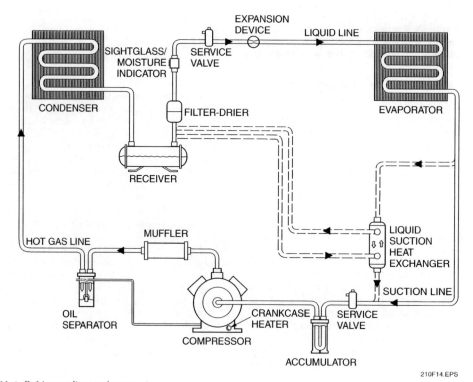

Figure 14 ♦ Refrigerant lines and accessories.

10.32   HVAC ♦ LEVEL TWO

Inside Track

Capillary Tubes

Capillary tubes, like the one shown here, were once widely used in systems up to five tons. With the introduction of the fixed-orifice metering device, the capillary tube is now mainly used in room air conditioners and packaged terminal air conditioners (PTACs). They also continue to be used in domestic refrigerators and freezers.

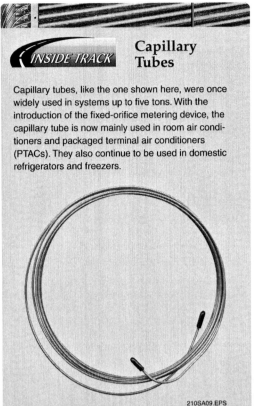

Inside Track

Refrigerant Line Restrictions

Split HVAC systems (evaporator and condenser in different locations) are susceptible to refrigerant line restrictions and damage because the field-installed interconnecting tubing may be routed through areas where they are subject to accidents, abuse, or vandalism. Packaged HVAC products have all refrigerant lines contained within the security of the package so they are much less susceptible to damage.

Liquid Line Restrictions

Liquid and suction line service valves on the condensing unit of a split system can be the source of restrictions. This is because at installation or during some service procedures, technicians may forget to fully open one or both of the valves, causing restriction. Some condensers also have a liquid line strainer built into the condenser. This strainer, which can be easily overlooked by the technician, can become restricted.

It is a good practice to consider any accessories that are installed in a system as a potential cause of a restriction and check them for an abnormal temperature or pressure drop when troubleshooting. Refrigerant tubing can become restricted as a result of outside physical damage that causes the tubing to be pinched or bent. This often occurs on systems that have been running normally for a long time. Physical damage to the tubing often goes undetected because it can be hidden under insulation or behind a fixture.

When a partial restriction occurs in a liquid line, it acts like a metering device and drops the pressure and temperature of the refrigerant at the point of the restriction. A restriction could be detected by measuring the pressure drop across the accessory or tubing suspected to have the restriction; however, this is not always practical because gauge ports are not usually available. The easiest way is to measure the temperature drop on each side of the accessory or tubing suspected to contain the restriction.

Depending on the severity of the restriction, you can sometimes feel a temperature change in the line or observe frost or sweat on the line. If the restriction is very slight, temperature measurement is usually the only way to find the location of the restriction.

A plugged or undersized filter-drier is a common cause of a restriction. On a newly installed system, the restriction may occur shortly after startup as a result of being plugged by solid contaminants that entered the system during installation. If a system is equipped with a sightglass installed downstream of the filter-drier, a restricted filter-drier will cause flash gas (bubbles) to appear in the sightglass. To confirm that a filter-drier is restricted, always measure the temperature drop across it, since a restricted filter-drier can go undetected using the touch method. The amount of pressure drop across a filter-drier can also be an indication of a restriction. The pressure drop should not exceed 5 psi or 2 degrees.

Define noncondensibles as air and water in a refrigeration system.

Identify the ways that noncondensibles enter the system.

List the symptoms of system contamination.

Explain how acids are formed inside a refrigeration system.

Restrictions can be caused by valves that do not open all the way. Normally, TXVs either work or do not work. However, a partial restriction can occur in the valve if it loses part of the charge in the thermal bulb, or the inlet strainer becomes dirty. Also, if moisture is present and circulating in a system that operates below freezing, it can freeze at the expansion device and restrict refrigerant flow.

You can test for loss of charge in the bulb by holding the bulb in your hand. If the valve fails to feed the evaporator more refrigerant, the bulb has probably lost its charge. Another potential restriction point in split-system products is the service valves. During normal operation, ensure that both the liquid line and suction line service valves are 100 percent open. During installation, technicians sometimes fail to open these valves fully, causing a restriction. During certain service procedures these valves must be closed, but make sure to open one or both valves 100 percent after the service procedure is complete.

12.0.0 ♦ NONCONDENSIBLES AND CONTAMINATION IN A SYSTEM

Refrigeration systems should contain only refrigerant and oil. Anything else is considered a contaminant.

12.1.0 Air and Moisture Contamination

Air and moisture are the most common contaminants in a refrigeration or air conditioning system. Air is a **noncondensible gas**, which means that it does not change into a liquid at operating temperatures and pressures. Air entering the system can also contain moisture in the form of water vapor. When air is in a system, it accumulates in the condenser, taking up space needed for condensing the refrigerant. This results in high head pressure, an increase in the condensing temperature, and high subcooling temperatures. Under the heat of compression, moisture will react with the refrigerant to form acids. These acids cause corrosion of metals and breakdown of the insulation on the motor windings and wiring. Moisture in the refrigerant can also cause oil sludge, which reduces the lubricating properties of the oil and plugs oil passages and screens in the compressor. Moisture can also freeze at the expansion device.

Air and moisture can enter a system by the following:

- Improper evacuation/dehydration at installation or after system servicing
- Refrigerant system leaks under a vacuum
- Failure to install a filter-drier at installation or after servicing
- Failure to purge hoses prior to charging refrigerant

Symptoms of system contamination include the following:

- Sealed-tube refrigerant test shows acid and/or moisture
- Discoloration of moisture indicator in liquid line sightglass
- Refrigeration oil acid test shows acid
- System valves stick
- Plugged metering device (ice)
- Poor compressor operation
- Compressor failure
- Overheating of compressor
- High compressor current draw

In a system that uses a TXV, the problem of noncondensibles in the system is often diagnosed as an overcharge of refrigerant because the symptoms are very similar.

NOTE
If a system using R-410a refrigerant is left open and exposed to the air, the compressor oil can absorb large amounts of moisture. This moisture cannot be removed with a vacuum pump. Instead, an R-410a liquid-line filter-drier must be used to remove the moisture.

12.2.0 Acid Contamination

Acid is not introduced into a system; it is formed inside an improperly operating system by the reaction of air or moisture with the refrigerant. Acid is produced in greater quantities in an overheated system. Acid creates sludge and varnish, which can plug oil passages and metering devices, and it can restrict the strainers in the compressor and filter-driers, as well as other refrigeration system components. Acid most likely exists in a system when the compressor has failed as the result of an electrical failure (**burnout**) involving shorted or grounded motor windings. Burnouts are classified as either mild or severe based on the amount of acid present in the system. A mild burnout tends to yield little acid in the system because it occurs suddenly, causing the motor to stop before the contaminants created by the burnout leave the compressor. A severe burnout usually occurs over a longer period of

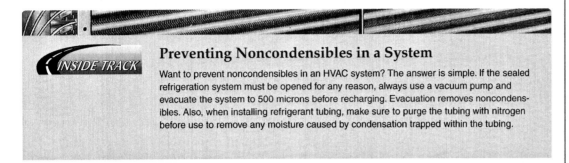

Inside Track

Preventing Noncondensibles in a System

Want to prevent noncondensibles in an HVAC system? The answer is simple. If the sealed refrigeration system must be opened for any reason, always use a vacuum pump and evacuate the system to 500 microns before recharging. Evacuation removes noncondensibles. Also, when installing refrigerant tubing, make sure to purge the tubing with nitrogen before use to remove any moisture caused by condensation trapped within the tubing.

Emphasize that gloves and eye protection must be worn when working on a system suspected of having a burnout.

Explain how to test and clean up a contaminated system.

time, allowing considerable acid and other contaminants to be produced and pumped through the system while the compressor is still running.

 WARNING!
When working on any system containing refrigerant, eye and hand protection must be worn at all times to prevent personal injury. This is especially true when working with acid-contaminated systems. Acid in contact with eyes or skin can cause severe burns.

Before replacing a compressor that has failed because of a burnout, you must test the system for acid contamination with a sealed-tube acid/moisture test kit or an oil test kit (*Figure 15*) to determine if the burnout was mild or severe. A sealed-tube acid/moisture test kit is connected to a system service port to obtain a sample of the refrigerant. Oil acid test kits require that the test be performed on a sample of the compressor oil. In either case, follow the test kit manufacturer's instructions for using the kit and for determining the amount of acid/moisture contamination in the system, if any.

Cleanup of a system to eliminate contaminants is relatively simple if the compressor failed because of a mild burnout. This is because in a mild burnout, few contaminants enter the system. Normally, the only thing that must be done after the compressor installation is completed is to remove the existing liquid line filter-drier and replace it with one that is one size larger. If the system is not equipped with a filter-drier, install one. Also, the system should be triple evacuated, with a final evacuation to 500 microns, before recharging the system with refrigerant. Cleanup of a system when the compressor failed because of a severe burnout can be quite extensive. The system must be properly cleaned or the result will most likely be another failed compressor.

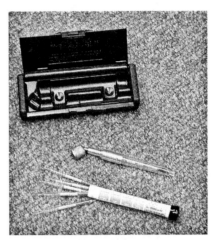
SEALED-TUBE ACID/MOISTURE TEST KIT

ACID TEST KIT

210F15.EPS

Figure 15 ◆ Acid and moisture test kits.

MODULE 03210-07 ◆ TROUBLESHOOTING COOLING 10.35

Discuss condensate water problems in refrigeration systems.

Show trainees how to develop a checklist for troubleshooting cooling systems.

Have trainees develop a checklist for troubleshooting cooling systems. Note the proficiency of each trainee. This laboratory corresponds to Performance Task 1.

Reinforce safety practices for troubleshooting equipment with power supplied to it. Emphasize the danger of rotating parts.

Show trainees how to test a system for moisture/acid contamination.

Use pre-faulted components, jumpers, or other means to insert malfunctions into a cooling system. Provide the trainees with the wiring diagram for the unit and have them isolate and correct the fault(s). Note the proficiency of each trainee. This laboratory corresponds to Performance Tasks 3 and 4.

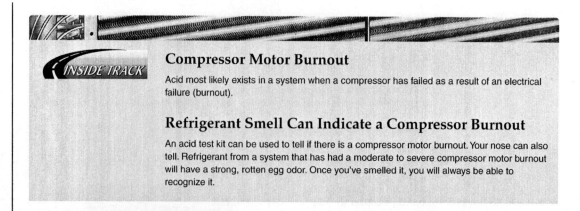

Compressor Motor Burnout

Acid most likely exists in a system when a compressor has failed as a result of an electrical failure (burnout).

Refrigerant Smell Can Indicate a Compressor Burnout

An acid test kit can be used to tell if there is a compressor motor burnout. Your nose can also tell. Refrigerant from a system that has had a moderate to severe compressor motor burnout will have a strong, rotten egg odor. Once you've smelled it, you will always be able to recognize it.

13.0.0 ◆ CONDENSATE WATER DISPOSAL PROBLEMS

Condensate water can back up and cause water damage due to failed condensate pumps, blocked drain lines, or algae growing in the drain pan(s). This is usually indicated by water around the furnace or air handler or dripping from the ceiling. The problem can be corrected by thorough cleaning of the evaporator coil and condensate pan, and a complete flushing of the drain line. Condensate pumps are used in systems where gravity drains are impractical. Good condensate pumps contain a float switch that will shut off the system if the water level in the pump gets too high.

Instructor's Notes:

Review Questions

Refer to the schematic shown in *Figure 4* to answer Questions 1 through 3.

1. The normal sequence of compressor operation for the cooling system shown in *Figure 4* is _____.
   a. thermostat calls for cooling, short cycle module times out, compressor contactor energizes, compressor and outdoor fan run
   b. compressor contactor energizes, thermostat calls for cooling, compressor and outdoor fan run, short cycle module times out
   c. thermostat calls for cooling, compressor contactor energizes, short cycle module times out, compressor and outdoor fan run
   d. compressor contactor energizes, short cycle module times out, thermostat calls for cooling, compressor and outdoor fan run

2. The purpose of the indoor fan relay in *Figure 4* is to _____.
   a. select high speed operation for the indoor fan during the heating mode
   b. select low speed operation for the indoor fan during the cooling mode
   c. select high speed operation for the indoor fan during the cooling mode
   d. turn off the induced-draft motor during the cooling mode

3. If there is a problem only with heating in the combined cooling/heating unit shown in *Figure 4*, which component can be eliminated as a possible cause of the problem?
   a. Time delay relay (TDR)
   b. Induced-draft relay (IDR)
   c. Pressure switch (PS)
   d. Start thermistor (ST)

4. The terminal of a standard heating/cooling thermostat associated with the cooling mode of unit operation is normally designated _____.
   a. R (red)
   b. Y (yellow)
   c. G (green)
   d. W (white)

5. When troubleshooting a mechanical refrigeration system cooling problem, you find high suction and low discharge pressures. The compressor current draw is considerably lower than normal. The most likely cause of the problem is _____.
   a. undercharge of refrigerant
   b. bad compressor valves and/or rings
   c. wrong oil charge
   d. restriction in the liquid line

6. You notice that heavy foaming is present in the compressor oil sightglass when the system is first started. This foaming should diminish _____.
   a. immediately after startup
   b. about an hour after startup
   c. about five to ten minutes after startup
   d. about thirty minutes after startup

7. When troubleshooting a mechanical refrigeration system cooling problem, you find that the suction and discharge pressures are lower than normal, superheat is high, and subcooling is low. The compressor current draw is also low. The most likely cause of the problem is _____.
   a. a dirty condenser
   b. low evaporator airflow
   c. an undercharge of refrigerant
   d. an overcharge of refrigerant

8. Poor cooling because of too little evaporator airflow is most likely caused by _____.
   a. a dirty evaporator blower wheel
   b. grass clippings on the surface of the coil
   c. shrubs blocking airflow
   d. air recirculation

9. Periodic knocking in the compressor can be an indication of _____.
   a. low refrigerant
   b. liquid refrigerant in the compressor
   c. undercharge of oil
   d. oil in the compressor crankcase absorbing refrigerant

Have trainees complete the Review Questions, and go over the answers prior to administering the Module Examination.

Review Questions

10. A system that uses a distributor to feed the evaporator refrigerant circuits is experiencing high superheat and liquid floodback to the compressor. One likely cause of this problem is that the cooling load is too low.
    a. True
    b. False

11. When troubleshooting a mechanical refrigeration system cooling problem, you find that the suction and discharge pressures are lower than normal, and superheat and subcooling are both high. The most likely cause of the problem is a _____.
    a. refrigerant undercharge
    b. loose TXV bulb
    c. high evaporator load
    d. liquid line restriction

12. When troubleshooting a mechanical refrigeration system cooling problem, you find that the suction and discharge pressures are higher than normal, with a high subcooling temperature. The compressor current draw is also high. The most likely cause of the problem is _____.
    a. a liquid line restriction
    b. noncondensibles in the system
    c. an undercharge of refrigerant
    d. an undersized expansion valve

Refer to the unit shown in *Appendix A* to answer Questions 13 through 15.

13. When troubleshooting the unit shown in *Appendix A*, the indoor blower motor runs continuously. The cause is most likely the _____.
    a. thermostat switch is set to the AUTO position
    b. thermostat switch is set to the ON position
    c. starter overload contacts are closed
    d. blower motor contactor is de-energized

14. When the cooling thermostat calls for cooling in the unit shown in *Appendix A*, all of the following occur *except* the _____.
    a. control relay energizes
    b. outdoor fan relay energizes
    c. compressor contactor energizes
    d. indoor blower relay de-energizes

15. The crankcase heater for the compressor used in the unit shown in *Appendix A* is _____.
    a. energized when the compressor is turned off
    b. always energized regardless of whether the compressor is turned on or off
    c. de-energized when the compressor is turned on
    d. always de-energized regardless of whether the compressor is turned on or off

Instructor's Notes:

Summary

Effective troubleshooting of cooling systems is a process by which, as an HVAC technician, you listen to a customer's complaint, perform an independent analysis of a problem, and then use a systematic, step-by-step approach to troubleshooting that results in the correction of the problem. You must understand the purpose and principles of operation of each component in the cooling equipment being serviced. You must be able to tell whether a given device is functioning properly and to recognize the symptoms arising from the improper operation of any part of the equipment. Based on the symptoms revealed by your analysis of system operation, you then use the troubleshooting aids provided by the manufacturer for the system being serviced to identify and repair the cause of the problem.

Summarize the major concepts presented in the module.

Administer the Module Examination. Record the results on Craft Training Report Form 200, and submit the results to the Training Program Sponsor.

Administer the Performance Test, and fill out Performance Profile Sheets for each trainee. If desired, trainee proficiency noted during laboratory sessions may be used to complete the Performance Test. Record the results on Craft Training Report Form 200, and submit the results to the Training Program Sponsor.

Trade Terms Introduced in This Module

Burnout: The condition in which the breakdown of the motor winding insulation causes the motor to short out or ground electrically.

Capacitance boost: A procedure used to start a stuck PSC compressor. It involves momentarily connecting a start (boost) capacitor across the run capacitor of the stuck compressor in an attempt to start the stuck compressor by increasing the starting torque.

Flooded starts: When a compressor is started, slugging, foaming, and inadequate lubrication occur as a result of the oil in the compressor crankcase having absorbed refrigerant during shutdown. Flooded starts can also occur as a result of rapid return of liquid refrigerant during startup.

Flooding: The condition in which there is a continuous return of liquid refrigerant or liquid droplets in the suction vapor being returned to the compressor during operation.

Hygroscopic: The characteristic of absorbing and retaining moisture. The word stems from *hygroscope*, which is an instrument showing changes in humidity.

Noncondensible gas: A gas (such as air) that does not change into a liquid at operating temperatures and pressures.

Seized (stuck) compressor: The condition in which a compressor is electrically or mechanically unable to start (locked up). It hums but will not start, and draws locked rotor current.

Short cycling: The condition in which the compressor is restarted immediately after it has been turned off.

Slugging: The condition that occurs when a compressor tries to compress liquid refrigerant, oil, or both, instead of superheated gas.

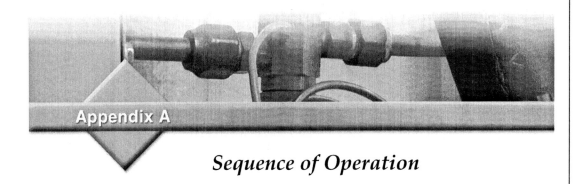

Appendix A

Sequence of Operation

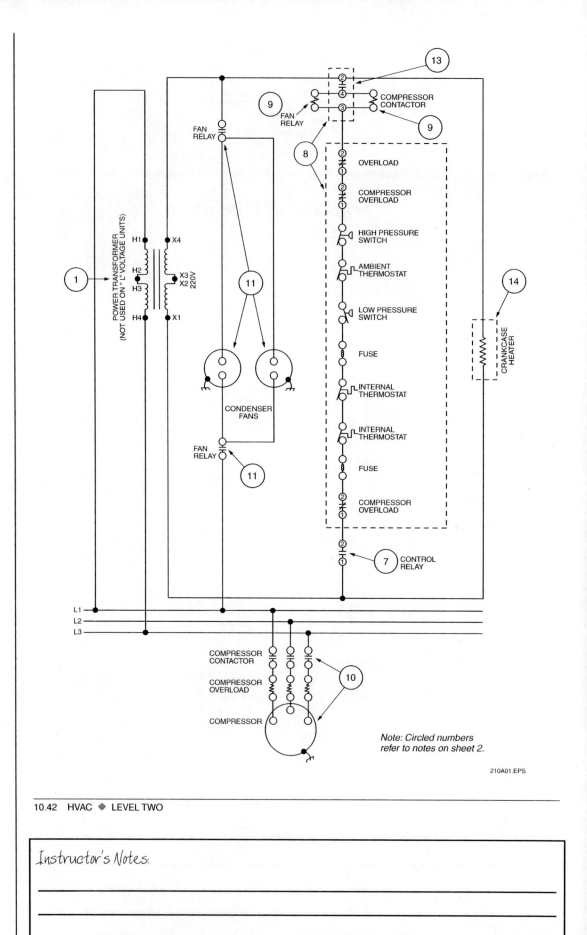

10.42 HVAC ♦ LEVEL TWO

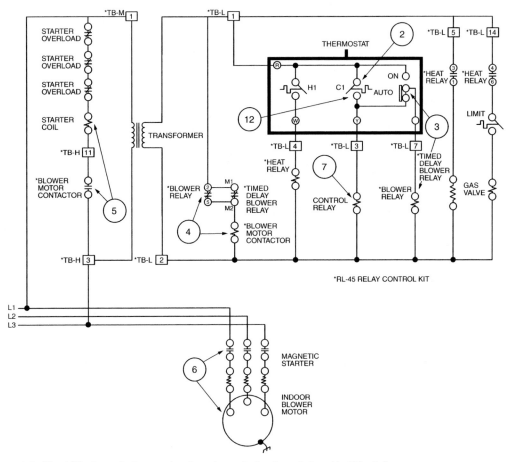

1. On "F" and "H" voltage units, the power transformer is energized continuously. It provides 220 volts for compressor safety circuit and condenser fan motors.
2. The thermostat makes on a cooling demand.
3. If the thermostat is set on "Auto", the Blower Relay is energized.
4. The Blower Relay closes its N.O. contacts to energize the Blower Motor Contactor.
5. In turn, the Blower Motor Contactor closes its N.O. contacts to energize the Magnetic Starter.
6. The Magnetic Starter then closes its N.O. contacts to power the Indoor Blower Motor.
7. As the thermostat makes, it also energizes the Control Relay.
8. This completes a circuit to the Timed Off Control through the L2 compressor protections circuit:

   Compressor Overloads   Ambient Thermostat   Fuses
   High Pressure Switch   Low Pressure Switch   Internal Thermostat

   Note: On " L" voltage units, the control circuit is protected by fuses located in the control box.

9. Twenty seconds after the Timed Off Control is powered, it energizes the No. 1 Compressor Contactor and Outdoor Fan Contactor ("G" and "F" voltage units only).
10. The Compressor Contactor closes its N.O. contacts to power the compressor. On "Y" voltage units it also powers the condenser fan motors.
11. On "G" and "F" voltage units, the Outdoor Fan contactor closes its N.O. contacts to power the condenser fan motors.
12. As the cooling demand is satisfied, the thermostat cycles off the compressor and condenser fan motors.
13. If the original cycle was less than five minutes, the timed off control will wait the remaining time plus twenty seconds on the next cooling demand.
14. Crankcase heater operation is continuous.

210A02.EPS

Appendix B

Cooling Troubleshooting Chart

GENERALLY THE CAUSE Make these checks first.	OCCASIONALLY THE CAUSE Make these checks only if the first checks failed to locate the trouble.	RARELY THE CAUSE Make these checks only if other checks failed to locate the trouble.
PROBLEM: Compressor and condenser fan motor will not start		
Power failure Open disconnect switch Blown fuses or tripped circuit breaker Faulty wiring, loose terminals Contactor coil Protective interlock circuit contacts open Short cycle device not timed out	Low line voltage Defective contacts in contactor Thermostat or setting Control transformer Refrigerant charge low	Single-phase failure of three-phase power Unbalanced power supply (three phase) Condenser motor Condenser fins dirty or plugged Condenser fan belt slipping Condenser air short circuiting Low condenser air volume High head pressure Low suction pressure Noncondensibles (air, etc.)
PROBLEM: Compressor will not start but condenser fan will run		
Faulty wiring or loose connection, load side of contactor to compressor motor terminal Run/start capacitor Start relay contacts stuck open Open motor windings or overload	Loose terminals Low line voltage Single-phase failure of three-phase power Unbalanced power supply (three phase)	Defective compressor bearings
PROBLEM: Condenser fan motor will not start		
Faulty wiring or loose connection, load side of contactor to condenser motor terminal Run capacitor Open condenser fan motor windings or overload Bearings seized		
PROBLEM: Compressor hums but will not start		
Faulty wiring Loose terminals Compressor overload Start relay Run or start capacitor Defective compressor bearings Unequalized pressures on PSC motor Low line voltage Seized compressor	Single-phase failure of three-phase power Unbalanced power supply (three phase) Defective contacts in contactor Shorted or grounded motor windings	Defective compressor valves Compressor oil level Condenser fins dirty or plugged Condenser fan belt slipping Low condenser air volume Condenser air temperature low Superheat setting TEV stuck open Loose TEV thermal bulb

210A03.EPS

GENERALLY THE CAUSE Make these checks first.	**OCCASIONALLY THE CAUSE** Make these checks only if the first checks failed to locate the trouble.	**RARELY THE CAUSE** Make these checks only if other checks failed to locate the trouble.
PROBLEM: Compressor short cycles on low pressure		
Power failure Refrigerant charge low Low suction pressure Dirty TEV Defective TEV power element TEV push rod packing loose Evaporator fins dirty or plugged Evaporator blower motor belt slipping Low evaporator air volume Dirty filters Restrictions	TEV valve too small Low outdoor air temperature	Stratified air in space
PROBLEM: Compressor runs continuously – no cooling/inadequate cooling		
Primary or safety controls Defective compressor valves Refrigerant charge low Evaporator fins dirty or plugged Evaporator blower motor belt slipping Low evaporator air volume Dirty filters	High suction pressure Noncondensibles (air, etc.) Superheat setting TEV dirty TEV push rod packing loose Restrictions Thermostat locations	Condenser fans dirty or plugged Condenser fan belt slipping Condenser air short circuiting Low condenser air volume Condenser air temperature low Defective TEV power element
PROBLEM: Compressor runs continuously – cooling		
Faulty wiring, control circuit Thermostat set too low Excessive load in space	System too small Thermostat location	
PROBLEM: Compressor noisy		
Loose hold-down bolts Defective compressor bearings Defective compressor valves Compressor oil level	Incorrect refrigerant piping Compressor internal mounts (hermetic)	Compressor/condensing unit not level
PROBLEM: Compressor loses oil		
Refrigerant charge low Very low suction pressure Oil trapping in system Compressor short cycling Low oil in system Refrigerant leak	TEV stuck Compressor oil pump Dirty oil pump screen	Superheat setting Valve too small Evaporator fins dirty or plugged Evaporator belt slipping Low evaporator air volume Dirty filters

Instructor's Notes:

GENERALLY THE CAUSE Make these checks first.	**OCCASIONALLY THE CAUSE** Make these checks only if the first checks failed to locate the trouble.	**RARELY THE CAUSE** Make these checks only if other checks failed to locate the trouble.
PROBLEM: Head pressure too high		
Overcharge of refrigerant Noncondensibles (air, etc.) Condenser fans dirty or plugged Condenser fan belt slipping Inoperative condenser fan Wrong condenser fan rotation	Excessive load in space Condenser air short circuiting Higher than ambient air entering condenser Oversized TEV valve	TEV stuck open Loose thermal bulb Restrictions in discharge line and/or suction line
PROBLEM: Head pressure too low		
Refrigerant charge low Low suction pressure Condenser input air temperature low	Defective compressor valves TEV valve too small	Restrictions in liquid line Open compressor internal pressure relief valve
PROBLEM: Liquid line frosting or sweating		
Liquid valve partially closed Restrictions		
PROBLEM: Suction pressure too high		
Defective compressor valves High head pressure Noncondensibles (air, etc.)	Overcharge of refrigerant High heat load Incorrect TEV Loose TEV thermal bulb Oversized TEV valve TEV stuck open Suction line insulation missing	High evaporator air volume
PROBLEM: Suction pressure too low		
Refrigerant charge low Evaporator belt slipping Evaporator fins dirty or plugged Low evaporator air flow Dirty filters	Low outdoor air temperature TEV dirty Defective TEV power element TEV push rod packing loose Restrictions in liquid line	TEV valve incorrect Incorrect refrigerant piping Recirculation of evaporator air Wrong evaporator fan rotation Inadequate duct system Low return air temperature
PROBLEM: Suction line frosting or sweating		
Overcharge of refrigerant TEV stuck open Oversized TEV valve Evaporator fins dirty or plugged Evaporator belt slipping Dirty filters	TEV dirty Low evaporator air volume	Low return air temperature

210A05.EPS

GENERALLY THE CAUSE Make these checks first.	OCCASIONALLY THE CAUSE Make these checks only if the first checks failed to locate the trouble.	RARELY THE CAUSE Make these checks only if other checks failed to locate the trouble.
PROBLEM: Evaporator blower will not start		
Power failure Blown fuses or tripped circuit breaker Faulty wiring or loose terminals Control transformer Evaporator fan relay coil or contacts Evaporator blower motor	Thermostat	Line voltage Single-phase failure of three-phase power
PROBLEM: Liquid refrigerant flooding back to compressor (TEV system)		
Superheat setting Loose thermal bulb Overcharge of refrigerant	Low heat load TEV stuck open Oversized TEV valve	Evaporator fins dirty or plugged Evaporator blower motor belt slipping Low evaporator air volume Dirty filters Thermostat setting
PROBLEM: Liquid refrigerant flooding back to compressor (cap tube system)		
Overcharge of refrigerant High head pressure Excessive subcooling	Evaporator fins dirty or plugged Evaporator blower motor belt slipping Low evaporator air volume	
PROBLEM: Space temperature too high		
Thermostat Refrigerant charge low Low suction pressure Evaporator fins dirty or plugged Evaporator blower motor belt slipping Low evaporator air volume Dirty filters Ductwork small or restricted Thermostat setting too high System too small for load	Defective compressor valves Superheat setting TEV dirty Defective TEV power element Thermostat location	TEV stuck TEV push rod packing loose Incorrect refrigerant piping Stratified air in space

Instructor's Notes:

Additional Resources and References

Additional Resources

This module is intended to be a thorough resource for task training. The following reference works are suggested for further study. These are optional materials for continued education rather than for task training.

Air Conditioning Systems, Principles, Equipment, and Service, 2001, Joseph Moravek. Upper Saddle River, NJ: Prentice Hall.

General Training Air Conditioning 2, GTAC. Syracuse, NY: Carrier Corporation.

HVAC Servicing Procedures, Syracuse, NY: Carrier Corporation.

Figure Credits

Carrier Corporation, 210SA01, 210SA04

Courtesy of Honeywell International Inc., 210SA03

Topaz Publications, Inc., 210F06, 210SA05, 210SA06, 210F07, 210SA07, 210SA08, 210SA09, 210SA10, 210F15 (top)

Highside Chemicals, 210F15 (bottom)

MODULE 03210-07 — TEACHING TIPS

The following are suggested activities or instructional methods to help you teach the material in this module.

General

When you call on someone to answer a question, the rest of the class relaxes or even tunes out because they expect that the question and answer will take place only between you and the trainee you called on. Instead, use this technique to involve more trainees in answering questions and to keep them on their toes.

1. Ask trainees to define a term or explain a concept.
2. After one trainee has answered, ask a trainee seated nearby if the answer is right. Then ask whether a trainee in the back of the room agrees.
3. Ask trainees to explain why they think an answer is right or wrong.
4. Use the session to clear up incorrect ideas and encourage trainees to learn from their mistakes.

Sections 1.0.0 through 4.0.0

Quick Quiz

This Quick Quiz will familiarize trainees with the terms and definitions that are commonly used when troubleshooting cooling systems. You will need photocopies of the quiz provided in the following page. Trainees will need pencils. If you allow trainees to use the Trainee Guide, decrease the amount of time you give them to complete the quiz.

1. Make a photocopy of the quiz for each trainee.
2. Give trainees between 5 and 10 minutes to complete the quiz.
3. Go over the answers to the quiz.
4. Ask trainees if they have questions.

Answers to Quick Quiz

1. d
2. g
3. f
4. a
5. h
6. c
7. e
8. b

Quick Quiz   *Troubleshooting Cooling*

For each description listed, identify the term that the text best describes. Write the corresponding letter in the blank provided.

_____ 1. The compressor oil in R-410a compressors is very _____.

_____ 2. The condition in which a compressor is restarted immediately after it has been turned off is called _____.

_____ 3. A compressor that hums but will not start is considered a(n) _____.

_____ 4. You can attempt to start a stuck PSC compressor with a(n) _____.

_____ 5. A compressor trying to compress liquid refrigerant, oil, or both instead of superheated gas, causes a condition known as _____.

_____ 6. Oil in the compressor crankcase absorbing refrigerant can cause _____.

_____ 7. Air does not change into a liquid at operating pressures and temperatures and is known as a _____.

_____ 8. An oversized TXV can cause _____.

a. capacitance boost
b. flooding
c. flooded starts
d. hygroscopic
e. noncondensible gas
f. seized (stuck) compressor
g. short cycling
h. slugging

Section 3.0.0 *Control Circuits*

This exercise will familiarize trainees with circuit diagrams for cooling equipment. Trainees will need pencils and paper. You will need to obtain the wiring diagrams for several types of units. Allow 20 to 30 minutes for this exercise. This exercise corresponds to Performance Task 3.

1. Divide the class into several groups.
2. Have each group study a portion of the circuit diagram, and then have them teach the operating sequence for that portion to the rest of the class. For example, Group 1 could teach the compressor problems and Group 2 the electrical controls.
3. Have trainees teach the rest of the class what they learned by studying the wiring diagram. Answer any questions the trainees may have.

Section 3.0.0 *Identifying Parts of a Cooling System*

This exercise will familiarize trainees with identifying various parts of a cooling system. Trainees will need pencils and paper. You will need an overhead projector, acetate sheets, and transparency markers. Allow 20 to 30 minutes for this exercise. This exercise corresponds to Performance Task 3.

1. Divide the class in half. On one sheet write "Team One" on one half and "Team Two" on the other half.
2. Call out the components in the legend in *Figure 4*. Have trainees locate the component on the circuit diagram in *Figure 4*.
3. To ensure complete participation, each team member can answer no more than two times (this can be adjusted according to the number of participants and items to be identified). Each correct answer gets a check on the acetate sheet. An incorrect answer means the other team has the opportunity to answer.
4. Keep score to encourage class participation. At the end of the game, answer any questions.

Section 4.0.0 *Troubleshooting Cooling Systems*

This exercise will familiarize trainees with troubleshooting various types of cooling systems. Trainees will need pencils and paper. You will need to arrange for a manufacturer's representative to give a presentation on cooling systems. Allow 20 to 30 minutes for this exercise.

1. Tell trainees that a guest speaker will be presenting information on cooling systems. Have trainees brainstorm questions for the guest speaker before the speaker arrives.
2. Introduce the speaker. Ask the presenter to speak about troubleshooting various types of cooling systems. Have them discuss various troubleshooting procedures.
3. Have the trainees take notes and write down questions during the presentation.
4. After the presentation, answer any questions trainees may have.

Section 7.1.0 *Think About It – Refrigerant Undercharge*

The superheat reading will be higher than normal because when a system is undercharged, there is less refrigerant in the evaporator to absorb the same amount of heat. Therefore, the refrigerant passing through each evaporator tube is heated to saturation more quickly. This allows superheating to begin earlier in the evaporator tubes. Superheating also occurs more rapidly, since there is less refrigerant. Thus, the refrigerant's superheat is higher than normal. It should be pointed out that this is true in systems with a fixed-orifice, capillary, or TXV metering device.

Section 8.2.0   *Think About It – Dirty Condenser Coil*

With a dirty condenser coil, the system's ability to reject heat is reduced. A dirty condenser coil retards the flow of heat from the refrigerant into the air flowing over the condenser coil. Therefore, the refrigerant's saturated condensing temperature climbs, causing the system discharge and suction pressures to climb with it. The compressor has to compress and discharge against these higher pressures.

Section 10.0.0   *Think About It – Compressor Current Draw*

Compressor current is lower than normal. The compressor current always tracks the discharge pressure. When the pressure is low, the compressor is pumping against a low head pressure, causing it to draw less current. Similarly, when the pressure is high, the compressor is pumping against a high pressure, causing it to draw more current.

Section 10.1.0   *Think About It – TXV Sensing Bulbs*

If the sensing bulb for a TXV is disconnected from the suction line, it will open the orifice size to increase refrigerant flow. This happens because instead of sensing the temperature of the refrigerant in the suction line, the sensing bulb will sense the much higher ambient temperature of the air surrounding the evaporator. The change in sensed temperature raises the pressure in the bulb, sending the signal to open the TXV orifice.

MODULE 03210-07 — ANSWERS TO REVIEW QUESTIONS

Answer	Section
1. a	3.0.0
2. c	3.0.0
3. d	3.0.0
4. b	5.3.2
5. b	5.4.2; 9.5.0
6. c	6.2.1
7. c	7.1.0
8. a	8.1.0
9. b	9.1.0
10. a	10.2.0
11. d	11.0.0
12. b	12.1.0
13. b	Appendix A
14. d	Appendix A
15. b	Appendix A

CONTREN® LEARNING SERIES — USER UPDATE

NCCER makes every effort to keep these textbooks up-to-date and free of technical errors. We appreciate your help in this process. If you have an idea for improving this textbook, or if you find an error, a typographical mistake, or an inaccuracy in NCCER's Contren® textbooks, please write us, using this form or a photocopy. Be sure to include the exact module number, page number, a detailed description, and the correction, if applicable. Your input will be brought to the attention of the Technical Review Committee. Thank you for your assistance.

Instructors – If you found that additional materials were necessary in order to teach this module effectively, please let us know so that we may include them in the Equipment/Materials list in the Annotated Instructor's Guide.

Write: Product Development and Revision
National Center for Construction Education and Research
3600 NW 43rd St, Bldg G, Gainesville, FL 32606

Fax: 352-334-0932

E-mail: curriculum@nccer.org

Craft Module Name

Copyright Date Module Number Page Number(s)

Description

(Optional) Correction

(Optional) Your Name and Address

Module 03211-07

Heat Pumps

NCCER STANDARDIZED CRAFT TRAINING PROGRAM

The National Center for Construction Education and Research (NCCER) provides a standardized national program of accredited craft training. Key features of the program include instructor certification, competency-based training, and performance testing. The program provides trainees, instructors, and companies with a standard form of recognition through a National Craft Training Registry. The program is described in full in the *Guidelines for Accreditation*, published by NCCER. For more information on standardized craft training, contact the NCCER by writing us at 3600 NW 43rd St., Bldg. G, Gainesville, FL 32606; calling 352-334-0911; or emailing info@nccer.org. More information may be found at our website, www.nccer.org.

HOW TO USE THIS ANNOTATED INSTRUCTOR'S GUIDE

Each page presents two sections of information. The larger section displays each page exactly as it appears in the Trainee Module. The narrow column ties suggested trainee and instructor actions to each page and provides icons (detailed below) to call your attention to material, safety, audiovisual, or testing requirements. The bottom of each page includes space for your notes.

The **Audiovisual** icon indicates an appropriate time to show a transparency or other audiovisual aid.

The **Classroom** icon prompts you to define a term, stress a point, ask trainees to explain a concept, or give examples.

The **Demonstration** icon directs you to show trainees how to perform tasks.

The **Examination** icon tells you to administer the written module examination.

The **Homework** icon is placed where you may wish to assign reading for the next class, assign a project, or advise trainees to prepare for an examination.

The **Laboratory** icon is used when trainees are to practice performing tasks.

The **Materials** icon is a reminder for you to gather materials needed for classes, labs, and testing.

The **Performance Testing** icon tells you to administer a performance test or a portion thereof.

The **Safety** icon is used to emphasize safety issues. It is often keyed to *Caution* and *Warning!* statements in the Trainee Module.

The **Teaching Tip** icon indicates additional guidance is available, such as how to conduct an exercise, get the most educational value from a field trip, or encourage class participation. Teaching Tips may expand on a feature (*Think About It*, *Did You Know?*) or provide *Quick Quizzes* or similar exercises. You will be referred to the Teaching Tips section at the back of the module if there is additional material.

The **Combination** icon indicates that the laboratory listed corresponds with a performance task. If desired, you can note the proficiency of the trainees during the laboratory, and use it to satisfy performance testing requirements.

PREPARATION

Before teaching this module, you should review the Objectives, Performance Tasks, Materials and Equipment List, and Module Outline. Be sure to allow ample time to prepare your own training or lesson plan and gather all required materials and equipment.

Heat Pumps
Annotated Instructor's Guide

Module 03211-07

MODULE OVERVIEW

This module introduces covers operation, installation, and control circuit analysis for heat pumps.

PREREQUISITES

Prior to training with this module, it is recommended that the trainee shall have successfully completed *Core Curriculum*; *HVAC Level One*; and *HVAC Level Two*, Modules 03201-07 through 03210-07.

OBJECTIVES

Upon completion of this module, the trainee will be able to do the following:

1. Describe the principles of reverse-cycle heating.
2. Identify heat pumps by type and general classification.
3. Describe various types of geothermal water loops and their application.
4. List the components of heat pump systems.
5. Describe the role and basic operation of electric heat in common heat pump systems.
6. Describe common heat pump ratings, such as Coefficient of Performance (COP), Heating Season Performance Factor (HSPF), and Seasonal Energy Efficiency Ratio (SEER).
7. Demonstrate heat pump installation and service procedures.
8. Identify and install refrigerant circuit accessories commonly associated with heat pumps.
9. Analyze a heat pump control circuit.
10. Isolate and correct malfunctions in a heat pump.

PERFORMANCE TASKS

Under the supervision of the instructor, the trainee should be able to do the following:

1. Identify components that are unique to heat pumps and explain the function of each.
2. Calculate the balance point of a heat pump.
3. Simulate the installation procedures for a heat pump.
4. Perform heat pump servicing procedures.
5. Analyze a heat pump circuit diagram and perform simulated troubleshooting exercises.

MATERIALS AND EQUIPMENT LIST

Overhead projector and screen
Transparencies
Blank acetate sheets
Transparency pens
Whiteboard/chalkboard
Markers/chalk
Pencils and scratch paper
Appropriate personal protective equipment
Copies of a heat pump wiring diagram
Reversing valve
Check valve
Metering device
Fan coil with electric heating elements
Applicable manufacturer's heat pump installation instructions
Transparency or sketch of structure in which heat pump will be installed
Manufacturers' heat pump balance point calculation sheets

Electrical wire, cable, connectors, conduit (if required)
Disconnect switch
Refrigerant
Videotape *Inside the Heat Pump* (optional)
Videotape *Troubleshooting Heat Pumps* (optional)
DVD/VCR/TV set (optional)
Disassembled air-to-air split-system heat pump with controls
TEV with built-in check valve
Refrigeration mechanic's tool set
Gauge manifold set
Electrical test instruments
Operational, properly wired and assembled air-to-air split-system heat pump
Air conditioning service tools and test equipment set
Copies of Quick Quiz*
Module Examinations**
Performance Profile Sheets**

* Located in the back of this module.
**Located in the Test Booklet.

SAFETY CONSIDERATIONS

Ensure that the trainees are equipped with appropriate personal protective equipment and know how to use it properly. This module requires trainees to work with appliances. Make sure that all trainees are briefed on appropriate safety procedures. Emphasize electrical safety and chemical hazards.

ADDITIONAL RESOURCES

This module is intended to present thorough resources for task training. The following reference work is suggested for both instructors and motivated trainees interested in further study. This is optional material for continued education rather than for task training.

Inside the Heat Pump, Videotape. Catalog No. 020-538. Syracuse, NY: Carrier Corporation.

TEACHING TIME FOR THIS MODULE

An outline for use in developing your lesson plan is presented below. Note that each Roman numeral in the outline equates to one session of instruction. Each session has a suggested time period of 2½ hours. This includes 10 minutes at the beginning of each session for administrative tasks and one 10-minute break during the session. Approximately 15 hours are suggested to cover *Heat Pumps*. You will need to adjust the time required for hands-on activity and testing based on your class size and resources. Because laboratories often correspond to Performance Tasks, the proficiency of the trainees may be noted during these exercises for Performance Testing purposes.

Topic   **Planned Time**

Session I. Introduction to Heat Pumps
- A. Introduction   _____
- B. Heat Pump Operation   _____
- C. Heat Pump Classification   _____
- D. Heat Pump Refrigeration Cycle   _____

Session II. Heat Pump Components and Controls
- A. Heat Pump Components   _____
- B. Laboratory   _____
  Trainees practice identifying the components of a heat pump. This laboratory corresponds to Performance Task 1.
- C. Supplemental Electric Heat   _____
- D. Heat Pump Performance   _____
- E. Heat Pump Balance Point   _____
- F. Laboratory   _____
  Trainees practice calculating the balance point of a heat pump. This laboratory corresponds to Performance Task 2.

Session III. Installation
- A. Installation   _____
- B. Laboratory   _____
  Trainees practice simulating or describing the installation procedures for a heat pump. This laboratory corresponds to Performance Task 3.

Sessions IV and V. Servicing and Troubleshooting
- A. Servicing   _____
- B. Laboratory   _____
  Trainees practice heat pump servicing procedures. This laboratory corresponds to Performance Task 4.
- C. Laboratory   _____
  Trainees practice troubleshooting heat pumps. This laboratory corresponds to Performance Task 5.
- D. Heat Pump Controls   _____

Session VI. Review and Testing
- A. Review   _____
- B. Module Examination   _____
  1. Trainees must score 70% or higher to receive recognition from NCCER.
  2. Record the testing results on Craft Training Report Form 200, and submit the results to the Training Program Sponsor.
- C. Performance Testing   _____
  1. Trainees must perform each task to the satisfaction of the instructor to receive recognition from NCCER. If applicable, proficiency noted during laboratory exercises can be used to satisfy the Performance Testing requirements.
  2. Record the testing results on Craft Training Report Form 200, and submit the results to the Training Program Sponsor.

HVAC Level Two

03211-07
Heat Pumps

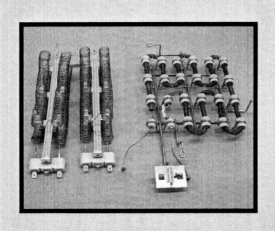

Assign reading of Module 03211-07.

03211-07
Heat Pumps

Topics to be presented in this module include:

1.0.0	Introduction	11.2
2.0.0	Heat Pump Operation	11.2
3.0.0	Heat Pump Classifications	11.2
4.0.0	Heat Pump Refrigeration Cycle	11.8
5.0.0	Heat Pump Components	11.12
6.0.0	Supplemental Electric heat	11.18
7.0.0	Heat Pump Performance	11.23
8.0.0	Balance Point	11.26
9.0.0	Installation	11.27
10.0.0	Service	11.31
11.0.0	Heat Pump Controls	11.32

Overview

Heat pumps provide both cooling and heating using the mechanical refrigeration cycle. The secret to this capability is that even cold air contains some heat. The heat pump can extract that heat by reversing the refrigeration cycle. The heat pump contains a few more control devices than the cooling-only system, and its heat exchangers are larger because they serve two functions. Other than that, the heat pump looks just like a common air conditioning system. Air-to-air heat pumps are the most common type, but some heat pumps use water from wells or ponds as a heat source. Others, called geothermal heat pumps, extract heat from coils buried in the earth.

Instructor's Notes:

Objectives

When you have completed this module, you will be able to do the following:

1. Describe the principles of reverse-cycle heating.
2. Identify heat pumps by type and general classification.
3. Describe various types of geothermal water loops and their application.
4. List the components of heat pump systems.
5. Describe the role and basic operation of electric heat in common heat pump systems.
6. Describe common heat pump ratings, such as Coefficient of Performance (COP), Heating Season Performance Factor (HSPF), and Seasonal Energy Efficiency Ratio (SEER).
7. Demonstrate heat pump installation and service procedures.
8. Identify and install refrigerant circuit accessories commonly associated with heat pumps.
9. Analyze a heat pump control circuit.
10. Isolate and correct malfunctions in a heat pump.

Trade Terms

Absolute zero
Balance point
Coefficient of Performance (COP)
Compression heat
Defrost
Dual-fuel or hybrid system
Fan coil unit
Four-way valve
Ground-source (geothermal) system
Heat sink
Heating Season Performance Factor (HSPF)
Indoor coil
Kelvin scale
Outdoor coil
Packaged unit
Reverse cycle heat
Reversing valve
Seasonal Energy Efficiency Ratio (SEER)
Split system

Required Trainee Materials

1. Pencil and paper
2. Appropriate personal protective equipment

Prerequisites

Before you begin this module, it is recommended that you successfully complete *Core Curriculum*; *HVAC Level One*; and *HVAC Level Two*, Modules 03201-07 through 03210-07.

This course map shows all of the modules in the second level of the HVAC curriculum. The suggested training order begins at the bottom and proceeds up. Skill levels increase as you advance on the course map. The local Training Program Sponsor may adjust the training order.

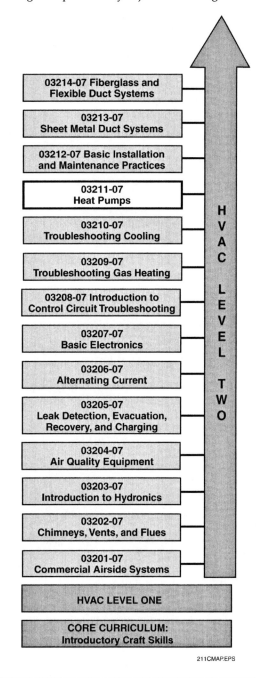

Ensure that you have everything required to teach the course. Check the Materials and Equipment list at the front of this module.

See the general Teaching Tip at the end of this module.

Explain that terms shown in bold are defined in the Glossary at the back of this module.

Show Transparency 1, Objectives, and Transparency 2, Performance Tasks. Review the goals of the module, and explain what will be expected of the trainee.

Review the modules covered in Level Two and explain how this module fits in.

Identify the differences between a heat pump and a conventional cooling unit.

Explain that heat pumps operate most efficiently when the outdoor temperature is 35°F or higher.

Point out that heat pumps used in cold climates are normally equipped with supplemental heat.

Explain that heat pumps are classified according to their method of heat source and heat sink.

See the Teaching Tip for Sections 1.0.0–8.0.0 at the end of this module.

1.0.0 ♦ INTRODUCTION

A heat pump is a combination heating and cooling unit. It produces cooling in the same manner as a conventional cooling unit, then reverses the cycle to produce heat. Both systems have the same components:

- Compressor
- Blower fan
- Condenser fan
- Service valves
- Condenser and evaporator coils
- Pressure and temperature controls
- Refrigerant and refrigerant lines

In addition to the typical air conditioning components listed, the heat pump requires a **reversing valve** and additional metering and control devices.

Since heat pumps perform both heating and cooling functions, the coils are identified by location: the evaporating coil and condensing coil on an air conditioner are identified as the **indoor coil** and **outdoor coil** on a heat pump in the cooling mode.

Why are heat pumps used? Heat pumps operate for about half the cost of electric resistance heat. Electric utilities offer customers, contractors, and developers rebates to install heat pumps. This is because electricity demand is low in the winter months, and heat pumps encourage customers to switch from oil or gas to electric utilities. Air Conditioning and Refrigeration Institute (ARI) statistics indicate that over one million heat pumps are sold each year.

Besides being efficient to operate in the heating mode, a heat pump provides cooling. For heating purposes, a heat pump may or may not be less costly to operate than gas or oil heat, depending on the climate and fuel costs. However, it is likely to be more cost-effective for combined heating and cooling than other methods.

2.0.0 ♦ HEAT PUMP OPERATION

To understand how the heat pump works, it is necessary to understand that, regardless of how cold it is outdoors, there is still some heat in the air. This is true as long as the temperature is above **absolute zero** (–460°F), which is 0°K on the **Kelvin scale**. At warmer temperatures, there is more heat available. As temperatures drop, there is less heat available for the heat pump to extract. Most heat pumps begin losing efficiency rapidly when the outdoor temperature falls below 35°F. When the temperature falls below 20°F, they can

Indoor and Outdoor Coils

When a heat pump switches from the cooling mode to the heating mode, the indoor and outdoor coils switch functions. Like cooling equipment, a heat pump can consist of a single packaged unit, or it can be a matched split system.

be quite ineffective. For that reason, air-source heat pumps are usually equipped with some form of auxiliary heat, such as electric or fossil fuel, that cycles on automatically when the heat pump alone can no longer meet the demand.

Water-source units are able to extract heat more effectively and dependably due to the consistent temperatures found beneath the earth's surface. Although their capacity also fluctuates as the temperature of the water source changes, these changes are generally within a very small range.

3.0.0 ♦ HEAT PUMP CLASSIFICATIONS

Heat pumps are classified according to their heat source and the medium to which the heat is transferred (**heat sink**). A water-to-air heat pump, for example, picks up heat from water flowing through a coil and transfers it to the air flowing over another coil.

The more common types of heat pumps are: air-to-air, water-to-water, air-to-water, and water-to-air. In the air-to-air system, the external heat source is air, and the medium (heat sink) that comes in contact with the indoor refrigerant coil is also air. Some larger heat pump systems may have more than one heat source and may also use both water and air as heat sinks.

An air-to-water system also uses air as the heat source, but the heating and cooling are provided by a water distribution system within the structure. The air-to-water system uses convection devices to transfer the heat between the water and the air.

Groundwater is an excellent heat source, but it is not available in all areas and in some areas is in limited supply during certain periods. Therefore, air is the predominant heat source in residential and small commercial installations. Quite often, heat pump systems use a coil of tubing buried in the earth as a heat source. These are known as **ground-source (geothermal) systems**. Using a

---

11.2 HVAC ♦ LEVEL TWO

Instructor's Notes:
_____
_____
_____

closed loop eliminates many water quality issues associated with using groundwater or lake water.

Other heat sources besides air and water are sometimes used. Some of these are waste heat from selected industrial processes, exhaust air from ventilation, solar energy, and heat extracted from refrigerated spaces. These sources are commonly used in addition to, rather than as replacements for, the basic heat source.

3.1.0 Air-to-Air Heat Pumps

The most common heat source is air. The refrigeration circuit schematic in *Figure 1* represents an air-to-air heat pump in the cooling cycle. Outdoor air is used as the heat source in air-to-air units. The heat is then delivered to, or removed from, the indoor air through the use of refrigerant flowing through a coil.

Air-to-air heat pumps are generally smaller-tonnage units. Packaged heat pumps are available for residential and commercial applications up to 30 tons. However, built-up systems of any size may also be installed. Current air-to-air models are either self-contained in one unit (**packaged unit**) or **split systems**, which are divided into two sections. Depending on the type of application, both kinds offer definite advantages.

When air is used as the heat source, efficiency improves in mild climates. A heat pump is generally selected to match the cooling load.

If the heating load exceeds the heating capacity of the selected heat pump unit, electric resistance heaters may be required to supply supplementary heat during periods of higher heat load.

Auxiliary heat is also required for heating during the **defrost** cycle. Auxiliary heat should be sized for 100 percent of the building's heat load. This ensures that there will be adequate heat in the event of a **compression heat** failure.

Air-to-air heat pumps offer several advantages:

- They do not use water, so the problems related to piping, corrosion, and condenser water disposal are nonexistent.
- They operate as in-space units supplying filtered, conditioned air at a nominal cost.
- They are available as packaged (integral) or remote (split) systems.
- They operate at half the operating cost of electric resistance heat.

Show Transparency 3 (Figure 1). Discuss the operation and applications of air-to-air heat pumps. List their advantages. Point out that they are available as packaged units or split systems.

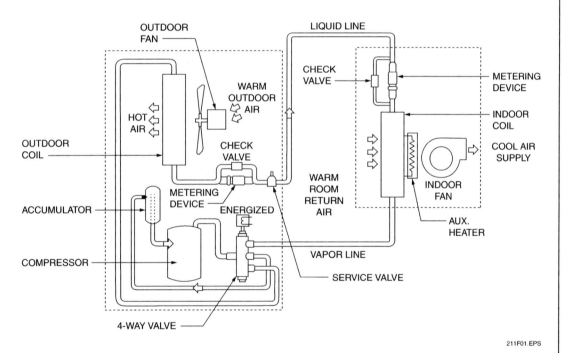

Figure 1 ◆ Schematic of an air-to-air heat pump (shown in cooling mode).

Show Transparencies 4 and 5 (Figures 2 and 3). Discuss the operation and applications of water-to-air heat pumps. List their advantages.

Discuss the operation and applications of water-to-water heat pumps.

See the Teaching Tip for Section 3.2.0 at the end of this module.

3.2.0 Water-to-Air Heat Pumps

Water-to-air heat pumps use well water, lake water, or another available water supply as a heat source and transfer the heat indoors through the use of a heat exchanger. These units are similar in size to the air-to-air type and are marketed in the range from 2 to 50 tons.

Figure 2 is an example of a water-to-air system using two wells. This is considered an open loop system. One well provides the supply water, while the other provides a means for the water to be returned to the ground. The water returned to the ground is at a different temperature once heat has been removed during the heating mode or added in the cooling mode. Therefore, the two wells must be placed at a sufficient distance from each other to prevent discharge water from being drawn back to the system before its temperature returns to normal. In some cases, wells drilled at significantly different depths are employed. Installation of the two wells can significantly increase the installed cost of a system, with the cost and drilling approach greatly dependent upon local geology.

Figure 3 shows a typical water-to-air system using a closed loop. Mechanical systems provide water heating and cooling for distribution to a variety of water-source heat pump equipment in a commercial setting. Water temperature in the loop is typically maintained between 70°F and 80°F, with a boiler as a heat source and a cooling tower providing cooling. Mixing valves are often used to precisely control water temperature in the loop.

Water-to-air heat pumps have some of the same basic problems as other systems associated with water, such as water availability, mineral content and quality, and disposal. Except under emergency conditions, using a treated domestic water supply is not cost-effective, due to both the initial water cost and the cost for disposal through a sewage treatment system. The system in *Figure 3*, however, can be chemically treated to circumvent some of these issues. Unlike water-to-water systems (discussed in the next section), the problem of water and its treatment is confined to one side of the system. The water-to-refrigerant heat exchanger coils in water-source systems of all types are generally made of special alloys such as copper-nickel for both good heat transfer and corrosion resistance.

The following are some of the advantages of water-to-air heat pumps:

- They generally operate more efficiently than air-source units because water temperatures are generally warmer than air in winter, while cooler in summer. The increased heat transfer characteristics of water is also a factor.
- Self-contained or integral water-to-air units have greater application flexibility than integral air-to-air units because they are not restricted to the use of an outdoor air supply.
- They may be installed as free-standing, in-space conditioning units.
- They do not require defrost.
- They generally do not require supplemental electric heat as their heating capacity remains stable throughout the winter months. The heating capacity of air-to-air heat pumps drops as outdoor temperatures become colder. However, electric or fossil fuel heating equipment may be added to provide an emergency source in the event of a heat pump failure.
- They operate at half the operating cost of electric resistance heat.

3.3.0 Water-to-Water Heat Pumps

Water-to-water heat pumps use water from wells, lakes, or closed-loop systems to transfer heat to a second circulating loop of water. This method can be used for domestic water heating or for comfort applications. Both methods can be integrated into a single system. For domestic water heating, water-to-water systems can generally produce water of sufficient temperature to meet common

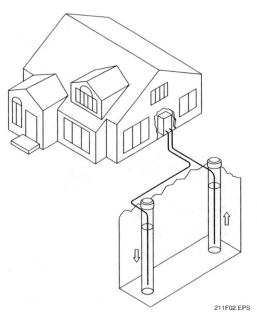

Figure 2 ♦ Open loop system using two wells.

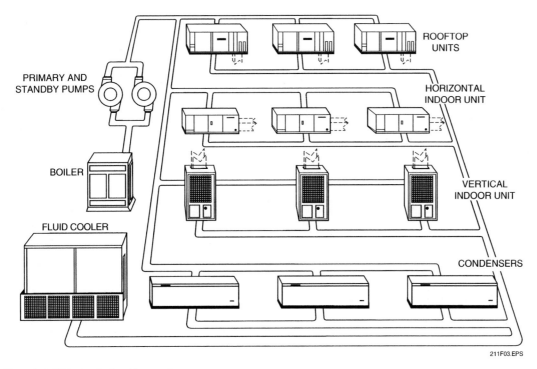

Figure 3 ♦ Water-to-air closed loop systems.

List the advantages of water-to-water heat pumps.

Discuss the operation and applications of air-to-water heat pumps.

needs. However, for comfort heating applications, these systems are generally unable to produce water at 170°F to 180°F, which is an older hydronic heating design standard. As a result, the overall heating system must be designed to provide sufficient heat transfer surface area to meet heating needs with cooler water. Although in use since the 1930s, growth has been slow due to the introduction of improved and more cost-effective air-source units.

Water quality and water-side maintenance issues are magnified in the water-to-water system, as two separate water loops must be addressed. Soluble minerals such as calcium salts, magnesium salts, and iron can form deposits on heat exchanger surfaces and impede heat transfer. Without proper maintenance and system design, mineral deposits and corrosion can quickly cripple these systems, and maintenance costs are often increased.

The water-to-water heat pump has three major cost advantages:

- When compared to fossil fuel units, the water-to-water heat pump has substantially lower yearly electrical operating costs.
- It has a more efficient refrigeration cycle than that of the air-source heat pump. As a result, less electricity is usually required.
- The initial cost of the dual system will generally be lower than that of a conventional heating system combined with a second system to provide summer cooling.

Water-to-water heat pumps are sometimes used in conjunction with closed circuit, flat-plate solar collectors. The solar collectors are used as a heat source, and a water storage tank is used as a heat sink in these types of applications.

3.4.0 Air-to-Water Heat Pumps

An air-to-water heat pump is a system that uses the outdoor air as a heat source. It delivers the heat to the conditioned space through the use of a secondary medium that, in this case, is water. These units are often used to heat domestic water and swimming pools. Air-to-water installations can be designed to compete with conventional heating and cooling systems of all sizes. Heat is generally supplemented by electric resistance heaters if the load exceeds the heating capacity of

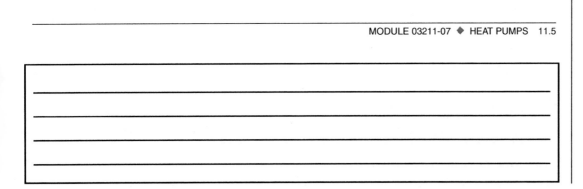

List the advantages of air-to-water heat pumps.

Show Transparency 6 (Figure 4). Discuss the operation and applications of ground-source heat pumps.

Show Transparency 7 (Figure 5). Describe horizontal loop systems.

Show Transparency 8 (Figure 6). Describe vertical loop systems.

the chosen heat pump. Air-to-water systems provide the following benefits:

- They can eliminate the need for large supply and return duct systems because warm and cold water are piped through the building.
- Rooms and areas may be thermostatically controlled more easily in large, central installations than with large supply and return air systems.
- They can be used as hot water heaters.

3.5.0 Ground-Source Heat Pumps

A ground-source or geothermal heat pump (*Figure 4*) uses coils of tubing buried beneath the ground in a variety of configurations to transfer heat to and from the earth. As is the case with most water-source systems, the consistency of ground temperatures provides a distinct advantage on days of peak demand, both in heating and cooling modes. Since the ground loop is closed, it is easily treated chemically to prevent corrosion and mineral deposits. Anti-freeze solutions are often added to prevent any potential for freezing, which can occur in locations where the

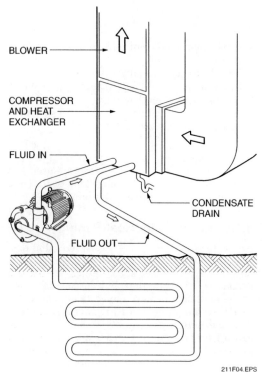

Figure 4 ◆ Ground-source heat pump.

INSIDE TRACK — Heat Pumps and Groundwater

Environmental laws are becoming very strict with regard to thermal pollution of any type. For this reason, it is important to always check the local codes concerning the use of groundwater for heat pump applications.

loop is exposed to the ambient air. When properly installed, freezing beneath the ground is unlikely. A variety of piping materials can be used below ground. Determining the amount of tubing to be buried depends upon the heat transfer characteristics of the selected materials, as well as geology. Although plastic tubing, such as high-density polyethylene, cannot generally match the heat transfer speed and efficiency of copper, its low cost often compensates for the higher volume of tubing to be buried. Plastic tubing also has the advantages of inherent corrosion resistance, flexibility, and ease of connection. Geothermal systems enjoy a continued rise in popularity due to both their efficiency and low maintenance costs.

3.5.1 Horizontal Loop Systems

Horizontal loop systems (*Figure 5*) use trenches dug to an average depth of four to six feet to access highly stable ground temperature zones. Generally more cost-effective to install than vertical loops, they are commonly found where land is readily available and accessible to the system owner. The tubing layouts can be installed in residential yards, open fields, landscaped areas, sports stadiums, and under parking lots.

Where a suitable body of water exists, geothermal loops can simply be submerged and placed at the bottom where water temperatures remain more consistent. This system can be an extremely cost-effective installation solution where circumstances allow. However, sediment and other debris naturally deposited on the outer surface of the tubing can impede heat transfer over time.

3.5.2 Vertical Loop Systems

Vertical loop systems (*Figure 6*) use holes drilled into the earth, and are often used when insufficient land is available for a wide-spread system such as the horizontal loop. The supply and

11.6 HVAC ◆ LEVEL TWO

Instructor's Notes:

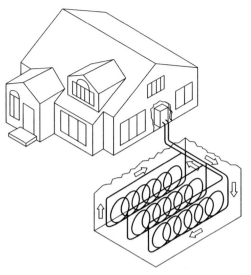

HORIZONTAL CLOSED LOOP SYSTEM

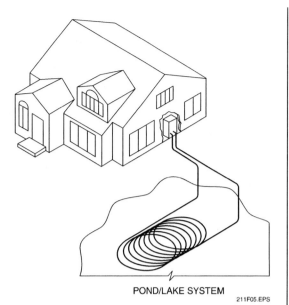

POND/LAKE SYSTEM

Figure 5 ♦ Horizontal loop systems.

Point out that heat pumps can also use nontraditional heat sources, such as heat from energy recovery systems and commercial processes.

Show the video *Inside the Heat Pump*.

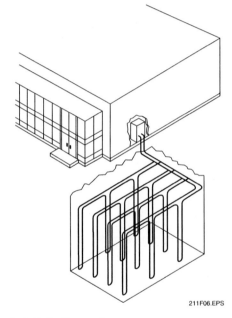

Figure 6 ♦ Vertical loop system.

return piping is inserted into each hole; as the return piping exits the hole, it is then routed into additional holes as needed for sufficient heat transfer. After the loop piping is inserted into the hole, it is grouted with materials such as Bentonite to ensure good thermal contact with the surrounding earth and improve heat transfer.

3.6.0 Special Heat Sources

Economical, special heat sources for heat pump systems may be available from operations or processes that exist in nearby areas.

On large commercial and industrial buildings, one potential source of extracted heat is the central-point exhaust unit. Direct-expansion coils located in the exhaust air stream may be used to remove this heat and return it to other portions of the building.

Another method of extracting heat is the coil energy recovery loop (runaround loop) system. The coil energy recovery loop system incorporates coils filled with an antifreeze solution located in the ventilation supply and exhaust air ducts. Heat removed from the exhaust air is absorbed by the incoming ventilation air, thereby reducing the heating requirements of the mechanical equipment. Coil energy recovery loop systems are covered in more detail in *Energy Conservation Equipment*.

A third method of extracting heat and reducing compressor motor electrical consumption is to place a refrigerant-liquid subcooling coil in the ventilation air plenum. The subcooling coil provides preheat to the incoming fresh air, while subcooling the liquid refrigerant as it flows to the evaporator.

Explain that the refrigeration cycle of air-source heat pump systems function much the same as a conventional air conditioning unit when in cooling mode.

Show Transparency 9 (Figure 7). Trace the refrigerant flow through the cooling, heating, and defrost cycles of a heat pump.

The use of a subcooling coil is limited to buildings where a large portion of the ventilation supply air is introduced to the building at a central point close to the compressor room. It is impractical to run long refrigerant lines to widely dispersed coils.

Other sources from which heat may be recovered include computer rooms, industrial plants where heat is removed from process work, and refrigeration condensing units or packaged air coolers in supermarkets and warehouses.

4.0.0 ◆ HEAT PUMP REFRIGERATION CYCLE

The refrigeration cycle of air-source (and most other) heat pump systems functions much the same as a conventional air conditioning unit when in the cooling mode. Hot refrigerant vapor is sent to the condenser to be cooled and condensed, and liquid refrigerant is sent through the metering device to the evaporator to provide cooling. In order to provide heating, however, the refrigerant must flow in the opposite direction. The indoor coil becomes the condenser, while the outdoor coil assumes the duties of the evaporator. As a result, the industry refers to the two coils in a heat pump system as the indoor and outdoor coils, since their duty changes as needed. In the heating mode, heat is added to the refrigerant in the outdoor coil, despite cold ambient conditions, because the refrigerant entering the outdoor coil is still colder than the surrounding air. After being compressed (adding further to the heat content), the hot refrigerant vapor is sent to the indoor coil where the heat is further transferred to the relatively cooler air circulating inside.

4.1.0 Cooling Cycle

Figure 7 shows a schematic diagram of the refrigerant cycle of a typical air-to-air heat pump. The reversing valve, also known as a **four-way valve**, is the key to the heat pump's dual nature. In most heat pumps, the reversing valve is energized when the thermostat calls for cooling and de-energized when it calls for heating. This guarantees that heat is available even if there is a failure in the reversing valve control circuits.

When the unit is in the cooling mode (*Figure 7, View A*), the refrigerant flow (see arrows) is the same as that of any cooling unit. Notice the path through the reversing valve. The cold, low-pressure refrigerant flowing through the indoor coil (evaporator) absorbs heat from the conditioned space and is boiled into a superheated vapor. Hot, high-pressure refrigerant gas leaving the compressor is pumped through the outdoor coil (condenser), where the heat is rejected. The refrigerant pressures and temperatures shown are typical of those we have seen in the discussion of cooling systems.

4.2.0 Heating Cycle

In the heating mode (*Figure 7, View B*), the reversing valve changes position. The refrigerant leaving the compressor is routed in the opposite direction from that of the cooling cycle. Instead of flowing through to the outdoor coil, the hot, high-pressure refrigerant vapor leaving the compressor flows to the indoor coil, which is now acting as a condenser. Heat is extracted from the refrigerant at that point because the air in the conditioned space is cool in relation to the refrigerant in the coil (heat transfer from warm to cool). The condensed liquid refrigerant then flows outdoors, through the metering device and into the outdoor coil, where it encounters cold air. Since the refrigerant here, by design, is colder than the air, heat is extracted and transferred to the refrigerant. This causes the refrigerant to begin boiling and returning to the vapor state. Heat provided in this way is known as **reverse cycle heat** or compression heat. Note the significant differences in the pressures and temperatures at key points in the system as compared with the cooling mode.

4.3.0 Defrost Cycle

When heat is transferred from the cold outdoor air to the refrigerant in the outdoor coil, moisture condenses on the coil. Because of the temperature relationships, this moisture will freeze on the coil, even at outdoor temperatures above 40°F. Over time, the frost will build up on the coil, blocking airflow and preventing effective heat transfer. This is more likely to be a problem at outdoor temperatures between 28°F and 40°F than it is at lower temperatures, because the colder the air, the less moisture it contains.

To eliminate ice buildup on the outdoor coil, most heat pumps have a defrost function. In air-to-air heat pumps, the heating cycle is reversed, placing the unit in the cooling mode (*Figure 7, View A*). Hot refrigerant then flows through the outdoor coil to melt the frost buildup. While in the defrost mode, the outdoor fan is cycled off to prevent cold air from being forced across the coil and impede or prevent the melting process. This usually lasts about 10 minutes. During that

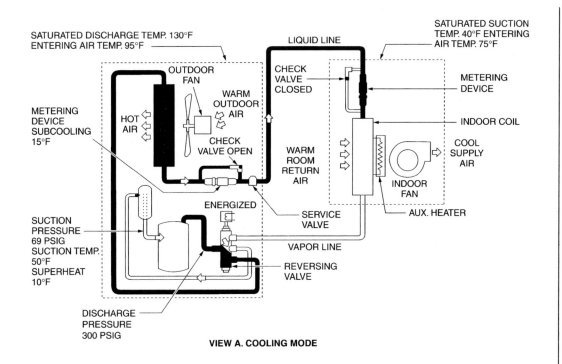

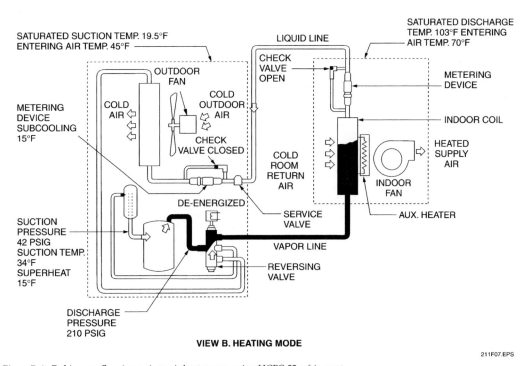

Figure 7 ♦ Refrigerant flow in an air-to-air heat pump, using HCFC-22 refrigerant.

Explain that the defrost cycle prevents frost buildup on the outdoor coil.

Show Transparency 10 (Figure 8). Discuss electromechanical defrost controls.

Show Transparency 11 (Figure 9). Discuss electronic defrost controls.

time, the heat pump is not providing any heat. In fact, cool air is blowing off the evaporator into the conditioned space, just like it would on a hot day. To prevent discomfort to building occupants during the defrost cycle, an electric heater located downstream of the indoor coil is automatically switched on. When an air-source heat pump is paired with a fossil fuel system, the heat pump is generally switched off when conditions that can allow significant frost to develop are present. The fossil fuel system then takes over.

There are many different ways to control the defrost function. One popular method uses a combination of time and temperature. The logic behind this method is that defrost is needed only if the unit has been operating long enough for frost to form on the coil (for example, 90 minutes), and the temperature is low enough for moisture to freeze on the coil. If both conditions don't exist, defrost is unnecessary. Most timers can be set for 45, 60, or 90 minutes of operation before defrost will occur.

This method requires a timer to keep track of the system's operating time and a means of sensing outdoor temperature. The latter is easy; a small thermostat or thermistor on the outdoor coil will do the job. Keeping time is easily accomplished with an electronic control.

4.3.1 Electromechanical Defrost Control

In some heat pumps, especially older models, the timing function is partly mechanical. One method uses a motor-driven timing device, which is switched on whenever the compressor is running. *Figure 8* shows the electrical schematic of such an arrangement. Whenever the compressor contactor is energized, power is supplied to the compressor motor and the defrost timer. As the defrost timer motor runs, it turns a cam that closes a set of contacts for 10 seconds every 90 minutes. Because the unit does not always run for 90 minutes at a time, it may take several on-cycles to accumulate 90 minutes.

The 10-second closure of the mechanical contacts is not enough to start defrost. The defrost thermostat (DFT) must also be closed, indicating that the outdoor temperature is below 45°F (or some other selected setting). When both are closed, the defrost relay (DFR) will energize. One set of normally open DFR contacts (1) closes to act as a holding circuit for the DFR coil. This is needed because the mechanical contacts will open after 10 seconds.

Another set of normally open DFR contacts (2) closes, causing the reversing valve to energize and placing the system in the cooling mode. After 10 minutes (or some other selected period) of defrost operation, the other contacts that are mechanically linked to the timer motor will open momentarily, stopping the defrost process and restarting the 90-minute timing cycle.

If the heating thermostat opens during the defrost cycle, defrost will be suspended. When the thermostat cycles the unit on again, it will pick up where it left off, as long as the defrost thermostat has remained closed.

Notice that a third set of normally open DFR contacts (3) controls the heater relay (HR). When defrost cycles on, HR energizes, and its contacts in the heater circuit close, bringing on the electric heater to prevent cold air from blowing into the conditioned space. Further, a set of normally closed contacts (4) opens during the defrost cycle to shut down the outdoor fan motor while defrost is in progress.

In addition to mechanical timers, pressure switches that sense airflow through the outdoor coil and thermal controls may also be used to start and stop defrost. These pressure switches are located in the liquid line, and open the defrost circuit when the pressure at the outdoor coil reaches 275 psi. This equates to a coil temperature of about 124°F, which is high enough to indicate that the coil is free of ice.

4.3.2 Electronic Defrost Controls

Most modern heat pumps use electronic defrost controls in which all the defrost circuitry, including the relays, is mounted on a printed circuit board. The timer is electronic, and therefore less likely to fail (see *Figure 9*). A temperature sensor, a defrost thermostat, or a thermistor sends temperature information from the outdoor coil to the board. Instead of keeping the operating time mechanically, an electronic clock in the timing logic receives a signal as long as the heating thermostat is closed. If defrost is interrupted by the opening of the heating thermostat, the timing logic will remember where it left off. If the coil temperature still indicates the presence of frost when the unit cycles on again, the unit will automatically go back into the defrost mode.

A jumper or other selection device on the board allows the installer to select the defrost frequency (30, 50, or 90 minutes, for example) based on local conditions.

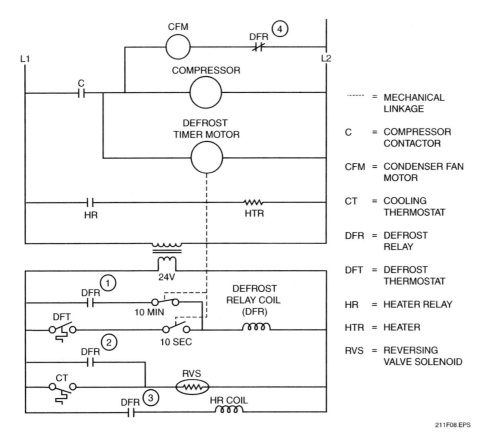

Figure 8 ♦ Electromechanical defrost control.

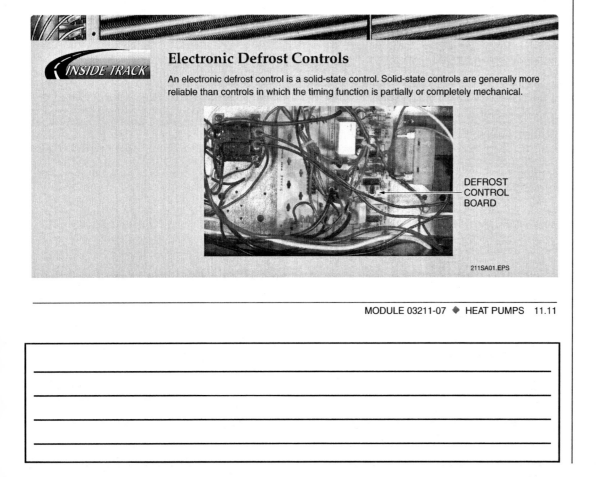

Keep Transparency 11 (Figure 9) showing. Explain how the built-in defrost test cycle functions. Explain how to optimize defrost cycles.

Have the trainees review Sections 5.0.0–8.0.0.

Ensure that you have everything required for teaching this session.

Explain that while heat pumps and cooling-only units are similar, the size and functions of their components differ.

Disassemble a heat pump and point out each of the components and controls.

See the Teaching Tip for Section 5.0.0 at the end of this module.

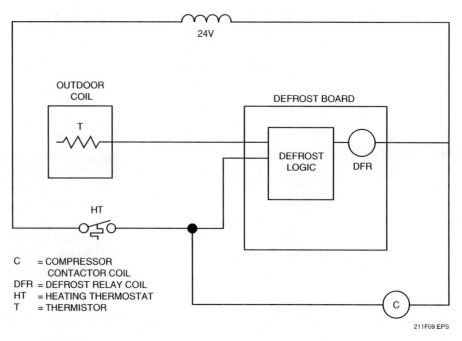

Figure 9 ♦ Electronic defrost control simplified schematic.

An important feature of many of these boards is a built-in defrost test cycle. On one such control, the technician can, by placing a jumper across two terminals on the board, artificially force a defrost cycle to determine whether the system is working properly in the defrost mode. The 90-minute defrost cycle is overridden and the usual 10-minute running cycle can be reduced to just a few seconds if desired. A major advantage of the electronic defrost control is that it is completely self-contained. If the technician finds that there is a defrost control problem, it is not necessary to isolate the failed component. Instead, the entire board is replaced.

As previously noted, the heat pump delivers cold air to the space while in a defrost cycle. Electric heat is typically energized during this period to temper the supply air temperature. As a result, the defrost cycle consumes far more energy than a typical heating or cooling cycle, and must be factored in when considering overall system efficiency. Obviously, minimizing the number of defrost cycles, as well as their length, is a great advantage. With the advent of newer, more versatile electronic controls, manufacturers have developed demand, or adaptive, defrost control strategies. These systems combine information relative to current operating conditions, such as the difference in temperature between the outdoor coil and the ambient temperature, and determine precisely when a defrost cycle is needed and when it can be terminated for maximum operating efficiency. Rather than respond solely to the closure of thermostats and other electromechanical devices, these defrost circuits use solid-state logic to precisely determine when defrost is necessary.

5.0.0 ♦ HEAT PUMP COMPONENTS

Although a heat pump and a cooling-only unit are schematically similar, they are physically quite different. The heat pump not only has more components, but the components it has in common with the cooling unit are different because they perform a dual role. In a cooling-only unit, the evaporator can be smaller than the condenser because it handles less heat. The condenser must also reject the heat of the compression cycle, which typically represents an additional 3,000 Btuh per ton more than the evaporator will absorb. In a heat pump, the evaporator must be able to reject the heat gained through the outdoor coil, plus the additional heat of compression, which is now working to the benefit of the user. The heat pump requires a heavy-duty compressor because it runs year-round. The heat pump compressor must be able to withstand compression ratios of 8:1, where the cooling-only compressor might reach only 3:1. Piping is also different. In a cooling-only unit, the vapor line carries only low-pressure suction gas from the evaporator. In a heat pump, this same line must also be able to carry hot, high-pressure gas from the compressor discharge.

5.1.0 Reversing Valves

There are several types of reversing valves, but they all operate in basically the same way. One particular type of valve has a main valve body containing a slide and a piston, plus a three-way pilot valve that is actuated by an electrically energized solenoid.

The reversing valve has four piping connections (*Figure 10*). Flow of refrigerant through two of these connections never changes. These two connections are the hot gas discharge from the compressor and the suction line back to the compressor. The remaining two ports connect to the indoor and outdoor coils.

The refrigerant gas is directed to either coil depending on the position of the piston in the valve. See *Figure 11*. When the piston shifts, the path of the discharge flow is changed from one coil to the other.

The valve shifts due to the pressure difference within the valve body. The pressure difference is controlled by a 24V solenoid-actuated pilot valve. As the pilot valve moves, it directs the flow of higher pressure refrigerant gas to one end of the reversing valve, pushing the piston toward the opposite end. It is important to note that a pressure difference of roughly 100 psig is required to move the piston. As a result, the valve cannot change position until the compressor has started and developed sufficient differential pressure to reposition the piston. The pilot valve voltage may also be 120V or 240V. During the cooling cycle, the valve operates as shown in *Figure 12, View A*:

- The solenoid is de-energized and the pilot solenoid pin carrier is at the far left of the pilot solenoid chamber (point 1).
- The compressor discharge gas bleeds behind the left side of the slide (point 2).
- At point 3, the discharge gas is trapped in the capillary tube. The pressure builds until it reaches the discharge pressure.
- At point 4 (*Figure 12, View B*), the pressure behind the right side of the slide equals the compressor suction pressure, which then passes through the pilot solenoid chamber, down through the center capillary to the suction line port at point 5.

The pressure behind the right side of the piston is less than the pressure behind the left side of the piston; therefore, the reversing valve piston is moved to the right. With the piston in this position, the flow of refrigerant is directed to the outdoor coil and the heat pump is operating in the cooling cycle (*Figure 13*).

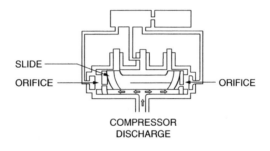

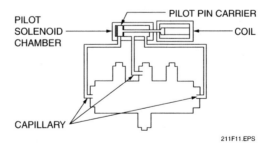

Figure 10 ◆ Reversing (four-way) valve.

Figure 11 ◆ Refrigerant flow in a reversing valve.

Show Transparency 12 (Figure 10). Discuss the function of the reversing valve.

Show Transparency 13 (Figure 11). Explain refrigerant flow in a reversing valve.

Show Transparencies 14 and 15 (Figures 12 and 13). Discuss the operation of the reversing valve during the cooling cycle.

Provide a reversing valve for the trainees to examine.

Show Transparencies 16 and 17 (Figures 14 and 15). Discuss the operation of the reversing valve during the heating cycle.

Explain that there are several methods for metering refrigerant in a heat pump.

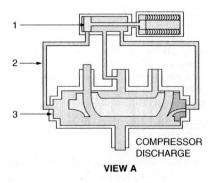

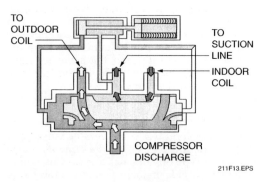

Figure 13 ♦ Refrigerant flow – cooling cycle.

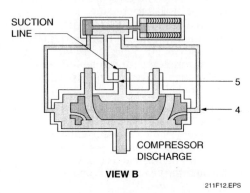

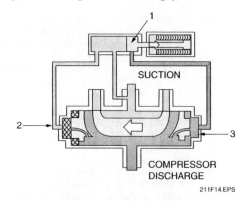

Figure 12 ♦ Reversing valve position – cooling cycle.

Figure 14 ♦ Reversing valve position – heating cycle.

In order to place the heat pump in the heating mode, the solenoid valve is energized and the action is as follows:

- Upon energizing the solenoid valve, the pilot pin carrier moves to the right to point 1 (*Figure 14*).
- Compressor suction pressure is now able to pass through the center capillary and the pilot chamber down to the left side of the piston at point 2.
- At point 3, the compressor discharge pressure bleeds through the orifice behind the right side of the piston, where it is trapped.
- In this position, the pressure behind the left side of the piston is less than the pressure behind the right side and the piston will move to the left.
- With the piston in this position (*Figure 15*), the hot gas from the compressor is now directed to the indoor coil and the unit is operating in the heating cycle.

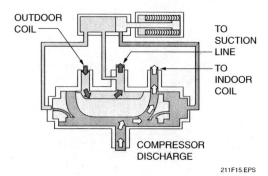

Figure 15 ♦ Refrigerant flow – heating cycle.

5.2.0 Metering Devices

There are a number of methods of metering refrigerant in a heat pump. Most of them involve the use of two metering devices.

11.14   HVAC ♦ LEVEL TWO

In one arrangement, a fixed-orifice metering device is installed at each coil. A check valve in parallel with each device either blocks or allows refrigerant flow, depending on the flow direction. The metering device piston and the check valve oppose each other. In the cooling mode, for example, the metering device at the output of the indoor coil passes the refrigerant flowing to the indoor coil. The check valve is positioned to prevent flow in this direction. The piston in the other metering device faces in the opposite direction, and is therefore forced into the orifice to block flow. The check valve in this location is positioned to pass refrigerant flowing toward the indoor coil, so the refrigerant flows through the check valve and is not metered. In the heating mode, the refrigerant flows in the other direction and the opposite occurs; that is, refrigerant is metered by the metering device at the input to the outdoor coil.

A similar arrangement can be made using thermostatic expansion valves (abbreviated as TEVs and TXVs) and check valves, as shown in *Figure 16*. TXVs with built-in check valves are also available. Systems using TXVs operate more efficiently than those using fixed metering devices, but are more expensive. However, with recent changes mandating increases in minimum energy efficiency standards, TXVs are appearing in more new systems to help achieve higher efficiency ratings

Point out that heat pumps typically use a metering device and check valve at each coil.

Show Transparency 18 (Figure 16). Explain that a TXV can also be used.

Provide a metering device and check valve for the trainees to examine.

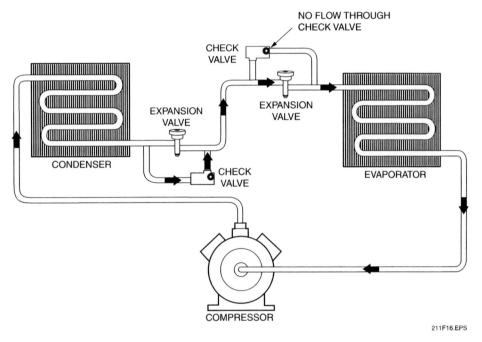

Figure 16 ◆ Use of thermostatic expansion valves in a heat pump.

Check Valves and Fixed-Orifice Metering Devices

Today, many fixed-orifice metering devices combine the check valve and expansion device valve functions in the same assembly. Special heat pump TXVs are also available with a built-in check valve and bypass circuit.

Explain that heat pumps use liquid accumulators to help prevent liquid slugging and balance the refrigerant charge between cycles.

Show Transparency 19 (Figure 17). Discuss the controls required on heat pump thermostats.

5.3.0 Liquid Accumulators

A liquid accumulator is a large cylinder or container in series in the refrigerant suction line and is often necessary on many heat pumps. It is usually placed close to the compressor to help eliminate compressor damage due to liquid slugging. It collects liquid refrigerant and prevents it from entering the compressor.

Another function of the accumulator is to adjust the charge between the two cycles. During the heating mode, more refrigerant is available than is required for operation, so it has to be stored. Without the accumulator, this liquid could spill into the compressor crankcase and cause mechanical failure. A metering device built into the accumulator feeds the refrigerant oil and refrigerant to the compressor at a controlled rate.

5.4.0 Thermostats

A two-stage heating, one-stage cooling thermostat is used to control most residential heat pumps. The first heating stage controls the compression heat (heat pump), while the second controls the supplemental electric heat. Many water-source units, not equipped with supplemental heat, use a simple one-stage heating, one-stage cooling thermostat. Although a number of mercury-filled, electromechanical units are still in service *(Figure 17)*, digital models now dominate the installation/repair market entirely. Heat pump thermostats are often equipped with lights to indicate the performance status of the system, display filter

Figure 18 ♦ Digital, programmable thermostat.

change reminders, and provide built-in, five minute delay-on-break compressor start protection. Today's advanced, programmable units *(Figure 18)* provide a wide variety of scheduling options and features. They use solid-state logic to more efficiently control and lock out supplemental heat while recovering from temperature setback, and have become quite user-friendly. Models able to also control humidification and dehumidification cycles and ventilation air equipment continue to be introduced, providing complete environmental and indoor air quality control.

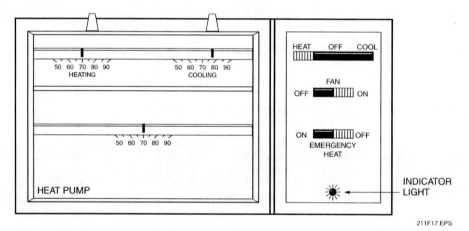

Figure 17 ♦ Heat pump thermostat.

5.5.0 Pressure Controls

Both high-pressure and low-pressure safety controls are frequently installed on heat pumps. These two pressure controls are in addition to the defrost termination pressure switch. Physically, the high-pressure control is located in the compressor discharge line between the compressor and the reversing valve.

The high-pressure control protects the system against an excessive pressure buildup in both the heating and cooling cycle. Electrically, this control is connected in series with the coil of the compressor contactor (*Figure 19*).

The low-pressure control is located in the liquid line and has the primary function of protecting the system against a loss of charge. Therefore, it is often referred to as the loss-of-charge switch. The low-pressure control is wired in series with the compressor contactor coil and will open on a drop in pressure.

5.6.0 Crankcase Heaters

Heat pump compressors are often equipped with crankcase heaters. The crankcase heater prevents compressor damage as a result of liquid slugging that can occur on cold start-up. It is usually wired into the circuit through a crankcase heater relay so it is on whenever the compressor is off and the main disconnect is on. A thermostat may also be used with the crankcase heater relay (*Figure 20*) to turn on the heater when the outdoor temperature falls below a preset level. As shown, energizing voltage is available to the heater whenever the compressor contactor is de-energized.

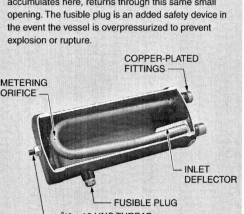

Liquid Accumulators

The liquid accumulator, also commonly referred to as a suction accumulator in reference to its installation location, captures liquid refrigerant before it can reach the compressor and cause damage. The compressor draws vapor from the top through a U-shaped tube, while liquid accumulates in the bottom, often creating a visible frost line on the outside of the vessel as the liquid inside boils to a vapor. A small hole at the bottom of the U-tube allows small amounts of liquid refrigerant to enter the suction vapor stream, where it is vaporized before reaching the compressor. Oil, which also accumulates here, returns through this same small opening. The fusible plug is an added safety device in the event the vessel is overpressurized to prevent explosion or rupture.

Show Transparency 20 (Figure 19). Discuss the pressure controls used in heat pumps.

Show Transparency 21 (Figure 20). Explain that heat pumps use crankcase heaters to help prevent liquid slugging on cold starts.

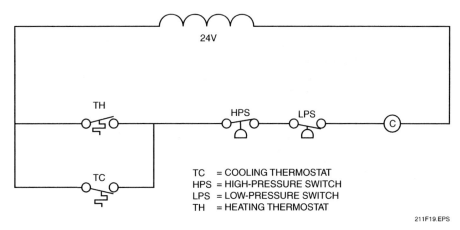

Figure 19 ♦ Pressure switches used to protect the compressor.

Point out that if the unit has been turned off for a long period, the crankcase heater must be energized for 24 hours before system start-up.

Show trainees how to identify the components on a heat pump.

Have trainees practice identifying the components on a heat pump. Note the proficiency of each trainee. This laboratory corresponds to Performance Task 1.

Explain that electric heating units are used as a supplemental heat source. Identify situations where supplemental heat is needed.

Show Transparency 22 (Figure 21). Describe a fan coil with electric heating elements.

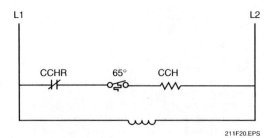

Figure 20 ♦ Crankcase heater control.

If power is disconnected for a long period of time, no attempt should be made to start the unit for 24 hours after power is restored. This allows all of the liquid refrigerant to be driven out of the compressor by the crankcase heater. A warning label pertaining to supplying power to the crankcase heater should be provided with the installation instructions for the outdoor unit. Attach this label to the disconnect switch where the owner or occupant can see it.

6.0.0 ♦ SUPPLEMENTAL ELECTRIC HEAT

Electric heating systems use electricity as an energy source, rather than combustible fuels. In all electric heat systems, electrical current flows through resistive heating elements, causing them to generate heat. The resistive heating elements can be positioned in the airstream of a fan coil or packaged HVAC unit where the heat is transferred into the supply air. In a majority of heat pump systems, electric heat is used to provide a source of supplemental heat in the following situations:

- When the structure is below the **balance point**
- When recovering from setback (unless disabled by thermostats with the capability to do so)
- When the thermostat setpoint is increased sufficiently to activate second-stage heating
- During defrost operation to temper supply air temperatures

Some control schemes disable all or a portion of the electric heat above a predetermined outdoor temperature to reduce unnecessary energy consumption. In addition, the electric heat provides an emergency backup in the event of a heat pump unit failure.

When emergency heat is selected at the wall thermostat, any controls linking the heating elements to outdoor temperature are generally bypassed.

6.1.0 General Description

The fan coil with electric heating elements (*Figure 21*) converts electrical energy to heat energy. This conversion takes place in resistance heaters that convert electrical energy into heat energy.

Electric forced-air heating systems differ from fuel furnaces in that no special heat exchanger is required. Because no fuel is burned, neither a chimney nor a vent is needed to carry the products of combustion outdoors. This feature allows for greater safety and more installation flexibility. The return air from the conditioned space passes directly over the resistance heaters and into the supply air plenum. The amount of heat supplied by an electric forced-air heating system depends upon the number and size of the resistance heaters used.

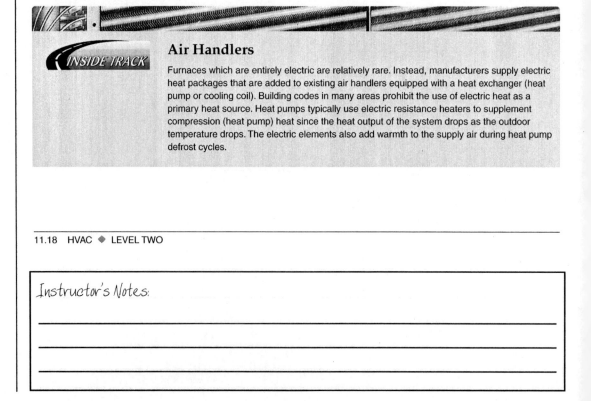

Air Handlers

Furnaces which are entirely electric are relatively rare. Instead, manufacturers supply electric heat packages that are added to existing air handlers equipped with a heat exchanger (heat pump or cooling coil). Building codes in many areas prohibit the use of electric heat as a primary heat source. Heat pumps typically use electric resistance heaters to supplement compression (heat pump) heat since the heat output of the system drops as the outdoor temperature drops. The electric elements also add warmth to the supply air during heat pump defrost cycles.

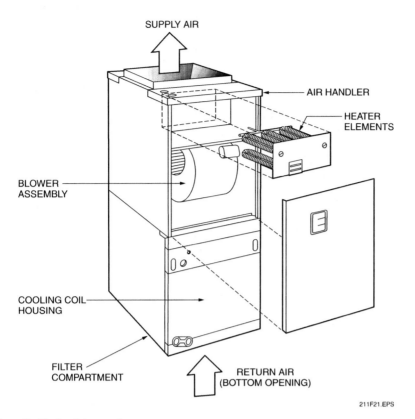

Figure 21 ♦ Fan coil with electric heat package.

6.2.0 Major Components

The major components of a fan coil with electric heat (*Figure 22*), excluding the controls, are the heating elements; the blower and motor assembly; the heat pump or indoor coil; the furnace enclosure or cabinet; and the filter.

6.2.1 Heating Element

The heating element provides the heat required for the conditioned space. Its wires are an alloy of nickel and chromium (Nichrome™). The heating element wire is spiraled and threaded through a metal holding rack, which has ceramic insulators that prevent the resistance wires from shorting out on the frame of the rack.

A forced-air electric heating system may have one or more banks of heating elements, depending on the required capacity. For example, assume there are four banks of heating elements. If each provides 5kW, the furnace capacity would be 20kW. This converts to 68,260 Btuh when multiplied by 3.413, which is the number of Btus per kilowatt. At 240V, each 5kW circuit draws about 21A (I = P/E). Because of the high current drawn by the resistive heating elements, banks of electric heat are often controlled independently in a heat pump system based on outdoor temperature, mode of operation, or other control input.

Show Transparency 23 (Figure 22). Identify the major components of a fan coil with electric heat.

Take apart a fan coil with electric heating elements and show trainees the components.

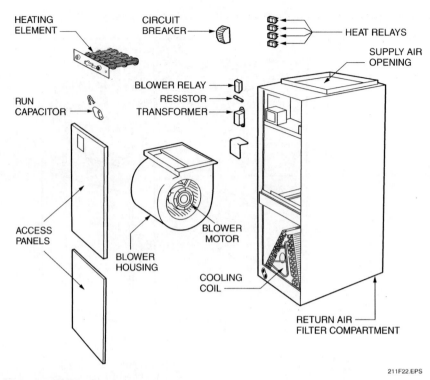

Figure 22 ♦ Fan coil components.

The Rise and Fall of Electric Heating

In the 1960s, electricity was plentiful and inexpensive. The all-electric home was the ideal. Electric baseboard heaters were relatively inexpensive to install in comparison to forced-air systems, which required the installation of ductwork. Electric baseboard heat gave the added advantages of zoned heating and clean operation. In addition, utility companies were offering incentives to people who built homes with electric heat. In some parts of the country, buildings were heated with electric furnaces, which are forced-air systems using banks of heating coils instead of a fuel-burning heat source.

Over the next decade, the cost of electricity gradually rose, while the cost of fossil fuels such as natural gas and heating oil remained steady or declined. Heat pumps came into common use in many parts of the country. Eventually, the cost of heating homes with electricity far outweighed the cost of other methods, and the owners of all-electric homes began having trouble selling their homes. Conversion to forced-air heat was prohibitively expensive because of the difficulty in installing ductwork in a finished home.

Today, electric heat serves as the benchmark for measuring the efficiency of all other heating methods. A heat pump, for example, is 1.5 to 3 times more efficient than electric heat.

Electric heat is still in use because it is ideal for certain applications. One very common use is placing electric heaters under windows in commercial buildings that use forced-air heat. The electric heaters deal with the puddle of cold air that forms under a window in cold climates. These heaters are usually controlled by outdoor thermostats, so they are only used when the outdoor temperature is low.

Electric heater packages (*Figure 23*) typically range in capacity from about 17,000 Btuh (5kW) to 120,000 Btuh (35kW). At 240V, the higher capacity packages will draw about 145A. The equipment manufacturer will supply all components necessary to install the pre-wired heater packages. In most fan coils or packaged units, an access panel is removed to insert the actual heater.

The sheet metal in the air handler cabinet is pre-drilled to accept the other components, such as the 24V transformer, relays, and terminal strip. New wiring diagrams and complete instructions are provided.

6.2.2 Blower and Motor Assembly

Like those used in fuel furnaces, the fans used in fan coils and packaged units equipped with electric heaters are usually direct-driven. Most are equipped with multi-speed blower motors and are sized to handle the airflow requirements of cooling, which tend to be greater than those of heating. Field adjustment of multi-speed blowers may be required.

6.2.3 Enclosure

The casing of an air handler is similar to that of a gas or oil furnace, but without the vent pipe connection. The interior of the cabinet is designed to permit the air to flow first over the cooling coil and then over the heating elements, which are usually insulated from the exterior casing by an air space.

6.2.4 Accessories

Filters and humidifiers are added to a fan coil in much the same manner as they are in gas and oil furnaces. Therefore, a fan coil with electric heaters can provide all of the climate-control features found in fuel-fired furnaces.

6.2.5 Power Supply

Fan coils with electric heaters usually require 208/240V single-phase, 60Hz AC. This type of heating unit is supplied by three wires: two hot and one grounded. The hot lines leading to the furnace contain fused disconnects. The connection at the air-handling unit is typically done at a single-point terminal block, which distributes power internally to the heating elements, blower, and control transformer.

All wiring should be enclosed in conduit with the proper connectors, as specified by the *National Electrical Code®* (*NEC®*). Because these furnaces use 240V, every possible protection must be provided. The *NEC®* also requires that the fan coil be grounded. A supply ground is provided for that purpose. Fuses or circuit breakers may be used at the cabinet terminal block. Check local codes for maximum temperature and clearance from combustible building components.

6.3.0 Safety Controls

Safety controls must be included in any forced-air electric heating system. These include limit switches and thermal fuses.

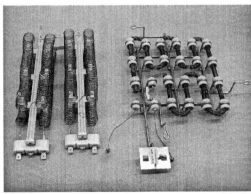

Figure 23 ♦ Electric heater packages.

Describe an electric heater package.

Describe the blower and motor assembly, the enclosure, accessories, and the power supply.

Identify the primary safety controls.

Electrical Disconnects

Packaged air conditioners and heat pumps equipped with electric heaters may require more than one electrical disconnect. For example, one disconnect would supply power to compressor and fan motor circuits while another disconnect would power the electric heaters. This is commonly seen in large-capacity equipment installed in commercial settings. On some commercial units, higher voltages and three-phase power circuits may be encountered. Exercise appropriate caution when working around energized electrical circuits.

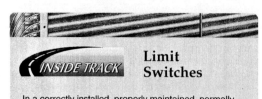

Limit Switches

In a correctly installed, properly maintained, normally operating system, a limit switch may never open during the life of the system. If you encounter a forced-air electric heating system that is cycling on the limit switch, there is probably something wrong. Restricted airflow, dirty filters, and an inadequate duct system are some problems that could cause a limit switch to open. Each of these conditions results in increased temperature due to reduced airflow.

6.3.1 Limit Switch

The function of the limit switch is to de-energize the heating element branch circuit when the ambient temperature surrounding the element exceeds the limit switch setting. The limit switch protects the heating element and surrounding materials.

The limit switches are wired in series with the heating elements, and have a disc-type, bimetal sensor placed in a small metal housing. The position of the switch is such that the metal shell absorbs radiant heat from the element. As the blower moves air across the metal shell of the limit switch, heat is dissipated from the switch and it remains closed. If the air cannot be moved for some reason, the heat is absorbed by the shell. The bimetal sensor reacts to the excessive heat and will snap the contacts open, taking the element off-line. When the element cools down, the limit switch automatically closes and the heating element will begin to function again.

There is a differential between the temperature at which the contacts open and the one at which they close. The close setpoint is usually 35°F cooler than the open setpoint. Limit switches have different open and close setpoints; therefore, only a limit switch with the same rating as the original may be used as a replacement for a defective switch.

6.3.2 Thermal Fuse

The thermal fuse is designed to open the line voltage circuit to the heating element if an abnormally high temperature exists around the element. It acts as a backup safety switch if the limit switch fails to operate. It has a higher temperature opening setting time than the limit switch. The thermal fuse must be replaced with one of the same rating. If both the limit switch and thermal fuse open, the cause should be determined and repaired before the unit is put back into service.

NOTE

A manual reset limit switch may be used in place of a thermal fuse. Since it must be manually reset, the manual reset device provides the same type of protection as the thermal fuse.

6.4.0 Determining Btuh Output

The heating capacity of electric heating elements is rated in thousands of watts or kilowatts (kW). As previously discussed, kilowatts can be translated into Btuh output. Recall that 1 watt equals 3.413 Btuh heat output. Thus, 1kW = 3,413 Btuh.

Most resistance elements are rated for full power at 240V input. This data is stamped directly on the frame of the heating element. As the voltage decreases, the kilowatt capacity also decreases. To determine the actual kilowatts at the unit for various voltages, use *Table 1* as a guide.

After using a VOM/DMM to find the resistance of an element in ohms (*Figure 24*), Ohm's law can be used to determine the actual kilowatt output of the element. *Figure 25* shows the various relationships between voltage, power, current, and resistance.

Table 1 Voltage Multipliers

Volts	Multiplier
208	Multiply output at 240 volts by 0.715
220	Multiply output at 240 volts by 0.839
230	Multiply output at 240 volts by 0.917

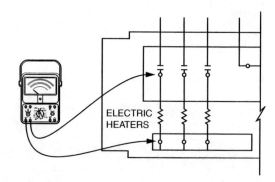

WITH THE POWER OFF, MAKE A RESISTANCE CHECK ON EACH HEATING ELEMENT WITH OHMMETER. IF INFINITY IS READ, REPLACE THE ELEMENT.

Figure 24 ♦ Resistance check.

Multi-Stage Heat Control

A heat pump with two stages of electric heat is really a three-stage heating system. The first stage is reversed cycle (compression) heat. When there are two stages of electric heat, the second stage is often controlled by an outdoor thermostat (ODT). This stage does not kick in until the outdoor temperature falls below a preset level (55°F is a common setting).

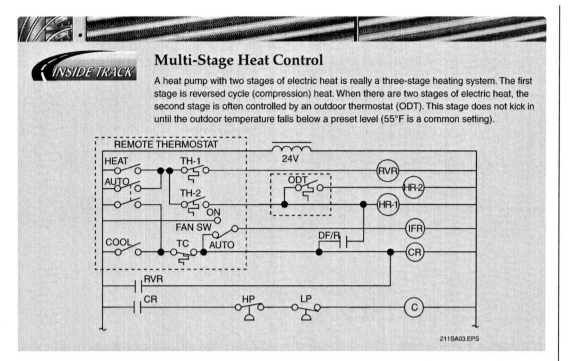

Work the sample problem on the whiteboard/chalkboard.

Provide sample problems for the trainees to solve.

Explain that heat pump performance can be measured in several ways.

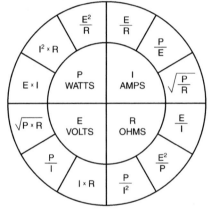

E = VOLTAGE
I = CURRENT
R = RESISTANCE
P = POWER

Figure 25 ◆ Ohm's law and derivative equations.

NOTE

The resistance changes as the coil heats up. Use a correction factor of 1.065 when making calculations on a cold coil.

The following formula can be used to calculate the output of an element with a known resistance:

$$P = \frac{E^2}{R}$$

Study Example

Find the kW value of a heating element with a cold resistance of 12.26 ohms (Ω).

Using the applicable power formula, find P (in watts):

$$P = \frac{E^2}{R}$$

$$P = \frac{240^2}{12.26} \times 1.065$$

$$P = 5{,}004W, \text{ or about } 5kW$$

7.0.0 ◆ HEAT PUMP PERFORMANCE

The electric heat pump, when operating as a cooling unit, removes more heat from the conditioned space than the electrical energy required to run it. Likewise, when operating as a heating unit, it provides more heat energy to the conditioned space than the electrical energy it consumes. Heat pump economy can be measured using one of several rating methods.

Show the trainees how to calculate the Coefficient of Performance (COP) of a heat pump for both cooling and heating.

Work the sample problem on the whiteboard/chalkboard.

Provide sample problems for the trainees to solve.

Explain how to calculate the Coefficient of performance and have trainees answer the sample problem. See the answers to the "Think About It" at the end of this module.

Describe the Performance Factor.

7.1.0 Coefficient of Performance

The ratio of useful heat delivered to the equivalent heat consumed in operating the entire system is known as the **Coefficient of Performance (COP)**. This ratio is measured in Btus per hour (Btuh) or kilowatts and is determined under specific operating conditions. Occasionally, the COP may be specified with respect to only the compressor or to any given component or portion of the system.

Heat pump efficiency for heating is defined as follows:

$$COP = \frac{Btuh\ output}{kW\ input \times 3,413\ Btu}$$

or

$$COP = \frac{Btuh\ output}{watts \times 3,413}$$

When electric heat is used, it generates 3,413 Btus of heat for each kilowatt of power consumed. Therefore, the COP of straight electric resistance heat is 1 and can never be any higher.

To calculate the COP of a three-ton heat pump delivering 36,000 Btuh, and with a total input including fan motor wattage of 4.8kW, proceed as follows:

$$Cooling\ COP = \frac{Btuh\ output}{kW\ input \times 3,413}$$

$$Cooling\ COP = \frac{36,000}{4.8 \times 3,413} = 2.197$$

Cooling COP = 2.2 (rounded off)

In order to calculate the COP of the heat pump in the heating mode for the same size unit, assume that the heating capacity of the same unit is 33,480 Btuh with a total input, including fan motor wattage, of 3.768kW.

$$Heating\ COP = \frac{Btuh\ output}{kW\ input \times 3,413}$$

$$Heating\ COP = \frac{33,489}{3.768 \times 3,413}$$

Heating COP = 2.6

Study Examples

1. Calculate the heating COP of a heat pump with a heating capacity of 55,800 Btuh, and with a total input of 6.28kW.

$$Heating\ COP = \frac{Btuh\ output}{kW\ input \times 3,413}$$

$$Heating\ COP = \frac{55,800}{6.28 \times 3,413}$$

Heating COP = 2.6

2. Calculate the cooling COP of a heat pump with a four-ton cooling capacity consuming 4.03kW.

$$Cooling\ COP = \frac{Btuh\ output}{kW\ input \times 3,413}$$

$$Cooling\ COP = \frac{4 \times 12,000\ Btuh\ per\ ton}{4.03 \times 3,413} = 3.489$$

Cooling COP = 3.5 (rounded off)

When COP figures are used to compare different heat pump systems or installations, care must be taken. The COPs should be based on the same operating conditions, the same complete systems, or the same components. The COP varies with operating conditions; therefore, it cannot be used to compare systems operating over a period of time. The advantage of COP is that it uses like units. The kilowatt input is multiplied by 3,413 to convert it to Btus.

7.2.0 Performance Factor

The Performance Factor (PF) is used to calculate the ratio of the total heating or cooling energy delivered during a given period (usually one season). When calculated for one season, it is called the **Heating Season Performance Factor (HSPF)**. The PF is more often associated with the heating mode of the heat pump system rather than with the cooling mode.

The HSPF is found by dividing the total heating output for a season by the total energy consumed during that season. For example, to

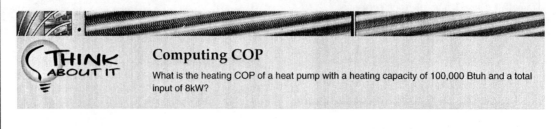

Computing COP

What is the heating COP of a heat pump with a heating capacity of 100,000 Btuh and a total input of 8kW?

11.24 HVAC ♦ LEVEL TWO

Instructor's Notes:

calculate the HSPF of a heat pump system that produces 102,000 Btus of heating energy while consuming 14,560 kW/hour (which includes power used to supply resistance heaters and drive all electrical components), use the formula:

$$HSPF = \frac{\text{total seasonal heat output (Btu)}}{\text{total power input (watt-hours)}}$$

$$HSPF = \frac{\text{total Btuh output}}{\text{total kW/hour}}$$

$$HSPF = \frac{102,000}{14,560}$$

$$HSPF = 7.0$$

Calculated performance factors are affected by several variables: climatic conditions; sizing and application of equipment; building characteristics and their internal loads; and the type of heat pump systems. The advantage of the HSPF is that it can be used to compare systems operating over a period of time. Unlike COP, however, HSPF uses mixed units (Btu output divided by kilowatt input).

Residential heat pump installations in the southern regions of the U.S. operate with higher HSPFs than similar installations in the northern regions because the average outdoor air heat source temperature is higher. Commercial and industrial installations usually operate with higher HSPFs than residential installations due to higher motor and compressor efficiency. Care must be exercised to make sure the same proportion of components are included when HSPFs are used to compare heat pump systems.

7.3.0 Seasonal Energy Efficiency Ratio

The Air Conditioning and Refrigeration Institute (ARI) established the **Seasonal Energy Efficiency Ratio (SEER)** as a standard by which all heat pumps are rated. The efficiency of central air conditioning and heat pump units is governed by U.S. law and regulated by the U.S. Department of Energy (DOE). Every air conditioning and heat pump unit is assigned an SEER value, defined as the total cooling output (in Btus) provided by the unit during its normal annual usage period, divided by its total energy input (in watt-hours) during the same period. Initially, units were rated by EER, rather than SEER. However, EER reflects only a snapshot of unit efficiency, reflecting its performance at only one outdoor temperature, usually 95°F. The SEER reflects performance over the entire operating season for a clearer picture of actual results. Higher values are preferred, as it indicates that a unit provides more capacity per watt-hour of power input than a lower SEER value. SEER values provide a clear picture to consumers of the anticipated energy cost for comparison of various units prior to purchase.

Heat Pump Ratings

SEER ratings are generally used to express the cooling efficiency of air conditioning/heating systems, while COP and HSPF ratings are used to express the heating efficiency of these systems. Equipment that is certified by the ARI has an ARI label like the one shown here. For such equipment, the certified capacities, SEER, COP, and HSPF ratings, as well as other performance data, can be found by looking up the equipment by manufacturer and model number in the appropriate ARI Performance Rating Directory. These directories are available in printed copy and CD-ROM versions. This information is also available on the ARI website (http://www.ari.org).

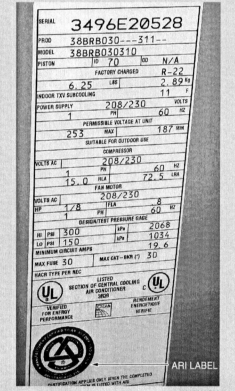

Show the trainees how to calculate the Heating Season Performance Factor (HSPF) of a heat pump.

Provide sample problems for the trainees to solve.

Show the trainees how to calculate the Seasonal Energy Efficiency Ratio (SEER) of a heat pump.

Provide sample problems for the trainees to solve.

Discuss the U.S. Department of Energy (DOE) recommendations for heat pump efficiency.

Show Transparency 27 (Figure 26). Define balance point.

Explain how to find the balance point for typical heat pumps using manufacturers' heat pump balance point calculation sheets.

Provide sample problems for the trainees to solve.

Show trainees how to calculate the balance point of a heat pump.

Have trainees practice calculating the balance point of a heat pump. Note the proficiency of each trainee. This laboratory corresponds to Performance Task 2.

As a result of new federal energy standards adopted in 2006, a SEER rating of 13.0 is the minimum rated efficiency level allowed by the DOE for new equipment. However, heat pumps with SEER levels in excess of 20 are now readily available, and increases in efficiency levels continue through engineering advances.

7.4.0 Heat Pump Efficiency Recommendations

The DOE Energy Star® program rates various appliances and residential heating and cooling products according to their efficiency. Citizens are encouraged to purchase Energy Star® products to help conserve energy. Many local utilities have rebate programs to encourage their customers to buy Energy Star® products. According to the DOE, the most efficient heat pumps available today have an HSPF of 11 and SEER ratings of 21. As of April, 2006, the minimum standard for an Energy Star® heat pump is an HSPF of 8.2 for split systems and 8.0 for packaged units. The minimum SEER rating to achieve the designation is 14 for both types of systems. The DOE publishes lists of products that qualify for Energy Star® ratings. This list is available at energystar.gov. Note that the higher the value for SEER, the higher the HSPF.

8.0.0 ◆ BALANCE POINT

The balance point of a system is the outdoor temperature at which the heating capacity of the heat pump is equal to the heat loss of the building. The balance point will usually vary depending on the climate, building design, type of construction, and other factors which affect heat loss or gain. Generally speaking, the lower the balance point, the more economical a heat pump system will be. The best method of lowering the balance point is to reduce the heat loss of the structure. It is important to note however, that system capacity for a given application is generally selected based on the cooling load of the structure. Once the equipment is chosen, the balance point can be calculated more accurately.

Figure 26 illustrates the effects of outdoor temperature on the operation of an air-to-air heat pump. Line A indicates that the heating capacity of the unit decreases with decreasing outdoor temperatures. At the same time, the heat loss and the heating load necessary to displace the heat loss increases. This is shown by line B. The balance point of the system is the temperature at which the heat loss is equal to the heating capacity of the heat pump. The balance point, therefore, is the intersection of lines A and B. As shown in the graph, it is about 23°F for this case.

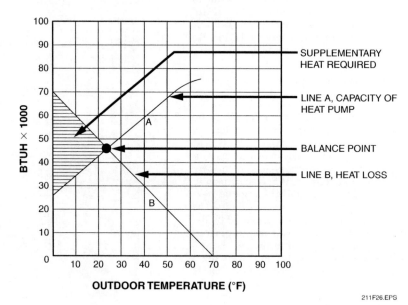

Figure 26 ◆ Heat pump balance point.

Inside Track: Outdoor Thermostats

In some parts of the country, the electric utility mandates the use of outdoor thermostats to control all stages of supplemental electric heaters.

Supplementary heaters are generally used to maintain the capacity of the unit when temperatures fall below the balance point. These heaters are usually electric resistance heaters. If heat staging is used (individual banks of electric heaters are staged on as temperature decreases), some of the heaters will be controlled by outdoor thermostats.

9.0.0 ♦ INSTALLATION

This section covers the installation of split systems, packaged units, and **dual-fuel or hybrid systems**.

9.1.0 Split Systems

A split system (*Figure 27*) is one that has two separate units. The outdoor unit contains the compressor, reversing valve, outdoor coil and fan, and controls. In a heat pump system, this unit would also contain one of the metering devices. The indoor unit, which is also known as the air handler or **fan coil unit**, contains the indoor fan (blower), metering device, and the indoor coil. Electric heaters, if installed, would also be located in this unit.

A variation on the split system is the triple-split system, in which the outdoor unit contains only the outdoor coil, outdoor fan, and a metering device. The compressor, reversing valve, and controls are located in a separate unit designed to be installed indoors. This arrangement is very efficient and is much easier to service in bad weather.

Another type of add-on system is the ductless split system (*Figure 28*), which is designed to serve a single room or area. The indoor coil in these applications is mounted on a wall or in a drop ceiling. These systems are especially good for add-on applications because they require no

Have the trainees review Sections 9.0.0–9.4.0.

Ensure that you have everything required for teaching this session.

Show Transparency 28 (Figure 27). Discuss the components and operation of a typical split system.

Show Transparency 29 (Figure 28). Point out that two variations of this system are the triple-split system and the ductless split system.

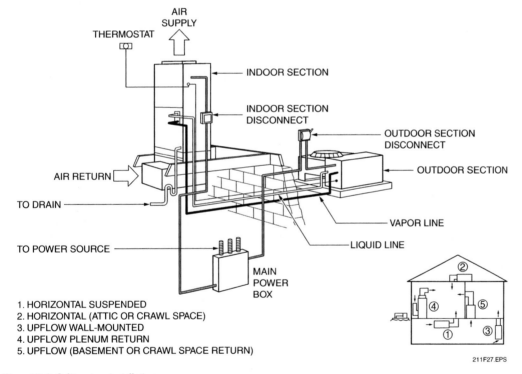

Figure 27 ♦ Split system installation.

MODULE 03211-07 ♦ HEAT PUMPS 11.27

Discuss the requirements for locating and installing the outdoor unit.

Discuss the requirements for locating and installing the indoor unit.

Show Transparency 30 (Figure 29). Discuss the components and operation of a typical packaged system.

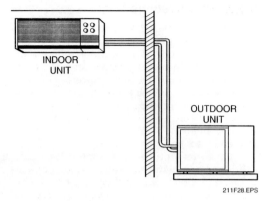

Figure 28 ♦ Ductless split system.

ductwork. A family room addition in a home is a good example. They may also be a good choice for historical buildings where the installation of ductwork would destroy the character of the building.

There are versions of these ductless systems in which a single outdoor unit serves two or three indoor units. These are known as multiplexed systems. One type of indoor unit has connections for two or three lengths of flexible duct that can distribute air to diffusers or rooms within 10' or 15' of the unit.

Several important factors must be considered before selecting the location of the indoor and outdoor units of a split-system heat pump. These are explained in the following paragraphs.

9.1.1 Outdoor Section Location

The location of the outdoor unit of the air-to-air heat pump should consider these factors: sound, wind direction, proximity to the structure, and the treatment of defrost and drainage. The recommended procedure is to select a site near the residence, but away from bedroom windows or rooms where the sound might be objectionable. The unit should not be located where discharge air will be directed toward windows of neighboring residences. Local ordinances often impose restrictions on the placement of outdoor units. It is very important, therefore, to consult local codes before starting the installation.

If the outdoor unit is to be placed in front of the building, it should be concealed by shrubbery so that the sound level will blend with traffic sounds. Ample clearance should be allowed around the unit for air movement and service access. Clearance dimensions are usually specified in the manufacturer's installation instructions. Local codes may also apply.

The outdoor unit should be located near enough to the building to eliminate lengthy piping runs, but it should not be placed directly below eaves or gutters. Prevailing winds can interfere with defrost. This can be dampened by placing the unit so that wind does not blow directly across the outdoor coil. Trees, shrubs, fences, or even corners of buildings will reduce capacity losses.

9.1.2 Outdoor Section Mounting

Once the location of the outdoor unit has been selected, the unit should be mounted in accordance with recommendations for existing climate conditions. Where there is little or no snowfall, a concrete pad is recommended. The thickness of the pad should provide about 6" between the base of the unit and the gravel bed in the ground. A recession formed in the slab under the coil will allow drainage of condensate and melting frost.

Where snowfall is heavy, it is recommended that the unit be mounted on an angle-iron frame with the supports imbedded in a concrete base. The unit should be 12" to 18" above the ground.

The unit should be level and rigidly mounted. Proper clearance for service access and airflow must be allowed. A 12" gravel bed extending out and away from the perimeter of the outdoor section should be provided to prevent mud splashing. Care must be taken to make sure that condensate does not drain directly onto areas that could become frozen and slippery and result in personal injury to pedestrian traffic.

9.1.3 Indoor Section Placement

The indoor section of a split system may be placed in the basement, utility room, attic, closet, or attached garage. The indoor location should be selected to minimize the length of piping runs between the indoor and outdoor units while allowing for efficient ductwork design. Wire and conduit lengths from the main service panel to the electric heaters must also be considered. The supply and return air distribution duct runs should have canvas flex connections on either side of the plenums. Internal acoustical duct lining and vibration isolators or sound-dampening pads should also be incorporated into the unit installation procedure.

9.2.0 Packaged Units

A packaged unit (*Figure 29*) contains all the components in a single package, which is located outside the building. The supply and return ducts of a packaged unit penetrate the building. This is

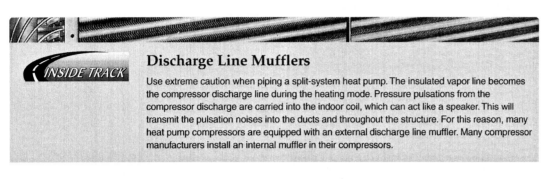

Discharge Line Mufflers

Use extreme caution when piping a split-system heat pump. The insulated vapor line becomes the compressor discharge line during the heating mode. Pressure pulsations from the compressor discharge are carried into the indoor coil, which can act like a speaker. This will transmit the pulsation noises into the ducts and throughout the structure. For this reason, many heat pump compressors are equipped with an external discharge line muffler. Many compressor manufacturers install an internal muffler in their compressors.

significantly different from a split system, in which only the refrigerant piping penetrates the wall or roof. For that reason, packaged units do not usually make good add-on systems. Packaged units are built in a variety of configurations, serving cooling loads ranging from a couple of tons to 75 tons or more. The larger units will have multiple compressors. Packaged units are available with gas heat for cold climates. Heat pump versions are popular in more moderate climates.

Packaged heat pump systems require the same installation considerations as the outdoor units of split systems. There are also several additional considerations. One pertains to the added weight and larger dimensions of the packaged units; since the packaged unit contains the indoor coil and blower in addition to the outdoor coil, blower and compressor, it will require sturdier mounting. This is especially true in the snow belt areas where 12" to 18" of ground clearance is recommended.

Figure 30 shows typical locations for packaged units; roof mounting is frequently preferred in the south and southwest. Supply and return ductwork must be insulated if it runs through an unconditioned attic or other spaces. Adequate weatherstripping and sealer must be applied where plenum or ductwork connections enter the roof or walls of the building.

9.3.0 Dual-Fuel or Hybrid Systems

Due to the cost of oil and gas, the dual-fuel or hybrid air-to-air heat pump (*Figure 31*) has some advantages for the homeowner. When it is more efficient or economical to heat with the heat pump (usually between 35°F and 65°F) the system automatically selects that mode. When it is more efficient or economical to heat with gas or oil, the system automatically selects that mode. Overall

Figure 29 ♦ Packaged unit installation.

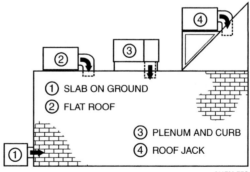

Figure 30 ♦ Packaged unit locations.

Keep Transparency 32 (Figure 31) showing. Describe the procedures for locating and installing an add-on system.

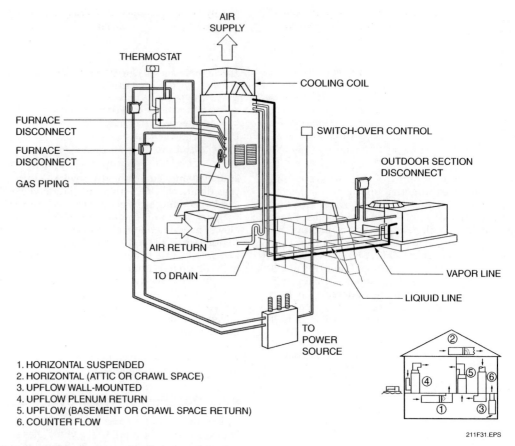

1. HORIZONTAL SUSPENDED
2. HORIZONTAL (ATTIC OR CRAWL SPACE)
3. UPFLOW WALL-MOUNTED
4. UPFLOW PLENUM RETURN
5. UPFLOW (BASEMENT OR CRAWL SPACE RETURN)
6. COUNTER FLOW

Figure 31 ♦ Add-on heat pump installation.

heating bills will be lower using this method than when heating with a single fuel. The add-on system shown in *Figure 31* features a cooling coil added to the top of the furnace and connected to a heat pump outdoor unit by refrigerant piping. This is a relatively inexpensive way to add central cooling to a building served by a furnace. The add-on heat pump uses the existing furnace as a backup heat source.

When the temperature falls below the heat pump's balance point, a dual-fuel heating system such as this is much more cost-effective and energy efficient than using electric heaters. An outdoor thermostat is used to control a switching circuit that turns off the heat pump and turns on the furnace when the temperature falls below a selected setpoint such as 35°F.

It is important to note that the heat pump and fossil fuel furnace cannot be operated simultaneously, unlike systems with electric heat. This is primarily due to the fact that the heat pump indoor coil is typically located in the supply air stream of the furnace. With the furnace in operation and high temperature air flowing across the indoor coil, the heat pump would rapidly develop excessive discharge pressures and shutdown due to opening of the high-pressure safety switch. If a high-pressure switch is not present or fails to function, the compressor internal relief valve would open.

When a heat pump is added to an existing heating system, the furnace blower must have sufficient static pressure to deliver the manufacturer's recommended cfm per ton across the evaporator coil. If the blower cannot produce this capacity, either the motor and blower assembly should be replaced with one of sufficient capacity, or the add-on unit should not be installed. Two-speed blowers are common in systems of this type.

Instructor's Notes:

A heat pump system generally requires 450 to 500 cfm per ton. The size of the ductwork is critical in the heating mode. The capacity of the ductwork must meet or exceed the required 450 to 500 cfm per ton.

CAUTION

The heat pump must be turned off before activating gas or oil heat. The hot air from the furnace will cause excessive high-side pressure and temperature that will cause the compressor to overheat and overload.

9.4.0 Installation Checklist

Before placing a heat pump in operation, do the following, as applicable:

- Check or change filter.
- Check fan motor (indoor/outdoor).
- Check supply ducts for leaks.
- Check return air for leaks.
- Check thermostat/cycle unit.
- Check all safety controls.
- Check crankcase heater.
- Check defrost cycle operation.
- Check auxiliary heat.
- Check emergency heat.
- Check that unit is correctly charged.
- Check electrical connections and tighten, if necessary.

The following data should be collected when a heat pump is placed in operation:

- Voltage and current balance on a three-phase system
- Compressor motor current draw within manufacturer's specification
- Voltage at indoor heater at full load operation (electrical resistance heat operational)
- Measured amperage draw of each electrical resistance heating element individually
- Voltage at outdoor contactor when compressor is operating
- Condensing unit pressures (psig) on both the high side and the low side
- Superheat at compressor after accumulator
- Temperature rise with heat pump only
- Outdoor temperature
- Outdoor weather conditions (cloud cover, precipitation)
- Make, model, and serial number of indoor and outdoor units
- Any general comments that may apply
- The date the data is taken
- The location of the site

Record values for all the parameters just listed, and leave a copy at the site for future reference.

10.0.0 ◆ SERVICE

A service call is recommended prior to each heating or cooling season. The check for cooling should be made when the outdoor temperature is above 70°F, and the heating check should be made when the outdoor temperature is below 55°F.

A check sheet should be developed prior to servicing the system for recording information and also as a record of operational functions. When servicing a heat pump, follow these steps:

Step 1   If a bimetal thermostat is used, make sure it is level.

Step 2   Check the thermostat for faulty wiring and loose connections.

Step 3   Check the supply voltage and verify that it is within the allowed tolerance.

Step 4   On a three-phase system, check the voltage and current balance.

Step 5   Check the compressor motor current draw and verify that it is within the manufacturer's specification.

Step 6   Change or clean the filter(s).

Step 7   Clean the blower and the blower compartment.

Step 8   Turn off the disconnect and check all wiring for damage and loose connections.

Step 9   If applicable, check the blower motor drive belt for damage and proper alignment.

Step 10 Clean the outdoor coil as necessary.

Step 11 Apply power, then turn the thermostat down to turn on cooling. Verify that the unit runs properly in the cooling mode.

Step 12 Set the thermostat to activate heating and verify that the heating function is working properly.

Step 13 Use the built-in servicing feature to activate defrost and verify that the defrost function is working.

Step 14 Check the charge and adjust as required.

Step 15 Check the auxiliary heater(s).

Emphasize that the heat pump must be turned off before the gas or oil heat is energized.

Review the installation checklist for a typical heat pump.

Show trainees how to install a heat pump.

Under your supervision, have the trainees simulate/describe the installation procedures for a heat pump. Note the proficiency of each trainee. This laboratory corresponds to Performance Task 3.

Have the trainees review Sections 10.0.0–11.0.0.

Ensure that you have everything required for teaching this session.

Review the pre-season service procedures for a typical heat pump.

Show trainees how to perform various service procedures for a typical heat pump.

Under your supervision, have the trainees perform various service procedures for a typical heat pump. Note the proficiency of each trainee. This laboratory corresponds to Performance Task 4.

See the Teaching Tip for Section 10.0.0 at the end of this module.

Show Transparency 33 (Figure 32). Trace the operation of the control circuits for a typical heat pump.

Explain how to solve troubleshooting problems through examination of the heat pump's circuit diagram.

11.0.0 ♦ HEAT PUMP CONTROLS

The control circuits of a heat pump are more complicated than those of a cooling-only unit. There are several reasons for this. Refer to *Figure 32* and locate the components or circuits as they are discussed. The numbers in parentheses in the text relate to the circled numbers on the diagram. Try to answer the questions as you progress through the discussion. They are designed to help you understand the operation of a heat pump.

The heat pump requires a defrost control circuit that changes the position of the reversing valve to switch the unit into the defrost mode. In this unit, an electronic defrost control board (1) is used. This board receives the coil temperature and outdoor air temperature from thermistors located on or near the outdoor coil.

Question – Is the reversing valve energized or de-energized in the defrost mode?

Answer – Energized. Normally open contacts of the defrost relay control the reversing valve. When the defrost board energizes the defrost relay, its contacts (2) complete the circuit to the reversing valve.

Supplementary electric heaters are usually required on a heat pump. These heaters (3) perform two functions: they augment reverse cycle heat when the temperature falls below the balance point, and they provide heat while the unit is in the defrost mode.

Question – How is the electric heater activated during the defrost mode?

Answer – A set of normally open defrost relay contacts energizes heater relay HR1 (4) when the unit goes into the defrost mode. Normally open contacts of HR1 (5) close to energize heating element #1.

Heat Pump Setback

For periods when homes or commercial buildings are unoccupied, heating temperature setback is a common means of conserving energy. However, setting the temperature back too far on heat pump systems can actually result in reduced energy savings, especially when done with non-programmable thermostats. The recovery process to the desired occupied temperature generally results in the supplemental electric heat being energized, and remaining on throughout the process, until the space temperature has recovered within 3° of the setpoint. Most thermostats will revert to the heat pump alone within that range. Selecting a more reasonable setback—5°F, for example—will significantly reduce the recovery time and energy cost.

The vast majority of today's digital programmable heat pump thermostats are designed with this problem in mind. Using information gathered in recent days of operation relative to cycling rates and run times, these thermostats determine when to begin the process of recovery to meet the desired occupied setpoint at the appropriate time. In addition, these calculations are often based on the recovery time required without the assistance of electric supplemental heat. This is commonly referred to as adaptive or intelligent recovery.

Another type of problem can occur when unoccupied temperatures are set too low on split heat pump systems. In cold weather, while the temperature is set back and the compressor is off for lengthy periods of time, refrigerant begins to migrate to the coldest part of the system and condense to a liquid. Obviously, the logical destination is the outdoor unit. This leaves very little refrigerant volume remaining in the indoor coil—thin vapor only. Once the compressor starts, the suction pressure often drops to an abnormally low value for a short period of time. If the unit is equipped with a low-pressure safety switch, the pressure may fall below its setpoint, forcing the unit to shut down within seconds of startup. One solution often employed is to add a bypass timer to the low-pressure switch circuit. This timer, with normally closed contacts in parallel with the low-pressure switch, provides a temporary bypass of the low pressure switch for a short period of time. After the timing period has lapsed, the timer contacts open, placing the low-pressure switch back in control of the circuit. This same strategy is often used in cooling-only split systems required to operate at low ambient conditions. This problem does not generally occur in packaged units, as the entire refrigerant circuit is located outdoors.

11.32 HVAC ♦ LEVEL TWO

Instructor's Notes:

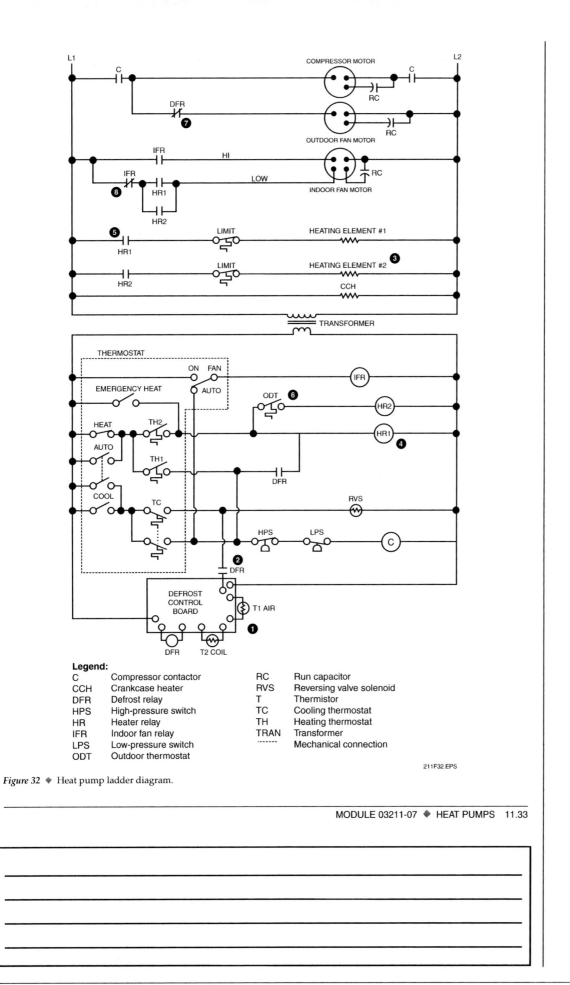

Figure 32 ◆ Heat pump ladder diagram.

Continue a discussion about troubleshooting.

Show the video *Troubleshooting Heat Pumps, Residential-Light Commercial*.

Show trainees how to troubleshoot a heat pump using a circuit diagram.

Under your supervision, have the trainees simulate/describe how to troubleshoot a heat pump using a circuit diagram. Note the proficiency of each trainee. This laboratory corresponds to Performance Task 5.

A multi-stage heat-cool thermostat is needed. It may be an automatic changeover thermostat. The first stage of the heating side of the thermostat (TH1) controls reverse cycle heat. It energizes the compressor contactor when closed. The second stage, which is the first stage of electric resistance heat, is controlled by TH2. The third stage (second stage of electric heat) is controlled by an outdoor thermostat (6).

Question – Does this circuit have any means of preventing heating element #2 from cycling on at the same time heating element #1 comes on?

Answer – Yes. Both heating elements are controlled by TH2. HR2 cannot be energized until TH2 and ODT are closed. Therefore, heating element #2 will never be on unless heating element #1 is also on.

Try to answer these additional questions. They will help you understand how heat pump control circuits are arranged.

Question – How could you modify this circuit so that the crankcase heater is on only when the compressor is off?

Answer – Add a crankcase heater relay with a set of normally closed contacts in series with the crankcase heater. The relay coil would be wired in parallel with the contactor coil.

Question – Is the outdoor fan motor on or off during defrost?

Answer – Off. Normally closed contacts of the defrost relay (7) are in series with the outdoor fan motor. When the defrost relay energizes, the motor is disabled. This is common in heat pump controls. If the fan is blowing cold air across the coil, it retards the defrost.

Question – What feature does this circuit have that makes sure the indoor blower runs whenever one of the electric resistance heaters is on?

Answer – Normally closed contacts of the indoor fan relay (8) will provide voltage to the low-speed side of the blower whenever either of the heater relays is energized. This feature is important when the thermostat has an emergency heat control that can be used to turn on the electric heaters if the reverse-cycle heating circuits fail. Without a fan to disperse the heat created by the heating elements, the unit could overheat.

Instructor's Notes:

Review Questions

1. In a water-to-air heat pump operating in the heating mode, _____.
   a. water is the heat source
   b. water is the heat sink
   c. air is the heat source
   d. water is both the heat source and heat sink

2. In a ground-source heat pump _____.
   a. refrigerant flows over a buried coil
   b. water flows over a buried coil
   c. nothing flows over the outdoor coil
   d. water flows from a well back to another well

3. During the heating cycle of a heat pump, _____.
   a. electric heaters are always energized
   b. air flows in the same direction as in the cooling mode, and refrigerant flow reverses
   c. refrigerant flows in the same direction as in the cooling mode, and airflow reverses
   d. refrigerant and air both flow in the opposite direction from cooling

4. When the unit is operating in the defrost mode, what usually happens in the conditioned space?
   a. The unit continues to provide compression heat.
   b. Occupants experience discomfort.
   c. The furnace takes over.
   d. An electric heater is cycled on to temper, or warm, the supply air.

5. A usual frequency and duration for a defrost cycle is every _____.
   a. 10 minutes for 30 seconds
   b. 90 minutes for 10 seconds
   c. 90 minutes for 10 minutes
   d. 15 minutes for 5 minutes

6. When a heat pump is in the reverse cycle heating mode, _____.
   a. the compressor discharge is routed to the indoor coil
   b. the compressor discharge is routed to the outdoor coil
   c. hot, high-pressure liquid flows through the compressor
   d. the outdoor coil is not used

7. Refrigerant metering in a heat pump _____.
   a. is done the same way as it is in cooling
   b. requires two metering devices
   c. requires two metering devices and two check valves
   d. is done with a single metering device and a check valve

8. Which of the following is *not* determined from the balance point?
   a. The setting of the outdoor thermostat that turns on supplementary heat.
   b. The temperature at which reverse cycle heat will no longer handle the heating load.
   c. The amount of supplementary heat needed.
   d. The SEER rating of the heat pump.

9. The term split system refers to _____.
   a. any system that can provide both heating and cooling
   b. a system that has two units, one indoors and the other outdoors
   c. any system that has two or more units in an outdoor unit
   d. a system in which a furnace shares the heating load with a heat pump when the outdoor temperature falls below 35°F

10. A ductless system is one in which _____.
    a. the condenser and evaporator are both indoors
    b. the ductwork is outside the building
    c. conditioned air is discharged directly into the conditioned space by the indoor unit
    d. the ductwork is both indoors and outdoors

Review Questions

Refer to *Figure 32* in the module to answer questions 11 through 15.

11. Which of the following is *not* true of the outdoor fan motor?
    a. It is controlled by the compressor contactor during cooling.
    b. It is on only when defrost is energized.
    c. It is off when defrost is energized.
    d. It is a single-phase motor.

12. When the unit is in the defrost mode, the _____.
    a. reversing valve solenoid is energized
    b. reversing valve solenoid is de-energized
    c. indoor fan turns off
    d. indoor fan operates at low speed

13. Which of the following is true about heating element #2?
    a. It turns on whenever TH1 is closed.
    b. It turns on whenever the outdoor temperature is below the ODT setpoint.
    c. It will not turn on unless the indoor fan is running.
    d. It will not turn on unless heater element #1 is on.

14. What would happen if the DFR contacts controlling HR1 stuck in the closed position?
    a. Heating element #1 would be on all the time.
    b. Heating element #1 would turn on whenever the thermostat called for heating or cooling.
    c. The unit would overheat.
    d. The unit would be in a continuous defrost mode.

15. The normally closed IFR contacts are designed to _____.
    a. keep the electric heaters from turning on if the indoor fan has failed
    b. turn the indoor fan on when both heaters are energized
    c. turn the fan on when either of the heaters is energized
    d. keep the indoor fan running when the unit is turned off

11.36 HVAC ◆ LEVEL TWO

Summary

Even on the coldest day, there is some heat in the air. Heat pumps take advantage of that heat by reversing the refrigerant cycle. The condensing coil becomes an evaporator and the evaporator becomes a condenser. Heat pumps provide much better heating efficiency than electric heat in most climates. In cold climates, they are often combined with furnaces in dual-fuel arrangements that use each type of heat to its maximum efficiency. Heat pumps are available in a wide variety of system configurations and capacities. Their greatest advantage is that one system provides both heating and cooling.

Notes

Trade Terms Introduced in This Module

Absolute zero: The temperature at which all molecular motion ceases. It is –460°F, –273°C, and 0°K (Kelvin).

Balance point: The outdoor temperature at which the heating capacity of the heat pump is equal to the heat loss of the building. The balance point varies depending on the climate, building design, type of construction, and other factors that affect heat loss or gain.

Coefficient of Performance (COP): The ratio of work performed in relation to energy used. A rating method for heat pumps. It is further defined as the Btuh output divided by the total electrical input (watts) required to produce this Btuh output times 3.413.

Compression heat: The heat produced by a heat pump when the refrigerant cycle is reversed. See *reverse cycle heat*.

Defrost: In a heat pump, the process of cycling hot refrigerant through the outdoor coil to melt accumulated frost due to condensation.

Dual-fuel or hybrid system: A system in which a heat pump is combined with a furnace.

Fan coil unit: A term often applied to the indoor unit of a split system. Also known as an air handler.

Four-way valve: A heat pump reversing valve. Also known as a switch-over or cross-over valve.

Ground-source (geothermal) system: A system in which the outdoor coil is buried in the ground and the heat exchange occurs between the earth and the refrigerant flowing through the coil.

Heat sink: A low-temperature surface to which heat can be transferred.

Heating Season Performance Factor (HSPF): A heat pump performance rating that has been adjusted for seasonal operation. It is the total heating output of a heat pump (in Btus) during its normal annual usage period for heating divided by the total electric power input in watt-hours during the same period.

Indoor coil: The designation given to the heat pump coil used to transfer heat to or from the conditioned space.

Kelvin scale: A temperature scale in which zero equals –460°F. See *absolute zero*.

Outdoor coil: The heat pump coil used to transfer heat to or from the outdoor air.

Packaged unit: A self-contained air conditioning system.

Reverse cycle heat: The heat produced by a heat pump when refrigerant flow is reversed. See *compression heat*.

Reversing valve: A valve that changes the direction of refrigerant flow in a heat pump. See *four-way valve*.

Seasonal Energy Efficiency Ratio (SEER): The ARI standard for measuring heat pump efficiency. It is the total cooling of a heat pump (in Btus) during its normal annual period for cooling divided by the total electric input in watt-hours during the same period.

Split system: An air conditioning system with an indoor coil and an outdoor coil connected with refrigerant lines.

Additional Resources and References

Additional Resources

This module is intended to be a thorough resource for task training. The following reference work is suggested for further study. This is optional material for continued education rather than for task training.

Inside the Heat Pump, Videotape. Catalog No. 020-538. Syracuse, NY: Carrier Corporation.

Figure Credits

Carrier Corporation, 211SA01

Courtesy of Honeywell International Inc., 211F18

Emerson Climate Technologies, 211SA02

Topaz Publications, Inc., 211F23, 211SA04

MODULE 03211-07 — TEACHING TIPS

The following are suggested activities or instructional methods to help you teach the material in this module.

General

When you call on someone to answer a question, the rest of the class relaxes or even tunes out because they expect that the question and answer will take place only between you and the trainee you called on. Instead, use this technique to involve more trainees in answering questions and to keep them on their toes.

1. Ask trainees to define a term or explain a concept.
2. After one trainee has answered, ask a trainee seated nearby if the answer is right. Then ask whether a trainee in the back of the room agrees.
3. Ask trainees to explain why they think an answer is right or wrong.
4. Use the session to clear up incorrect ideas and encourage trainees to learn from their mistakes.

Sections 1.0.0 through 8.0.0 *Quick Quiz*

This Quick Quiz will familiarize trainees with the terms and definitions that are commonly used when discussing heat pumps. You will need photocopies of the quiz provided in the following page. Trainees will need pencils. If you allow trainees to use the Trainee Guide, decrease the amount of time you give them to complete the quiz.

1. Make a photocopy of the quiz for each trainee.
2. Give trainees between 5 and 10 minutes to complete the quiz.
3. Go over the answers to the quiz.
4. Ask trainees if they have questions.

Answers to Quick Quiz

1. i
2. a
3. f
4. h
5. e
6. d
7. c
8. j
9. g
10. b

Quick Quiz *Heat Pumps*

For each description listed, identify the term that the text best describes. Write the corresponding letter in the blank provided.

_____ 1. The refrigerant gas is directed to either coil depending on the position of the piston in the _____.

_____ 2. If the temperature is above _____, there is still some heat in the air.

_____ 3. Heat pumps that use a coil of tubing buried in the earth as a heat source are known as _____.

_____ 4. An air-to-air heat pump that is self-contained in one unit is known as a _____.

_____ 5. Auxiliary heat is also required for heating during the _____ cycle.

_____ 6. Auxiliary heat should be sized for 100 percent of the building's heat load to ensure that there is adequate heat in the event of a(n) _____ failure.

_____ 7. The ratio of useful heat delivered to the equivalent heat consumed in operating the entire system is known as the _____.

_____ 8. The _____ is the standard by which all heat pumps are rated.

_____ 9. To obtain the _____, divide the total heating output for the season by the total energy consumed for the season.

_____ 10. The outdoor temperature at which the heating capacity of the heat pump is equal to the heat loss of the building is known as the _____.

a. absolute zero
b. balance point
c. Coefficient of Performance
d. compression heat
e. defrost
f. ground-source (geothermal) systems
g. Heating Season Performance Factor
h. packaged unit
i. reversing valve
j. Seasonal Energy Efficiency Ratio

Section 3.2.0   *Water-to-Air Heat Pumps*

This exercise will familiarize trainees with water-to-air heat pumps. Trainees will need appropriate personal protective equipment, pencils, and paper. You will need to arrange for a tour of a commercial building that uses a water-to-air heat pump. Allow 45 to 60 minutes for this exercise.

1. Discuss water-to-air heat pumps. Have trainees brainstorm questions for the tour before embarking.
2. Have the safety or facility manger give a tour of the water-to-air heat pump system. Have them point out the various controls and components.
3. Have the trainees take notes and write down questions during the tour.
4. After the presentation, answer any questions trainees may have.

Section 5.0.0   *Identifying Heat Pump Components*

This exercise will familiarize trainees with identifying various heat pump components. Trainees will need pencils and paper. You will need an overhead projector, acetate sheets, transparency markers, and a heat pump system or diagram. Allow 15 to 20 minutes for this exercise. This exercise corresponds to Performance Task 1.

1. Divide the class in half. On one sheet write "Team One" on one half and "Team Two" on the other half.
2. Point to a component in the system. Have one trainee identify it. If desired, add a bonus point if the trainee can describe the function of the component.
3. To ensure complete participation, each team member can answer no more than two times (adjust according to the number of participants and items to be identified). Each correct answer gets a check on the acetate sheet. An incorrect answer means the other team has the opportunity to answer.
4. Keep score to encourage class participation. At the end of the game, answer any questions.

Section 5.1.0   *Think About It – Computing COP*

Heating COP = $100{,}000/(8 \times 3{,}413)$ = 3.7 (rounded off)

Section 5.2.0   *Think About It – Computing HSPFs*

HSPF = $200{,}000/35{,}000$ = 5.7 (rounded off)

Section 10.0.0   *Servicing Procedures*

This exercise will familiarize trainees with basic maintenance procedures. Trainees will need appropriate personal protective equipment, pencils, and paper. You will need to arrange workstations with different types of equipment each one. Allow 20 to 30 minutes for this exercise.

1. Describe basic maintenance procedures.
2. Have trainees rotate between the various workstations. Have trainees perform one service procedure on each piece of equipment, or an entire set of service checks on one piece.
3. Answer any questions trainees may have.

MODULE 03211-07 — ANSWERS TO REVIEW QUESTIONS

Answer	Section
1. a	3.2.0
2. c	3.5.0
3. b	4.2.0
4. d	4.3.0
5. c	4.3.1
6. a	5.1.0
7. c	5.2.0
8. d	8.0.0
9. b	9.1.0
10. c	9.1.0
11. b	11.0.0
12. a	11.0.0
13. d	11.0.0
14. b	11.0.0
15. c	11.0.0

CONTREN® LEARNING SERIES — USER UPDATE

NCCER makes every effort to keep these textbooks up-to-date and free of technical errors. We appreciate your help in this process. If you have an idea for improving this textbook, or if you find an error, a typographical mistake, or an inaccuracy in NCCER's Contren® textbooks, please write us, using this form or a photocopy. Be sure to include the exact module number, page number, a detailed description, and the correction, if applicable. Your input will be brought to the attention of the Technical Review Committee. Thank you for your assistance.

Instructors – If you found that additional materials were necessary in order to teach this module effectively, please let us know so that we may include them in the Equipment/Materials list in the Annotated Instructor's Guide.

Write: Product Development and Revision
National Center for Construction Education and Research
3600 NW 43rd St, Bldg G, Gainesville, FL 32606

Fax: 352-334-0932

E-mail: curriculum@nccer.org

Craft _____ Module Name _____

Copyright Date _____ Module Number _____ Page Number(s) _____

Description
_____
_____
_____
_____

(Optional) Correction
_____
_____
_____

(Optional) Your Name and Address
_____
_____
_____

Module 03212-07

Basic Installation and Maintenance Practices

NCCER STANDARDIZED CRAFT TRAINING PROGRAM

The National Center for Construction Education and Research (NCCER) provides a standardized national program of accredited craft training. Key features of the program include instructor certification, competency-based training, and performance testing. The program provides trainees, instructors, and companies with a standard form of recognition through a National Craft Training Registry. The program is described in full in the *Guidelines for Accreditation*, published by NCCER. For more information on standardized craft training, contact the NCCER by writing us at 3600 NW 43rd St., Bldg. G, Gainesville, FL 32606; calling 352-334-0911; or emailing info@nccer.org. More information may be found at our website, www.nccer.org.

HOW TO USE THIS ANNOTATED INSTRUCTOR'S GUIDE

Each page presents two sections of information. The larger section displays each page exactly as it appears in the Trainee Module. The narrow column ties suggested trainee and instructor actions to each page and provides icons (detailed below) to call your attention to material, safety, audiovisual, or testing requirements. The bottom of each page includes space for your notes.

 The **Audiovisual** icon indicates an appropriate time to show a transparency or other audiovisual aid.

 The **Classroom** icon prompts you to define a term, stress a point, ask trainees to explain a concept, or give examples.

 The **Demonstration** icon directs you to show trainees how to perform tasks.

 The **Examination** icon tells you to administer the written module examination.

 The **Homework** icon is placed where you may wish to assign reading for the next class, assign a project, or advise trainees to prepare for an examination.

 The **Laboratory** icon is used when trainees are to practice performing tasks.

 The **Materials** icon is a reminder for you to gather materials needed for classes, labs, and testing.

 The **Performance Testing** icon tells you to administer a performance test or a portion thereof.

 The **Safety** icon is used to emphasize safety issues. It is often keyed to *Caution* and *Warning!* statements in the Trainee Module.

 The **Teaching Tip** icon indicates additional guidance is available, such as how to conduct an exercise, get the most educational value from a field trip, or encourage class participation. Teaching Tips may expand on a feature (*Think About It*, *Did You Know?*) or provide *Quick Quizzes* or similar exercises. You will be referred to the Teaching Tips section at the back of the module if there is additional material.

 The **Combination** icon indicates that the laboratory listed corresponds with a performance task. If desired, you can note the proficiency of the trainees during the laboratory, and use it to satisfy performance testing requirements.

PREPARATION

Before teaching this module, you should review the Objectives, Performance Tasks, Materials and Equipment List, and Module Outline. Be sure to allow ample time to prepare your own training or lesson plan and gather all required materials and equipment.

Basic Installation and Maintenance Practices
Annotated Instructor's Guide

Module 03212-07

MODULE OVERVIEW

This module introduces the trainee to the basic mechanical procedures commonly performed in HVAC servicing work. Basic maintenance procedures, documentation, and customer relations are also covered.

PREREQUISITES

Prior to training with this module, it is recommended that the trainee shall have successfully completed *Core Curriculum*; *HVAC Level One*; and *HVAC Level Two*, Modules 03201-07 through 03211-07.

OBJECTIVES

Upon completion of this module, the trainee will be able to do the following:

1. Identify, explain, and install threaded and non-threaded fasteners.
2. Identify, explain, remove, and install types of gaskets, packings, and seals.
3. Identify types of lubricants, and explain their uses.
4. Use lubrication equipment to lubricate motor bearings.
5. Identify the types of belt drives, explain their uses, and demonstrate procedures used to install or adjust them.
6. Identify and explain types of couplings.
7. Demonstrate procedures used to remove, install, and align couplings.
8. Identify types of bearings, and explain their uses.
9. Explain causes of bearing failures.
10. Demonstrate procedures used to remove and install bearings.
11. Perform basic preventive maintenance inspection and cleaning procedures.
12. List ways to develop and maintain good customer relations.

PERFORMANCE TASKS

Under the supervision of the instructor, the trainee should be able to do the following:

1. Identify different types of threaded fasteners.
2. Identify non-threaded fasteners.
3. Identify different types of gaskets.
4. Identify mechanical seal parts.
5. Install an oil seal.
6. Align and properly adjust V-belts.
7. Identify different types of drive couplings.
8. Tighten a four-bolt flange.
9. Install an expandable anchor bolt.
10. Identify different types of bearings.
11. Recognize and use a manual bearing puller to remove a bearing.
12. Recognize and use a feeler gauge to measure bearing clearances.
13. Lubricate a bearing using a lever-type grease gun.
14. Fill out typical forms used for installation and service calls.

MATERIALS AND EQUIPMENT LIST

Overhead projector and screen

Transparencies

Blank acetate sheets

Transparency pens

Whiteboard/chalkboard

Markers/chalk

Pencils and scratch paper

Appropriate personal protective equipment

Assortment of machine bolts, machine screws, cap screws, and stud bolts

Various types of the following:
   Set screws
   Flat and lock washers
   Nuts
   Thread-forming and thread-cutting screws
   Toggle and anchor bolts
   Thread repair inserts
   Retainer rings
   Pin fasteners
   Keys
   Rivets
   Preformed gaskets and gasket materials
   Packing materials
   Non-mechanical seals
   Lubricants
   Hydraulic fittings
   V-belts
   Couplings
   Mechanical seals

Measuring devices, including:
   Circumference rules
   Calipers and dividers
   Micrometers
   Feeler gauges
   Thread pitch gauges

Assortment of flange sets used to practice different bolt tightening sequences

Manufacturers' installation instructions for selected mechanical seals

Assortment of different types of sleeve, ball, and roller bearings

Examples of bearings that show the common types of bearing damage

Metal plates with holes drilled through and/or with threaded holes

Several pieces of sheet metal of different sizes and weights

Rags

Examples of job forms

Wrenches and pliers of the type and size used to fit selected fasteners

Torque wrenches for use with the size of selected fasteners

Tap and die set

Electric drill and drill bits

Blind rivet tool

Packing and seal removal/installation tools

Equipment to demonstrate and practice removing and installing packing and seals

Equipment to demonstrate and practice removing and installing mechanical seals

Equipment to demonstrate and practice removing and installing bearings

Various types of bearing/coupling pullers

Arbor press and/or hydraulic press

Feeler gauge

Equipment to demonstrate and practice the lubrication of motor bearings

Lever-type grease gun

Equipment to demonstrate and practice V-belt installation and pulley alignment

V-belt tension gauge

Equipment to demonstrate and practice removing and installing couplings

Straightedge

Copies of Quick Quiz*

Module Examinations**

Performance Profile Sheets**

* Located in the back of this module.
**Located in the Test Booklet.

SAFETY CONSIDERATIONS

Ensure that the trainees are equipped with appropriate personal protective equipment and know how to use it properly. This module requires trainees to work with appliances. Make sure that all trainees are briefed on appropriate safety procedures. Emphasize electrical safety and chemical hazards.

ADDITIONAL RESOURCES

This module is intended to present thorough resources for task training. The following reference work is suggested for both instructors and motivated trainees interested in further study. This is optional material for continued education rather than for task training.

Air Conditioning Systems, Principles, Equipment, and Service, Latest Edition. Upper Saddle River, NJ: Prentice Hall.

TEACHING TIME FOR THIS MODULE

An outline for use in developing your lesson plan is presented below. Note that each Roman numeral in the outline equates to one session of instruction. Each session has a suggested time period of 2½ hours. This includes 10 minutes at the beginning of each session for administrative tasks and one 10-minute break during the session. Approximately 17½ hours are suggested to cover *Basic Installation and Maintenance Practices*. You will need to adjust the time required for hands-on activity and testing based on your class size and resources. Because laboratories often correspond to Performance Tasks, the proficiency of the trainees may be noted during these exercises for Performance Testing purposes.

Topic   **Planned Time**

Session I. Introduction and Mechanical Fasteners

    A. Introduction

    B. Mechanical Fasteners

    C. Laboratory

    Trainees practice identifying various threaded and non-threaded fasteners. This laboratory corresponds to Performance Tasks 1 and 2.

    D. Installing Threaded Fasteners

    E. Laboratory

    Trainees practice tightening a four-bolt flange. This laboratory corresponds to Performance Task 8.

    F. Installing Anchor Bolts

    G. Laboratory

    Trainees practice installing anchor bolts. This laboratory corresponds to Performance Task 9.

Session II. Gaskets, Packing, and Seals

    A. Gaskets

    B. Laboratory

    Trainees practice identifying various types of gaskets. This laboratory corresponds to Performance Task 3.

    C. Packing

    D. Identifying Seals

    E. Laboratory

    Trainees practice identifying mechanical seal parts. This laboratory corresponds to Performance Task 4.

    F. Installing and Removing Seals

    G. Laboratory

    Trainees practice installing an oil seal. This laboratory corresponds to Performance Task 5.

Session III. Bearings
A. Identifying Bearings

B. Laboratory

Trainees practice identifying different types of bearings. This laboratory corresponds to Performance Task 10.

C. Removing Bearings

D. Laboratory

Trainees practice using a manual bearing puller to remove a bearing. This laboratory corresponds to Performance Task 11.

E. Installing Bearings

F. Laboratory

Trainees practice using a feeler gauge to measure bearing clearances. This laboratory corresponds to Performance Task 12.

Session IV. Lubrication, Belts, and Belt Drives
A. Lubricating Bearings

B. Laboratory

Trainees practice lubricating a bearing using a lever-type grease gun. This laboratory corresponds to Performance Task 13.

C. Belts and Belt Drives

D. Laboratory

Trainees practice aligning and adjusting a V-belt. This laboratory corresponds to Performance Task 6.

Session V. Couplings and Direct Drives
A. Couplings and Direct Drives

B. Laboratory

Trainees practice identifying different types of drive couplings. This laboratory corresponds to Performance Task 7.

C. General Coupling Removal and Installation Methods

D. Coupling Alignment

E. Basic Maintenance Procedures

Session VI. Documentation and Customer Relations
A. Documentation

B. Laboratory

Trainees practice filling out forms used for installation and service calls. This laboratory corresponds to Performance Task 14.

C. Customer Relations

D. Customer Communications

Session VII. Review and Testing
A. Review

B. Module Examination

  1. Trainees must score 70% or higher to receive recognition from NCCER.
  2. Record the testing results on Craft Training Report Form 200, and submit the results to the Training Program Sponsor.

C. Performance Testing

  1. Trainees must perform each task to the satisfaction of the instructor to receive recognition from NCCER. If applicable, proficiency noted during laboratory exercises can be used to satisfy the Performance Testing requirements.
  2. Record the testing results on Craft Training Report Form 200, and submit the results to the Training Program Sponsor.

HVAC Level Two

03212-07

Basic Installation and Maintenance Practices

Assign reading of Module 03212-07.

03212-07

Basic Installation and Maintenance Practices

Topics to be presented in this module include:

1.0.0	Introduction	12.2
2.0.0	Mechanical Fasteners	12.2
3.0.0	Gaskets	12.13
4.0.0	Packing	12.15
5.0.0	Seals	12.18
6.0.0	Bearings	12.21
7.0.0	Lubrication	12.27
8.0.0	Belts and Belt Drives	12.29
9.0.0	Couplings and Direct Drives	12.31
10.0.0	Basic Maintenance Procedures	12.34
11.0.0	Documentation	12.43
12.0.0	Customer Relations	12.47
13.0.0	Customer Communication	12.51

Overview

Anyone performing installation work must be familiar with mechanical devices such as fasteners and anchors, which are used to attach system components to mounting surfaces. These surfaces can be concrete, wood, or metal. There are many types of fasteners and anchors, and the installer must know which type to select for a given application. Large systems use more complex mechanical devices such as belt-driven fans, serviceable compressors, direct-drive compressors, and mechanically operated dampers. All of these devices require knowledge of special tools, materials, and methods.

Instructor's Notes:

Objectives

When you have completed this module, you will be able to do the following:

1. Identify, explain, and install threaded and non-threaded fasteners.
2. Identify, explain, remove, and install types of gaskets, packings, and seals.
3. Identify types of lubricants, and explain their uses.
4. Use lubrication equipment to lubricate motor bearings.
5. Identify the types of belt drives, explain their uses, and demonstrate procedures used to install or adjust them.
6. Identify and explain types of couplings.
7. Demonstrate procedures used to remove, install, and align couplings.
8. Identify types of bearings, and explain their uses.
9. Explain causes of bearing failures.
10. Demonstrate procedures used to remove and install bearings.
11. Perform basic preventive maintenance inspection and cleaning procedures.
12. List ways to develop and maintain good customer relations.

Trade Terms

Axial load	Oxidation	Thrust
Break-away torque	Pour point	Tolerance
	Radial load	Torque
Dropping point	Run-down resistance	Viscosity
Dynamic seal		Viscosity index (VI)
Fire point	Set or seizure	
Flash point	Static seal	
Journal	Stuffing box	

Required Trainee Materials

1. Pencil and paper
2. Appropriate personal protective equipment

Prerequisites

Before you begin this module, it is recommended that you successfully complete *Core Curriculum*; *HVAC Level One*; and *HVAC Level Two*, Modules 03201-07 through 03211-07.

This course map shows all of the modules in the second level of the HVAC curriculum. The suggested training order begins at the bottom and proceeds up. Skill levels increase as you advance on the course map. The local Training Program Sponsor may adjust the training order.

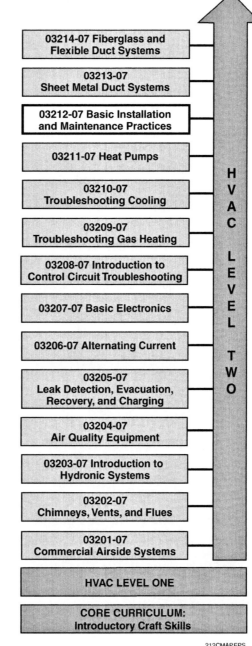

Ensure that you have everything required to teach the course. Check the Materials and Equipment list at the front of this module.

See the general Teaching Tip at the end of this module.

Explain that terms shown in bold are defined in the Glossary at the back of this module.

Show Transparency 1, Objectives, and Transparency 2, Performance Tasks. Review the goals of the module, and explain what will be expected of the trainee.

Review the modules covered in Level Two and explain how this module fits in.

MODULE 03212-07 ♦ BASIC INSTALLATION AND MAINTENANCE PRACTICES 12.1

List some of the maintenance tasks required of an HVAC technician.

Explain that HVAC equipment requires the use of many types of fasteners.

Explain how to identify fasteners based on thread designations.

Show Transparency 3 (Figure 1). List the standard designations for threaded fasteners commonly used by HVAC technicians.

See the Teaching Tips for Sections 1.0.0–13.0.0 and Section 2.0.0 at the end of this module.

1.0.0 ♦ INTRODUCTION

The HVAC technician must be able to properly install and maintain the mechanical components used in HVAC systems. This module describes the mechanical maintenance skills required to install and work with the following components:

- Mechanical fasteners
- Gaskets, packings, and seals
- Lubricants
- Belts and belt drives
- Couplings and direct drives
- Bearings

Mechanical maintenance tasks commonly performed during the scheduled maintenance of most types of HVAC equipment are also covered in this module, including procedures for motor lubrication, air filter and screen maintenance, coil maintenance, and damper maintenance.

In addition to performing the actual installation and maintenance field service tasks, the technician is often required to fill out job reports and checklists used by both the technician and the employer to track the progress of a job and/or record the performance of the equipment. In addition, the technician must also interface with the customer before, during, and after performing the work. For this reason, the last part of this module describes some common types of documentation used in the field, then focuses on guidelines for promoting and maintaining good customer relations.

2.0.0 ♦ MECHANICAL FASTENERS

Fasteners are used to install many types of parts and equipment. Common fasteners include bolts, screws, pins, and retainers.

2.1.0 Thread and Grade Designations

This section covers thread and grade designations. Fasteners, bolts, and screws are distinguished from one another through the use of standard designations.

2.1.1 Thread Designations

Fastener threads are made to established standards. The most common standard is the unified or American National Standard. There are three series or classes of threads defined by the unified standard. These series are based on the number of threads per inch for a fastener of a certain diameter. The three series are:

- *Unified National Coarse (UNC) Thread* – UNC thread is used for bolts, screws, nuts, and other general applications. Fasteners with UNC threads are used for rapid assembly and disassembly of parts where corrosion or slight damage may occur.
- *Unified National Fine (UNF) Thread* – UNF thread is used for bolts, screws, nuts, and other uses where a finer thread than UNC is required.
- *Unified National Extra Fine (UNEF) Thread* – UNEF thread is used for thin-walled tubes, nuts, ferrules, and couplings.

Bolt and screw threads are designated by a standard method (*Figure 1*). The standard designations include:

- Nominal size (diameter)
- Number of threads per inch
- Thread series symbol
- Thread class (Classes 1A, 2A, and 3A are external threads, such as used with a bolt. Classes 1B, 2B, and 3B are internal threads, such as used with a nut.)
- Left-hand thread symbol (Unless shown, the threads are right hand. This symbol is used only for left-hand fasteners.)

Metric screw threads based on the American National Standard are also in common use. Metric M-profile threaded screws are a coarse-thread series of fasteners used for general fastening purposes. Metric MJ-profile threaded screws are used with aircraft parts and in other high-stress uses requiring extra strength.

2.1.2 Fastener Grade Designations

The strength and quality of a fastener can be determined by special grade markings on the head of the fastener. These markings are standardized by

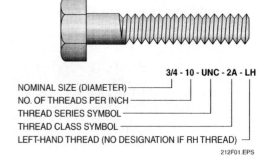

Figure 1 ♦ Thread designations.

the Society of Automotive Engineers (SAE) and ASTM International. Grade markings are sometimes called line markings. *Figure 2* shows the SAE and ASTM markings for steel bolts and screws.

NOTE

Always use bolts that have grade markings. Bolts that do not may be of inferior quality.

ASTM AND SAE GRADE MARKINGS FOR STEEL BOLTS & SCREWS

GRADE MARKING	SPECIFICATION	MATERIAL
(plain hex)	SAE-GRADE 0	STEEL
	SAE-GRADE 1 ASTM-A 307	LOW CARBON STEEL
	SAE-GRADE 2	LOW CARBON STEEL
(one line)	SAE-GRADE 3	MEDIUM CARBON STEEL, COLD WORKED
(A 449)	SAE-GRADE 5	MEDIUM CARBON STEEL, QUENCHED AND TEMPERED
	ASTM-A 449	
(A 325)	ASTM-A 325	MEDIUM CARBON STEEL, QUENCHED AND TEMPERED
(BB)	ASTM-A 354 GRADE BB	LOW ALLOY STEEL, QUENCHED AND TEMPERED
(BC)	ASTM-A 354 GRADE BC	LOW ALLOY STEEL, QUENCHED AND TEMPERED
(five lines)	SAE-GRADE 7	MEDIUM CARBON ALLOY STEEL, QUENCHED AND TEMPERED ROLL THREADED AFTER HEAT TREATMENT
(six lines)	SAE-GRADE 8	MEDIUM CARBON ALLOY STEEL, QUENCHED AND TEMPERED
	ASTM-A 354 GRADE BD	ALLOY STEEL, QUENCHED AND TEMPERED
(A 490)	ASTM-A 490	ALLOY STEEL, QUENCHED AND TEMPERED

ASTM SPECIFICATIONS
- A 307 – LOW CARBON STEEL EXTERNALLY AND INTERNALLY THREADED STANDARD FASTENERS.
- A 325 – HIGH STRENGTH STEEL BOLTS FOR STRUCTURAL STEEL JOINTS, INCLUDING SUITABLE NUTS AND PLAIN HARDENED WASHERS.
- A 449 – QUENCHED AND TEMPERED STEEL BOLTS AND STUDS.
- A 354 – QUENCHED AND TEMPERED ALLOY STEEL BOLTS AND STUDS WITH SUITABLE NUTS.
- A 490 – HIGH STRENGTH ALLOY STEEL BOLTS FOR STRUCTURAL STEEL JOINTS, INCLUDING SUITABLE NUTS AND PLAIN HARDENED WASHERS.

SAE SPECIFICATION
- J 429 – MECHANICAL AND QUALITY REQUIREMENTS FOR THREADED FASTENERS.

212F02.EPS

Figure 2 ♦ Grade markings for steel bolts and screws.

List the types of threaded fasteners commonly used by HVAC technicians.

Show Transparency 5 (Figure 3). Discuss machine bolts, machine screws, stud bolts, and cap screws, and applications for each.

Provide machine bolts, machine screws, stud bolts, and cap screws for the trainees to examine.

2.2.0 Threaded Fasteners

Threaded fasteners are one of the most common types of fasteners. Many are assembled with nuts and washers. Others are installed in threaded holes. Types of threaded fasteners include the following:

- Set screws
- Machine bolts, machine screws, stud bolts, and cap screws
- Flat and lock washers
- Nuts
- Thread-forming and thread-cutting screws
- Toggle and anchor bolts
- Inserts

2.2.1 Machine Bolts, Machine Screws, Stud Bolts, and Cap Screws

Machine bolts (*Figure 3*) are used to assemble parts that do not require close **tolerances**. The tolerance is the amount of variation allowed from a standard. Machine bolts are made with diameters ranging from ¼" to 3" and with lengths from ½" to 30". A machine bolt is tightened and released by turning its mating nut that is usually furnished along with the bolt.

Machine screws are used for general assembly. They have slotted or recessed heads. Machine screws are available in diameters from #6 through #12 (0.060" to 0.2160") and from ¼" to ½", and in lengths from ⅛" to 3". They are also made in metric sizes. A machine screw normally mates with an

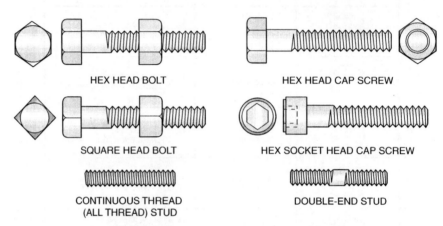

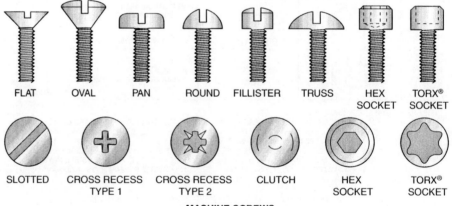

Figure 3 ♦ Machine bolts, machine screws, cap screws, and stud bolts.

internally threaded hole into which it is tightened or released. They can also be used with nuts.

Cap screws are generally used on assemblies that need a finished appearance. They pass through a clearance hole in one part of the assembly and are screwed into a threaded hole in the other part. This clamps the parts together when the cap screw is tightened.

Cap screws are made to close tolerances with machined or semifinished bearing surfaces under the head. Cap screws come in coarse and fine threads, and in diameters from ¼" to 2". Lengths from ⅜" to 10" are available. They are also made in metric sizes.

Stud bolts are headless bolts, threaded either along the entire length or on both ends. One end can be screwed into a threaded hole. The part to be clamped is fitted over the other end of the stud, and a nut and washer are screwed on to fasten the two parts together.

2.2.2 Set Screws

Set screws are usually made of heat-treated steel. They are used to fasten pulleys and fan blades on shafts, and to hold collars in place. Set screws are classified by head styles and point styles. *Figure 4* shows several set screw head and point styles.

2.2.3 Flat and Lock Washers

Flat washers (*Figure 5*) provide an enlarged surface used to distribute the load from bolt heads and nuts. Flat washers are made in light, medium, heavy-duty, and extra heavy-duty series. Fender washers have a wide surface area to bridge oversized holes or other wide clearance requirements. Lock washers are used to keep bolts or nuts from working loose. They are placed between the flat washers and the bolts or nuts. Some common types of lock washers include the following:

- *External* – Provides the greatest resistance
- *Internal* – Used on small screws
- *Internal-external* – Used for oversized mounting holes
- *Countersunk* – Used with flat or oval-head screws
- *Split ring* – Commonly used with bolts and cap screws

2.2.4 Nuts

Nuts used with most threaded fasteners have hex (hexagonal) or square shapes and are used with bolts having the same shaped head. *Figure 6* shows different types of nuts. Some special-purpose nuts include the following:

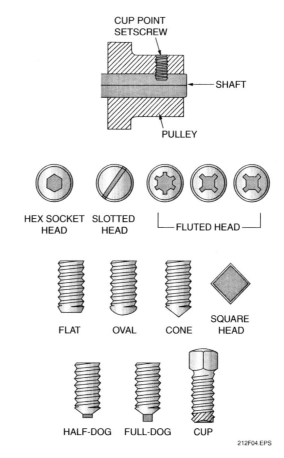

Figure 4 ♦ Set screws.

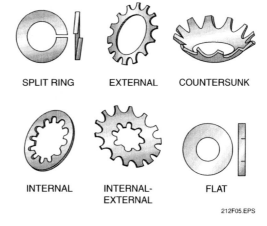

Figure 5 ♦ Flat and lock washers.

Show Transparency 6 (Figure 4). Discuss set screws and their applications.

Show Transparency 7 (Figure 5). Discuss washers and their applications.

Show Transparency 8 (Figure 6). Discuss types of nuts and their applications.

Provide set screws, washers, and nuts for trainees to examine.

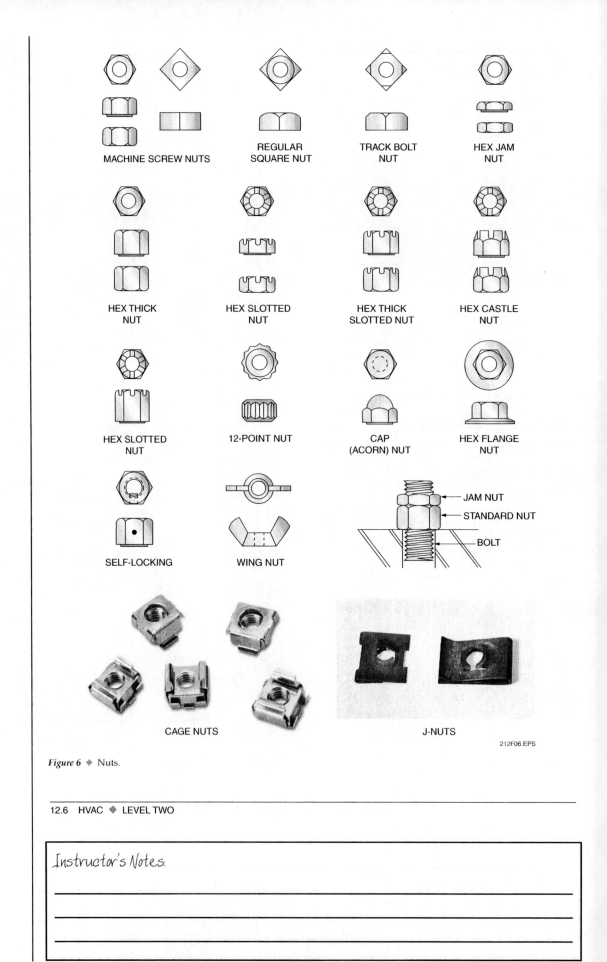

Figure 6 ♦ Nuts.

- *Acorn nut* – These are used when appearance is important or when there are exposed sharp threads on the fastener.
- *Castellated (or castle) and slotted nuts* – After the nut is tightened, a cotter pin is fitted into one set of slots and through a hole in the bolt. The cotter pin keeps the nut from loosening.
- *Self-locking nut* – This has a nylon insert or is slightly deformed so it cannot work loose. Once a self-locking nut is used and removed, it should be thrown away and replaced with a new one.
- *Wing nut* – These are used where frequent adjustments and service are necessary. Wing nuts allow for loosening and tightening without the use of a wrench.
- *Jam nut* – This is a thin nut used to lock a standard nut in place.
- *U-nuts and J-nuts* – These are used with soft, thin panel materials, or with thin sheet metal. The U-nut or J-nut has a threaded hole and is slid onto the edge of thin, relatively soft material and clamps it onto a sheet metal edge. The metal and the spring tension of the nut help to keep the screw from tearing out of the material.
- *Cage nut* – These are square nuts enclosed in a stamped sheet metal box. The cage has a spring clip at the bottom which can be clipped into a square hole in a rack. The nut is loose enough inside the cage to allow for some small misalignment when mounting equipment.

2.2.5 Thread-Forming and Thread-Cutting Screws

Thread-forming screws (*Figure 7*) are used mainly to fasten light-gauge metal parts. They form a thread as they are driven. This eliminates tapping. Some also drill their own hole. This eliminates drilling, punching, and aligning of parts.

Thread-cutting screws cut threads into the metal as they are driven into a pilot hole. They are made of hardened steel and are used to join heavy-gauge sheet metal and nonferrous metal parts.

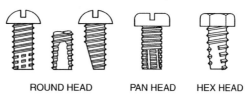

Figure 7 ♦ Thread-forming and thread-cutting screws.

2.2.6 Toggle and Anchor Bolts

Toggle bolts are used to fasten a part to a hollow wall or panel. *Figure 8* shows some common toggle bolts.

Anchor bolts are used to fasten parts and equipment to concrete and masonry. There are several types of non-expansion anchor bolts designed for installation in both wet and hardened concrete and other surfaces. Expansion-type anchor bolts are installed in holes drilled in hardened concrete. When the nut on the anchor bolt is tightened, the bolt base expands to provide the holding force. *Figure 8* shows typical non-expansion and expansion anchor bolts installed in concrete.

Show Transparency 9 (Figure 7). Discuss thread-forming and thread-cutting screws and their applications.

Show Transparency 10 (Figure 8). Discuss toggle bolts and anchor bolts and their applications.

Provide thread-forming screws, thread-cutting screws, toggle bolts, and anchor bolts for trainees to examine.

Self-Drilling Sheet Metal Screws

If you are installing a sheet metal duct system, self-drilling sheet metal screws can be real timesavers. Use a cordless drill equipped with a ¼" or ⁵⁄₁₆" magnetized drive socket. Insert the self-drilling screw in the socket and drive the screw in. Be careful not to overtorque the screws as they will strip out.

MODULE 03212-07 ♦ BASIC INSTALLATION AND MAINTENANCE PRACTICES 12.7

Show Transparency 11 (Figure 9). Discuss thread repair inserts and their applications.

Provide thread repair inserts for the trainees to examine.

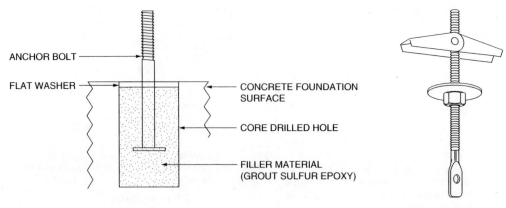

Figure 8 ♦ Toggle bolts and anchor bolts.

2.2.7 Thread Repair Inserts

Thread repair inserts, or heli-coils (*Figure 9*), are a special kind of fastener used to provide high-strength threads in soft metals and plastics. They are also used to replace damaged or stripped threads in a tapped hole. Inserts are made in standard sizes and forms including metric sizes. The insert, which is larger in diameter than the tapped hole, is compressed during installation and allowed to spring back, permanently anchoring the insert in the tapped hole.

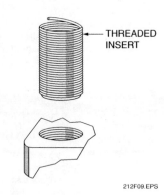

Figure 9 ♦ Thread repair insert.

2.3.0 Non-Threaded Fasteners

Non-threaded fasteners have many uses. This section describes the following types of non-threaded fasteners:

- Retainer rings
- Pins
- Keys
- Rivets

2.3.1 Retainer Rings

Retainer rings are used for both internal and external fastening. Some retainer rings are seated in grooves. Others are self-locking and do not require a groove. Special pliers are used to remove internal and external rings. *Figure 10* shows typical retainer rings.

2.3.2 Pins

Pin fasteners (*Figure 11*) are used to align mating parts, to hold gears and pulleys on shafts, and to secure slotted nuts. Some common pins and their uses are as follows:

- *Dowel pins* – These are fitted into reamed holes to position mating parts. They also support a portion of the load placed on the parts.
- *Taper and spring pins* – These are used to fasten gears, pulleys, and collars to a shaft.
- *Cotter pins* – These are fitted into a hole drilled through a shaft. Cotter pins are used to prevent parts from slipping on or off the shaft, and also to keep slotted nuts from working loose.

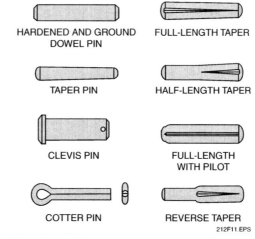

Figure 11 ♦ Pin fasteners.

2.3.3 Keys

Keys are metal parts used to prevent a gear or pulley from rotating on a shaft. One half of the key fits into a keyseat on the shaft. The other half fits into a keyway in the hub of a gear or pulley. *Figure 12* shows some common keys and their trade names.

2.3.4 Rivets

Rivets are used to permanently join two pieces of material. Two common types of rivets are the tinner's rivet and the blind rivet (*Figure 13*). To install a tinner's rivet, the rivet is inserted into a hole that

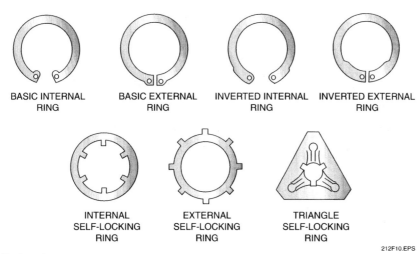

Figure 10 ♦ Retainer rings.

Show trainees how to identify various threaded and non-threaded fasteners.

Have trainees practice identifying various threaded and non-threaded fasteners. Note the proficiency of each trainee. This laboratory corresponds to Performance Tasks 1 and 2.

Explain that to install threaded fasteners you need to be able to select the correct fastener and use the correct method to install it.

Emphasize the importance of tightening fasteners to the proper torque.

Show Transparency 16 (Figure 14). Explain that flanges also follow a tightening sequence and must not be over-tightened.

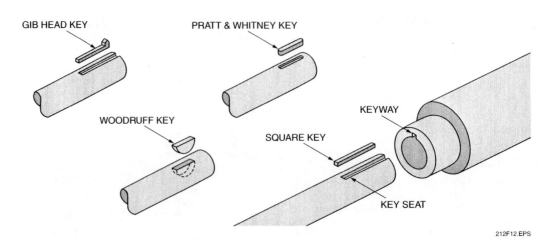

Figure 12 ◆ Keys.

has been drilled or punched. The small end of a tinner's rivet is hammered or set in the form of a head. Hollow tinner's rivets are clinched at the small end with a special tool. Tinner's rivets are sized in ounces or pounds per 1,000 rivets. For example, a 6-ounce rivet means that 1,000 of these rivets will weigh 6 ounces. As the weight increases, so does the diameter and length of the rivet.

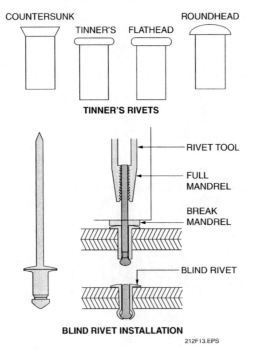

Figure 13 ◆ Rivets.

Blind rivets are used to fasten sheet metal, fiberglass, and plastics. They are available in various lengths and diameters. Blind rivets are used when the joint can only be reached from one side. To install a blind rivet, the rivet is inserted into a previously drilled or punched hole, then the rivet is set, or popped, using a special tool.

2.4.0 Installing Threaded Fasteners

To install threaded fasteners, you must know the type of fastener to use and the correct method of installation. You must also understand the tightening sequence and **torque** specifications. This section describes the guidelines you should follow when installing common fasteners.

2.4.1 Torquing Steel Fasteners

When torquing steel fasteners, you must first select the proper type and grade of fastener for the job, then torque (tighten) it to the recommended specifications. *Figure 14* shows typical torque specifications for fasteners.

INSIDE TRACK — Removing Rivets

Chisel or grind off the heads of tinner's rivets or steel blind rivets; remove the shaft by striking through with a punch. Drill off blind or countersunk rivet heads using a high-speed electric drill, then punch out the shaft.

12.10   HVAC ◆ LEVEL TWO

Instructor's Notes:

Discuss inspection and maintenance procedures when replacing fasteners.

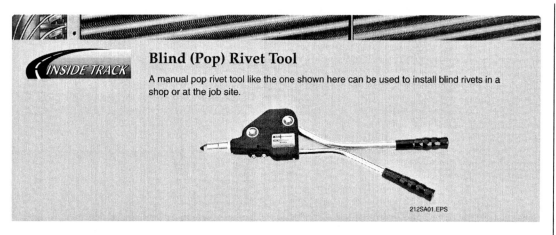

Blind (Pop) Rivet Tool

A manual pop rivet tool like the one shown here can be used to install blind rivets in a shop or at the job site.

TORQUE IN FOOT POUNDS

FASTENER DIAMETER	THREADS PER INCH	MILD STEEL	STAINLESS STEEL 18-8	ALLOY STEEL
1/4	20	4	6	8
5/16	18	8	11	16
3/8	16	12	18	24
7/16	14	20	32	40
1/2	13	30	43	60
5/8	11	60	92	120
3/4	10	100	128	200
7/8	9	160	180	320
1	8	245	285	490

SUGGESTED TORQUE VALUES FOR GRADED STEEL BOLTS

GRADE	SAE 1 OR 2	SAE 5	SAE 6	SAE 8
TENSILE STRENGTH	64,000 PSI	105,000 PSI	130,000 PSI	150,000 PSI
GRADE MARK				

BOLT DIAMETER	THREADS PER INCH	FOOT POUNDS TORQUE			
1/4	20	5	7	10	10
5/16	18	9	14	19	22
3/8	16	15	25	34	37
7/16	14	24	40	55	60
1/2	13	37	60	85	92
9/16	12	53	88	120	132
5/8	11	74	120	169	180
3/4	10	120	200	280	296
7/8	9	190	302	440	473
1	8	282	466	660	714

Figure 14 ♦ Torque specifications.

Emphasize the importance of following manufacturer's instructions when istalling fasteners.

Define torque and other common terms associated with the use of torque wrenches.

Explain why fasteners must not be overtightened.

Show Transparency 17 (Figure 15). Explain that the proper tightening sequence must be followed when installing bolts or other fasteners.

Discuss methods used to express torque values. See the answer to the "Think About It" at the end of this module.

Show trainees how to tighten a four-bolt flange.

Under your supervision, have the trainees practice tightening a four-bolt flange. Note the proficiency of each trainee. This laboratory corresponds to Performance Task 8.

Always follow the manufacturer's instructions and recommended torque values when installing fasteners on equipment. You must use a suitable torque wrench to tighten fasteners to their correct torque specifications. You might want to review the torque wrenches covered in the *Core Curriculum*.

Torque is the resistance to a turning or twisting force. The correct size torque wrench for a job is one that will read between 25 and 75 percent of the scale when the required torque is applied. This allows for adequate capacity and provides satisfactory accuracy. Avoid using an oversized torque wrench because the scale divisions are too coarse, making it difficult to get an accurate reading. Using a wrench that is too small will not allow for extra capacity in the event of seizure or **run-down resistance**. All threaded fasteners must be clean and undamaged in order to get accurate readings. The calibration of torque wrenches must be checked periodically to guarantee accuracy. The following terms, sometimes used in manufacturers' service literature, must be understood when using a torque wrench:

- *Break-away torque* – The torque required to loosen a fastener. This is generally lower than the torque to which it has been tightened. For a given size fastener, there is a direct relationship between tightening torque and break-away torque. This relationship is determined by actual test. Once known, the tightening torque can be checked by loosening and checking breakaway torque.
- *Set or seizure* – In the last stages of rotation in reaching a final torque, seizing or set of the fastener may occur. When this happens, there is usually a noticeable popping sound and vibration. To break the set, back off and then again apply the tightening torque. Accurate torque settings cannot be made if the fastener is seized.

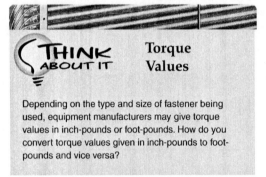

Torque Values

Depending on the type and size of fastener being used, equipment manufacturers may give torque values in inch-pounds or foot-pounds. How do you convert torque values given in inch-pounds to foot-pounds and vice versa?

- *Run-down resistance* – The torque required to overcome the resistance of associated hardware, such as locknuts and lockwashers, when tightening a fastener. To obtain the proper torque value where tight threads on locknuts produce a run-down resistance, add the resistance to the required torque value. Run-down resistance must be measured on the last rotation or as close to the makeup point as possible.

CAUTION

Tighten fasteners only small amounts at a time, following the proper tightening sequence for the bolt pattern and fastener type before reaching the final torque. Failure to follow the correct tightening sequence or overtightening can result in damage to the fasteners, any sealing gaskets, or the object being fastened.

2.4.2 Flange Tightening Sequences

When tightening bolts on flanges and similar surfaces, the bolts must be tightened to the proper torque and in the proper sequence. Tighten each fastener only a small amount at a time until snug, following the proper tightening sequence for the bolt pattern and fastener type, then tighten to the final torque. This prevents warping or damaging the flange or machine part and also prevents leaks. The numbers in *Figure 15* show the proper tightening sequences for some common bolt patterns.

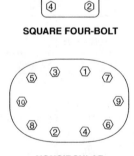

SQUARE FOUR-BOLT   **CIRCULAR FOUR-BOLT**

NONCIRCULAR MULTI-BOLT   **CIRCULAR MULTI-BOLT**

212F15.EPS

Figure 15 ◆ Common fastener tightening sequences.

Instructor's Notes:

Fastening Steel Fasteners

Steel fasteners are commonly fastened using a torque wrench like the one shown here. Whenever possible, apply force to the torque wrench by pulling rather than pushing. This reduces the chance of injury to the fingers or knuckles should the wrench slip. Be sure to grip the handle. Do not use extensions.

2.4.3 Thread Tapping

When assembling equipment using a threaded fastener, each fastener should be inspected for damaged threads. If damaged, replace the fastener with a new one. If the threads are dirty or rusty, clean and lightly lubricate them. If the fastener is being installed in a threaded hole, inspect the hole threads for damage. If the threads are damaged, the hole should be retapped. Be sure to use the correct size tap for the fastener being used.

2.4.4 Installing Anchor Bolts

Anchor bolts come in either expandable or non-expandable forms. When installing either type in hardened concrete, make sure the area where the equipment is to be fastened is smooth so that the equipment will have solid footing. Uneven footing might cause the equipment to twist, warp, not tighten properly, or vibrate during operation.

WARNING!
Drilling in concrete generates noise, dust, and possible flying objects. Always wear safety glasses, ear protectors, respiratory protection, and gloves. Ensure that others in the area also wear appropriate personal protective equipment.

Carefully inspect the drill and bit to make sure they are in good operating condition. If they are not, you could be injured by parts that chip and fly off during operation.

When installing a non-expansion anchor bolt in hardened concrete, it is installed in a drilled hole filled with a filler material (grout). When installing this type of anchor, the drill bit must be slightly larger in diameter than the head of the fastener and the flat washer used with the fastener. The flat washer must be able to fit down inside the hole, just below surface level. Also, the anchor bolt should extend out of the hole far enough for the threads and a little of the unthreaded bolt to be above the surface level. During the drilling process, the drill may be lubricated with water.

After the hole is drilled, the bolt installed, and the hole filled with grout, make sure the bolt is left straight after working the anchor around in the hole and installing the washer. The washer centers the bolt and holds it until the grout hardens. If the grout sets and the bolt is not straight, the bolt will be unusable, and the job will have to be repeated after the bolt is removed. Allow the grout to fully dry before mounting anything to the anchor bolt.

3.0.0 ◆ GASKETS

The basic function of a gasket is to create a seal between two fixed parts, such as joints in pipe systems, flanges in pumps, and compressor cylinder heads. Generally, the thinnest gasket that provides the seal is the most efficient and lasts the longest. The type of gasket used depends on:

- The operating temperature of the system or equipment
- The type of connection being made
- The type of fluid the system or equipment handles
- The system or equipment pressure range

3.1.0 Gasket Types

Pre-formed gaskets in standard sizes and patterns, and those used in HVAC equipment and components, are readily available from HVAC equipment suppliers and manufacturers. These

Note that both fasteners and threaded holes must be inspected before use and replaced or retapped, as appropriate.

Discuss the installation considerations for anchor bolts.

Emphasize the importance of using appropriate personal protective equipment while installing fasteners in concrete.

Show trainees how to install expandable anchor bolts.

Have trainees practice installing expandable anchor bolts. Note the proficiency of each trainee. This laboratory corresponds to Performance Task 9.

Have the trainees review Sections 3.0.0–5.3.0.

Ensure that you have everything required for teaching this session.

Discuss factors to consider when selecting gaskets.

Explain that pre-formed gaskets are widely available for HVAC applications.

factory-made gaskets are cut to the required pattern, with the holes already punched for the specific application. When manufactured gaskets are not available, gaskets must be laid out and cut from the proper material as needed. Depending on the intended use, gaskets are made from natural or man-made materials, or both. *Figure 16* shows some common gasket types, including:

- Flat
- Ring
- Spiral-wound
- Full-face
- Jacketed
- Envelope

3.1.1 Flat Gaskets

Flat gaskets are made of various materials and have many uses. They are typically pre-formed and made by equipment manufacturers in patterns for use with specific equipment.

3.1.2 Ring Gaskets

Ring gaskets are flat flange gaskets made of various materials for different uses. They are made to fit inside the bolt circle of a flange; therefore, they have no bolt holes. They usually have a pressure range of 150 to 200 psig.

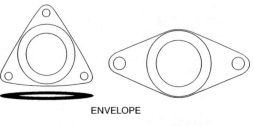
ENVELOPE

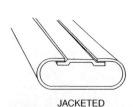
JACKETED

SPIRAL-WOUND

FULL-FACE

FLAT

RING

Figure 16 ◆ Common gasket types.

3.1.3 Spiral-Wound Gaskets

Spiral-wound gaskets are flat flange gaskets, commonly made of stainless steel with a graphite insert. They are used on high-temperature, high-pressure systems. They are crushable gaskets with high elasticity that allows them to adjust automatically to changes in line pressure, thermal shocks, vibration, and minor flange separation.

3.1.4 Full-Face Gaskets

Full-face gaskets are flat flange gaskets. They are similar to ring gaskets but are made to fit the flange and have holes for the flange bolts. They are made of various materials for different uses.

3.1.5 Jacketed Gaskets

Jacketed gaskets have a metal exterior cover with an internal filler. The filler can be of various materials. The outer jacket prevents the fluid in a system from contacting the inner filler material, preventing the filler from contaminating the liquid.

3.1.6 Envelope Gaskets

Envelope gaskets are similar in construction to the jacketed gasket. The outer cover is usually Teflon® with various types of fillers. Envelope gaskets have a limited temperature range. Because they have a superior sealing ability, envelope gaskets are frequently used on flanges with imperfections.

3.2.0 Installing and Removing Gaskets

The methods used to install and remove gaskets vary depending on the type of gasket material and the type of joint. Before a new gasket can be installed, the old gasket must be removed and the flanges thoroughly cleaned, using solvent if needed. The flanges and old gasket should be inspected for irregularities that might indicate a damaged flange.

WARNING!
Always replace a gasket with one specified by the equipment manufacturer, or an approved equivalent. Substituting the wrong gasket or material may cause equipment failure and possible personal injury.

When installing the new gasket, do not overtighten the bolts because this could damage the gasket, the flange or mating surfaces, or the fasteners. Adhesives, sealers, or lubricants should not be used with gaskets unless specified by the manufacturer. In some cases, adhesives can be used to hold the gasket in place during installation. If used, make sure the adhesive is compatible with the gasket material and will not cause breakdown or contaminate the fluid in the system.

CAUTION
Tighten fasteners only small amounts at a time, following the proper tightening sequence for the bolt pattern and fastener type, before reaching the final torque. An incorrect tightening sequence or overtightening can result in damage to the fasteners, any sealing gaskets, or the object being fastened.

4.0.0 ◆ PACKING

Packing is a rope-like material that is impregnated with a lubricant or rubber-like material and is pre-formed into special shapes. Packing is used to control or prevent leakage in equipment, such as valves and pumps that handle fluids. *Figure 17* shows an example of a valve packing.

4.1.0 Packing Shapes and Materials

The most common types of packing are solid, braided, and granulated fibers. Packing also takes many different shapes. It may be square,

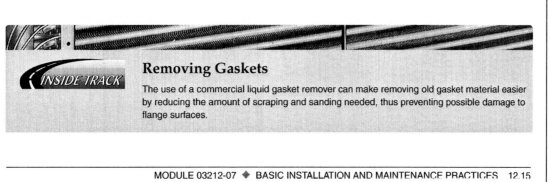
Removing Gaskets
The use of a commercial liquid gasket remover can make removing old gasket material easier by reducing the amount of scraping and sanding needed, thus preventing possible damage to flange surfaces.

Describe spiral-wound, full-face, jacketed, and envelope gaskets and their applications.

Show trainees how to identify various types of gaskets.

Have trainees practice identifying various types of gaskets. Note the proficiency of each trainee. This laboratory corresponds to Performance Task 3.

Note that flanges must be inspected and cleaned before a new gasket is installed.

Emphasize the importance of using only the gasket specified by the equipment manufacturer.

Remind trainees to follow the proper tightening sequence and avoid overtightening when installing gaskets.

Show Transparency 19 (Figure 17). Explain the purpose of packing.

List the common types of packing.

Discuss different types of packing materials and their applications.

Show trainees how to identify various types of packing.

Provide common types of packing for the trainees to examine.

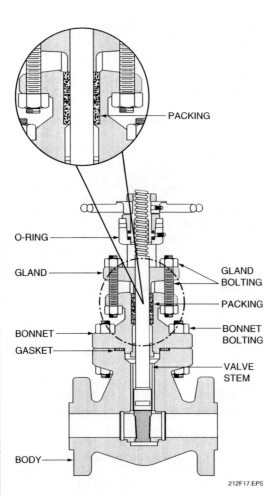

Figure 17 ◆ Valve packing.

4.1.1 Teflon® Yarn Packing

Teflon® yarn packing is a cross-braided yarn impregnated with Teflon®. Teflon®-impregnated packing resists concentrated acids, such as sulfuric acid and nitric acid, sodium hydroxide, gases, alkalis, and most solvents. It is good for use in applications of up to approximately 550°F.

4.1.2 Teflon® Filament Packing

Teflon® filament (cord) packing is a braided packing made from TFE filament. It is impregnated with Teflon® and an inert softener lubricant, or sometimes with graphite. It is often used on rotating pumps, mixers, agitators, kettles, and other equipment.

4.1.3 Lubricated Graphite Yarn Packing

Lubricated graphite yarn packing is an intertwined braid of pure graphite yarn impregnated with inorganic graphite particles. The graphite particles dissipate heat. Lubricated graphite yarn packing also contains a special lubricant that provides a film to prevent wicking and reduce friction. It is good for use in high-temperature applications.

4.1.4 Lubricated Carbon Yarn Packing

Lubricated carbon yarn (graphite impregnated) packing is made from an intertwined braid of carbon fibers impregnated with graphite particles and lubricants to fill voids and block leakage. Lubricated carbon yarn packing is used in systems containing water, steam, and solutions of acids and alkalis. It is considered suitable for steam applications with temperatures up to approximately 1,200°F, and up to 600°F where oxygen is present. It is reactive to oxygen atmospheres.

4.1.5 TFE/Synthetic Fiber Packing

TFE/synthetic fiber packing is made from braided yarn fibers saturated and sealed with TFE particles before being woven into a multi-lock braided packing. TFE/synthetic fiber packing protects against a variety of chemical actions. It is used in applications where caustics, mild acids, gases, and many chemicals and solvents are present. It is often used in general service for rotating and reciprocating pumps, agitators, and valves. Some variations also use TFE yarns, sometimes combined with Inconel, and a filling of graphite or other material, held in place with the yarn exterior.

wedge-shaped, or ring-shaped. Commonly used types of packing include Teflon® yarn and filament, lubricated graphite yarn, lubricated carbon yarn (graphite impregnated), and tetrafluoroethylene (TFE/synthetic fiber). Types of packing are shown in *Figure 18*. Though most packing materials will normally be specified, the following factors must be considered when choosing the type of packing material to use:

- Material flowing through the line
- Operating pressures
- Operating temperatures
- Minimum temperature of the piping system
- Composition of the valve stem

INCONEL® AND CARBON WIRE YARN TEFLON® YARN GRAPHITE YARN GRAPHITE RIBBON

PACKING IN PLACE SYNTHETIC FIBER EXPANDED PTFE

Figure 18 ♦ Types of packing.

4.2.0 Installing and Removing Packing

This section covers installing and removing packing. Knowing how to properly install and remove packing is critical due to the great impact these actions may have on system performance.

4.2.1 Removing Packing

The life of a packing depends on the type of packing used and its application. The most common symptom of packing failure is excessive leakage. When packing fails, it must be replaced.

Before the new packing can be installed, the old packing must be removed. Removal is usually simple, but it is important that the procedure be performed properly. The procedure for removing packing varies slightly from one piece of equipment to another. Always refer to the manufacturer's manual for the specific piece of equipment.

Regardless of the procedure, a packing puller or packing hook is typically used to remove the packing from a shaft installed in a **stuffing box**. *Figure 19* shows a stuffing box with the gland removed and packing puller screwed into the packing. If the puller is properly screwed into the packing ring, the ring usually comes out in one piece. If the ring breaks during removal, all of the smaller pieces must be removed. To help prevent breakage, two pullers may be used at the same time to pull the packing on both sides.

4.2.2 Installing Compression Packing

Packing is a major item of concern in the maintenance of equipment. Selecting the proper type and size of packing and installing it correctly is critical to maintaining optimum equipment operation. Incorrect selection or installation of

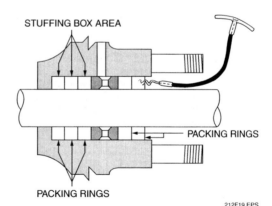

Figure 19 ♦ Packing puller screwed into packing.

Show Transparency 21 (Figure 20). Explain how to install various types of packing.

Explain that cutting packing on the shaft may result in equipment damage.

Discuss the purpose of seals. Explain that there are two kinds of seals: mechanical and nonmechanical.

Using a Packing Puller

When using a packing puller, be careful not to scratch or score the shaft. To prevent damage to the shaft, keep the puller angled away from the shaft at all times. Even minor damage to the shaft will cause accelerated packing wear and failure. If there are any burrs or nicks on the shaft and/or stuffing box, they must be removed before the new packing can be installed. If there are deep scratches or excessive wear, an evaluation must be made as to the need for repairs before new packing is installed.

packing can result in premature failure of the packing or in equipment damage. *Figure 20* shows important points related to installing compression packing. As shown in *Figure 20(A)*, the packing should be wrapped tightly around the shaft. The number of turns should equal the number of packing rings needed for the job. *Figure 20(B)* shows the method used to mark the packing while still wrapped around the shaft in preparation for cutting.

Figure 20(C) shows the use of an S-twist method for placing the cut packing rings on the shaft. Do not force the rings open because this might break the packing. When installing the rings, install them one at a time, tamping (pressing) each one firmly into position. The cut in each ring should be rotated 180 degrees from the cut in the ring installed before it. Staggering the cuts as shown in *Figure 20(D)* provides a better seal than stacking the cuts of the rings in line with one another.

CAUTION
Do not cut packing on the shaft because the cutting tool may scratch the shaft. Cut the packing on a round steel shaft or pipe of the approximate size of the shaft being packed. Some packing manufacturers recommend a straight cut; others recommend a cut made on an angle (skive cut). Check the manufacturer's recommendations before cutting and packing.

5.0.0 ♦ SEALS

Seals are devices used to prevent or control leakage between moving and fixed parts or between two fixed parts. There are two classes of seals: nonmechanical and mechanical.

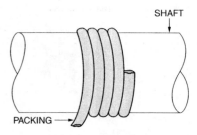
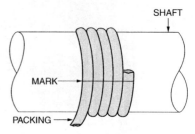
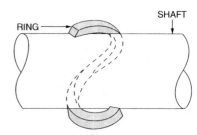
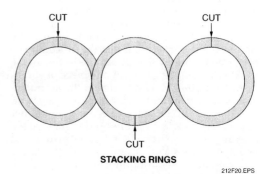

Figure 20 ♦ Installing packing. (A) Packing on shaft. (B) Mark on packing. (C) Installing packing. (D) Stacking rings.

Instructor's Notes:

5.1.0 Nonmechanical Seals

Nonmechanical seals are used as both **static seals** and **dynamic seals**. A static seal is one in which there is no movement between the two joining parts or between the seal and the mating part. O-rings are examples of seals often used as a static seal. A dynamic seal is one in which there is movement between two mating parts or between one of the parts and the seal. Lip and oil-type seals are examples of seals used as dynamic seals.

5.1.1 O-Rings

O-rings are circular seals used as either static or dynamic seals. They come in a variety of sizes and are made from various materials for use in different applications. When used as a static seal, an O-ring is installed in a slightly wider O-ring groove (*Figure 21*). A static seal needs to be a complete seal, so the O-ring should almost fill the O-ring groove. When used as a dynamic seal, an O-ring is usually placed in a groove or joint that is wider than the diameter of the O-ring. This is because a dynamic seal requires a running (friction) fit.

When an O-ring comes into contact with the areas to be sealed, it is slightly distorted in a motion called mechanical squeeze. The pressure caused by mechanical squeeze holds the O-ring in contact with the surfaces to be sealed. This pressure also causes the O-ring to roll and slide to the side of the groove away from the pressure.

To allow for initial mechanical squeeze, general purpose O-rings are made with a cross-section that is 10 percent larger than the nominal size. The use of O-rings with too much squeeze wears them out quickly, while insufficient squeeze can cause leaks.

5.1.2 Lip and Oil Seals

Lip seals are low-pressure, positive-contact seals used with rotating shafts. Pressure on the lip or a vacuum behind the lip pushes it against the shaft

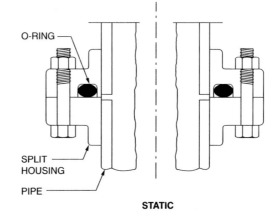
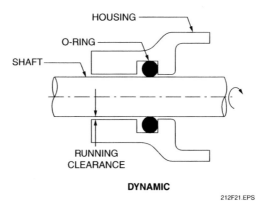

Figure 21 ♦ Static and dynamic seals using an O-ring.

for a tighter seal. Lip seals are usually installed with the lip facing in to contain lubricant in a housing where the outside conditions are relatively clean. In dirty conditions, the seal is installed with the lip facing out to prevent foreign matter from getting into the housing.

O-Ring Lubrication

All O-rings require lubrication. Keep in mind, however, that the lubricant and the O-ring material must be compatible. If they are not properly matched, the seal could be damaged. Because new refrigerants and oils are constantly being introduced, it is increasingly important to check the manufacturer's requirements to ensure a match between the O-ring and the lubricant.

Show Transparency 23 (Figure 22). Describe how an oil seal works.

Show Transparency 24 (Figure 23). Point out the components of a typical mechanical seal.

Provide mechanical seals for trainees to examine.

In cases where the area being sealed changes from pressure to vacuum conditions, double-lip seals are used to prevent air or dirt from getting in and lubricant from getting out.

The oil seal, or radial lip seal, is a positive-contact seal used on rotating or reciprocating shafts. It is used to keep fluids in or dirt and other foreign matter out. Oil seals are most often used to keep fluids in and are installed with the lip facing in. When an oil seal is used to keep contaminants out, it is installed with the lip facing away from the housing. Double-lip seals that perform both functions are also available. *Figure 22* shows a typical oil seal.

5.2.0 Mechanical Seals

When used with rotating shafts, mechanical seals provide a far superior dynamic seal to compression packings and other nonmechanical seals. They are typically used on shafts of open-type compressors and similar devices. Mechanical seals are made in many varieties, but they all basically function in the same way. As shown in *Figure 23*, mechanical seals usually include the following components:

- A set of primary faces is used, one that rotates and one that is stationary (such as the seal ring and insert shown in *Figure 23*).

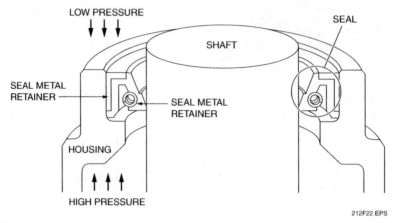

Figure 22 ◆ Oil seal.

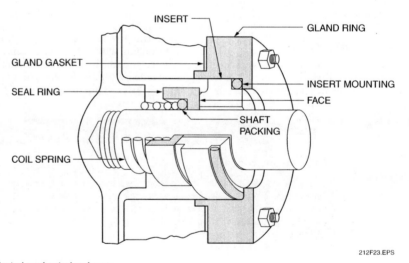

Figure 23 ◆ Typical mechanical seal parts.

12.20   HVAC ◆ LEVEL TWO

- A closing mechanism provides the load required to keep the seal faces in contact with each other. These are usually multiple springs, a single spring, or a metal bellows.
- A set of secondary seals is used, such as shaft packing and insert mounting, which can be O-rings, V-rings, wedges, or U-cups.
- The mechanical seal hardware includes gland rings, collars, compression rings, pins, and springs.

The primary seal is achieved by the compression of two very flat, lapped faces that are held tightly together by springs, metal bellows, or similar devices. Operation of the mechanical seal depends on an extremely thin film of lubricant being applied between the sealed faces. Since no seal is perfect, this lubricant is furnished by virtue of a minuscule level of leakage that occurs at the seal surfaces. Maintenance of this extremely thin film is made possible by machining the seal faces to very high tolerances in respect to flatness and surface finish. To preserve this precise surface finish requires that seal parts be carefully handled and protected. Mating surfaces should never be placed in contact without lubrication. Each seal application determines the best type of mechanical seal to use.

5.3.0 Installing and Removing Seals

The life of a seal depends on the type of seal used and its application. The most common symptom of seal failure is leakage. When a seal fails it must be replaced. Before the new seal can be installed, the old seal must be removed. The removal and installation of an O-ring seal is a relatively simple procedure. The procedures for removing lip, oil, and mechanical seals vary widely from one piece of equipment to another. These seals should be removed by following the instructions given in the manufacturer's service manual for the equipment you are servicing. Inspect the parts of a mechanical seal after they are removed for possible clues as to why the seal failed.

To achieve proper operation of lip, oil, and mechanical seals, it is important that the installation procedures be performed properly, according to the following guidelines:

- The equipment must be properly prepared to receive the new seal. All parts must be clean and free of sharp edges.
- The seal used must be right for the application.
- It is important to install the seal according to the equipment and/or seal manufacturer's instructions, or both. Study the manufacturer's data so that you are familiar with the procedure before attempting to install the seal.
- Lubricate the shaft, O-rings, and seal faces using the lubricant specified by the manufacturer or in the seal catalog.
- Protect precise seal parts from dirt and damage during installation. Do not touch seal faces with your hands. Handle seals by the edges. Inspect seal faces to ensure that there are no nicks or scratches. During assembly, keep the seal centered on the shaft.
- During assembly, protect all O-ring and/or other static seals from damage on sharp edges such as threads, keyways, or the end of the shaft.
- All parts must fit properly without binding.
- Follow the proper bolt-tightening procedure.

6.0.0 ♦ BEARINGS

Bearings of one type or another are used in practically every piece of equipment that has rotating parts. Bearings are used to reduce friction between parts in motion. In HVAC equipment, they are commonly used to support shafts in motors, fans, and compressors. Bearings can be divided into two broad classifications: plain and anti-friction.

The following terms are commonly used when describing the operation of bearings:

- *Axial load* – This is an external load that acts lengthwise along a shaft, such as that applied in a direct-drive coupled compressor and motor.
- *Journal* – This is the part of a shaft, axle, or spindle that is supported by and revolves in a bearing.
- *Radial load* – This is the side or radial force applied at right angles to a bearing and shaft, such as that applied from a pulley.
- *Thrust* – This is the force acting lengthwise along the axis of a shaft, either toward it or away from it.

6.1.0 Plain Bearings

Plain bearings (*Figure 24*) are simple in construction, operate efficiently, and can support heavy loads. During operation they develop an oil film between the shaft journal and bearing surfaces that overcomes the friction of the sliding motion. There are three general classifications of plain bearings: radial, thrust, and guide bearings. Radial bearings are used to support radial loads.

Show trainees how to identify mechanical seal parts.

Have trainees identify mechanical seal parts. Note the proficiency of each trainee. This laboratory corresponds to Performance Task 4.

Provide general guidelines for installing and removing seals.

Show trainees how to install an oil seal.

Have trainees practice installing an oil seal. Note the proficiency of each trainee. This laboratory corresponds to Performance Task 5.

Have the trainees review Sections 6.0.0–6.4.2.

Ensure that you have everything required for teaching this session.

Discuss the purpose of bearings, and define common terms associated with their use.

Show Transparency 25 (Figure 24). Discuss plain bearings and their applications.

Describe thrust and sleeve bearings and their applications.

Describe anti-friction bearings and their applications.

Show Transparency 26 (Figure 25). Describe the function and applications for ball bearings.

Provide various types of bearings for trainees to examine.

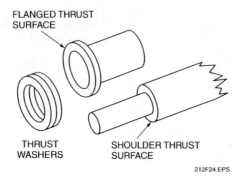

Figure 24 ♦ Plain bearings.

Thrust bearings are used to support axial loads on rotating members. Both radial and thrust bearings are also called rotational bearings. Guide bearings are used to guide moving parts in a straight line, such as in machine tools. The most common plain bearings used in HVAC equipment are sleeve and thrust bearings.

6.1.1 Sleeve Bearings

The sleeve bearing provides the bearing surface for the shaft journal and is usually press-fitted into a supporting member. Sleeve bearings are quiet in operation and have a good radial load capacity, such as needed with a blower motor used in a residential furnace. If made of bronze, sleeve bearings have good resistance to humidity, dirt infiltration, and corrosion.

The simplest and most widely used types of sleeve bearings are cast-bronze and porous bronze. Lubrication in a sleeve bearing is important because the bearing must have an oil film between the shaft and bearing surface. During rotation, the shaft actually floats on this oil film and never touches the bearing surface. Cast-bronze bearings are oil or grease lubricated. This type usually has a reservoir that is filled from the outside by means of an access port. Cast-bronze bearings usually need to be lubricated on a regular basis with the proper type of oil or grease.

Porous bearings are impregnated with oil and often have an oil reservoir with a wick to gradually feed the oil to the bearing. Under normal operating conditions, this type of sleeve bearing will operate in a motor for years without the need for lubrication.

6.1.2 Thrust Bearings

Thrust bearings support axial loads and/or restrain lengthwise movement. The simplest type is a thin disc-type thrust bearing called a thrust washer. Two or more thrust washers made of low-friction materials are often combined. It is also common to support thrust loads on the end surface of journal bearings. If the area is too small for the applied load, a flange may be provided. A shoulder is usually cut on the mating surface of the shaft.

6.2.0 Anti-Friction Bearings

Anti-friction bearings are so named because they operate on the principle of rolling motion, using either balls or rollers between rotating and fixed surfaces. Because of this, friction is reduced to a fraction of that in plain bearings. The rolling action of anti-friction bearings normally makes them noisier than plain bearings. Under certain conditions, anti-friction bearings can rust because they are usually made of steel.

Compared to plain bearings, anti-friction bearings normally require much less maintenance, especially if they are grease packed. Lubricant is used in anti-friction bearings mainly to keep out dirt and moisture. It also helps dissipate the heat that builds up in the bearing, but it does not provide an oil film to reduce bearing friction. Many types are permanently lubricated. There are two general types of anti-friction bearings: ball bearings and roller bearings.

6.2.1 Ball Bearings

Ball bearings (*Figure 25*) consist of outer and inner rings, also called races, with balls in between. The inner race has a groove around its outside circumference for the balls to roll in. The outer race usually has a similar groove on its inside circumference. A separator or retainer spaces the balls around the track and guides them through the load zone. Ball bearings fall into three classes: radial, thrust, and angular. Angular ball bearings are used in applications that have combined radial and thrust loads, and where precise shaft location is needed. Radial and thrust ball bearings are used in applications with radial loads and axial thrust loads, respectively.

12.22 HVAC ♦ LEVEL TWO

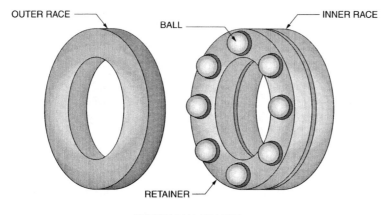

 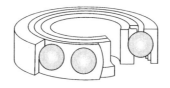

THRUST BALL BEARING

RADIAL BALL BEARING **ANGULAR BALL BEARING**

212F25.EPS

Figure 25 ♦ Typical ball bearings.

6.2.2 Roller Bearings

In general, roller bearings (*Figure 26*) have higher capacities than ball bearings. They are used in heavy-duty, moderate-speed applications. The construction of roller bearings is similar to ball bearings except that rollers are used instead of balls. Also, the outer race is called the cup, and the inner race is called a cone. The separator or retainer is called the cage. Roller bearings fall into three classes:

- *Cylindrical bearing* – These are used for radial loads.
- *Tapered bearing* – These are used for radial loads, axial loads, or both, depending on the design.
- *Spherical bearing* – These are used in applications requiring high load capacity and/or those with a shock load, such as a conveyer or speed reducer.

Another bearing considered to be a roller bearing is called a needle bearing. Needle bearings do not have cages to separate the rollers. Also, the rollers are longer and more of them are used than in the other roller bearings. This gives the needle bearing more contact area. Variations of this type of bearing allow it to be used for the same applications as other roller bearings. The advantage is that they give more support in less space.

6.3.0 Identifying Bearing Failures

Touch and sound are two effective methods used to detect bearing failures. Touching or feeling the bearing housing when a motor or other rotating device is operating can give an indication of a bearing's condition. A bearing is probably in the process of failing or has failed if the bearing housing feels overly hot to the touch.

Listening for the sound of foreign noises coming from a motor or other rotating device will often help you detect a bearing problem.

Another method used to detect a bearing problem is to place one end of a steel rod on the bearing housing while the other end is held close to the ear. The rod acts as an amplifier, transmitting unusual sounds such as thumping or grinding, which would indicate bearing failure. Special listening devices, such as a stethoscope, can be used for this purpose.

Describe the functions and applications for roller bearings.

Show trainees how to identify different types of bearings.

Have trainees practice identifying different types of bearings. Note the proficiency of each trainee. This laboratory corresponds to Performance Task 10.

Describe the motor checks used to determine the condition of bearings.

Provide examples of failed bearings for trainees to examine.

See the Teaching Tip for Section 6.3.0 at the end of this module.

Show Transparency 27 (Table 1). Explain how an examination of the lubricant can help in identifying the cause of bearing failure.

NEEDLE ROLLER BEARING

TAPERED ROLLER BEARING

SPHERICAL ROLLER BEARING

CYLINDRICAL ROLLER BEARING

Figure 26 ◆ Typical roller bearings.

Excessive end play of a motor shaft can also be an indication of bearing wear. Ball-bearing motors should typically have an end play of about $\frac{1}{32}$" to $\frac{1}{16}$". Sleeve-bearing motors may have an end play of up to $\frac{1}{2}$".

Examining the lubricant on a damaged bearing can also be a help in identifying a cause of a bearing failure. *Table 1* summarizes some common problems. Lubricants are discussed in detail later in this module.

Table 1 Lubricant Symptoms Related to Common Bearing Failures

Lubricant	Symptom	Cause
Oil or grease	Reddish deposit at ball contact	Indicates chafing; look for oscillatory or vibratory motion during operation
Oil or grease	Metallic particles accompanied by indications of wear on bearing components	Lubrication failure or external contamination
Oil	Dry surface with little evidence of residue, or slight brownish haze on ball or raceway surfaces	Too little lubricant, either initially or because of migration
Grease	Grease darkened but still oily	High loads or contaminants
Grease	Grease darkened and dry	Extremely high loads or contaminants
Grease	Grease surface dry and hard, soft interior	High external heat

6.4.0 Removing and Installing Bearings

When bearings wear out, they must be removed and replacements installed. They must also be removed to disassemble a piece of equipment for repair or maintenance.

6.4.1 Removing Bearings

The removal of bearings can vary widely from one piece of equipment to another, depending on the type of bearing used and its housing. Bearings should be removed by following the instructions given in the manufacturer's service manual for the equipment. This is important in order to prevent damaging the bearing or the shaft. If the bearing is being removed because it has failed, inspect the bearing after it has been removed for possible clues as to the reason the bearing failed.

The type of bearing removal tool and the method used to remove a bearing generally depend on the size, type, and fit of the bearing to be removed. Bearing pullers, arbor presses, and hydraulic presses can be used. The manual bearing puller is the most common tool for removing bearings in the field. The puller can usually be used to remove the bearing from the shaft while the shaft associated with the bearing is still in the equipment. The manual puller has a bolt that is turned using a wrench to provide the pressure to pull the bearing. *Figure 27* shows a manual bearing puller being used to remove a bearing.

When removing a bearing using a puller, use the following guidelines:

- The puller jaws must apply pressure only to the inner race of the bearing. If pressure is applied to the outer race, the bearing will be damaged and may come apart.
- Be sure that the puller is pulling straight. If it is misaligned, the bearing will become cocked and may damage the shaft.
- Do not let the bearing fall on the floor when it comes off the shaft because it could be damaged and get dirty.

6.4.2 Installing Bearings

The tools and methods used to install a bearing generally depend on the size, type, and fit of the bearing. Three kinds of fits for bearings are the slip fit, press fit, and interference fit. The slip fit is the simplest to install because the bearing fits fairly loosely and can usually be pushed into place by hand. The press fit is much tighter and

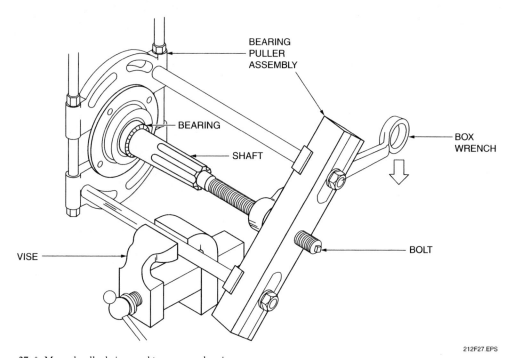

Figure 27 ♦ Manual puller being used to remove a bearing.

Discuss common methods of installing bearings.

Explain that heated bearings must be handled with appropriate gloves to avoid burns.

requires more effort to press the bearing into place. Bearings usually have a slip fit on one ring and a press fit on the other. The ring that rotates is usually press-fitted. In most cases, the inner ring of the bearing rotates. Interference fit bearings must be heated before they are pressed on because the inside diameter is smaller than the shaft. Bearings are usually installed using either a temperature mounting method, press mounting method, or locknut method, depending on the type of bearing. The method used should be the one recommended by the instructions given in the manufacturer's service manual for the equipment that you are servicing.

The temperature mounting method used for interference bearings can usually be performed in the field or shop while the shaft associated with the bearing is still in the equipment. This method uses a bearing heater (*Figure 28*) to heat the bearing to a temperature level specified by the manufacturer. Heating the bearing causes it to expand enough so that it can be slipped onto the shaft. Once on the shaft, it is quickly moved to its proper mounted position and held there to prevent it from moving. When it cools, it shrinks to fit the shaft. When using this method, care must be taken not to overheat the bearing, because overheating can adversely affect the hardness of the bearing steel. The maximum temperature to which a bearing should be heated is 250°F.

> **WARNING!**
> Heated bearings can burn your hands. Wear clean, protective gloves and handle the heated bearing in a manner that prevents your skin from coming in contact with the heated metal. To avoid being burned, wait until the bearing has cooled before touching it.

The press mounting method uses an arbor press (*Figure 29*) to press the shaft into a bearing. This method is typically used when installing bearings where the bearing shaft race is press-fitted and the housing race is slip-fitted, such as with a thrust bearing. In this method, the bearing is placed on the arbor so that the bearing shaft race is well supported. The shaft is then positioned in the bearing bore and pressed into place. The assembled shaft and bearing are then installed in the equipment. When pressing the shaft into the bearing, the shaft must be kept square with the bearing at all times. If the shaft is cocked during the pressing operation, the bearing will gouge the shaft.

Figure 28 ♦ Bearing heater.

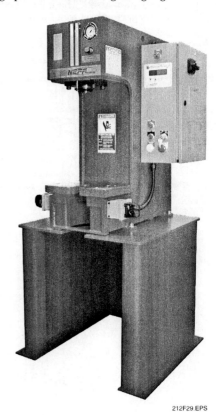

Figure 29 ♦ Arbor press.

The locknut method is typically used when installing tapered-bore bearings, such as a spherical roller bearing. A tapered-bore bearing is mounted on a tapered shaft, or a tapered sleeve. In this method, the bearing is placed on the shaft and positioned in place by tightening a locknut. As the locknut is turned, it causes the bearing to be forced onto the shaft, and the clearance between the races and the rolling elements is reduced. This clearance must be controlled. To do this, the clearance is measured with a feeler gauge before installation and during the tightening process (*Figure 30*). The locknut is turned as needed until the bearing is at the proper position on the shaft. This position is attained when the amount of clearance initially measured for the bearing is reduced to an amount specified in tables supplied by the bearing manufacturer. When no bearing clearance tables are available, a rule of thumb is to reduce the clearance by about 50 percent.

7.0.0 ♦ LUBRICATION

Any piece of equipment or machinery that has rotating or moving parts produces friction. Friction is the resistance to motion that takes place when two parts move or rub against one another. Lubrication is required whenever friction is a problem and must be controlled. Lubrication is the application of any substance (lubricant) that reduces friction by creating a slippery film between two surfaces. Lubricants can be made from animal fats, vegetable oils, mineral oils, or synthetics. However, over 90 percent of the total lubricants used are made from mineral oils which come from refined crude oil. Lubricants can be in the form of a liquid (oil), semisolid (grease), or solid films. All three are used to lubricate rolling and sliding bearings and gears.

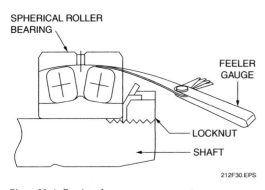

Figure 30 ♦ Bearing clearance measurement.

7.1.0 Oils

Oils are a broad class of fluid lubricants. Basically, lubricating oils are made from two types of crude oil: naphthenic and paraffinic. Naphthenic oils contain very little wax and are good lubricants for almost any use. Paraffinic oils are very waxy, and lubricants made from them are used mainly in hydraulic equipment and other machinery. All lubricants possess certain properties that govern their suitability for a particular use. These properties include the following:

- **Viscosity**
- **Viscosity index (VI)**
- **Pour point**
- **Flash point**
- **Fire point**
- **Oxidation** resistance

7.1.1 Viscosity

Viscosity is the most important property of oil. Viscosity is the thickness of a liquid or the ability of the liquid to flow at a specific temperature. Low-viscosity oils are light oils that flow freely when poured. High-viscosity oils are heavy oils that flow slowly when poured. Medium viscosity oils range somewhere in between.

The viscosity rating of oils can be expressed in two ways. Automotive and gear-lubricating oils are expressed in Society of Automotive Engineers (SAE) ratings. The other viscosity rating system, which applies to industrial lubricants, is the Saybolt Universal Seconds (SSU or SUS). This rating system was developed by ASTM International. Generally, low-viscosity oils are more applicable for use with bearings. Higher-viscosity oils are more applicable for use with gears.

7.1.2 Viscosity Index

The viscosity index (VI) is a measure of how viscosity varies with temperature. Not all oils change viscosity at the same rate when subjected to the same temperature changes. If an oil has a low VI, the viscosity changes rapidly with temperature. A high VI represents a lower change in viscosity with temperature changes. Typically, naphthenic oils have a low VI and paraffinic oils have a high VI.

7.1.3 Pour Point

The pour point of an oil refers to the lowest temperature at which an oil will flow freely. This must be considered in cold weather applications

Show Transparency 29 (Figure 30). Describe the locknut method for installing bearings. Explain how to use a feeler gauge to measure bearing clearances.

Show trainees how to use a feeler gauge to measure bearing clearances.

Have trainees practice using a feeler gauge to measure bearing clearances. Note the proficiency of each trainee. This laboratory corresponds to Performance Task 12.

Have the trainees review Sections 7.0.0–8.2.0.

Ensure that you have everything required for teaching this session.

Explain that proper lubrication is essential to prevent damage caused by friction.

Discuss oil-based lubricants and their applications.

Define common terms associated with oils.

See the Teaching Tip for Section 7.0.0 at the end of this module.

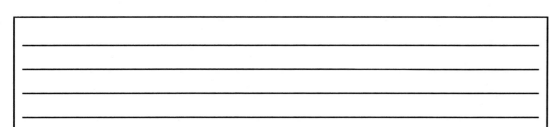

Discuss greases and their applications.

Explain that most HVAC equipment is lubricated manually.

Show Transparency 30 (Figure 31). Discuss lever-type grease guns and fittings.

Note that motors using large amounts of grease normally incorporate lubrication fittings.

Provide a lever-type grease gun for trainees to examine.

and refrigeration systems. The lower the pour point of an oil, the more freely it flows at very low temperatures. A high pour point means that the oil stops flowing freely at low temperatures. At the same viscosities, naphthenic oils have lower pour points than paraffinic oils.

7.1.4 Flash Point and Fire Point

The flash point is the temperature at which oil gives off ignitable vapors. This temperature is not high enough for the oil to support combustion. When the flash point is reached, the lubricant film between the mating surfaces is destroyed, and damage of the surfaces can occur.

The fire point is the temperature at which oil will burn if ignited. The flash and fire points of an oil are only a consideration when using an oil in high-temperature applications.

7.1.5 Oxidation Resistance

Oxidation resistance is related to the service life of an oil. As oils are exposed to air and heat, they take on oxygen from the air. This process is known as oxidation. All lubricants tend to oxidize over time. Oxidation breaks down the chemical structure of oil and eventually destroys its lubricating qualities. Oxidation is a slow process, but if the oil splashes around excessively or is exposed to high temperatures, the process speeds up.

7.2.0 Greases

Greases are a solid or semisolid lubricant formed by adding a thickening agent, usually soap, and an additive to oil. An additive (such as a rust inhibitor or antifoam agent) is an extra ingredient used to improve the qualities of the lubricant. Grease produces the same lubricating action as oil, only in a different way. Oil forms a film that keeps surfaces apart to reduce friction and heat. Grease is not a liquid and cannot form a liquid film when it is initially applied. Grease forms the lubricating film when the surfaces begin to move and exert pressure on the grease. When pressure is applied, grease releases some of its oil, allowing the lubrication action to begin. When the motion is stopped and the pressure is reduced, grease tends to solidify again.

Greases are usually rated by their relative hardness or consistency on a scale developed by the National Lubricating Grease Institute (NLGI). The softest greases are rated at 000, with higher numbers indicating harder greases. Most greases fall in the range from 1 to 6. The consistency of grease varies with temperature, and there is generally an increase in the softening of a grease as the temperature increases.

Another common rating method for greases is their **dropping point.** The dropping point is the temperature at which grease is soft enough for a drop of oil to fall away or flow from the bulk of the grease. As a rule of thumb, a temperature that is 50°F less than the dropping point is the maximum operating temperature for the grease.

Normally, NLGI Grade 2 greases are used in most roller-type bearings because they combine good lubricant feeding with resistance to mechanical churning. Frequently, stiffer Grade 3 greases are used for double-sealed bearings and large bearings. Typically, the use of conventional, multi-purpose grease is adequate for temperatures between –20°F and +200°F. When choosing a grease, follow the grease, bearing, or equipment manufacturer's recommendations.

7.3.0 Lubrication Equipment

Several types of equipment and methods are used to apply lubricants. Some equipment requires manual lubrication; others use mechanical lubricating devices. The lubrication of most HVAC equipment involves the manual method. Manual lubrication equipment is hand-operated and is used to apply oil and grease. The operator must know how to refill and maintain this equipment.

7.3.1 Lever-Type Grease Gun

The most common manual lubrication tool used in the field is the lever-type grease gun (*Figure 31*). Lever guns are hand-operated and develop high pumping pressure with little effort. Grease guns are available in low-pressure models, high-pressure models, or both.

Generally, lever guns can be filled with grease in three ways: standard cartridge, bulk fill, and with a grease transfer pump. The standard cartridge is the most common method for HVAC work.

Adapters, such as flexible extension hoses and adjustable swivel fittings, can be used with the lever gun to gain easier access to lubrication fittings in hard to reach or tight places. Also, special grease injectors can be used when lubricating sealed bearings or for other special applications.

7.3.2 Lubrication Fittings

Equipment that is lubricated using grease or other semisolid lubricants normally has fittings that connect with the coupling on the lever gun

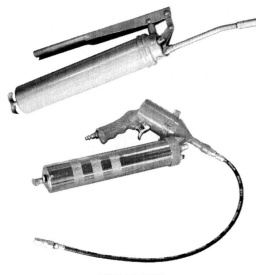

GREASE GUNS

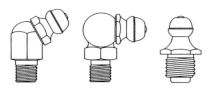

GREASE FITTINGS

HYDRAULIC COUPLERS

Figure 31 ♦ Lever-type grease gun and hydraulic grease fittings.

to form a pressure-tight connection. The most widely used is the hydraulic type, which mates with the jaws in the lever gun coupling. The lubricant pressure seats the coupling. Other types of fittings include button-head and flush types. These fittings are typically found in equipment using large quantities of grease. High pressures are required to pump grease through a fitting and into the grease cavity around a bearing. Lever guns can develop pressures up to 10,000 psi. Overpumping the handle can cause excessive grease pressure that could blow out the bearing seals. Some fittings and grease cavities are equipped with a vent groove or plug that is removed to relieve pressure and allow the old grease to be forced out.

8.0.0 ♦ BELTS AND BELT DRIVES

Belt drives are a quiet, smooth, and economical form of power transmission. They are commonly used in commercial HVAC equipment. The drive mechanism transfers a motor's rotating power to the driven device. A belt drive consists of a drive pulley with one or more sheaves, a driven pulley with a matching number of sheaves, and belts to match the sheaves (*Figure 32*). The pulleys or sheaves are available in many shaft sizes, diameters, and types of construction. Multiple-groove pulleys are made for use with equipment that uses two or more belts in the drive. Some air conditioning equipment uses step pulleys for driving the air movement fan. By changing the belt from one groove to another, the speed of the fan can be changed. Special variable-pitch pulleys are also available. These are made with half of the pulley threaded on the hub of the other half. A setscrew locks the variable half in place when it is properly adjusted. By turning the variable half, the V-groove can be widened to let the belt ride lower in the hub. This reduces the speed of the driven fan or other device. Belt drives can be divided into two basic types: V-belts and synchronous belts. HVAC equipment typically uses V-belt drives.

8.1.0 V-Belts

V-belts are made of a combination of fabric, cord, and/or metal reinforcement vulcanized with natural rubber compounds. The V-belt has a tapered shape that causes it to wedge firmly into the grooves of the sheave when it is under load. A V-belt works through frictional contact between the sides of the belt and the tapered sheave groove. Most V-belts fall into three general classifications:

- Fractional horsepower belts
- Standard multiple belts
- Wedge belts

8.1.1 Fractional Horsepower Belts

Fractional horsepower (FHP) belts are light-duty belts, usually used singularly. The size of FHP belts is indicated by a code marked on the outside of the belt. The first number and letter in the code tell the width of the belt in eighths of an inch. The next three numbers in the code tell the

Explain how a lever-type grease gun is used.

Show trainees how to lubricate a bearing using a lever-type grease gun.

Under your supervision, have the trainees practice lubricating a bearing using a lever-type grease gun. Note the proficiency of each trainee. This laboratory corresponds to Performance Task 13.

Identify the advantages of belt drives and explain that they are commonly used in commercial systems.

Show Transparency 31 (Figure 32). Explain how a V-belt drive operates.

Discuss various types of V-belts, including their alphanumeric designations and applications.

Provide examples of V-belts for trainees to examine.

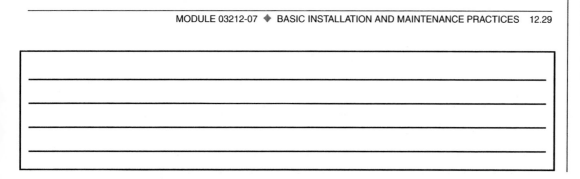

Explain that belt-drive failures can be the result of improper tension and that all belts should be checked using a tension gauge.

Explain that belts must never be run on the edges of the pulleys.

Show Transparency 32 (Figure 33). Discuss the three types of belt misalignment: angular, parallel, and sheave groove.

Show Transparency 33 (Figure 34). Explain how to check for proper belt alignment using a straightedge.

Show trainees how to align and adjust a V-belt.

Under your supervision, have the trainees practice aligning and adjusting a V-belt. Note the proficiency of each trainee. This laboratory corresponds to Performance Task 6.

Have the trainees review Sections 9.0.0–10.7.0.

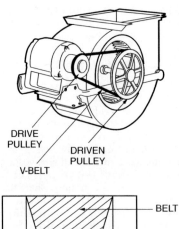

Figure 32 ♦ V-belt drive.

length of the belt in inches, with the last number indicating tenths of an inch. For example, a belt marked 4L300 is ½" wide and 30" long. FHP belts are measured on the outside surface of the belt. FHP belts come in the following standard widths and heights:

2L – ⅜" wide × ⅛" high
3L – ⅜" wide × ⁷⁄₃₂" high
4L – ½" wide × ⁵⁄₁₆" high
5L – ⅝" wide × ⅜" high

8.1.2 Standard Multiple Belts

Standard multiple belts are used for continuous service. They are used in sets of two or more. The size of standard multiple belts is indicated by a code marked on the belt, with a letter indicating the width and a number indicating the length. For example, a belt marked A42 is ½" wide and 42" long. The length of standard belts is measured on the inside surface of the belt. They are available in various lengths for each width size and in the following standard widths and heights:

A – ½" wide × ¹¹⁄₃₂" high
B – ⅝" wide × ⁷⁄₁₆" high
C – ⅞" wide × ⁹⁄₁₆" high
D – 1¼" wide × ½" high
E – 1½" wide × 1" high

8.1.3 Wedge Belts

The wedge belt is a multiple belt that has a smaller cross-section per horsepower than the standard multiple V-belt. Also, it is used on smaller diameter sheaves with shorter center distances than the standard belt. Wedge belts are not interchangeable with standard multiple belts and should not be run on sheaves made for standard belts. The size of wedge belts is indicated by a code marked on the outside of the belt. The first number and letter in the code tell the width and cross-section of the belt in eighths of an inch. The next three numbers in the code tell the length of the belt in inches. For example, a belt marked 3V500 has a 3V cross-section (⅜" wide × ⁵⁄₁₆" high) and is 50" long. The length of the wedge belt is measured along the pitch line, which runs along the center of the belt thickness. Wedge belts come in the following standard widths and heights:

3V – ⅜" wide × ⁵⁄₁₆" high
5V – ⅝" wide × ¹⁷⁄₃₂" high
8V – 1" wide × ⅞" high

8.2.0 Belt Drive System Maintenance

Belt drives are intended to give many hours of service in their particular use. Belt-drive failure can often be traced to improper belt tension. All belts should be tightened according to the manufacturer's instructions using a tension gauge. When installing the belts, they should be slipped loosely over the pulleys.

> **WARNING!**
> Do not try to run the belts on the pulleys. This will place excessive stress on the cords of the belt and can cause it to flop under load and possibly turn over in the sheaves.

Belt-drive failure can also often be traced to pulley misalignment. There are basically three kinds of alignment problems that can occur: angular, parallel, and sheave groove (*Figure 33*).

A simple way to check alignment using a straightedge ruler is shown in *Figure 34*, with arrows indicating the four check points. When the pulleys are properly aligned, the straightedge will be flat on the faces of both pulleys and no light should show at these four points.

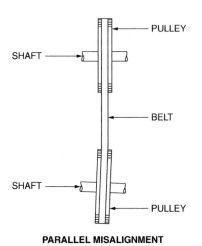

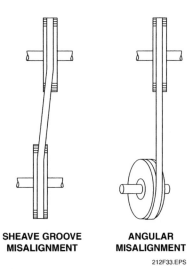

Figure 33 ◆ Belt-drive misalignment.

Variable-Pitch Pulley

The photo below shows a variable-pitch pulley being adjusted to change the speed of a blower in a belt-drive system.

9.0.0 ◆ COUPLINGS AND DIRECT DRIVES

Couplings are typically used to connect the shaft of a motor (driver) to the shaft of a compressor (driven). This method of coupling is called direct drive.

9.1.0 Coupling Types

Couplings are made in a variety of types and for many uses. Some couplings allow for slight misalignment and end play between the rotating shafts. Other couplings reduce or absorb vibrations or torque. Three common categories of couplings include rigid, flexible, and soft-start.

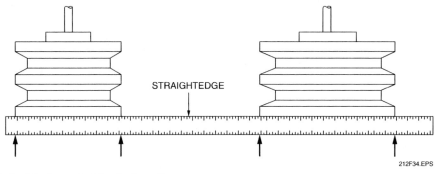

Figure 34 ◆ Straightedge method for aligning pulleys.

Ensure that you have everything required for teaching this session.

Discuss the purpose of a coupling and list the types of couplings in common use.

Provide various types of couplings for the trainees to examine.

Show Transparency 34 (Figure 35). Discuss rigid and flexible couplings and their applications. Explain that rigid couplings require precise alignment, while flexible couplings do not require the same degree of alignment.

9.1.1 Rigid Couplings

Rigid couplings provide a nonflexible connection between the driver and driven shafts. Rigid couplings do not compensate for misalignment and require precise alignment during installation. If a rigid coupling is misaligned and forced together, the drive will be damaged. Even slight misalignment can cause vibration and operating problems. There are three common types of rigid couplings: flanged, sleeve, and ribbed.

Flanged couplings (*Figure 35A*) join the driver and driven shafts using two mating flanges bolted together. Flanged couplings require keys to prevent them from rotating on the shafts.

Sleeve couplings, also called compression couplings, are similar to flanged couplings, except that they are taper-bored and have tapered sleeves that fit on the shafts. The wedge principle is used to tighten the coupling on the shafts. As the two halves of the coupling are pulled together over the tapered sleeve by the flange bolts, the coupling halves are tightened on the shafts. Sleeve couplings do not require keys and are normally used on small-diameter shafts. They are not suitable for use with heavy loads.

Ribbed couplings, also called clamp couplings, are made in two pieces. They are used when sleeve couplings are difficult to install. One advantage of using ribbed couplings is that they can be installed on shafts that are already in place without moving one of the shafts. They are used when the shafts are the same size and are also used for low-speed drives because of their unbalanced design and weight distribution.

9.1.2 Flexible Couplings

Flexible couplings are much more common than rigid couplings because they are usually easier to install and maintain and do not require precise alignment. Although flexible couplings allow for some misalignment, they should be aligned as close as possible, using the same methods employed with rigid couplings. Flexible couplings should not be used when major angular misalignment is known to exist. Deliberate misalignment requires the use of universal joints. Flexible couplings are divided into two categories: mechanical and material. Depending on the application, there are many types of couplings made in both categories.

Mechanical flexible couplings have metal components that may or may not need lubrication. They use the play or clearance in a mechanical device, such as chains or gears, to compensate for misalignment.

Material flexible couplings (*Figure 35B*) are made to allow parts of the coupling to flex to compensate for misalignment. These flexing elements can be made of various materials, such as metal, rubber, or plastic. The life of a coupling depends on the life of the flexible material. As the material flexes, it begins to wear. The more the coupling is misaligned, the more the material is flexed and the faster it wears.

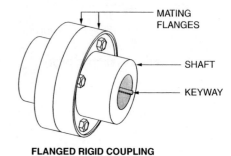

FLANGED RIGID COUPLING

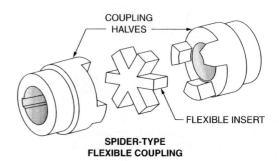

SPIDER-TYPE FLEXIBLE COUPLING

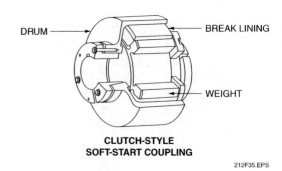

CLUTCH-STYLE SOFT-START COUPLING

212F35.EPS

Figure 35 ♦ Typical couplings.

9.1.3 Soft-Start Couplings

Soft-start couplings (*Figure 35C*) are used in applications where smooth, even starts are needed. Soft-start couplings allow the driving motor to pick up speed before the load is engaged, allow the driven device to start slowly and smoothly, and prevent stalls during overload conditions.

9.2.0 General Coupling Removal and Installation Methods

The removal of couplings can vary from one piece of equipment to another, depending on the type of coupling used. The manual coupling puller is the most common field method for removing couplings that are press-fitted on the shafts. This puller is similar to a bearing puller (see *Figure 27*).

The tools and methods used to install couplings generally depend on the type of coupling. Many types of couplings are relatively simple to install. Press-fitted couplings are usually installed using either the press-fit mounting method or the interference fit method. The method used should be the one recommended by the manufacturer's service manual for the equipment. The press-fit method of installation uses an arbor press to mount the coupling on the shaft. This requires that the shaft be removed from the equipment. The shaft is securely mounted on the press, then the coupling is positioned on the shaft and pressed into place.

In the interference fit method, the coupling has a bore that is slightly smaller than the shaft diameter. The coupling must be preheated to expand the bore before it can be installed on the shaft. As the coupling cools, it shrinks to grip the shaft securely. This method is basically the same as that previously described for heating and installing a bearing.

9.3.0 Coupling Alignment

Rotating shafts are usually aligned with the couplings in place. The measurements are taken from the couplings, and adjustments are made accordingly.

> **CAUTION**
> Alignment is critical to a few thousandths of an inch. Vibration, coupling wear, and bearing failure may occur if proper alignment is not maintained.

When aligning two coupling halves so that the shafts will be aligned, there are two ways that they must line up. First, the outer diameters, also called the rims, of the couplings must be lined up all the way around; then the faces of the couplings must be lined up. This results in four basic ways the coupling must be aligned:

- Outer diameter alignment, side view
- Outer diameter alignment, top view
- Face alignment, side view
- Face alignment, top view

If the couplings are misaligned, the alignment must be corrected to avoid poor equipment operation or damage.

> **NOTE**
> Motor mounts can weaken and sag over time, which will cause misalignment.

9.3.1 Correcting Outer Diameter Alignment, Side View

Outer diameter (OD) misalignment, side view is also called parallel misalignment. This type of misalignment occurs when one of the coupling halves is not in line with the other when viewed from the side (*Figure 36*). In OD misalignment, one coupling is higher than the other, and one of the units must be raised or lowered to align the couplings.

9.3.2 Correcting Outer Diameter Alignment, Top View

Outer diameter misalignment, top view (*Figure 37*) occurs when the outer diameters of the couplings are misaligned from side to side as viewed from the top. In this type of misalignment, one of the units must be moved to one side or the other to align the couplings.

9.3.3 Correcting Face Alignment, Side View

Face misalignment is also called angular misalignment. Face misalignment, side view (*Figure 38*) means that the faces of the couplings are not square with one another when viewed from the side. In this type of misalignment, one of the units must be tilted on its base to align the couplings.

Keep Transparency 34 (Figure 35) showing. Discuss soft-start couplings and their applications. Explain that these couplings allow the motor to pick up speed before engaging the load.

Show trainees how to identify various types of couplings.

Have trainees practice identifying various types of drive couplings. Note the proficiency of each trainee. This laboratory corresponds to Performance Task 7.

Provide general guidelines for removing couplings.

Emphasize the importance of proper coupling alignment.

Show Transparencies 35–38 (Figure 36–39). Explain how to correct outer diameter (OD) misalignment and face misalignment.

List some of the basic maintenance procedures required of HVAC technicians.

Emphasize the importance of using accurate measuring devices.

Show Transparency 39 (Figure 40). Describe a circumference rule and explain how it is used.

Describe calipers and dividers and explain how they are used.

Provide various measuring devices for the trainees to examine.

See the Teaching Tip for Section 10.0.0 at the end of this module.

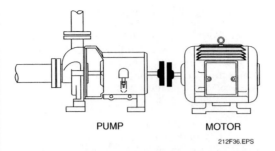

Figure 36 ♦ OD misalignment, side view.

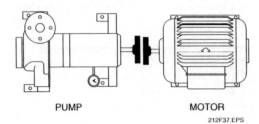

Figure 37 ♦ OD misalignment, top view.

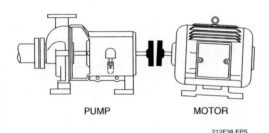

Figure 38 ♦ Face misalignment, side view.

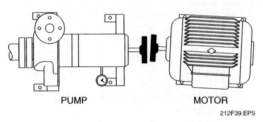

Figure 39 ♦ Face misalignment, top view.

9.3.4 Correcting Face Alignment, Top View

Face misalignment, top view (*Figure 39*) means that the faces of the couplings are not square with one another when viewed from the top. In this type of misalignment, one of the units must be rotated on its base to align the couplings.

10.0.0 ♦ BASIC MAINTENANCE PROCEDURES

This section covers some of the basic maintenance procedures commonly used in the scheduled maintenance of several types of HVAC equipment. These procedures include motor lubrication, system air filter and screen inspection and cleaning, cooling system/heat pump coil inspection and cleaning procedures, and damper inspection and cleaning.

10.1.0 Measuring Devices

Accurate measurements are vital in the HVAC trade. Some measuring devices that will provide precise measurements in a variety of applications are discussed in this section.

10.1.1 Circumference Rule

The upper edge of the circumference rule can be used for the same general-purpose measuring as the steel rule described in the *Core Curriculum* module, *Introduction to Hand Tools*. The lower edge, however, indicates at a glance the circumference of any given cylinder. The circumference rule in *Figure 40* illustrates that a cylinder with a diameter of 5" would be 15⅞" in circumference.

The reverse side of the rule contains various tables indicating container sizes, dry and liquid measures, and container capacity computing rules.

10.1.2 Calipers and Dividers

Calipers (*Figure 41*) are used to measure the thickness or diameter of a piece of work, to measure distances between or on surfaces, and to compare and transfer work dimensions.

Inside calipers are used to take inside measurements of work such as pipes, screw holes, and similar positions. Outside calipers are for measuring outside diameters of work such as pipes, tubes, and other round objects. Vernier calipers are used for precision work.

Dividers are used to measure the distance between lines or points; to transfer dimensions from measurements to the work; and to scribe lines, arcs, and circles.

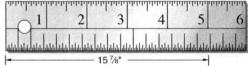

Figure 40 ♦ Circumference rule.

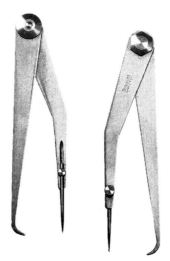

HERMAPHRODITE CALIPERS

OUTSIDE CALIPER

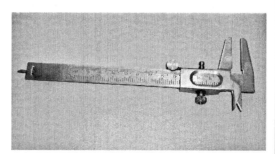

VERNIER CALIPER

DIVIDER

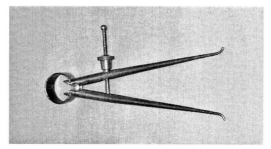

INSIDE CALIPER

212F41.EPS

Figure 41 ♦ Dividers and calipers.

Explain how to use various types of calipers.

Describe micrometers, thickness (feeler) gauges, and thread pitch gauges. Explain the use and applications of each.

Show trainees how to use various types of measuring devices.

Identify supplemental tools used for servicing HVAC equipment.

Describe a drill press.

Hermaphrodite and keyhole calipers are variations of dividers. Their principal use is in scribing parallel lines from an edge or for locating the center of cylindrical work. The keyhole caliper differs from the hermaphrodite caliper in that the leg does not taper and has a blunt end.

The vernier caliper shown in *Figure 41* is made up of a graduated steel rule or beam with two jaws, one fixed and the other movable. The movable jaw has a graduated vernier scale and a mechanism for making fine adjustments. One side of the scale is graduated for reading outside measurements, and the other side is for reading inside measurements.

Making accurate measurements with the vernier caliper requires the same care and sensitive touch as when measuring with a micrometer. Also, reading a vernier caliper is very similar to reading a micrometer. Each inch of the beam, like each inch on the sleeve of a micrometer, is graduated into 40 equal parts with the decimal equivalent of $\frac{1}{40}$ of an inch equal to 0.025". As with the micrometer, every fourth division line is a little longer and is marked 1, 2, 3, etc., denoting tenths of an inch. The vernier plate correlates with the micrometer thimble and is graduated into twenty-five equal divisions, representing thousandths of an inch. Vernier calipers are available with dial and digital readouts.

10.1.3 Micrometer

A micrometer is an instrument that measures in thousandths of an inch (*Figure 42*). Metric micrometers are also available that measure in hundredths of a millimeter. An outside micrometer is most commonly used. It is used to measure outside diameters of round objects and thicknesses of flat stock and materials.

The micrometer is a precision tool and should be handled with extreme care. Dust, dirt, and grease may damage its internal mechanisms and render it useless.

All micrometers are calibrated alike and operate on the principle that a screw with a pitch of 40 threads per inch will advance $\frac{1}{40}$ of an inch or 0.025" (twenty-five one-thousands of an inch) with each complete turn. There are also 40 lines to the inch on the sleeve, the same number of threads as on the screw. Each turn, as indicated above, equals 0.025". If one turn equals 0.025", then $\frac{1}{25}$ of a turn will equal 0.001" ($\frac{1}{25}$ of 0.025"), making it possible to measure in thousandths of an inch. One-half and lesser parts of a thousandth can be judged by sight as nearly as possible, or a ten-thousandth (vernier) micrometer may be used. Micrometers are available with digital readouts that eliminate complicated conversions.

10.1.4 Thickness (Feeler) Gauge

Feeler gauges (*Figure 43*) are used to measure small distances or clearances between objects that cannot be measured in any other way. The feeler gauge is made up of a number of thin steel blades that fold into a handle. To determine a distance, feeler gauges of various thicknesses are tried until one fits snugly between the objects. The thickness of each blade is marked by a number. The number on each blade indicates the thickness of that blade in thousandths of an inch and/or in millimeters. A modification of the feeler gauge is a stepped feeler gauge, which is often called a go/no-go gauge.

10.1.5 Thread Pitch Gauge

Nuts, bolts, and screws come in many sizes, from very small to very large. These fasteners can be measured in terms of pitch. Not only can a rule be used to count the number of threads per inch, but one can also use a thread pitch gauge (*Figure 43*). A thread gauge has many leaves. Each leaf measures 1" across and has a specific pitch marked on the side. Proper use of the gauge involves finding the leaf that has the proper number of teeth to fit the number of threads per inch (pitch) on the fastener, and then reading the number imprinted on the leaf.

10.2.0 Supplemental Tools

Selecting the proper tool for the job is an important HVAC technician skill. The information in this section on tools and their uses will assist you in making the best possible choices.

10.2.1 Drill Press

The two types of drill presses are the bench type and the floor model (*Figure 44*). The bench drill press can either be fastened to a standard work

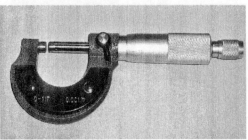

Figure 42 ♦ Micrometer.

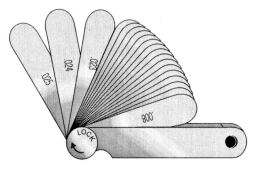

THICKNESS GAUGE

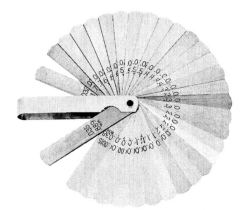

THREAD PITCH GAUGE

Figure 43 ♦ Thickness (feeler) gauge and thread pitch gauge.

Figure 44 ♦ Floor drill press.

bench or may be held in place solely by the weight or shape of the machine. The floor drill press is usually heavy enough so that it need not be secured to the floor, but can be set where desired. The table of the press upon which the workpiece is held is usually adjustable over the entire range of the support column.

The speed of the drill press can be varied by the placement of drive belts or through the use of gearing. Speed can be regulated for the metal being drilled. The feed of the drill is measured in fractions of an inch per revolution and should vary with the type of workpiece material and size of the drill. The correct speed and feed are dependent upon many conditions but can generally be determined by judgment based on experience.

When using electric drill presses, the following safety and maintenance considerations should be observed:

 WARNING!
Always wear proper eye protection and safety gloves. Do not operate a drill press without proper ground fault protection. Before connecting to the power source, make sure the POWER ON/OFF switch is in the OFF position.

- Make sure to use the right drill bit for the job.
- Always use a sharp drill bit.
- Make sure the drill bit is securely tightened in the chuck and the chuck key is removed before starting.
- Let the drill do the work. Never force the drill while drilling; this dulls the drill bit and can damage the bearings of the drill press.

Explain how drill presses are used.

Review safety guidelines for using a drill press. Emphasize the importance of eye protection.

Discuss torque and emphasize the importance of securing the workpiece.

Show trainees how to safely use a drill press.

Identify various power tools used for servicing HVAC equipment.

Review rules for the care and safe use of hand tools.

- Drill presses require relatively little maintenance. Follow the manufacturer's instructions.
- Keep the drill press and ventilation passages clean.

WARNING!
A drill press develops a great deal of torque. The drill can dig into the work as it penetrates through the material. This will nearly always occur when drilling thin materials. Do not hold the material with your hands when using a drill press to bore a hole, especially through sheet metal. When the drill grabs the material, it will spin it out of your hands and may cause severe injury. Always secure the work with a vise or clamps.

10.2.2 Portable Power Tools

A number of portable power tools, including hand drills and saws of various types and sizes, are used in HVAC work. Cordless versions of most of the corded power tools, including drills and reciprocating saws, are also available. *Figure 45* illustrates a set of cordless power shears that can be used instead of tin snips to cut sheet metal. The professional cordless tools are usually powered by rechargeable 24V batteries. These batteries can provide continuous use for an hour or more before recharging is required. Cordless tools are extremely convenient when a plug-in power source is not readily available.

10.3.0 General Guidelines for the Care and Safe Use of Hand and Power Tools

Using hand and power tools for their intended purpose, using them safely, and properly caring for tools are characteristics of a professional. Follow OSHA and manufacturers' safety rules and regulations.

Figure 45 ◆ Cordless portable power shears.

10.3.1 Rules for the Care and Safe Use of Hand Tools

The following are some rules for the care and safe use of hand tools:

- Use the correct tool for the job.
- Keep tools clean.
- Maintain tools properly as directed by the tool manufacturer.
- Keep tools sharp (as applicable).
- Inspect tools frequently to make sure they are in good condition.
- Repair damaged tools promptly; dispose of tools that cannot be repaired.
- Do not use broken tools.
- Do not throw or drop tools.
- Protect the cutting edge of tools when carrying and storing them.
- Store tools properly when not in use. Do not carry tools in your pocket; do not place tools where they can roll and fall; lightly oil tools before storing; store tools in a dry place.
- Wear eye, ear, and respiratory protection when appropriate.
- Stay alert when using tools.
- Keep fingers away from cutting edges.
- Work away from your body when using cutting tools.
- Be sure the area is clear before you swing a hammer.
- Use tools with insulated or wood handles when working near or with electrical equipment.

10.3.2 Rules for the Safe Use of Power Tools

The following are some rules for the care and safe use of all power tools:

- Do not attempt to operate any power tool before being cleared by your instructor or supervisor on that particular tool.
- Always wear eye protection and a hard hat when operating all power tools.
- Wear face protection when necessary.
- Wear proper respiratory equipment when necessary.
- Wear appropriate clothing for the job. Never wear clothing that can become caught in the moving tool. Roll up long sleeves, tuck in shirttails, and tie back long hair.
- Do not distract others or let anyone distract you while operating a power tool.
- Consider the safety of others, as well as yourself.

- Do not leave a power tool running while it is unattended.
- Assume a safe and comfortable position when using a power tool.
- Be sure that a power tool is properly grounded before using it.
- Be sure that a power tool is disconnected before performing maintenance or changing accessories.
- Do not use a dull or broken tool or accessory.
- Use a power tool only for its intended use.
- Keep your feet, fingers, and hair away from the blade and/or other moving parts of a power tool.
- Do not use a power tool with the guards or safety devices removed.
- Do not operate a power tool if your hands or feet are wet.
- Keep the work area clean at all times.
- Become familiar with the correct operation and adjustment of a power tool before attempting to use it.
- Keep a firm grip on the power tool at all times.
- Use electric extension cords of sufficient size to service the particular power tool you are using.
- Report unsafe conditions to your instructor or supervisor.

10.3.3 Guidelines for the Care of Power Tools

The following are some guidelines for the proper care of power tools:

- Keep all tools clean and in good working order.
- Keep all machine surfaces clean and waxed.
- Follow the manufacturer's maintenance procedures.
- Protect cutting edges.
- Keep all tool accessories (such as blades and bits) sharp.
- Always use the appropriate blade for the arbor size.
- Report any unusual noises, sounds, or vibrations to your instructor or supervisor.
- Regularly inspect all tools and accessories.
- Keep all tools in their proper place when not in use.
- Use the proper blade for the job being done.

10.4.0 Motor Lubrication

To reduce friction and prevent wear, some motor bearings and other rotating and moving components need to be lubricated on a regular basis using a hand oiler (*Figure 46*) or a lever-type grease gun.

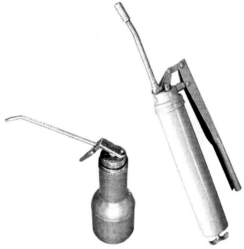

Figure 46 ♦ Oiler and lever-type grease gun.

Review rules for the care and safe use of power tools.

Explain that some motors require periodic lubrication using a hand oiler or grease gun.

Power Tool Safety

Two HVAC technicians were installing air conditioning ductwork in a residence. One technician was in a crawl space installing aluminum straps around the new duct to hold it in place. As he drilled a hole, the drill bit of the ungrounded tool contacted the house wiring above him, causing the drill to become energized. This provided a path to ground through the victim, who happened to be leaning on a metal cold water pipe. The second technician heard a noise and went to investigate, instinctively grabbing for the victim instead of de-energizing the power. He also contacted the metal water pipe. Both technicians received fatal injuries.

The Bottom Line: Use grounded power tools and be aware of all possible hazards before proceeding with an installation. All craftworkers should be aware of proper rescue procedures.

Discuss the effects of over-lubrication.

Stress the importance of following manufacturers' instructions for lubricating any equipment.

Provide an overview of the procedure for lubricating motors fitted with oil ports.

Provide an overview of the procedure for lubricating motors equipped with grease fittings.

Discuss the benefits of proper filter maintenance.

The bearings in many motors used in HVAC equipment are permanently lubricated at the factory and require no lubrication in the field. The service literature for the equipment being serviced normally will state when this is the case. Other motors still require periodic lubrication. These motors are normally equipped with oil ports or grease fittings.

When lubricating motor bearings, it is important to follow the lubrication interval and use the type and quantity of lubricant recommended in the equipment manufacturer's service literature.

The most common problem with lubricating motor bearings is over-lubrication. It can cause an increase in operating temperature and a decrease in viscosity. If the lubricant becomes too thin, it cannot carry the load inside the bearing, and the bearing will fail and have to be replaced. Too much grease being forced into a bearing may burst the bearing seals, causing the lubricant to escape from the bearing and allowing contaminants to enter.

To lubricate motors with oil ports, proceed as follows:

Step 1 Shut off power to the equipment, then lock out and tag the disconnect.

Step 2 Remove the dust caps from the oil ports on both ends of the motor.

Step 3 Oil the motor using the type and quantity of oil specified by the manufacturer. Typically, this is about 16 to 25 drops of nondetergent SAE 20 motor oil applied in each oil port. Rotate the shaft while lubricating the bearings to evenly distribute the lubricant.

Step 4 Wipe up any excess oil, turn on the power, and observe the motor operation.

To lubricate motors or bearings equipped with grease fittings, proceed as follows:

Step 1 Shut off power to the equipment, then lock out and tag the disconnect.

Step 2 Wipe away any old or hardened grease from the grease fittings. Remove the relief plug if so equipped.

Step 3 Using a grease gun, add the grease type and amount specified by the equipment manufacturer. Rotate the shaft while lubricating the bearings to evenly distribute the lubricant. *Do not* over-lubricate.

Step 4 Wipe up any excess grease, turn on the power, and observe the motor operation.

10.5.0 Air Filter and Screen Maintenance

Maintaining the cleanliness of air filters and screens in an HVAC system provides many benefits including maintaining equipment efficiency and improving indoor air quality.

There are several types of air filters and it is important to use the type specified by the equipment manufacturer. Conventional disposable filters (commonly called dust stop filters) are typically constructed of fiberglass, hog hair, polyester, or open-cell foam. Sometimes a coating is applied to the filter media to trap additional particles. This type of filter is very common and is often seen in residential and light commercial HVAC equipment. The main disadvantage of this type of filter is that it only removes the larger particles in the air.

Better filtration can be achieved with an extended surface filter or an electrostatic filter. Extended surface filters have the material folded accordion-style which creates more filtration surface in a given area. Electrostatic filters have an inherent electrostatic charge which polarizes particles in the air stream, causing them to become trapped in the filter. Some of these types of filters can be cleaned and reused.

The best mechanical filter *(Figure 47)* is the high-efficiency mechanical air cleaner. The filter media is several inches thick and folded accordion-style to get a tremendous amount of filtration surface in a small area. This filter type can remove extremely small particles from the air.

Figure 47 ♦ Typical mechanical air filter.

12.40 HVAC ♦ LEVEL TWO

Instructor's Notes:

Sizing an Air Filter

Air must flow through the filter, not around it. If the supply air can bypass the filter, the purpose of the filter is defeated. It is import to use a correctly sized air filter and to install it properly. Also, check gaskets, liners, insulation, slide tracks, filter baffles, blank-off plates, as well as access doors and panels to ensure that all are in place and secure.

A highly efficient filter is the electronic air cleaner (*Figure 48*). This device uses high voltage to impart a charge on all particles that enter the device. A plate with an opposite charge then attracts these charged particles before they have a chance to exit the air cleaner. The electronic air cleaner can remove smaller particles than any other type of filter.

Regardless of type, all filters should be cleaned and/or replaced at the intervals recommended by the HVAC equipment manufacturer. Since different filters provide different resistance to air flowing through them, it is important to replace the filter with a similar or identical unit. Failure to do this could affect the operation of the system. If a filter is dirt clogged, the frame is bent, or the filter media is torn, the filter should be replaced. Disposable filters should be replaced and washable filters should be cleaned per the filter manufacturer's instructions.

To clean permanent filters and screens, proceed as follows:

Step 1 Shut off all power to the unit.

Step 2 Remove the filter and/or screen.

Step 3 Remove heavy accumulations of dirt by tapping the filter and/or vacuuming it.

Step 4 Wash the filter with a mild detergent solution until clean.

Step 5 Dry the filter before re-installing. Do not oil or coat the filter unless directed to do so by the filter manufacturer. If the filter has an airflow arrow, make sure it points in the direction of airflow (usually toward the blower motor).

Step 6 Replace all panels, turn the power on, and observe the equipment for correct operation.

The media core of high-efficiency mechanical filters can be removed and replaced with a new media core when necessary (*Figure 49*). Do this in accordance with the manufacturer's instructions, being careful not to tear or damage the filter media during replacement.

Because there are many different manufacturers and models of electronic air cleaners, and because of the high voltages involved in their operation, maintenance of these devices, including cleaning, must always be done in strict accordance with the manufacturer's instructions.

List some of the common types of air filters.

Provide an overview of the procedure for cleaning permanent filters and screens.

Show trainees how to clean permanent filters and screens.

Figure 48 ◆ Electronic air cleaner.

Figure 49 ◆ Replacing filter media in a high-efficiency mechanical filter.

Emphasize the importance of keeping the indoor and outdoor coils clean.

Provide an overview of the procedure for inspecting and cleaning evaporator and condenser coils.

Provide an overview of the procedure for inspecting and cleaning outdoor air dampers.

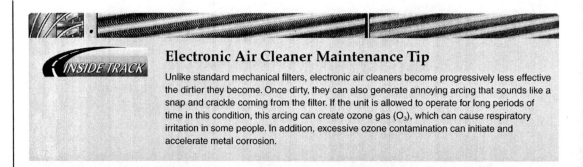

Electronic Air Cleaner Maintenance Tip

Unlike standard mechanical filters, electronic air cleaners become progressively less effective the dirtier they become. Once dirty, they can also generate annoying arcing that sounds like a snap and crackle coming from the filter. If the unit is allowed to operate for long periods of time in this condition, this arcing can create ozone gas (O_3), which can cause respiratory irritation in some people. In addition, excessive ozone contamination can initiate and accelerate metal corrosion.

10.6.0 Coil and Condensate System Maintenance

The condenser and evaporator coils of a cooling system need to be inspected on a yearly basis and cleaned if necessary. Similarly, the indoor and outdoor coils of a heat pump system should be cleaned before each heating and cooling season. Dirt on coil surfaces cuts down on the ability of the coil to transfer heat, causing system inefficiency, poor indoor comfort, and even compressor failure.

To inspect and clean the evaporator/indoor coil, proceed as follows:

Step 1 Shut off power to the equipment, then lock out and tag the disconnect.

Step 2 Remove the coil access panel and inspect the coil for dirt accumulation on the coil surface.

Step 3 Check for debris and/or dirt in the condensate drain pan. Remove any algae by washing the pan with an approved commercial solution.

NOTE

Be sure the condensate drain has a trap as called for by the installation instructions (some units have built-in traps and do not require a separate trap). The trap prevents condensate from flowing back to the unit.

Step 4 Use a vacuum cleaner brush attachment to remove light dust or lint. For stubborn dirt, spray the coil with a mild detergent solution. Rinse the coil and condensate pan with clear water. Make sure water runs freely from the condensate pan drain.

Step 5 If local codes permit, tablets or chemicals can be placed in the condensate drain pan to inhibit the growth of slime or algae.

Step 6 Straighten any bent coil fins with a fin comb before replacing the coil access panel, then restore power and observe the equipment operation.

To inspect and clean the condenser coil/outdoor coil, proceed as follows:

Step 1 Shut off power to the unit, then lock out and tag the disconnect.

Step 2 Gain access to the coil by removing the appropriate panels.

Step 3 Vacuum or brush away any light dust or lint. Then spray the coil with a mild detergent solution. Use a low-pressure garden hose to rinse dirt and debris from the coil. Multi-row or multi-section coils may have to be separated to get at hidden dirt.

Step 4 Replace all access panels, restore power, and observe the equipment operation.

Step 5 Make sure that plants and shrubs in the vicinity of the unit have not grown to the point that they block or restrict airflow to the condenser/outdoor coil. If they have, recommend that the property owner trim back the plants.

10.7.0 Damper Inspection and Cleaning

Commercial HVAC units like economizers and similar equipment are equipped with outdoor air dampers. To inspect and clean these dampers, proceed as follows:

Step 1 Shut off power to the unit, then lock out and tag the disconnect.

12.42 HVAC ♦ LEVEL TWO

Inside Track

Cleaning Coils
Do not use a high-pressure washer to clean any coil surface. The high-pressure jet of water will flatten the aluminum fin stock on the coil, blocking airflow through the coil. Also, when washing a coil, be sure to prevent water from splashing in any nearby motors or electrical components. A simple plastic bag over a motor can provide a very effective shield. Some units have multi-layer coils. In order to properly clean the coils, carefully separate the layers by 3" to 4" and clean between them.

Coil Cleaners
Many chemical coil cleaners are available to HVAC service technicians. Some of these chemicals are alkaline or acidic and can corrode metal if not thoroughly rinsed off. They also pose a safety hazard if splashed on your skin or in your eyes. If you choose to use these chemicals, first check to see if the equipment manufacturer approves of their use. Follow the application directions for the coil cleaner, and always wear appropriate personal protective equipment.

Step 2 Remove and clean the mesh screen(s) as previously described above under filter and screen maintenance.

Step 3 Clean the damper blades of any dirt, soot, etc., and make sure none are damaged.

Step 4 Check any pins, straps, and bushings/bearings for wear, rust, or corrosion.

Step 5 Check that the seal strips on the damper hood top and sides are not damaged.

Step 6 With the unit in operation, check to make sure that the damper moves freely, with no binding.

Step 7 Replace the screen(s).

11.0.0 ♦ DOCUMENTATION

One of the most important tasks concerning the installation and maintenance of HVAC equipment involves the completion of various forms and/or reports. Proper completion of forms and/or reports does not end with simply adding the data to the form. The forms used in the HVAC industry are primarily used as communication tools, but they serve many purposes. They must be legible and understandable. In written or oral communications with your customers, remember not to use trade jargon. Customers need to understand what you have done during your service call, how their unit is currently functioning, and what to expect from it in the future. Using trade terms makes this much more difficult. This aspect of your job responsibilities cannot be overemphasized. If you cannot clearly communicate with the customer and your service dispatcher, and leave accurate documentation of what you have done, your potential for success in this industry will be limited.

Forms and reports serve as historical records about the equipment. For this reason, it is important to fill them out completely and accurately. When troubleshooting at some later date, recorded data is useful because it can be compared to current system readings in order to determine areas of possible system degradation. Unfortunately, the popularity of civil and criminal litigation is slipping rapidly into the HVAC industry as well. This trend is forcing the use of the documentation you are generating to be used in court cases. In the event of a lawsuit, these forms or reports are often used in court to prove that specific tasks have been performed and/or that the installed equipment has met design and customer specifications. Forms and paperwork are commonly used to do the following:

- Record the results of system performance tests.
- Verify that required inspection or quality control milestones have been met.
- Verify that installed systems meet design and customer specifications.
- Notify equipment manufacturers of the start of their equipment warranties.
- Record important facts about service calls or other field maintenance activities.
- Communicate to all concerned parties what you have done.

Have the trainees review Sections 11.0.0–13.4.8.

Ensure that you have everything required for teaching this session.

List the functions served by job documentation.

Discuss the types of forms commonly used by the HVAC technician.

Emphasize the importance of recording the make, model, and serial number on each document associated with the unit being serviced.

Show Transparencies 40 and 41 (Figures 50 and 51). Explain how to properly complete a typical service ticket and work order form.

If all of this talk about documentation seems like overkill, you should be reminded that accurate and legible documentation is the key to successfully communicating with all concerned parties about work you and your employer perform. Documentation is not filled out only for the benefit of your service manager or dispatcher. It is done primarily for the benefit of the customer. If improper communication is allowed, then customer dissatisfaction is sure to follow. Without customers willing to pay for your services, there will be no need to worry over any paperwork at all.

Some types of documentation you routinely will be required to fill out include:

- Service ticket/invoice
- Commissioning job report
- Start-up report
- Warranty ticket

11.1.0 The Importance of Make, Model Number, and Serial Number

The three most critical items that should always appear on everything you document are the make, model number, and serial number for the unit being serviced. These three pieces of information can be found on the unit nameplate. The make identifies the name of the manufacturer such as Carrier, Trane, Lennox, York, etc. The model number identifies the specific model of the unit. It describes its style, type, capacity, electrical characteristics, and application. The serial number is the sequential manufacturing number assigned by the manufacturer for the unit. Its use is critical when you need to get replacement parts for the unit. This is because many units are manufactured as a series or group of numbers over the years, and they may even share the same model numbers over time.

Unfortunately, manufacturers often change critical parts in the life of a particular model series, and the serial number is needed to make sure that you get the correct part for use in the unit being serviced. This reduces callbacks and customer dissatisfaction.

11.2.0 Service Ticket/Invoice

The service ticket/invoice is the most basic form used by the HVAC technician. Many firms have incorporated the use of computer dispatching and data retrieval. This reinforces the need for accurate and clear communication. The data from the service ticket is then entered into the computer and used for preparing the formal customer invoice or bill. Some systems allow for a computer log-in from a remote site so that the technician can type in the data, then receive an automatic printout of an invoice at the site.

Figure 50 shows a typical service ticket/invoice form. A service ticket/invoice form is filled out by the HVAC technician for each job. Initially, the form may contain information provided by the shop supervisor/service dispatcher that gives the technician the customer's name, location, and details about the nature of the job or service call. Sometimes this information is initially provided on a work order form (*Figure 51*).

> **NOTE**
> Many of the newer computer-managed dispatch and service systems do not allow the processing of payroll until all forms and required inputs are correctly entered into the system. This should be an ample incentive for completing accurate paperwork.

At the completion of the service call, the technician fills in all applicable portions of the form to provide a specific description of the work that was done, the quantity and types of materials used, the labor hours expended, and so on. After the service ticket is completed and signed by the technician, it is then given to the customer who also signs the ticket to acknowledge that the service performed was requested and authorized by

---

INSIDE TRACK

Roof or Campus Maps

In larger commercial projects where multiple units are in service, you may find that the use of a roof or campus map is useful. In these cases, the addition of a mark or unit tag number (such as RTU-I or AC-I) will be required in addition to the standard make, model, and serial number information that you will always provide.

LOCATION OF WORK

NAME

ADDRESS

WORK TO BE DONE

JOB NUMBER

PHONE NO.

☐ SERVICE CONT. ☐ REGULAR ☐ WARRANTY ☐ CALL BACK

Source	Cost	Quan.	Item	Price
			MILEAGE	

Customer No.	Tech No.	Invoice No.	Date Filled	Total
			/ /	

Service or Repair Recommended	Amount $	Authorized Y or N	Parts Ordered Y or N

XYZ
Company

P.O. Box 0105
Street Address
City, State 28277

BILL TO
NAME
STREET
CITY STATE ZIP

MAKE	MODEL	SERIAL NUMBER

SERVICE INVOICE **083065**

SERVICE DATE ____/____/____

This invoice is subject to the maximum finance charge as allowed by law or 1½% per month which is an annual percentage rate of 18%.

I HAVE AUTHORITY TO ORDER THE WORK, WHICH HAS BEEN SATISFACTORILY PERFORMED, AS OUTLINED BELOW. IT IS AGREED THAT THE SELLER WILL RETAIN TITLE TO ANY EQUIPMENT OR MATERIAL THAT MAY BE FURNISHED UNTIL FINAL PAYMENT IS MADE AND IF PAYMENT IS NOT MADE AS AGREED, THE SELLER SHALL HAVE THE RIGHT TO REMOVE SAME AND THE SELLER WILL BE HELD HARMLESS FOR ANY DAMAGES RESULTING FROM THE REMOVAL THEREOF, IF THIS ACCOUNT IS PLACED IN THE HANDS OF A COLLECTION AGENCY OF ATTORNEY. I AGREE TO PAY ALL REASONABLE COSTS AND FEES INCURRED IN THE COLLECTION.

CUSTOMERS SIGNATURE

DATE	DESCRIPTION OF WORK PERFORMED

MINIMUM 1 HOUR LABOR CHARGE

WARRANTY
PARTS WARRANTY – 1 YEAR
LABOR WARRANTY – 30 DAYS

TERMS
COD ☐ CHARGE ☐
PLEASE PAY FROM THIS INVOICE

Thank You!
FOR THIS OPPORTUNITY TO SERVE YOU. PLEASE KNOW THAT WE'LL BE READY FOR YOUR NEXT SERVICE NEED.

MECHANIC

SERVICE SALES	
ITEM	AMOUNT
MATERIALS	
LABOR HOURS	
MILAGE	
SUB TOTAL	
SALES TAX	
TOTAL AMOUNT	

212F50.EPS

Figure 50 ◆ Example of a typical repair order/service report form.

the customer and that all materials and services have been received. The signed ticket usually serves as a customer billing invoice and a copy is given to the customer.

A typical service ticket is a multi-sheet form consisting of duplicate white, yellow, and pink sheets. Normally, the white sheet is retained by the technician for office use, the yellow sheet is given to the customer, and the pink sheet is retained for the office historical files.

11.3.0 Commissioning Job Report

At the completion of most large commercial and industrial HVAC installations, a commissioning process is used to document and verify the performance of the HVAC systems in order to ensure that they operate in conformance with the design intent. Detailed checklists and reports are filled out during the commissioning process to record HVAC equipment readiness, start-up, and performance. At the completion of the process, these checklists and reports are turned over to the building owner or other designated authority. The specific checklists used during a commissioning process are determined by the specific types of installed equipment.

The National Environmental Balancing Bureau (NEBB) has developed a series of forms and check sheets widely used by its members and other NEBB-certified firms for certifying building system commissioning. These forms are grouped into the following categories:

- Administrative forms
- General equipment forms
- Hydronic equipment forms
- Cooling equipment forms
- Air handling equipment forms
- Control system forms

Explain what happens during the commissioning process of a large installation.

MODULE 03212-07 ◆ BASIC INSTALLATION AND MAINTENANCE PRACTICES 12.45

WORK ORDER

XYZ Company
P.O. Box 0105
Street Address
City, State 28277

Date:
Summary:
Reference #:
Tech:
Start Time:

Bill To:

Job Name:

Description of Work

Material	Labor	Other	Subtotal	Tax 1	Tax 2	Total

All material is guaranteed to be as specified. All work to be completed in a professional manner according to standard practices. Any alteration or deviation from above specifications involving extra costs will be executed only upon written orders and will become an extra charge over and above the estimate. All agreements contingent upon delays beyond our control. Purchaser agrees to pay all costs of collection, including attorney's fees.

212F51.EPS

Figure 51 ◆ Example of a typical work order form.

Instructor's Notes:

11.4.0 Start-Up Report

As the name implies, start-up reports or forms are used to record the specific operating conditions and parameters that exist at the time of initial startup and operation for all types of HVAC systems and/or individual components of an HVAC system. These reports not only provide a record to verify proper operation at startup, but they also serve as a record for troubleshooting at some later date where the data can be compared to current system readings in order to determine areas of possible system degradation. *Figures 52* and *53* show examples of start-up reports used for recording the startup operating conditions for a heat pump and a furnace, respectively.

11.5.0 Warranty Ticket

At the completion of a job or HVAC system commissioning process, HVAC equipment warranty forms or tickets should be filled out and given to manufacturers to notify them that their equipment has been put into operation. Copies of all warranties should also be given to the building owner or designated person.

12.0.0 ♦ CUSTOMER RELATIONS

Good customer relations are essential to the success of most businesses, including those in the HVAC industry, where each service technician represents the company and influences what customers think about the business. Keeping these impressions positive requires that all employees work toward generating customer good will. Providing good service is a part of this, but so are good personal habits, good work practices, and customer-pleasing attitudes. Customer good will often results in more business through referrals and repeat servicing of existing customers.

12.1.0 Why Customer Relations are Important

Good customer relations consist of the habits, behaviors, and attitudes which guarantee that the customer's first impression of you and your company is a good one; that the customer's needs are clearly understood; and that appropriate customer service is provided to meet these needs.

When you provide this level of effective service to your customers, you can increase your company's share of business. Your ability to relate well to your customers is critical. As a service technician, you're likely to be the only person from your company that the customer ever sees. In the customer's eyes, you are the company.

In this section, personal habits, handling service calls, and handling difficult situations will be explained. Mastering these concepts will not only make your job easier, it will also make it more likely that a customer will call your company the next time HVAC service is needed.

12.2.0 Personal Habits, Behaviors, and Attitudes

Appearances do count. First impressions happen only once, in the first sixty seconds. You don't get a second chance to make a first impression.

Because customers don't have technical skills, they call a servicing company for help with heating and cooling problems. Although they can't judge the servicing technician's professional competence, they can, and do, form an impression of the technician's appearance and attitudes. The customer's first glimpse of a technician can determine that customer's opinion of the worker and the company the technician represents. Examine your first impression. Do you:

- Practice good personal hygiene?
- Get enough sleep and look alert?
- Wear a neat, clean uniform?
- Wear shoe covers or remove shoes when in the home?
- Carry a pencil and pad to take notes?

Once the customer has formed a first impression of you and your company, your on-site work habits will confirm or change that impression.

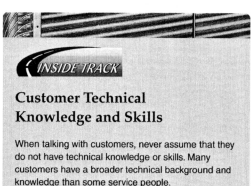

Customer Technical Knowledge and Skills

When talking with customers, never assume that they do not have technical knowledge or skills. Many customers have a broader technical background and knowledge than some service people.

XYZ Company
P.O. Box 0105
Street Address
City, State 28277

START-UP REPORT

JOB/WORK ORDER #

NAME:	DATE:	TECH:
ADDRESS:		
TIME – IN:	TIME – OUT:	OSA TEMP: DB: WB

OUTDOOR SECTION

MAKE:	MODEL:	SERIAL:	
RATED: VOLTS	ACTUAL: VOLTS	RATED: AMPS	ACTUAL: AMPS
FUSE OR BREAKER SIZE:		CONNECTIONS:	
DEFROST TIME INTERVAL:		ANTI-CYCLE TIMER:	
REFRIGERANT PRESSURES:		TYPE:	
COOL: HIGH LBS	LOW LBS	HEAT: HIGH LBS	LOW LBS
COIL CONDITION:			
OUTDOOR THERMOSTAT LOCATION:		SETTING:	

INDOOR SECTION

MAKE:	MODEL:	SERIAL:	
MOTOR: VOLTS	ACTUAL: VOLTS	RATED: AMPS	ACTUAL: AMPS
MOTOR TYPE:	BELT SIZE:	ADJUSTMENT:	
FILTER TYPE:	SIZE:	CONDITION:	
FILTER LOCATION:			
ELECTRIC HEAT RATED KW:		OPERATION:	
COIL CONDITION:		CONDENSATE DRAIN:	
AIR TEMPERATURES – COOLING IN:		COOLING OUT:	
HEAT PUMP – IN:	OUT:	H/P & ELECT – IN:	OUT:
NOTES:			

TECHNICIAN:	CUSTOMER SIGNATURE:

212F52.EPS

Figure 52 ♦ Example of a typical start-up report for a heat pump.

XYZ Company
P.O. Box 0105
Street Address
City, State 28277

START-UP REPORT

CUST. NAME ______________________ DATE ______________________

ADDRESS ______________________ TIME ______________________

CITY ______________________

PHONE # ______________________ DATE INSTALLED ______________________

UNIT MODEL NO. ______________________ SERIAL NO. ______________________

CHECK LIST

____ GAS LINE PRESSURE

____ MANIFOLD PRESSURE

____ FAN ON TEMP

____ TEMPERATURE RISE

____ TEMP LIMIT OPENS

____ TEMP LIMIT CLOSES

____ FAN OFF TEMP

____ CARBON MONOXIDE TEST

____ FILTER SIZE

____ BLOWER WHEEL

____ FURNACE GROUND

____ LINE VOLTAGE

____ ALL ELECTRICAL CONNECTIONS TIGHT

____ GAS LEAK SEARCH

____ DAMPERS OPEN

____ FRESH AIR DAMPER ADJUSTED

____ REGISTERS ON AND OPEN

____ FLOOR NOISE

____ VENTING COMPLETE

____ CAULKED GAS LINE & VENTING

____ STICKERS

____ DRAIN COMPLETE

____ INSULATION IN BOX SILLS

NOTES: ______________________

____

____

WHICH BLOWER SPEED USED?

HEATING ______ COOLING ______

START-UP PERFORMED BY:

____

TECHNICIAN

212F53.EPS

Figure 53 ◆ Example of a typical start-up report for a furnace.

MODULE 03212-07 ◆ BASIC INSTALLATION AND MAINTENANCE PRACTICES 12.49

Discuss the personal habits, behaviors, and attitudes needed by a service technician.
See the answers to the "Think About It" at the end of this module.

List the three parts of a service call: the opening, servicing, and the closing.

Discuss how to handle the opening of a service call.

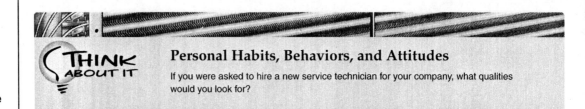

Personal Habits, Behaviors, and Attitudes

If you were asked to hire a new service technician for your company, what qualities would you look for?

- Are you on time?
- Is your tool set complete and neatly packed?
- Do you show concern, courtesy, and respect to the customer by listening carefully while the customer identifies the problem?
- Do you tackle the problem promptly and quietly?
- Do you remember to protect the work area? Do you carry rags for cleaning up and dropcloths to protect floors and carpets?
- Do you take the time to explain carefully to the customer what's wrong and how the unit will perform satisfactorily after you've repaired it?
- Do you respect the job site by not tracking in dirt and by cleaning up after yourself?
- Do you refrain from smoking on the premises, avoid smelling of smoke, or spitting?
- Do you respect company equipment, tools, and vehicles? Trucks and vehicles displaying the company name are rolling advertisements. Is your vehicle clean and in good repair? Are your driving habits courteous?
- Do you avoid using alcohol and drugs while on the job or around vehicles?
- Do you avoid profanity and horseplay? Customers, especially those with children on the premises, are likely to be offended by inappropriate language.

Every hour of every working day, you are an advertisement for your company. Your appearance and behavior must be consistent with the positive image your company wants you to reflect.

Off the job, you can increase word-of-mouth advertising (free advertising created by satisfied customers who recommend your company to friends) by having a positive attitude about your employers, co-workers, and customers. Positive statements about your job are free advertising for your company.

12.3.0 Customer Relations: Handling Service Calls

During each service call, you can do several things to enhance the customer's image of you and your company.

The typical service call has three parts:
- *The opening* – Identifying the problem
- *Servicing* – Solving the problem
- *The closing* – Leaving the customer with a positive impression

Remember that your objective in servicing your customer's equipment goes beyond technical competence. Your objective is to win the customer's repeat business.

12.3.1 The Opening

As you open the service call, you are already making your first impression. Your personal appearance is good. Your vehicle looks good. You show respect for the customer by appearing promptly and politely at the door. Now what? You should:

- Smile and display confidence and polite respect for the customer.
- Promptly identify yourself. Many customers waiting for a service worker are naturally cautious about admitting a stranger. Identify yourself clearly and show an appropriate ID, if you have one. Give the customer a company business card with your name on it. If your company does not have business cards made with the names of individual service technicians, encourage the company to at least have company business cards available so that you can give one to the customer in case they need to contact the company if they have any follow up questions or concerns.
- Understand your customer's needs. Listen to your customer to learn what the problem is. Ask questions about what the customer has seen, heard, smelled, and felt. Ask when things occurred and how often they happened. Often, the customer can help speed your diagnosis by providing clues to the equipment problem. When you listen attentively, everyone wins.

It's important to ask open-ended questions, the kinds that result in specific, information-gathering answers. You don't want to be led to the equipment without any information from the customer. Ask as many questions as you need to get

Instructor's Notes:

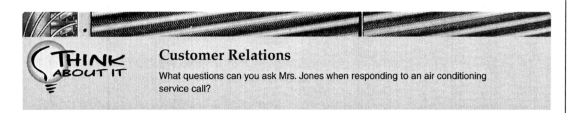

Customer Relations

What questions can you ask Mrs. Jones when responding to an air conditioning service call?

the exact symptoms. Don't say, "I'm here to fix your furnace. Where is it?" Instead, say, "I came to repair your furnace. What seems to be the problem?"

12.3.2 Servicing

While servicing the equipment, you should be adding to your customer's positive impression of you and your company. The customer has an equipment problem. You're there to solve it. Build customer confidence by getting right to work to determine the probable cause of the malfunction. Practice good work habits. Avoid general conversation while working. Remember, many service technicians are paid by the hour; that's how the customer pays for their time. The customer must feel that you are filling the time with productive work, not idle chatter.

During this part of the service call, you'll need your professional skills to:

- Analyze the symptoms.
- Isolate the problem.
- Eliminate probable causes that do not apply.
- Isolate probable causes.
- Determine the solution.
- Explain the solution to the customer, including your best estimate of what it will cost.
- If the customer approves, remedy the situation.
- Follow up on the service visit.

Following up is important, but it's often overlooked. If time permits, call the customer a day or two later to ask if your repairs were satisfactory. If you made adjustments, stop by to see if they were acceptable. Often, a simple phone call from you can be the touch that creates a repeat customer. In your company, if the service technicians don't have the time to follow up on service visits, perhaps someone from the office staff can handle these calls.

12.3.3 The Closing

This very important step helps set the tone for your customer's image of you and your company. When you finish the work:

- Neatly pack your tools.
- Return the premises to its original condition. (Replace covers, wipe off dirty fingerprints, clean up dropcloths, etc.)
- Explain to the customer that the problem has been solved. Explain what parts you replaced and offer to show the customer the defective parts.
- Demonstrate that the equipment works.
- Wrap up the call by relating your service to the customer, not the equipment. Don't say: "I fixed the thermostat; your furnace will turn on now." Rather, say: "I repaired your thermostat so you can be warm and comfortable again." A positive closing like this builds confidence in you and your company.

13.0.0 ♦ CUSTOMER COMMUNICATION

Having covered the basic customer relations requirements for handling a service call (opening, servicing, and closing), you are ready to examine the customer relations topics that can give your company an extra edge over the competition. These are: keeping communications positive, taking the positive approach, showing concern for customers, and handling difficult customers.

Discuss how to handle servicing and the closing aspects of a service call.

Have the trainees enact one or more service calls, then discuss their approaches.

Emphasize the importance of positive customer communications.

Discuss customer communication skills needed by a service technician. See the answers to the "Think About It" at the end of this module.

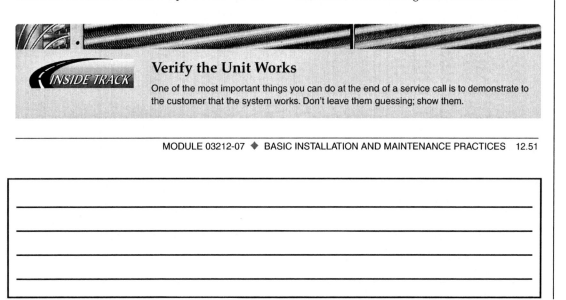

Verify the Unit Works

One of the most important things you can do at the end of a service call is to demonstrate to the customer that the system works. Don't leave them guessing; show them.

MODULE 03212-07 ♦ BASIC INSTALLATION AND MAINTENANCE PRACTICES 12.51

Discuss closing a service call. See the answers to the "Think About It" at the end of this module.

List the elements of positive customer communication.

Explain that customers respond well to friendly, knowledgeable service technicians.

Ask the trainees to recall their most positive and negative interactions with service organizations. Encourage them to isolate the attitudes and behaviors that left these impressions.

Explain that one of the first steps in solving a customer's problem is to show concern for it.

List some of the ways to express concern for the customer.

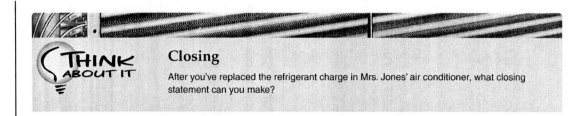

Closing
After you've replaced the refrigerant charge in Mrs. Jones' air conditioner, what closing statement can you make?

It's important to know the technical requirements of your job. Knowing the steps in completing a typical call is important, too. However, you can gain a little extra leverage by learning how to interact well with customers. This may make all the difference for your company and may determine whether you or your competitor is called the next time. When you practice good public relations throughout the service call, you give the customer confidence in your service and your company and predispose that customer to call you again.

13.1.0 Keeping Communications Positive

On every service call, you should strive to keep the communications positive. This means paying close attention to verbal and nonverbal elements. This includes communicating:

- The technical elements of the job
- What the customer needs
- Your problem-solving expertise
- What you've done to make the customer happy

Here are some simple but effective tips:

- Do the best technical job you can, since actions speak louder than words.
- Always treat the customer with courtesy, concern, and respect. Treating others as you'd like to be treated yourself is particularly important, especially in unpleasant situations. Remember, it is always easy to be courteous when things are going well. A truly mature person is courteous even when things aren't going well.
- Treat each service call as if it were an emergency. To the owner, it is. Usually, customers do not call a service technician unless they are without heat in cold weather or without cooling in hot weather. When comfort is at stake, repair becomes a top priority in the customer's mind.
- If your schedule is running late, report this to your supervisor or the customer. If you're running late, the customer must often make other arrangements to fulfill later commitments.

13.2.0 The Positive Approach

Here are some important tips on keeping things positive:

- *Smile* – Be genuinely interested in the customer. You may think you were called to repair a furnace. You were really called to make the customer's home comfortable again.
- *Use positive statements with the customer* – Compliment the equipment if you know it is good. Compliment the cleanliness of the furnace area if that contributes to the operation of the equipment. Speak of your employer in positive terms, because that builds customer confidence in your company.
- *Don't criticize or condemn* – Whenever possible, avoid negative statements. Be careful of what you say about the brand of equipment, the design of the system, or the quality of the workmanship, for you may offend the customer. Homeowners do not want to hear that you think they own inferior equipment. Even if it is inferior, fix what is possible or explain respectfully to the customer what must be replaced.

13.3.0 Showing Concern for Customers

Another part of good, positive, on-the-job communications is showing concern for your customers. Do this by:

- *Being a good listener* – Listen carefully when the customer explains what went wrong, what was observed, etc. Ask questions as appropriate to gather the information you need to solve the problem. Remember that although the customer may not have the technical vocabulary you do, the customer's own words can supply you with important clues.
- *Talking in terms of the customer's interests* – Always try to look at the problem from the customer's point of view. For example, Mrs. Jones may be upset that you are late. Maybe the emergency you fixed for your last customer caused this. However, to Mrs. Jones, the important thing may be that she's running late in picking up her child from school. If this

12.52 HVAC ♦ LEVEL TWO

Instructor's Notes:
___
___
___

understandable concern for her child is added to concern over a furnace that isn't heating, Mrs. Jones might be short-tempered. If she speaks to you abruptly, try turning the conversation around by saying, "I know you must be concerned about my being late. I apologize. Please show me your furnace so we can get you back on schedule quickly."
- *Keeping your personal problems to yourself* – Remember, you're on the job to solve the customer's problem.
- *Answering questions honestly, but positively* – If Mr. Brown asks if his equipment is worn out, you may reply, "It needs to be replaced," but you should avoid negative comments such as, "This pile of junk needs to be replaced."
- *Respecting your customer's opinions* – Suggest alternatives that fall short of telling a customer, "You're wrong!" Show the customer his opinion may have value. For example, if Mr. Smith announces that the thermostat is no good and you suspect there is a wiring problem, don't say, "No, you're wrong. It's in the wiring." Instead, say, "I think it's the wiring, but I'll check out the thermostat too."
- *Not socializing while on the call* – Generally, service technicians are paid by the customer by the hour. If you explain pleasantly that you appreciate the offer of a cup of coffee, but that you're sure they'll understand if you decline, the customer will not be offended by your trying to keep the charges to minimum.

Further, you continue to show concern for your customers when you make it easy for them to contact you for repeat business. When possible:

- Leave a business card.
- Put a sticker with your company's phone number on the equipment itself or leave one with the customer for placement near the phone.
- Show the customer some simple maintenance techniques that will prolong equipment life, such as changing filters frequently. Tell the customer why this maintenance is helpful.
- Teach the owner steps for more efficient use of the equipment, if applicable.
- Discuss billing accurately and honestly.
- Perform customer follow-up, if possible.
- If the opportunity arises, discuss additional equipment to improve comfort and efficiency that the customer can purchase from your company. Ask for the customer's permission to notify the sales staff. Follow up by providing this lead to the sales staff.

13.4.0 Handling Difficult Customers

When you deal with customers, remember: the customer wants safety, comfort, and convenience. If your habits, behaviors, and attitudes inspire that customer to believe your company will provide for their safety and comfort in a convenient manner, they will return to your company rather than seeking out a competitor next time.

Sometimes, the service technician encounters problems that go beyond failed machinery. For example:

- A customer may be unduly worried about safety.
- A customer may be upset about something totally unrelated to the problem, but is taking out that irritation on the service technician.
- A customer may be angry over equipment performance and your company's bill.

In cases like these, the service technician's actions and attitudes will often determine whether the company loses a customer.

Explain that all technicians will encounter difficult customers and that the best approach is to remain calm and positive.

Explain that a good approach to an opinionated or argumentative customer is to divert the conversation to the technical problem you are there to solve.

Discuss how to show concern for customers. See the answers to the "Think About It" at the end of this module.

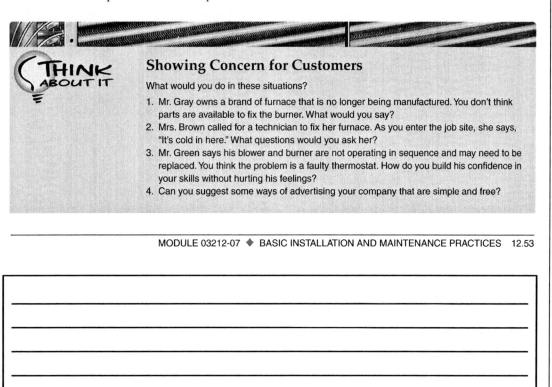

Showing Concern for Customers

What would you do in these situations?

1. Mr. Gray owns a brand of furnace that is no longer being manufactured. You don't think parts are available to fix the burner. What would you say?
2. Mrs. Brown called for a technician to fix her furnace. As you enter the job site, she says, "It's cold in here." What questions would you ask her?
3. Mr. Green says his blower and burner are not operating in sequence and may need to be replaced. You think the problem is a faulty thermostat. How do you build his confidence in your skills without hurting his feelings?
4. Can you suggest some ways of advertising your company that are simple and free?

Explain that fearful customers can often be reassured by providing them with a complete explanation of the system problem and how it has been corrected.

Have trainees practice solving the exercise questions and discuss the solutions.

In this section, we will discuss several situations that require more advanced customer relations skills. Decide how you'd handle these customers and compare your ideas with those of your classmates.

Remember, there may be no single right way out of an angry confrontation, but your attitudes and reactions can go a long way to help resolve even the most difficult situations.

13.4.1 Fearful Customers

Sometimes customer fears are based on experience, rumors, or misunderstandings. As the technician, you have the technical information that can calm your fearful customer. There is always a reason why your customer has a fear. Find it and work from the facts to ease the customer's mind. If you listen with concern and take customers' fears seriously, they will feel reassured and be more willing to listen to your explanation.

Exercise Question:

Mr. Fearful thinks his furnace is leaking gas because he smelled some when the pilot light blew out. You have re-ignited the pilot and are closing the call. What can you say to calm his fear that the house will explode from a leak at the pilot light?

Exercise Solution:

Fear is real. Often you can discover the reason for it by asking questions:

- When does this happen? When did it start?
- When it happens, what does it look like?
- What does it sound like?
- What does it smell like?
- Before it happens, are you doing something different from what you used to do?

After you have found the answers, don't just say, "Yes, when the pilot light goes out there is a smell. It is fixed." This has not addressed Mr. Fearful's concerns for the next time it happens. Instead, explain: "When a pilot goes out, there is a smell of gas. A minute quantity escapes but will not ignite. It is also mixed with a definite odor that alerts you to re-light the burner. The small amount of gas will not cause a fire. The vent pipe will exhaust it. Your family is not in any danger." Possibly suggest that a gas sniffer alarm and carbon monoxide sensor be installed to give the customer peace of mind.

Since this explanation not only addresses the technical problem, but also the fear associated with it, the customer will be more satisfied.

13.4.2 Opinionated Customers

Opinions are beliefs based not entirely on facts, but also on what seems probable in a person's mind. Everyone has opinions. Everyone also likes to get approval for opinions held. If our opinions are ridiculed, we feel hurt and offended.

It is important that customers do not feel their opinions have been challenged or belittled. With a customer, it's best to turn the conversation to the technical problem at hand (even when you agree with the opinion).

Exercise Question:

Mrs. Opinionated has had a bad week, including the failure of her air conditioner. She's trying to draw you into her earlier problems and her opinions about them. You're trying to finish quickly because you have two other service calls before lunch. What do you say?

Exercise Solution:

Taking what seems to be a negative approach might offend her. So don't say: "Yeah, we all have problems. My problem is fixing air conditioners. Where is it?" Instead, say: "I've had times like that myself. I can help you with one of your problems, though. Please show me your air conditioner so I can help you solve that problem." This offers sympathy while at the same time diverting the customer to the task at hand.

13.4.3 Argumentative Customers

Some customers are just the argumentative type. They want you to challenge their opinions, just to get an argument going. You're sensitive to the customer, and you don't want to seem aloof. You don't want to express negative opinions either. You also know that some topics are really argument-prone, such as politics, religion, children, and in-laws. Here, too, the safest approach is to avoid controversy by diverting the conversation to the task at hand.

Exercise Question:

Mr. Argumentative has no airflow from his furnace. You've discovered a broken fan belt. He wants to pull you into a discussion of local politics. He thinks the mayor is a crook. You agree. How would you handle this?

Exercise Solution:

The best way to handle this is to state a general fact that will not offend, and to avoid prolonging or expanding the political discussion while you gently turn the conversation to the problem you're there to solve. (Even if you disagreed with the customer, the technique would still be same: shifting attention to the fan belt problem.) Say, "Some people feel that way. But my feeling is that you'll be a lot more comfortable after I replace this fan belt."

13.4.4 Sloppy Customers

As you make service calls, you'll see all kinds of housekeeping: some spotless; some cluttered; some downright dirty. Always, it's best to avoid offending your customer. When admitted to the job site, avoid being judgmental in your words and gestures. Since you're a visitor, the owner will often offer an excuse or reason if things are messy. In some homes, the owner's sloppy habits can adversely affect the operation of the equipment.

You can handle such sensitive situations in these ways:

- *Focus on the task at hand* – Don't look like you're judging the job site. Avoid comments like "What a mess!" and raised eyebrows or rolled eyes that say the same thing. Assume that the owner is the unfortunate victim of several small domestic catastrophes that will be remedied soon.
- *Put the owner at ease* – If you must reply, avoid degrading and comparing them unfavorably. Say: "This happens at our home, too."
- *Be diplomatic* – If it's cluttered in front of the furnace, move only what's necessary and try to restore things when you're finished. Don't be obvious in moving things. As you clear space, avoid grunts, groans, and comments.
- *Don't sound like an accuser* – Avoid "you" statements. Talk about what's best for the health of the furnace or make up a similar situation that you solved, using your own family as an example.

13.4.5 Angry Customers

In the last section, you were advised to avoid "you" statements to a customer when it might suggest a negative judgment; for example, accusing them of bad housekeeping. There is, however, a good use of a "you" statement, and that's when the customer has a right to be angry.

When equipment breaks down twice in three days, or when the customer has tried to meet the service technician two days in a row without success, the customer may feel justified in venting anger at the first available target: the technician who appears at the door. What should you do?

First, let the angry customer vent his or her emotions. Next, acknowledge the customer's feelings. Sometimes, this means saying something sympathetic like, "I'd feel that way, too." Finally, assure the customer that you'll do your best to solve the problem.

Exercise Question:

Mrs. Smith's furnace was repaired just last week. It has broken down again, leaving the family without heat for two days. The children have colds. The Smiths have already paid a sizable sum for the repair that did not last. As you walk in the door, Mrs. Smith snarls at you. What do you say?

Exercise Solution:

Here is an appropriate answer: "Hello, Mrs. Smith. I hear your furnace isn't working again. I'm sure you're very concerned about your children's comfort. I know just how you feel. My air conditioner quit twice on me last summer during a heat wave. I'll get to the problem as quickly as I can. Let's see what the matter is." You have shown concern for her feelings and worries; you have acknowledged that she is justified in her concern; and you have assured her that you will address the problem quickly and to her satisfaction.

Emphasize that a technician should never be judgmental about the appearance of a customer's home.

Explain that when a customer is unhappy with the service or overly critical, it is best to allow the customer to voice their complaint and then, if possible, assure the customer that you will rectify it.

Discuss how to handle sloppy customers. See the answers to the "Think About It" at the end of this module.

Have trainees practice solving the exercise questions and discuss the solutions.

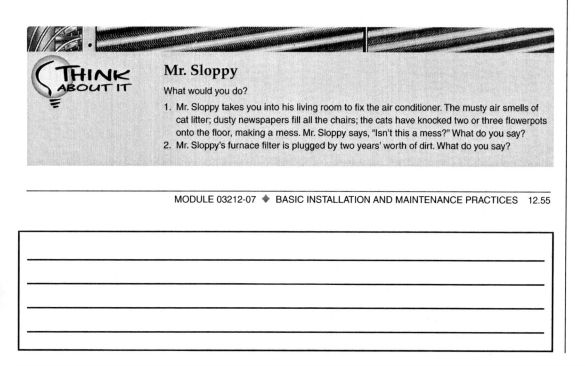

Mr. Sloppy

What would you do?

1. Mr. Sloppy takes you into his living room to fix the air conditioner. The musty air smells of cat litter; dusty newspapers fill all the chairs; the cats have knocked two or three flowerpots onto the floor, making a mess. Mr. Sloppy says, "Isn't this a mess?" What do you say?
2. Mr. Sloppy's furnace filter is plugged by two years' worth of dirt. What do you say?

Explain that when a customer has a complaint that cannot be solved by the technician, the best approach is to offer some direction that will help the customer to solve the problem.

Discuss a positive approach to dealing with customers who want help with odd jobs during a service call.

Have trainees practice solving the exercise questions and discuss the solutions.

13.4.6 Critical Customers

Sometimes you'll encounter a customer who can't be satisfied, no matter what you say or do. Often this happens because the customer is under a lot of stress or has had a bad experience that hasn't been dealt with properly. When this customer explodes, you may be the nearest target. How can you handle these situations with sensitivity and tact?

We've all heard of cases where a parent, angry about a bad day at work, comes home and yells at the kids. This displacement of anger can also happen when a person has an unresolved conflict that is upsetting.

- A man who refused to pay extra for a weekend call now has a mess on his hands because the pipes in his poorly insulated laundry room are frozen.
- A woman having trouble with her sales accounts doesn't really want to be home waiting for a service technician when she should be dealing with her office problems.

The first step in handling these situations is to remember that all of us have bad days. Approach the customer with the attitude that some day you will feel out of sorts. On these days, we all hope those around us will make allowances and be tolerant. Give your customer the same understanding you'd like to have at these times.

If the customer explodes, don't take it personally. After all, you may have just been a convenient target. You may not be responsible for the problem. Review what you have done. If your work was correct, shrug off the comments silently.

If your customer is justifiably upset, let him vent some anger. Empathize with the feelings, explaining that you'd probably feel the same under the same circumstances. Reassure the customer that you'll fix the HVAC technical problem as quickly as you can, so at least part of the day will be a positive experience.

Get the job done as quickly as you can. Get out of the situation as soon as possible without offending the customer. Stick to the facts. Focus on the mechanical problem as you solve it.

13.4.7 Customers With Unresolvable Problems

Once in a while, you'll have a customer whose problem you cannot resolve. For example, the wiring may be too outdated to support the air conditioner they want installed, or the existing furnace may be inadequate to heat a newly expanded space. What should you do?

Here is a poor response: "Sorry, sir. Your fuses are too small to handle this air conditioner. I can't do electrical work."

A better response might be: "I found that your electrical service was too small to handle the new air conditioner. We can't do this kind of work, but here is the name of an electrical company we often deal with. Or, you can check the phone book for the names of other electricians. If you prefer, we can subcontract a licensed electrical contractor to do the job, then after the electrical service is fixed, we'll be glad to come back to install the air conditioner."

A typical homeowner might prefer to contact a company that is recommended, rather than searching out several competing companies and taking a chance on the quality of their work. Your positive, helpful attitude can leave a favorable impression, even if the customer already has an electrician. You have identified the problem and proposed a good solution, then offered to complete the job when the customer is ready.

13.4.8 Customers Who Request Help With Odd Jobs

Sometimes a customer asks a service technician to perform a task that is outside company regulations. In these situations, explain why you can't help. Then suggest another way to solve the problem.

Exercise Question:

While working on her furnace, Mrs. Oddjob asks you to move her freezer. What do you say?

Exercise Solution:

A good answer is: "I'm sorry, Mrs. Oddjob, but company regulations won't let me handle appliances we don't service. I know you're anxious to move that freezer. Perhaps a neighbor can help you. It wouldn't be fair to my later customers if I spent too much time at any one place. As you know, you're billed by the hour for my time. I know you want to keep your costs down." This response recognizes that the customer has a problem, but explains why you can't help her solve it.

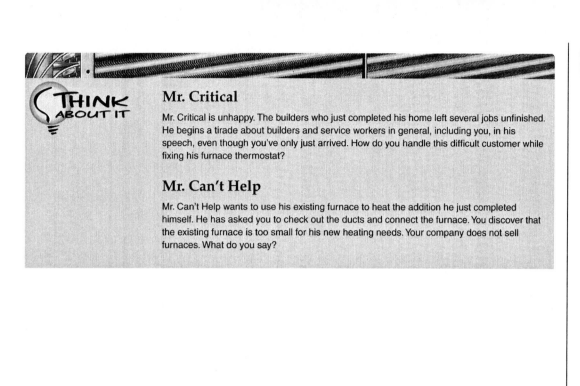

Mr. Critical

Mr. Critical is unhappy. The builders who just completed his home left several jobs unfinished. He begins a tirade about builders and service workers in general, including you, in his speech, even though you've only just arrived. How do you handle this difficult customer while fixing his furnace thermostat?

Mr. Can't Help

Mr. Can't Help wants to use his existing furnace to heat the addition he just completed himself. He has asked you to check out the ducts and connect the furnace. You discover that the existing furnace is too small for his new heating needs. Your company does not sell furnaces. What do you say?

Discuss how to handle critical customers or customers you cannot help. See the answers to the "Think About It" at the end of this module.

Have trainees complete the Review Questions, and go over the answers prior to administering the Module Examination.

Review Questions

1. The bolts and screws most likely used on assemblies that must have a finished appearance are called _____.
   a. machine bolts
   b. machine screws
   c. cap screws
   d. stud bolts

2. A jam nut is used _____.
   a. where frequent removal of the nut is required
   b. to lock a standard nut in place
   c. when appearance is important
   d. to lock a split washer in place

3. When installing a ½" Grade 8 steel bolt, it should be tightened to _____.
   a. 37 foot-pounds
   b. 60 foot-pounds
   c. 85 foot-pounds
   d. 92 foot-pounds

4. When installing gaskets at a flange, tighten the _____.
   a. flange bolts only a small amount at a time
   b. bolts one at a time to final torque, before tightening the next bolt in sequence
   c. bolts in the proper sequence
   d. flange bolts only a small amount at a time and in the proper sequence

5. Lip-type packings are used primarily _____.
   a. with shafts
   b. with hydraulic cylinders
   c. with valve stems
   d. as a packing gland

6. O-rings can be used _____.
   a. as a static seal
   b. as a dynamic seal
   c. as both static and dynamic seals
   d. in place of packings

7. A force acting either toward or away from and lengthwise along the axis of a shaft is called _____.
   a. an axial load
   b. thrust
   c. a radial load
   d. a journal load

8. A type of anti-friction bearing is a ____ bearing.
   a. sleeve
   b. ball
   c. thrust
   d. glide

9. When installing a bearing using the temperature method, you should heat the bearing to _____.
   a. the temperature recommended by the bearing manufacturer
   b. a minimum of 275°F
   c. a temperature lower than that recommended by the manufacturer, if the bearing has expanded enough to mount on the shaft
   d. a maximum of 275°F

10. The viscosity of oil refers to the _____.
    a. lowest temperature at which the oil will flow freely
    b. ability of the oil to flow at a specific temperature
    c. temperature at which the oil gives off ignitable vapors
    d. highest temperature at which the oil will flow freely

11. On initial movement, grease lubricates _____.
    a. by forming a liquid film that keeps the surfaces apart to reduce friction and heat
    b. when the oil in the grease flows away from the bulk of the grease
    c. when the surfaces begin to move and exert pressure on the grease
    d. when the surfaces begin to move and friction heats the grease

Instructor's Notes:

Review Questions

12. A V-belt is marked 3L405. Its size is _____.
    a. ¼" wide × ⅛" high × 405" long
    b. ⅜" wide × ⁷⁄₃₂" high × 40½" long
    c. ½" wide × ⁵⁄₁₆" high × 40½" long
    d. ⅝" wide × ⅜" high × 405" long

13. Flexible couplings _____.
    a. require precise alignment
    b. are generally easier to install than rigid couplings
    c. are used where smooth, even starts are needed
    d. are used where a major angular misalignment exists

14. When looking at the faces of a coupling from a side view, the faces are not square. To achieve correct alignment, one of the units must be _____.
    a. raised or lowered
    b. tilted on its base
    c. moved to one side or the other
    d. rotated on its base

15. Overlubrication of a bearing can cause _____.
    a. contaminants to enter the bearing
    b. a decrease in operating temperature
    c. an increase in viscosity
    d. no problems

16. A document that a technician fills out to indicate that a manufacturer's equipment has been put into service is called a _____.
    a. commissioning job report
    b. start-up job report
    c. warranty ticket
    d. service ticket

17. The customer begins forming a first impression of you and your company _____.
    a. after you've fixed his air conditioner
    b. after you've submitted your bill
    c. as soon as you appear at the door
    d. as soon as you speak

18. When speaking to a customer, always _____.
    a. expect that the customer understands technical jargon
    b. show courtesy, concern, and respect
    c. break the ice by chatting about local politics
    d. tell the customer if you think the equipment is low-quality

19. Your customer is worried about furnace safety. You should _____.
    a. tell him to relax
    b. tell him you have everything under control
    c. calm him and work from facts to ease his mind
    d. call your supervisor

20. If you encounter a problem you can't fix because your company doesn't provide the type of service required, you should _____.
    a. tell the customer you fixed part of the problem
    b. call someone you know who can do the job
    c. help the customer find someone else who can do the job
    d. criticize the customer for wasting your time

Summarize the major concepts presented in the module.

Administer the Module Examination. Record the results on Craft Training Report Form 200, and submit the results to the Training Program Sponsor.

Administer the Performance Test, and fill out Performance Profile Sheets for each trainee. If desired, trainee proficiency noted during laboratory sessions may be used to complete the Performance Test. Record the results on Craft Training Report Form 200, and submit the results to the Training Program Sponsor.

Summary

Mechanical repairs account for a high percentage of HVAC work. Most of these repairs can be traced to improper installation procedures or failure to perform preventative maintenance. The HVAC technician must be able to properly install and maintain the mechanical components used in HVAC systems.

The service technician who helps a company acquire and retain happy customers practices good personal and work habits, behaviors, and attitudes. In addition, the technician who learns how to approach problems and work with customers perpetuates the company's positive image. The customer's impression of the technician, who represents the company at the job site, determines if the customer returns to the company for future service calls. Practicing good customer relations increases your value to your company and ensures that you will always have customers to serve.

Notes

Instructor's Notes:

Trade Terms Introduced in This Module

Axial load: An external load that acts lengthwise along a shaft.

Break-away torque: The torque required to loosen a fastener. This is usually lower than the torque to which the fastener has been tightened.

Dropping point: The temperature at which grease is soft enough for a drop of oil to fall away or flow from the bulk of the grease.

Dynamic seal: A seal made where there is movement between two mating parts, or between one of the parts and the seal.

Fire point: The temperature at which oil will burn if ignited.

Flash point: The temperature at which oil gives off ignitable vapors.

Journal: The part of a shaft, axle, spindle, etc., which is supported by and revolves in a bearing.

Oxidation: The process of combining with oxygen. All petroleum products react with oxygen to some degree, and this increases as the temperature increases.

Packing gland: A part used to compress the packing in a stuffing box.

Pour point: Refers to the lowest temperature at which an oil will flow freely.

Radial load: The side or radial force applied at right angles to a bearing and shaft.

Run-down resistance: The torque required to overcome the resistance of associated hardware, such as locknuts and lockwashers, when tightening a fastener.

Set or seizure: In the last stages of rotation in reaching a final torque, the fastener may lock up; this is known as seizing or set. This is usually accompanied by a noticeable popping effect.

Static seal: A seal made where there is no movement between the two joining parts or between the seal and the mating part.

Stuffing box: The housing used to control leaking along a shaft or rod. Typically composed of three parts: the packing chamber (also called the box); the packing rings; and the gland follower (also called the stuffing gland).

Thrust: The force acting lengthwise along the axis of a shaft, either toward it or away from it.

Tolerance: The amount of variation allowed from a standard.

Torque: The resistance to a turning or twisting force.

Viscosity: The thickness of a liquid or its ability to flow at a specific temperature.

Viscosity index (VI): A measure of how an oil's viscosity varies with temperature.

Additional Resources and References

Additional Resources

This module is intended to be a thorough resource for task training. The following reference work is suggested for further study. This is optional material for continued education rather than for task training.

Air Conditioning Systems, Principles, Equipment, and Service, Latest Edition. Upper Saddle River, NJ: Prentice Hall.

Figure Credits

Alcoa Fastening Systems (Marson Corporation), 212SA01

Topaz Publications, Inc., 212SA02, 212F41 (outside caliper, vernier caliper, divider, inside caliper), 212F42, 212F44–212F48

Photo courtesy of Garlock Sealing Technologies, 212F16 (bottom left, top right)

Inertech, Inc., 212F16 (bottom right)

RBC Bearings Incorporated, 212F26 (top left)

The Timken Company, 212F26 (top right, bottom left, bottom right)

Pruftechnik AG, 212F28

Neff Press Inc. – St. Louis, MO, 212F29

Carrier Corporation, 212SA03, 212F49

L.S. Starrett Company, 212F41 (hermaphrodite calipers)

Kastar Hand Tools, 212F43 (thread pitch gauge)

MODULE 03212-07 — TEACHING TIPS

The following are suggested activities or instructional methods to help you teach the material in this module.

General

When you call on someone to answer a question, the rest of the class relaxes or even tunes out because they expect that the question and answer will take place only between you and the trainee you called on. Instead, use this technique to involve more trainees in answering questions and to keep them on their toes.

1. Ask trainees to define a term or explain a concept.
2. After one trainee has answered, ask a trainee seated nearby if the answer is right. Then ask whether a trainee in the back of the room agrees.
3. Ask trainees to explain why they think an answer is right or wrong.
4. Use the session to clear up incorrect ideas and encourage trainees to learn from their mistakes.

Sections 1.0.0 through 13.0.0 *Quick Quiz*

This Quick Quiz will familiarize trainees with the terms and definitions that are commonly used in basic maintenance and installation practices. You will need photocopies of the quiz provided in the following page. Trainees will need pencils. If you allow trainees to use the Trainee Guide, decrease the amount of time you give them to complete the quiz.

1. Make a photocopy of the quiz for each trainee.
2. Give trainees between 5 and 10 minutes to complete the quiz.
3. Go over the answers to the quiz.
4. Ask trainees if they have questions.

Answers to Quick Quiz

1. h
2. b
3. f
4. a
5. g
6. k
7. c
8. j
9. l
10. e
11. d
12. i

Quick Quiz   *Basic Installation and Maintenance Practices*

For each description listed, identify the term that the text best describes. Write the corresponding letter in the blank provided.

_____ 1. O-rings are examples of _____.

_____ 2. Lip and oil-type seals are examples of _____.

_____ 3. In cold weather applications and refrigeration systems, the _____ of the oil must be considered.

_____ 4. The _____ is the torque required to loosen a fastener.

_____ 5. Locknuts and lockwashers produce _____.

_____ 6. There are two ways to express the _____ rating of oils.

_____ 7. The _____ is the temperature at which an oil will burn if ignited.

_____ 8. When installing equipment, always follow the manufacturer's instructions and recommended _____ values.

_____ 9. Typically, paraffinic oils have a high _____.

_____ 10. The service life of an oil depends on its _____.

_____ 11. The _____ is the temperature at which oil gives off ignitable vapors.

_____ 12. A packing puller is typically used to remove the packing from a shaft installed in a(n) _____.

   a. break-away torque
   b. dynamic seals
   c. fire point
   d. flash point
   e. oxidation resistance
   f. pour point
   g. run-down resistance
   h. static seals
   i. stuffing box
   j. torque
   k. viscosity
   l. viscosity index

Section 2.0.0   *Identifying Fasteners*

This exercise will familiarize trainees with identifying various types of fasteners. Trainees will need pencils and paper. You will need an overhead projector, acetate sheets, transparency markers, and various fasteners. Allow 15 to 20 minutes for this exercise. This exercise corresponds to Performance Tasks 1 and 2.

1. Divide the class in half. On one sheet write "Team One" on one half and "Team Two" on the other half.
2. Hold up a fastener and have trainees identify it. If desired, add a bonus point if the trainee can name a typical application for the fastener.
3. To ensure complete participation, each team member can answer no more than two times (adjust according to the number of participants and items to be identified). Each correct answer gets a check on the acetate sheet. An incorrect answer means the other team has the opportunity to answer.
4. Keep score to encourage class participation. At the end of the game, answer any questions.

Section 2.4.1   *Think About It – Torque Values*

To convert inch-pounds to foot-pounds, divide the inch-pounds by 12. For example, 84 inch-pounds ÷ 12 = 7 foot-pounds. To convert foot-pounds to inch-pounds, multiply the foot-pounds by 12. For example, 7 foot-pounds × 12 = 84 inch-pounds.

Section 3.1.0   *Identifying Gaskets*

HVAC technicians must be able to identify different types of gaskets. Trainees will need appropriate personal protective equipment, pencils, and paper. You will need to arrange workstations with different types of gaskets at each one. Allow 20 to 30 minutes for this exercise. This exercise corresponds to Performance Task 3.

Obtain old or broken equipment, motors, pumps, or appliances from local plants, a disposal facility, an equipment recycler, or the local scrap yard. Take the equipment apart to expose the gasket.

This equipment may be used to demonstrate and practice service techniques.

1. Discuss various types of gaskets and their applications.
2. Have trainees rotate between the various workstations and identify the type of gasket at each station.
3. After all trainees have visited each station, have one trainee identify the type of gasket at their station. Answer any questions trainees may have.

Section 6.3.0   *Bearing Failures*

This exercise will familiarize trainees with bearing failures. Trainees will need appropriate personal protective equipment, pencils, and paper. You will need to arrange workstations with different piece of equipment each one. Allow 20 to 30 minutes for this exercise.

Obtain old or broken equipment, motors, pumps, or appliances from local plants, a disposal facility, an equipment recycler, or the local scrap yard. Find equipment with different types of bearing failures.

1. Describe different types of bearing failures.
2. Have trainees rotate between the various workstations. Have trainees identify the type of bearing failure and possible causes.
3. Answer any questions trainees may have.

Section 7.0.0 *Lubricants*

This exercise will familiarize trainees with lubricants. Trainees will need pencils and paper. You will need to arrange for a manufacturer's representative to give a presentation on lubricants. Allow 20 to 30 minutes for this exercise.

1. Tell trainees that a guest speaker will be presenting information on lubricants. Have trainees brainstorm questions for the guest speaker before the speaker arrives.
2. Introduce the speaker. Ask the presenter to speak about lubricants. Have them cover typical applications, lubricant selection, safety precautions, personal protective equipment needed, hazards, storage, handling, and disposal requirements.
3. Have the trainees take notes and write down questions during the presentation.
4. After the presentation, answer any questions trainees may have.

Section 10.0.0 *Basic Maintenance Procedures*

This exercise will familiarize trainees with basic maintenance procedures. Trainees will need appropriate personal protective equipment, pencils, and paper. You will need to arrange workstations with different type of equipment each one. Allow 20 to 30 minutes for this exercise.

Obtain old or broken equipment, motors, pumps, or appliances from local plants, a disposal facility, an equipment recycler, or the local scrap yard that can be used to perform basic maintenance procedures.

1. Describe basic maintenance procedures.
2. Have trainees rotate between the various workstations. Have trainees perform one service procedure on each piece of equipment or an entire set of service check on one piece.
3. Answer any questions trainees may have.

Section 12.0.0 *Think About It – Customer Relations*

There are no specific answers to this "Think About It." It is intended to generate class discussion among the trainees. Ideally, the trainees will arrive at answers via discussion among themselves.

Section 13.0.0 *Think About It – Customer Communication*

There are no specific answers to this "Think About It." It is intended to generate class discussion among the trainees. Ideally, the trainees will arrive at answers via discussion among themselves.

MODULE 03212-07 — ANSWERS TO REVIEW QUESTIONS

Answer	Section Reference
1. c	2.2.1
2. b	2.2.4
3. d	2.4.1; Figure 14
4. d	3.2.0
5. b	4.1.0
6. c	5.1.1
7. b	6.0.0
8. b	6.2.0
9. a	6.4.2
10. b	7.1.1
11. c	7.2.0
12. b	8.1.1
13. b	9.1.2
14. b	9.3.3
15. a	10.4.0
16. c	11.5.0
17. c	12.2.0
18. b	13.1.0
19. c	13.4.1
20. c	13.4.7

CONTREN® LEARNING SERIES — USER UPDATE

NCCER makes every effort to keep these textbooks up-to-date and free of technical errors. We appreciate your help in this process. If you have an idea for improving this textbook, or if you find an error, a typographical mistake, or an inaccuracy in NCCER's Contren® textbooks, please write us, using this form or a photocopy. Be sure to include the exact module number, page number, a detailed description, and the correction, if applicable. Your input will be brought to the attention of the Technical Review Committee. Thank you for your assistance.

Instructors – If you found that additional materials were necessary in order to teach this module effectively, please let us know so that we may include them in the Equipment/Materials list in the Annotated Instructor's Guide.

Write: Product Development and Revision
National Center for Construction Education and Research
3600 NW 43rd St, Bldg G, Gainesville, FL 32606

Fax: 352-334-0932

E-mail: curriculum@nccer.org

Craft _____ Module Name _____

Copyright Date _____ Module Number _____ Page Number(s) _____

Description
_____
_____
_____
_____

(Optional) Correction
_____
_____
_____

(Optional) Your Name and Address
_____
_____
_____

Module 03213-07

Sheet Metal Duct Systems

NCCER STANDARDIZED CRAFT TRAINING PROGRAM

The National Center for Construction Education and Research (NCCER) provides a standardized national program of accredited craft training. Key features of the program include instructor certification, competency-based training, and performance testing. The program provides trainees, instructors, and companies with a standard form of recognition through a National Craft Training Registry. The program is described in full in the *Guidelines for Accreditation*, published by NCCER. For more information on standardized craft training, contact the NCCER by writing us at 3600 NW 43rd St., Bldg. G, Gainesville, FL 32606; calling 352-334-0911; or emailing info@nccer.org. More information may be found at our website, www.nccer.org.

HOW TO USE THIS ANNOTATED INSTRUCTOR'S GUIDE

Each page presents two sections of information. The larger section displays each page exactly as it appears in the Trainee Module. The narrow column ties suggested trainee and instructor actions to each page and provides icons (detailed below) to call your attention to material, safety, audiovisual, or testing requirements. The bottom of each page includes space for your notes.

The **Audiovisual** icon indicates an appropriate time to show a transparency or other audiovisual aid.

The **Classroom** icon prompts you to define a term, stress a point, ask trainees to explain a concept, or give examples.

The **Demonstration** icon directs you to show trainees how to perform tasks.

The **Examination** icon tells you to administer the written module examination.

The **Homework** icon is placed where you may wish to assign reading for the next class, assign a project, or advise trainees to prepare for an examination.

The **Laboratory** icon is used when trainees are to practice performing tasks.

The **Materials** icon is a reminder for you to gather materials needed for classes, labs, and testing.

The **Performance Testing** icon tells you to administer a performance test or a portion thereof.

The **Safety** icon is used to emphasize safety issues. It is often keyed to *Caution* and *Warning!* statements in the Trainee Module.

The **Teaching Tip** icon indicates additional guidance is available, such as how to conduct an exercise, get the most educational value from a field trip, or encourage class participation. Teaching Tips may expand on a feature (*Think About It, Did You Know?*) or provide *Quick Quizzes* or similar exercises. You will be referred to the Teaching Tips section at the back of the module if there is additional material.

The **Combination** icon indicates that the laboratory listed corresponds with a performance task. If desired, you can note the proficiency of the trainees during the laboratory, and use it to satisfy performance testing requirements.

PREPARATION

Before teaching this module, you should review the Objectives, Performance Tasks, Materials and Equipment List, and Module Outline. Be sure to allow ample time to prepare your own training or lesson plan and gather all required materials and equipment.

Sheet Metal Duct Systems
Annotated Instructor's Guide

Module 03213-07

MODULE OVERVIEW

This module introduces sheet metal duct systems and explains how to lay out and install sheet metal and flexible ducts.

PREREQUISITES

Prior to training with this module, it is recommended that the trainee shall have successfully completed *Core Curriculum*; *HVAC Level One*; and *HVAC Level Two*, Modules 03201-07 through 03212-07.

OBJECTIVES

Upon completion of this module, the trainee will be able to do the following:

1. Identify and describe the basic types of sheet metal.
2. Define properties of steel and aluminum alloys.
3. Describe a basic layout method and perform proper cutting.
4. Join sheet metal duct sections using proper seams and connectors.
5. Describe proper hanging and support methods for sheet metal duct.
6. Describe thermal and acoustic insulation principles.
7. Select, apply, and seal the proper insulation for sheet metal ductwork.
8. Describe guidelines for installing components such as registers, diffusers, grilles, dampers, access doors, and zoning accessories.
9. Install takeoffs and attach flexible duct to a sheet metal duct.

PERFORMANCE TASKS

Under the supervision of the instructor, the trainee should be able to do the following:

1. Join duct sections and fittings.
2. Install takeoffs and attach flexible duct.

MATERIALS AND EQUIPMENT

Overhead projector and screen
Transparencies
Blank acetate sheets
Transparency pens
Whiteboard/chalkboard
Markers/chalk
Pencils and scratch paper
Appropriate personal protective equipment
Sheet metal gauge
Several gauges of sheet metal
Samples of stainless steel and aluminum
Samples of various types of seams
Connectors
Hangers and supports
Grilles, registers, and diffusers
Sections of duct

Tools for joining ductwork
Insulation materials
Fibrous glass duct liner
Flexible blanket insulation
Measuring tape
Utility knife
Straightedge
Mechanical fasteners
Dampers
Takeoffs
Flexible duct
Tin snips and other sheet metalworking tools
Copies of Quick Quiz*
Module Examinations**
Performance Profile Sheets**

* Located in the back of this module.
**Located in the Test Booklet.

SAFETY CONSIDERATIONS

Ensure that the trainees are equipped with appropriate personal protective equipment and know how to use it properly. This module requires trainees to work with sheet metal. Make sure that all trainees are briefed on appropriate safety procedures. Emphasize the dangers posed by sharp metal edges and cutting tools, and appropriate safety precautions.

ADDITIONAL RESOURCES

This module is intended to present thorough resources for task training. The following reference works are suggested for both instructors and motivated trainees interested in further study. These are optional materials for continued education rather than for task training.

Air Distribution Basics for Residential and Small Commercial Buildings, 2000. Hank Rutkowski. Arlington, VA: Air Conditioning Contractors of America.

Air Distribution in Rooms: Ventilation for Health and Sustainable Environment, 2000. Hazim B. Awbi, editor. New York, NY: Elsevier Science.

Fibrous Glass Duct Liner Standard: Design, Fabrication, and Installation Guidelines, Third Edition, 2002. NAIMA. Alexandria, VA: North American Insulation Manufacturers Association.

HVAC Duct Construction Standard—Metal and Flexible. Chantilly, VA: Sheet Metal and Air Conditioning Contractors' National Association.

Standard for the Installation of Air Conditioning and Ventilating Systems, 1999. Quincy, MA: National Fire Protection Association.

Thermal Insulation Building Guide, 1990. Edin F. Strother and William C. Turner. Melbourne, FL: Krieger Publishing.

Ultimate Sheet Metal Fabrication, 1999. Tim Remus. Osceola, WI: Motorbooks International.

Working With Fiber Glass, Rock Wool, and Slag Wool Products, 2001. Alexandria, VA: North American Insulation Manufacturers Association.

TEACHING TIME FOR THIS MODULE

An outline for use in developing your lesson plan is presented below. Note that each Roman numeral in the outline equates to one session of instruction. Each session has a suggested time period of 2½ hours. This includes 10 minutes at the beginning of each session for administrative tasks and one 10-minute break during the session. Approximately 5 hours are suggested to cover *Sheet Metal Duct Systems*. You will need to adjust the time required for hands-on activity and testing based on your class size and resources. Because laboratories often correspond to Performance Tasks, the proficiency of the trainees may be noted during these exercises for Performance Testing purposes.

Topic **Planned Time**

Session I. Introduction to Sheet Metal Duct Systems
    A. Introduction   _____
    B. Steel and Other Metals   _____
    C. Seams (Locks)   _____
    D. Connectors   _____
    E. Hangers and Supports for Sheet Metal Ducts   _____
    F. Installing Registers, Grilles, and Diffusers   _____
    G. Laboratory   _____
        Trainees practice joining duct sections and fittings. This laboratory corresponds to Performance Task 1.

Session II. Accessories, Review, and Testing
   A. Insulation
   B. Dampers and Access Doors
   C. Takeoffs
   D. Laboratory

   Trainees practice installing takeoffs and attaching flexible duct. This laboratory corresponds to Performance Task 2.

   E. Zoning Accessories and Coils
   F. Review
   G. Module Examination
      1. Trainees must score 70% or higher to receive recognition from NCCER.
      2. Record the testing results on Craft Training Report Form 200, and submit the results to the Training Program Sponsor.
   H. Performance Testing
      1. Trainees must perform each task to the satisfaction of the instructor to receive recognition from NCCER. If applicable, proficiency noted during laboratory exercises can be used to satisfy the Performance Testing requirements.
      2. Record the testing results on Craft Training Report Form 200, and submit the results to the Training Program Sponsor.

HVAC Level Two

03213-07
Sheet Metal Duct Systems

Assign reading of Module 03213-07.

03213-07
Sheet Metal Duct Systems

Topics to be presented in this module include:

1.0.0	Introduction	13.2
2.0.0	Steel and Other Metals	13.2
3.0.0	Seams (Locks)	13.4
4.0.0	Connectors	13.5
5.0.0	Hangers and Supports for Sheet Metal Duct	13.6
6.0.0	Installing Registers, Grilles, and Diffusers	13.9
7.0.0	Insulation	13.10
8.0.0	Dampers and Access Doors	13.24
9.0.0	Takeoffs	13.30
10.0.0	Zoning Accessories and Coils	13.31

Overview

HVAC ductwork can be made of sheet metal or fiberglass ductboard. In addition, flexible metal duct can be used for branch runs in many applications. Because the interior surfaces of sheet metal ducts are smooth, these ducts offer less resistance to airflow than fiberglass ductboard. However, in many instances they must be insulated to reduce noise transmission and heat transfer. Therefore, working with sheet metal ductwork requires knowledge of various types of metals, insulation practices, as well as unique methods for joining and supporting sheet metal ductwork. Sheet metal ductwork systems also include additional components for safe and efficient air distribution, such as dampers, takeoffs, and zoning accessories.

Instructor's Notes:

Objectives

When you have completed this module, you will be able to do the following:

1. Identify and describe the basic types of sheet metal.
2. Define properties of steel and aluminum alloys.
3. Describe a basic layout method and perform proper cutting.
4. Join sheet metal duct sections using proper seams and connectors.
5. Describe proper hanging and support methods for sheet metal duct.
6. Describe thermal and acoustic insulation principles.
7. Select, apply, and seal the proper insulation for sheet metal ductwork.
8. Describe guidelines for installing components such as registers, diffusers, grilles, dampers, access doors, and zoning accessories.
9. Install takeoffs and attach flexible duct to a sheet metal duct.

Trade Terms

Alloy
Angle bracket
Carbon
Channel
Cold-rolled steel
Lap-joined
Mastic
Non-ferrous
Opposed-blade damper
Runout
Shear load
Sheet metal gauge
Sheet steel
Sound attenuation
Stainless steel
Tensile load
Tensile strength
Vapor barrier

Required Trainee Materials

1. Appropriate personal protective equipment
2. Pencil and paper
3. Sheet Metal and Air Conditioning Contractors' National Association (SMACNA) Manual: *HVAC Duct Construction Standards—Metal and Flexible*, Third Edition, 2005.

Prerequisites

Before you begin this module, it is recommended that you successfully complete *Core Curriculum*; *HVAC Level One*; and *HVAC Level Two*, Modules 03201-07 through 03212-07.

This course map shows all of the modules in the second level of the HVAC curriculum. The suggested training order begins at the bottom and proceeds up. Skill levels increase as you advance on the course map. The local Training Program Sponsor may adjust the training order.

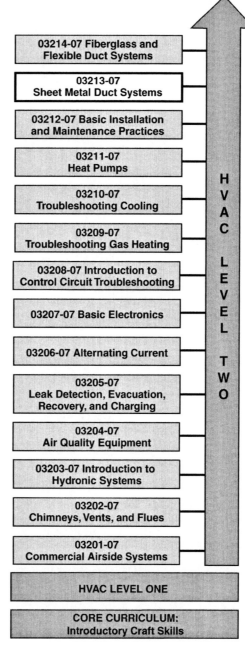

Ensure that you have everything required to teach the course. Check the Materials and Equipment list at the front of this module.

See the general Teaching Tip at the end of this module.

Explain that terms shown in bold are defined in the Glossary at the back of this module.

Show Transparency 1, Objectives, and Transparency 2, Performance Tasks. Review the goals of the module, and explain what will be expected of the trainee.

Review the modules covered in Level Two and explain how this module fits in.

Identify the advantages of sheet metal ductwork. Emphasize the importance of insulating sheet metal ductwork.

Explain that steel is a combination of iron and carbon. Discuss the advantages of alloys over pure metals.

Identify the three categories of sheet metals: basic metals, coated metals, and alloy metals.

Discuss the basic metals used in sheet metal ductwork. Describe cold-rolled steel.

Show Transparency 3 (Figure 1). Explain how to measure sheet metal using a sheet metal gauge.

Show trainees how to measure sheet metal using a sheet metal gauge.

Explain that sheet metal is often coated with zinc.

See the Teaching Tip for Sections 1.0.0–6.0.0 at the end of this module.

1.0.0 ◆ INTRODUCTION

Sheet metal ductwork and fiberglass ductboard can serve the same purpose in most applications. However, there are instances when sheet metal is preferred. For example, sheet metal ductwork offers a longer duct life, and is less likely to be damaged in locations where service personnel will be working in and around the ducts. Sheet metal with outer insulation is used when it is necessary to clean the inside of the ductwork. Sheet metal ductwork is also used in locations where the ductwork is exposed to outdoor ambient conditions, and it may be required in healthcare facilities and in buildings where specified by fire safety codes. Standards published by the Sheet Metal and Air Conditioning Contractors National Association (SMACNA) are designed to guide HVAC installers in proper construction, support, and insulation for ductwork.

Sheet metal ductwork must often be insulated in order to reduce noise transmission and prevent heat transfer. Therefore, it is important for anyone installing sheet metal ductwork to understand the proper techniques for insulating this type of ductwork.

2.0.0 ◆ STEEL AND OTHER METALS

Steel is a combination of a metallic element, iron, and a non-metallic element, **carbon**. Though it is the most abundant of all metals, iron is rarely found in its pure state in nature. It is usually found in combination with oxygen and other elements in the form of an ore mixed with rocks, clay, and sand. About five percent of the Earth's crust is composed of iron compounds.

Metals are classified as pure metals and as **alloys**. A pure metal is one that is not combined with any other metal. Iron, aluminum, copper, lead, tin, zinc, and gold are examples of pure metals. Pure metals are often too soft or lack sufficient strength to be used in most construction application, so other elements are added to create alloys. Alloys are stronger and generally have other structural advantages necessary in certain construction applications. Later in this section you will learn how other elements are added to pure metals to produce alloys.

2.1.0 Types of Sheet Metal

Sheet metal can be divided into three basic groups:

- Basic metals
- Coated metals
- Alloy metals

The basic metals used in sheet metal ductwork are **sheet steel**, **cold-rolled steel**, hot-rolled steel, and copper. Sheet steel is an uncoated sheet with a bluish-black surface. It is commonly used to make stovepipes. Cold-rolled steel is silver-gray and has a smoother surface than regular sheet steel. The name refers to the steel mill process used to make it. The steel is compressed and rolled into the desired shape and thickness after is has cooled, giving it an improved surface finish and/or higher **tensile strength**. Hot-rolled steel is formed as it poured out of a smelting furnace. Copper sheets usually contain a small amount of another metal as an alloy. Copper sheets are most commonly used as roofing and guttering materials.

Sheet metal comes in various widths of thickness, otherwise identified as gauges. Typical thicknesses range from 1/64 inch (30 gauge) to 1/8 inch (11 gauge); the lower the gauge number, the thicker the sheet. Thickness is measured and verified using a **sheet metal gauge** (*Figure 1*). Be careful not to confuse this with a wire gauge. When measuring coated metals, only account for the base metal, and not the coating. The actual gauge of coated metal is one gauge thinner than that indicated by the measuring tool. Refer to *Appendix A* for a list of common sheet metal gauges.

2.1.1 Coated Metals

Steel is often coated with zinc to protect it from rust and corrosion. Steel treated in this way is called galvanized steel. The thickness of the coating on sheet metal may be G60 (0.60 ounce per square foot) or G90 (0.90 ounce per square foot). In the United States, galvanized sheets were first produced in the 1850s. Other metals now com-

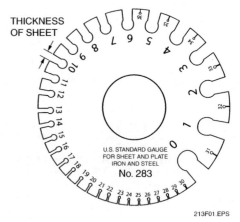

Figure 1 ◆ Sheet metal gauge.

How Metal is Galvanized

The zinc that protects may be applied by one of two methods: hot dipping or electroplating.

In the hot dipping method, clean, oxide-free iron or steel is placed into hot, melted zinc. This process bonds the zinc to the surface of the iron or steel.

In electroplating, a pure coating of a controlled thickness of zinc is applied without heat. This process does not have the soldering effect of the hot-dip process, so it is not suitable for articles that must be made watertight. However, with this method, the zinc coating has greater adherence.

How does corrosion take place? Here is a simple example. When an uncoated metal is placed in a corrosive atmosphere such as water, its atoms turn into ions that move into the water. This process eats away at the metal. If you place a barrier, like zinc, between the metal and the water, the corrosive process is stopped.

Zinc protects the base metal because the coating is more electronegative (charged with negative electricity) than the base metal. The zinc gives the metal what is called sacrificial protection. This means that if the base metal is scratched, the zinc is slowly consumed, or sacrificed, so that the iron or steel remains protected from corrosion. This protection continues as long as the zinc and base metal are in contact with one another.

monly used include copper, aluminum, and **stainless steel**. Some sheet metal applications in areas exposed to corrosive fumes or high amounts of moisture require the use of plastics and fiberglass.

When the zinc coating is properly applied, galvanizing may protect the base metal from corrosion for 15 to 30 years or more. The most common uses for galvanized sheet metals are in heating and air conditioning ducts and in gutters and downspouts—applications that are exposed to atmospheric corrosion. In addition, the zinc coating provides a surface that helps stop the growth of microbes in air distribution systems.

2.1.2 Alloy Metals

The properties of a pure metal can be changed by melting it and mixing in other elements. The critical melting and burning temperatures are shown in *Figure 2*. This process produces an alloy, an entirely new metal than can have characteristics very different from those of the original elements that went into it.

An alloy is named after the principal metal in its composition. When metals are added to steel, the resulting product is called alloy steel. When metals are added to aluminum, the result is an aluminum-base alloy. There are many other alloys; some common ones are shown in *Table 1*. Other non-metallic elements can be alloyed with a basic metal to change their properties. The most common of these elements is carbon. When carbon is added to iron, the result is steel. Thus, steel itself is an alloy of iron and carbon.

2.2.0 Steel Alloys

Steel alloys are produced when certain other metallic elements are added to plain steel. These other metals include nickel, chromium, manganese, tungsten, and vanadium.

°CELSIUS			°FAHRENHEIT
6,440°	Tungsten	Arc Flame	11,624°
3,500°	Oxyacetylene	Flame	6,332°
3,410°	Tungsten	Melts	6,170°
1,961°	Natural Gas	Flame	3,562°
1,535°	Iron	Melts	2,795°
1,083°	Copper	Melts	1,981°
660°	Aluminum	Melts	1,218°
419.5°	Zinc	Melts	787°
232°	Tin	Melts	449°
100°	Water	Boils	212°
0°	Ice	Melts	32°
−38.87°	Mercury	Melts	−38°
−78°	Dry Ice	Vaporizes	−110°
−273.16°	Absolute	Zero	−459.69°
°CELSIUS			°FAHRENHEIT

Figure 2 ♦ Critical melt temperatures.

Table 1 Common Alloys

Alloy	Components
Brass	Copper and zinc
Bronze	Copper and tin
Stainless steel	Iron, nickel, and chromium
Steel	Iron and carbon
Solder	Lead and tin

Discuss the applications of coated sheet metal.

Show Transparency 4 (Figure 2). Explain how the properties of a metal can be changed by mixing it with other metals.

Show Transparency 5 (Table 1). Describe common alloys and discuss their uses.

Identify other metals that are often added to plain steel.

Describe stainless steel and discuss its advantages.

Describe aluminum alloys and their advantages.

Provide samples of stainless steel and aluminum for trainees to examine.

Explain that seams are used to join the edges of the metal.

Show Transparency 6 (Figure 3). Describe a grooved lock seam and explain how it is formed.

Describe a snap lock seam.

Provide samples of various types of seams for trainees to examine.

DID YOU KNOW?
The High Cost of Rust

Rust is expensive. Experts estimate that rust and corrosion on buildings, bridges, vehicles, and aircraft cost about $300 billion nationally.

One of the most common alloy steels is stainless steel, which contains between 11 percent and 26 percent chromium. The chromium hardens and toughens steel and makes the grain finer. Chromium also makes the steel less likely to stain (thus stainless). Sometimes nickel is also added to the chromium and steel to produce a type of nickel-chromium stainless steel.

Stainless steel is valuable for its strength, toughness, and corrosion resistance. It is often used to make counters and cabinets for hospitals and restaurants. It is also used to make ductwork, particularly in industrial applications.

2.3.0 Aluminum Alloys

Aluminum is the most plentiful metallic element in the Earth's crust. It is the most widely used **non-ferrous** metal, but it is never found in metallic form in nature.

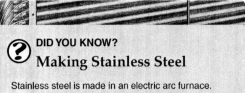

DID YOU KNOW?
Making Stainless Steel

Stainless steel is made in an electric arc furnace. Carbon electrodes make contact with recycled stainless scrap and alloys of chromium or other elements. A current passes through the electrodes, which increases the temperature and melts the scrap and alloys. This molten material then moves to an argon oxygen decarbonization (AOD) vessel. Here, carbon levels are reduced and final alloys are added to make the exact chemical mix wanted. Some of the material is hot-rolled or forged into its final form. Some material is cold-rolled to reduce the sheet thickness or to be made into rods and wire. Most stainless steels receive a final annealing (a heat treatment that soften the structure) and pickling (an acid wash that removes furnace scale).

Pure aluminum is soft and lacks strength, but it can be combined with small amounts of copper, magnesium, silicon, manganese, and other elements to form aluminum alloys. These lightweight, but strong, metals are easily bent and formed, so they are commonly used for ductwork. The following are additional advantages of aluminum alloys:

- Ductile and malleable
- Corrosion resistant
- Easy to install and maintain
- Excellent conductor of heat and electricity
- Recyclable

3.0.0 ◆ SEAMS (LOCKS)

Seams (or locks) are used to join the edges of the metal. They may be made by mechanical methods or by welding. The best method for joining the seams depends on the thickness of the metal, the kind of metal, the cost of fabrication, and the equipment available. Common seams include the grooved lock seam, snap lock seam, and the Pittsburgh lock.

3.1.0 Grooved Lock Seam

The grooved lock seam is also sometimes called the acme, or pipe lock (see *Figure 3*). The seam consists of two folded edges that are hooked together and locked with a grooving machine or hand groover.

The hand groover is a hardened steel tool with one end recessed to offset the grooved lock. It has a range of grooves ranging from $\frac{3}{32}$ to $\frac{19}{32}$ of an inch wide. Select a groover with a groove that is approximately $\frac{1}{16}$ of an inch wider than the width of the seam. Grooved seams are rarely used in metals heavier than 20 gauge.

3.2.0 Snap Lock Seam

The snap lock seam consists of a pocket lock and another edge that is inserted into the pocket. This edge is formed using a snap lock punch to raise buttons or bumps that produce a tighter fit inside the pocket.

Figure 3 ◆ Cross section of a grooved lock seam.

3.3.0 Pittsburgh Lock

The Pittsburgh lock is the most commonly used seam in the sheet metal shop. This seam consists of one edge formed into a pocket and a second edge that is bent and inserted into the pocket. The important feature of this seam is that after the bent edge is inserted into the pocket, the pocket edge is folded over to permanently lock the seam shut (see *Figure 4*).

This seam is used so often that a roll-forming machine called the Pittsburgh lock machine has been developed. The metal is inserted in one end of the machine, runs through a series of rolls, and comes out at the other end with the pocket lock completely formed. If no roll-forming machine is available, you must form the Pittsburgh lock on a brake. The allowance for the Pittsburgh lock formed on a brake is 1¼ inches.

4.0.0 ♦ CONNECTORS

Connectors are used to join individual pieces of duct. Connectors include the drive connector, S-slip, standing seam, standing S, standing drive slip, and Ductmate®.

4.1.0 Drive Connector

The drive connector is made in a two-step process. First, the edges of two pieces of metal to be joined are turned to form two pockets, each about a ½ inch wide. The drive is formed from a separate piece of metal, the edges of which are also turned, forming a sleeve. You will pull this sleeve over the folded edges of the pieces to be connected and drive it on with a hammer (see *Figure 5*). Drives are used to connect the sides of ductwork.

4.2.0 S-Slip

The S-slip is an S-shaped connector that forms two pocket locks for metal edges to slip into (see *Figure 6*). S-slips are used to connect the top and bottom edges of ductwork. The S-slip is sometimes also called a flat slip.

4.3.0 Standing Seam

The standing seam eliminates the need for additionally reinforcing a run of duct because this type of connector supplies its own reinforcement. This lock is easily made and is also used as a cross-seam on larger ducts (see *Figure 7*).

4.4.0 Transverse Duct Connector

The transverse duct connector is used to connect two sections of duct. Sometimes called a pocket lock, it is used in the same way as the S-slips and drive slips. The difference is that the finished transverse duct connector is not flush with the duct.

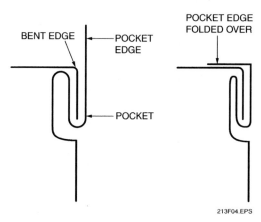

Figure 4 ♦ Cross section of a Pittsburgh lock.

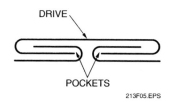

Figure 5 ♦ Cross section of a drive connector.

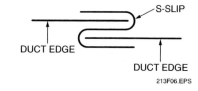

Figure 6 ♦ Cross section of an S-slip connector.

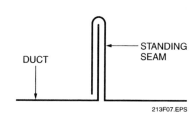

Figure 7 ♦ Cross section of a standing seam.

Show Transparencies 11 and 12 (Figures 8 and 9). Describe other commonly used connectors.

Identify the factors that cause air leakage in metal duct systems. Explain how to minimize air leakage.

Show Transparency 13 (Table 2). Discuss duct sealing requirements.

Identify the three main elements to the hanging system for HVAC ductwork.

Refer to *Appendices A* through *D* and discuss various requirements for hanging ductwork.

Discuss hanger spacing for straight ducts.

Provide various types of hangers and supports for trainees to examine.

See the Teaching Tip for Section 5.0.0 at the end of this module.

4.5.0 Other Connectors

Three other commonly used duct connectors include the standing S, standing drive slip, and Ductmate®. The standing S is an S-shaped connector with an additional hem reinforcement (*Figure 8*). The standing drive slip is a drive connector with this same type of additional reinforcement (*Figure 9*). Ductmate® is a patented duct connection system for sealing connecting duct sections. This system is used primarily to prevent air leaks in medium-pressure air systems. It also gives further reinforcement to larger ducts.

4.6.0 Air Leakage in Ductwork

An airtight duct system may be a desirable goal, but absolute air tightness is not possible. Seams and joints all allow some air to enter and exit the duct; this is taken into account when the system is designed. A certain amount of air leakage is expected and is directly related to the following factors:

- Operating pressure
- Square footage of the duct system
- Duct construction
- Sealing

To help keep air leakage at acceptable levels, the duct must be properly sealed at connections, joints, and seams. Installers use approved adhesives, gaskets, or tape systems either alone or in combination to seal the duct. Check the manufacturer's guidelines or follow your shop's standards for choosing and applying the appropriate sealing material.

Duct sealing requirements are placed into categories by pressure class (see *Table 2*). The duct system designer is responsible for selecting the pressure class and the seal class. This information is usually contained in the job specifications and engineering drawings.

5.0.0 ♦ HANGERS AND SUPPORTS FOR SHEET METAL DUCT

Once the ductwork is fabricated, joined, and sealed to specifications, it must be properly supported. There are three main elements to the hanging system for HVAC ductwork:

- The upper attachment to the structure
- The hanger
- The lower attachment to the duct

When planning the installation of hangers and supports, you must consider such things as proper spacing, material, and installation practices. The following sections present some construction code specifications for hanging straight ducts and riser supports. There are some specifications for minimum fasteners for **lap-joined** straps in *Appendix B*, maximum loads for single metal hangers in *Appendix C*, and strap sizes and spacings in *Appendix D*.

5.1.0 Hanger Spacing for Straight Duct

In straight duct sections, the joints are the weakest points, so support must be provided. SMACNA standards specify the allowable loads and maximum spacing for hangers, so be sure to consult these recommendations. These specifications vary

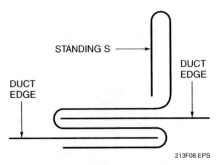

Figure 8 ♦ Cross section of a standing S connector.

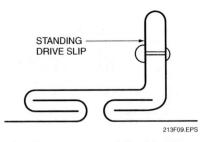

Figure 9 ♦ Cross section of a standing drive slip.

Table 2 Pressure Seal Requirements

Class	Requirements
Class C	**C**onnector joint sealed for 2 inches of water gauge pressure class and, when specified, for 0.5 inches of water gauge pressure class or for 1 inch of water gauge pressure class.
Class B	**B**oth connector joints and longitudinal seams sealed for 3 inches of water gauge pressure class.
Class A	**A**ll connector joints, longitudinal seams, and duct wall penetrations sealed for 4 inches of water gauge pressure class and greater.

according to hanger type, duct type and size, and type of installation.

5.2.0 Riser Supports

Rectangular risers (vertically running ducts) must be supported with angle irons or **channels** secured to the sides of the ducts with welds, bolts, sheet metal screws, or blind rivets. Place riser supports at one- or two-story height intervals (every 12 to 24 feet). Some risers are supported from the floor (see *Figure 10*). Reinforcing for the riser support is located below the duct joint. It may be necessary to install vibration isolators to eliminate duct-borne noise. These isolators keep vibration noise from transferring through the riser supports to surrounding walls, floors, or ceilings, which might act like sounding boards and amplify the noise.

Some risers are supported from the wall (*Figure 11*) and may be held in place by a band or by **angle brackets**. The allowable load per fastener and the number of fasteners to be used depend on the duct gauge and duct size. Ducts should be located against the wall or a maximum of 2 inches away from the wall. Each wall anchor must satisfy both the **tensile load** and the **shear load** of the supported member.

5.3.0 Rectangular Duct Hangers

Rectangular duct hangers include strap hangers (*Figure 12*) and trapeze hangers (*Figure 13*). Duct up to 60 inches wide must be supported with strap hangers. Duct more than 60 inches wide must be supported with trapeze hangers. Stacked trapeze hangers can also be used for stacked ducts (*Figure 14*). However, be sure that the hanger system chosen does not exceed the load rating for the installation. Minimum recommended sizes of hangers for rectangular duct are shown in *Table 3*.

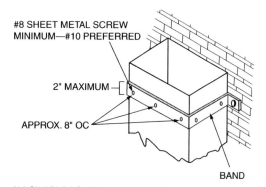

ALLOWABLE LOAD PER FASTENER:
25 Lbs – 26 to 28 Gauge
35 Lbs – 20, 22, 24 Gauge
50 Lbs – 16, 18 Gauge

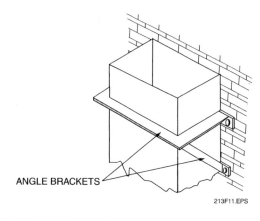

Figure 11 ◆ Wall-supported risers.

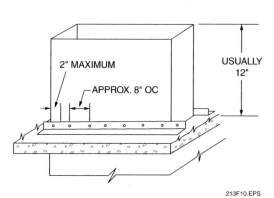

Figure 10 ◆ Floor-supported risers.

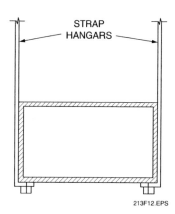

Figure 12 ◆ Strap hangers.

Show Transparencies 14 and 15 (Figures 10 and 11). Explain how risers are supported.

Show Transparencies 16-19 (Figures 12-14 and Table 3). Describe rectangular duct hangers and explain how they are used.

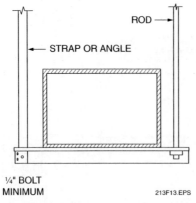

Figure 13 ♦ Trapeze hanger.

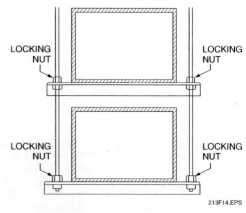

Figure 14 ♦ Stacked trapeze hangers.

Table 3 Types, Sizes, and Spacing of Hangers

P/2*	Hanger Pair Spacing	Strap Width Inches × Thickness	Wire or Rod
\multicolumn{4}{c}{Rectangular Metal Duct at 16-Gauge Maximum With Allowance for Insulation}			
30" maximum	10' apart	1" wide × 22 gauge	10 gauge (0.135")
30" maximum	8' apart	1" wide × 22 gauge	10 gauge (0.135")
30" maximum	5' apart	1" wide × 22 gauge	12 gauge (0.106")
30" maximum	4' apart	1" wide × 22 gauge	12 gauge (0.106")
72" maximum	10' apart	1" wide × 18 gauge	⅜" diameter
72" maximum	8' apart	1" wide × 20 gauge	¼" diameter
72" maximum	5' apart	1" wide × 22 gauge	¼" diameter
72" maximum	4' apart	1" wide × 22 gauge	¼" diameter
96" maximum	10' apart	1" wide × 16 gauge	⅜" diameter
96" maximum	8' apart	1" wide × 18 gauge	⅜" diameter
96" maximum	5' apart	1" wide × 20 gauge	⅜" diameter
96" maximum	4' apart	1" wide × 22 gauge	¼" diameter
120" maximum	10' apart	1½" wide × 16 gauge	½" diameter
120" maximum	8' apart	1" wide × 16 gauge	⅜" diameter
120" maximum	5' apart	1" wide × 18 gauge	⅜" diameter
120" maximum	4' apart	1" wide × 20 gauge	¼" diameter
168" maximum	10' apart	1½" wide × 16 gauge	½" diameter
168" maximum	8' apart	1½" wide × 16 gauge	½" diameter
168" maximum	5' apart	1" wide × 16 gauge	⅜" diameter
168" maximum	4' apart	1" wide × 18 gauge	⅜" diameter
192" maximum	10' apart	No straps allowed	½" diameter
192" maximum	8' apart	1½" wide × 16 gauge	½" diameter
192" maximum	5' apart	1" wide × 16 gauge	⅜" diameter
192" maximum	4' apart	1" wide × 16 gauge	⅜" diameter
193" and greater	Special analysis required		

The limits are for duct, insulation, and normal reinforcement and trapeze weights. No external loads are allowed in addition to this total weight.
* P = Total perimeter of the rectangular duct. P/2 = ½ of the duct perimeter.

6.0.0 ♦ INSTALLING REGISTERS, GRILLES, AND DIFFUSERS

Registers, grilles, and diffusers are the most visible part of an HVAC installation (see *Figure 15*). They are used as air inlets and outlets on walls or ceilings. Registers and grilles have **opposed-blade dampers** that control the volume of air. Diffusers are most often used as ceiling outlets for supply air; however, special diffusers can be used as wall outlets. Diffusers consist of vanes (blades) that discharge supply air in various directions. The vanes mix supply air with air that is already in the room. Vanes may be fixed or movable.

Before installing registers, grilles, or diffusers, review the project specifications for proper placement and any architectural or engineering details. Because these units must be installed flush with the finished walls, measure carefully to ensure that the ductwork does not extend beyond the finished surfaces. Follow these guidelines to install registers, grilles, and diffusers:

- For walls and ceilings made of drywall, complete the duct rough-in before the drywall is installed. Extend the duct beyond the framing studs to ⅛ inch less than the drywall width.
- Make sure the duct rough-in is level and square with the building.
- The duct or sleeve must have a flange (see *Figure 16*). The flange will hold the sheet metal screws used to fasten registers, grilles, and diffusers to the wall or ceiling. The flange must be wide enough to hold the screws, but not so wide that it is exposed on the finished side. If the flange is too wide, you will have to cut it. Cutting a flange against a wall or ceiling is very difficult.
- Install a wider flange on the raw end of the duct and against the studs to ensure a flush fitting and a solid surface for fastening the screws.
- Attach registers, grilles, and diffusers to the wall or ceiling with sheet metal screws. (The manufacturer supplies the screws for each unit.)
- Install the registers, grilles, and diffusers level and square with the building and flush with the finished wall or ceiling.

Special attention is required when installing diffusers in a rigid ceiling (a drywall ceiling with limited or no access). Follow the manufacturer's instructions for this type of installation. Note the following general guidelines:

- The ceiling opening is usually larger than the duct connection.

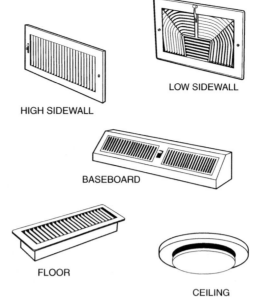

Figure 15 ♦ Registers, grills, and diffusers.

- The duct connection may be above the ceiling.
- The diffuser might have to be supported separately from the duct connection if the duct connection cannot support the weight of the diffuser.
- Support ceiling diffusers with ceiling radiation dampers from the structure.

6.1.0 Testing

To ensure that registers, grilles, and diffusers are properly installed, use the following checklist:

- Is the rough-in of the duct level and square with the building?
- Is the duct or sleeve flush with the finished wall?
- Does the duct have a flange for installing registers and diffusers with exposed sheet metal screws? Is the flange the correct width?
- At drywall construction, is the flange fastened to the studs and extended the width of the drywall sheet? (This helps prevent the duct from buckling in or bowing out.)
- Is the extra weight of the ceiling diffuser separately supported and not dependent for support on the duct alone?
- Is the opening in a rigid ceiling made according to the manufacturer's instructions?

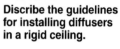

Show Transparency 20 (Figure 15). Explain the function of grilles, registers, and diffusers.

Show Transparency 21 (Figure 16). Review the installation guidelines for grilles, registers, and diffusers.

Discribe the guidelines for installing diffusers in a rigid ceiling.

Review the checklist of items that will ensure proper installation of grilles, registers, and diffusers.

Provide various types of grilles, registers, and diffusers for trainees to examine.

Show trainees how to join duct sections and fittings.

Have trainees practice joining duct sections and fittings. Note the proficiency of each trainee. This laboratory corresponds to Performance Task 1.

Have the trainees review Sections 7.0.0–10.1.1.

MODULE 03213-07 ♦ SHEET METAL DUCT SYSTEMS 13.9

Ensure that you have everything required for teaching this session.

Discuss the importance and advantages of insulating ductwork.

Define R-value.

Describe different insulation materials.

See the Teaching Tip for Section 7.0.0 at the end of this module.

Provide different types of insulation materials for trainees to examine.

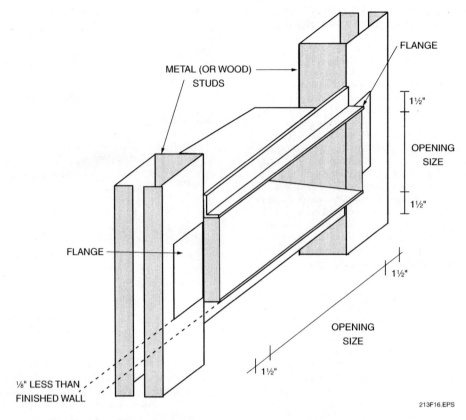

Figure 16 ♦ Installing registers, grilles, and diffusers.

7.0.0 ♦ INSULATION

The conditioned air and water that make living and working spaces comfortable may travel quite a distance from the air conditioning equipment to get to the room. The farther away from the equipment, the greater the chance of heat transfer. Heat naturally moves toward colder areas. Insulation is required to slow or minimize this heat transfer.

Insulation is also required for cold air and water systems. However, in addition to insulation, these cold systems also require a **vapor barrier** over porous insulation to prevent condensation and protect the insulation from moisture. Insulation also provides **sound attenuation;** it reduces noise generated in an HVAC system. Other characteristics and benefits of insulation are covered in the following sections that discuss the principles of thermal insulation and acoustic insulation.

The thermal resistance value, or R-value, is a measurement of the ability of insulation to retard the flow of heat. The American Society of Heating, Refrigerating, and Air-Conditioning Engineers (ASHRAE) has developed a standard that sets minimum recommended R-values for both commercial and residential construction. The standard, which has been published by ASHRAE, is called *ASHRAE/IES 90.1-1989 Energy Efficient Design of New Buildings Except Low-Rise Residential Buildings*.

Insulation is made of a variety of porous materials such as glass, rock or slag wool, calcium silicate, bonded perlite, vermiculite, and ceramics. It may also be made of closed-cell materials such as cork, foam rubber, polystyrene, and polyurethane. There is also reflective insulation, which has a smooth metallic or metallized surface.

13.10   HVAC ♦ LEVEL TWO

Instructor's Notes:

Insulation may be flexible, semi-rigid, rigid, or formed in place. Flexible and semi-rigid insulation comes in blanket or batt form and is available in sheets or rolls. A covering or facing material may be fastened to one or both sides. This facing may serve as a reinforcing vapor barrier, reflective surface, or surface finish.

Rigid insulation is available in blocks, boards, or sheets. Rigid insulation for pipes or curved surfaces is supplied in half sections designed to fit standard pipe sizes.

Insulation that is formed in place is available as a liquid or expandable pellets. This type of insulation is poured or sprayed in place, where it hardens to form rigid or semi-rigid foam insulation. Fibrous materials mixed with binders may also be sprayed in place.

7.1.0 Principles of Thermal Insulation

Thermal insulation is designed to slow the flow of heat energy. It is commonly used to control heat transfer from ductwork to the surrounding area in temperatures ranging from 0°F through 3,000°F and higher. Thermal insulation has the following characteristics:

- Conserves energy by reducing heat loss or heat gain through ducts, pipes, and buildings
- Controls surface temperatures of equipment for safety and personal comfort
- Allows for temperature control of a piece of equipment or a building
- Prevents vapor condensation at surfaces that have a temperature below the dew point of the surrounding environment
- Helps to control hazardous fire conditions that can be created by grease-laden air in a kitchen exhaust system

Depending on the type, thermal insulation may also have some additional advantages:

- Adds structural strength to a wall or ceiling
- Provides support for a surface finish
- Prevents condensation
- Reduces damage to equipment or buildings if they are exposed to fire or freezing temperatures
- Reduces noise and vibration

7.2.0 Principles of Acoustic Insulation

An HVAC system not only transmits air, but can also transmit noise. There are two types of potential noise in an HVAC system: system-generated noise (noise produced by HVAC equipment, air flowing through the system, and equipment vibration); and occupant-generated noise (people talking, televisions and radios playing, and so on).

The fan, or blower, is the main source of system-generated noise in the system. Part of the horsepower supplied to the fan radiates out as sound. The more powerful the fan, the noisier it will be. Manufacturers provide a rating, called the sound power level, that tells how much noise a fan will produce.

As air flows through the ductwork, it produces noise when it changes direction, when it flows through a damper, or when there is a change in the duct size. Under certain conditions, air can also produce noise as it flows through straight sections of duct.

Some motors in an HVAC system transmit vibrations and low-frequency noise to the structure to which they are attached. Anti-vibration mountings, therefore, are usually needed between such motors and the structure. These mountings, which may be made of springs or rubber, help to absorb the vibration and reduce the noise level.

Sheet metal ductwork can act as a speaker tube. For example, if one duct serves two rooms, sound can pass from one room through the duct into the other room. So without proper sound attenuation, this occupant-generated noise, called cross-talk, can get into the ductwork as easily as system-generated noise can get out. The direction of the airflow has little to do with the direction of the noise transmitted. Sound can be transmitted in either direction.

7.3.0 Fibrous Glass Duct Liner

Fibrous glass duct liner is made of glass fibers bonded with a thermosetting (heat-setting) resin.

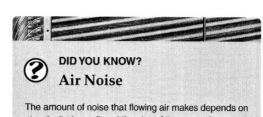

DID YOU KNOW?
Air Noise

The amount of noise that flowing air makes depends on air velocity (speed) and the size of the space through which the air flows. For instance, rolling the window up halfway or almost to the top while driving on the highway makes the wind sound higher and louder. In an HVAC system, air will produce sound in the same way. The more the airflow is restricted, the higher the sound is.

Describe different types of insulation.

Identify the characteristics of thermal insulation.

Discuss acoustic insulation.

Describe fibrous glass duct liner.

Provide a sample of fibrous glass duct liner for trainees to examine.

Identify the advantages of fibrous glass duct liner.

Explain applications where fibrous glass duct liner is not appropriate.

Identify requirements for the properties of fibrous glass duct liner.

Explain how fibrous glass duct liner is cut.

Two forms are available: flexible blankets supplied in rolls (*Figure 17*) and rigid boards supplied as sheets (*Figure 18*).

Fibrous glass duct liner is designed to do the following:

- Lessen noise generated by the HVAC system and by building occupants
- Reduce heat loss or gain through sheet metal duct surfaces
- Prevent water vapor condensation on the inside and outside of the duct

However, fibrous glass duct liner is not recommended for the following applications:

- With equipment that does not include automatic temperature controls
- In any application where an operating temperature of 250°F may be exceeded
- In kitchens or fume exhaust ducts
- In ducts carrying solids or corrosive gases
- With coal- or wood-fueled equipment

Figure 17 ♦ Flexible fibrous glass duct liner.

Figure 18 ♦ Rigid fibrous glass duct liner.

- In any application where the duct may come in contact with water (such as cooling coils, humidifiers, or evaporative coolers), unless the duct is protected from the water source
- Inside fire damper sleeves
- Near high-temperature heating coils unless the duct is protected from heat radiation
- In systems supplying operating rooms, delivery rooms, recovery rooms, nurseries, isolation rooms, and intensive care units

7.3.1 Property Requirements

ASTM International has established property requirements (*ASTM C1071*) for fibrous glass duct liner in the following areas:

- *Acoustical performance* – A standard for how well sound is attenuated.
- *Corrosiveness* – A standard for limiting corrosion where duct liner contacts sheet metal.
- *Moisture vapor absorption* – A standard to limit moisture absorption.
- *Fungi resistance* – A standard to limit the growth of fungi and bacteria.
- *Temperature resistance* – A standard to prevent burning, glowing, smoking, smoldering, or delamination of the liner material.
- *Erosion resistance* – A standard to prevent the insulation from breaking away, flaking off, or delaminating.
- *Odor emission* – A standard to prevent the transmission of objectionable odors from the insulation.
- *Flame and smoke rating* – A standard that defines how quickly flames and smoke will spread when the materials are ignited.
- *Thermal conductivity* – A standard that defines the rate at which heat is conducted.

7.3.2 Cutting

Fibrous glass duct liner may be cut using one of the following methods:

- By hand, using a utility knife or insulation knife (see *Figure 19*). Knives that are specially designed to cut duct liner are available from several manufacturers.
- By machine, using automatic coil line equipment (see *Figure 20*). Many different types of coil line equipment exist. Be sure to follow the manufacturer's cutting recommendations for the specific equipment you are using.
- By computer, which can be programmed to cut rectangular shapes for straight duct sections or special shapes for fittings.

Figure 19 ♦ Cutting fibrous duct liner by hand.

Figure 20 ♦ Cutting fibrous duct liner on an automatic coil line.

CAUTION
When using mechanical or computer-operated equipment, be sure to follow the manufacturer's operating instructions and safety guidelines.

7.3.3 Adhesives and Mechanical Fasteners

Fibrous glass duct liners are fastened inside sheet metal ductwork with special adhesives, which must meet the property requirements of *ASTM C916*. These adhesives are also used to repair minor damage to the liner surface and to coat exposed edges of the liner.

These adhesives are either solvent-based or water-based and are classified by ASTM according to their flammability when wet and dry. Vapors from solvent-based adhesives can be dangerous. However, solvent-based adhesives generally become tacky and bond more quickly than water-based adhesives. Adhesives can be applied by roller coating, spraying, or brushing.

WARNING!
Vapors from some solvent-based adhesives may be explosive. Never apply solvent-based adhesives near open flames, welding operations, or other potential ignition sources. You must always follow the manufacturer's recommendations regarding fire hazards, proper ventilation, and storage.

Mechanical fasteners are required in addition to adhesives when installing fibrous duct glass liners. Three types of fastener are commonly used:

- *Mechanically secured* – These are hardened steel fasteners that are driven into the duct (see *Figure 21*).
- *Weld-secured* – These fasteners either have integral (*Figure 22*) or press-on (*Figure 23*) heads. The welding equipment used to secure these fasteners must be carefully adjusted to get a solid weld without burning through the liner material.

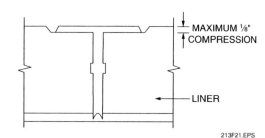

Figure 21 ♦ Hardened steel fastener.

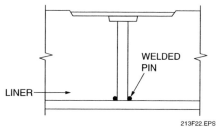

Figure 22 ♦ Weld-secured fastener with integral head.

Emphasize the importance of following the manufacturer's safety guidelines when using equipment.

Explain how adhesives are used to install fibrous glass duct liner.

Discuss hazards and safety precautions needed when using adhesives.

Show Transparencies 22-24 (Figures 21-23). Explain how mechanically secured and weld-secured fasteners are used to install fibrous glass duct liner.

Show Transparency 25 (Figure 24). Describe adhesive-secured fasteners.

Show Transparency 26 (Figure 25). Discuss spacing requirements for mechanical fasteners.

Review guidelines for inspecting the completed duct system.

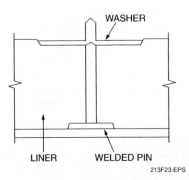

Figure 23 ♦ Weld-secured fastener with press-on head.

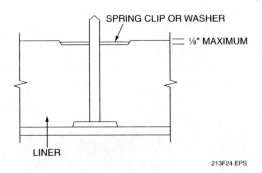

Figure 24 ♦ Adhesive-secured fastener.

- *Adhesive secured* – These fasteners have a large base to hold the adhesive (*Figure 24*). However, some adhesives lose their gripping power as they age, so this type of fastener may not be appropriate for all applications. Follow the manufacturer's recommendations.

Fastener heads should have a minimum area of 0.75 inches and a minimum thickness of 0.01 inches. They should also have either cupped or beveled heads. Fastener heads must not compress the insulation more than ⅛ inch.

Correct spacing of mechanical fasteners is very important. Mechanical fasteners must have the following characteristics:

- Be as corrosion-resistant as G60 galvanized steel when installed
- Indefinitely sustain a 50-pound tensile dead-load test perpendicular to the duct wall
- Not affect the fire hazard classification of the duct liner and adhesive
- Not damage the duct liner when applied as recommended
- Not cause leakage in the duct
- Be able to be installed perpendicular to the duct surface
- Project only nominally into the airflow

- Be the correct length for the specified duct liner thickness
- Not compress the duct liner more than ⅛ inch based on nominal insulation thickness

In addition, mechanical fasteners must be located and spaced regardless of airflow direction as shown in *Figure 25*. The expected velocity of air moving through the duct affects the spacing of mechanical fasteners.

7.3.4 Inspecting

After the lined duct system is completely installed, conduct a final inspection. This inspection should include the following:

- Check all registers, grilles, and diffusers to make sure that they are clean and free of construction debris.
- Using the manufacturer's instructions, check all filters.
- Cover air supply openings with temporary filters before starting up the system to catch any loose material that may still be in the ductwork.
- Turn the HVAC system on and allow it to run until steady operation is reached.
- Remove the temporary filters and any loose material caught by the filters.

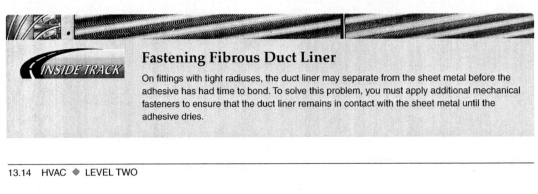

Fastening Fibrous Duct Liner

On fittings with tight radiuses, the duct liner may separate from the sheet metal before the adhesive has had time to bond. To solve this problem, you must apply additional mechanical fasteners to ensure that the duct liner remains in contact with the sheet metal until the adhesive dries.

13.14   HVAC ♦ LEVEL TWO

Instructor's Notes:

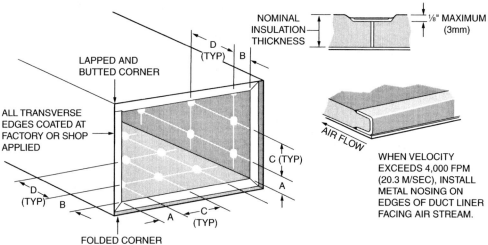

Dimension		Velocity, fpm (m/sec)	
		0–2,500 (0–12.7)	2,501–6,000 (12.7–30.5)
A	From corners of duct	4" (100mm)	4" (100mm)
B	From transverse end of duct liner	3" (75mm)	3" (75mm)
C	Across duct width, OC, minimum 1 per side	12" (300mm)	6" (150mm)
D	Along duct length, OC, minimum 1 per side	18" (450mm)	16" (400mm)

Figure 25 ◆ Mechanical fastener spacing.

Describe flexible blanket insulation.

Review safety guidelines for working with fiberglass.

Show Transparency 27 (Figure 26). Explain how the staple flap is used to install blanket insulation.

Provide flexible blanket insulation for the trainees to examine.

7.4.0 Fiberglass Blanket Insulation (Wrap)

Fiberglass blanket insulation is available in two forms: flexible blankets sold in rolls and rigid boards sold as sheets. Fiberglass blanket insulation is also called duct wrap, flexible insulation, or flexible blanket insulation. It is used to insulate HVAC duct systems in buildings and to insulate the exterior of plenums, fittings and valves, and surfaces that require temperature control. Blanket insulation is installed mainly above ceilings or behind finished walls.

CAUTION

To prevent fiberglass fibers from transferring to your other clothes, wash work clothing separately, then rinse your washing machine thoroughly before using it again. If you have a lot of glass fibers on your clothes, soak them first, then wash them.

Fiberglass blanket insulation is made of fine, inorganic glass fibers that are bonded with a heat-setting resin. It is sold in large rolls that are tightly compressed in a plastic bag, so it may not appear to be as thick as the package indicates. However, the insulation will return to the stated thickness when it is unrolled. It comes in ¾-pound, 1-pound, and 1½-pound densities (sometimes called Type 75, Type 100, and Type 150). Density is measured in cubic feet. For example, a density of 1 pound means that 1 cubic foot of insulation weighs 1 pound. Thickness ranges from 1 to 3 inches with 1½ inches by ¾-pound density as a common selection. The roll length depends on the thickness. The most common lengths are 100 feet for a 1½-inch thickness and 75 feet for a 2-inch thickness. The standard width is 48 inches.

Manufacturers also apply a facing material on one side of the insulation blanket. This facing material is designed to meet heat, sound, condensation, and fire-resistance requirements. Facings include Foil Scrim Kraft (FSK), vinyl, and All Service Jacket (ASJ). They usually have a 2-inch staple flap along the edge to assist in installation (see *Figure 26*).

Show Transparency 28 (Figure 27). Explain how blanket insulation is installed.

Explain how to cut blanket insulation.

Emphasize the importance of wearing appropriate personal protective equipment when working with fiberglass.

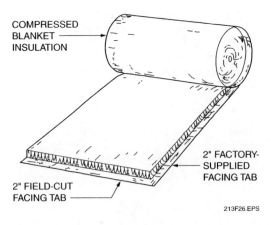

Figure 26 ◆ Fiberglass blanket insulation with staple flaps.

In a typical commercial HVAC system, the ducts are insulated after the ductwork is installed, tested, and inspected, and before the ceiling grid is in place. Insulate the trunk line or main line of duct first and then move on to the **runouts** or branches. A simplified drawing of a typical duct system is shown in *Figure 27*. Parts of some duct systems may be lined internally with insulation. Lined duct may or may not require external insulation.

7.4.1 Measuring and Cutting

To measure and cut fiberglass blanket insulation, use a tape measure, a utility knife and blades, a straightedge, and a large piece of heavy cardboard. Because fiberglass may irritate your eyes and skin, wear safety glasses, gloves, long pants, a long-sleeved shirt, and a hat. In addition, NIOSH recommends that when the level of dust is not known a NIOSH-rated and approved respirator be worn.

CAUTION

Always wear proper personal protective equipment (PPE), long pants, and long sleeves when working with fiberglass to prevent irritation to the eyes and skin.

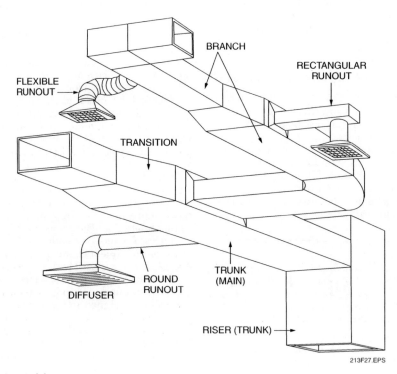

Figure 27 ◆ A typical duct system.

The following instructions contain information for rectangular, round, and oval duct. To measure and cut fiberglass blanket insulation, follow these steps:

Step 1 Choose an area in which to measure and cut the insulation. It should be out of the way of foot traffic and near the installation. Place heavy cardboard or plywood under the insulation to prevent floor damage.

Step 2 Roll out several feet with the facing side down. Save the plastic bag to dispose of scraps.

Step 3 Measure the distance around the straight section of rectangular duct. If using a 1½" thick blanket, add an 8" stretchout to this measurement. This new figure is the length of insulation you will cut to wrap around the duct. The additional 8" will allow for the thickness of the insulation at each corner of the duct plus a 2" facing (staple flap) tab. If the insulation is 2" thick, add a 10" stretchout to the length. An example of how to measure a 1½" thick blanket to fit a 12" × 6" rectangular duct is shown in *Figure 28*.

For round or oval duct, first measure the circumference. If the insulating material is 1½" thick, add 10" to the circumference to find the proper length to cut. If the insulation is 2" thick, add 12" to the circumference.

Roof Openings

During construction, some parts of a building may still be unfinished when you install the insulation. For example, openings may have been left in the roof for equipment installation. Be alert to any roof openings over the ducts you install. Because these areas are exposed to the weather, do not install insulation until the roof openings are closed.

Step 4 Check the end of the roll to ensure that it is straight and square. Some ends may become damaged or separated in packing. Use the utility knife and straightedge to cut off any damaged portion.

Step 5 Measure and mark both edges of the unrolled insulation (see *Figure 29*).

Step 6 Align your straightedge with the marks and cut from the far edge to the near edge. Make a clean cut through both fiberglass and facing (see *Figure 30*).

DID YOU KNOW?
How Fiberglass Insulation Works

Heat naturally moves toward cold by conduction or convection. For example, a stainless steel pan is an excellent conductor. It allows heat to travel freely from its source (a stove-top burner) to the food in the pan.

The glass fibers in fiberglass insulation, on the other hand, are poor conductors, so fiberglass insulation is an excellent material for controlling heat loss. Fiberglass insulation is made up of fine strands of glass randomly held together in a mat. Tiny pockets of air lie between each of the fibers. Heat can travel through these air pockets by convection. But because the glass fibers are randomly placed, no direct path exists. The longer it takes the heat to travel through the mat, the more effective the insulation is at controlling heat loss. This is why you must not overly compress the insulation blanket—you will shorten the convection path and reduce the insulation's effectiveness.

Show Transparencies 29-31 (Figures 28-30). Review the procedure for measuring and cutting blanket insulation.

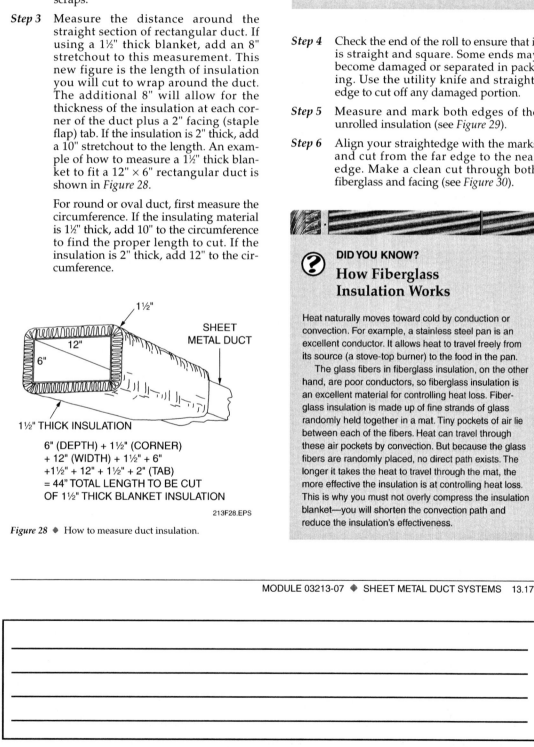

6" (DEPTH) + 1½" (CORNER)
+ 12" (WIDTH) + 1½" + 6"
+1½" + 12" + 1½" + 2" (TAB)
= 44" TOTAL LENGTH TO BE CUT
OF 1½" THICK BLANKET INSULATION

Figure 28 ♦ How to measure duct insulation.

Show Transparency 32 (Figure 31). Explain how to field cut a tab on blanket insulation.

Show trainees how to measure and cut blanket insulation.

Describe different methods for installing insulation.

Show Transparency 33 (Figure 32). Explain the procedure for installing insulation.

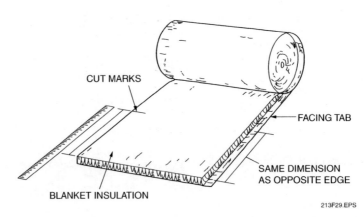

Figure 29 ◆ Measuring and marking the cut.

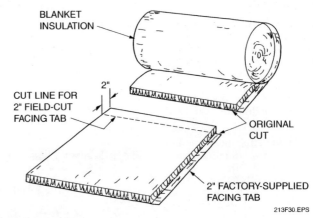

Figure 30 ◆ Fiberglass blanket after cutting.

Step 7 Create a facing tab. Move the straightedge 2" in from the cut. Cut through the fiberglass but not through the facing material. Remove the 2" wide strip of fiberglass. Use your knife to scrape the fiberglass loose from the facing (see *Figure 31*).

Step 8 Place scraps in the plastic bag for proper disposal. The piece is now ready for installation.

7.4.2 Installing Fiberglass Blanket Insulation-Butt Joint With Stapled Tab

There are several acceptable methods for attaching fiberglass blanket insulation to ductwork, including the butt joint with stapled tab method, the adhesive method, and the stitch method.

Another method, the wire method, may be required for some installations. For this method, wire is tied around the insulation blanket on 8" centers. However, this method is not recommended because the wire may puncture the vapor barrier and crush the insulation. Always follow the method required in the job specifications.

When attaching fiberglass blanket using the butt joint with stapled tab method, use flare-head staples and an outward-clinch staple gun to fasten the ends of the insulation. Follow these steps:

Step 1 Place the cut piece of insulation over the top of the duct with the facing tab toward you (see *Figure 32*).

Step 2 Hold the top end of the blanket and reach under the duct for the other end.

Instructor's Notes:

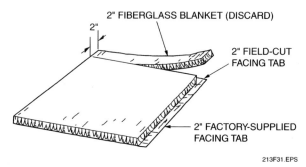

Figure 31 ♦ Field-cut facing tab.

Step 3 Pull the end of the blanket snugly around the duct. Do not compress the blanket more than 50 percent at the corners. Check for proper fit.

Step 4 Hold the facing tab down snugly on the blanket underneath. Staple in the middle of the facing tab. Check the underside of the duct to ensure the insulation is snug.

Step 5 Staple every 2" to 4" as required to securely fasten the facing tab. Staple from the center of the piece out to the edges (see *Figure 33*). You must install the staples with a flared head or outward clinch so that they do not penetrate the facing (see *Figure 34*).

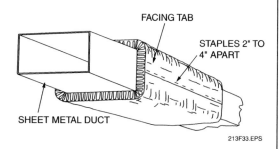

Figure 33 ♦ Insulation with staples.

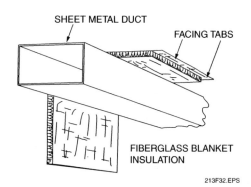

Figure 32 ♦ Applying the insulation to the duct.

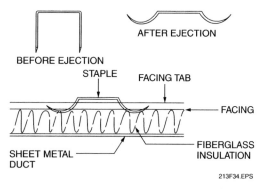

Figure 34 ♦ Proper staple application.

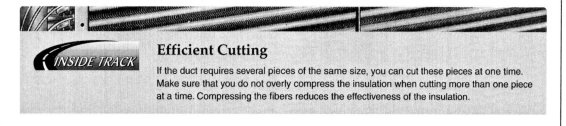

Efficient Cutting

If the duct requires several pieces of the same size, you can cut these pieces at one time. Make sure that you do not overly compress the insulation when cutting more than one piece at a time. Compressing the fibers reduces the effectiveness of the insulation.

MODULE 03213-07 ♦ SHEET METAL DUCT SYSTEMS 13.19

Show Transparencies 36 and 37 (Figures 35 and 36). Explain how to install insulation around projections and hangers.

Show Transparency 38 (Figure 37). Explain how to split insulation around branches.

Step 6 Install the next piece of insulation in the same way, butting it snugly against the first piece. After fastening the insulation with staples, pull the facing tab over the first piece of insulation and staple this tab around the duct. The staple ends should not penetrate through the facing.

NOTE
You can use tape tabs to hold down the nonflap side of the insulation against the ductwork. This will enable you to use both hands to control the flapped edged of the insulation and the staple gun.

When installing blanket insulation around duct hangers, split the butt end of the blanket at the hanger to allow the two split pieces to be pulled over the duct. Secure with staples. Seal the facing split with manufacturer-approved tape (see *Figure 35*).

Ductwork may have projections that stand out 1½" to 2" from the surface of the duct. These projections include standing seams, reinforcement flanges, and ribs (see *Figure 36*).

CAUTION
The corners of projections are sharp. Avoid puncturing the facing on these corners. If you puncture the facing, apply a patch made of insulation and facing over the puncture and seal it.

Install the insulation blanket so that it blouses over the duct and is taut but not compressed at these projections.

Step 7 Continue installing sections of blanket until you reach a branch connection, a reducer connection, or a tee. Do not cut a circular or rectangular piece from the blanket. Instead, split the blanket around the branch to allow it to flare out (see *Figure 37*).

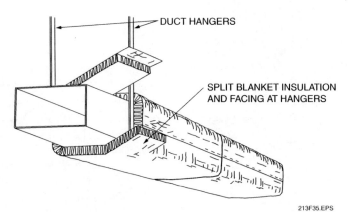

Figure 35 ♦ Installing the blanket around hangers.

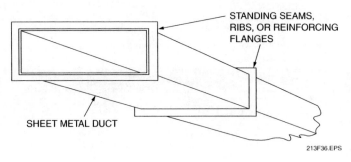

Figure 36 ♦ Projections on ductwork.

13.20 HVAC ♦ LEVEL TWO

Instructor's Notes:

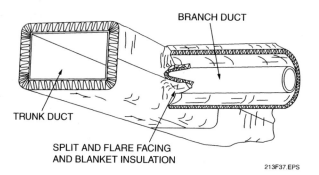

Figure 37 ♦ Splitting the blanket around a branch.

Install the blanket around fittings (offsets, elbows, transitions, reducers, and so on) by working from the largest duct to the smaller runouts. Wrap the insulation around the largest section of duct first, split it at the corners, tuck it, and secure it by pulling snugly on the smaller sections (see *Figure 38*).

Install blanket insulation on the branch lines and runouts beginning at the trunk line. The insulation must snugly butt or overlap the trunk line insulation. Split the facing tab at the corners to allow the tab to flare over the trunk line insulation. Staple the facing tab to the trunk line insulation (see *Figure 39*).

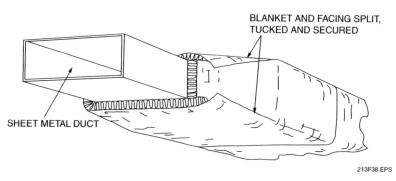

Figure 38 ♦ The blanket facing is split, tucked, and secured.

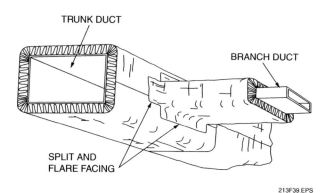

Figure 39 ♦ Insulating a branch duct.

Show Transparencies 39 and 40 (Figures 38 and 39). Explain how to install insulation around fittings and branches.

Explain how to complete the installation.

Show Transparency 41 (Figure 40). Review the procedure for using the adhesive method to install insulation.

Show Transparency 42 (Figure 41). Review the procedures for using the stitch method to install insulation.

Explain how to seal the insulation using insulation tape.

Overlap any pre-installed flexible duct insulation. Continue the insulation to the diffuser. Cut, fit, and seal the insulation to the back of the diffuser. Seal all insulation ends at fire dampers, connections, and so forth with manufacturer-approved tape. Make sure the tape overlaps to a metal surface. Rub the tape firmly to ensure a complete seal.

> **CAUTION**
> Do not overly compress the insulation during installation. Compression reduces the insulation's effectiveness.

7.4.3 Adhesive Method

Job specifications may require that the blanket insulation be cemented to the duct with a solvent-based adhesive. Follow these steps:

Step 1 Read the safety precautions on the adhesive label. They may require you to wear a respirator, to work in a well-ventilated area, or both.

Step 2 Make sure no one in your work area is using equipment that produces sparks or flames. Solvent-based adhesives are flammable when wet.

Step 3 Cover finished floors to protect the finish from adhesive drips and splatters.

Step 4 Brush a coat of adhesive on the top edge of the duct.

Step 5 Place the blanket insulation on the top edge of the duct and press it into place.

Step 6 Brush on strips of adhesive spaced about 6 inches apart around the sides and bottom of the duct (see *Figure 40*).

Step 7 Pull the blanket snugly around the duct and press into place.

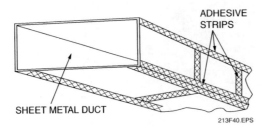

Figure 40 ♦ Brushed-on adhesive strips.

> **WARNING!**
> Solvent-based adhesives are flammable when wet. Make sure that there are no nearby sparks or flames when applying them.

7.4.4 Stitch Method

With this method, you will measure the duct as described earlier in this section, but you must add an extra 2 inches to the measurement. Follow these steps:

Step 1 Pull the insulation blanket snugly around the duct.

Step 2 Fold the two horizontal edges outward and staple along the edge every 2 to 4 inches (see *Figure 41*).

7.4.5 Sealing

You must seal the seams, joints, or facing tabs on blanket insulation with manufacturer-approved tape or adhesive. Some applications require that you also apply vapor barrier **mastic** to joints, seams, and staple penetrations. Review the job specifications for sealing requirements and recommended materials.

To use insulation tape, follow these steps:

Step 1 Measure and cut the amount of tape needed.

Step 2 Peel the protective paper contact strip from the tape.

Step 3 Press the tape down over the staples and the facing joint. Seal any punctures or tears in the facing with tape.

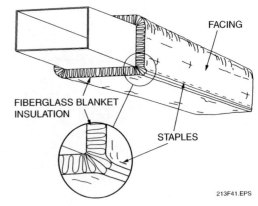

Figure 41 ♦ The stitch method.

13.22   HVAC ♦ LEVEL TWO

Step 4 Seal the facing-to-metal surfaces at termination points, such as air handling units, diffuser backs, and fire dampers. Rub the tape firmly to ensure a complete seal.

To use adhesive, follow these steps:

Step 1 Read the label for the manufacturer's instructions regarding application, drying time, and safety precautions. Wear recommended appropriate personal protective equipment.

Step 2 Brush the adhesive under the facing tab.

Step 3 Rub the tabs firmly to ensure a complete seal.

Step 4 Staple the blanket insulation or apply a quick-drying contact adhesive.

To use vapor barrier mastic, follow these steps:

Step 1 Read and follow the manufacturer's instructions and safety precautions. Wear recommended appropriate personal protective equipment.

Step 2 Protect finished floors and equipment from drops and splatters of mastic.

Step 3 If the manufacturer specifies glass-fiber mesh as reinforcement, fasten the mesh in place by stapling or gluing according to the manufacturer's instructions.

Step 4 Brush or trowel the mastic over the seams and staples. Apply the mastic to the hardest-to-reach areas first and work from the furthest point to the nearest. Do this to avoid reaching over wet mastic. Cover the mesh completely with the mastic.

Step 5 Seal all penetrations of the facing (such as duct hangers) tightly.

7.4.6 Mechanical Fasteners

Mechanical fasteners are used on the bottom surface of large ducts to keep fiberglass blanket insulation from sagging. Fasteners include self-adhesive stickpins, cemented stickpins, weld pins, and cupped-head weld pins.

Install cupped-head weld pins as you apply the insulation to the duct or after you have applied several pieces. When using cupped-head weld pins, follow these steps:

Step 1 Press the pins through the insulation and facing.

Step 2 Seal over the head with vapor barrier mastic or tape to ensure a complete vapor seal.

Install cemented stickpins and weld pins before applying the insulation to the duct. First cement the pins to the duct. Use a heavy-bodied, general-purpose adhesive. Follow these steps:

Step 1 Cut a piece of cardboard about 18" × 18".

Step 2 Stick several fastening pins into the cardboard with the anchor bases up (see *Figure 42*).

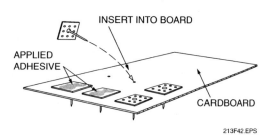

Figure 42 ◆ Fasteners set in cardboard.

Explain how to seal the insulation using adhesive and vapor barrier mastic.

Explain how mechanical fasteners are used to install insulation.

Describe the procedures for using cupped-head weld pins.

Show Transparency 43 (Figure 42). Review the procedure for using cemented stickpins and weld pins.

Provide mechanical fasteners for trainees to examine.

Insulation Tape

Manufacturers supply tape that has been specially designed for use with insulation and that meets the temperature requirements of the installation. The adhesive part of the tape is covered with a protective paper contact strip. Always read the manufacturer's specifications for any application restrictions (such as temperature).

MODULE 03213-07 ◆ SHEET METAL DUCT SYSTEMS 13.23

Show Transparencies 44 and 45 (Figures 43 and 44). Explain how to secure the fasteners to the duct surface.

Explain how to apply the insulation to the fasteners.

Show Transparency 46 (Figure 45). Discuss fastener placement.

Show trainees how to install insulation on ductwork.

Describe dampers and explain their functions.

Provide dampers for the trainees to examine.

Step 3 Apply adhesive to the entire base of each fastener. Cover each base with an even coat about 3/16 of an inch thick.

Step 4 Place the fastener on the duct surface and press firmly into position (see *Figures 43* and *44*).

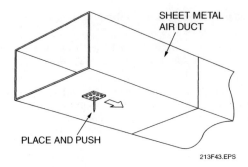

Figure 43 ♦ Fastener applied to sheet metal duct.

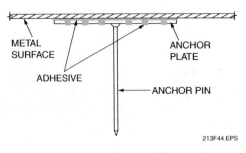

Figure 44 ♦ Close-up view of cemented fastener.

Wiggle the fastener on the surface slightly to set the adhesive. Some adhesive will press out through the holes in the base. Next, apply the insulation. Follow these steps:

Step 1 Press the pins into place. They should penetrate both the insulation and the facing.

Step 2 Place a self-locking washer or clip on the pin. Place the washer snug against the insulation. Do not press the washer completely to the duct surface.

Step 3 Cut the protruding pin flush with the washer using end-cutting nippers.

Step 4 Place vapor barrier mastic or tape over the washer to ensure a complete vapor seal.

Job specifications may require you to install these fasteners on the bottom surface of all ducts more than 24 inches wide. Generally, a row of fasteners spaced every 18 inches in the middle of a duct that is 24 to 36 inches wide is adequate (see *Figure 45*).

If the duct is more than 36 inches wide, two or more rows of fasteners are required. Note that if you use adhesive to cement the insulation blanket to the duct, mechanical fasteners usually are not required. Always check the job specifications.

8.0.0 ♦ DAMPERS AND ACCESS DOORS

Dampers are made up of a frame and movable blades, and they act like gates within the HVAC system. They regulate air distribution. Specially designed safety dampers control smoke and flames. Access doors must be installed wherever access to dampers or other components within the ductwork is needed.

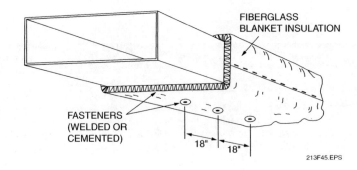

Figure 45 ♦ Fastener placement.

Instructor's Notes:
___
___
___

8.1.0 Manual Volume Dampers

Manual volume dampers (MVDs) may be round, oval, rectangular, or square. They may be made of galvanized steel, aluminum, or stainless steel. Depending on the overall size, an MVD may have a single blade or multiple blades.

Multiple-blade dampers are made in two basic types, opposed blade and parallel blade. In the opposed-blade damper, each blade moves in a direction opposite to the blade next to it (see *Figure 46*). This design allows for more precise airflow control than can be achieved with the parallel-blade damper. The parallel-blade damper (*Figure 47*) works like a venetian blind. This type of damper is a good choice where two-position operation—fully opened or fully closed—is necessary.

8.1.1 Installing Manual Volume Dampers

Before installing MVDs, review the project specifications and the architectural or engineering drawings to determine the proper location. The specifications may state, "MVDs are required at each branch duct or at each reheat coil." Generally, you must locate MVDs in branch ducts as close to the main trunk duct as possible. Sometimes, a wall or another duct may block one side of the duct where you must install an MVD. In this case, always locate the shaft for the damper quadrant on the unblocked side. To install an MVD, follow these steps:

Step 1 Insert the damper into the duct so that you can view the operation of the blades.

Step 2 Install the screws in the frame behind the travel of the blade while it is opening or closing. The blade must travel freely from the open to the closed position and vice versa.

Step 3 Extend the shaft through the ductwork to mount the damper quadrant with either pop rivets or sheet metal screws.

CAUTION

During installation it is important not to block any blade on the damper. If one blade is blocked, all the blades will be blocked.

Step 4 Using a hacksaw, cut a slot in the end of the damper. The slot should be parallel to (facing the same direction as) the damper blade.

Step 5 Leave the damper in the open position.

DAMPER QUADRANT

Figure 46 ♦ Opposed-blade volume damper with damper quadrant.

Figure 47 ♦ Parallel-blade damper.

8.1.2 Testing

To ensure that the damper is properly installed, use the following checklist:

- Can the damper rotate from fully open to fully closed without interference from mounting screws or rivets?
- Can the damper rotate from fully open to fully closed without interference from duct liner insulation?
- Does the slot in the damper shaft match the position of the damper blades?
- Is the damper in the open position with the damper handle locked?

Describe manual volume dampers.

Explain the procedures for installing manual volume dampers.

Review the checklist of items that will ensure dampers are properly installed.

Describe fire dampers and fire/smoke dampers and explain their functions.

Show Transparency 47 (Figure 48). Explain how fire dampers close automatically when triggered.

Show Transparency 48 (Figure 49). Discuss mounting considerations for fire dampers.

8.2.0 Fire Dampers and Fire/Smoke Dampers

Openings for ducts in walls and floors with fire-resistance ratings must be protected by fire dampers as required by local codes. Air transfer openings should also be protected. Smoke dampers are used for smoke management (smoke containment) or for smoke control.

At locations requiring both fire and smoke dampers, combination dampers that meet the requirements can be used.

Smoke detection systems monitor the smoke density that exists in a chimney or duct system. One method uses a photoelectric cell (photocell) and beam tube installed in the chimney or duct. A signal generated by the photocell indicates the amount of smoke. It is amplified for display on a local indicator or can be recorded using a recording device. If the amount of smoke detected exceeds a preset level, an alarm may be given. In addition to the alarm, the signal may also be applied to a control circuit where it initiates the shutdown of the system. Photocells and other electronic sensors are also used to control smoke dampers in duct runs. If the amount of smoke detected exceeds a preset level, a damper-holding device is tripped, causing the smoke damper to close.

Unless specially designed dampers are installed, an air delivery system can also carry fire or smoke. These special dampers include fire dampers, combination fire/smoke dampers, and ceiling radiation dampers. Usually these dampers are located near fire-rated walls and fire-rated ceiling assemblies. All fire and smoke dampers must have a UL label attached. This label certifies that the damper has been tested and approved for the specified use.

A fire damper closes automatically with a spring-loaded action (see *Figure 48*). By closing, it restricts the airflow that a fire needs to continue burning. It also restricts the passage of flames.

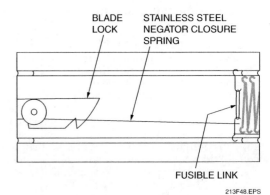

Figure 48 ◆ Spring-loaded fire damper.

The damper is controlled by a fusible link, which holds the damper open until the heat of a fire causes it to melt and close the damper. The fire rating for fire dampers is specified in hours and may not necessarily match the fire rating of the wall or partition.

A fire damper must stay in the wall, even if ducts or ceiling components on either side of the wall fall down. Therefore, breakaway connections are placed between the duct sleeve containing the damper and the ductwork on either side. Retaining angles are attached to the sleeve or damper frame, not to the wall. The connectors shown in *Figure 49* are acceptable for fire damper installation. To properly seal the joints, follow the manufacturer's written instructions and the local code.

A fire/smoke damper is controlled by a smoke detector and is designed to close automatically when smoke is present. This damper must be fail-safe; in other words, if it fails, it must fail in the closed position. This type of damper may accomplish other tasks in the duct system and still serve as an approved smoke damper.

The UL Mark

INSIDE TRACK

All fire and smoke dampers must have an Underwriters Laboratories, Inc. (UL) label. UL is a not-for-profit organization that tests and certifies a wide variety of products for public safety. Since its founding in 1894, it has certified nearly 16.1 billion products.

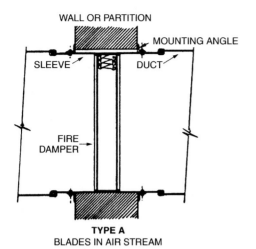
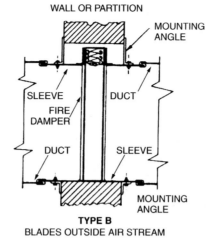

Discuss combination fire/smoke dampers.

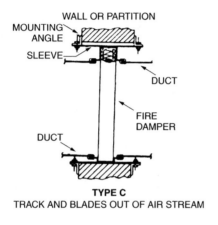

Figure 49 ◆ Mounting angles for fire dampers.

DID YOU KNOW?
NFPA

The National Fire Protection Association (NFPA) publishes standards for fire prevention in residential and commercial structures. The standards are revised annually. HVAC designers must use the latest edition of the standards.

Some HVAC installations require combination fire/smoke dampers (see *Figures 50* and *51*). These dampers combine fire- and smoke-detection capabilities. A ceiling radiation damper is designed for use in horizontal assemblies that are required to have a fire-resistance rating. It has a protective ceiling membrane that closes off a duct opening through the membrane in the event of a fire. This damper can detect heat radiating from the floor below.

Explain how to install and test fire and fire/smoke dampers.

Show Transparencies 49 and 50 (Figures 52 and 53). Describe duct access doors and explain their functions.

Figure 50 ♦ Combination fire/smoke damper rough-in at wall opening.

Figure 51 ♦ Combination fire/smoke damper installed in wall.

8.2.1 Installing Fire Dampers and Fire/Smoke Dampers

The wall contractor usually provides the damper openings. Floor and wall openings must be sized according to the manufacturer's written instructions and local codes.

Before installing these dampers, review the project specifications and the architectural or engineering drawings to determine the proper location. In addition, follow the manufacturer's instructions for each type of damper. Because these instructions vary from manufacturer to manufacturer, no specific installation steps are presented here. Never assume that one manufacturer's instructions will apply to a damper made by another manufacturer.

8.2.2 Testing

Follow the manufacturer's instructions for testing the fire damper. You can test the closing ability of most fire dampers by removing the fusible link that holds the damper in the open position. The local building inspector will examine each damper installation. The building inspector will also review the manufacturer's installation instructions to verify that the installation has been done correctly.

8.3.0 Duct Access Doors

Once fire and smoke dampers have been installed, access to them is necessary to test the damper installation and to replace the fusible link when needed. A duct access door must be installed in the duct at most fire damper and combination fire/damper locations. Duct access doors are available from a number of manufacturers in a variety of styles. Approved view ports (window-like panels) may be part of the door construction so that damper controls and fusible links can be inspected without opening the access door (see *Figure 52*). Framing, fastening, and hinge recommendations are shown in *Figure 53*.

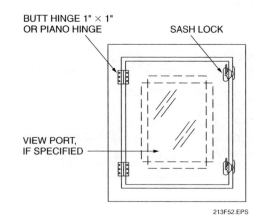

Figure 52 ♦ Access door.

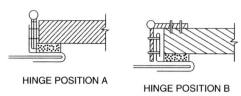
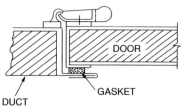
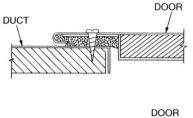
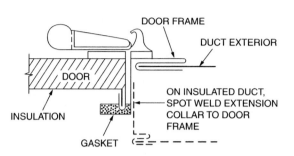
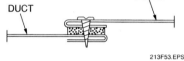

Explain how to install and test duct access doors.

Describe motorized control dampers.

Figure 53 ◆ Access door standards.

8.3.1 Installing Access Doors

Follow the manufacturer's installation instructions when installing access doors. Observe these general guidelines for installing access doors:

- Measure and cut the opening to the proper size.
- Use the hinge approved for the specific application.
- Make sure that the doors open fully and that they open out, not in.
- Ensure that nothing interferes with the door swing. Structural elements, pipe, electrical lines, and conduits cannot be in the way. Note that some of these elements may not yet be installed.
- Make sure the door fits into the opening snugly and that it is level and square with the duct.
- If the access door has a view port, make sure that it is aligned properly so that you can see the necessary parts of the damper through it.
- Lock the access door.

8.3.2 Testing

To ensure that you have properly installed the access doors, use this checklist:

- Do the doors fit snugly and are they level and square?
- Do the doors swing out? Are they clear of any structural elements that are in place or that will be in place in the future?
- Is the access door locked?

8.4.0 Motorized Control Dampers

Motorized control dampers are used in HVAC systems to control, or regulate, the mixture of the air. They are usually located near the air handler in the mechanical room to modulate the return air, outside air, and exhaust air.

This type of damper may be round, oval, rectangular, or square and may be made of galvanized steel, aluminum, or stainless steel. Depending on the overall size, a motorized control damper may have a single blade or multiple blades. In the multiple-blade type, the blades may be either opposed or parallel.

Explain how to install and test motorized control dampers.

Define takeoffs.

Show Transparency 51 (Figure 54). Explain how to install takeoffs.

Explain how to test takeoffs.

8.4.1 Installing Motorized Dampers

Before installing motorized dampers, review the project specifications and the architectural or engineering drawings to determine the proper location. Sometimes, a wall or another duct may block one side of the duct where you must install a motorized damper. In this case, locate the shaft for the motor on the unblocked side. To install a motorized damper, follow these steps:

Step 1 Insert the damper into the duct so that you can view the operation of the blades.

Step 2 Install the screws into the frame behind the travel of the blade while it is opening or closing.

Step 3 Make sure the duct liner does not interfere with the blade rotation.

Step 4 Note that the motor may be mounted in the airstream or on the outside of the duct. If the motor is mounted on the outside of the duct, you must extend the motor's shaft through the ductwork. The opening for the shaft should be just large enough to allow the shaft to rotate.

Step 5 Install an access door so that you can install the damper linkage or the motor. This door will also allow for inspection of the damper.

8.4.2 Testing

To ensure that the damper is properly installed, use the following checklist:

- Do the damper blades open and close properly?
- Do the blades rotate freely without interference from mounting screws?
- Do the blades rotate freely without interference from the duct liner?

9.0.0 ◆ TAKEOFFS

Takeoffs include tap-ins, hammerlocks, and spin-ins. Tap-ins or hammerlocks are square or rectangular fittings that connect branch ducts to the main trunk duct. Spin-ins are round fittings, usually with dampers, that connect ceiling diffusers to the main duct.

 CAUTION
You must remove the duct liner when installing a spin-in. If you don't, no air will flow through the branch line.

9.1.0 Installing Takeoffs

To install a takeoff, cut the opening so that the fitting fits into the duct snugly and smoothly. There must be no holes at the corners. Use a drill to make a pilot hole and then cut the opening with aviation snips.

A tap-in may be installed as an inside knockover or an outside knockover. These terms refer to the direction in which the tabs are hammered. To ensure an airtight, professional installation, hammer the tabs smoothly and tightly to the duct.

Tap-ins must be installed with a transition type takeoff, or clinch tee, with the tapered end upstream of the airflow (see *Figure 54*). Extractors may be required at tap-ins with the hinged end upstream of the airflow. The extractor arm must extend out of the duct on the downstream side of the airflow.

After installing the spin-in, ensure that the damper is open and locked in position. When installing spin-ins with scoops, ensure that the scoop is installed on the downstream side of the airflow. It is a scoop-shaped piece of metal attached to the spin-in. When properly installed, it directs the airflow down the branch line from the main line.

9.1.1 Testing

To ensure that tap-ins and spin-ins are properly installed, use the following checklist:

- Is the hole cut to the proper size?
- Are the corners neat and free of holes?
- Can the fitting be set easily and snugly into the duct?
- Are the tabs hammered over neatly, smoothly, and tight to the duct?

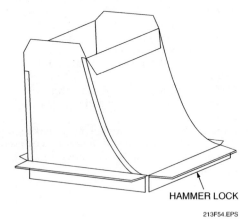

Figure 54 ◆ Clinch tee.

- Is the fitting installed with the proper airflow direction?
- Is the spin-in damper open and locked in position?
- Has the duct liner been removed?

10.0.0 ♦ ZONING ACCESSORIES AND COILS

Some air distribution accessories provide separate zones in the HVAC air distribution system. These accessories include variable air volume boxes (VAVs), constant air volume boxes (CAVs), fan power terminal units (FPTUs), and inline duct fans.

VAVs are designed to save operating costs by varying the amount of air instead of varying the temperature. They save fan energy and reheat energy and allow for the use of smaller air handlers and ductwork. These units may be mounted on ceiling support members or suspended from the ceiling. A VAV box with a hot water reheat coil is shown in *Figure 55*. Some models are designed for use in fire-rated ceiling assemblies.

CAVs provide a constant volume of air within an air distribution system. They are less expensive to install than VAVs and are suitable for applications where variable air volume is not necessary.

FPTUs redistribute excess heat from interior zones in a building to perimeter zones where it is needed. They also can respond to a thermostat demand for heat when the central fan system is shut down.

Inline duct fans are designed to correct air distribution problems. For example, a building may have rooms that are too hot in the summer and too cold in the winter or rooms whose temperature varies greatly depending on the time of the day. The fans boost airflow so that a greater volume of heated or cooled air can reach the problem area and produce a comfortable temperature. These fans are designed to fit into rigid round or square duct as well as into flexible duct.

Each of these accessories has a top side, a bottom side, an inlet opening, and a duct outlet opening. Units with electric or hot water coils also include locations for the electrical or pipe connections. An installation project may have many similar pieces of equipment, but the connection locations for each unit may be on different sides. Pay attention to details such as the location of the top side and the location of the connections to ensure proper installation and operation.

Show trainees how to install takeoffs and flexible duct.

Have the trainees practice installing takeoffs and flexible duct. Note the proficiency of each trainee. This laboratory corresponds to Performance Task 2.

Describe accessories used in zoned systems.

Figure 55 ♦ VAV box with hot water reheat coil.

Explain how to install and test zoning accessories.

10.1.0 Installing Zoning Accessories

Before installing zoning accessories, review the project specifications and the architectural or engineering drawings to determine the proper location. You must also read and follow the manufacturer's instructions for each type of unit. These instructions specify the number and type of hangers to use. Because these instructions vary from manufacturer to manufacturer, no specific installation steps are presented here. Never assume that one manufacturer's instructions will apply to a unit made by another manufacturer.

Observe the following general guidelines for installation of these units:

- Ensure that nothing interferes with the duct or with the pipe or electrical connections. Structural elements, pipe, electrical lines, and conduits cannot be in the way. Note that some of these elements may not yet be installed.
- Allow enough room for access doors. The doors for electric reheat coils are very large and require additional space.
- Select and use properly sized hangers. Follow the manufacturer's recommendations for type and size of hanger and proper spacing.
- Ensure that hangers do not interfere with the unit's connection points.
- Ensure that hangers do not interfere with access doors.
- Install vibration isolators (cushioned supports) on motor-driven equipment to minimize noise.
- Ensure that the filters installed on the units draw return air from the plenum (the space between a suspended ceiling and the structure above).

10.1.1 Testing

To ensure that these units are properly installed, use the following checklist:

- Are the hangers sized correctly?
- Is there proper clearance of connections at all points?
- Is the equipment level?
- Is the installation in proper alignment with the airflow?
- Will there be enough room, now and in the future, for the access doors to open fully?
- Are the filters installed on the units correctly and are they drawing return air from the plenum?

Instructor's Notes:

Review Questions

1. A metal that is not combined with any other metal is called a(n) _____.
   a. base metal
   b. alloy
   c. pure metal
   d. natural metal

2. Steel that is coated with zinc is said to be _____.
   a. stainless
   b. galvanized
   c. cold-rolled
   d. hot-rolled

3. The metal that gives stainless steel the ability to resist staining is _____.
   a. copper
   b. chromium
   c. brass
   d. iron

4. A seam consisting of one edge formed into a pocket and a second edge that is bent and inserted into the pocket is called a _____.
   a. Pittsburgh lock
   b. snap lock seam
   c. S-slip
   d. grooved lock

5. Which of the following connectors is used to connect two sections of duct?
   a. Drive
   b. Standing seam
   c. S-slip
   d. Transverse

6. The duct seal class that requires the tightest seal is Class _____.
   a. A
   b. B
   c. C
   d. D

7. Thermal insulation is designed to _____.
   a. transfer heat by convection
   b. transfer heat by conduction
   c. slow the flow of heat energy
   d. increase the flow of heat energy

8. An appropriate application for fibrous glass duct liner is _____.
   a. in kitchens or fume exhaust ducts
   b. inside fire damper sleeves
   c. in moist environments
   d. in operating temperatures under 250°F

9. Fire dampers are controlled by a(n) _____.
   a. spring-loaded lock
   b. fusible link
   c. heat-sensing coil
   d. internal thermostat

10. When installing tap-ins and spin-ins, you must cut the opening so that _____.
    a. the fitting can extend above the opening by ⅛ inch
    b. there are no holes at the corners
    c. there is no need for tabs or other fasteners
    d. the damper can stay in a closed position

Have trainees complete the Review Questions, and go over the answers prior to administering the Module Examination.

Summarize the major concepts presented in the module.

Administer the Module Examination. Record the results on Craft Training Report Form 200, and submit the results to the Training Program Sponsor.

Administer the Performance Test, and fill out Performance Profile Sheets for each trainee. If desired, trainee proficiency noted during laboratory sessions may be used to complete the Performance Test. Record the results on Craft Training Report Form 200, and submit the results to the Training Program Sponsor.

Summary

HVAC installers work with sheet metal in fabricating and assembling ductwork. Installers must understand basic layout and cutting techniques, as well as seam fabrication and the use of various connectors. Spacing hangers properly to support the system will ensure safe and efficient air distribution.

Ductwork must be properly insulated to provide comfort and energy savings. Insulation also prevents unwanted noise from moving through the system. Manufacturers design and make different types and thicknesses of insulation to suit a wide variety of applications. The installation and operating instructions for each accessory vary by manufacturer. Whether applying insulation with adhesives or mechanical fasteners, be sure to follow all manufacturer instructions and adhere to safety guidelines.

In addition to the duct and insulation, additional components may be installed in an HVAC system to perform various functions: regulating the air distribution, allowing access for service, safeguarding against fire or smoke, providing zoning, connecting the various parts in the system, and regulating the airflow. A variety of accessories for these functions is available, and installation and testing instructions will vary from manufacturer to manufacturer. Always adhere to recommendations for the specific installation at hand.

Notes

Instructor's Notes:

Trade Terms Introduced in This Module

Alloy: A metal mixed with other metals or elements to create strength and durability.

Angle bracket: An L-shaped metal supporting member used to support vertical risers. Also called angle iron.

Carbon: A non-metallic element found in nature, such as in diamonds and graphite, or a part of coal or petroleum. Carbon combines with iron to make steel.

Channel: A U-shaped piece of structural steel used as a supporting device.

Cold-rolled steel: Metal that has been formed by rolling at room temperature, usually to obtain an improved surface or higher tensile strength.

Lap-joined: A condition in which one piece is joined to another by partly covering one piece with the other.

Mastic: A protective coating applied by trowel or spray on the surface of thermal insulation to prevent its deterioration and to weatherproof it.

Non-ferrous: A metal that contains little or no iron.

Opposed-blade damper: A type of damper in which each blade moves in a direction opposite the blade next to it.

Runout: A term describing the connection between the main trunk line of duct and the diffuser, grille, or register.

Shear load: The amount of weight or pressure that causes an exposed piece of metal to break or shear off.

Sheet metal gauge: A measuring tool used to find the thickness, or gauge, of sheet metal.

Sheet steel: An uncoated sheet of steel with a bluish-black surface.

Sound attenuation: The reduction in the level of sound that is transmitted from one point to another.

Stainless steel: A high-strength, tough, corrosion- and rust-resistant alloy that contains chromium and sometimes nickel.

Tensile load: The weight required to cause metal to stretch or compress.

Tensile strength: The resistance of a material against rupture when placed under tension.

Vapor barrier: A barrier, such as a mastic, placed between insulation and the surrounding air to control condensation.

Appendix A

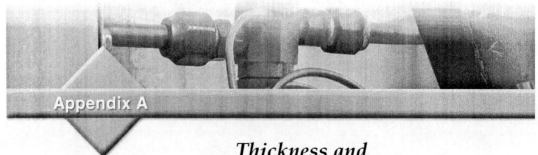

Thickness and Weights of Metals

Galvanized			Stainless				Aluminum			Copper		
Gauge	Mfr's Thickness for Steel + .0037"	Weight per Square Foot, Pounds	Gauge	U.S. Standard Approximate Decimal Parts of an Inch Thickness	Pounds per Square Foot		Gauge		Approximate Thickness in Inches	Ounces per Square Foot		Approximate Thickness in Inches
					Chrome	Nickel						
										96		—
							—		0.0403	88	⅛	0.1250
							9		0.0359	72	⁷⁄₆₄	0.1080
							—		0.0320	64	³⁄₃₂	0.0972
10	0.1382	5.781	10	0.140	5.793	5.906	10	¹⁄₃₂	0.0313	56		0.0863
							—		0.0253	48	⁵⁄₆₄	0.0647
12	0.1084	4.531	12	0.109	4.506	4.593	12		0.0226	44	¹⁄₁₆	0.0647
							—		0.0201	40		0.0593
							13		0.0179	36		0.0539
14	0.0785	3.281	14	0.078	3.218	2.625	14		0.0159	32	³⁄₆₄	0.0485
							—	¹⁄₆₄	0.0156	28		0.0431
							15		0.0142	24		0.0377
16	0.0635	2.656	16	0.062	2.575	2.625	16		0.0126	20	¹⁄₃₂	0.0323
							—		0.0113	18		0.0270
							—		0.0100	16		0.0243
18	0.0516	2.156	18	0.050	2.060	2.100				15		0.0216
20	0.0396	1.656	20	0.037	1.545	1.575				14		0.0202
22	0.0336	1.406	22	0.310	1.287	1.312				13		0.0189
24	0.0276	1.156	24	0.250	1.030	1.050				12		0.0175
26	0.0217	0.906	26	0.018	0.772	0.787				11	¹⁄₆₄	0.0162
28	0.0187	0.781	28	0.015	0.643	0.656				10		0.0148
30	0.0157	0.656	30	0.012	0.515	0.525				9		0.0135
										8		0.0121

Instructor's Notes:

Appendix B

Minimum Fasteners for Lap-Joined Straps

When straps are lap-joined for hanging rectangular metal duct, you must place the fasteners along the length of the strap in series, not side by side at the width of the strap. The minimum recommended sizes of fasteners used for lap joining are as follows:

1" × 22-, 20-, or 18-gauge	Two #10 screws or one ½" diameter bolt
1" × 16-gauge	Two ½" diameter bolts
1½" × 16-gauge	Two ⅜" diameter bolts

Appendix C

Maximum Loads for Single Metal Hangers

The maximum allowable loads for single metal hangers are as follows:

Size	Hanger Material	Maximum Load
1-inch	22-gauge strap	260 lbs.
1-inch	20-gauge strap	320 lbs.
1-inch	18-gauge strap	420 lbs.
1-inch	16-gauge strap	700 lbs.
1½-inch	16-gauge strap	1,100 lbs.
106-inch diameter	rod	80 lbs.
135-inch diameter	rod	120 lbs.
162-inch diameter	rod	160 lbs.
½ inch diameter	rod	270 lbs.
⅜ inch diameter	rod	680 lbs.
½ inch diameter	rod	1,250 lbs.
⅝ inch diameter	rod	1 ton
¾ inch diameter	rod	1½ tons

213A03.EPS

Instructor's Notes:

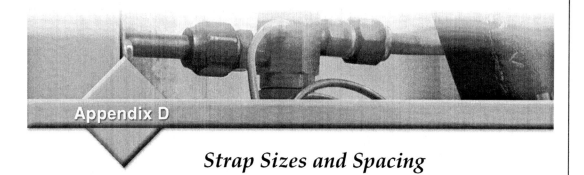

Appendix D

Strap Sizes and Spacing

Minimum metal hanger sizes recommended for round sheet metal duct (conventional wall thickness with allowance for insulation) with maximum spacing of 12 feet are as follows:

Duct Diameter	Strap Hanger(s)	Rod Hanger(s)	Wire Hanger(s)
Up through 24"	(1) 1" × 22-gauge	(1) ¼ inch diameter	(2) 10 gauge
25" through 36"	(1) 1" × 20-gauge	(1) ⅜ inch diameter	(2) 8 gauge
37" through 50"	(2) 1" × 20-gauge	(2) ⅜ inch diameter	No wires allowed
51" through 60"	(2) 1" × 18-gauge	(2) ⅜ inch diameter	No wires allowed
61" through 84"	(2) 1" × 16-gauge	(2) ⅜ inch diameter	No wires allowed

213A04.EPS

Additional Resources

This module is intended to be a thorough resource for task training. The following reference works are suggested for further study. These are optional materials for continued education rather than for task training.

Air Distribution Basics for Residential and Small Commercial Buildings, 2000. Hank Rutkowski. Arlington, VA: Air Conditioning Contractors of America.

Air Distribution in Rooms: Ventilation for Health and Sustainable Environment, 2000. Hazim B. Awbi, editor. New York, NY: Elsevier Science.

Fibrous Glass Duct Liner Standard: Design, Fabrication, and Installation Guidelines, Third edition, 2002. NAIMA. Alexandria, VA: North American Insulation Manufacturers Association.

HVAC Duct Construction Standards—Metal and Flexible. Chantilly, VA: Sheet Metal and Air Conditioning Contractors' National Association.

Standard for the Installation of Air Conditioning and Ventilating Systems, 1999. Quincy, MA: National Fire Protection Association.

Thermal Insulation Building Guide, 1990. Edin F. Strother and William C. Turner. Melbourne, FL: Krieger Publishing.

Ultimate Sheet Metal Fabrication, 1999. Tim Remus. Osceola, WI: Motorbooks International.

Working With Fiber Glass, Rock Wool, and Slag Wool Products, 2001. Alexandria, VA: North American Insulation Manufacturers Association.

Acknowledgements and References

Acknowledgments

The material on fibrous glass duct liner in this module has been adapted with the permission of the North American Insulation Manufacturers Association (NAIMA), which holds the copyright on *Fibrous Glass Duct Liner Standard, Design, Fabrication, and Installation Guidelines*. For more information about NAIMA and its publications, contact NAIMA, 44 Canal Center Plaza, Suite 310, Alexandria, VA 22314.

NAIMA assumes no responsibility and accepts no liability for the application of principles or techniques contained in the material on fibrous glass duct liner. NAIMA makes no warranty of any kind, express or implied or regarding merchantability or fitness for any particular purpose in connection with the information supplied.

Authorities considering adoption of NAIMA standards should review all federal, state, local, and contractual regulations applicable to specific installations. These materials are not intended to address issues relating to thermal or acoustical insulation within and furnished as integral parts of HVAC equipment such as air handling units, coils, air cleaners, silencers, humidifiers, and terminal devices. Manufacturers of such equipment are responsible for design, specification, and installation of appropriate insulation components in their products so that thermal, acoustical, and indoor air quality requirements are met.

Figure Credits

Courtesy North American Insulation Manufacturers Association, 213F17–213F20, 213F25

Action Air Inc., 213F46, 213F47, 213F50, 213F51, 213F55

Vent Products Company, Inc., 213F48

MODULE 03213-07 — TEACHING TIPS

The following are suggested activities or instructional methods to help you teach the material in this module.

General

When you call on someone to answer a question, the rest of the class relaxes or even tunes out because they expect that the question and answer will take place only between you and the trainee you called on. Instead, use this technique to involve more trainees in answering questions and to keep them on their toes.

1. Ask trainees to define a term or explain a concept.
2. After one trainee has answered, ask a trainee seated nearby if the answer is right. Then ask whether a trainee in the back of the room agrees.
3. Ask trainees to explain why they think an answer is right or wrong.
4. Use the session to clear up incorrect ideas and encourage trainees to learn from their mistakes.

Sections 1.0.0 through 6.0.0 *Quick Quiz*

This Quick Quiz will familiarize trainees with the terms and definitions that are commonly used when discussing sheet metal duct systems. You will need photocopies of the quiz provided in the following page. Trainees will need pencils. If you allow trainees to use the Trainee Guide, decrease the amount of time you give them to complete the quiz.

1. Make a photocopy of the quiz for each trainee.
2. Give trainees between 5 and 10 minutes to complete the quiz.
3. Go over the answers to the quiz.
4. Ask trainees if they have questions.

Answers to Quick Quiz

1. g
2. j
3. a
4. c
5. i
6. b
7. d
8. h
9. e
10. f

Quick Quiz *Sheet Metal Duct Systems*

For each description listed, identify the term that the text best describes. Write the corresponding letter in the blank provided.

_____ 1. To measure and verify the thickness of sheet metal, a(n) _____ is used.
_____ 2. Cold-rolled steel has a higher _____.
_____ 3. Pure metals are often too soft to be used in most construction applications, so _____ are often used.
_____ 4. Aluminum is the most widely used _____ metal.
_____ 5. One of the most common alloy steels is _____, which contains chromium.
_____ 6. Some risers are supported by the wall and are held in place by a band or by _____.
_____ 7. Registers and grilles have _____ that control the volume of air.
_____ 8. In addition to minimizing heat transfer, insulation also provides _____.
_____ 9. Insulate the trunk line of the duct first and then move on to the _____ or branches.
_____ 10. Each wall anchor must satisfy both the tensile load and the _____ of the supported member.

a. alloys
b. angle brackets
c. non-ferrous
d. opposed-blade dampers
e. runouts
f. shear load
g. sheet metal gauge
h. sound attenuation
i. stainless steel
j. tensile strength

Section 5.0.0 *Hangers and Supports for Sheet Metal Duct*

This exercise will familiarize trainees with sheet metal duct systems. Trainees will need appropriate personal protective equipment, pencils, and paper. You will need to arrange to tour a commercial building with a sheet metal duct system. Allow 20 to 30 minutes for this exercise.

1. Describe sheet metal duct systems and the hangers and supports used to install them.
2. Tour the utility area of a commercial building or the building in which you are holding class. Point out various aspects of a sheet metal duct system, including the hangers and supports used to support the duct system.
3. Answer any questions trainees may have.

Section 7.0.0 *Insulation*

This exercise will familiarize trainees with various types of insulation used with sheet metal duct systems. Trainees will need appropriate pencils and paper. You will need to arrange for a manufacturer's representative to give a presentation on duct insulation. Allow 45 to 60 minutes for this exercise.

1. Discuss insulation for sheet metal duct systems. Have trainees brainstorm questions before the presentation.
2. Have the speaker discuss different types of materials used for insulation and various installation techniques.
3. Have the trainees take notes and write down questions during the presentation.
4. After the presentation, answer any questions trainees may have.

MODULE 03213-07 — ANSWERS TO REVIEW QUESTIONS

Answer	Section Reference
1. c	2.0.0
2. b	2.1.1
3. b	2.2.0
4. a	3.3.0
5. d	4.4.0
6. a	4.6.0; Table 2
7. c	7.1.0
8. d	7.3.0
9. b	8.2.0
10. b	9.1.0

CONTREN® LEARNING SERIES — USER UPDATE

NCCER makes every effort to keep these textbooks up-to-date and free of technical errors. We appreciate your help in this process. If you have an idea for improving this textbook, or if you find an error, a typographical mistake, or an inaccuracy in NCCER's Contren® textbooks, please write us, using this form or a photocopy. Be sure to include the exact module number, page number, a detailed description, and the correction, if applicable. Your input will be brought to the attention of the Technical Review Committee. Thank you for your assistance.

Instructors – If you found that additional materials were necessary in order to teach this module effectively, please let us know so that we may include them in the Equipment/Materials list in the Annotated Instructor's Guide.

Write: Product Development and Revision
National Center for Construction Education and Research
3600 NW 43rd St, Bldg G, Gainesville, FL 32606

Fax: 352-334-0932

E-mail: curriculum@nccer.org

Craft _____ Module Name _____

Copyright Date _____ Module Number _____ Page Number(s) _____

Description _____

_____

_____

_____

_____

(Optional) Correction _____

_____

_____

_____

(Optional) Your Name and Address _____

_____

_____

_____

Module 03214-07

Fiberglass and Flexible Duct Systems

NCCER STANDARDIZED CRAFT TRAINING PROGRAM

The National Center for Construction Education and Research (NCCER) provides a standardized national program of accredited craft training. Key features of the program include instructor certification, competency-based training, and performance testing. The program provides trainees, instructors, and companies with a standard form of recognition through a National Craft Training Registry. The program is described in full in the *Guidelines for Accreditation*, published by NCCER. For more information on standardized craft training, contact the NCCER by writing us at 3600 NW 43rd St., Bldg. G, Gainesville, FL 32606; calling 352-334-0911; or emailing info@nccer.org. More information may be found at our website, www.nccer.org.

HOW TO USE THIS ANNOTATED INSTRUCTOR'S GUIDE

Each page presents two sections of information. The larger section displays each page exactly as it appears in the Trainee Module. The narrow column ties suggested trainee and instructor actions to each page and provides icons (detailed below) to call your attention to material, safety, audiovisual, or testing requirements. The bottom of each page includes space for your notes.

The **Audiovisual** icon indicates an appropriate time to show a transparency or other audiovisual aid.

The **Classroom** icon prompts you to define a term, stress a point, ask trainees to explain a concept, or give examples.

The **Demonstration** icon directs you to show trainees how to perform tasks.

The **Examination** icon tells you to administer the written module examination.

The **Homework** icon is placed where you may wish to assign reading for the next class, assign a project, or advise trainees to prepare for an examination.

The **Laboratory** icon is used when trainees are to practice performing tasks.

The **Materials** icon is a reminder for you to gather materials needed for classes, labs, and testing.

The **Performance Testing** icon tells you to administer a performance test or a portion thereof.

The **Safety** icon is used to emphasize safety issues. It is often keyed to *Caution* and *Warning!* statements in the Trainee Module.

The **Teaching Tip** icon indicates additional guidance is available, such as how to conduct an exercise, get the most educational value from a field trip, or encourage class participation. Teaching Tips may expand on a feature (*Think About It*, *Did You Know?*) or provide *Quick Quizzes* or similar exercises. You will be referred to the Teaching Tips section at the back of the module if there is additional material.

The **Combination** icon indicates that the laboratory listed corresponds with a performance task. If desired, you can note the proficiency of the trainees during the laboratory, and use it to satisfy performance testing requirements.

PREPARATION

Before teaching this module, you should review the Objectives, Performance Tasks, Materials and Equipment List, and Module Outline. Be sure to allow ample time to prepare your own training or lesson plan and gather all required materials and equipment.

Fiberglass and Flexible Duct Systems
Annotated Instructor's Guide

Module 03214-07

MODULE OVERVIEW

This module introduces fiberglass and flexible duct systems and explains how to lay out and install them.

PREREQUISITES

Prior to training with this module, it is recommended that the trainee shall have successfully completed *Core Curriculum*; *HVAC Level One*; and *HVAC Level Two*, Modules 03201-07 through 03213-07.

OBJECTIVES

Upon completion of this module, the trainee will be able to do the following:

1. Identify types of fiberglass duct, including flexible duct.
2. Describe fiberglass duct layout and some basic fabrication methods.
3. Describe the various closure methods for sealing fiberglass duct.
4. Fabricate selected duct modules and fittings using the appropriate tools.
5. Describe hanging and support methods for fiberglass duct.
6. Describe how to repair major and minor damage to fiberglass duct.
7. Install takeoffs and attach flexible duct to a fiberglass duct.

PERFORMANCE TASKS

Under the supervision of the instructor, the trainee should be able to do the following:

1. Fabricate and assemble fiberglass duct fittings and sections.
2. Install takeoffs and attach flexible duct.

MATERIALS AND EQUIPMENT

Overhead projector and screen
Transparencies
Blank acetate sheets
Transparency pens
Whiteboard/chalkboard
Markers/chalk
Pencils and scratch paper
Appropriate personal protective equipment
Rigid ductboard
Rigid round duct
Flexible round duct
Manufacturer's literature on ductboard
Copies of NFPA and UL standards for the fabrication and installation of fiberglass and flexible duct
Tapes and mastics and manufacturer's instructions
Iron, stapler, rubbing tool, and other tools to apply tape
Insulation knife and other tools to cut and fabricate ductwork
Wire, metal straps, and channel reinforcements
Copies of Quick Quiz*
Module Examinations**
Performance Profile Sheets**

* Located in the back of this module.
**Located in the Test Booklet.

SAFETY CONSIDERATIONS

Ensure that the trainees are equipped with appropriate personal protective equipment and know how to use it properly. This module requires trainees to work with fiberglass ductboard. Make sure that all trainees are briefed on appropriate safety procedures. Emphasize chemical hazards and skin protection.

ADDITIONAL RESOURCES

This module is intended to present thorough resources for task training. The following reference works are suggested for both instructors and motivated trainees interested in further study. These are optional materials for continued education rather than for task training.

"Energy Efficient Design of New Low-Rise Residential Buildings," 1992. *In National Voluntary Consensus Standard*. ANSI/ASHRAE.

Environmental Protection Agency Website, www.epa.gov, "Biocontaminant Control," reviewed July 2002.

Environmental Protection Agency Website, www.epa.gov, "Should You Have the Air Ducts in Your Home Cleaned," Indoor Environment Division, October 1997, reviewed July 2002.

Fibrous Glass Duct Construction Standard, Fourth Edition, 2001. Alexandria, VA: North American Insulation Manufacturers Association.

Flexible Duct Performance and Installation Standards®, Fourth Edition, 2007. Schaumberg, Illinois: Air Diffusion Council.

Fibrous Glass Duct Construction Standard, Third Edition, 1993. Alexandria, VA: North American Insulation Manufacturers Association.

Fibrous Glass Residential Duct Construction Standard, Second Edition, 1998. Alexandria, VA: North American Insulation Manufacturers Association.

UL 181, Standard for Factory-Made Air Ducts and Air Connectors, 2005. Northbrook, IL: Underwriters Laboratories, Inc.

UL 181A, Standard for Closure Systems for Use with Rigid Air Ducts and Air Connectors, 2005. Northbrook, IL: Underwriters Laboratories, Inc.

UL 181B, Standard for Closure Systems for Use with Flexible Air Ducts and Air Connectors, 2005. Northbrook, IL: Underwriters Laboratories, Inc.

NFPA 90A, Standard for the Installing of Air Conditioning and Ventilating Systems, 2002. Quincy, MA: National Fire Protection Association.

Working With Fiber Glass, Rock Wool, and Slag Wool Products. 2001. Alexandria, VA: North American Insulation Manufacturers Association.

TEACHING TIME FOR THIS MODULE

An outline for use in developing your lesson plan is presented below. Note that each Roman numeral in the outline equates to one session of instruction. Each session has a suggested time period of 2½ hours. This includes 10 minutes at the beginning of each session for administrative tasks and one 10-minute break during the session. Approximately 7½ hours are suggested to cover *Fiberglass and Flexible Duct Systems*. You will need to adjust the time required for hands-on activity and testing based on your class size and resources. Because laboratories often correspond to Performance Tasks, the proficiency of the trainees may be noted during these exercises for Performance Testing purposes.

Topic   **Planned Time**

Session I. Introduction to Fiberglass and Flexible Duct Systems

    A. Introduction   _____

    B. Types and Standards of Fiberglass Duct   _____

    C. Advantages of Modular Duct Construction   _____

    D. Extended Plenum Supply System   _____

    E. Closure Systems for Fiberglass Duct   _____

    F. Fabricating and Joining a Duct Module   _____

    G. Laboratory   _____

        Trainees practice fabricating and assembling fiberglass duct and sections.
        This laboratory corresponds to Performance Task 1.

Session II. Installation
   A. Connecting Ductboard to Sheet Metal
   B. Flexible Round Duct Connections
   C. Laboratory

   Trainees practice attaching flexible duct. This laboratory corresponds to Performance Task 2.

   D. Hanging and Supporting Fiberglass Duct
   E. Repairing Damage
   F. Laboratory

   Trainees practice installing takeoffs. This laboratory corresponds to Performance Task 2.

Session III. Review and Testing
   A. Review
   B. Module Examination
      1. Trainees must score 70% or higher to receive recognition from NCCER.
      2. Record the testing results on Craft Training Report Form 200, and submit the results to the Training Program Sponsor.
   C. Performance Testing
      1. Trainees must perform each task to the satisfaction of the instructor to receive recognition from NCCER. If applicable, proficiency noted during laboratory exercises can be used to satisfy the Performance Testing requirements.
      2. Record the testing results on Craft Training Report Form 200, and submit the results to the Training Program Sponsor.

HVAC Level Two

03214-07

Fiberglass and Flexible Duct Systems

Assign reading of Module 03214-07.

03214-07
Fiberglass and Flexible Duct Systems

Topics to be presented in this module include:

1.0.0	Introduction	14.2
2.0.0	Types and Standards of Fiberglass Duct	14.2
3.0.0	Advantages of Modular Duct Construction	14.5
4.0.0	Extended Plenum Supply System	14.6
5.0.0	Closure Systems for Fiberglass Duct	14.6
6.0.0	Fabricating and Joining a Duct Module	14.9
7.0.0	Connecting Ductboard to Sheet Metal	14.11
8.0.0	Flexible Round Duct Connections	14.11
9.0.0	Hanging and Supporting Fiberglass Duct	14.14
10.0.0	Repairing Damage to Fiberglass Duct	14.20

Overview

Fiberglass ductboard and flexible ducts (flex-duct) have become the standard for residential HVAC systems. Both fiberglass ductboard and flex-duct are lightweight and easy to handle. They are pre-insulated, so it is not necessary to add insulation as it is with sheet metal duct. Fiberglass ductboard is supplied in sheets that are easily assembled in the shop or on the job site. Flex-duct is preassembled, so it simply needs to be cut to the required length. These two types of ductwork are often combined in a system, with ductboard serving as plenums and trunk ducts, and flex-ducts serving as branch ducts. Anyone doing HVAC installation work needs to know how to assemble, join, and support these types of ductwork.

Instructor's Notes:
_____
_____
_____

Objectives

When you have completed this module, you will be able to do the following:

1. Identify types of fiberglass duct, including flexible duct.
2. Describe fiberglass duct layout and some basic fabrication methods.
3. Describe the various closure methods for sealing fiberglass duct.
4. Fabricate selected duct modules and fittings using the appropriate tools.
5. Describe hanging and support methods for fiberglass duct.
6. Describe how to repair major and minor damage to fiberglass duct.
7. Install takeoffs and attach flexible duct to a fiberglass duct.

Trade Terms

Closure system
EI rating
Fatigue test
Foil-scrim-kraft (FSK)
Plenum
R-value
Shiplap
Staple flap
Tap-ins
Torsion
Vertical riser
Young's modulus of elasticity (E)

Required Trainee Materials

1. Pencil and paper
2. Appropriate personal protective equipment

Prerequisites

Before you begin this module, it is recommended that you successfully complete *Core Curriculum*; *HVAC Level One*; and *HVAC Level Two*, Modules 03201-07 through 03213-07.

This course map shows all of the modules in the second level of the HVAC curriculum. The suggested training order begins at the bottom and proceeds up. Skill levels increase as you advance on the course map. The local Training Program Sponsor may adjust the training order.

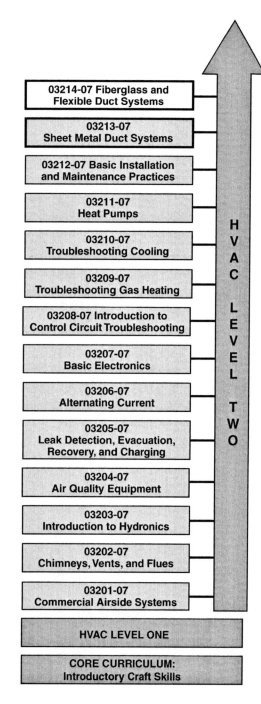

Ensure that you have everything required to teach the course. Check the Materials and Equipment list at the front of this module.

See the general Teaching Tip at the end of this module.

Explain that terms shown in bold are defined in the Glossary at the back of this module.

Show Transparency 1, Objectives, and Transparency 2, Performance Tasks. Review the goals of the module, and explain what will be expected of the trainee.

Review the modules covered in Level Two and explain how this module fits in.

Identify the advantages of fiberglass ductwork.

Identify the standards that cover the construction and installation of fiberglass duct.

Identify the three types of fiberglass duct and describe each type.

See the Teaching Tips for Section 1.0.0–10.0.0 and Section 2.0.0 at the end of this module.

Provide samples of fiberglass duct for the trainees to examine.

1.0.0 ♦ INTRODUCTION

Fiberglass ductboard, like sheet metal, is used to make the main trunk lines in an air distribution system. Flex-ducts are a type of fiberglass duct that are used for branch ducts, or runouts, to the diffuser or grille. Because fiberglass duct is already insulated, it is not necessary to wrap or line it with insulation as you must with sheet metal duct. In addition, it is lightweight and easy to transport and handle. The following are some additional benefits of fiberglass duct:

- Provides consistent heating and cooling performance through controlled insulation thickness
- Absorbs noise caused by air handling equipment and turbulence
- Reduces expansion, contraction, and vibration noise
- Exhibits a lower air leakage rate than sheet metal duct

In this module, you will learn how fiberglass duct is fabricated into straight duct sections and fittings. You will get practice in joining straight duct modules and flexible round duct connections. You also will learn how to properly hang and support fiberglass duct and how to identify and repair different types of damage.

The standards in this module apply to air distribution systems designed for use in residential situations, including constant air volume systems operating at 0.5 inches water gauge. For systems operating between 0.5 inches and 2 inches water gauge, refer to *Fibrous Glass Duct Construction Standard* for additional reinforcing requirements and details. You will find a complete reference to this publication in the Acknowledgments and References section of this module.

The material and figures in this module have been adapted from *Fibrous Glass Duct Construction Standard* (third and fourth editions) and *Fibrous Glass Residential Duct Construction Standard* (second edition) with the permission of the North American Insulation Manufacturers Association (NAIMA). Be sure to refer to the most recent editions of these standards when installing fiberglass duct.

 CAUTION
Take extra precautions when working with fiberglass duct. Always wear the appropriate personal protective equipment. Gloves, dust masks, and goggles will protect your hands, lungs, and eyes.

DID YOU KNOW?
Fiberglass

Fiberglass has been commercially manufactured and marketed for more than 50 years. It is used to make energy-conserving products that help reduce pollution and protect the environment. Fiberglass can absorb sound, control heat flow, remove impurities from liquids and gases, reinforce other materials, and, with a vapor barrier, help control condensation. It does not support the growth of mold or bacteria. These qualities make fiberglass an important material in the heating, ventilation, and air conditioning (HVAC) industry.

2.0.0 ♦ TYPES AND STANDARDS OF FIBERGLASS DUCT

Fiberglass duct, sometimes called fibrous glass duct, is made from flame-resistant glass fibers that are bonded with a heat-setting resin. There are three types of fiberglass duct:

- Rigid ductboard
- Rigid round duct
- Flexible round duct

Rigid ductboard comes in three thicknesses: 1 inch, 1½ inches, and 2 inches. Fabricators use these rigid boards to make rectangular and ten-sided duct sections and air distribution boxes.

Rigid round duct is no longer made, but may be found in existing systems. The exterior is covered with a tough, abuse-resistant **foil-scrim-kraft (FSK)** vapor barrier. FSK is a flame-retardant barrier made from a sandwich of aluminum foil, fiberglass yarn (called scrim), and kraft paper. The smooth interior surface offers minimal resistance to airflow.

Flexible duct is round duct that may or may not be insulated. A wire coil that runs around the inside of the duct gives flexible duct some stiffness and makes it easier to handle. Flexible duct is used as a connection between the rigid duct and heating/cooling equipment and ceiling diffusers. Standard practice permits the installation of flexible duct and connectors only for indoor comfort HVAC systems.

Insulated flexible round duct has a reinforced inner air barrier core and an outer jacket that is usually made from polyvinyl chloride (PVC) plastic or foil. Between the inner and outer jackets is a layer of insulation. Flexible round duct is available in standard 25-foot lengths that can be cut to

Instructor's Notes:

The Health and Safety Aspects of Fiberglass

In October 2001, the International Agency for Research on Cancer (IARC) removed fiberglass from its list of materials that may cause cancer. The U.S. National Academy of Sciences reached a similar decision in 2000. This means that the fiberglass materials used in thermal and acoustical installation are believed to pose no risk of cancer in people who are exposed to them on the job.

Discuss the hazards and safety precautions needed when working with fiberglass duct in confined spaces. Explain how to determine if the duct is safe to touch.

the required length. This type of duct gets its flexibility from a Mylar®-coated inner steel coil. It is used to connect the main duct to registers and diffusers.

CAUTION

Fiberglass duct is often installed in tight spaces. If you are working in a small or crowded space, pay special attention to your surroundings. Wear appropriate personal protective equipment and protect yourself against falls and cuts. Before touching uninsulated duct, check its temperature to avoid burns. Hold your hand about 1 inch away from the duct to make sure it is not too hot to touch.

2.1.0 Requirements and Standards for Rigid Ductboard

Rigid ductboard used in HVAC work must be manufactured in accordance with industry standards and must be tested to ensure that it meets those standards. Underwriters Laboratories (UL) tests the materials that are manufactured for use in air distribution systems. The North American Insulation Manufacturers Association (NAIMA) recommends standards for their fabrication and installation. These organizations work to ensure that the materials used are safe and durable.

2.1.1 Strength, Deflection, and Fatigue

Rigid ductboard must be strong enough to withstand building stress. It must also be able to span certain distances without deflecting (deforming), and it must stand up under the pressure of varying air cycles.

The strength and stiffness of rigid ductboard are identified by its **EI rating.** This rating is the result of a calculation based on two tests: **Young's modulus of elasticity (E)** and the moment of inertia (I). Engineers and testing organizations use this calculation to determine the level of stress the ductboard can withstand. There are three EI ratings for fiberglass duct: 475-EI, 800-EI, and 1400-EI. These ratings indicate the increasing stiffness of the duct.

Organizations such as NAIMA test duct systems under normal service conditions. These **fatigue tests** gauge the ability of rigid duct to withstand normal operating pressures. The tests are important in establishing the strength and performance of rigid ductboard over time.

2.1.2 Moisture Control

Avoid exposing ductboard to water, ice, or water vapor. If the insulation gets wet, it will become ineffective. Therefore, when working with fiberglass duct, follow these guidelines:

- Use a drip pan and sheet metal sleeve to protect the immediate area near evaporative coolers or humidifiers.
- Tightly close cooling duct systems that run through non-conditioned space. This will prevent water vapor from accumulating in the duct system during the heating season.
- Do not install wet fiberglass duct.
- Replace any sections of fiberglass duct that get wet during installation.
- Find and correct the source of any water that appears in installed duct sections during normal service. Contact the manufacturer for help or additional information if needed.

2.1.3 Fabrication and Installation Requirements

The National Fire Protection Association (NFPA) and UL have set the following fabrication and installation requirements for fiberglass duct:

- Fiberglass duct shall be constructed of Class 1 duct materials as tested in accordance with *UL Standard for Factory-Made Air Ducts and Air Connectors (UL 181).*

Explain that rigid ductboard must be manufactured in accordance with industry standards.

Explain that the strength of ductboard is identified by its EI rating. Explain how the EI rating is calculated.

Show trainees the EI rating on a typical piece of ductboard, or provide manufacturer's literature on ductboard for trainees to examine.

Discuss guidelines for controlling moisture when working with fiberglass duct.

Explain that both NFPA and UL have set standards for the fabrication and installation of fiberglass duct.

Provide copies of NFPA and UL standards for the fabrication and installation of fiberglass duct for the trainees to examine.

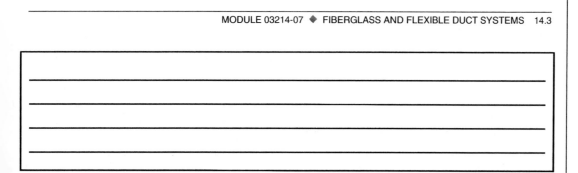

Discuss NFPA and UL standards for the fabrication and installation of fiberglass duct.

Discuss the thermal and acoustical performance of fiberglass duct.

Discuss industry standards for air leakage and identify the benefits of minimizing air leakage.

Explain that flexible duct is covered by UL standards.

Provide a copy of *UL 181* for trainees to examine.

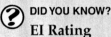 **DID YOU KNOW?**
EI Rating

Young's modulus of elasticity—the E in the EI rating—is a test. During this test, the ductboard is placed under stress and strain to test its rigidity. The result is a ratio of stress to strain. The I part of the rating is based on a mathematical calculation. The results of the test and the calculation are combined to assign the appropriate numerical rating to rigid ductboard.

- Such duct shall be installed in accordance with the conditions of its listing. This means that the installation recommendations of the testing organization must be followed.
- Such duct may not be used in air duct systems that operate with an air temperature higher than 250°F (121°C) entering the ducts.
- Such duct shall not be used as a **vertical riser** in air duct systems serving buildings that are more than two stories high.

UL tests rigid fiberglass duct for its ability to withstand a wide variety of conditions. The UL testing program covers the following areas:

- Surface burning characteristics
- Flame penetration
- Burning (other than surface burning)
- Corrosion
- Mold growth and humidity
- Temperature
- Puncture
- Static load
- Impact
- Leakage
- Erosion
- Pressure
- Collapse

2.1.4 Thermal and Acoustical Performance

The American Society of Heating, Refrigerating, and Air Conditioning Engineers (ASHRAE) has set standards for thermal performance of fiberglass duct. These standards, which have been made part of many local building codes, recommend **R-values** for ductboard based on climate zones. The recommended R-values also depend on where in a building the ductboard is installed.

For example, the recommended R-value for a roof or building exterior is higher than that for an attic or garage.

Fiberglass insulation absorbs fan and air turbulence noise and reduces the popping noises caused by expansion and contraction. ASTM International has established typical acoustical performance standards for fiberglass duct. Because acoustical performance is a fairly complex subject, it is best to contact the fiberglass manufacturer for detailed acoustical performance information.

2.1.5 Air Leakage

The industry provides standards for air leakage, which is the expected amount of leakage along duct system seams and joints. For example, a system fabricated to the Leak Class 6 standard will have an expected rate of leakage of no more than 6 cubic feet per minute (cfm) for every 100 square feet of fiberglass duct that is put under 1 inch water gauge pressure. The same amount of unsealed sheet metal duct under the same pressure is expected to leak up to 48 cfm. The lower the Leak Class number, the more airtight the system should be.

According to NAIMA, air tightness is not an absolute requirement. However, limiting the amount of air leakage has the following benefits:

- Better noise reduction
- More economical performance
- More efficient operation

2.1.6 Other Standards

In addition to the standards just discussed, standards exist for connections between duct and runouts and between fiberglass duct and sheet metal, for sealing seams and joints, for hangers and supports, and for repairs. These recommendations will be described later in this module.

2.2.0 Requirements and Standards for Flexible Duct

Flexible duct is also tested in accordance with *UL 181*. In addition to all the tests conducted for rigid duct, flexible duct is tested for tension, **torsion,** and bending. The Air Diffusion Council (ADC) tests other properties of flexible duct. These properties include thermal performance, friction loss, acoustical performance, static pressure performance, temperature performance, and air leakage. An ADC publication that discusses this topic in more detail is listed at the end of this module.

2.3.0 Restrictions on the Use of Fiberglass Duct

Fiberglass duct systems have many benefits. However, fiberglass duct is not suitable for some applications, including, but not limited to, the following:

- In kitchen or fume exhaust ducts
- For solids or corrosive gases
- For installations in concrete or buried below grade
- For outdoor installations
- Near high-temperature electric heating coils without radiation protection
- For vertical risers in air duct systems serving more than two stories
- With equipment of any type that does not include automatic maximum temperature controls
- With equipment fueled by coal or wood
- As penetrations in construction where fire dampers are required, except when the fire damper is installed in a sheet metal sleeve extending through the wall

3.0.0 ♦ ADVANTAGES OF MODULAR DUCT CONSTRUCTION

Modular duct construction allows installers to take advantage of the standard-length modules for each straight and fitted section in a duct system. This method of constructing a duct system can be used only with fiberglass duct. To understand this concept, compare a section of a sheet metal duct with a similar section of fiberglass duct (see *Figure 1*). When working with sheet metal, fittings such as elbows and reducers of various lengths must be added to the straight, nominal length duct sections. Nominal lengths for straight sections of duct are 4 feet (from metal sheets) and 5 feet (from metal coil). Note that nominal sizes are not precise measurements. A 4-foot nominal section of duct actually has a finished length of 47 inches. In the sheet metal section in *Figure 1*, a 2-foot reducer is installed after the two straight 47-inch sections. Farther down the air stream, the sections vary in length from 47 inches (with different side dimensions than those used in the first two sections) to 34 inches, to the elbow with its 4-inch throat.

Discuss the limitations on the use of fiberglass duct.

Show Transparency 3 (Figure 1). Explain the advantages of modular construction.

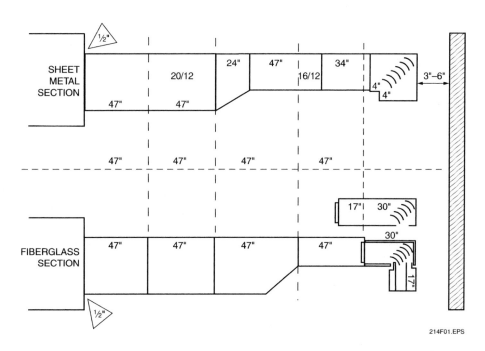

Figure 1 ♦ Modular duct construction compared to sheet metal installation.

Describe modular fabrication.

Describe the extended plenum supply system.

Show Transparency 4 (Figure 2). Identify the components in an extended plenum system.

Show Transparencies 5 and 6 (Figures 3 and 4). Explain how to seal the joints on fiberglass duct.

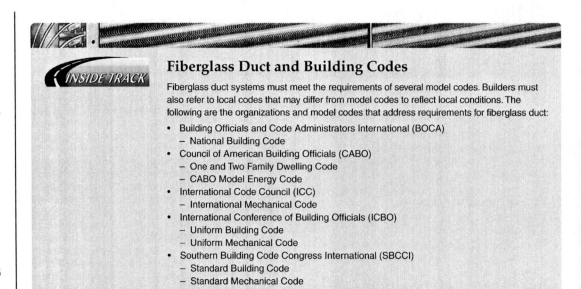

Fiberglass Duct and Building Codes

Fiberglass duct systems must meet the requirements of several model codes. Builders must also refer to local codes that may differ from model codes to reflect local conditions. The following are the organizations and model codes that address requirements for fiberglass duct:

- Building Officials and Code Administrators International (BOCA)
  - National Building Code
- Council of American Building Officials (CABO)
  - One and Two Family Dwelling Code
  - CABO Model Energy Code
- International Code Council (ICC)
  - International Mechanical Code
- International Conference of Building Officials (ICBO)
  - Uniform Building Code
  - Uniform Mechanical Code
- Southern Building Code Congress International (SBCCI)
  - Standard Building Code
  - Standard Mechanical Code

When working with fiberglass duct, there is the advantage of fabricating all the pieces using the same standard-length sections, whether they are straight ducts or fittings. Using a throatless elbow in a fiberglass duct system allows for easy adjustment to the prescribed layout. Any portion cut off in locating the elbow can be plugged in downstream, maintaining the 47-inch module. Therefore, modular duct construction is fairly easy. With a sheet metal layout, the throats of the elbows and the lengths of the fittings are sized to get a precise layout. With fiberglass, the pieces can be field-modified to make up for minor changes in equipment location or obstacles, and the same precision results. Learning how to pick off the duct system pieces for modular construction and mastering the fabrication information in this module are critical to success in working with fiberglass ductboard.

4.0.0 ♦ EXTENDED PLENUM SUPPLY SYSTEM

The extended **plenum** supply system is standard for residential applications. It is the ideal system for basement installations. It is sturdy and efficient and provides excellent acoustical and thermal performance. The size does not need to be changed after each runout or branch.

Figure 2 represents a sample extended plenum supply system. The plenum is installed on the air handler. Runouts extend from the plenum to the registers or diffusers. The runouts may be made from rigid ductboard, rigid round duct, 10-sided duct, flexible round duct, or insulated metal duct. Note that an actual installation would probably not combine all the types of duct shown in the figure. This module focuses on flexible round duct joining for the branches. A complete illustration of the system would also include the required hangers and supports. These elements have been left out for the sake of simplicity. The following items are identified in the figure:

1. Rigid ductboard (plenum extension)
2. Connection of the plenum to the central air equipment
3. Plenum
4. **Tap-ins** for rectangular, round, and ten-sided duct runouts
5. Registers or diffusers
6. Connections to registers or diffusers using flexible duct runouts

In this module, you will learn how some of the system components of an extended plenum supply system work. You will also learn how to join duct modules and fabricate flex-duct fittings that make up the system.

5.0.0 ♦ CLOSURE SYSTEMS FOR FIBERGLASS DUCT

Whether joining straight duct modules, connecting fiberglass duct to sheet metal, or fabricating flex-duct connections, the components must be sealed properly. For maximum efficiency and performance, the correct **closure system** must be used. Tapes are used either to reinforce **staple flap** joints or are laid in cross-tabs to seal joints without staple flaps (see *Figures 3* and *4*). Only closure sys-

Instructor's Notes:

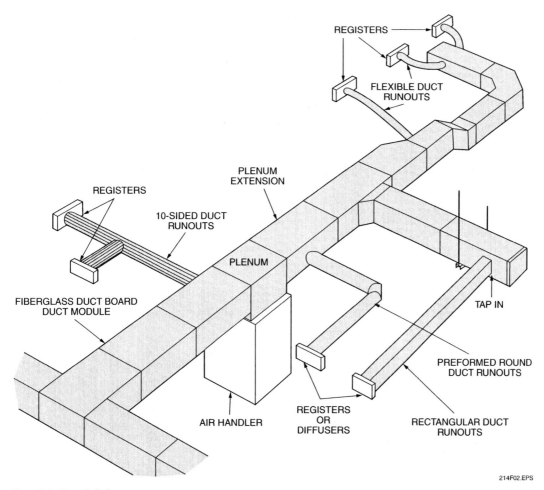

Figure 2 ♦ Extended plenum system components.

tems that comply with *UL Standard 181A: Closure Systems for Use with Rigid Air Ducts and Air Connectors* are acceptable for use in rigid fiberglass duct systems. Closure machines are also used in cases of high-volume production and installation, but only authorized personnel may use these. Basic closure systems for fiberglass duct include the following:

- Pressure-sensitive tape
- Heat-activated tape
- Mastic and glass fabric tape

Preparing the surface is an important step in working with closure systems. The surfaces to be bonded must be clean and dry. If there is any dust, dirt, oil, grease, or moisture present, the adhesives won't stick. In most cases, the area to be bonded can be cleaned with an oil-free, lint-free cloth or paper towel. For best results, follow the manufacturer's cleaning recommendations.

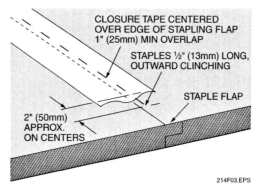

Figure 3 ♦ Tape closure with staple flap.

Identify the three types of closure systems for fiberglass duct.

Emphasize the importance of preparing the surface.

Explain that tapes and mastics have storage limitations that must be followed.

Discuss the hazards and safety precautions needed when working with fiberglass duct.

Provide tapes and mastics and manufacturer's instructions for the trainees to examine.

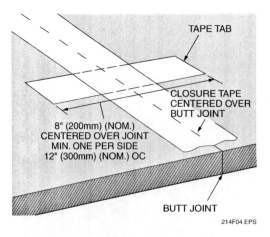

Figure 4 ◆ Tape closure without staple flap or staples.

Most tapes and mastics have storage requirements and a limited shelf life. It is important to make sure that these materials have been stored according to the manufacturer's instructions. No product that is past its shelf life should be used. The adhesive material may have dried out so much that it will not make an effective bond.

NOTE

For rigid fiberglass ductboard systems, only use closure systems that comply with *UL 181A*.

Pressure-sensitive tape is used to join fiberglass duct and also to connect fiberglass fittings to sheet metal. Be sure to read the manufacturer's metal cleaning instructions before using pressure-sensitive tape on sheet metal. Pressure-sensitive tape is

Working Safely with Fiberglass

According to NAIMA, which has done a great deal of research, fiberglass is safe to manufacture, install, and use when recommended work practices are followed. Recommended work practices and safety guidelines vary from one manufacturer to another. Therefore, you must always read the material safety data sheet (MSDS) for the latest information and recommendations regarding the fiberglass product being used. In general, however, NAIMA recommends the following:

- If there is a question as to the fiber count in the air, use a respirator. Because it is important to use a respirator properly, you must read and follow the manufacturer's recommendations for proper fit and use. The manufacturer may also recommend a program for training and testing.
- Wear a cap, gloves, a long-sleeved shirt, and long pants. To help keep glass fibers from being trapped between your clothing and your skin, it is best if your work clothes fit loosely. Fiberglass can irritate your skin. This is not an allergic reaction but a temporary mechanical response of the skin when the glass fibers rub against or become embedded in it.
- Wear safety glasses with side shields, safety goggles, or a face shield to protect your eyes.
- Do not rub or scratch your skin. Remove fiberglass particles from your skin by washing the area gently and thoroughly with warm water and mild soap. Using a skin lotion also may help.
- Wash work clothing separately from other household laundry to prevent the transfer of glass fibers to other clothing. Rinse the washing machine thoroughly before using it again. If your work clothes have lots of fibers on them, it is best to soak and rinse them separately before placing them in the washing machine.
- Keep your work area clean. Avoid handling scrap fiberglass by keeping disposal bins as close to the working area as possible. Don't let scrap material accumulate on the floor or other surfaces.
- Prevent airborne dust. Do not dry sweep or use a compressed air line to clean your work area. Use a filtered vacuum or wet sweeping technique.

Generally, workers engaged in such operations as sawing, machining, and blowing fiberglass products may be exposed to more airborne glass fibers than installers. Therefore, NAIMA recommends using dust collection systems whenever exposure to glass fibers may exceed recommended levels. Currently, NAIMA recommends that airborne exposures be kept below 1 fiber per cubic centimeter (f/cc). A pamphlet and video describing recommended work practices are available from NAIMA.

Instructor's Notes:

Inside Track: Effective Cleaning

When cleaning a surface to be sealed, move a clean cloth across the area in one smooth sweep. If necessary, fold the cloth and make another smooth sweep. Apply even pressure on the cloth because too little pressure is ineffective and too much pressure could damage the ductboard. Don't rub the cloth back and forth or scrub at the surface. Doing that can redeposit some of the dirt or grease you just removed.

Inside Track: Handling Tape

Tape is a convenient and easy way to seal joints and seams and to make repairs. However, the tape can fold back on itself or wrinkle as you apply it. To better control the tape, pull out a small section and firmly press it into place while holding the roll with your other hand. Hold the adhered section in place briefly with one hand as you steadily pull the tape roll with the other. Smooth the tape into place, moving your free hand down the tape as it comes off the roll.

normally pressed and rubbed into place, but in colder conditions (below 50°F) it may need to be heated before being applied to the duct seams.

CAUTION

When rubbing pressure-sensitive tape, avoid placing too much pressure on the sealing tool. Too much pressure could cause punctures wherever the tape covers the staples.

Heat-activated tape is applied using a heating iron with a temperature between 550°F and 600°F (288°C and 326°C). After ironing to apply and bond the tape, all joints and seams must be allowed to cool below 150°F (66°C) before putting any stress on them.

WARNING!

Both the iron and the heated surfaces get very hot. To avoid serious burns, pay close attention to what you are doing. To protect your hands, wear KEVLAR® gloves. When positioning the tape, set the iron down on a heat-resistant surface. Work so that the hot iron is always in front of the electric wire. Work carefully to keep your feet from getting tangled in the wire.

Do not use a blowtorch with heat-activated tape. Both heat and pressure are required for an effective bond.

Mastic and glass fabric tape is a two-part closure system consisting of liquid mastic that comes in a can and a glass fabric tape that looks like a screen. Although there are some similarities among mastics, the application rate, safety precautions, shelf life limits, and minimum setup time will vary from one manufacturer to another. Therefore, you must always read the manufacturer's instructions and recommendations. Depending on the application, either pressure-sensitive or heat-activated tape can be used in combination with this closure system.

When using any type of closure system, follow all manufacturer recommendations for proper application of the tape.

6.0.0 ♦ FABRICATING AND JOINING A DUCT MODULE

To fabricate a duct module of standard straight duct, **shiplaps** and corner grooves should be cut using either hand tools or machine tools. In either case, you can use the procedures outlined earlier in this module for using these tools. Refer to *Figure 5* for the following steps:

Step 1 Cut shiplaps and corner grooves with hand or machine tools. Remove scrap from the grooves.

Step 2 Fold the panels to form the duct. Make sure that the ends are flush and properly seated in the shoulder of the shiplap edges.

Step 3 Hold the duct at about a 30-degree angle. Staple the long flap with ½-inch outward clinching staples approximately 2 inches on center.

Step 4 Close the long seam using one of the closure methods discussed earlier in this module.

Once the duct modules are constructed, they must be joined together. This is a fairly simple task; however, it must be done carefully to ensure that the joints fit tightly and evenly. The joints

Explain how to apply tape to fiberglass duct.

Discuss the hazards and safety precautions needed when applying tape to fiberglass duct.

Explain how to apply mastics to fiberglass duct.

Review the procedure for fabricating a duct module.

Show trainees how to fabricate a duct module.

Emphasize the importance of properly joining duct modules.

Review the procedure for joining duct modules.

Show trainees how to join duct modules.

Have trainees practice fabricating and joining duct modules. Note the proficiency of each trainee. This laboratory corresponds to Performance Task 1.

Have the trainees review Sections 7.0.0–10.2.2.

STEP 1

STEP 2

STEP 3

STEP 4

214F05.EPS

Figure 5 ♦ Fabricating a straight duct module.

must be sealed properly with the approved tape. When standing on end, the large duct sections tend to wobble. Prop the sections against a wall to steady them while joining the components together. Refer to *Figure 6* for the following steps:

Step 1 Using an insulation knife, slit the facing tabs at each corner. Do not cut below the male shiplap shoulder or into the inside duct surfaces. The best way to make these slits is to rest the end of the knife blade on the shoulder of the shiplap and pull the blade outward to cut the facing.

Step 2 Push the two sections together. Make sure the male and female shiplaps fit tightly together. This step is easiest if you place one duct section on the floor and set the other section on top of it.

Step 3 Using a staple gun, staple the flaps on all four sides. Use ½-inch outward clinching staples about 2 inches on center.

Step 4 Make sure the ductboard surface is clean and dry. Then tape the joint with pressure-sensitive tape. Using a plastic rubbing tool, rub the tape firmly until the facing screen can be seen clearly through the tape.

14.10   HVAC ♦ LEVEL TWO

Instructor's Notes:

STEP 1

STEP 2

STEP 4

Figure 6 ♦ Joining duct modules.

7.0.0 ♦ CONNECTING DUCTBOARD TO SHEET METAL

In a duct system, there may be areas where fiberglass ductboard must be connected to sheet metal. Refer to *Figure 7* to see how to connect ductboard to sheet metal flanges on air distribution equipment. The four possible connections are as follows:

- Pressure-sensitive tape over sheet metal screws on a 26-gauge sheet metal U-channel (*Figure 7A*)
- Pressure-sensitive tape over sheet metal screws on a 26-gauge pocket lock (*Figure 7B*)
- Pressure-sensitive tape over sheet metal screws on a 22-gauge sheet metal sleeve (*Figure 7C*)
- Pressure-sensitive tape over sheet metal screws on a 22-gauge sheet metal flange (*Figure 7D*)

The sheet metal must be cleaned carefully and the connections sealed properly with pressure-sensitive tape. The mechanical connections (screws or screws and washers) must be placed at a maximum of 12 inches on center. Washers must be 0.028 inches thick with turned edges to prevent them from cutting into the ductboard or the facing.

Apply the pressure-sensitive tape using the same procedures outlined earlier in this module. Avoid putting too much pressure on the sealing tool when rubbing the tape over mechanical connections.

8.0.0 ♦ FLEXIBLE ROUND DUCT CONNECTIONS

Flexible round duct is used for runouts from low-pressure ducts, mixing boxes, diffusers, or other low-air velocity units. Flexible duct is not recommended for use on the return air side of an air distribution system. Flexible duct is connected to rigid ductboard using two methods: the closure strap method and the insulated collar method. In both methods you must pull back the insulation that covers the inner core of the flexible duct to properly make the connection. Once the connection is made, push the insulation back into place so that the entire assembly butts tightly against the ductboard.

8.1.0 Closure Strap Method

Panduit® straps are a commonly used brand of closure straps. To use the closure strap method, follow these steps (*Figure 8*):

Step 1 Cut a hole in the ductboard using a hole cutter sized to accept a sheet metal collar. Use either a spin-in collar or a dovetail collar. If using a spin-in collar, also cut a 1-inch slit radial to the hole (*Figure 8A*).

Ensure that you have everything required for teaching this session.

Show Transparency 7 (Figure 7). Explain how to join fiberglass ductboard to sheet metal.

Show trainees how to join fiberglass ductboard to sheet metal.

Describe flexible round duct and explain its applications.

Explain that there are two methods to join flexible duct to rigid ductboard.

Show Transparency 8 (Figure 8). Explain the closure strap method for joining flexible round duct to fiberglass ductboard.

Show trainees how to join flexible round duct to fiberglass ductboard using the closure strap method.

Explain that for the insulated collar method a ring must be fabricated.

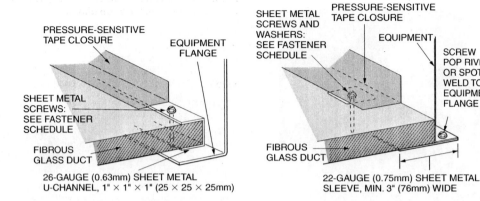

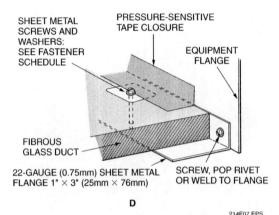

Figure 7 ♦ Connecting fiberglass ductboard to sheet metal.

Step 2 To reduce air leakage, coat the collar flange with mastic to seal the collar to the ductboard.

Step 3 Insert either a spin-in collar (*Figure 8B*) or a dovetail collar (*Figure 8C*) made for fiberglass duct. If using a spin-in collar, bend the leading edge of the inner flange down at an angle and slip it through the slit in the ductboard. Screw the collar into the hole in the ductboard until the flange is snug against the ductboard facing. If using a dovetail collar, push it into the hole in the ductboard until the flange is snug against the ductboard facing. Bend all tabs 90 degrees to lock the collar into place.

Step 4 Push the insulation back and slide 1 inch of the duct core over the collar. Seal the core to the collar if required by the manufacturer or by the job specifications (*Figure 8D*).

Step 5 Pull the insulation back over the core so that it butts firmly against the duct wall (*Figure 8E*). Complete the installation in accordance with the manufacturer's instructions.

8.2.0 Insulated Collar Method

In the insulated collar method, a ring must be fabricated from the fiberglass ductboard to insulate a sheet metal collar. Depending on the job

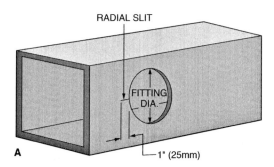

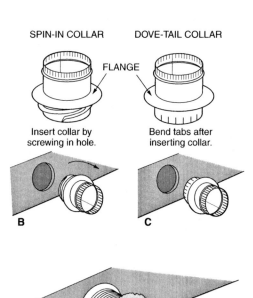

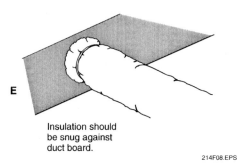

Figure 8 ♦ Closure strap method.

requirements, you may use either a spin-in collar or a dovetail collar. The insulated collar method maintains full insulation thickness across the flexible duct connection and helps prevent condensation. To use the insulated collar method, follow these steps (*Figure 9*):

Step 1 Cut a ductboard ring to the following specifications:
- An inside diameter equal to the diameter of the sheet metal collar being installed.
- An outside diameter with a thickness corresponding to the insulation's R-value (see *Table 1*).

Step 2 Slide the ring onto the sheet metal collar with the foil facing of the ductboard away from the flange. Tape the insulation to the back of the flange with pressure-sensitive tape (*Figure 9A*).

Step 3 Cut a hole in the ductboard sized to accept the sheet metal collar. Use either a spin-in collar or a dovetail collar. If using a spin-in collar, cut a 1-inch slit radial to the hole. Bend the leading edge of the inner flange down at an angle and slip it through the slit in the ductboard. Screw the collar into place with the outer flange snug against the foil facing and the inner ring fully visible inside the duct. To reduce air leakage, coat the collar of the flange with approved mastic. If using a dovetail collar, push the collar into the hole until the outer flange is snug against the facing. Bend the dovetails 90 degrees outward to lock them into place.

Step 4 Slide the insulation back from the core of the flexible duct and pull the duct core over the collar. Apply sealant if required by the manufacturer or the job specifications. Secure with a closure strap placed between the bead on the collar and the ductboard (*Figure 9B*).

Step 5 Butt the flexible duct and insulation firmly against the ductboard ring (*Figure 9C*). Pull the vapor barrier over the ring so that it covers about one-half of the width of the ring.

Step 6 Tape the vapor barrier to the ring using tape that is compatible with the jacket of the flexible duct (*Figure 9D*). Complete the installation in accordance with the manufacturer's instructions.

Show Transparencies 9 and 10 (Figure 9 and Table 1). Explain the insulated collar method for joining flexible round duct to fiberglass ductboard.

Show trainees how to join flexible round duct to fiberglass ductboard using the insulated collar method.

Have trainees practice joining flexible round duct to fiberglass ductboard. Note the proficiency of each trainee. This laboratory corresponds to Performance Task 2.

Review safety guidelines for using foam insulating sealant.

Discuss general considerations for hanging fiberglass duct.

Show Transparency 11 (Figure 10). Explain how to support fiberglass duct up to 48 inches wide.

Show Transparency 12 (Figure 11). Explain how to determine the hanger extension.

Provide wire, metal straps, and channel reinforcements for the trainees to examine.

See the Teaching Tip for Section 9.0.0 at the end of this module.

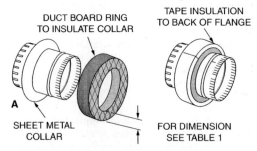

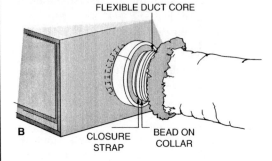

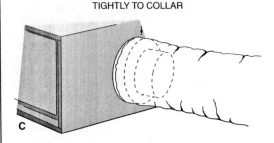

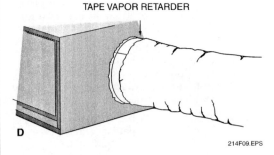

Figure 9 ◆ Insulated collar method.

Table 1 Inside Dimension and R-Value

Inside Diameter in Inches	R-Value
2	4.2
3	6.0
4	8.0

CAUTION

When using foam insulating sealant, remember to wear gloves and safety goggles. Apply the foam sealant in small amounts. The sealant is messy and expands when it is released from its can.

9.0.0 ◆ HANGING AND SUPPORTING FIBERGLASS DUCT

Fiberglass duct is lightweight and, for the most part, self-supporting. A minimum of carefully placed hangers can be used for support. Ensure that the hanging method used has adequate load-bearing capability for the installation, without placing too much stress on either the hanger or the fiberglass duct system.

Remember to install enough hangers to support additional accessories, including heaters and dampers. Hanger treatment and spacing requirements depend on duct dimensions. Channel gauge and profile vary with duct size. Always refer to your local code for specifications regarding the method and materials.

9.1.0 Hanging and Supporting Rectangular Fiberglass Duct

Fiberglass duct up to 48 inches wide can be suspended using either 12-gauge (minimum) wire or 22-gauge metal straps that are a minimum of 1 inch wide (*Figure 10*). Use of a 12-gauge wire is the preferred method. When using channel reinforcement members, consider the maximum unreinforced duct dimension to determine spacing. The gauge and profile of the channel will vary with the size of the duct. Sheet metal straps may be bolted to channel reinforcement.

9.1.1 Determining Hanger Extension

To determine hanger extension, add the distance between the hanging wires and the duct walls (*Figure 11*). For rectangular fiberglass duct, the supporting channel should never be less than 2

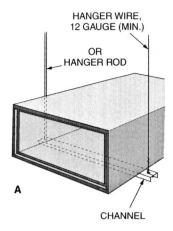
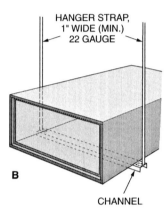
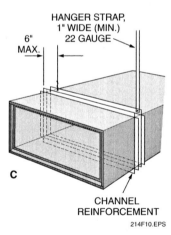

Figure 10 ◆ Suspending fiberglass duct with wire, metal straps, and channel reinforcements.

inches wide. If the hanger extension is less than 6 inches, use 24-gauge (minimum) wire and 3-inch by 1.5-inch-wide channels. If the hanger extension is less than 18 inches, use 22-gauge (minimum) wire and 3-inch by 2-inch-wide channels. If the hanger extension is less than 30 inches, use 18-gauge (minimum) wire and 3-inch by 2-inch-wide channels.

For ducts less than 48 inches wide and 24 inches high, 2-inch-wide hangers can be used instead of 3-inch-wide hangers. Use 22-gauge wire and space the 2-inch hangers at maximum 4 feet apart (see *Figure 12*). Refer to *Figure 13* to determine the maximum hanger spacing for straight duct using 3-inch channels.

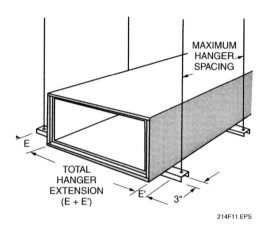

Figure 11 ◆ Hanger spacing and extension.

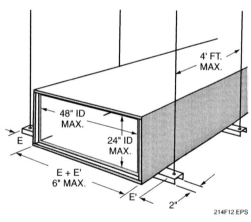

Figure 12 ◆ Two-inch-wide hangers.

Show Transparencies 13 and 14 (Figures 12 and 13). Discuss hanger spacing.

Give trainees sample problems on hanger spacing.

Show Transparencies 15–20 (Figures 14–19). Explain how to support fiberglass duct fittings.

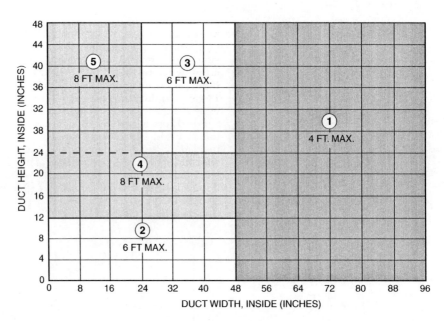

Figure 13 ♦ Maximum hanger spacing.

9.1.2 Fiberglass Duct Fittings

Fiberglass duct fittings up to 48 inches wide will require additional support. If duct is wider than 18 inches, support the elbows with a channel reinforcement located two-thirds of the diagonal distance from the throat to the heel (*Figure 14*). If a trunk duct hanger should be located where a branch duct connects with the trunk, install hangers on either side of the branch duct (*Figure 15*). If a tee runout hanger should be located where a branch duct connects with the trunk duct, install run-out hangers on either side of the trunk duct (*Figure 16*). Do not exceed maximum hanger spacing when adding branch and tee supports.

Offset supports are needed when the angled portion of offset is longer than 48 inches (*Figure 17*). For transition support, locate hangers as you would for straight duct (*Figure 18*). If there is an inclined bottom surface on a duct with a width greater than 48 inches, support the offsets and transitions (*Figure 19*). When supporting offsets and transitions, add hangers to comply with hanger spacing requirements.

14.16   HVAC ♦ LEVEL TWO

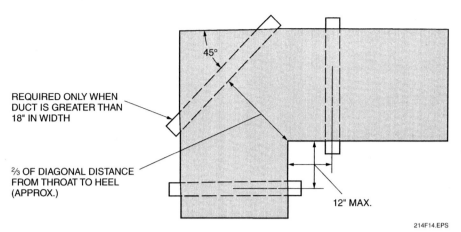

Figure 14 ◆ Elbow support.

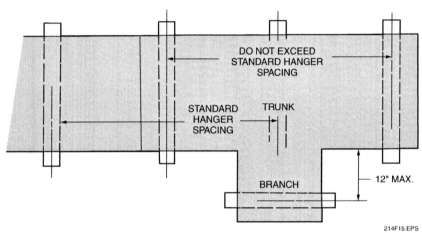

Figure 15 ◆ Branch support.

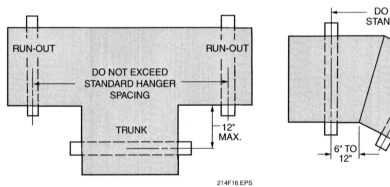

Figure 16 ◆ Tee support.

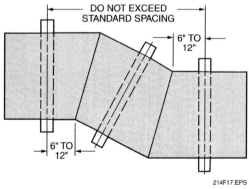

Figure 17 ◆ Offset support, flat-bottom surface.

Show Transparencies 21–23 (Figures 20–22). Explain the procedures for hanging and supporting flexible duct and fittings.

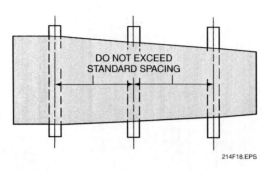

Figure 18 ♦ Transition support, flat-bottom surface.

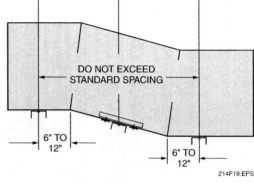

Figure 19 ♦ Offset and transition support, inclined bottom surfaces.

9.2.0 Hanging and Supporting Flexible Duct

Before installing flexible duct, review the project plans for proper placement. To install flexible duct, follow these steps:

Step 1 Measure to determine the length of duct required. Be sure to allow the correct amount for any turns.

Step 2 Cut completely around and through the duct with a sharp utility knife.

Step 3 Trim the interior wire with lineman pliers (side cutters).

Step 4 Pull back the insulation (if any) from the core and slide at least 1 inch of the core over the collar, pipe, or fitting (*Figure 20*). Note that sheet metal collars must be at least 2 inches long.

Step 5 Tape with at least two wraps of approved duct tape (*Figure 21*). Note that the tape must be plenum-rated if the ceiling space is to be used as a return air plenum. You may use an approved clamp in place of or with the tape.

Step 6 Fully extend the duct for installation. Do not compress the duct. Radius bends at the duct center line should not be less than the diameter of the duct (*Figure 22*).

Step 7 Support the duct at the manufacturer's recommended intervals, but at least every 4 feet. The maximum permissible sag is ½ inch per foot of spacing between supports. For long horizontal runs that have sharp bends, use additional supports before and after the bend.

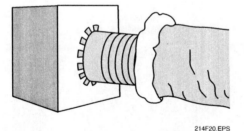

Figure 20 ♦ Pull back vapor barrier jacket and insulation.

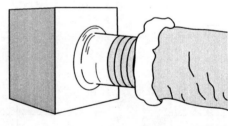

Figure 21 ♦ Joint wrapped with tape.

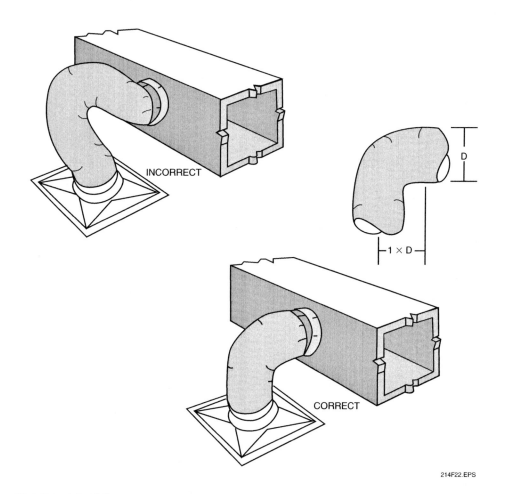

Figure 22 ♦ Extend duct fully.

Step 8 Splice flexible duct together with a sheet metal sleeve at least 4 inches long. Tape splices with at least two wraps of approved duct tape (*Figure 23*).

Step 9 Place a hanger at connections to rigid ducting, splice connections, or equipment supports. Hangers or saddles used to support flexible duct must be at least 1½ inches wide (*Figure 24*). Other types of approved supports for horizontal duct on ceiling joists, for angled duct, and for vertical duct are shown in *Figure 24*.

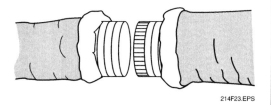

Figure 23 ♦ Splicing.

Show Transparency 24 (Figure 23). Explain how to splice flexible duct.

Show Transparency 25 (Figure 24). Explain how to support flexible duct.

Explain how to check to determine if the duct is installed properly.

Explain that fiberglass duct can become damaged during installation.

Describe minor damage that can be repaired with tape.

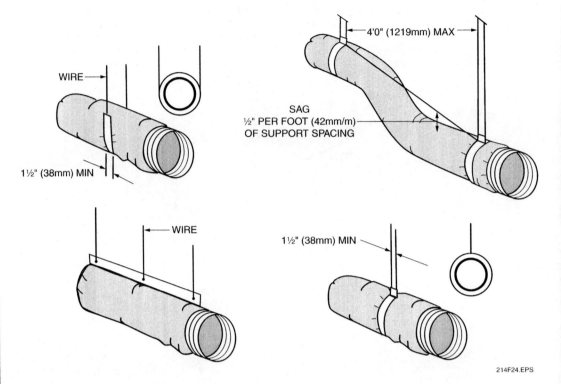

Figure 24 ♦ Flexible duct supports.

Step 10 Repair any tears or damage to the vapor barrier jacket with approved duct tape or according to the manufacturer's instructions. If the internal core is penetrated, replace flexible duct or splice and seal it as in Steps 4 and 5.

To ensure that flexible duct is properly installed, use the following checklist:

- Do the hangers support the duct without cutting into it, especially at the bends?
- Have tears or damage to the vapor barrier jacket been repaired with approved duct tape or according to the manufacturer's instructions?
- Have you checked to ensure that the duct does not sag more than ½ inch between supports?

10.0.0 ♦ REPAIRING DAMAGE TO FIBERGLASS DUCT

Fiberglass duct can be damaged during installation or by workers from other trades on the construction site as they do their own installations. In this section, you will learn how to identify and repair different types of damage.

10.1.0 Repairing Minor Damage

Minor damage includes small, straight slits in the facing material that have not damaged the insulation. Use the appropriate closure material (pressure-sensitive tape, heat-activated tape, or mastic and glass fabric tape). Clean and dry the surface before applying the closure material to ensure a good bond.

Instructor's Notes:

If the facing damage is more than just a straight slit but is not more than ½ inch wide, make the repair as shown in *Figure 25*. The closure material must extend at least 1 inch beyond all sides of the tear.

If the facing damage is wider than ½ inch but is less than the width of the closure material, smooth the facing and make the repair as shown in *Figure 26*. Apply two layers of closure material. The first layer is a single piece of closure material centered over the tear. The second layer is two pieces of closure material butted side by side so that the second layer covers the first layer with a 1-inch minimum overlap.

10.2.0 Repairing Major Damage

Major damage occurs when the fiberglass insulation is damaged or displaced. When this happens, the damaged section must be removed and a replacement plug fabricated (*Figure 27*). To replace a damaged panel, follow these steps:

Step 1 Using a shiplap tool, cut out the damaged area and discard it. Cut a square or rectangle that is slightly larger than the damaged area. Doing this will allow the replacement plug to be fabricated more easily.

Step 2 Measure and cut a replacement plug from fiberglass ductboard. Shiplap the edges and leave a 1¾-inch staple flap on all four sides. Be sure the replacement board thickness matches the thickness of the duct panel being repaired.

Show Transparencies 26 and 27 (Figures 25 and 26). Explain how to repair minor damage.

Show trainees how to repair minor damage.

Show Transparency 28 (Figure 27). Explain how to repair major damage.

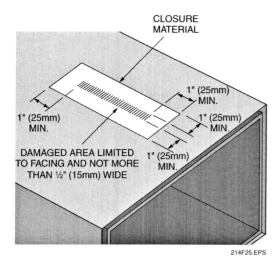

Figure 25 ◆ Repair of minor facing damage less than ½ inch wide.

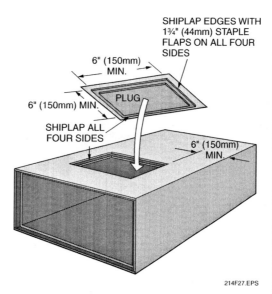

Figure 26 ◆ Repair of minor facing damage ½ inch wide or greater.

Figure 27 ◆ Repairing major damage to one panel.

Show Transparency 29 (Figure 28). Explain when and how to replace the width of the panel.

Show Transparency 30 (Figure 29). Explain when and how to replace an entire shiplapped panel.

Show Transparency 31 (Figure 30). Explain when and how to replace an entire square-edged panel.

Show trainees how to repair damaged ductwork.

Have trainees practice repairing damaged ductwork.

Step 3 Insert the plug and test the fit. The plug should fit smoothly and snugly.

Step 4 Staple the flaps and seal with the approved closure system.

If the damaged area extends to within 6 inches of the edge of the panel, cut out and repair the entire width of the damaged panel, using the previous steps. Make end cuts on the plug to match the ends of the damaged panel (*Figure 28*).

> **CAUTION**
> To ensure a strong bond, use only approved closure materials and methods. Always read and follow the manufacturer's instructions. Do not use heat-activated tape over either pressure-sensitive tape or mastic and glass fabric closure systems. The heat may cause these other closure systems to fail.

10.2.1 Replacing an Entire Shiplapped Panel

Sometimes the damage to a duct panel is so great that the entire panel must be removed. This usually happens when something falls on the panel, causing it to crack or crease. To make this repair, follow these steps (see *Figure 29*):

Step 1 Using a shiplap tool, cut away and discard the damaged panel.

Step 2 Measure and fabricate a new panel. Shiplap the edges of the new panel.

Step 3 Remove insulation from the new panel to form a staple flap on two sides as shown. See the detail drawings in *Figure 29* for 1⅜ inch and 1¹¹⁄₁₆ inch minimum staple flaps.

Step 4 Clean out any debris from inside the duct panel and insert the replacement panel.

Step 5 Test the fit. It should be smooth and snug.

Step 6 Staple the flaps and seal with the approved closure system.

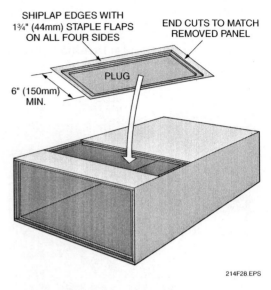

Figure 28 ◆ Repairing an entire panel.

10.2.2 Replacing an Entire Square-Edged Panel

The process for replacing an entire square-edged panel is similar to that for replacing an entire shiplapped panel. Cut the replacement panel with square, instead of shiplapped, edges, using the following steps (*Figure 30*):

Step 1 Using a shiplap tool, cut away and discard the damaged panel.

Step 2 Measure and fabricate a new panel with square edges.

Step 3 Remove insulation from the new panel to form a staple flap on two sides as shown.

Step 4 Clean out any debris from inside the duct panel and insert the replacement panel.

Step 5 Test the fit. It should be smooth and snug.

Step 6 Staple the flaps and seal with the approved closure system.

14.22   HVAC ◆ LEVEL TWO

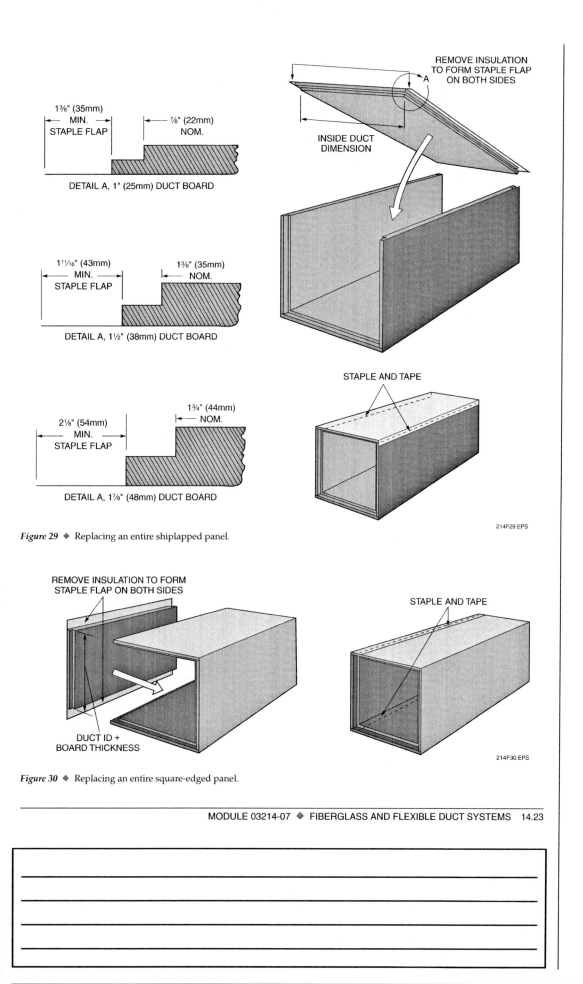

Figure 29 ◆ Replacing an entire shiplapped panel.

Figure 30 ◆ Replacing an entire square-edged panel.

Have trainees complete the Review Questions, and go over the answers prior to administering the Module Examination.

Review Questions

1. The three types of fiberglass duct are _____.
   a. 475, 800, 1400
   b. 375, 700, 1300
   c. shiplapped, square-edged, and grooved
   d. rigid, rigid round, and flexible round

2. The strength and stiffness of rigid ductboard are identified by its _____ rating.
   a. IE
   b. EI
   c. IS
   d. SI

3. Fiberglass duct may be used for all of the following applications *except* _____.
   a. residential
   b. commercial
   c. indoor
   d. outdoor

4. A 4-foot nominally sized section of duct actually measures _____ inches.
   a. 44
   b. 47
   c. 49
   d. 51

5. In an extended plenum supply system, flexible round duct can be used for _____.
   a. the plenum
   b. runouts
   c. hangers and supports
   d. joining duct modules

6. All of the following are acceptable finished closure systems for fiberglass duct *except* _____.
   a. heat-activated tape
   b. soldered duct joints
   c. pressure-sensitive tape
   d. mastic and glass fabric tape

7. A blowtorch should not be used with heat-activated tape because _____.
   a. the blowtorch is too large a tool for the purpose
   b. the tape will melt
   c. the chemicals in the tape will not bond under an open flame
   d. both heat and pressure are required for an effective bond

8. When joining duct modules, ensure a steady, airtight fit by _____.
   a. hanging the duct modules before sealing them together
   b. joining the modules together temporarily with duct tape
   c. gluing inside the shiplaps
   d. steadying the modules against a wall while joining them

9. When fabricating a duct module, staple the long flap with _____.
   a. 1-inch staples approximately 2½ inches on center
   b. 1-inch outward clinching staples approximately 2 inches on center
   c. ½-inch outward clinching staples approximately 2 inches on center
   d. ½-inch stainless steel staples approximately 2½ inches on center

10. The two methods for joining flexible round duct to sheet metal are _____.
    a. closure strap and insulated collar
    b. mastic tape and glass fabric tape
    c. spin-in and dovetail
    d. tape-over-screw and tape-over-flange

11. To properly hang rectangular fiberglass duct, the width of the supporting channel must be at least _____ inch(es).
    a. ½
    b. 1
    c. 2
    d. 3

12. Two methods for suspending rectangular fiberglass duct are _____.
    a. channels suspended from 12-gauge (minimum) wire or 22-gauge 1-inch wide (minimum) metal straps
    b. channels suspended from 12-gauge (minimum) wire or 1 inch wide (minimum) flexible fiberglass straps
    c. channels suspended from 10-gauge (minimum) wire or 1½ inch wide (minimum) metal straps
    d. channels suspended from 10-gauge (minimum) wire or ½-inch wide (minimum) metal straps

Instructor's Notes:

Review Questions

13. After flexible round duct is installed, it should not sag more than _____ inch(es) between supports.
    a. ¼
    b. ½
    c. 1
    d. 2

14. When repairing minor facing damage with tape, the closure material must extend beyond all sides of the tear by _____ inch(es).
    a. 1
    b. 1½
    c. 2
    d. 3

15. An entire duct panel should be removed if a damaged area extends to within _____ inches of the duct panel edges.
    a. 2
    b. 4
    c. 6
    d. 8

Summarize the major concepts presented in the module.

Administer the Module Examination. Record the results on Craft Training Report Form 200, and submit the results to the Training Program Sponsor.

Administer the Performance Test, and fill out Performance Profile Sheets for each trainee. If desired, trainee proficiency noted during laboratory sessions may be used to complete the Performance Test. Record the results on Craft Training Report Form 200, and submit the results to the Training Program Sponsor.

Summary

Fiberglass duct, like sheet metal duct, is used to fabricate the straight duct sections and fittings that make up an air distribution system. Fiberglass is strong, lightweight, easy to handle, and increases the efficiency of an air distribution system. Types of fiberglass duct include rigid ductboard, rigid round duct, and flexible round duct (flex-duct).

Fiberglass must be properly measured, laid out, cut, and assembled before installation. Assembling fiberglass duct sections is somewhat easier than assembling sheet metal duct sections. Instead of rivets and soldering and welding techniques, fiberglass duct assembly uses staples and closure systems made up of tape or tape and mastic. Closure systems are also used to connect fiberglass ductboard and flexible round duct to sheet metal.

Fiberglass and flex-duct also require special care during installation and hanging to ensure that the system is properly supported and protected from damage. Depending on the extent of the damage, fiberglass components and closure systems can be easily fabricated and assembled for repair on the job site.

Notes

14.26 HVAC ◆ LEVEL TWO

Instructor's Notes:

Trade Terms Introduced in This Module

Closure system: Any one of the three types of sealing products used with fiberglass duct: pressure-sensitive tape, heat-activated tape, and mastic and glass fabric tape.

EI rating: A strength and rigidity rating assigned to rigid ductboard that is based on a calculation called Young's modulus of elasticity.

Fatigue test: A test in which a material is subjected to normal wear and tear to gauge how long it will remain intact and effective before beginning to break down.

Foil-scrim-kraft (FSK): A vapor barrier made from layers of aluminum foil, fiberglass yarn or scrim, and kraft paper. FSK is also flame retardant.

Plenum: In an air distribution system, an enclosed volume of air that is at a slightly higher pressure than the atmosphere and is connected to several branch supply ducts. In a return air system, the air in the plenum is at a slightly lower pressure than the atmosphere and is connected to several return air grilles or registers.

R-value: A measure of the ability of a material such as insulation to slow heat transfer, expressed as a numerical rating. Also known as thermal-resistance value. The higher the rating number, the better the insulating properties.

Shiplap: A groove cut into ductboard that allows interior edges to overlap when the ductboard is folded or fit to another piece. A special shiplap cutting tool is used.

Staple flap: A piece of ductboard facing from which the insulation has been removed that is pulled over a joint or seam and then stapled into place.

Tap-ins: Square or round fittings made of fiberglass ductboard, flexible duct, or sheet metal that tap into the main trunk duct to connect the main duct to a runout.

Torsion: The twisting of a building's structural member along its length by two equal and opposite forceful rotations at each end.

Vertical riser: In an air distribution system, a duct that extends vertically one full story or more to deliver air to branch ducts on different floors of a building.

Young's modulus of elasticity (E): A mathematical ratio that describes tensile elasticity, or the tendency of an object to deform along an axis when opposing forces are applied along that axis.

Additional Resources

This module is intended to be a thorough resource for task training. The following reference works are suggested for further study. These are optional materials for continued education rather than for task training.

"Energy Efficient Design of New Low-Rise Residential Buildings," 1992. In *National Voluntary Consensus Standard*. ANSI/ASHRAE.

Environmental Protection Agency website, www.epa.gov, "Biocontaminant Control," reviewed July 2002.

Environmental Protection Agency website, www.epa.gov, "Should You Have the Air Ducts in Your Home Cleaned," Indoor Environments Division, October 1997, www.epa.gov, reviewed July 2002.

Fibrous Glass Duct Construction Standard, Fourth Edition, 2001. Alexandria, VA: North American Insulation Manufacturers Association.

Flexible Duct Performance & Installation Standards®, Fourth Edition, 2007. Schaumburg, Illinois: Air Diffusion Council.

Fibrous Glass Duct Construction Standard, Third Edition, 1993. Alexandria, VA: North American Insulation Manufacturers Association.

Fibrous Glass Residential Duct Construction Standard, Second Edition, 1998. Alexandria, VA: North American Insulation Manufacturers Association.

UL 181, Standard for Factory-Made Air Ducts and Air Connectors, 2005. Northbrook, IL: Underwriters Laboratories, Inc.

UL 181A, Standard for Closure Systems for Use with Rigid Air Ducts and Air Connectors, 2005. Northbrook, IL: Underwriters Laboratories, Inc.

UL 181B, Standard for Closure Systems for Use with Flexible Air Ducts and Air Connectors, 2005. Northbrook, IL: Underwriters Laboratories, Inc.

NFPA 90A, Standard for the Installation of Air Conditioning and Ventilating Systems, 2002. Quincy, MA: National Fire Protection Association.

Working With Fiber Glass, Rock Wool, and Slag Wool Products, 2001. Alexandria, VA: North American Insulation Manufacturers Association.

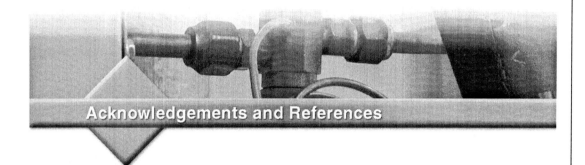

Acknowledgements and References

Acknowledgments

The material and figures in this module have been adapted from *Fibrous Glass Duct Construction Standard* (fourth edition), *Fibrous Glass Duct Construction Standard* (third edition), and *Fibrous Glass Residential Duct Construction Standard* (second edition) with the permission of the North American Insulation Manufacturers Association (NAIMA). For more information about NAIMA and its publications, contact NAIMA, 44 Canal Center Plaza, Suite 310, Alexandria, VA 22314.

NAIMA assumes no responsibility and accepts no liability for the application of principles or techniques covered in this module. NAIMA makes no warranty of any kind, express or implied, or regarding merchantability or fitness for any particular purpose in connection with the information supplied herein. Authorities considering adoption of NAIMA standards should review all federal, state, local, and contractual regulations applicable to specific installations. These materials are not intended to address issues relating to thermal or acoustical insulation within and furnished as integral parts of HVAC equipment such as air handling units, coils, air cleaners, silencers, humidifiers, and terminal devices. Manufacturers of such equipment are responsible for design, specification, and installation of appropriate insulation components in their products so that thermal, acoustical, and indoor air quality requirements are met.

Figure Credits

Michael Bergen, 214F01

Courtesy of North American Insulation Manufacturers Association (NAIMA), 214F02–214F19, 214F25–214F30

MODULE 03214-07 — TEACHING TIPS

The following are suggested activities or instructional methods to help you teach the material in this module.

General

When you call on someone to answer a question, the rest of the class relaxes or even tunes out because they expect that the question and answer will take place only between you and the trainee you called on. Instead, use this technique to involve more trainees in answering questions and to keep them on their toes.

1. Ask trainees to define a term or explain a concept.
2. After one trainee has answered, ask a trainee seated nearby if the answer is right. Then ask whether a trainee in the back of the room agrees.
3. Ask trainees to explain why they think an answer is right or wrong.
4. Use the session to clear up incorrect ideas and encourage trainees to learn from their mistakes.

Sections 1.0.0 through 10.0.0 *Quick Quiz*

This Quick Quiz will familiarize trainees with the terms and definitions that are commonly used when discussing fiberglass and flexible duct systems. You will need photocopies of the quiz provided in the following page. Trainees will need pencils. If you allow trainees to use the Trainee Guide, decrease the amount of time you give them to complete the quiz.

1. Make a photocopy of the quiz for each trainee.
2. Give trainees between 5 and 10 minutes to complete the quiz.
3. Go over the answers to the quiz.
4. Ask trainees if they have questions.

Answers to Quick Quiz

1. d
2. b
3. j
4. c
5. i
6. a
7. g
8. e
9. h
10. f

Quick Quiz *Fiberglass and Flexible Duct Systems*

For each description listed, identify the term that the text best describes. Write the corresponding letter in the blank provided.

_____ 1. The exterior of rigid round duct is covered with _____ vapor barriers.

_____ 2. The strength and stiffness of rigid ductboard is identified by its _____.

_____ 3. The EI rating is the result of a calculation based on the moment of inertia and _____.

_____ 4. Tests used to gauge the ability of ducts to withstand normal operating pressure are known as _____.

_____ 5. In air duct systems servicing buildings that are more than two stories high, fiberglass duct shall not be used as a(n) _____.

_____ 6. When joining fiberglass duct, the correct _____ must be used to ensure maximum efficiency and performance.

_____ 7. To fabricate a duct module of standard straight duct, _____ and corner grooves should be cut using either hand tools or machine tools.

_____ 8. Many residential basement installations use an extended _____ supply system.

_____ 9. Tapes are used to reinforce _____ joints.

_____ 10. Many building codes include recommended _____ for ductboard based on climate zones.

a. closure system
b. EI rating
c. fatigue tests
d. foil-scrim-kraft (FSK)
e. plenum
f. R-value
g. shiplaps
h. staple flap
i. vertical riser
j. Young's modulus of elasticity

Section 2.0.0   *Types and Standards of Fiberglass Duct*

This exercise will familiarize trainees with various types and standards of fiberglass duct. Trainees will need appropriate pencils and paper. You will need to arrange for a manufacturer's representative to give a presentation on fiberglass duct. Allow 45 to 60 minutes for this exercise.

1. Discuss types and standards for fiberglass duct systems. Have trainees brainstorm questions before the presentation.
2. Have the speaker discuss various types of fiberglass duct systems and installation techniques.
3. Have the trainees take notes and write down questions during the presentation.
4. After the presentation, answer any questions trainees may have.

Section 9.0.0   *Hangers and Supports for Fiberglass and Flexible Duct*

This exercise will familiarize trainees with fiberglass and flexible duct systems. Trainees will need appropriate personal protective equipment, pencils, and paper. You will need to arrange to tour a commercial building with a fiberglass and flexible duct system. Allow 20 to 30 minutes for this exercise.

1. Describe fiberglass and flexible duct systems and the hangers and supports used to install them.
2. Tour the utility area of a commercial building or the building in which you are holding class. Point out various aspects of a fiberglass and flexible duct system including the hangers and supports used to support the duct system.
3. Answer any questions trainees may have.

MODULE 03214-07 — ANSWERS TO REVIEW QUESTIONS

Answer	Section
1. d	2.0.0
2. b	2.1.1
3. d	2.3.0
4. b	3.0.0
5. b	4.0.0
6. b	5.0.0
7. d	5.0.0
8. d	6.0.0
9. c	6.0.0
10. a	8.0.0
11. c	9.1.1
12. a	9.1.0
13. b	9.2.0
14. a	10.1.0
15. c	10.2.0

CONTREN® LEARNING SERIES — USER UPDATE

NCCER makes every effort to keep these textbooks up-to-date and free of technical errors. We appreciate your help in this process. If you have an idea for improving this textbook, or if you find an error, a typographical mistake, or an inaccuracy in NCCER's Contren® textbooks, please write us, using this form or a photocopy. Be sure to include the exact module number, page number, a detailed description, and the correction, if applicable. Your input will be brought to the attention of the Technical Review Committee. Thank you for your assistance.

Instructors – If you found that additional materials were necessary in order to teach this module effectively, please let us know so that we may include them in the Equipment/Materials list in the Annotated Instructor's Guide.

Write: Product Development and Revision
National Center for Construction Education and Research
3600 NW 43rd St, Bldg G, Gainesville, FL 32606

Fax: 352-334-0932

E-mail: curriculum@nccer.org

Craft _____ Module Name _____

Copyright Date _____ Module Number _____ Page Number(s) _____

Description
_____
_____
_____
_____

(Optional) Correction
_____
_____
_____

(Optional) Your Name and Address
_____
_____
_____

Glossary of Trade Terms

Absolute zero: The temperature at which all molecular motion ceases. It is -460°F, -273°C, and 0°K (Kelvin).

Actuator: The portion of a regulating valve that converts one type of energy, such as pneumatic pressure, into mechanical energy (for example, opening or closing a valve).

Adiabatic: A term used to describe a thermodynamic process that happens without loss or gain of heat.

Air handler: A commercial air handler is a packaged unit containing a cooling coil and usually several other components such as a filter, heating coil or element, dampers, and fans connected to ducts. Conditioned air from the air handler leaves through ducts and is delivered to air terminals for distribution around the conditioned spaces.

Alloy: A metal mixed with other metals or elements to create strength and durability.

Alternator: A device that generates alternating current by means of conductors rotated in a magnetic field.

Analog-to-digital converter: A device designed to convert analog signals such as temperature and humidity to a digital form that can be processed by logic circuits.

Angle bracket: An L-shaped metal supporting member used to support vertical risers. Also called angle iron.

Anode: The positive terminal of a diode.

Aquastat: A control that works basically the same way as a thermostat with the exception that it is designed to control water temperature instead of air temperature.

Armature: The rotating component of a generator.

Automatic changeover thermostat: A thermostat that automatically selects heating or cooling.

Axial load: An external load that acts lengthwise along a shaft.

Balance point: The outdoor temperature at which the heating capacity of the heat pump is equal to the heat loss of the building. The balance point varies depending on the climate, building design, type of construction, and other factors that affect heat loss or gain.

Basic input/output system (BIOS): The basic method by which a computer exchanges information.

Bleed control: A valve with a tiny opening that permits a small amount of fluid to pass.

Break-away torque: The torque required to loosen a fastener. This is usually lower than the torque to which the fastener has been tightened.

Bridge rectifier: A rectifier circuit that uses four diodes, two of which conduct current on each half-cycle. Has the advantage of not needing a center-tapped transformer.

Burnout: The condition in which the breakdown of the motor winding insulation causes the motor to short out or ground electrically.

Capacitance boost: A procedure used to start a stuck PSC compressor. It involves momentarily connecting a start (boost) capacitor across the run capacitor of the stuck compressor in an attempt to start the stuck compressor by increasing the starting torque.

Capacitor: An electrical storage device containing two metal plates separated by an insulating (dielectric) material.

Carbon: A non-metallic element found in nature, such as in diamonds and graphite, or a part of coal or petroleum. Carbon combines with iron to make steel.

Cathode: The negative terminal of a diode.

Cavitation: The result of air formed due to a drop in pressure in a pumping system.

Centrifugal force: The force that makes rotating objects tend to move away from the center of rotation.

Channel: A U-shaped piece of structural steel used as a supporting device.

Chip: A common term used to describe an integrated circuit.

Closure system: Any one of the three types of sealing products used with fiberglass duct: pressure-sensitive tape, heat-activated tape, and mastic and glass fabric tape.

Coefficient of Performance (COP): The ratio of work performed in relation to energy used. A rating method for heat pumps. It is further defined as the Btuh output divided by the total electrical input (watts) required to produce this Btuh output times 3.413.

Glossary of Trade Terms

Cold-rolled steel: Metal that has been formed by rolling at room temperature, usually to obtain an improved surface or higher tensile strength.

Commutator: The movable contact surface on an electric generator or motor.

Complete combustion: Burning in which there is enough oxygen to prevent the formation of carbon monoxide.

Compression heat: The heat produced by a heat pump when the refrigerant cycle is reversed. See *reverse cycle heat*.

Condensing furnace: A high-efficiency furnace containing a secondary heat exchanger that extracts additional heat from the flue gases.

Constant-volume system: A constant-volume system maintains a constant airflow while varying the air temperature in response to the space load.

Control zone: In HVAC, a control zone is a building, group of rooms, single room, or part of a room controlled by its own thermostat.

Cooling compensator: A fixed resistor installed in a thermostat to act as a cooling anticipator.

Corrosion: The breaking down or destruction of a material, especially a metal, through chemical reactions. The most common form of corrosion is rusting, which occurs when iron combines with oxygen and water.

Deadband: A temperature band, usually 3°F, that separates heating and cooling in an automatic changeover thermostat.

Defrost: In a heat pump, the process of cycling hot refrigerant through the outdoor coil to melt accumulated frost due to condensation.

Dehumidifier: A device used to remove moisture from the air.

Diac: A three-layer diode designed for use as a trigger in AC power control units.

Dielectric: A material that strongly resists the passage of current.

Differential: The difference between the cut-in and cut-out points of a thermostat.

Dilution air: Air added to the flue gases in a natural-draft furnace to aid flue gas removal.

Droop: A mechanical condition caused by heat that affects the accuracy of a bimetal thermostat.

Dropping point: The temperature at which grease is soft enough for a drop of oil to fall away or flow from the bulk of the grease.

Dual-duct systems: Dual-duct systems condition all the air in a central unit and distribute it to the conditioned spaces through two parallel ducts, one duct carrying cold air and the other warm air. In each conditioned zone, a device mixes the warm and cold air in proper proportions to satisfy the load of the zone.

Dual-fuel or hybrid system: A system in which a heat pump is combined with a furnace.

Dynamic seal: A seal made where there is movement between two mating parts, or between one of the parts and the seal.

Economizer: An HVAC device that substitutes outdoor air for the cooled air produced by the air conditioning system, when outdoor air conditions permit. It also controls the amount of outdoor air used to ventilate a building.

Effective voltage: See *root-mean-square (rms) voltage*.

EI rating: A strength and rigidity rating assigned to rigid ductboard that is based on a calculation called Young's modulus of elasticity.

Electric-pneumatic (E-P) relay: A three-way pneumatic solenoid valve.

Electromechanical components: Electrical devices that contain moving parts.

Electronics: The science that deals with the behavior and effects of electron movement in conductors, insulators, and semiconductors.

Energy recovery ventilator (ERV): HVAC equipment used to supply fresh air and recover both heating and cooling energy year-round.

Enthalpy: The total heat content of a substance. In HVAC, the total heat content of the air and water vapor mixture as measured from a predetermined base or point.

Evaporation: The condition in which the heat absorbed by a liquid causes it to change into a vapor.

Fan coil unit: A term often applied to the indoor unit of a split system. Also known as an air handler.

Fatigue test: A test in which a material is subjected to normal wear and tear to gauge how long it will remain intact and effective before beginning to break down.

Glossary of Trade Terms

Fault isolation diagram: A troubleshooting aid usually contained in the manufacturer's installation, startup, and service instructions for a particular product. Fault isolation diagrams are also called troubleshooting trees. Normally, fault isolation diagrams begin with a failure symptom then guide the technician through a logical decision-action process to isolate the cause of the failure.

Fire point: The temperature at which oil will burn if ignited.

Flame rectification: A process in which exposure of a sensing rod to a flame produces a tiny current that can be sensed and used to control the gas valve.

Flash point: The temperature at which oil gives off ignitable vapors.

Flooded starts: When a compressor is started, slugging, foaming, and inadequate lubrication occur as a result of the oil in the compressor crankcase having absorbed refrigerant during shutdown. Flooded starts can also occur as a result of rapid return of liquid refrigerant during startup.

Flooding: The condition in which there is a continuous return of liquid refrigerant or liquid droplets in the suction vapor being returned to the compressor during operation.

Foil-scrim-kraft (FSK): A vapor barrier made from layers of aluminum foil, fiberglass yarn or scrim, and kraft paper. FSK is also flame retardant.

Four-way valve: A heat pump reversing valve. Also known as a switch-over or cross-over valve.

Fractionation: A term related to refrigerant blends; refers to the process by which each of the refrigerants in the blend leaks at a different rate if released into the atmosphere.

Free cooling: A mode of economizer operation. It is the cooling provided by outside air rather than the compressor.

Free electrons: Valence electrons that can easily be knocked out of orbit.

Frequency: The number of complete cycles of an alternating current, sound wave, or vibrating object that occur in a period of time.

Full-wave rectifier: A rectifier circuit that uses two diodes.

Fusible link: A circuit protective device that melts, opening the circuit, when the current is excessive.

Gravity hot-water system: A hot-water heating system in which the circulation of the hot water through the system results from thermal conduction. No system circulating pump is used.

Ground-source (geothermal) system: A system in which the outdoor coil is buried in the ground and the heat exchange occurs between the earth and the refrigerant flowing through the coil.

Half-wave rectifier: A rectifier circuit that uses a single diode.

Head pressure: A measure of pressure drop, expressed in feet of water or psig. It is normally used to describe the capacity of circulating pumps. It indicates the height of a column of water that can be lifted by the pump, neglecting friction losses in piping. Commonly referred to as head.

Heat anticipator: A resistive heating element in a thermostat that shuts off the furnace before the space temperature reaches the setpoint. It prevents the system from exceeding the desired temperature.

Heat recovery ventilator (HRV): HVAC equipment that saves energy by using a heat exchanger to transfer heat from the heating system exhaust air to the cold ventilation air that is entering the building.

Heat sink: A low-temperature surface to which heat can be transferred.

Heating Season Performance Factor (HSPF): A heat pump performance rating that has been adjusted for seasonal operation. It is the total heating output of a heat pump (in Btus) during its normal annual usage period for heating divided by the total electric power input in watt-hours during the same period.

Hertz (Hz): The unit of measure for the frequency of alternating current. One Hertz equals one cycle per second.

High/low pump head: Trade terms used to indicate the relative magnitude of the height of a column of water that a circulating pump is moving, or must move, in a water system. See head pressure.

Humidifier: A device used to control humidity.

Glossary of Trade Terms

Hydronic system: A system that uses water or water-based solutions as the medium to transport heat or cold from the point of generation to the point of use.

Hygroscopic: The characteristic of absorbing and retaining moisture. The word stems from *hygroscope*, which is an instrument showing changes in humidity.

Incomplete combustion: Burning in which there is not enough oxygen to prevent the formation of carbon monoxide.

Indoor coil: The designation given to the heat pump coil used to transfer heat to or from the conditioned space.

Induced-draft furnace: A fan-assisted furnace with an AFUE rating of 78 to 85 percent.

Induction motor: An AC motor.

Induction: To generate a current in a conductor by placing it in a moving magnetic field.

Inertia: The tendency of a body in motion to remain in motion and a body at rest to remain at rest.

Integrated circuit: A plug-in circuit containing microminiature electronic circuits. Sometimes called a chip.

Invar®: An alloy of steel containing 36 percent nickel. It is one of the two metals in a bimetal device.

Isolation transformer: A transformer with a one-to-one turns ratio. It is used for personnel safety and to prevent electrical interference.

Isothermal: The relationship between variables, especially pressure and volume, at a constant temperature.

Journal: The part of a shaft, axle, spindle, etc., which is supported by and revolves in a bearing.

Kelvin scale: A temperature scale in which zero equals -460°F. See *absolute zero*.

Label diagram: A troubleshooting aid usually placed in a convenient location inside the equipment. It normally depicts a wiring diagram, a component arrangement diagram, a legend, and notes pertaining to the equipment.

Ladder diagram: A troubleshooting aid that depicts a simplified schematic diagram of the equipment. The load lines are arranged like the rungs of a ladder between vertical lines representing the voltage source. Normally, all the wire color and physical connection information is eliminated from the diagram to make it easier to use by focusing on the functional, not the physical, aspects of the equipment.

Lap-joined: A condition in which one piece is joined to another by partly covering one piece with the other.

Light-emitting diode (LED): A diode that gives off light when current flows through it.

Mastic: A protective coating applied by trowel or spray on the surface of thermal insulation to prevent its deterioration and to weatherproof it.

MBh: One MBh equals 1,000 Btus per hour.

Mechanical cooling: A mode of economizer operation. It is the cooling provided in the conventional manner by the compressor.

Megohmmeter (megger): A test instrument used to test high-resistance circuits.

Microfarad: One-millionth of a farad. Used to rate capacitors.

Microminiaturization: The technology that allows the manufacture of microscopic electronic circuits.

Micron: One-millionth of a meter (about $1/25,400$ of an inch). It is also a precise measurement of pressure used with electronic vacuum measuring instruments and vacuum pumps. One inch of mercury equals 25,400 microns.

Microprocessor: An integrated circuit chip designed to perform computing functions. The microprocessor is the heart of a personal computer.

Modulating gas valve: A gas valve that provides precise control of burner capacity by modulating the flow of gas to the burners in small increments.

Monochrome: Able to display a single color.

Natural-draft furnace: A furnace that depends on the pressure created by the heat in the flue gases to force them out through the vent system.

Glossary of Trade Terms

Noncondensible gas: A gas (such as air) that does not change into a liquid at operating temperatures and pressures.

Non-ferrous: A metal that contains little or no iron.

Opposed-blade damper: A type of damper in which each blade moves in a direction opposite the blade next to it.

Outdoor coil: The heat pump coil used to transfer heat to or from the outdoor air.

Outgassing: The slow release of a gas that was trapped or absorbed by a material.

Oxidation: The process of combining with oxygen. All petroleum products react with oxygen to some degree, and this increases as the temperature increases.

Packaged air conditioner (PAC): Packaged air conditioners provide cooling only or can provide both cooling and heating when equipped with electric resistance heaters.

Packaged rooftop unit: A self-contained air conditioning unit that is installed outdoors on a rooftop with connections through a roof opening to the internal duct system. They are commonly used on flat roof commercial structures such as office buildings and shopping malls.

Packaged unit: Packaged units are factory-assembled units that contain all the components needed to support an HVAC function, such as cooling, heating, or air handling.

Packaged unit: A self-contained air conditioning system.

Packing gland: A part used to compress the packing in a stuffing box.

Photo diode: A diode that conducts current when exposed to light.

Pixel: An abbreviation for picture element. A pixel is a single dot in a graphic image on a computer screen.

Plenum: In an air distribution system, an enclosed volume of air that is at a slightly higher pressure than the atmosphere and is connected to several branch supply ducts. In a return air system, the air in the plenum is at a slightly lower pressure than the atmosphere and is connected to several return air grilles or registers.

Pneumatic-electric (P-E) relay: A pressure switch in which a pneumatic signal causes an electrical change.

Pour point: Refers to the lowest temperature at which an oil will flow freely.

Pressure drop: The difference in pressure between two points. In a water system, it is the result of power being consumed as the water moves through pipes, heating units, and fittings. It is caused by the friction created between the inner walls of the pipe or device and the moving water.

Primary air: Air that is added to the fuel before it goes to the burner.

Radial load: The side or radial force applied at right angles to a bearing and shaft.

Reclamation: The remanufacture of used refrigerant to bring it up to the standards required of new refrigerant. Reclamation is not a field service procedure. It is a complicated process done only at reprocessing or manufacturing facilities.

Recovery: The removal and temporary storage of refrigerant in containers approved for that purpose. Recovery does not provide for any cleaning or filtration of the refrigerant.

Rectification: The conversion of AC into DC using diodes.

Recycle: To circulate recovered refrigerant through filtering devices that remove moisture, acid, and other contaminants. This does not mean that it meets the purity standards for new refrigerants.

Redundancy: In HVAC systems, designs that provide a back-up of primary equipment such as boilers or pumps, allowing for system operation to continue in spite of a failed unit. With 100 percent redundancy, for example, a system may have two boilers installed, each sized to handle the complete heating needs of the structure alone

Reverberator: A wire screen used to cover the ceramic grids in a gas-fired heater that operates by submerged flame.

Reverse cycle heat: The heat produced by a heat pump when refrigerant flow is reversed. See *compression heat.*

Reversing valve: A valve that changes the direction of refrigerant flow in a heat pump. See *four-way valve.*

Glossary of Trade Terms

Root-mean-square (rms) voltage: The value of AC voltage that will produce as much power when connected across a load as an equivalent amount of DC voltage. Also known as effective voltage.

Rotor: The rotating component of an induction motor.

Run capacitor: A capacitor that remains in the motor circuit while the motor is running to improve running efficiency.

Run winding: The stator winding of a motor that draws current during the entire running cycle of the motor.

Run-down resistance: The torque required to overcome the resistance of associated hardware, such as locknuts and lockwashers, when tightening a fastener.

Runout: A term describing the connection between the main trunk line of duct and the diffuser, grille, or register.

R-value: A measure of the ability of a material such as insulation to slow heat transfer, expressed as a numerical rating. Also known as thermal-resistance value. The higher the rating number, the better the insulating properties.

Seasonal Energy Efficiency Ratio (SEER): The ARI standard for measuring heat pump efficiency. It is the total cooling of a heat pump (in Btus) during its normal annual period for cooling divided by the total electric input in watt-hours during the same period.

Secondary air: Air that is added during combustion.

Seized (stuck) compressor: The condition in which a compressor is electrically or mechanically unable to start (locked up). It hums but will not start, and draws locked rotor current.

Semiconductor: A material that contains four valence electrons and is used in the manufacture of integrated circuits.

Set or seizure: In the last stages of rotation in reaching a final torque, the fastener may lock up; this is known as seizing or set. This is usually accompanied by a noticeable popping effect.

Shear load: The amount of weight or pressure that causes an exposed piece of metal to break or shear off.

Short cycling: The condition in which the compressor is restarted immediately after it has been turned off.

Sheet metal gauge: A measuring tool used to find the thickness, or gauge, of sheet metal.

Sheet steel: An uncoated sheet of steel with a bluish-black surface.

Shiplap: A groove cut into ductboard that allows interior edges to overlap when the ductboard is folded or fit to another piece. A special shiplap cutting tool is used.

Sick building syndrome: A combination of symptoms (headache, nausea, eye, nose and throat irritation) that are attributed to flaws in the HVAC systems. Symptoms can be cured by boosting the overall turn-over rate in fresh air exchange with the outside air. Other causes have been attributed to contaminants produced by out-gassing of some types of building materials, or improper exhaust ventilation of light industrial chemicals.

Silicon-controlled rectifier (SCR): A device that is used mainly to convert AC voltage into DC voltage. To do so, however, the gate of the SCR must be triggered before the device will conduct current.

Sinusoidal (sine) wave: The waveform created by an AC generator.

Slugging: The condition that occurs when a compressor tries to compress liquid refrigerant, oil, or both, instead of superheated gas.

Sound attenuation: The reduction in the level of sound that is transmitted from one point to another.

Specific heat: The amount of heat required to raise the temperature of one pound of a substance one degree Fahrenheit. Expressed as Btu/lb/°F. At sea level, water has a specific heat of 1 Btu/lb/°F. At sea level, air has a specific heat of 0.24 Btu/lb/°F.

Split system: An air conditioning system with an indoor coil and an outdoor coil connected with refrigerant lines.

Stainless steel: A high-strength, tough, corrosion- and rust-resistant alloy that contains chromium and sometimes nickel.

Staple flap: A piece of ductboard facing from which the insulation has been removed that is pulled over a joint or seam and then stapled into place.

Glossary of Trade Terms

Start winding: The stator winding of a motor that is used to provide starting torque.

Static pressure: In a water system, static pressure is created by the weight of the water in the system. It is referenced to a point such as a boiler gauge. Static pressure is equal to 0.43 pounds per square inch, per foot of water height.

Static seal: A seal made where there is no movement between the two joining parts or between the seal and the mating part.

Stator: The stationary windings of a motor.

Stuffing box: The housing used to control leaking along a shaft or rod. Typically composed of three parts: the packing chamber (also called the box); the packing rings; and the gland follower (also called the stuffing gland).

Sub-base: The portion of a two-part thermostat that contains the wiring terminals and control switches.

Submerged flame: A method of gas-fired heater operation that uses burning gas within two-layer porous ceramic grids.

Synchronous speed: The maximum rated speed of a motor.

Tap-ins: Square or round fittings made of fiberglass ductboard, flexible duct, or sheet metal that tap into the main trunk duct to connect the main duct to a runout.

Temperature glide: A range of temperatures in which a zeotrope refrigerant will evaporate and condense for a given pressure.

Tensile load: The weight required to cause metal to stretch or compress.

Tensile strength: The resistance of a material against rupture when placed under tension.

Thermostat base: The portion of a two-part thermostat that contains the heating and cooling thermostats.

Thrust: The force acting lengthwise along the axis of a shaft, either toward it or away from it.

Tolerance: The amount of variation allowed from a standard.

Torque: The force that must be generated to turn a motor.

Torque: The resistance to a turning or twisting force.

Torsion: The twisting of a building's structural member along its length by two equal and opposite forceful rotations at each end.

Triac: A bi-directional triode thyristor that functions as an electrically controlled switch for AC loads.

Troubleshooting table: A troubleshooting aid usually contained in the manufacturer's installation, startup, and service instructions for a particular product. Troubleshooting tables are intended to guide the technician to a corrective action based on observations of system operation.

Troubleshooting: A procedure by which the technician locates the source of a problem, then makes the repairs and/or adjustments to correct the cause of a problem so that it will not recur.

Turns ratio: The ratio between the number of turns in the primary and secondary windings of a transformer.

Two-stage gas valve: A gas valve body containing two separate valves that supply fuel gas to the burners based on the amount of heat demand.

Valence electrons: Electrons located in the outer orbit of an atom.

Vane switch: A switch that is actuated by a vane that reacts to the flow of water or air.

Vapor barrier: A barrier, such as a mastic, placed between insulation and the surrounding air to control condensation.

Variable air volume (VAV) system: A VAV system is one that controls the temperature within a control zone by varying the quantity of supply air rather than by varying the supply air temperature. Dual-duct VAV systems blend cold and warm air in various volume combinations.

Variable volume, variable temperature (VVT) system: A VVT system is one that delivers a variable volume of air to each controlled zone, as the load dictates. The temperature of the air supplied by the central unit varies with time.

Vent connector: The horizontal section of the vent system that connects the appliance(s) to the vent pipe or chimney.

Vent: The vertical section of the vent pipe.

Glossary of Trade Terms

Vertical riser: In an air distribution system, a duct that extends vertically one full story or more to deliver air to branch ducts on different floors of a building.

Viscosity: The thickness of a liquid or its ability to flow at a specific temperature.

Viscosity index (VI): A measure of how an oil's viscosity varies with temperature.

Wiring diagram: A troubleshooting aid, sometimes called a schematic, that provides a picture of what the unit does electrically and shows the actual external and internal wiring of the unit.

Year-round air conditioner (YAC): The year-round air conditioner provides both cooling and heating. It differs from a packaged air conditioner in that its heating capability is provided by a natural or LP gas heating section, instead of electric resistance heaters.

Young's modulus of elasticity (E): A mathematical ratio that describes tensile elasticity, or the tendency of an object to deform along an axis when opposing forces are applied along that axis.

Zoned system: A system that has more than one thermostat used to control the areas (zones) it conditions.

HVAC Level Two

Index

Index

A

Absolute zero, 11.2, 11.38, 13.3
AC. *See* Alternating current
ACCA. *See* Air Conditioning Contractors of America
Accidents. *See also* Explosion; Fire
   burns, 3.8, 6.29, 10.35, 12.26, 14.3, 14.9
   cuts, 12.38, 12.39, 14.3
   with drill press, 12.38
   electric shock, 6.29–6.30, 8.10, 8.30, 10.17, 12.39
   falls, 14.3
Accumulator, liquid (suction), 10.18, 11.16, 11.17
Acids, 5.9, 5.21, 9.10, 10.34–10.36
Actuator, 1.27, 3.22, 6.20, 8.53, 8.54, 8.65
Adhesives
   for blanket fiberglass, 13.22, 13.23, 13.24
   for fiberglass duct, 14.7, 14.8
   for fibrous glass duct liner, 13.13, 13.14
   for gaskets, 12.15
Adiabatic process, 1.45, 1.50
AFUE. *See* Annual Fuel Utilization Efficiency
Air
   common particulates, 4.11
   components, 2.4
   compressed, 5.6, 5.15, 8.53
   as contaminant in refrigeration system, 10.34
   dilution, 1.5, 2.4, 2.20
   excess, 2.4
   in hot-water heating system, 3.15–3.16, 3.18
   maximum allowable velocity, 1.19–1.20
   primary, 2.3, 2.20
   return. *See* Exhaust air
   secondary, 2.3, 2.4, 2.20
   supply. *See* Supply air
   techniques for varying velocity, 1.39
   temperature and circulation dynamics, 1.23, 1.25, 2.8–2.9, 4.4, 11.20
Air cleaner, electronic, 4.13–4.14, 12.41, 12.42
Air conditioners
   branch circuits, 6.12
   packaged (PAC), 1.31–1.32, 1.35, 1.50
   packaged terminal (PTAC), 10.33
   vertical packaged (VPAC), 1.32–1.33
   year-round (YAC), 1.32, 1.35, 1.50, 9.5

Air Conditioning and Refrigeration Institute (ARI), 5.2, 11.2, 11.25
Air Conditioning Contractors of America (ACCA), 2.9, 5.2
Air Diffusion Council (ADC), 14.4
Air exchange rate, 4.10, 4.15
Air handlers, 1.3, 1.4, 1.39–1.46, 1.50, 11.18
Airport, 4.14
Air purification system, ultraviolet light, 4.21
Air quality, indoor. *See also* Air conditioners; Dehumidifiers; Filters; Humidifiers
   ASHRAE standards, 1.44, 4.10
   carbon dioxide from respiration, 1.6
   chemical contaminants, 1.5
   common pollutants, 1.5, 4.10
   contaminated humidifier, 1.45
   contaminated supply air, 2.6
   dust, 1.45, 4.8, 4.11, 14.8
   equipment, 4.10–4.14
   maintaining body comfort, 4.2–4.4, 4.24, 9.2
   odors. *See* Odors and odor control
   sick building syndrome, 1.5, 1.7, 1.50, 4.2
   smoke from cigarettes, 4.2, 4.10, 4.11
   smoke from fire, 1.8
   ultraviolet light air purification system, 4.21
Airshaft, 1.21
Air source equipment, 1.30–1.39
Air terminals
   in all-air chilled-water system, 1.5
   in all-air refrigeration system, 1.3, 1.4
   dual-duct variable air volume system, 1.18, 1.19, 1.29–1.30
   mixing box, 1.28–1.29
   single-duct variable air volume system, 1.17, 1.18, 1.27–1.28
   supply outlets, 1.23–1.26
   types, 1.22, 1.26–1.30
Alcohol, 12.50
All-air system
   components, 1.3–1.5, 1.6
   regulation of supply and exhaust air, 1.5–1.10
   types, 1.10–1.19
Alloy, 13.2, 13.3–13.4, 13.35

Alternating current (AC)
　current draw, 8.9–8.10
　effects on human body, 6.30
　historical background, 6.2
　induction motors, 6.18–6.24
　overview, 6.2, 6.18, 6.34
　power generation and distribution, 6.5–6.16, 7.5
　safety, 6.29–6.30
　sine wave, 6.2, 6.6–6.7, 6.35
　testing components, 6.24–6.29
　transformers, 6.2–6.5
　using AC power, 6.16–6.18
　voltage on circuit diagrams, 6.30–6.31
Alternator, 6.6, 6.35
Aluminum, 1.21, 13.3, 13.4, 13.36
American Gas Association, 2.5
American National Standard, 12.2
American Society of Heating, Refrigerating and Air-Conditioning Engineers (ASHRAE)
　building ventilation, 1.6, 1.10, 4.2, 4.15
　fiberglass duct, 14.4
　indoor air quality, 1.44, 4.10
　prohibition of reheat, 1.14
　R-value, 13.10
American Society of Testing and Materials International (ASTM), 12.3, 12.27, 13.12
Ammeter, clamp-on, 8.10, 8.35, 8.40
Amperage draw, 8.10–8.11
Analog signal, 8.58, 8.59
Annual Fuel Utilization Efficiency (AFUE), 2.2, 9.7, 9.8, 9.10
Anode, 7.5, 7.10, 7.30
Antifreeze, 11.6, 11.7
Apartment, 1.20
Aquastat, 3.13, 3.39, 8.50–8.51
ARI. *See* Air Conditioning and Refrigeration Institute
ARI Performance Rating Directory, 11.25
Armature, 6.6, 6.8–6.9, 6.35
ASHRAE. *See* American Society of Heating, Refrigerating and Air-Conditioning Engineers
ASTM. *See* American Society of Testing and Materials International
Atomic structure, 7.2–7.3, 7.4
Attic, 2.7, 11.28
Auditorium, 1.20

B

Backdraft, 2.9
Backflow preventer, 3.11, 3.20–3.21, 3.29, 3.30
Balance point, 11.18, 11.26–11.27, 11.38
Bank, 1.14, 1.16, 1.18, 1.20
Basement, 2.5
Basic input/output system (BIOS), 7.20, 7.30. *See also* Input/output function
Battery, 6.2, 8.7, 8.67, 9.14
Bearings
　anti-friction, 12.22–12.23, 12.24
　common terms, 12.21
　identification of failures, 12.23–12.24
　lubrication, 12.22, 12.24, 12.40
　plain, 12.21–12.22
　removal and installation, 12.25–12.27

Behavior on the job, 12.47, 12.50
Bellows, 8.54
Belt and belt drive, 12.29–12.31
Bentonite, 11.7
Bias, forward and reverse, 7.5, 7.6
Bimetal element, 8.2, 8.11, 8.21, 8.22, 8.54
BIOS. *See* Basic input/output system
Bladder, 3.14, 3.15
Bleed control, 8.54, 8.65
Blowers
　air conditioner, 9.5
　electric heating system, 11.20, 11.21
　furnace, 2.10, 9.5, 9.9–9.10, 9.21, 9.22, 11.30. *See also* Fans, furnace
　refrigeration system evaporator, 10.26, 10.48
BOCA. *See* Building Officials and Code Administrators International
Boilers
　condensing, 3.9, 3.12
　control circuit, 9.11
　in domestic water heater, 3.24, 3.25
　dual-temperature water system, 3.31
　forced hot-water system, 3.6
　gas-fired, 3.2, 9.10–9.12
　low water cutoff, 8.51–8.52
　operation and types, 3.7–3.10, 3.38
　safety controls and accessories, 3.10–3.13
　steam, 3.8
　troubleshooting, 8.51–8.52, 9.10–9.12
Bolts
　anchor, 12.7–12.8, 12.13
　installation, 12.12, 12.13
　threaded, 12.3, 12.4–12.5, 12.7–12.8
　tightening sequence, 12.12, 12.15
　toggle, 12.7, 12.8
　torque, 12.11
Boxes. *See* Burner box; Duct box; Firebox; Fuse box; Junction box; Mixing box; Outlet box; Stuffing box
Bracket, angle, 13.7, 13.35
Brass, 13.3
Brazing, 5.7
British thermal unit (Btu), 6.16–6.17
Bronze, 12.22, 13.3
Building code, 14.6
Building Officials and Code Administrators International (BOCA), 14.6
Building-related illness, 1.7. *See also* Sick building syndrome
Burner box, 9.7, 9.10, 9.17
Burners
　boiler, 3.7, 3.9–3.10, 8.11, 9.12
　flame detector to monitor, 7.15–7.16
　flames from, 2.3–2.4, 9.7
　infrared gas-fired heater, 9.12
　input adjustment, 2.10
　troubleshooting, 8.51, 9.30–9.31
Burnout, 10.18, 10.34–10.35, 10.36, 10.40
Business card, 12.50

C

CABO. *See* Council of American Building Officials
Cadmium sulfide, 7.15–7.16

Cafeteria, 1.20
Calculator, subcooling charging, 5.29, 5.30
Calipers, 12.34–12.36
Cap, weather, 2.7
Capacitance boost, 10.17, 10.40
Capacitor analyzer, 6.24
Capacitors
   electric heating system, 11.20
   overview, 6.17–6.18, 6.35
   rectifier, 7.7
   single-phase motor, 6.20–6.22
   symbol, 8.67
   troubleshooting, 6.28, 8.46, 10.11, 10.19
Carbon, 12.16, 12.17, 13.2, 13.4, 13.35
Carbon dioxide
   from combustion, 2.2, 2.4
   in flue gases, 2.4
   from respiration, 1.6
   sensing equipment, 1.10, 1.11, 1.14, 4.22
Carbon monoxide, 2.3, 2.4, 2.17, 4.10, 4.22, 9.10
Carborundum, 9.18
Carpeting, 1.5, 4.10
Casing, air handler, 1.42
Cathode, 7.5, 7.6, 7.10, 7.30
CAV. *See* Constant air volume box
Cavitation, 3.17, 3.18, 3.39
CD. *See* Compact disc
Ceiling, 1.6, 2.7, 13.9, 13.26, 13.27
Central units
   constant volume system, 1.11, 1.12, 1.13
   exhaust air, 1.7–1.8, 11.7
   variable air volume system, 1.17, 1.18, 1.30
   variable volume, variable temperature system, 1.15, 1.16
Centrifugal force, 7.2, 7.3, 7.30
Certification, technician, 5.2
CFC. *See* Chlorofluorocarbon
Channels, 13.7, 13.35, 14.12, 14.14
Charcoal, 4.14
Charging process, refrigeration system, 5.18–5.32
Chattering (bounce), 8.7, 10.13
Checklist, 12.45
Chiller, 1.5, 1.13, 3.31, 5.24
Chimney, 2.7, 2.12–2.13
Chimney liner, 2.12–2.13
Chips
   integrated circuit (IC), 7.18, 7.19, 7.30
   microprocessor, 7.2, 7.3, 7.18–7.19, 7.30
Chlorofluorocarbon (CFC), 5.3, 5.6, 5.8
Chromium, 13.3, 13.4
Church, 1.20
Circuit breaker, 6.4, 6.10, 8.39–8.40, 11.20, 11.21
Circuit diagram, 6.10, 6.30–6.31, 8.26, 10.8
Circuits
   branch, 6.10–6.12
   bridge, 7.14
   capacitive, 6.18, 6.19
   compressor, 8.20
   control
      compressor, 8.20–8.21
      gas heating system, 9.2–9.15
      heat pump, 11.32–11.34
      HVAC, 8.13–8.26, 8.34, 8.64

      overview, 7.18, 8.2
      refrigeration system, 10.7–10.10
   delta arrangement, 6.13, 6.14, 6.24
   ECM, 7.16
   and electronic components, 7.4–7.17
   furnace, 8.22
   gas heating system, 9.2–9.15
   glow coil, 9.18
   HVAC, 8.34
   inductive, 6.17, 6.19, 8.41–8.42
   integrated microprocessor, 7.2, 7.18–7.19, 7.30, 10.15
   printed circuit boards (PC board), 7.4, 7.17–7.19,
      8.16–8.17, 10.11, 11.11, 11.12
   recording instruments, 6.26–6.27
   refrigeration system, 10.11–10.19
   resistive, 6.16–6.17, 8.41–8.42
   troubleshooting, 8.35–8.37
   wye arrangement, 6.13–6.14, 6.24
Clean Air Act, 5.2, 5.7
Cleaning products, 12.43
Clearance, 2.5, 11.28, 12.27
Climate control system, integrated programmable, 7.2, 7.20,
   8.6–8.7, 8.13, 10.15
Closure strap method, 14.11–14.12, 14.13
Closure system, for fiberglass duct, 14.6–14.9, 14.21, 14.22,
   14.27
Clothing, 12.38, 12.47, 13.15, 14.8
Coal, 2.4, 3.7, 4.11, 13.12, 14.5
Codes. *See* Building code; National Electrical Code®;
   National Fuel Gas Code
Coefficient of performance, 11.24, 11.38
Coefficient of resistance, 7.13
Coil guard, 1.38
Coils
   chilled-water, 1.4, 1.5
   condenser, 1.34–1.35, 10.26. *See also* Coils, indoor
   cooling, 1.3, 1.4, 1.42, 1.43, 9.22. *See also* Coils, outdoor
   direct-expansion (DX; evaporator), 1.3, 1.4, 11.7
   evaporator, 1.34, 10.27, 12.42. *See also* Coils, indoor
   fan, with electric heating element, 11.18, 11.19, 11.20
   glow, 9.15, 9.17–9.18
   heating, 1.10, 1.25, 1.40, 1.41–1.42, 1.42–1.43, 13.31
   indoor, in heat pump, 11.2, 11.8, 11.38
   maintenance, 12.42, 12.43
   multi-zone constant volume system, 1.12, 1.13
   outdoor, in heat pump, 11.2, 11.8, 11.12, 11.38
   in packaged equipment, 1.34–1.35
   relay, 8.15, 8.16, 8.19
   solenoid, 8.41
   subcooling, 11.7–11.8
   symbols, 8.66, 8.67
   tankless, 8.51
   two-deck multi-zone system, 1.17
   variable volume, variable temperature system, 1.15
Cold conditions
   coil freeze-up, 1.39
   furnace vent termination above snow level, 2.14
   and heat pump operation/installation, 11.2, 11.6, 11.8,
      11.28
   low-ambient control, 1.37
Collar, duct-connection, 14.11–14.14
Combustion process, 2.2–2.4, 2.20

Comfort zone, 4.2, 4.3
Commercial airside system
　air handler, 1.39–1.46
　air source equipment, 1.30–1.39
　air terminal, 1.22–1.30
　duct system, 1.19–1.22
　indoor all-air components, 1.2–1.5, 1.6
　outdoor air and air system, 1.5–1.10
　overview, 1.2
　types, 1.10–1.19
　zones, 1.2
Commissioning process, 12.45
Communication bus, 1.19
Communication with the customer, 8.27, 8.64, 12.43, 12.44, 12.50–12.56
Commutator, 6.5, 6.6, 6.35
Compact disc (CD), 7.22–7.23, 7.25–7.27
Compensator, cooling, 8.3, 8.65
Compression heat, 11.3, 11.12, 11.38
Compressors
　air, for pneumatic system, 8.53, 8.55–8.56
　all-air chilled water system, 1.5
　capacity control, 10.19
　control circuit, 8.20–8.21
　economizer, 1.10
　heat pump, 8.5, 11.12–11.13
　hermetic, 7.16, 8.46, 8.50, 10.21
　lubrication, 10.4
　packaged equipment, 1.33–1.34, 5.22
　programmable thermostat, 8.11, 8.12
　reciprocating, 10.17
　refrigeration system, 5.22, 10.2, 10.3–10.5, 10.10, 10.15–10.20, 10.26–10.29
　rotary, 10.17, 10.18
　scroll, 10.17, 10.20, 10.29
　seized (stuck), 10.16, 10.17, 10.40
　short-cycle timer, 8.20
　single-zone constant volume system, 1.10, 1.11
　start assist, 6.22, 6.23, 7.14, 10.18, 10.20
　symbol, 8.67
　three-phase circuit, 6.15
　troubleshooting, 10.15–10.20, 10.21, 10.26–10.29, 10.45–10.46
Computers
　to control HVAC system, 7.9, 7.19–7.20, 8.57–8.60
　to cut fibrous glass duct liner, 13.12
　the first, 7.22
　heat recovery from rooms, 11.8
　mainframe, 7.21
　payroll function, 12.44
　personal, 7.21–7.25
　recordkeeping and service ticket/invoice, 12.44
　special terms, 7.20–7.21
　storage media, 7.22–7.23, 7.25–7.27
Concrete, 1.21, 3.26, 12.13
Condensate
　boiler, 3.9
　condensing gas furnace, 9.10
　definition, 1.3
　drainage and cleanup, 1.42, 12.42
　handler cooling coil, 1.42
　heat pump, 11.28
　refrigeration system, 10.36
　steam heating system, 3.2
Condensation process
　definition, 1.3
　in the furnace room, 2.17
　inside chimney, 2.4
　inside ducts, 1.21
　inside vent system and furnace, 2.9, 2.11
　in the refrigeration system, 10.34, 10.35
Condensers
　heat pump, 11.12
　maintenance, 12.42
　refrigeration system
　　airflow, 10.25–10.26
　　definition, 10.2
　　line restrictions, 10.33
　　operation, 10.3, 10.4, 10.5, 10.7, 10.23
　　troubleshooting, 10.33, 10.45
Condensing units
　air handler, 1.39, 1.40
　all-air chilled water system, 1.5
　heat recovery from, 11.8
　refrigeration system, 1.3, 1.4, 5.27
　split-system air conditioning system, 8.24
Conduction, 4.4, 13.17
Conductors, 6.6, 6.13, 7.3–7.4, 8.66
Connector, duct, 13.4–13.5
Constant air volume box (CAV), 13.31
Constant volume system, 1.10–1.15, 1.50
Contactor, 8.17, 8.18, 8.43, 10.11, 11.17
Control circuit. See Circuits, control
Controllers
　dual-duct variable air volume system, 1.30
　economizer, 4.17
　fan, 1.37, 1.39
　gas-fired systems, 9.13, 9.21
　HVAC, 8.58–8.59
　reset, 8.51
　residential zone, 4.20
　variable air volume system, 1.17, 1.27, 1.28, 1.29
　variable frequency drive, 7.17
　variable volume, variable temperature system, 1.15, 1.16
　ventilation, 4.22
Convection, 3.6, 4.4, 11.2, 13.17
Convector, 3.22–3.23
Converter, analog-to-digital, 8.59, 8.65
Cooler. See Evaporators, all-air chilled-water system
Cooling systems. See Air conditioners; Refrigeration system
Cooling tower, 1.5
Copper
　in bimetal element, 8.2
　as a conductor, 7.4
　in heat exchanger coil, 11.4
　melting temperature, 13.3
　piping for Edison hookup, 6.10
　sheet metal, 13.2, 13.3, 13.36
　thickness and weight, 13.36
Council of American Building Officials (CABO), 14.6
Couplings and direct drives, 12.31–12.34
Crankcase, 11.16. See also Heaters, for crankcase
Cross-talk, 13.11
Crystal, piezoelectric, 4.8

Current, electrical, 11.23. *See also* Alternating current; Diagrams, circuit; Direct current
Customer relations
    callbacks, 12.44
    communication skills, 8.27, 8.64, 12.43, 12.44, 12.50–12.56
    difficult customers, 12.53–12.56
    follow up phone call, 12.51
    handling service calls, 12.43, 12.44, 12.50–12.51
    importance, 12.47
    personal habits, behaviors, and attitudes, 12.47, 12.50
Cutter, tubing, 5.7
Cylinders
    charging, 5.19–5.20, 5.23
    gas, 5.6, 5.16, 5.18–5.24

D

Dampers
    all-air chilled-water system, 1.5
    all-air refrigeration system, 1.3, 1.4
    balancing, 1.20
    barometric relief, 1.7, 1.8, 1.36
    bypass, 1.15, 1.16, 4.21
    central unit, 1.11
    control, 8.60, 13.29–13.30
    economizer, 1.8, 1.9, 4.17–4.18
    fire and fire/smoke, 13.26–13.28, 14.5
    gravity relief, 1.7
    maintenance, 12.42–12.43
    manual volume (MVD), 13.25
    mixing box or air terminal, 1.26, 1.44
    multi-shutter or multiple blade, 1.24, 13.25
    multi-zone constant volume system, 1.12, 1.13, 1.14
    opposed-blade, 13.9, 13.25, 13.35
    pneumatic system, 8.54–8.55
    potentiometer, 8.18
    residential zone, 4.20–4.21
    two-deck multi-zone system, 1.17
    variable volume, variable temperature system, 1.27
    vent, 2.12, 2.16–2.17
Data
    processing room, 1.44
    storage, 7.22–7.23, 7.25–7.27
    transfer via high-speed telecommunications, 7.20, 7.21, 7.24–7.25
DC. *See* Direct current
DCV. *See* Ventilation, demand-controlled
DDC. *See* Direct digital control
Deadband, 8.5, 8.65
Defrost cycle, 11.3, 11.8, 11.10–11.12, 11.38
Dehumidifiers, 4.6, 4.25, 11.16
Delta arrangement, 6.13, 6.14, 6.24
Diac, 7.9, 7.10, 7.11, 7.30
Diagnostic capability, 7.9, 7.19
Diagnostic equipment and testers, 8.28–8.29, 9.13
Diagrams
    circuit, 6.10, 6.30–6.31, 8.26, 10.8
    fault isolation, 8.28, 8.30, 8.65
    label, 8.28, 8.29, 8.65
    ladder, 8.16, 8.28, 8.29, 8.65, 11.33
    wiring (schematic), 8.28, 8.59, 8.65, 9.4, 9.8
Diaphragm, expansion tank, 3.14, 3.15
Dielectric material, 6.17, 6.35

Differential, 8.5, 8.65
Diffusers
    ceiling, 1.24, 1.26, 13.9
    concentric duct, 1.26
    flexible duct runout from, 14.11
    installation, 13.9–13.10
    location selection, 1.3–1.4
    multi-zone constant volume system, 1.12, 1.13
    overview, 13.9
    perforated panel, 1.24
    single-zone constant volume system, 1.11
    slot, 1.24–1.25
    variable air volume system, 1.18
Digital signal, 8.58, 8.59
Diodes
    and current flow, 7.5–7.8
    light-emitting (LED), 5.12, 5.21, 7.8–7.9, 7.30, 8.16–8.17, 9.13
    photo (light-sensing), 7.8, 7.30
Direct current (DC), 6.2, 6.5–6.6, 6.11, 7.5, 7.16
Direct digital control (DDC), 1.19, 1.27, 8.57–8.58
Distributor, 10.31
Diverter, draft, 2.17, 3.7
Divider, 12.34–12.35
DMM. *See* Multimeter, digital
Documentation, installation and maintenance, 12.43–12.51
DOE. *See* U.S. Department of Energy
Door, duct access, 13.28–13.29
Doping, 7.4
Downdraft, 2.17
Draft (air circulation in a room), 1.3, 8.27
Draft (drawing action of a chimney), 2.8–2.9, 2.17
Draft control, air feeding a furnace flame, 2.16–2.17
Draperies, 1.5
Drill and drill bits, 12.7, 12.13, 12.36–12.38
Drives
    belt, 12.29, 12.30–12.31
    direct, 12.31–12.34
    variable frequency (VFD), 7.17
Droop, 8.7, 8.65
Dropping point, 12.28, 12.61
Drugs, 12.50
Dual-duct system, 1.18–1.19, 1.50
Dual-fuel system, heat pump, 11.27, 11.29–11.31, 11.38
Ductboard, rigid fibrous glass. *See* Ducts, rigid fibrous glass
Duct box, 1.18, 1.28–1.29, 13.31
Duct liner, fibrous glass, 1.21, 13.11–13.15, 13.30
Ductmate®, 13.5, 13.6
Ducts
    access door, 13.28–13.29, 13.30
    airflow control and sail switch, 8.56
    construction, 1.21–1.22
    dimensions, 14.16
    fabricating and joining, 14.9–14.11
    flange, 13.9
    flexible fiberglass, 14.2, 14.4–14.5, 14.11–14.14
    header, 1.20
    installation, 14.3–14.4
    materials, overview, 1.21
    repair, 14.20–14.23
    rigid fibrous glass, 1.21, 13.11–13.15, 14.2–14.11, 14.14–14.23
    roof opening for, 1.21

Ducts (*continued*)
    and roof opening for equipment, 13.17
    size selection, 9.22
Duct system
    in all-air chilled-water system, 1.5
    in all-air refrigeration system, 1.3
    classification, 1.19–1.20
    closure system, 14.6–14.9
    condensation, 1.21
    dual-duct variable air volume, 1.18, 1.19, 1.29
    extended plenum supply system, 14.6, 14.7
    fiberglass blanket insulation (wrap), 13.16–13.24
    gas heating system, 9.21
    hangers and supports, 13.6–13.8, 13.37–13.39, 14.14–14.20
    layout, 1.20–1.21
    leaks, 13.6, 14.2, 14.4
    modular, 14.5–14.6
    multi-zone constant volume system, 1.12, 1.13
    plenum ceilings, 1.6, 1.21
    purpose, 1.19
    runout or branch lines, 1.20, 13.16, 13.30–13.31, 13.35, 14.6, 14.7
    seams and connectors, 13.4–13.6
    single-zone constant volume system, 1.11
    trunk line, 1.20, 13.16
    typical, 13.16
    variable air volume system, 1.17, 1.28
Dust, 1.45, 4.8, 4.11, 14.8

E

ECM. *See* Motors, electronically commutated
Economizers
    all-air system, 1.6, 1.36
    definition, 4.25
    multi-zone constant volume system, 1.12, 1.13
    overview of function, 4.16–4.18
    three types, 1.8–1.10
Edison hookup, 6.10
EI rating, 14.3, 14.4, 14.27
Elasticity, 14.3, 14.27
Electrical system
    current. *See* Alternating current; Circuit; Circuit diagram; Direct current
    safety. *See* Safety, electrical
    testing, 6.24–6.29
    troubleshooting, 8.27–8.30
    using AC power, 6.16–6.18
Electricity, static, 7.18, 7.19, 8.8
Electromagnetic field, 8.14
Electromechanical components, 7.2, 7.30, 10.13–10.15, 11.10, 11.16
Electromotive force (EMF), 6.21–6.22
Electron, 7.2–7.3, 7.4, 7.30
Electronics
    components and circuits, 7.4–7.17
    computer, 7.19–7.27
    overview, 6.19, 7.30
    printed circuit board, 7.4, 7.17–7.19, 10.11
    and semiconductors, 7.3–7.4
    theory, 7.2–7.3, 11.23
Electroplating, 13.3
EMF. *See* Electromotive force

Energy conservation
    air conditioning system, 4.14–4.20
    energy recovery loop, 11.7
    energy recovery ventilator, 4.11, 4.14–4.16
    Energy Star® program, 11.26
    government-mandated, 9.2
    heat pump, 11.2, 11.15, 11.20, 11.25–11.26
    hydronic system, 8.51
    insulation, 13.11
    radiant floor heating, 3.26
    Seasonal Energy Efficiency Ratio, 11.25–11.26, 11.38
    thermostat setting, 8.7
    variable air volume box, 13.31
Energy Star® program, 11.26
Enthalpy, 1.9, 4.17, 4.25
EPA. *See* U.S. Environmental Protection Agency
Equipment
    closure machine, 14.7
    diagnostic, 8.28–8.29, 9.13
    duct liner cutter, 13.13
    importance of make, model number, and serial number, 12.44
    lubrication basics, 12.28–12.29
    packaged. *See* Packaged equipment
    personal protective. *See* Protective equipment
ERV. *See* Ventilators, energy recovery
Evacuation process, refrigeration system, 5.9–5.17, 5.38, 10.35
Evaporation, 4.5, 4.6, 4.18–4.19, 4.25
Evaporators
    airflow, 10.25–10.26
    all-air chilled-water system (cooler), 1.4, 1.5, 5.24
    definition, 10.2
    heat pump, 11.12
    refrigeration system, 1.3, 1.4, 10.3, 10.4–10.5, 10.23, 10.25–10.26, 10.27
    troubleshooting, 10.23, 10.25–10.26, 10.27, 10.48
Exfiltration, 1.7
Exhaust air (return air)
    central equipment, 1.7–1.8, 11.7
    furnace, 2.6
    outlets. *See* Diffusers; Grilles; Registers
    powered exhaust, 1.7, 1.8, 1.37
    three methods, 1.7
    through plenum ceiling, 1.6, 13.32
Expansion device, 10.2, 10.3, 10.5, 10.7, 10.23
Explosion
    from adhesive, 13.13, 13.22
    boiler, 3.7, 3.12
    from electronic leak detector, 5.3
    prohibited gases to pressurize a system, 5.6, 5.15
Extension, hanger, 14.14–14.15

F

Fan coil unit, 11.27
Fan power terminal unit (FPTU), 13.31
Fans. *See also* Blowers
    air handler, 1.40, 1.41, 1.43–1.44
    all-air chilled-water system, 1.5
    all-air refrigeration system, 1.3
    boiler, 3.10
    condenser, 1.35, 1.37, 10.7, 10.26
    dual-duct variable air volume system, 1.19

evaporator (indoor; supply), 1.35, 10.7
fan-powered mixing box location, 1.28–1.29
forced-air heating system, 3.2, 11.18, 11.19
furnace, 8.21–8.22. *See also* Blowers, furnace
humidifier, 4.7
inline duct, 13.31
motor, 6.19
multi-zone constant volume system, 1.12, 1.13
packaged equipment, 1.34–1.35
for powered exhaust, 1.7, 1.8, 1.37
rating curves, 1.35
return, 1.8
single-zone constant volume system, 1.10, 1.11
unit heater and unit ventilator, 3.24
variable air volume system, 1.17, 1.39
Fasteners, mechanical
for blanket fiberglass, 13.23–13.24
for fibrous glass duct liner, 13.13–13.14
grade designations, 12.2–12.3
for hanger strap, 13.37
installation, 12.10–12.13
nonthreaded, 12.9–12.10
replacement, 12.8, 12.13
threaded, 12.2–12.8, 12.10–12.13, 12.27, 12.29
threads, 12.2
Fatigue, 14.3, 14.27
Fault isolation diagram, 8.28, 8.30, 8.65
Fiberglass
air filter, 4.11–4.12, 12.40
blanket insulation for duct system, 13.15–13.24
and clothing, 13.15
cutting, 13.15, 13.16, 13.17, 13.19
flexible duct, 14.2, 14.4–14.5, 14.11–14.14
how it works, 13.17, 13.19
properties, 14.2
rigid fibrous glass duct, 1.21, 13.11–13.15, 14.2–14.11, 14.14–14.23
safety, 13.15, 13.16, 13.19, 14.2, 14.8
Filter-drier, 10.23, 10.33
Filters
activated carbon, 1.4, 4.12, 4.14
air handler, 1.40
all-air system, 1.3, 1.4
bag-type, 1.36, 4.12
cartridge, 1.36
check-filter switch, 1.38
disposable (dust stop), 12.40, 12.41
electric heating system, 11.20, 11.21
electronic air cleaner, 4.13–4.14, 12.41, 12.42
electrostatic, 4.12–4.13, 12.40
fiberglass, 4.11–4.12, 12.40
gas-phase, 4.14
high-efficiency particulate arresting (HEPA), 1.36, 12.40, 12.41
humidifier, 1.45
maintenance, 4.11, 4.13, 4.14, 12.40–12.42
mechanical, 4.11–4.13, 4.14, 12.40
packaged equipment, 1.36
pneumatic system, 8.53
sizing, 12.41
steel/aluminum mesh, 4.13
Fire
chimney-related, 2.8

damper to control, 13.26–13.28
electronic equipment near flammable vapor, 5.3
fire-rated walls and ceilings, 13.26
kitchen, 13.11
smoke control, 1.8
warning on oil and oxygen mix, 5.6, 5.15
Firebox, 3.10
Fire panel, 1.8
Fire point, 12.28, 12.61
Fire protection system, 13.26
Firestop, 2.7
Fittings
duct, 13.30–13.31, 14.11, 14.12–14.14, 14.16–14.18
lubrication, 12.28–12.29
modular duct system, 14.5–14.6
water piping system, 3.27, 3.34
Flame
from a burner, 2.3–2.4, 9.7, 9.9
from a pilot, 9.3, 9.15, 9.17–9.19
submerged, 9.12, 9.26
Flame sensor, 7.15–7.16, 9.15, 9.19–9.21
Flange, 12.15, 12.32, 14.12, 14.13
Flash point, 12.28, 12.61
Flooded start, 10.28, 10.40
Flooding, 10.27–10.28, 10.29, 10.40, 10.48
Flow chart, troubleshooting, 8.30, 9.27–9.28, 10.13
Fluorescent light, 5.3–5.5, 6.14
Flux, 5.7
Foil scrim kraft (FSK), 13.15, 14.2, 14.27
Food processing area, 1.44, 10.2
Forms and checksheets, 12.45
FPTU. *See* Fan power terminal unit
Fractionation, 5.31, 5.37
Free cooling, 4.17, 4.25
Frequency, 6.2, 6.8–6.9, 6.35
Friction, on bearings, 12.22
Friction loss, 3.34
FSK. *See* Foil Scrim Kraft
Fuel oil, 3.7
Furnaces
amperage, 8.11
combustion efficiency, 2.3
controls, 8.21–8.22
electric, 11.20, 13.4
gas
condensing, 2.2, 2.4, 2.13–2.15, 2.20, 9.10
induced-draft, 2.2, 2.9–2.13, 2.20, 9.5–9.9, 9.17
natural-draft, 2.2, 2.8–2.9, 2.20, 9.2–9.5
installation, 2.5, 2.10
troubleshooting, 7.19
venting process, 2.4–2.6
Furniture, 1.5
Fuse box, 6.10
Fuses
symbol, 8.66
testing, 6.28–6.29, 8.39
thermal, 11.22
transformer, 6.4
troubleshooting, 8.39–8.40
Fusible link, 6.4, 6.35

G

Galvanization process, 13.3

GAMA. *See* Gas Appliance Manufacturers Association
Gas. *See also* Carbon dioxide; Carbon monoxide; Oxygen
   corrosive, 14.5
   flue, 2.2, 2.4, 2.14
   natural gas, 1.10, 1.14, 1.35–1.36, 2.8, 13.3
   noncondensible, 10.34, 10.35, 10.40
   ozone, 4.14, 5.2, 12.42
   volatile organic compounds, 4.10
Gas Appliance Manufacturers Association (GAMA), 2.3
Gaskets, 12.13–12.15
Gauges. *See under* Instrumentation
   feeler, 12.27, 12.36, 12.37
   sheet metal, 13.2, 13.35
   thread pitch, 12.36, 12.37
Generator, 6.5, 6.9
Geothermal system, 11.2–11.3, 11.6–11.7, 11.38
Germanium, 7.18
GFCI. *See* Ground fault circuit interrupter
Gland, packing, 12.16, 12.61
Glass, 7.4, 14.7–14.9. *See also* Fiberglass
Glow bar. *See* Ignitors, hot-surface
Glow coil. *See* Coils, glow
Graphite, 12.15, 12.16, 12.17
Gravity hot-water system, 3.5–3.6
Grease, properties, 12.28
Grease gun, 12.28–12.29, 12.39
Grilles
   installation, 13.9–13.10
   on packaged rooftop unit, 1.8
   for return air leaving room
      in all-air chilled-water system, 1.5
      in all-air refrigeration system, 1.3, 1.4
      dual-duct variable air volume system, 1.18
      multi-zone constant volume system, 1.12, 1.13
      single-zone constant volume system, 1.11
      variable air volume system, 1.17
   for supply air entering room, 1.3–1.4, 1.23–1.24
   troubleshooting, 8.27
Groover, 13.4
Ground (electrical), 6.28, 6.30, 10.18, 10.34, 12.37, 12.39
Ground (mechanical), 8.67
Ground fault circuit interrupter (GFCI), 6.30
Ground-source system, 11.2–11.3, 11.6–11.7, 11.38
Groundwater, 11.2–11.3, 11.6
Gypsum board, 1.21

H
HACR. *See* Heating, air conditioning, and refrigeration
Hair, 12.38
Hammerlock, 13.30
Hangar, aircraft, 9.12
Hangers and supports, duct, 13.6–13.8, 13.37–13.39, 14.14–14.20
Hard drive, 7.25, 7.26, 7.27
HCFC. *See* Hydrochlorofluorocarbon
Health issues
   and air exchange rate, 4.10, 4.15
   carbon monoxide, 2.3, 2.5, 2.17
   fiberglass, 13.15, 14.3
   low humidity, 4.5
   mercury, 8.3
   microorganisms, 1.21, 1.46, 4.21
   mold and mildew, 4.9, 4.10, 4.15, 4.21

   polychlorinated biphenyl, 6.28
   sick building syndrome, 1.5, 1.7, 1.50, 4.2
   volatile organic compounds, 4.10
Heat anticipators, 2.9, 2.10–2.11, 2.20, 7.12, 8.10–8.11
Heaters
   baseboard, 1.27, 3.23, 11.20
   bearing, 12.26
   for crankcase, 1.38, 8.11, 10.28, 11.17–11.18, 11.34
   dual-duct variable air volume system, 1.18, 1.19, 1.29, 1.30
   electric, 11.20, 11.21
   electric resistance, 1.10, 1.14, 1.28, 1.35, 11.3
   low- and high-intensity infrared gas-fired, 9.12, 9.14–9.15
   multi-zone constant volume system, 1.12, 1.13, 1.14
   natural gas, 1.10, 1.14, 1.35–1.36
   in packaged equipment, 1.35–1.36
   propane, 1.10, 1.14, 1.35–1.36
   radiator, 3.23
   single-zone constant volume system, 1.10, 1.14
   unit heater, 3.24
   variable volume, variable temperature system, 1.27
   vertical panel, 9.12, 9.14
   water, 1.10, 1.14, 1.28, 3.24–3.25, 11.6
Heat exchangers
   airflow and temperature rise, 2.10
   boiler, 3.8, 3.9
   domestic hot water heater, 3.25
   furnace, 2.6, 8.21, 9.10, 9.21
   heat pump, 11.4
   heat supply from process steam, 3.2
   residual heat, 8.3
   warranty and supply air, 2.6
Heating, air conditioning, and refrigeration system (HACR), 8.40
Heating, ventilation, and air conditioning system (HVAC)
   commercial. *See* Commercial airside system
   control system
      basic characteristics, 8.13
      digital control, 8.57–8.60
      programmable, 7.9, 7.20, 8.6–8.7, 8.13
      sequence of operation, 8.23–8.26
   cooling. *See* Refrigerant; Refrigeration system
   energy conservation equipment, 4.14–4.20
   motors, 6.18, 6.20, 6.21, 6.22
   occupied *vs.* unoccupied period, 1.10–1.11, 1.14
   overview of zoned control, 4.19–4.21
   purpose, 1.2, 4.2
   simultaneous heating and cooling, 1.1, 1.12
   three functional circuit areas, 8.34
   troubleshooting. *See* Troubleshooting, HVAC
   voltage, 6.3, 6.8
Heating element, in electric heating systems, 11.19–11.21, 11.22–11.23, 11.34
Heating Season Performance Factor (HSPF), 11.24–11.25, 11.38
Heating system
   electric, 11.18–11.23, 11.32
   forced-air, 3.2, 11.18–11.23
   gas. *See* Furnaces, gas; Troubleshooting, gas heating system
   hot-water or steam
      components, 1.10, 3.6–3.26
      dual-temperature, 3.31, 3.32
      overview, 3.2, 3.38

piping, 3.26–3.30
types, 3.5–3.6
overhead, 1.25
packaged gas heating and cooling system, 9.2–9.7
radiant floor, 3.25–3.26, 3.38
as supplemental heat with heat pump, 11.18–11.23
terminals, 3.22–3.24, 3.38
Heat sink, 7.11, 11.2, 11.38
Heat transfer. *See also* Insulation
and comfort air conditioner system, 4.4
in heat pump, 11.2–11.3, 11.4, 11.5, 11.8
in hot-water heating system, 3.6
and insulation, 13.11
Heli-coil, 12.8
HEPA. *See* Filters, high-efficiency particulate arresting
Hertz, 6.2, 6.35
HFC. *See* Hydrofluorocarbon
High-pressure systems, 1.22
Hog hair, 4.11, 4.12, 12.40
Hood, draft, 2.17
Hospital, 3.2, 4.12, 4.14, 13.12
Hot dipping, 13.3
Hotel, 1.18, 1.20, 4.13, 4.14, 8.7
Hot spot, 1.3
HRV. *See* Ventilators, heat recovery
HSI. *See* Ignitors, hot-surface
HSPF. *See* Heating Season Performance Factor
Humidifiers, 1.44–1.46, 4.5, 4.6–4.9, 4.25, 8.60, 11.16
Humidistat, 8.53
Humidity
in furnace room, 2.17
humidification of a building, 4.5–4.6
relative
definition, 1.2
and evaporation, 4.5
maximum allowable, 1.9
most common range, 1.4, 4.2, 4.3, 4.6
Hunting, 10.30
HVAC. *See* Heating, ventilation, and air conditioning system
Hybrid system, heat pump, 11.27, 11.29–11.31, 11.38
Hydrochlorofluorocarbon (HCFC), 5.3, 5.5, 5.6, 5.8. *See also* Refrigerant, R-22
Hydrofluorocarbon (HFC), 5.3, 5.8. *See also* Refrigerant, R-22
Hydronic system
calculation of water flow rate, 3.31, 3.34
components, 3.6–3.26
definition, 3.2, 3.38, 3.39
dual-temperature water system, 3.31, 3.32
overview, 3.2, 3.38
piping system, 3.26–3.30
troubleshooting, 8.50–8.53
types, 3.5–3.6
water balance, 3.31, 3.33–3.35
water system terms, 3.3–3.5
Hygroscopic, 10.4, 10.40

I

IAQ. *See* Indoor air quality routine
IC. *See* Chips, integrated circuit
ICBO. *See* International Conference on Building Officials
ICC. *See* International Code Council
Ignition system, 7.10, 9.7, 9.13, 9.15, 9.17–9.21, 9.30–9.31

Ignitors
glow coil, 9.15, 9.17–9.18
hot-surface (HSI; glow bar), 3.9, 9.7, 9.15, 9.17, 9.18–9.19, 9.20–9.21
nitride, 9.15, 9.18
spark, 3.7, 3.9, 9.17
Impeller, pump, 3.17, 3.18
Inconel®, 12.17
Indoor air quality routine (IAQ), 1.44
Induction, 6.2, 6.35
Industrial facility and factory, 1.15, 3.2, 3.24, 9.12, 11.3, 11.8
Inertia, 6.19, 6.35, 14.3, 14.27
Infiltration, 1.6, 2.5, 2.9
Infrared system, 4.8, 9.12, 9.14–9.15
Input/output function (I/O), 7.20, 7.30, 8.58, 9.2, 10.12
Installation/removal
air filter and air cleaner, 4.14, 12.41
bearings, 12.25–12.27
belt, 12.30
blind rivet, 12.10
boiler, 3.13
coupling, 12.33–12.34
damper, 13.25, 13.28, 13.30
duct, 14.3–14.4
duct access door, 13.29, 13.30
duct liner, 13.12–13.15, 13.30, 14.10
fiberglass blanket insulation, 13.15–13.24
furnace, 2.5, 2.10
gasket, 12.15
heat pump, 11.27–11.31
packing, 12.17–12.18
register, grille, and diffuser, 13.9–13.10
seal, 12.21
takeoff, 13.30
thermostat, 8.7–8.12, 10.13
threaded fastener, 12.10–12.13
zoning accessories, 13.31, 13.32
Instrumentation. *See also* Thermostat
black bulb radiant sensor, 9.14, 9.15
boiler sensor, 3.10, 3.11, 3.13, 9.11
capacitor analyzer, 6.24
carbon dioxide sensor, 1.10, 1.11, 1.14, 4.22
carbon monoxide sensor, 2.3, 2.5, 4.22
cylinder pressure gauge, 5.6, 5.16
differential pressure gauge, 3.31, 3.33
digital multimeter, 6.8, 6.28
dual-duct variable air volume system sensor, 1.18–1.19, 1.29–1.30
electrical circuit recording, 6.26–6.27
flame sensor, 7.15, 9.15, 9.19–9.21, 9.26
megohmmeter (megger), 6.25–6.26, 6.27, 6.35, 7.6
pressure/temperature gauge, 3.11, 3.29, 3.30
refrigerant analyzer, 5.28
refrigerant leak detector, 5.2–5.5, 10.4
single-duct variable air volume system sensor, 1.27
smoke detector, 13.26
space sensor, 1.15, 8.59–8.60
static pressure sensor, 1.16
vacuum gauge, 5.11–5.16
variable air volume system sensor, 1.17
voltmeter, 7.14
water flow measuring and flow-control device, 3.31, 3.33
wattmeter, 6.24–6.25

Insulation
    acoustic, 13.10, 13.11, 14.4
    and air exchange rate of a building, 4.10
    attic, 2.7
    door and window, 2.4, 4.10
    duct, 1.3, 1.22, 11.29, 13.11–13.24, 14.4, 14.18
    electrical, 6.26
    fiberglass wrap, 13.15–13.24
    foam, 14.15
    overview, 13.10–13.11
    thermal, 13.10–13.11, 14.4
Insulator, 7.4
International Code Council (ICC), 14.6
International Conference on Building Officials (ICBO), 14.6
Invar®, 8.2, 8.65
Inverter, 7.17
Invoice, billing, 12.44–12.45, 12.46
Iron, 13.2, 13.3
Isothermal process, 1.45, 1.46, 1.50

J
Job report, 12.45
Joints
    staple flap in duct system, 14.6–14.8, 14.21, 14.22
    universal, 12.32
Journal, 12.21, 12.61
Jumper, 11.10, 11.12
Junction box, 6.19

K
Kelvin scale, 11.2, 11.38
KEVLAR®, 14.9
Keys, 12.9, 12.10, 12.32
Kitchen, 13.11, 13.12, 14.5

L
Label diagram, 8.28, 8.29, 8.65
Laboratory, 1.14
Ladder diagram, 8.16, 8.28, 8.29, 8.65, 11.33
Lamp, 8.67
Lap-joining, 13.6, 13.35, 13.37
Lawsuit. See Liability issues
LCD. See Liquid crystal display
Leaks
    capacitor, 6.28
    duct system, 13.6, 14.2, 14.4
    expansion tank, 3.15
    hydronic system, 3.3
    refrigerant, 5.2–5.7, 5.36, 8.28, 10.4, 10.25
LED. See Diodes, light-emitting
Liability issues, 2.4, 12.43
Library, 1.14, 1.18, 1.20
Lighting, 6.14, 7.10, 7.11
Liquid crystal display (LCD), 5.19, 7.8
Liquid (bubble) detector, 5.5
Load (electrical), 6.27, 8.33–8.35, 8.41
Load (force)
    axial, 12.21, 12.61
    maximum for metal hangers, 13.38
    radial, 12.21, 12.61
    shear, 13.7, 13.35
    tensile, 13.7, 13.35

Lock (metal edge connector), 13.4–13.5, 14.12
Locknut, 12.27
Lockout/tagout, 8.32–8.33, 10.17, 10.20
Log, system operating, 8.28
Louver, 1.24
Lubricant properties, 12.27–12.28
Lubrication
    bearings, 12.22, 12.24, 12.40
    compressor, 10.4, 10.23, 10.27, 10.28, 10.34, 10.46
    equipment basics, 12.28–12.29
    motor, 12.39–12.40
    O-ring, 12.19

M
Magnetic field, 6.5–6.6, 7.16, 9.16
Maintenance and repair
    basic procedures, 12.34–12.43
    belt drive system, 12.30–12.31
    coil and condensate system, 9.22, 12.42, 12.43
    damper, 12.42–12.43
    fiberglass duct, 14.20–14.23
    filter, 4.11, 4.13, 4.14, 9.9, 9.21, 11.22, 12.40–12.42
    motor, 12.39–12.40
    tool, 12.38–12.39
Mall, shopping, 1.12, 8.7
Manganese, 13.3
Manifold, 2.3, 3.26, 5.12, 5.13, 5.14
Manometer, 3.31, 4.14
Manufacturers' aids for troubleshooting, 8.28–8.29
Map, roof or campus, 12.44
Mastic, 13.22, 13.35, 14.7, 14.8
MAT. See Thermostats, mixed air
Material safety data sheet (MSDS), 14.8
MBh, 3.10, 3.39
Measuring device, 12.27, 12.31, 12.34–12.36, 12.37, 13.16.
    See also Straightedge
Medical clinic, 1.14–1.15, 1.16, 1.18
Megohmmeter (megger), 6.25–6.26, 6.27, 6.35
Memory, computer, 7.20, 7.21, 7.22, 7.25, 8.58
Mercury, 8.3, 11.16
Metering device, 1.4, 1.37, 10.29–10.32, 11.9, 11.14–11.15
Metric-based fasteners and tools, 12.2, 12.36
Microfarad, 6.17, 6.35
Micrometer, 12.36
Microminiaturization, 7.2, 7.18, 7.30
Micron, 4.11, 4.25
Microorganisms
    in distribution system, 1.21, 4.21
    galvanization to prevent, 13.3
    in humidifiers, 1.46
    mold and mildew, 4.1, 4.9, 4.10, 4.15, 4.21
Microprocessors, 7.2, 7.3, 7.18–7.19, 7.30, 8.57, 8.58, 10.15
Mineral deposits and salts
    and heat pumps, 11.5
    and humidifiers, 1.45, 1.46, 4.7, 4.9
    and pressure-relief valves, 3.13
    removal, 3.3
    and water system piping, 3.3
Mixing box, 1.18, 1.28–1.29, 1.40, 13.31, 14.11
Model number, 12.44
Mold and mildew, 4.9, 4.10, 4.15, 4.21
Moment of inertia, 14.3

Monitor, computer, 7.23–7.24
Monochrome, 7.23, 7.30
Monoflow® fitting, 3.27
Motel, 1.18
Motors
   blower, 9.21, 10.26, 11.20, 11.21, 11.34, 12.22
   checking for grounded windings, 6.28
   compressor, 10.7
   control damper, 13.30
   control system. *See* Drives, variable frequency
   electronically commutated (ECM), 7.16, 8.48, 10.10
   induction (AC), 6.18–6.24, 6.34, 6.35
   insulation tests, 6.26
   lubrication procedure, 12.39–12.40
   mount, 12.33, 13.11
   pneumatic, 8.54–8.55
   single-phase, 6.18–6.23, 8.48, 8.49
   speed control, 8.17–8.19, 8.45
   starter, 8.17, 8.18
   symbols, 8.67
   three-phase, 6.12, 6.23–6.24, 8.45–8.46, 8.49, 10.19
   troubleshooting, 8.22–8.23, 8.44–8.50
MSDS. *See* Material safety data sheet
Muffler, discharge line, 11.29
Multimeter, digital (DMM)
   overview, 6.8, 8.35
   testing procedures with
      control transformer, 8.43
      flame sensor, 9.20
      fuse, 6.28–6.29
      heating element, 11.22–11.23
      hot surface ignitor, 9.19, 9.20
      refrigeration system module/board, 10.11
      thermocouple, 9.16
   troubleshooting overview, 8.35–8.41
Multi-zone constant volume system, 1.12–1.15
Museum, 1.14, 1.44, 4.14
MVD. *See* Dampers, manual volume
Mylar®, 14.3

N
NAIMA. *See* North American Insulation Manufacturers Association
Nameplate, 2.10, 5.22, 6.8, 6.27
National Electrical Code®, 2.5, 6.12, 6.30, 8.32, 11.21
National Environmental Balancing Bureau (NEBB), 12.45
National Fire Protection Association (NFPA), 1.27, 2.5, 13.27, 14.3
National Fuel Gas Code, 2.5, 2.6, 2.10, 2.14
National Institute for Occupational Safety and Health (NIOSH), 13.16
National Lubricating Grease Institute (NLGI), 12.28
Natural gas, 1.10, 1.14, 1.35–1.36, 2.8, 13.3
NEBB. *See* National Environmental Balancing Bureau
Net positive suction head required (NPSHR), 3.18
Neutron, 7.2
NFPA. *See* National Fire Protection Association
Nichrome™, 11.19
Nickel, 7.12, 8.2, 11.4, 11.19, 13.3, 13.4
NIOSH. *See* National Institute for Occupational Safety and Health
Nitrogen, 2.4, 3.3, 5.5, 5.6, 5.15

Nitrogen dioxide, 4.10
NLGI. *See* National Lubricating Grease Institute
Noise
   acoustic insulation, 13.10, 13.11, 14.4
   air terminal location, 1.30
   bearing, 12.23
   compressor, 10.26, 10.46
   duct system, 1.22, 2.5, 13.2, 13.7, 13.11
   electronic air cleaner, 12.42
   fan (blower), 1.28, 13.11, 14.4
   flexible material on duct, 1.20
   gas heating system, 9.31
   heat pump, 11.28, 11.29
   HVAC equipment, 13.11
   hydronic system, 3.22
   occupant-generated, 13.11
   pump cavitation, 3.18
   vacuum pump, 5.15
   vibration-isolation roof curb, 1.36
Non-ferrous metal, 13.4, 13.35
North American Insulation Manufacturers Association (NAIMA), 14.2, 14.3, 14.4, 14.8
NPSHR. *See* Net positive suction head required
Nut, 12.5–12.7

O
Occupational Safety and Health Administration (OSHA), 8.32
Odors and odor control, 1.4, 4.12, 4.14, 9.31, 10.36, 12.54
Office building
   air cleaner, 4.14
   air conditioner, 1.31
   control zones, 1.2, 1.16
   maximum allowable air velocity, 1.20
   thermostat, 8.7
   variable air volume system, 1.18
Ohmmeter, 6.25–6.26, 6.27, 6.35, 7.6, 7.9, 10.19
Ohm's law, 11.23
Oil, properties, 12.27–12.28
Oil lock, 10.18
OSHA. *See* Occupational Safety and Health Administration
Outdoor air. *See* Pollution, outdoor; Supply air
Outgassing, 1.5, 4.10, 4.25
Outlet, heating, 1.23
Outlet box, 8.8
Overheating, 10.29, 10.47
Oxidation, 12.28, 12.61. *See also* Rust
Oxygen
   combustion process, 2.2, 2.4
   compressed, 5.6
   explosion from mixing with oil, 5.6, 5.15
   in hydronic system, 3.3
   respiration process, 1.6
Ozone, 4.14, 5.2, 12.42

P
PAC. *See* Air conditioners, packaged
Packaged equipment
   accessories, 1.36–1.38
   advantages and disadvantages, 1.33
   air conditioner, 1.31–1.33
   air handler, 1.39–1.40
   boiler, 3.10

Packaged equipment (*continued*)
  components, 1.33–1.36
  definition, 11.38
  electric heating system, 11.21
  gas heating and cooling system, 9.2–9.7
  heat pump, 1.33, 11.3, 11.28–11.29
  overview, 1.31
  refrigeration system, 10.24
  rooftop units. *See* Rooftop units
Packaged unit, 1.8, 1.14, 1.50
Packing, 12.15–12.18
Panduit®, 14.11–14.12, 14.13
Payment, 12.44, 12.51
PC. *See* Computers, personal
PCB. *See* Polychlorinated biphenyl
PC board. *See* Circuits, printed circuit boards
PDA. *See* Personal digital assistant
Performance, heat pump, 11.16, 11.23–11.26
Performance curve (chart), circulating pump, 3.17–3.18, 3.33, 3.35
Performance Factor (PF), 11.24–11.25
Personal digital assistant (PDA), 7.24
PF. *See* Performance Factor
Pharmaceutical process rooms, 4.12
Photoelectric cell (photocell), 7.11, 13.26
Pilot, 9.3, 9.15, 9.17–9.19, 12.54
Pin, 12.9, 13.23, 13.24
Pipe and piping system, 2.14–2.15, 3.26–3.30, 5.7, 6.10, 11.6–11.7, 11.28
Pixel, 7.23, 7.30
Plastic, 7.4, 12.8, 12.10, 12.32
Plenum, 1.6, 1.21, 11.28, 13.32, 14.6, 14.27
Pneumatic controls, troubleshooting, 8.53–8.56
Pole. *See* Relays, types
Pollution
  indoor. *See* Air quality, indoor
  outdoor, 4.10, 4.11, 5.7, 6.28
Polychlorinated biphenyl (PCB), 6.28
Polyester, 12.40
Polyethylene, 11.6
Polystyrene, 13.10
Polyurethane, 13.10
Polyvinyl chloride (PVC), 14.2
Potentiometer, 8.18, 8.55
Pour point, 12.27–12.28, 12.61
Power, 6.5–6.16, 7.5, 11.2, 11.23
Power tools, 12.38–12.39
Pre-cooler, evaporative, 4.18–4.19
Press, 12.25–12.26, 12.33, 12.36–12.38
Pressure
  within boiler, 3.8, 3.12
  building, slight positive, 1.6, 1.7
  depressurization of venting system, 2.9
  discharge, 11.13, 11.14
  and duct seal, 13.6
  gasket values, 12.13
  head, 1.37, 3.4, 3.17, 3.34–3.35, 3.39, 10.22–10.23, 10.46
  within heat pump, 11.9, 11.13, 11.14, 11.17
  heat pump values, 11.9
  overpressurization, 1.7
  for refrigerant evacuation, 5.11, 5.14, 5.15
  refrigeration system values, 10.23, 10.24
  of return duct system, 1.8
  static, 1.16, 1.19, 3.5, 3.17, 3.39, 4.21
  suction, 10.22, 10.23, 10.47, 11.9, 11.13, 11.14
Pressure chart, 5.30–5.31
Pressure drop, 1.20, 3.3–3.4, 3.17, 3.39, 4.14
Printing plant, 1.44
Protective equipment
  bearing installation, 12.26
  cleaning products, 12.43
  compressor oil, 10.4
  drilling in concrete, 12.13
  drill press, 12.37
  electricity, 6.30, 8.32
  fiberglass, 13.15, 14.2, 14.8
  foam insulation, 14.15
  heat-activated tape, 14.9
  refrigerant, 10.35
  tools, 12.38
Proton, 7.2, 7.3
Psychrometric chart, 4.2, 4.3
Public building, 1.18
Pullers
  bearing or coupling, 12.25, 12.33
  packing, 12.17, 12.18
Pulley, 12.29, 12.30, 12.31
Pulse width modulation (PWM), 7.17
Pump head, high/low, 3.17, 3.39
Pumps
  chilled water, 1.4, 1.5
  circulating (booster)
    in hot-water heating system, 3.11, 3.16–3.19, 3.28
    in hydronic system, 3.4, 3.6, 3.31, 3.33–3.35, 3.38, 8.52
    in radiant floor heating system, 3.26
  condensate, 10.36
  condenser water, 1.5
  heat
    balance point, 11.26–11.27
    components, 11.12–11.18
    controls, 8.5, 11.32–11.34
    installation, 11.27–11.31
    operation, 11.2
    overview, 11.2
    packaged, 1.33
    performance, 11.16, 11.23–11.26
    refrigeration cycle, 11.8–11.12
    service, 11.31
    supplemental electric heat, 11.18–11.23
    thermostat, 8.5
    types, 11.2–11.8
  performance curve (chart), 3.17–3.18, 3.33, 3.35
  troubleshooting in hydronic system, 8.52
  vacuum, for refrigeration system, 5.8, 5.10–5.16, 10.35
Pushbutton, 8.66
PVC. *See* Polyvinyl chloride
PWM. *See* Pulse width modulation

R

Race, bearing, 12.22, 12.23, 12.26
Radiation, 3.6, 3.25, 4.4
Radiator, 3.23
Radio studio, 1.15
Reclamation of refrigerant, 5.7, 5.37
Recorder, strip chart, 6.27
Recordkeeping. *See* Documentation

Recovery
   energy, 4.11, 4.14–4.16, 4.25, 11.7
   refrigerant, 5.7, 5.10, 5.20, 5.31, 5.37
Rectification, 7.5, 9.15, 9.26
Rectifier, 7.7–7.8, 7.9–7.10, 7.17, 7.30, 8.67
Recycle
   mercury, 8.3
   refrigerant, 5.7, 5.20, 5.37
Redundancy, 3.8, 3.39
Reflector, 9.12
Refrigerant
   all-air system, 1.3, 1.4
   backflow, 1.39
   effects on compressor testing, 8.50
   heat pump, 11.8, 11.9, 11.13, 11.15, 11.17
   leaks, 5.2–5.7, 5.36, 8.28, 10.4, 10.25
   lines and accessories, 10.2, 10.26, 10.32–10.34, 10.47
   low charge or overcharge, 10.24–10.25, 10.28
   odor, 10.36
   recovery, recycling, reclamation, 5.7–5.9, 5.10, 5.20, 5.31, 5.37
   R-410a (HCFC), 10.3–10.4, 10.5, 10.8, 10.23, 10.25, 10.34
   R-22 (HFC), 10.3–10.4, 10.5, 10.8, 10.23, 11.9
   zeotrope, 5.31–5.33
Refrigeration cycle
   basic, in refrigeration system, 10.3, 10.5–10.7
   heat pump, 11.8–11.12
   mechanical, in refrigeration system, 10.20–10.24
Refrigeration system
   basic operation, 10.2–10.3
   basic refrigeration cycle, 10.3, 10.5–10.7
   charging, 5.18–5.32
   chilled-water, 1.4–1.5
   contamination, 10.4, 10.34–10.36
   evacuation, 5.9–5.17, 5.38
   leak detection, 5.2–5.7, 5.36, 10.4
   lines and accessories, 10.2, 10.26, 10.32–10.34, 10.47
   mechanical refrigeration cycle, 10.20–10.24
   packaged unit, 9.2–9.7, 10.24
   replacement parts, 10.5
   sequence of operation, 10.41–10.43
   troubleshooting. *See* Troubleshooting, refrigeration system
   typical all-air, 1.3–1.4
Registers, 13.9–13.10
Regulator, draft, 2.16
Reheat, 1.14, 1.38
Relays
   current-sensing, capacitor, 6.21, 6.22
   electric heating system, 11.20, 11.32
   heat pump, 11.10, 11.11
   induced-gas heat/electric cool unit, 9.6
   lockout, 8.19
   motor start, 8.46–8.47
   pneumatic system (P-E; E-P), 8.54, 8.55, 8.56, 8.65
   refrigeration system, 10.12
   replacement, 8.18
   symbols, 8.66
   time-delay, 8.20, 9.5, 10.10, 10.20
   troubleshooting, 8.43, 8.46–8.47, 10.11
   types, 8.13–8.17
Removal procedures. *See* Installation/removal
Repair. *See* Maintenance and repair

Reports, 12.45, 12.47, 12.48, 12.49
Residence, domestic, 1.20, 3.8, 4.20–4.21
Resistance
   electrical, 8.42, 9.17–9.18, 10.18, 11.23
   fire, 13.15, 13.26, 13.27
   oxidation, 12.28
   run-down, 12.12, 12.61
   thermal, 13.10
Resistors, 7.9, 7.11–7.13, 8.67. *See also* Thermistors
Restaurant, 1.20, 4.13
Return air. *See* Exhaust air
Reverberator, 9.12, 9.26
Reverse cycle heat, 11.8, 11.38
R-410a. *See* Refrigerant, R-410a
Rings
   compressor, 10.28, 10.29
   O-, 12.19, 12.21
   packing, 12.17, 12.18
   retainer, 12.9
Riser, vertical, 13.7, 14.4, 14.5, 14.27
Rivet, 12.9–12.10, 12.11
rms. *See* Voltage, root-mean-square
Roof curb, 1.36
Rooftop units
   for all-air system, 1.8
   concentric duct diffuser on, 1.26
   definition, 1.50
   location selection, 1.21
   multi-zone constant volume system, 1.12, 1.13, 1.14
   single-zone constant volume system, 1.10
   variable air volume system, 1.18
   variable volume, variable temperature system, 1.16
Rotors, 6.18, 6.19–6.20, 6.22, 6.35
R-22. *See* Refrigerant, R-22
Rubber, 7.4, 12.29, 12.32
Rule, circumference, 12.34. *See also* Straightedge
Rust, 3.3, 12.22, 13.4
R-value, 13.10, 14.4, 14.14, 14.27

S
SAE. *See* Society of Automotive Engineers
Safety
   adhesive, 13.13, 13.22
   boiler, 3.7, 3.8, 3.12
   compressor oil, 10.4
   electrical
      capacitor, 6.28, 10.17
      megohmmeter, 6.26
      overview, 6.29–6.30. *See also* Lockout/tagout; Protective equipment
      power tools, 12.38, 12.39
      static electricity, 7.18
      thermostat wiring, 8.10
   electric heating system, 11.21
   electronic refrigerant leak detector, 5.3
   fiberglass, 13.15, 13.16, 14.2, 14.8
   gas cylinder, 5.18
   motor testing, 8.46
   overview, 6.34
   sheet metal, 12.38
   space heater, 4.10
   tool, 12.38–12.39
Saybolt Universal Seconds (SSU; SUS), 12.27

SBCCI. *See* Southern Building Code Congress International
Scale (contaminant). *See* Mineral deposits
Scale (weighing tool), for refrigerant charging, 5.19, 5.23
Schematic (wiring diagram), 8.28, 8.59, 8.65, 9.4, 9.8
School
   air conditioner, 1.31
   air filter, 4.13
   control zones, 1.14, 1.15, 1.16
   indoor air quality, 4.13
   variable air volume system, 1.18
Screws
   self-drilling sheet metal, 12.7
   threaded, 12.3, 12.4–12.5, 12.7, 12.29, 12.36
   thread-forming and thread-cutting, 12.7
   torque, 12.11
SCR package (silicon-controlled rectifier), 7.10
Seal, 12.18–12.21, 12.61
Seams
   fiberglass, 13.22–13.23, 14.2, 14.9, 14.11
   sheet metal (locks), 13.4–13.5
Seasonal Energy Efficiency Ratio (SEER), 11.25–11.26, 11.38
SEER. *See* Seasonal Energy Efficiency Ratio
Semiconductor, 7.3, 7.4, 7.18, 7.30
Sensors. *See* Instrumentation; Thermostat
Serial number, 12.44
Service calls, 12.43, 12.44, 12.50–12.51. *See also*
   Communication with the customer
Service entrance panel, 6.10, 8.40
Service ticket, 12.44–12.45, 12.46
Set (seizure) of fastener, 12.12, 12.61
Sheet metal
   connecting fiberglass ductboard to, 14.11, 14.12
   safety, 12.38
   thickness, 13.2, 13.36
   types, 13.2–13.3
   weights, 13.36
Sheet Metal and Air Conditioning Contractor's National
   Association (SMACNA), 1.19, 13.2, 13.6
Shiplap, 14.9, 14.21, 14.22, 14.27
Shoes, 12.47
Short cycling, 10.10, 10.40
Sick building syndrome, 1.5, 1.7, 1.50, 4.2
Sight glass, 5.20–5.21
Silicon, 7.10, 7.18
Silicon carbide, 9.18
Sine wave, 6.2, 6.6–6.7, 6.35
Single-zone constant volume system, 1.10–1.12, 1.15, 8.59
Slugging, 10.26–10.27, 10.28, 10.40
SMACNA. *See* Sheet Metal and Air Conditioning
   Contractor's National Association
Smart technology. *See* Climate control system
Smoke
   cigarette, 4.2, 4.10, 4.11, 12.50
   fire/smoke damper, 13.26–13.28
   from vacuum pump, 5.15
Smoke detector, 13.26
Society of Automotive Engineers (SAE), 12.3, 12.27
Solar collector, 11.5
Solder, 13.3
Solenoid, 8.41, 11.11, 11.13
Solid state. *See* Circuits, printed circuit boards
Soot, 2.2, 2.4, 9.16
Sound attenuation, 13.10, 13.35

Southern Building Code Congress International (SBCCI),
   14.6
Specific heat, 3.2, 3.39
Spillage, 2.17
Spin-in, 13.30
Split systems, 8.24, 11.3, 11.27–11.28, 11.38
Sports arena, 9.12
S-slip, 13.5
SSU. *See* Saybolt Universal Seconds
Staple flap, 14.6–14.8, 14.21, 14.22, 14.27
Starter, 8.17
Start-up report, 12.47, 12.48, 12.49
Starving, 10.30
Stator, 6.18–6.19, 6.35
Steam, 3.8, 4.8–4.9. *See also* Heating system, hot-water or
   steam
Steel
   alloys, 13.2, 13.3–13.4
   cold-rolled, 13.2, 13.4, 13.35
   ducts, 1.21
   fasteners, 12.3
   galvanized, 13.2, 13.36
   hot-rolled, 13.2, 13.4
   overview, 13.2
   sheet, 13.2
   stainless, 13.3, 13.4, 13.35, 13.36
Stethoscope, 12.23
Store, 1.15, 1.20, 1.31
Straightedge, 12.31, 13.16
Stuffing box, 12.17, 12.18, 12.61
Sub-base, 8.5–8.6, 8.8, 8.65
Subcooling
   heat pump, 11.7–11.8, 11.9
   refrigeration system, 5.29–5.30, 5.32, 5.33, 10.24, 10.31
Superheat
   heat pump, 11.9
   refrigeration system, 5.24, 5.26–5.29, 5.32, 5.33, 10.24,
     10.30
Supermarket, 11.8
Supply air (outdoor air; ventilation air)
   and air systems, 1.5–1.10
   flushing the system with, 1.11
   furnace, 2.2–2.3, 2.5–2.6
   heat pump, 11.3, 11.5, 11.7
   location of entry into room, 1.3–1.4, 1.23
   most common temperature, 1.4
   outlets for, 1.23–1.26. *See also* Diffusers; Grilles
   purpose in commercial building, 1.5–1.6
   reheat option, 1.38
   smooth entry into duct system, 1.21
   troubleshooting, 8.27
   varying airflow, 1.39
Surge suppression, 7.14
SUS. *See* Saybolt Universal Seconds
Switches
   aquastat, 3.13, 3.39, 8.50–8.51
   bimetal, 8.2, 8.3, 8.4
   centrifugal start winding, 6.20, 6.21
   check-filter, 1.38
   control circuit safety, 8.20–8.21
   disconnect, 6.12, 6.28, 11.18
   door interlock, 9.9
   emergency or auxiliary heat, 8.5, 8.6

flame rollout sensor, 9.10
freeze-stat, 8.21
inducer fan-proving, 8.22–8.23
limit, 8.22, 9.9, 9.21, 11.22
mode, 9.14
pilot safety, 9.18
pressure, 8.20–8.21, 11.10, 11.17
sail, 8.56
symbols, 8.66
troubleshooting, 8.43, 8.44
vane, 9.5, 9.26
Symbols
coil, 8.66, 8.67
diac, 7.11
light-emitting diode, 7.8
photo diode, 7.8
rectifier, 7.10, 8.67
relay, 8.15, 8.66
schematic, 8.66–8.67
switch, 8.66
triac, 7.11
variable resistor, 7.12
Synchronous speed, 6.23, 6.35
System pilot. *See* User interface module

T
Takeoff, 13.30–13.31
Tanks
expansion/compression, 3.11, 3.14–3.15, 3.17, 3.29, 3.30, 3.38
receiver, 3.2
Tap, speed, 6.23
Tape
for fiberglass duct, 14.6, 14.7–14.9, 14.10, 14.11, 14.23
insulation, 13.23
Tap-in, 13.30, 14.6, 14.27
Teflon®, 12.15, 12.16
Telecommunications, 7.20, 7.21, 7.24–7.25
Television studio, 1.15
Temperature
air, and circulation dynamics, 1.23, 1.25, 2.8–2.9, 4.4, 11.20
balance point of a heat pump, 11.18, 11.26–11.27, 11.38
comfort zone, 1.4, 4.2, 4.3
difference between supply air and return air, 2.10
dry-bulb, 1.2
freezing, boiling, and vaporization points of water, 13.3
maximum bearing, 12.26
measured by economizer, 1.9, 4.17
measured by pre-cooler, 4.19
melting points of metals, 13.3
optimum for grease, 12.28
refrigerant, change during cycle, 10.4, 10.5, 10.7
relationship with density, 3.34
setback, 1.11, 11.18, 11.32
setpoint, 1.10, 1.14, 4.17, 8.3
typical values in heat pump, 11.9
typical values in refrigeration system, 10.23, 10.24
in variable volume, variable temperature system, 1.15–1.16
Temperature glide, 5.31, 5.32, 5.37
Tensile strength, 13.2, 13.35
Terminal. *See* Air terminals; Computers, personal; Heating system, terminals

Testing. *See also* Troubleshooting
acid test kit, 10.35, 10.36
Btuh output of electrical heating element, 11.22–11.23
capacitor, 6.28
control circuits. *See* Troubleshooting, control circuits
damper, 13.25, 13.28, 13.30
diagnostic, 8.28–8.29
duct access door, 13.29
electrical
AC components, 6.24–6.29
capacitor, 6.28, 10.11, 10.19
compressor, 10.17
fuse, 6.28–6.29
inductive load, 6.27, 8.41
motor for grounded wirings, 6.28
resistive load, 8.14–8.42
fatigue, 14.3, 14.27
hot surface ignitor, 9.18
motor insulation tests, 6.26
motor protection thermistor, 7.15
motors, 8.46–8.50
R-410a refrigerant, 10.3–10.4
for refrigerant leak, 5.5–5.7, 5.36, 10.4
register, grille, and diffuser installation, 13.9
rigid ductboard, 14.3
takeoff, 13.30–13.31
thermal-electric expansion valve sensor, 7.14
Tetrafluoroethylene (TFE), 12.16, 12.17
TEV. *See* Valves, thermostatic expansion
Textile facility, 1.44
TFE. *See* Tetrafluoroethylene
Theater, 1.20
Thermal expansion, 3.6, 14.2
Thermal offset (droop), 8.7
Thermistors
in heat pump, 11.10, 11.11, 11.12
motor start, 8.47–8.48
overview, 7.13–7.15
troubleshooting, 8.7, 8.21, 8.42, 8.47–8.48
Thermocouple, 8.22, 8.67, 9.15–9.17
Thermostat base, 8.6
Thermostats
adjustment, 8.12
aquastat, 3.13, 3.39, 8.50–8.51
bulb in, 1.2, 8.2, 8.4, 8.5, 8.7
crankcase heater, 11.17
furnace, 2.10–2.11
heat anticipator in, 2.9, 2.10–2.11, 2.20, 7.12
heat pump, 11.10–11.12, 11.16, 11.17, 11.26–11.27, 11.32
hot water zoning, 3.28
induced-draft gas furnace, 2.9, 9.5–9.7
infrared heater, 9.14
installation, 8.7–8.12
mixed air (MAT), 4.17, 4.18
multi-zone constant volume system, 1.12, 1.13
outdoor, 8.21
pneumatic system, 8.53–8.54
principles of operation, 8.2–8.3
programmable, 1.10, 7.20, 8.6–8.7, 8.11, 10.15, 11.16, 11.32
refrigeration system, 10.7, 10.13–10.15
single-zone constant volume system, 1.11
troubleshooting, 8.2–8.12, 8.43–8.44, 10.13–10.15
two-deck multi-zone system, 1.17

Thermostats (*continued*)
　　types, 8.3–8.7, 8.65
　　variable volume, variable temperature system, 1.15
　　vs. thermistor, 8.21
Threaded insert, 12.8
Threads, 12.2, 12.8, 12.36
Throw. *See* Relays, types
Ticket, 12.44–12.45, 12.46, 12.47
Time clock, programmable, 1.13
Timer, 8.20, 8.66, 9.17, 10.10, 11.10
Timken, 12.24
Tolerance, 12.4, 12.61
Tools
　　bearing-removal, 12.25
　　blind (pop) rivet, 12.11
　　care and safe use, 12.38–12.39
　　coupling puller, 12.33
　　to cut fibrous glass duct liner, 13.12–13.13
　　drilling, 12.7, 12.13, 12.36–12.38
　　groover, 13.4
　　lubrication, 12.28–12.29
　　measuring devices, 12.27, 12.31, 12.34–12.36, 12.37
　　packing puller, 12.17, 12.18
　　power, 12.38–12.39
　　shiplap, 14.9, 14.21, 14.22, 14.27
　　stethoscope, 12.23
　　wrench, 12.12, 12.13, 12.25
Torch, 5.7
Torque
　　break-away, 12.12, 12.61
　　definition, 6.35, 12.61
　　drill press, 12.38
　　fastener installation, 12.10–12.12
　　to start motor, 6.19, 6.23, 6.24, 7.14
Torsion, 14.27
Tranducer, electronic, 1.45
Transformer, 6.2–6.5, 6.10, 6.13–6.14, 6.35, 7.7, 8.43
Transmitter, 8.6
Triac, 7.9, 7.10–7.11, 7.30, 8.16–8.17, 8.18
Trip, nuisance, 9.21
Troubleshooting
　　bearings, 12.23–12.24
　　burner, 8.51, 9.30–9.31
　　control circuits, overview, 7.18, 8.2
　　definition, 8.65
　　diagnostic capability, 7.9, 7.19
　　flame sensor, 9.15, 9.19–9.21
　　furnace control, 7.19
　　gas heating system
　　　　air system, 9.21–9.22
　　　　combustion system, 9.15–9.21
　　　　control circuit, 9.2–9.15
　　　　flow chart, 9.27–9.28
　　　　summary sheet, 9.30–9.31
　　historical records, 12.43
　　HVAC
　　　　control circuit, 8.13–8.26, 8.34, 8.64
　　　　control transformer checks, 8.43
　　　　digital control system, 8.57–8.60
　　　　electrical, common to all equipment, 8.37–8.44
　　　　fuse/circuit breaker checks, 8.39–8.40
　　　　input power and load, 8.33–8.34
　　　　isolating to a faulty circuit, 8.35
　　　　isolating to a faulty circuit component, 8.36–8.37
　　　　mechanical problems, 8.33
　　　　motors and motor circuit, 8.44–8.50
　　　　resistive and inductive load checks, 8.41–8.42
　　　　similarities among equipment, 10.10
　　　　switch and contactor/relay contact checks, 8.43, 10.11
　　　　thermostat checks, 8.43–8.44
　　hydronic controls, 8.50–8.53
　　ignition system, 9.7, 9.13, 9.15, 9.17–9.21, 9.30–9.31
　　manufacturers' aids, 8.28–8.29
　　organized approach, 8.27–8.30
　　refrigeration system
　　　　basic approach, 10.2, 10.10, 10.39
　　　　basic refrigerant cycle, 10.3, 10.5–10.7
　　　　compressor, 10.15–10.20, 10.21, 10.26–10.29, 10.45–10.46
　　　　condensate water disposal problems, 10.36
　　　　control circuit, 10.7–10.10
　　　　electrical circuit, 10.11–10.19
　　　　evaporator and condenser airflow, 10.25–10.26
　　　　flow chart, 9.27–9.28, 10.13
　　　　low charge or overcharge of refrigerant, 10.24–10.25, 10.28
　　　　mechanical refrigeration cycle, 10.20–10.24
　　　　metering device, 10.29–10.32
　　　　noncondensibles and system contamination, 10.4, 10.34–10.36
　　　　refrigerant lines and accessories, 10.2, 10.26, 10.32–10.34, 10.47
　　　　special tools and test equipment with new refrigerants, 10.3–10.4
　　　　summary sheet, 10.45–10.48
　　thermocouple, 9.15–9.17
　　thermostat, 8.2–8.12, 8.43–8.44
Troubleshooting table, 8.28, 8.65
Troubleshooting tree (fault isolation diagram), 8.28, 8.30, 8.65
Tubes
　　capillary, 10.31, 10.33, 11.13
　　finned, 3.8, 3.23
　　in ground-source heat pump, 11.6–11.7
　　in radiant floor heating system, 3.26
　　refrigerant. *See* Refrigerant, lines and accessories
Tungsten, 13.3
Turns ratio, 6.3, 6.35
Two-deck multi-zone system, 1.17
Two-zone constant volume system, 1.11–1.12
TXV. *See* Valves, thermostatic expansion

U

UL. *See* Underwriters Laboratories
Ultrasound, 5.3, 5.6
Ultraviolet light, 4.21, 5.3–5.5
Underwriters Laboratories (UL), 2.7, 13.26, 14.3, 14.4, 14.8
Unified National thread designations, 12.2
UNIVAC®, 7.22
U.S. Department of Energy (DOE), 4.16, 11.26
U.S. Environmental Protection Agency (EPA), 5.2, 5.5, 5.6, 5.7, 10.25
User interface module (system pilot), 1.15

V

Valence, 7.3, 7.30
Valves
  angle, 3.19, 3.20
  backflow-preventer, 3.20–3.21
  balancing, 3.6, 3.28
  ball, 3.19, 3.29
  check, 3.6, 3.20, 3.29, 3.30, 11.9, 11.15
  components, 12.16
  compressor, 10.28, 10.29
  diverting, 3.22
  drain, 3.11
  flow-control, 3.31
  gas cylinder safety, 5.6, 5.16
  gate, 3.19
  globe, 3.19–3.20
  mixing, 3.29, 3.30
  modulating gas, 9.26
  packing for, 12.15–12.18
  pressure-reducing, 3.20, 3.29, 3.30
  pressure-relief (safety)
    boiler, 3.8, 3.11–3.13, 3.15
    factory settings, 3.21
    hot water system, 3.29, 3.30
    pneumatic system, 8.53
    refrigeration system, 5.6, 5.16
  purging process, 3.29, 3.30
  reversing (four-way), 11.2, 11.8, 11.9, 11.13–11.14, 11.32
  Schrader core, 5.9, 5.11
  shutoff, 3.16
  solenoid, 8.41, 11.13, 11.14
  thermostatic expansion (TEV; TXV)
    heat pump, 11.15
    no field adjustment, 10.30
    refrigeration system, 5.20–5.21
    rooftop unit, 1.34
    testing thermistor for, 7.14
    troubleshooting, 10.24, 10.27, 10.29–10.31, 10.34
  two-stage gas, 9.5, 9.6, 9.26
  two-way and three-way, 3.22
  zone control, 3.18, 3.21–3.22, 3.25, 3.28, 8.52
Vanadium, 13.3
Vandalism, 1.38, 10.33
Vane (airflow guide), 1.18, 1.24, 1.39
Vane (indicator for switch), 9.5, 9.26
Vane (on a compressor), 10.18
Vapor barrier, 13.10, 13.35
Variable air volume system (VAV), 1.8, 1.17–1.18, 1.27–1.28, 1.38–1.39, 1.50, 13.31
Variable volume, variable temperature system (VVT), 1.15–1.16, 1.26–1.27, 1.50
VAV. *See* Variable air volume system
Vehicle, 12.50
Vent connector, 2.4, 2.11–2.12, 2.20
Ventilation
  air. *See* Supply air
  demand-controlled (DCV), 1.10, 1.11, 1.14, 4.22
  volume per person (rate), 1.6, 4.2
Ventilators
  energy recovery (ERV), 4.11, 4.14–4.16, 4.25
  heat recovery (HRV), 4.11, 4.15–4.16, 4.25
  unit, 3.24
Venting
  boiler, 3.9, 9.12
  furnace, 2.4–2.6, 2.11–2.13, 2.14–2.15, 2.16
  overview, 2.2, 2.19
  power, 2.16
Venting system
  for furnace
    components, 2.6–2.8
    draft controls, 2.16–2.17
    draft dynamics, 2.8–2.9
    piping considerations, 2.14–2.15
  for hot-water heating system, 3.16
Venturi tube, 3.31
Vertical packaged unit, 1.10, 1.16
VFD. *See* Drives, variable frequency
VI. *See* Viscosity index
Vibration
  coupling misalignment, 12.32, 12.33
  duct system, 13.7
  piezoelectric crystal, 4.8
  pump cavitation, 3.18
  and thermostat installation, 8.7, 10.13
  vibration isolator, 1.36, 11.28, 13.11, 13.32
Viscosity, 12.27, 12.61
Viscosity index (VI), 12.27, 12.61
Vise, 12.25
Volatile organic compounds (VOC), 4.10
Voltage
  capacitive circuit, 6.19
  capacitive current, 6.18
  on circuit diagrams, 6.30–6.31
  effective, 6.7, 6.8, 6.35
  glow coil and hot surface ignitor, 9.18–9.19, 9.20
  heating element, 11.22–11.23
  HVAC system, 6.3, 6.8
  imbalance in a three-phase system, 6.16, 10.22
  inductive circuit, 6.17, 6.18, 6.19
  measurement to troubleshoot HVAC, 8.36–8.38
  Ohm's law, 11.23
  peak, 6.7
  polarity, 6.7
  relay coil, 8.14
  resistive circuit, 6.16, 6.18
  root-mean-square (rms), 6.7, 6.35
  spark ignitor, 9.17
  thermostat, 8.7
  three-phase, 6.13, 6.16, 10.22
  transformer output, 6.3, 6.5, 6.10
Voltmeter, 7.14, 8.32
VPAC. *See* Air conditioners, vertical packaged
VVT. *See* Variable volume, variable temperature system

W

Wall, fire-rated, 13.26, 13.27
Wall covering, 1.5
Warehouse, 3.24, 9.12, 11.8
Warranty, 12.43, 12.47
Washer, 12.5, 12.13, 12.22
Waste disposal
  mercury, 8.3
  refrigerant recovery, 5.7, 5.10, 5.20, 5.31, 5.37, 10.25

Water. *See also* Condensate
  boiler, 3.13, 3.14, 8.51–8.52
  density, 3.34
  freezing, boiling, and vaporization temperature, 13.3
  groundwater, 11.2–11.3, 11.6
  high-temperature applications, 3.8
  makeup, 3.15
  physical and chemical properties, 3.3
  return, 11.4
  specific heat, 3.2
  supply, 1.45, 1.46, 11.2–11.3, 11.4, 11.5
Water balance, 3.31, 3.33–3.35
Water hammer, 3.22
Water system
  dual-temperature, 3.31, 3.32
  hot-water heating, 3.5–3.26
  terms, 3.3–3.5
  water flow and pressure in, 3.3–3.4
Watt, 6.17
Wattmeter, 6.24–6.25
Well, 11.4
Wind, 11.28
Windings
  on electronically commutated motors, 7.16, 8.48
  on inductive motors
    capacitor-start, 6.21
    definition, 6.35
    run, 6.19, 6.20, 6.21
    split-phase, 6.19–6.20
    start, 6.19, 6.20, 6.21
    testing, 6.26–6.28, 7.15
    thermistor on, 7.15
    troubleshooting, 8.48–8.50, 10.18, 10.34
  on transformer, 6.2–6.3, 6.5, 6.10
Wiring, electrical
  color codes, 8.8, 8.9
  copper, 7.4
  electric heating system, 11.21
  natural-gas heating system, 9.3–9.4
  nickel on resistor, 7.12
  splicing, 8.9
  thermostat, 8.8–8.9, 8.10, 8.44, 10.15
Wiring diagram, 8.28, 8.59, 8.65, 9.4, 9.8
Work area, 12.51, 12.55, 14.3
Work order, 12.44, 12.46
Wrap, duct, 13.16–13.24
Wrench, 12.12, 12.13, 12.25
Wye arrangement, 6.13–6.14, 6.24

Y
YAC. *See* Air conditioners, year-round
Young's modulus of elasticity, 14.3, 14.4, 14.27

Z
Zeotrope, 5.31–5.32
Zinc, 13.3
Zone, control, 1.2
Zoned system
  constant volume, 1.11–11.15, 1.33, 1.50
    definition, 1.50
    dual-duct, 1.18–1.19, 1.50
  hot-water heating system, 3.6, 3.18, 3.19, 3.21–3.22, 3.28–3.30
  multi-zone constant volume, 1.12–1.15
  overview of HVAC control, 4.19–4.21
  with packaged equipment, 1.33
  single-zone constant volume, 1.10–1.12, 1.15
  two-deck multi-zone, 1.17
  two-zone constant volume, 1.11–1.12
  variable air volume (VAV), 1.17–1.19, 1.38–1.39, 1.50, 13.31
  variable volume, variable temperature (VVT), 1.15–1.16, 1.26–1.27, 1.50
  zoning accessories and coils, 13.31–13.32